The Birds of Northern Melanesia

The Birds of Northern Melanesia

Speciation, Ecology, & Biogeography

ERNST MAYR
JARED DIAMOND

Color Plates by H. Douglas Pratt

2001

OXFORD
UNIVERSITY PRESS

Oxford New York
Athens Auckland Bangkok Bogotá Buenos Aires Cape Town
Chennai Dar es Salaam Delhi Florence Hong Kong Istanbul Karachi
Kolkata Kuala Lumpur Madrid Melbourne Mexico City Mumbai Nairobi
Paris São Paulo Shanghai Singapore Taipei Tokyo Toronto Warsaw

and associated companies in
Berlin Ibadan

Published by Oxford University Press, Inc.
198 Madison Avenue, New York, New York 10016

Library of Congress Cataloging-in-Publication Data
Mayr, Ernst, 1904–
The birds of northern Melanesia : speciation, ecology, and biogeography /
by Ernst Mayr and Jared Diamond.
p. cm.
Includes bibliographical references
ISBN 0-19-514170-9
1. Birds—Papua New Guinea—Bismarck Archipelago. 2. Birds—Solomon Islands.
I. Diamond, Jared II. Title
QL691.B52 M38 2001
598'.09958—dc21 00-061148

9 8 7 6 5 4 3 2 1

Printed in the United States of America
on acid-free paper

TO THE PEOPLE OF NORTHERN MELANESIA,

guardians of the future of Northern Melanesian birds,
in gratitude for their help and friendship.

Origins and Acknowledgments

This book, about the proximate and ultimate origins of species, has had its own proximate and ultimate origins. Its proximate origins lie in two letters, composed 42 years apart. More than 70 years ago (in 1929), the distinguished ornithologist Ernst Hartert wrote the first of those two letters to the then 25-year-old Ernst Mayr at the beginning of Mayr's professional career in ornithology. Hartert, who had studied birds from all parts of the world, told Mayr, "There is no other place in the world more favorable for the study of speciation in birds than the Solomon Islands." That suitability represented an irresistible challenge to Ernst Mayr then, and again 42 years later to Jared Diamond. The ultimate origins of this book lie in that irresistible challenge posed by the Solomons and the adjacent Bismarck Archipelago, which together constitute Northern Melanesia, to anyone interested in the origins of species.

Mayr remained in the Solomons from 1929 to 1930 as a member of the famous Whitney South Sea Expedition, exploring Buka, Bougainville, Shortland, Choiseul, San Cristobal, and Malaita Islands. Beginning in 1931 and continuing for more than 25 years, Mayr went on to study the taxonomy of the birds of Northern Melanesia, New Guinea, and adjacent islands, describing more than 100 new species and subspecies and writing the first field guide to the birds of the Solomons and other tropical Southwest Pacific islands (Mayr 1945a). He drafted some chapters of this book on Northern Melanesian birds in the 1950s and published a preliminary analysis in 1969 (Mayr 1969). By 1971 Mayr had compiled a set of island-by-species distributional tables, which he sent to Diamond with a cover letter asking, "Do you think that this could be the beginning of a joint paper?" In that letter lay the second proximate origin of this book, which in the following decades expanded far beyond the dimensions of the initially envisioned "joint paper."

Just as Hartert's letter had stimulated Mayr, Mayr's letter stimulated Diamond, who in 1964 had embarked on the first of his 19 expeditions to study the avifaunas of New Guinea and other Pacific archipelagoes from Java east to Samoa. Diamond carried out three expeditions to Northern Melanesia to fill in gaps in our knowledge, including ornithological surveys of New Britain's west end and the Long Island group, ecological studies in the mountains of Bougainville and Umboi, resurveys of almost all ornithologically significant Solomon Islands, and surveys of a hundred small islands. Our first joint analyses of our database were published in 1975 and 1976 (Diamond 1975a, Diamond & Mayr 1976, Diamond *et al.* 1976, Mayr & Diamond 1976, Gilpin & Diamond 1976). The first draft of this book was completed in 1984 and was followed by five more drafts, of which the last was finished in June 2000.

The two of us bring different perspectives to this book. We belong to different generations of scientists—one of us (E.M.) born in 1904, and the other (J.D.) in 1937, and represent different perspectives—evolutionary biology, systematics, and speciation (E.M.) and ecology and biogeography (J.D.). Thus, we have been able to contribute different insights. Despite our differences in perspective, though, we agreed on the basic principle of this book: to steer a middle course between overgeneralization and mere descriptive detail, and to seek to understand the rich diversity of solutions attained by different species. Too much of the literature of population biology consists either of detailed species-by-species descriptions that fail to provide a general framework, or else of generalizations that are derived from case studies of certain species but that actually do not apply to all species. The goal of our book is to recognize and understand differences among species and among islands: differences in the outcomes of geographic isolation, in the distributional strategies that have evolved, in the levels of endemism attained, and in the courses that speciation has taken.

While this book has only two authors, it rests on the contributions of literally hundreds of people who have helped us. Space limitations preclude giving a complete list, because so many people contributed their time and knowledge so generously. Some of the principal contributors are acknowledged below.

Alisasa Bisili of New Georgia, Teu Zinghite of Kulambangra, David Akoitai of Bougainville, Charles Tetuha of Rennell, and many other residents of Northern Melanesia helped J.D. with field work and shared their knowledge of birds.

People who shared their own observations or examined museum specimens of birds included Dean Amadon, Bruce Beehler, Harry Bell, K. David Bishop, Stephen Blaber, David Buckingham, Henry Childs, Brian Coates, Francis Crome, Richard Donaghey, Guy Dutson, John Engbring, Edmund Fellows, Brian Finch, Chris Filardi, Jon Fjeldså, Ian Galbraith, David Gibbs, Don Hadden, Warren King, Geoff Kuper, Mary LeCroy, Annette Lees, Eric Lindgren, Richard Loyn, Roy Mackay, Jonathan Newman, H.R. Officer, Shane Parker, William Peckover, Finn Salomonsen, Keith Sanders, Richard Schodde, Paul Scofield, Charles Sibley, David Steadman, James Tedder, Michael Walters, H. Price Webb, Torben Wolff, and Shane Wright.

For supplying information and/or scrutinizing our draft chapters on the following subjects, we owe big debts to T.C. Whitmore and Robert Thorne (vegetation); Eldon Ball, Robert Heming, R.W. Johnson, C.O. McKee, James Mori, and Eli Silver (geology); Timothy Flannery, J.E. Hill, Dieter Kock, Karl Koopman, James Smith, Hobart Van Deusen, and Alan Ziegler (mammals); Walter Brown, Harold Cogger, Allen Greer, Samuel McDowell, B. Mys, Ernest Williams, and Richard Zweifel (amphibians and reptiles); and Timothy Flannery and David Steadman (fossils and prehistory). Walter Bock heroically read the entire manuscript and made valuable suggestions.

The curators of the American Museum of Natural History, Bishop Museum (Honolulu), Los Angeles County Museum of Natural History, and University of Copenhagen Zoological Museum cooperated whole heartedly in enabling us to study their collections of birds.

The taxonomic conclusions of this book depend on bird collections amassed by collectors who worked under enormous difficulties and personal dangers. We especially acknowledge Albert Meek and Albert Eichhorn, who worked in Northern Melanesia for Lord Rothschild between 1900 and 1925; and E.M.'s colleagues in the Whitney South Sea Expedition, Rollo H. Beck, William Coultas, Frederick Drowne, Walter Eyerdam, Hannibal Hamlin, and G. Richards, who thoroughly explored the whole of the Solomons and many of the Bismarcks between 1927 and 1935.

E.M.'s field work was supported by the Whitney South Sea Expedition of the American Museum of Natural History; J.D.'s field work was supported by the National Geographic Society, the South Pacific Commission, and the Alpha Helix Expedition of the National Science Foundation.

Research for this book was generously supported by the Eve and Harvey Masonek and Samuel F. Heyman and Eve Gruber Heyman 1981 Trust Undergraduate Research Scholars Fund.

H. Douglas Pratt poured his artistic skills and scientific insight into painting the nine color plates. Ellen Modecki prepared the maps and illustrations.

Margaret Farrell-Ross and Jeffrey Gornbein carried out the statistical analyses of the tables and figures, with advice from Elliot Landaw. Michael Gilpin, Stuart Pimm, and James Sanderson collaborated, and Thomas Schoener gave advice, on statistical analyses of community ecology.

Walter Borawski, Barbara Burgeson, Michelle

Fisher-Casey, Lori Iversen, Laura Kim, Lori Rosen, Paula Stoessel, and Barbara Troyan devoted many years of their lives to manuscript and data preparation.

Kirk Jensen, our editor at Oxford University Press, patiently shepherded and supported this book through its long gestation; our copyeditor Karla Pace meticulously polished it; and Lisa Stallings oversaw its production.

To these and many others, we are profoundly grateful. This book could not have been written without them.

Ernst Mayr
Museum of Comparative Zoology
Harvard University
Cambridge, Massachusetts 02138

Jared Diamond
Department of Physiology
UCLA Medical School
Los Angeles, California 90095-1751

Contents

Part VI. Geographic Analysis: Differences Among Islands

Part VII. Synthesis, Conclusions, and Prospects

Appendices

Introduction

In this book we present a comprehensive, detailed study of speciation for all resident land and freshwater bird species of Northern Melanesia. Our database, set out in seven appendices, 52 distributional maps, and many tables, consists of lists of island occurrences, taxonomic affinities, and ecological attributes of every species. In addition to analyzing speciation, we use this database to illuminate other problems of population biology, such as extinction, colonization, and the evolution of dispersal ability.

We begin this introduction by addressing the obvious question: What is the need for a new book about speciation? This introduction then goes on to consider the advantages of Northern Melanesian birds for studying speciation, provide an outline of the book's contents, and describe the scope of this book.

What Is the Need for a New Book about Speciation?

As is well known, Darwin's (1859) book *The Origin of Species* established the fact and main mechanisms of evolutionary change with time. However, despite the book's title, Darwin did not solve the problem of *speciation*—namely, the problem of how multiple daughter species originate from a single ancestral species (fig. 0.1). Indeed, Darwin vacillated in his view about whether species were real units or just arbitrarily delineated sets of individuals (Mayr 1991, pp. 29–30).

Resolution of those two uncertainties was not achieved until after 1930. It then became recognized that, in most animal and plant groups, species do possess an underlying reality, termed the "biological species concept" and defined by reproductive cohesion within the species and reproductive isolation from other species. As part of that evolutionary synthesis, Mayr (1942) emphasized that most speciation in sexually reproducing taxa proceeds via geographic isolation, which may lead to reproductive isolation. That is, some populations belonging initially to the same species become geographically isolated from each other, either due to a barrier arising that divides the range of the species, or due to colonists founding a new population disjunct from the species' original range. The daughter populations are then said to live in allopatry (meaning in mutually exclusive geographic areas). The daughter populations may diverge in their reproductive behavior. Completion of speciation—that is, the achievement of sympatry (geographic coexistence) between the two daughter populations, now no longer interbreeding with each other—requires, in addition to the development of reproductive isolation, the development of ecological segregation between them (ecological differences permitting coexistence).

While this brief account of speciation is accepted today by many biologists, others continue to dis-

Evolutionary change with time, but without speciation or extinction

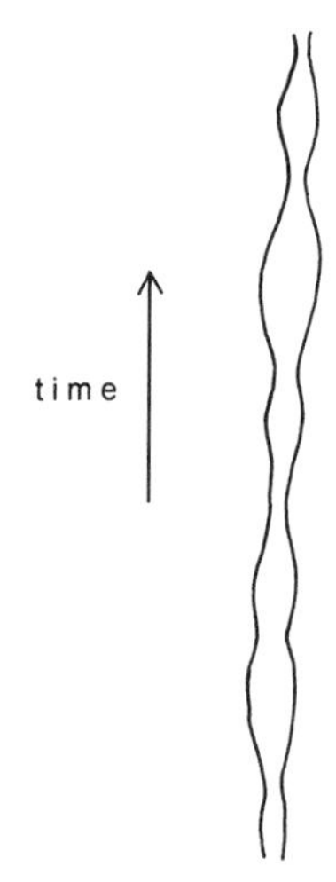

Evolutionary change with time, including speciation and extinction

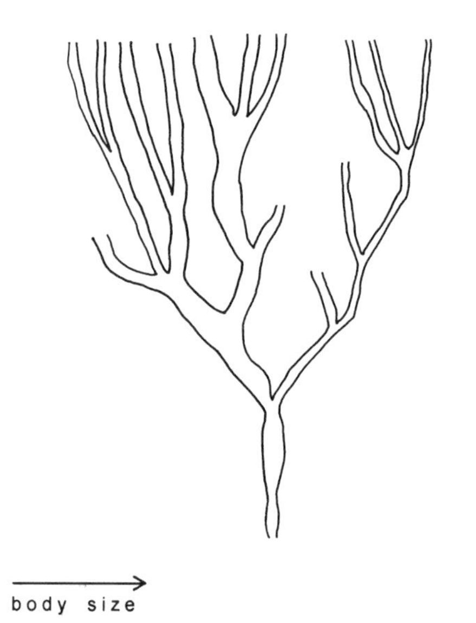

Fig. 0.1. Evolutionary changes in body size (or any other property) with time. Left: evolutionary change proceeds in a single species, without any speciation events, in the process termed "phyletic evolution." In effect, this is the problem illuminated by Darwin's book *The Origin of Species*. Right: evolutionary change, accompanied by speciation (i.e., repeated fission of ancestral species into daughter species), with subsequent extinction of some lines. Speciation is the main subject of this book.

pute it. Debates persist concerning the reality of species and concerning whether speciation proceeds allopatrically (i.e., via geographic isolation) or sympatrically (i.e., via reproductive isolation developing between members of the same population, while they co-occur at the same locality). Even for those biologists who would agree with our brief account of speciation, major questions remain unanswered. Examples of those questions include: does ecological segregation develop before reproductive isolation, or vice versa?; is there any preferred sequence to the development of different types of ecological segregation during speciation?; why are certain species highly variable geographically and actively speciating, while other species are widely distributed without any geographic variation?; what is the role of peripheral isolates in speciation?; how rapid are intermediate stages in speciation?; and might species remain "stuck" for a long time in allopatry, unable to achieve sympatry?

These debates about speciation have persisted for several reasons. First, answers to these specific questions, as well as to the basic questions about the reality of species and about the relative roles of allopatric and sympatric speciation, probably differ among different groups of organisms. Second, most speciation appears to be a slow historical process, unfolding over times far longer than a human lifetime. As a result, a biologist is unlikely to observe the entire course of natural speciation. This is not meant to deny the possibility that certain stages of speciation in at least some cases may unfold rapidly (see Mayr 1954, Gould & Eldredge 1977). Even in such cases, though, the same lineage may go through other, much slower, stages as well. Finally, in addition to proceeding much too slowly to be observed from start to finish in the field, speciation in most organisms cannot be replicated experimentally in the laboratory.

The student of speciation must instead attempt to reconstruct a chain of developments that occurred in the past. This is possible because the millions of species of existing plants and animals are at any time, including today, in various stages of speciation. Hence the course of speciation can be reconstructed by considering each modern species as an instantaneous snapshot of one of these stages in a long process (see fig. 13.1 on p. 98)—actually, in multiple parallel long processes, since speciation may follow various alternative courses. This procedure is analagous to that of someone who is presented with an incomplete, scrambled set of cut-up frames of a motion picture and who must try to reassemble the frames in the correct sequence so as to reconstruct the movie.

Similar methodological problems confront scientists studying numerous other slow phenomena, such as astronomers attempting to reconstruct the life cycles of stars, paleontologists attempting to reconstruct evolutionary change with time, archaeologists attempting to seriate individual cultural layers found in excavations of different sites, or cytologists of the 1870s seeking to understand the process of mitosis. The cytologists, for instance, arranged hundreds of microscope slides in sequence and thus inferred the course of mitosis. Similarly, no astronomer has watched for millions of years as one star formed from a condensing gas cloud and evolved first into a red giant and then into a white dwarf. Neither have cultural anthropologists watched hunter-gatherer bands evolve over the course of thousands of years into a literate empire, nor have paleontologists born in the Eocene watched *Eohippus* evolve into modern horses. Each of these disciplines, like speciation studies and all other historical sciences, makes use of the methodology of inferring temporal processes from sets of static snapshots.

Naturally, the success of this methodology requires an adequate database. That database must comprise a large enough set of snapshots, distributed over all stages of the process, to reconstruct the entire course. The number of available snapshots must be especially large if, as is true of speciation, the process has branch points leading to multiple alternative pathways. Hence the student of speciation requires a large database provided by a rich fauna whose component species are well understood in their distribution and taxonomy. The database should consist of the entire fauna, lest conclusions be biased by reliance on selected examples. Furthermore, some geographic configurations make it easier to reconstruct the course of speciation than do other configurations. Of course, such a database would be invaluable for understanding not only speciation but also many other major problems in ecology, biogeography, and evolutionary biology, some of which we explore here.

What Are the Advantages of Northern Melanesian Birds for Studying Speciation?

Birds in general offer the advantage of being conspicuous and relatively easy to observe and identify in the field. As a result, geographical distributions are better known, and taxonomic relationships have been studied more intensively and comprehensively, for birds than for any other group of organisms. Islands in general offer the advantage of being circumscribed units with sharp boundaries. As a result, island regions have been favorite sites for studying bird speciation, from the nineteenth century studies of Galapagos and Indonesian birds by Darwin and Wallace, respectively, to the modern studies of Galapagos, Hawaiian, and Caribbean birds by Lack (1945), Grant & Grant (1996), Amadon (1950), James & Olson (1991), and Rickleffs and colleagues (e.g., Rickleffs & Bermingham 1999).

As for Northern Melanesia, it consists of the islands of the Bismarck and Solomon archipelagoes, the two archipelagoes lying in the tropical Southwest Pacific Ocean east of New Guinea, northeast of Australia, and just south of the equator (figs. 0.2 and 0.3). Northern Melanesia and its birds possess decisive advantages for a study of speciation:

- Northern Melanesia contains approximately 200 ornithologically well-surveyed islands within a small area. The surveyed islands include every island of any individual importance within the region (most of them surveyed repeatedly), in addition to a hundred very small islands that are collectively significant. Thus, no further discoveries of modern distributions that might change our major conclusions are to be expected. Altitudinal and ecological distributions of Northern Melanesian birds are also well known. No other area of the tropics has been surveyed ornithologically as thoroughly.
- All islands of Northern Melanesia share similar climates and habitats. Not even the largest Northern Melanesian islands have climate gradients as marked as those on the larger islands of the Lesser Sundas and West Indies, within which one can find gradients from rainforest to desert scrub.
- Areas of Northern Melanesian islands span seven orders of magnitude in an evenly graded size series from the smallest (0.002 km^2) to

Fig. 0.2. Overview map depicting locations of the major land masses discussed in this book. The Bismarck Archipelago and the Solomon Archipelago, which together constitute Northern Melanesia, lie immediately east of New Guinea and south of the equator. Southern Melanesia consists of the Santa Cruz and New Hebrides archipelagoes and New Caledonia. Micronesia encompasses the Marianas, Carolines, Marshalls, and Palau, while Polynesia extends from Fiji eastward (that is, Polynesia as defined ornithologically but not anthropologically). The Greater Sundas are the islands of Borneo, Java, and Sumatra, which constitute the easternmost extension of the Southeast Asian continental shelf. Wallacea is the group of Indonesian islands (Celebes, Lesser Sundas, and Moluccas) lying between the Asian continental shelf and the Australian/New Guinean continental shelf.

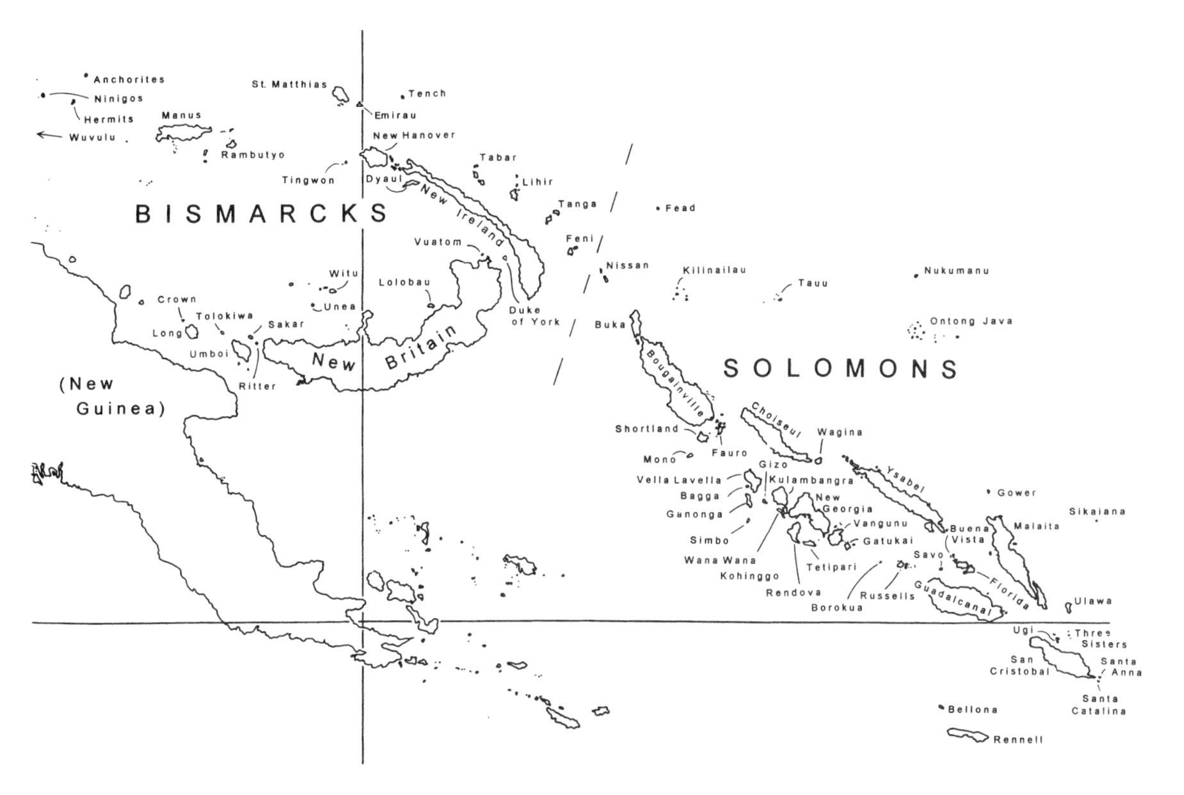

Fig. 0.3. Map depicting locations of the principal ornithologically significant islands of Northern Melanesia discussed in the text. The dashed line separates the Bismarcks from the Solomons, which together make up Northern Melanesia. Water gaps and bird distributions coincide in delineating Northern Melanesia, the Bismarcks, and the Solomons as unequivocal ornithogeographic regions, as discussed quantitatively at the end of chapter 29.

the largest (35,742 km^2) island, unlike the situation in the Greater Sundas and West Indies, which comprise a few huge islands with much smaller satellites.

- The islands are of several different types, including ones with recent land connections to each other, unconnected central islands, unconnected peripheral islands, mountainous islands up to 2591 m in elevation, flat low islands, old islands, young volcanic islands, and young, uplifted coral atolls.
- There are only three potential major colonization routes from distinct directions (west, southwest, and east), permitting easy identification of the source area for most colonists.
- The resident land and freshwater avifauna is rich enough to be interesting (195 species), but not so rich as to be overwhelming.
- The avifauna is diverse, comprising highly sedentary species as well as highly vagile ones, old endemics plus undifferentiated recent invaders, and large nonpasserines plus small songbirds.
- Only a few of those native bird species have become extinct in modern times, and only one nonnative bird species has established itself. In contrast, in many other island regions, such as Hawaii and New Zealand, most of the original native avifauna has become extinct since the arrival of humans, and in most habitats except remnant native forest it was replaced by individuals belonging to one or another of dozens of introduced species.
- The avifauna is well understood taxonomically. Virtually every bird genus and every species group represented in Northern Melanesia have been recently revised taxonomically, and we have re-revised all Northern Melanesian bird taxa.
- The information available for each species thus consists not only of number and size range of islands occupied, habitat preference,

colonization source, and degree of endemism, but also several independent measures of overwater colonizing ability. Species differences in overwater dispersal have received far too little attention in the past; they prove to be a major explanatory factor throughout this book.
- Numerous examples are available to illustrate every intermediate stage in speciation, including striking examples that have become classics and are cited in every textbook of biology (the white-eyes of Rendova and Tetipari [Plate 1] and the golden whistlers [Plate 2]).

For all these reasons, the birds of Northern Melanesia have played a disproportionate role in the development of current understanding of species and speciation.

There have been numerous previous studies of speciation among birds and other taxa, in Northern Melanesia and elsewhere in the world. Those previous studies have been based on selected examples. This book instead provides a quantitative analysis of every stage of speciation in every species of a rich fauna. This fact enables us to avoid trying to generalize from arbitrarily chosen anecdotes. We can instead calculate what percentage of all species in the fauna is in any given stage of speciation.

Naturally, the same advantages that make Northern Melanesian birds especially suitable for studying speciation also make them especially suitable for studying numerous other central problems of population biology. Hence, in addition to being a study of speciation, this book also mines the Northern Melanesian avifauna for the insights that it affords into numerous other ecological, evolutionary, and biogeographic problems.

Outline of the Book

The first four parts of our book present background material relevant to analyzing bird speciation in Northern Melanesia. Part I briefly summarizes the physical and biological environment in which Northern Melanesia's birds have evolved. We describe in turn Northern Melanesia's geology and geological history, climate, habitats and vegetation, and terrestrial vertebrates other than birds.

Part II surveys the history of human occupation, from the first human arrivals at least 30,000 years ago to the modern era. The history of ornithological exploration and of scientific description of Northern Melanesian birds is then described. We summarize modern and known prehistoric extinctions of Northern Melanesian birds. By evaluating the susceptibility of Northern Melanesian birds to known risk factors for island bird extinctions elsewhere in the world, we consider the extent to which not-yet-recognized extinctions may have distorted our knowledge of the Northern Melanesian avifauna.

Part III introduces the Northern Melanesian avifauna itself. We compare its family composition with that of its principal source avifauna, that of New Guinea. We examine the factors on which the number of bird species resident on a Northern Melanesian island depends: the island's area, isolation, elevation, altitudinal distribution of area, and geological history. We then tabulate for each species its level of endemism within Northern Melanesia, its habitat preference, its abundance, and its overwater dispersal ability. The evolutionary trade-offs selecting for high or low dispersal ability are considered. These tabulated ecological attributes are used to interpret each species' incidence—that is, the number and size range of islands that it occupies.

Part IV concludes the presentation of background material with an analysis of the colonization routes by which birds have reached Northern Melanesia. We identify both proximate and ultimate sources of colonists. This material permits us to test for Northern Melanesia the concept of "faunal dominance": Do adjacent biogeographic regions exchange colonists unequally, and, if so, why?

Parts V and VI, the longest parts of the book, then use this database to analyze speciation from two alternative perspectives: by species or by island. Part V, the analysis by species, examines why some clades seem to be speciating much more profusely than other clades. Our analysis categorizes each Northern Melanesian species according to the stage in speciation that it represents within Northern Melanesia: from species exhibiting no geographic variation, to species differentiated within Northern Melanesia at the level of weak or strong subspecies or allospecies, to cases of recently completed speciation, to cases of stalled or failed speciation, and finally to endemic full species or genera. At each level we seek to interpret species differences in speciation patterns in terms of species differences in ecological attributes. In the course of this analysis we also use the Northern Melanesian avifauna to address the debates about whether ecological segregation usually develops in allopatry or in sympatry, and whether different modes of ecological segregation predominate at different stages of speciation.

Whereas part V asks why some clades have speciated more than others, part VI asks why some interisland water gaps have been more effective at promoting differentiation than have other gaps. Some islands support no endemic taxa, while other islands support many. Some adjacent pairs of islands support avifaunas that are similar in species composition as well as in subspecific affiliation, while other island pairs have highly differentiated avifaunas. We use several numerical indices to express these differences in island distinctness and in barrier potency. We relate these indices to island properties such as area, isolation, geographic location, and geological history. We also analyze the effects of former land connections that existed at Pleistocene times of low sea level, when some sets of modern islands were joined into larger islands. Those former land connections not only made some water barriers intermittent but also left behind stranded populations of bird species that are poor at overwater dispersal.

Part VII synthesizes our conclusions with respect to two sets of problems. We first assemble the individual species snapshots that we have analyzed to reconstruct the motion picture of speciation and to assess the importance of primary isolation, secondary isolation, and peripheral isolation. We then review species differences as expressed in the evolution of dispersal, the so-called taxon cycle, and the heterogeneity of the montane avifauna. Our final chapter sketches many promising directions for future research.

Nine color plates of 129 birds, painted by H. Douglas Pratt, depict 88 species and allospecies, more than one-third of Northern Melanesia's total. They encompass 17 sets of taxa illustrating geographic variation and speciation, plus all of Northern Melanesia's endemic genera and the majority of its endemic species and allospecies. The distributions of nearly half of Northern Melanesia's species (80 out of the 195) are depicted on 52 maps (pp. 313–360), which convey patterns of geographic variation, colonization routes, and speciation more succinctly and in more detail than words can. Because completed speciation in sexually reproducing taxa involves changes in geographic distributions of related populations, maps are central to speciation studies. We do not provide plumage descriptions and field identification marks, which can instead be found in three published field guides or handbooks: Mayr (1945a), for the Solomons plus other tropical Southwest Pacific islands (but not the Bismarcks); Coates (1985, 1990), for the Bismarcks and Northwest Solomons plus New Guinea (but not those Solomon Islands outside the nation of Papua New Guinea); and Beehler et al. (1986), for New Guinea (hence for the many Northern Melanesian species shared with New Guinea).

Seven appendices at the end of the book summarize our database. Appendix 1, a systematic list, gives for each species its island-by-island distribution by subspecies and allospecies, our conclusions concerning its geographic variation and relatives and taxonomy, and references to key literature. Appendices 2–4 list Northern Melanesia's nonbreeding species, its very few introduced bird species, and the principal ornithological studies carried out on each island. Finally, appendices 5–7 present, for each species, a series of ecological and distributional attributes (level of endemism, habitat preference, abundance, dispersal ability, incidence category in each of the two archipelagoes, number of islands occupied, and number of subspecies and allospecies), 11 types of evidence used to assess overwater dispersal ability, and six types of evidence used to assess proximate and ultimate colonization routes.

The Scope of This Book

The subjects of this book draw on many disciplines and types of evidence, and they merge into many other subjects. Nevertheless, to keep the book's length, focus, and cost manageable, we have had to be selective. Inevitably, some readers would have selected differently and will feel that we treated some topics in inordinate detail while omitting other indispensible topics or approaches. So that readers will at least know what to expect, we now explain our selection. We have omitted some subjects by choice, and regretfully have had to treat other subjects scantily because of limited available information. We have dwelt in detail on information that will be of lasting value, that will be difficult for readers to synthesize from other sources, and that we feel uniquely qualified to summarize.

By choice, we have confined ourselves to a study of speciation in Northern Melanesian birds. We believe that the best use of space in this book is for as complete and detailed an account of the Northern Melanesian avifauna as possible. This will be the first time that a quantitative analysis of speciation is available for a rich fauna in its entirety, as a yardstick for comparative studies. The detailed factual material that we present will make it possible for scientists unfamiliar with our database and specializing in other biotas to test our conclusions, to develop and test new theories, and to compare and contrast the biotas that they study with ours.

This book is not a comparative study of speciation in general, for other taxa, or for other regions of the world. We restate explicitly that we expect findings about speciation to vary among taxa of plants and animals and among regions. Indeed, one of the major conclusions of this book is that, even within the Northern Melanesian avifauna, conclusions vary among bird taxa and among islands. In particular, it remains a task for the future to compare our study of the Northern Melanesian avifauna with the detailed studies of Galapagos finches by Lack (1940, 1945, 1947) and by Grant and Grant (e.g., Grant & Grant 1989, 1996, 1997; Grant 1981, 1984, 1999); of Hawaiian honeycreepers by Amadon (1950), Sibley and Ahlquist (1982), and James and Olson (1991); and of taxon cycles in Caribbean birds by Rickleffs and colleagues (Rickleffs & Cox 1972, 1978; Ricklefs & Bermingham 1997, 1999; Seutin et al. 1994). We explore such comparisons only briefly (e.g., in table 4.5 and chapter 36).

In addition, this book is a study of speciation resulting from geographic isolation by water barriers. Some readers may wonder why we do not begin by posing alternative competing hypotheses about the mechanism of speciation (e.g., is it sympatric or allopatric?) and proceed to test them. We would have begun in that way if we were analyzing speciation in some other taxa, for which the mode of speciation has been uncertain and might prove to be different. We would also have begun in that way if we had been writing this book about Northern Melanesian birds 80 years ago, when their mode of speciation was unresolved. Now, however, the evidence is overwhelming that speciation in Northern Melanesian birds results from geographic isolation by water barriers. The avifauna offers many examples of many stages in the whole process and thereby permits one to reconstruct the process convincingly. The avifauna provides only a couple of known examples even of low-level geographic variation resulting from land barriers within a single island. We cannot recognize any case that could plausibly be interpreted as a stage in sympatric speciation.

Given the modest areas of Northern Melanesian islands, the water barriers between them, and current knowledge of dispersal, genetics, and mate selection in birds, it is not surprising that speciation in Northern Melanesian birds results from geographic isolation of daughter populations by water barriers. In the face of this overwhelming evidence, to have begun by posing alternative hypotheses would have been a pretense. However, that does not mean that our detailed conclusions were obvious at the outset and that a lengthy book was unnecessary. As we already noted, many major questions about how speciation by geographic isolation operates have remained unresolved. In addition, geographic speciation in Northern Melanesian birds shows great diversity. Our treatment had to be comprehensive to test for potential contributions of numerous intrinsic factors (such as abundance and dispersal ability) and extrinsic factors (such as island isolation and Pleistocene sea-level changes) to this diversity.

Readers will find only scanty discussion of molecular and fossil evidence for evolution of Northern Melanesian birds. This relative neglect should not be misconstrued to imply that we place low value on these approaches—quite the contrary. We recognize that both approaches have been extraordinarily fruitful wherever they could be applied, and both of us have written extensively about their application to other faunas. Alas, we cannot devote much space to them because they as yet have been little applied to Northern Melanesian birds.

For instance, molecular evidence has yielded major insights into family-level evolutionary relationships of the New Guinea avifauna that served as the major source for Northern Melanesia's avifauna. We discuss those insights in chapter 8. Molecular evidence of relationships among West Indian bananaquit and yellow warbler populations has revealed that the history of those populations is considerably more complex than was apparent from classical morphological evidence (Klein & Brown 1994, Seutin et al. 1994). We assume that molecular studies of Northern Melanesian bird populations may similarly yield new insights. However, no such studies have been carried out yet.

Similarly, bird fossils excavated at archaeological sites on many Pacific islands outside Northern Melanesia have revealed the former existence of species exterminated prehistorically by the first human colonists of those islands (e.g., Olson & James 1982; Steadman & Olson 1985; Steadman 1989, 1993, 1995; James & Olson 1991). Fossils have also revealed extirpated populations of species still surviving elsewhere, and thus show that geographic ranges defined by extant populations can be very different from ranges occupied before human arrival. Such fossil evidence has just started to become available for Northern Melanesia (Steadman & Kirch 1998, Steadman et al. 1999). We devote chapter 7 to presenting that limited available evidence, summarizing other evidence bearing on possible recent extinctions of bird populations, and assessing the likely magnitude of as-yet-undocumented prehistoric human impacts on Northern Melanesian bird distributions.

Some readers may have expected still another discussion of the already much-discussed controversy (e.g., Mayr 1997, pp. 143–146) between the two prevalent systems of biological classification—namely, Darwinian classification based on both genealogy and similarity (similarity classes) and Hennigian cladification based exclusively on phylogeny (branches of the phylogenetic tree). Actually, whether to define higher taxa by Darwinian or Hennigian principles rarely has any consequences for the species-level problems that are the focus of this book. Furthermore, all published taxonomic studies of Northern Melanesian birds to date have used the Darwinian approach that we also use. Most Hennigians do not recognize the biological species concept and instead recognize species on the basis of degree of difference. Our treatment incorporates degree of difference by consistently allocating taxa to different levels of distinctions: zoogeographic species, superspecies, allospecies, megasubspecies, and weak subspecies (see chapters 16, 20, and 21 for details).

Other readers will object, "How can you possibly hope to draw any evolutionary conclusions if you don't have an adequate fossil record, or evidence obtained by modern molecular methods, or if your taxonomy is not based on cladistic methods?" Similar dilemmas recur throughout all fields of science. It is true that new evidence of any sort may reveal hitherto-unsuspected complexities and may radically alter interpretations. Scientific conclusions are never conclusions, just tentative formulations. The question then arises at what point should one's tentative formulations be published.

Should we wait until more molecular, paleontological, or cladistic studies of Northern Melanesian birds become available? Certainly not. Few such studies are currently underway, so they are unlikely to reach a stage that would permit a comprehensive new synthesis for the whole avifauna in the near future. In contrast, a vast body of distributional, ecological, and classical-taxonomic information about Northern Melanesian birds is already available. In particular, knowledge of modern distributions is sufficiently complete that future field work is unlikely to change it significantly. This diverse body of information has not been synthesized. Much of it is unpublished, and the rest has appeared in widely scattered and often obscure publications.

To select the most fruitful problems for future molecular, paleontological, and cladistic studies of Northern Melanesian birds, a synthesis of existing knowledge is required. Furthermore, it will not be possible to decide whether new information alters conclusions based on existing knowledge, unless those conclusions based on existing knowledge have been synthesized. Under present-day political, practical, and bureaucratic conditions prevailing in Northern Melanesia, exploration for fossils and acquisition of tissue specimens are difficult. Opportunities for collecting or even observing birds are now far less favorable than those that presented themselves to Mayr during the Whitney South Sea Expedition of the 1920s and 1930s, and to Diamond during his large-scale collecting expeditions of the 1960s and 1970s. Thus, the experience that both of us have gained over decades of studying Northern Melanesian birds provides a unique basis for synthesizing present knowledge.

We anticipate that much of the audience for this book may consist of biologists who are not interested in our conclusions, nor even in the questions that we ask, but who are instead interested in mining our raw data for their own purposes. We have seen again and again that ecologists and evolutionary biologists continue to reanalyze the same few data sets on insular species distributions. There are only a few published data sets complete enough to warrant detailed analysis, such as the data sets for Galapagos birds, West Indian birds and bats, Great Basin montane mammals, and New Hebridean birds. The data that we summarize for Northern Melanesian birds constitute by far the most detailed data set now available. Our appendices list individually the distributions of 195 bird species on 200 islands, provide a uniform taxonomic treatment for all species, state the degree of taxonomic distinction of every species and every island population, and provide coded annotations of ecological and biogeographic attributes and colonization sources of every species. As Peterson (1998, p. 557) has noted, "Very few geographic areas can count on a broad consideration of bird taxonomy that simultaneously places all species present in a consistent taxonomic context. This fault makes the consideration, comparison, and evaluation of alternative species concepts, or even documentation of patterns of avian biodiversity, extremely difficult."

Some readers will find this presentation of raw data inordinately detailed. We provide it because population biologists have so often analyzed much smaller published data sets, as the best ones hitherto available; and because we know how difficult the information would be for the nonspecialist to gather, and how unlikely it is that anyone else will find themselves in a position to synthesize the available information again. Most scientific publications of interpretations eventually become superseded. In contrast, at least in population biology, publications that continue to be used 50 years after their appearance are so used mainly because of the raw data that they provide.

PART I

NORTHERN MELANESIA'S PHYSICAL AND BIOLOGICAL ENVIRONMENT

1

Geology and Geological History

Northern Melanesia's avifauna has been molded by the physical and biological environment in which it has evolved. Part I introduces that environment, beginning in this chapter with a brief account of Northern Melanesia's geology; Chapters 2–4 describe in turn its climate, its habitats and vegetation, and its terrestrial vertebrates other than birds. To understand how birds reached and evolved on Northern Melanesian islands, we need first to understand how and when the islands formed, whether their configurations and locations in the past differed from those at present, and whether the islands have ever been joined to continents or to each other. Hence this chapter will summarize aspects of Northern Melanesian geology important for interpreting the avifauna—especially plate tectonic history, volcanism and rocks, and recent land bridges.[1]

1. References for Northern Melanesian geology in general include Bain et al. (1972), Dow (1977), Löffler (1977, 1982), Johnson (1979), and Pieters (1982); for plate tectonics, Johnson and Molnar (1972), Jaques and Robinson (1977), Crook and Belbin (1978), Johnson and Jaques (1980), Cooper and Taylor (1987), Davies et al. (1987), Honza et al. (1987), Pigram and Davies (1987), Silver et al. (1991), Abbott et al. (1994), and Kroenke (1996); for volcanism, Hammer (1907), Fisher (1957), Johnson (1976, 1981), Palfreyman et al. (1986), Zielinski et al. 1994, and Briffa et al. (1998); for the Solomons, Guppy (1887a, 1887b), Lever (1937), Grover (1955), Grover et al. (1959–1962), and Coleman (1966, 1970); for Bougainville and Buka, Blake and Miezitis (1967) and Scott et al. (1967); for New Ireland, Hohnen (1978); for New Ireland and Manus, Marlow et al. (1988); for New Britain, Fisher and Noakes (1942); and for the western half of the Bismarck Volcanic Arc, Johnson et al. (1972).

Plate Tectonic History

New Guinea and Northern Melanesia resulted from interaction between the north-moving Indo-Australian Plate and the west-moving Pacific Plate (fig. 1.1). The Indo-Australian Plate bears the Australian continental platform, which became detached from Antarctica between about 95 and 55 million years ago and included what is now the Fly River bulge of South New Guinea. The interaction between the Indo-Australian and Pacific plates in the Tertiary produced island arcs, which collided with the northern margin of the Australian continental platform, resulting in the fusion (possibly during the Miocene) of some of the islands into the platform's northern edge that was to become New Guinea. The collision also caused the uplift of New Guinea's Central Dividing Range and of parts of New Britain, New Ireland, and the Solomons.

Former islands thus incorporated into New Guinea's northern margin have contributed to what are now the Bewani, Torricelli, Prince Alexander, Adelbert, and Huon Peninsula mountains along New Guinea's north coast. New Britain is also moving toward New Guinea and is now closer to New Guinea than it ever has been. Thus, the modern Bismarck and Solomon islands have never had a land connection to New Guinea nor been nearer New Guinea than they are now. The original biota of Northern Melanesia must therefore have arrived over water.

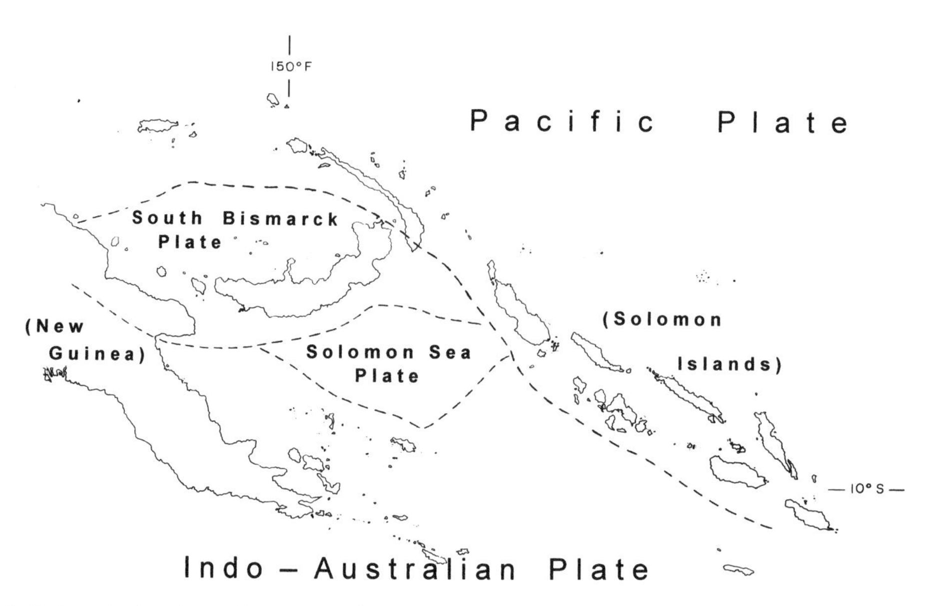

Fig. 1.1. Inferred plate boundaries in and near Northern Melanesia (from Cooper & Taylor 1987, Silver et al. 1991). The Bismarck and Solomon islands arose from the interaction between two major plates, the Pacific Plate and the Indo-Australian Plate. There are also two well-established minor plates, the Solomon Sea Plate and the South Bismarck Plate, plus a disputed minor plate (the North Bismarck Plate, not shown) immediately north of the South Bismarck Plate. Plate names are indicated without parentheses; geographic names for lands are in parentheses.

Two or possibly three minor plates lie between the Indo-Australian and Pacific Plates (fig. 1.1). The distinctness of the Solomon Sea Plate and South Bismarck Plate is well established, but that of a North Bismarck Plate is doubtful. Parts of the plate boundaries are marked by Late Cenozoic volcanism and by epicenters of recorded earthquakes, lying along lines displaced to the rear of the plate boundaries. The Solomon Sea Plate and the eastern part of the South Bismarck Plate meet at a submarine trench south of New Britain, where the former plate dives under the latter plate. The western part of the South Bismarck Plate meets the Indo-Australian Plate near the mountain ranges along the north coast of Papua New Guinea. Associated with these boundaries is the line of Quaternary volcanoes off the north coasts of New Britain and New Guinea, as well as a zone of earthquake epicenters. The Pacific Plate meets the Solomon Sea Plate in the southeast and the South Bismarck Plate in the northwest along a line that runs southwest of the Solomons and New Ireland and that is also associated with Quaternary volcanism and earthquake epicenters. The boundary between the South Bismarck Plate and postulated North Bismarck Plate runs through the Bismarck Sea and is marked by earthquake epicenters, which also mark the boundary between the Solomon Sea Plate and Indo-Australian Plate.

Northern Melanesian islands fall into two island arcs, each of which in turn includes both a younger and an older arc (fig. 1.2). The Bismarck Volcanic Arc is the chain of Quaternary volcanic islands and volcanoes that extends in the west from the Schouten Islands along New Guinea's north coast and along New Britain's north coast to Rabaul Volcano on New Britain's Gazelle Peninsula in the east. This young arc is paralleled by the remains of a much older (Paleocene) arc from New Guinea's Huon Peninsula extending east along south New Britain. Similarly, the Outer Melanesian Arc from Manus and the Admiralties in the west through New Ireland to the Solomons comprises both a Quaternary volcanic arc (including the Admiralty volcanoes, the islands east of New Ireland, Bougainville, and the New Georgia group of the Solomons) and a much older arc (including New Ireland, parts of Buka and Bougainville, and the Choiseul-Ysabel-Malaita-San Cristobal[2] chain of the Solomons). The younger and older arcs lie close to each other in places and are virtually superimposed in Bougainville.

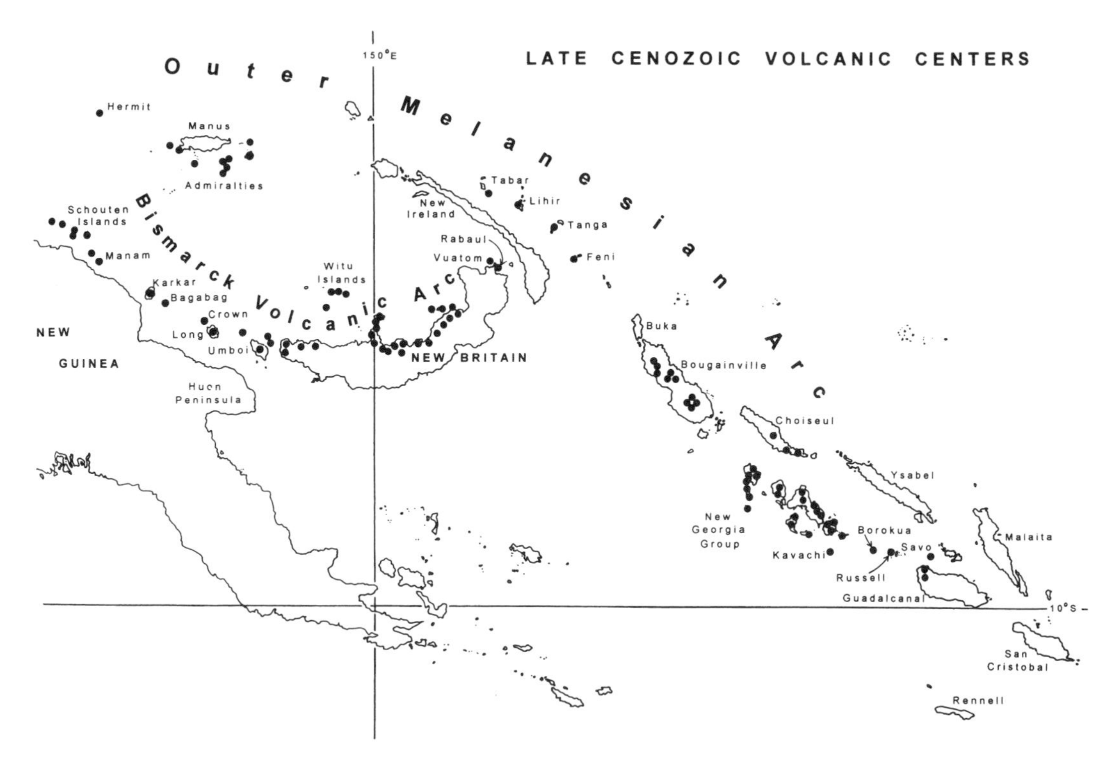

Fig. 1.2. Late Cenozoic volcanic centers of Northern Melanesia (●). The chain of centers from the Schouten Islands to Rabaul is called the Bismarck Volcanic Arc, whose western portion from the Schouten Islands to Bagabag belongs faunistically to the Papuan Region rather than to the Bismarcks. See text for further details (from Grover et al. 1959–62, Bain et al. 1972, Johnson & Davies 1972, Johnson et al. 1972, and Johnson 1976, 1979, 1981).

Volcanism and Rocks

Northern Melanesian islands consist mainly of volcanic products and limestone in varying proportions. At the one extreme, Rennell, Bellona, Gower, Three Sisters, Nissan, and Emirau are formed entirely of coral. At the opposite extreme, Long, Vuatom, Kulambangra, and Savo consist almost entirely of young, active or only recently extinct volcanoes. Among other islands, Buka is an especially simple combination, consisting of a raised Pleistocene coral reef in the northeast and Oligocene volcanic mountains in the southwest.

The oldest rocks in Northern Melanesia are Eocene. This does not necessarily mean that land has existed continuously since the Eocene, because Eocene islands may have been subsequently submerged, and because the Eocene rocks are submarine and have been subsequently raised to sea level. Volcanic activity has been concentrated in two phases separated by a Miocene period of limestone deposition: an older phase from the Eocene and Oligocene to the Early Miocene, and a Late Cenozoic phase from the Pliocene to the present. Islands with volcanism only in the older phase include New Ireland, New Hanover, Dyaul, Buka, Ysabel, Malaita, San Cristobal, and possibly St. Matthias. Both pre-Miocene and Late Cenozoic volcanism occurred on Bougainville, Manus, New Britain, Tabar, Choiseul, and Guadalcanal (but Late Cenozoic activity on the latter two islands was minor). The remaining volcanic islands noted in figure 1.2 experienced only Late Cenozoic volcanism.

2. Most of the distributional and taxonomic literature on Northern Melanesian birds was published before 1975. To facilitate use of the literature, throughout this book we refer to islands by their older names used in that literature. Since 1975, some European names of islands have become largely replaced by indigenous names—especially Lavongai, Makira, Mussau, Nggela, and Vanuatu instead of our older names New Hanover, San Cristobal, St. Matthias, Florida, and the New Hebrides, respectively. We also use older spellings for some islands that have retained their traditional names but with altered spellings—for example, Ghizo, Kolombangara, and Ranongga instead of our older spellings Gizo, Kulambangra, and Ganonga.

Figure 1.2 depicts Late Cenozoic volcanic centers in Northern Melanesia, which fall into three main groups. The most numerous group is the Bismarck Volcanic Arc, which includes volcanoes on the north coast of New Britain from Rabaul at the east end to Mt. Talawe and Mt. Tangi at the west end; islands north of New Britain such as Vuatom, Lolobau, and the Witu group; the islands between New Britain's west end and New Guinea (Ritter, Umboi, Sakar, Tolokiwa, Long, and Crown); and the westward continuation of this chain as islands along the north coast of New Guinea, belonging faunally to the Papuan region rather than to Northern Melanesia (islands from Karkar and Bagabag west to Vokeo in the Schouten Islands). A second group extends from the islands off the east coast of New Ireland (Tabar, Lihir, Tanga, and Feni) through several volcanoes on Bougainville to all the high islands of the New Georgia group, Borokua, the Russell group, Savo, parts of Choiseul, and western Guadalcanal. The third group includes the west end of Manus, islands south and southeast of Manus such as Rambutyo and Lou, and the Hermits west of Manus.

Volcanism has been a significant force in the evolution of habitat preferences and dispersal abilities of Northern Melanesian birds. Today at least a dozen volcanoes in the Bismarcks and three in the Solomons are considered active, and over 100 have been active in the Late Cenozoic. The creation of new habitats by formation of new land, and the defaunation of parts or the whole of existing islands (like the famous defaunation of Krakatau; Thornton 1996), are frequent. For example, the seventeenth-century explosion of Long Island defaunated Long, Crown, and possibly Tolokiwa (Diamond et al. 1989); the 1888 explosion of Ritter defaunated that island (Diamond 1974); the sixteenth-century explosion of Billy Mitchell defaunated about 200 km^2 of central Bougainville (Blake & Miezitis 1967, Zielinski et al. 1994, Briffa et al. 1998); the Rabaul explosion of around A.D. 600 defaunated Vuatom and Duke of York; and the Mt. Witori eruption of around 1500 B.C. defaunated several thousand square kilometers of central New Britain (Spriggs 1997). Hence at any given moment there is much habitat in a subclimax condition, providing opportunities for bird species of such habitats.

In addition to rapid bird succession dependent on such habitat succession, there is also more slowly continuing bird succession once the forest attains climax structure, as forest bird species specialized for dispersal first occupy forested volcanic islands and are eventually joined or replaced by forest bird species of lower dispersal ability. Bird succession is well illustrated by differences between the avifaunas of the recently defaunated Long and the older Umboi, two nearby forested mountainous Bismarck islands of similar area. Long's forest contains some highly vagile bird species lacking or occurring only as vagrants on Umboi (e.g., the monarch flycatcher *Monarcha cinerascens*, whistler *Pachycephala melanura*, honey-eater *Myzomela pammelaena*, and white-eye *Zosterops griseotinctus*), while Umboi's forest contains many poorly vagile species not present on Long (e.g., the pigeon *Henicophaps foersteri*, cuckoo *Centropus ateralbus*, monarch flycatcher *Monarcha verticalis*, and honey-eater *Philemon cockerelli*; Diamond et al. 1989). The extreme among vagile species favored by volcanic eruptions is the honey-eater *Myzomela sclateri*, which is confined to recently defaunated volcanoes and some nearby coral islets in the Bismarck Volcanic Arc.

Another potential significance of volcanism is for dating the origins and estimating rates of evolution of some Northern Melanesian endemic taxa confined to Late Cenozoic volcanic islands. The two *Myzomela* honey-eater species of Long, *M. sclateri* and *M. pammelaena*, have diverged significantly in size since they first met on Long following its defaunation three centuries ago (Diamond et al. 1989). Late Cenozoic volcanic islands with quite distinct endemic subspecies of birds include Lihir, Feni, and islands of the New Georgia group. The latter group has four endemic allospecies of the white-eye superspecies *Zosterops* [*griseotinctus*], two endemic allospecies of *Monarcha* flycatchers, and one of a *Gallirallus* rail. (Throughout this book, square brackets denote a superspecies name). Within the New Georgia group, Kulambangra, a stratovolcano whose form suggests that it is relatively young, has two endemic species, the flycatcher *Phylloscopus amoenus* and the white-eye *Zosterops murphyi*. It will be of interest to date the origins of these islands within the Late Cenozoic.

While this discussion has emphasized volcanism because it has contributed so much of the land area of Northern Melanesia, there is also much limestone. Miocene limestone deposits in the form of karst are extensive on northwestern and central New Ireland, large parts of New Britain, and Bougainville's Keriaka Plateau. Low-lying coral atolls that have not been uplifted include Ninigo,

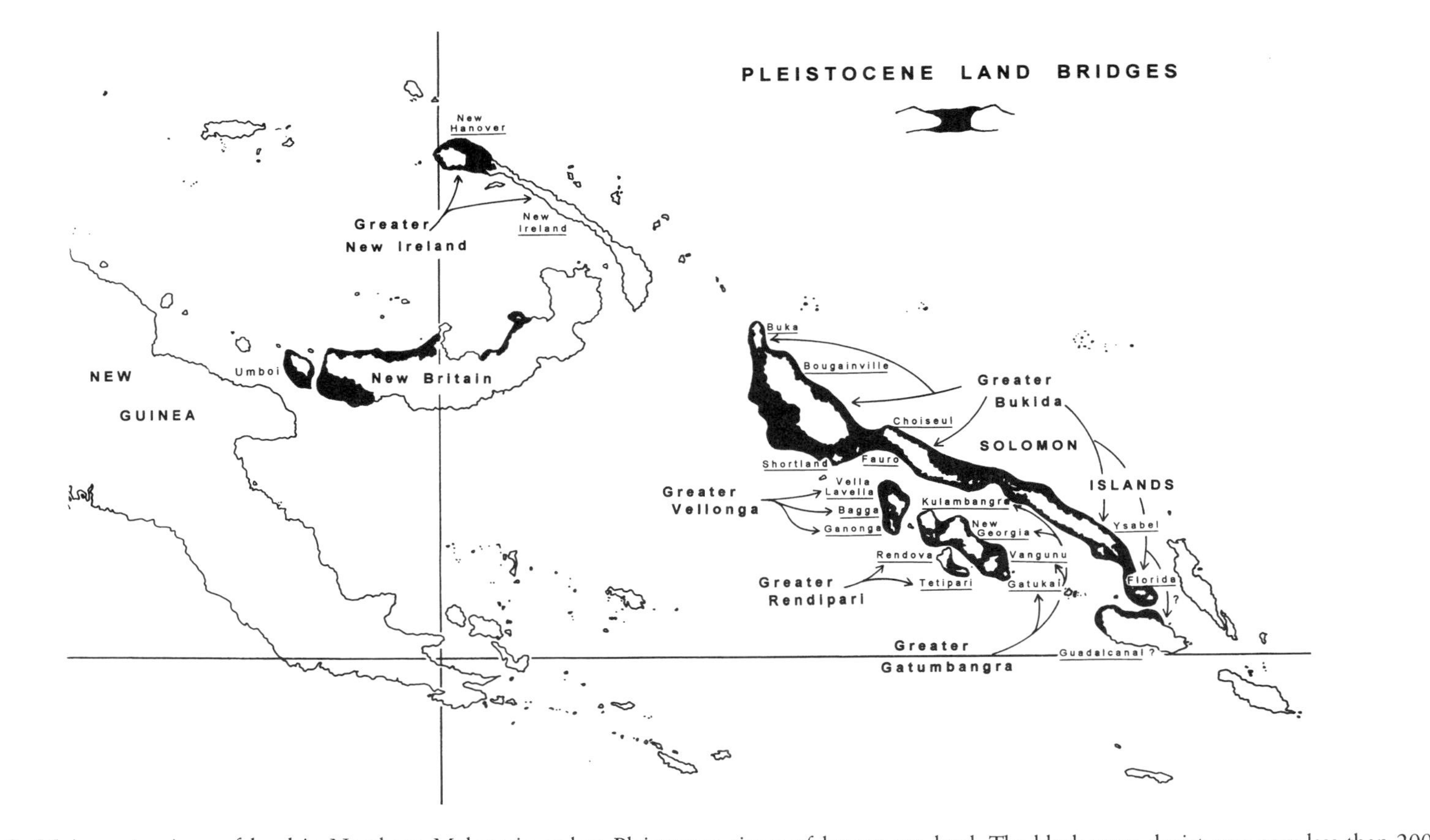

Fig. 1.3. Major extensions of land in Northern Melanesia at late-Pleistocene times of lowest sea level. The black areas depict seas now less than 200 m deep, and hence are areas that would have been converted into land by maximal late-Pleistocene lowering of sea level. Enlarged Pleistocene islands are named "Greater . . .", and the names of their modern island fragments produced by the post-Pleistocene rise of sea level are underlined. Note that the island chain from Buka and Bougainville through Shortland, Fauro, Choiseul, and Ysabel to Florida and possibly Guadalcanal would have been joined as Greater Bukida; the chain from Kulambangra, Wana Wana, and Kohinggo through New Georgia to Vangunu and Gatukai would have been joined as Greater Gatumbangra; Rendova would have been joined to Tetipari as Greater Rendipari; Vella Lavella would have been joined to Bagga and Ganonga as Greater Vellonga; New Hanover would have been joined to New Ireland as Greater New Ireland; and Umboi would almost have been joined to New Britain.

Sikaiana, Ontong Java, Kilinailau, Tauu, and Nukumanu. Islands consisting wholly of uplifted coral platforms include Rennell, Bellona, Nissan, Emirau, and Three Sisters. Rennell is a notable example of a former coral atoll that has been uplifted over 100 m vertically to form an island with steep cliffs and with a large lake now occupying the site of the former lagoon. Other wholly coral islands are Wuvulu, Tench, Tingwon, and Gower, plus Masahet and Mahur in the Lihir group and Boang in the Tanga group. Buka, Duke of York, Dyaul, St. Matthias, Santa Anna, Santa Catalina, and Ugi are mostly coral but have small residues of old volcanoes. Most of the large islands have significant areas of limestone.

Land Bridges

Figure 1.3 depicts for Northern Melanesia the 200-m bathymetric contour, which is close to the 100-m contour. In the late Pleistocene, when much ocean water was locked up in glaciers and polar ice caps, sea level is variously estimated to have dipped to between 120 and 150 m below its present level (Chappell & Shackleton 1986, Chappell & Polach 1991, Gallup et al. 1994). Hence modern islands lying within the same 100-m bathymetric contour are likely to have been joined in a single, late-Pleistocene island, while islands now separated by straits more than 200 m deep would not have been so joined.

Four factors complicate this interpretation. First, there has also been tectonic change, so that it is not the case that sea level simply rose and fell over an unchanging land. For instance, Tetipari Island has been rising at an estimated rate of nearly 4 m per 1000 years in the late Pleistocene. Second, deposition of volcanic ejecta, and redeposition of volcanic material from mountains, have widened some islands in the late Pleistocene, such as southern Bougainville. Third, there is some uncertainty about the lowest value of sea level reached within the last 200,000 years; a widely quoted estimate is 130 m below present sea level (Chappell & Shackleton 1986, Chappell & Polach 1991). Finally, that low was not maintained constant for long times. Instead, after sea level rose from its previous low at around 135,000 years ago to reach a high (6 m above the present stand) around 125,000 years ago, it fell again to fluctuate between 30 and 65 m below present level for most of the next 100,000 years, dropped to its low of −130 m around 18,000 years ago, then quickly rose to its present level (fig. 1.4).

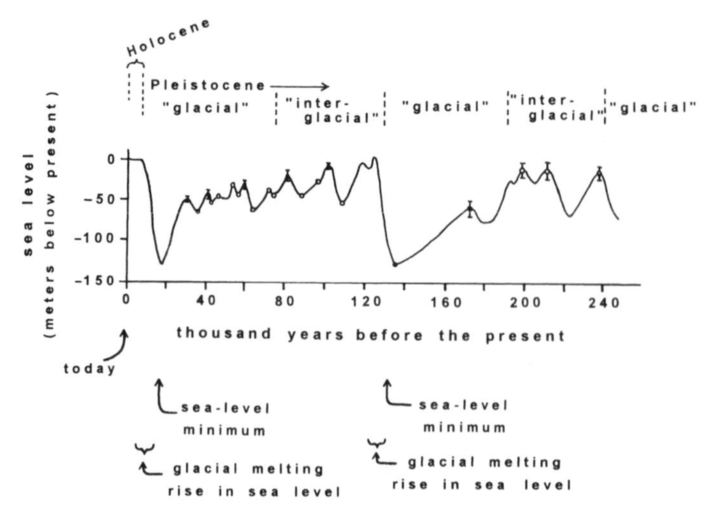

Fig. 1.4. Calculated fluctuations in sea level for the last 250,000 years. Based on fig. 1C of Chappell's & Shackleton's (1986) analysis of dated raised coral reef terraces on New Guinea's Huon Peninsula. An ordinate value of 0 indicates sea level identical to the present; a positive value indicates a level higher than the present (e.g., the brief value of + 6 m around 125,000 years ago); and a negative value indicates a level lower than the present (e.g., the values through most of the Pleistocene).

The largest effect of this late-Pleistocene sea-level drop was in the Solomon Islands, where Buka, Bougainville, Shortland, Fauro, Choiseul, Ysabel, and Florida were joined in a single large island that we call "Greater Bukida." In the absence of tectonic changes, Greater Bukida would have been separated from Guadalcanal by a narrow channel (Sealark Channel) up to 2 km wide, but this channel might have been erased by tectonic changes. Other coalescences in the Solomons were that Kulambangra, Wana Wana, Kohinggo, New Georgia, Vangunu, and Gatukai were joined in an island that we call "Greater Gatumbangra"; Rendova was joined to Tetipari (Greater Rendipari); Vella Lavella, Bagga, and Ganonga were joined (Greater Vellonga); and Greater Rendipari and Greater Gatumbangra were either joined or else separated by only a narrow channel. The main changes in the Bismarcks were that New Hanover was joined to New Ireland, and Umboi was either joined to New Britain or else separated from it by only a narrow channel.

All these islands that we have mentioned as having been joined are separated today by waters only 10–60 m deep. Thus, they would have been joined not only during the brief sea-level low around 18,000 years ago, but for most of the last glacial period as well. Similar cycles of alternate island joining and fission would have been repeated during the many previous Pleistocene cycles of sea-level rise and fall. However, other islands are surrounded by seas much deeper than 200 m—for example, more than 3000 m in the case of Rennell and Bellona Islands, and more than 2000 m in the case of the channel between New Britain and New Ireland. These deep-water islands—including not only Rennell and Bellona, but also Mono, Savo, Malaita, San Cristobal, Ugi, Simbo, the Russells, Dyaul, Tabar, Tanga, Lihir, Feni, and St. Matthias—probably remained as separate islands, at least during the late Pleistocene.

Elsewhere in the world, such late-Pleistocene land bridges had obvious effects on the distributions of flightless mammals, as Wallace noticed more than a century ago in discussing the distributions of rhinoceroses, tigers, and orang-utans on islands of the Sunda Shelf. Many tropical bird species are poor at dispersing over water, and so one might expect many of them, as well as flightless mammals, to be confined to modern fragments of a single Pleistocene island. In fact, such "land-bridge relict" distributions have been noted for bird species of the New Guinea region (Diamond 1972a), the Neotropics (MacArthur et al. 1972), the Sunda region (Diamond & Gilpin 1983), and the New Zealand region (Diamond 1984e). We previously noted that 13 bird species endemic to the Solomons are now confined to the modern island fragments of Pleistocene Greater Bukida (Diamond 1983). In chapter 32 we note similar examples for each of the other Pleistocene coalesced islands of Northern Melanesia (e.g., species 24, 70, 97C, and 111B on maps 3, 14, 22, and 23 respectively). In addition, modern island fragments of expanded Pleistocene islands share many subspecies and have few endemic taxa, whereas islands that have always been separate tend to have many more endemic taxa. Before the last phase of the Pleistocene, other fluctuations in sea level may have joined and severed the same islands repeatedly and may have joined other islands, but it is only the most recent set of land bridges whose effects can be deduced with confidence from modern bird distributions (chapters 32 and 33).

Summary

Northern Melanesian islands have never had a land connection to New Guinea or any continent and have never been closer to New Guinea than they are now. Hence the biota must have arrived over water. The oldest rocks are Eocene, but many of the islands are Late Cenozoic volcanoes, and others are coral or uplifted coral. Volcanism has defaunated islands and selected for some highly vagile bird species to recolonize such islands. Pleistocene drops in sea level periodically joined islands now separated by shallow seas. Those Pleistocene land bridges account not only for such islands now sharing subspecies, but also for shared populations of bird species unable to disperse across water.

2

Climate

To a casual observer accustomed to the marked and predictable seasonality of the temperate zones—for example, to American soldiers fighting in Melanesian jungles during World War II—it might at first seem that Northern Melanesia's climate could be described as "equatorial, tropical, hot, wet, and aseasonal." By comparison with the temperate zones, this statement is approximately correct, but the climate is actually more complex, and the complexities are relevant to understanding bird distributions. Our account of climate is derived from Brookfield and Hart (1966), Scott et al. (1967), Whitmore (1974), and McAlpine et al. (1975).

All of Northern Melanesia lies between latitudes 1°S and 11°S (fig. 0.2 on p. xiv). Thus, seasonal variation in temperature and in daily hours of sunlight is modest. Monthly means for daily temperature minima or maxima vary seasonally at any given site by only up to 1°C. At all available weather stations, all of which are located in the coastal lowlands, the mean maxima fall between 29 and 31°C, and the mean minima fall between 22 and 24°C. Presumably, temperatures decrease with elevation by about 2°C for every 300 m, as elsewhere in the equatorial tropics.

Mean annual rainfall for all available stations falls between 1730 and 6650 mm, with most stations having values between 3000 and 4000 mm. Again, all stations are in the lowlands, and rainfall may be higher in the mountains. The wettest sites are the south coasts of New Britain (5000–6650 mm) and of Guadalcanal, San Cristobal, and Malaita (4600–5000 mm), while the driest sites (1730–2500 mm) are the northeastern part of New Britain's Gazelle Peninsula, the north coast of Guadalcanal, and possibly Long.

Seasonality of rainfall varies among sites. A widespread pattern is for a single maximum of rainfall from around December to March and a minimum from around May to September. This pattern is especially marked at the driest sites. The pattern is reversed at the wettest sites, where the rainfall maximum is from May to September (see fig. 2.1 for an example). There is a gentle double maximum of rainfall on Manus and the New Georgia group, while some other sites show little seasonal variation in rainfall. However, "dry season" in Northern Melanesia does not mean no rain but just less rain. For example, mean rainfall for the driest month at any Bismarck site is 74 mm (September at Rapopo on New Britain's Gazelle Peninsula), and at any Solomon site, it is 61 mm (July at Kukum on the north coast of Guadalcanal).

Not only is seasonality of rainfall modest by world standards, but so is annual variability. Coefficients of variation for annual rainfall at all sites fall between only 7 and 31%, with the most seasonal sites (Honiara and Rabaul) having the highest coefficients of variation (31 and 29% respectively) and most other sites having values of 13–21%.

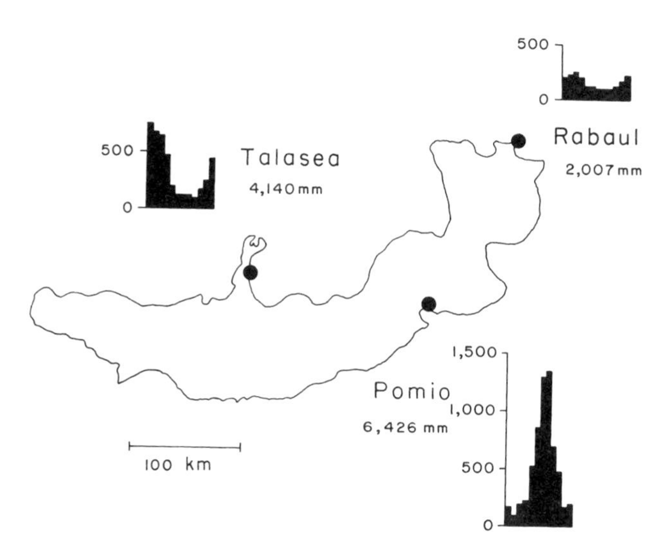

Fig. 2.1. Geographic variation in rainfall patterns within New Britain. For each of three stations, the histogram shows the monthly rainfall in millimeters from January (left) to December (right). Total annual rainfall is given below each station's name. The southern watershed (e.g., Pomio) is wetter than the northern watershed (e.g., Talasea), while the Gazelle Peninsula (e.g., Rabaul) is especially dry. Seasons are reversed between the watersheds: May–August are the wettest months in the south but the driest months in the north. Data are from Brookfield & Hart (1966).

This geographic variation in annual rainfall and in its seasonality mainly reflects geographic variation in rain during the months May–September. At the wettest sites those are the wettest months, with up to 1370 mm monthly of rain, while at the driest sites they are the driest months, with as little as 61 mm. Geographic variation in December—March rainfall is less marked.

These rainfall patterns may be understood in terms of the prevailing wind patterns. From May to September the strong and persistent trade winds blow constantly from the southeast, while more variable winds from the northwest (the so-called northwest monsoon) prevail from December to March. Geographic variation in exposure to the trades as a result of local topography is viewed as a major cause of geographic variation in May–September rainfall, total annual rainfall, and monthly distribution of rain.

This interpretation accounts for the striking reversal of rainfall patterns between the northern and southern watersheds on the largest Northern Melanesian island, New Britain, and also on the second largest Solomon island, Guadalcanal. Both of these islands have an east–west principal axis, are directly exposed to the south, and have high east–west mountains that shield the north coasts from the trade winds and create a rainshadow. Hence on both islands the windward south coast receives more than double the annual rainfall of the leeward north coast. During May–September the south coasts of these two islands are relatively wet, the north coasts dry, while the reverse is true in December–March (fig. 2.1).

Two anecdotal examples illustrate these striking effects. Between 1 and 13 July 1965 a rainstorm brought 3124 mm of rain (this number is not a misprint!), which equals 3.1 m or 123 inches, to Avuavu on the south coast of Guadalcanal, while Honiara, only a few dozen kilometers away on the north coast, received "only" 177 mm or 7 inches. In July 1969, one of us (J.D.) walked inland from New Britain's north coast, which was parched at the peak of the dry season. It began to rain constantly as Diamond moved into the southern watershed, and the rains did not stop until he returned to the north coast. The most marked north/south (or northwest/southeast) differences in rainfall arising from local topography are those just noted for New Britain and Guadalcanal, but such differences exist in more modest form for New Ireland, Bougainville, San Cristobal, and probably other mountainous islands of Northern Melanesia. On these islands those rainshadow effects have created local areas with natural nonforest vegetation, in which (as we discuss in the next chapter) Northern Melanesia's few endemic taxa of open-country birds may have evolved.

The remaining climatic feature to mention is cyclones, of which 15 were recorded in the Solomons between 1966 and 1972 during the months of November, December, and January. Most cyclones in Northern Melanesia have been confined to the southeastern Solomons, but some have extended as far northwest as Bougainville. For example, Cyclone Annie in November 1967 wrecked much of Ontong Java, Kulambangra, and Vella Lavella (Whitmore 1974); forests in areas of Florida, Savo, and Bougainville were flattened in 1972 by Cyclone Ida; and a cyclone destroyed gardens and forest on Rennell around 1905.

Cyclones are significant to the Northern Melanesian avifauna in several respects. First, cyclone damage is the most important ecological factor affecting intersite variation in forest composition throughout much of the Solomons (Whitmore 1974). Second, the immediate removal of tall trees and the subsequent development of dense secondary growth may cause local extinctions of some bird populations. For example, villagers on Florida interviewed in 1974 (Diamond, unpublished data) attributed the recent disappearance of the pigeon *Reinwardtoena crassirostris* to cyclone destruction of trees on whose fruits it depended. Finally, cyclones and other storms have brought birds to the Solomons from Australia and also driven birds from one Solomon island to another. Examples include large flocks of the cormorant *Phalacrocorax melanoleucos* (Diamond 2001) and pelican *Pelecanus conspicillatus* (Laird 1954, Cain & Galbraith 1956) driven to many islands from Australia by a storm in 1952, the cormorant *Phalacrocorax carbo* driven to Rennell Island by Cyclone Nina in 1993, and flocks of the parrot *Chalcopsitta cardinalis* driven to the outlying atoll of Ontong Java from the central Solomons by Cyclone Ida in 1972 (Bayliss-Smith 1973).

Summary

Since Northern Melanesia lies near the equator, seasonal variation in temperature, day-length, and rainfall is modest. Annual rainfall at most reported lowland sites is 3000–4000 mm, with only modest variation between years. Geographic variation in rainfall and in its seasonality depends especially on the interaction between local topography (e.g., mountains creating rainshadows) and two opposite and alternating wind systems. The southeastern Solomons are exposed to frequent cyclones, which have brought bird vagrants.

3

Habitats and Vegetation

In this chapter we introduce Northern Melanesia's major habitat types and the taxonomic composition of its vegetation.

Habitats

Northern Melanesian vegetation is discussed in detail by Whitmore (1966, 1969a, 1974, 1981, 1984), Scott et al. (1967), Percival and Womersley (1975), Cain and Galbraith (1956), Schodde (1977), and Lees et al. (1991), from whom the following account is derived. The climax vegetation in most of Northern Melanesia is rainforest. There is no natural savanna like that in South New Guinea, and no extensive areas of natural grassland, which is confined to small marshes, edges of streams and lakes, and (on New Britain and Guadalcanal) rainshadow areas. However, even under natural conditions much of the forest is in a subclimax state because of frequent deforestation by earthquakes, landslides, volcanism, and, especially in the Solomons, cyclones.

Most areas of nonforest habitat existing today result from human activity: coconut (and, on New Britain, oil-palm) plantations, secondary growth on abandoned garden and logging sites, and *Imperata–Themeda* grassland maintained by fires. The open grassy plains in the rainshadow area of northern Guadalcanal may also be maintained by human-lit fires, since shrublands develop there where fire is excluded. There is an endemic subspecies of grassland quail (*Turnix maculosa salomonis*) confined to those Guadalcanal plains, and six endemic allospecies or subspecies of grassland finches (*Lonchura [castaneothorax] melaena, L. [spectabilis] s. spectabilis, L. [s.] forbesi, L. [s.] hunsteini*) and quail (*Coturnix chinensis lepida, Turnix maculosa saturata*) are virtually confined to New Britain and New Ireland. These possibly old taxa suggest that at least small areas of natural grassland may have existed in rainshadow areas of those islands for a long time, as also discussed for New Guinea by Rand (1941b).

Many Northern Melanesian islands are high enough to exhibit significant changes in forest structure and composition with altitude. Ten Solomon and eight Bismarck islands exceed 900 m in elevation, and Bougainville, Guadalcanal, New Britain, and New Ireland reach 2,590, 2,450, 2,410, and 2,400 m, respectively (see tables 9.1 and 9.2). In the lowlands primary rainforest is 30–45 m tall (compared to 45–60 m in Malaysia and Borneo) and has large buttresses, large leaves, and large, woody climbers. With increasing elevation the rainforest decreases in height (e.g., to 25–30 m at an elevation of 700–1,200 m on Bougainville) and has smaller leaves, more moss, and few woody climbers. At higher elevations, within the zone shrouded by clouds for part of most days, one enters cloud forest, also called "mossy forest" or "upper montane forest." This forest is stunted (12–17 m tall), and its trunks and branches are gnarled and heavily draped with moss and ferns. Some areas of mountain forest have abundant bamboo, pandanus, or emergent palms (*Gulubia*). Unlike New Guinea, no Northern Melanesian island is high enough to reach the tree line or support alpine grassland, hence there is no alpine avifauna at all.

The lower limit of cloud forest is considerably lower on Northern Melanesian islands than on New Guinea, and is lower on lower islands or on mountains near the coast than on higher islands or on mountains far inland. For example, cloud forest descends to only 2,100–2,400 m on the central range of New Guinea (which is 5,000 m high) but descends to about 1,200 m on the highest Solomon islands (Bougainville [2,590 m high], Guadalcanal [2,450 m high], and Kulambangra [1,770 m high]), to about 650 m on Vangunu (1,125 m high) and San Cristobal (1,040 m high), and to about 600 m on Gatukai (890 m high). These differences are part of the usual compression of altitudinal zones on lower or more exposed mountains, well known in the Alps and elsewhere in the world and called the "Massenerhebung effect."

In addition to this altitudinal variation in forest structure, there is horizontal variation. The coasts of main islands, as well as coral atolls and raised reefs, support strand forest whose composition varies little throughout the Indo-Pacific region but which is better preserved and more extensive in Northern Melanesia than in much of Indonesia and Southeast Asia. Dominant tree species include *Barringtonia asiatica, Calophyllum inophyllum, Casuaria equisetifolia,* and *Terminalia catappa*. Some coasts support mangrove, dominated by species of *Bruguiera* and *Rhizophora,* and *Ceriops togal, Lumnitzera littorea*, and the palm *Nypa fruticans*. Freshwater swamp forest near the coast is dominated by *Terminalia brassii*. Ultrabasic soils on southern Ysabel and southeastern Choiseul support a low, species-poor forest in which *Casuarina papuana* and *Dillenia crenata* are abundant but which has few endemics, unlike the highly distinctive forests on ultrabasic soils of New Caledonia. The tallest and most species-rich lowland forests occur inland on well-drained soils.

The largest lake in Northern Melanesia, and the one supporting the greatest diversity of water birds, is Rennell's Lake Tegano. Next richest in water birds are Lake Buan of Umboi and lakes Dakataua, Hargy, and Namo of New Britain. Volcanic crater lakes include the large Lake Wisdom of Long and other much smaller ones: lakes Apalong, Bono, Molonpot, Pung, and Yokon on Umboi; lakes Billy Mitchell and Loloru on Bougainville; lakes Waipiapia and Wairafa on Santa Ana; and the unnamed and unexplored lake on Sakar's summit. The larger islands (especially New Britain) have rivers and marshes.

Taxonomic Composition of the Vegetation

Northern Melanesian floristics have been analyzed by van Balgooy (1960, 1971), Whitmore (1969b, 1981), Thorne (1969), and van Balgooy et al. (1996). The flora is an impoverished sample of the rich tropical flora of Malaysia, Indonesia, and New Guinea, without a strongly differentiated old-endemic element of its own. For example, van Balgooy (1971) listed about 1,400 phanerogam genera known from New Guinea but only 654 from the Solomons and 632 from the Bismarcks. Whitmore (1969b) ascribed 710 genera and 1,750 species of phanerogams and 290 species of pteridophytes to the Solomons; more recent studies have increased those numbers only slightly. Census plots of 0.625 ha support only 30–53 species of trees of girth greater than 0.3 m on the Solomon island of Kulambangra, less than half of the corresponding number for Borneo or the Malay Peninsula (compare fig. 2.7 of Whitmore 1974 with fig. 2.28 of Whitmore 1990).

The Bismarcks and Solomons lack or nearly lack many important plant genera or families of New Guinea and/or Indonesia, such as the entire family Dipterocarpaceae. Among the dominant tree genera of the New Guinea mountains, many (e.g., *Araucaria, Castanopsis, Engelhardtia, Papuacedrus, Phyllocladus,* and *Quercus*) are absent from Northern Melanesia, while *Agathis* and *Nothofagus* are confined within Northern Melanesia to New Britain. The huge genus *Eucalyptus,* with hundreds of species in Australia and many in New Guinea, is also confined in Northern Melanesia to New Britain. Northern Melanesia's relative poverty in plants parallels that in birds and presumably results from limitations imposed by island area and geographic isolation.

Northern Melanesia has few endemic plant genera: about three phanerogam genera for the Solomons and one for the Bismarcks, each consisting of only a single species. Instead, the overwhelming majority of the phanerogam genera are shared with New Guinea: 627 of the 632 genera in the Bismarcks, 637 of the 654 in the Solomons, according to van Balgooy's (1971) tabulation. (Remember that New Guinea has about 800 additional genera lacking in the Bismarcks and Solomons.) The Bismarcks and Solomons share most of their genera (501) with each other. About 31 or 29% of Bismarck or Solomon genera, respectively, are pantropical in distribution; 30 or 29% are paleotropical (including Australia but excluding the Americas), with or without ranges extending into Africa; and 30 or 31% have various distributions in tropical Asia, Indonesia, and New Guinea but not Australia or Africa. Genera centered on Australia and New Guinea constitute only 6% of the Bismarck or Solomon flora; genera centered on Australia, less than 1%; and Subantarctic-Pacific genera (e.g., genera such as Nothofagus, centered on extratropical South America, New Zealand, Australia, and New Guinea), also less than 1%.

Thus, to botanists Northern Melanesia is an impoverished extension of the Indomalayan region, which is known to botanists as Malesia and which botanically includes New Guinea. The sharp distributional break for plants lies to the south, between New Guinea and Australia. To ornithologists and mammalogists, however, New Guinea and Northern Melanesia belong not to the Indomalayan region but to the Australian region, and the distributional break lies instead to the west in Wallacea, the islands between the Asian and Australian/New Guinean continental shelves. For example, species centered on Australia and New Guinea, or on Australia, constitute 27% of the Northern Melanesian avifauna, much higher than the corresponding figure of only about 7% for plants. That is, there is a discordance in Northern Melanesia's as well as New Guinea's biogeographic affinities: plants (and most herbivorous insect groups; Gressitt 1982) have their affinities with the Indomalayan region to the west, while birds and mammals (and some freshwater insect groups such as water-striders; Polhemus & Polhemus 1993) have their affinities with Australia to the south. This discordance may have arisen because plant distributions and those of most terrestrial insects are more limited by climate than are distributions of homeothermic vertebrates (Mayr 1944c, 1953, Diamond 1984c). Australia lies much closer to Northern Melanesia and New Guinea than does tropical Southeast Asia, but wet tropical Southeast Asia is climatically more similar to Northern Melanesia and New Guinea than is the largely arid Australian continent. Hence Northern Melanesia and New Guinea ended up with mostly Indomalayan plants but with Australian birds.

Summary

Northern Melanesia's climax vegetation consists mostly of rainforest, whose structure changes with altitude. Other habitats include strand forest, mangrove, anthropogenic grassland and secondary growth, small areas of seemingly natural grasslands, and limited numbers of lakes, rivers, and swamps. Floristically, Northern Melanesia is an impoverished extension of the Indomalayan (Malesian) region, with low endemism and the absence of many plant groups dominant in New Guinea. Whereas New Guinea's and Northern Melanesia's birds and mammals are most closely related to those of Australia, their plants are instead most closely related to those of tropical Southeast Asia.

4

Terrestrial Vertebrates Other Than Birds

Northern Melanesia's nonavian, terrestrial vertebrates exhibit striking parallels as well as contrasts with its avifauna. In this chapter we briefly summarize these other classes for two reasons: to permit comparisons with conclusions drawn from the avifauna; and because diversity of native predators is one of the important factors determining the susceptibility of native bird species to extinction upon arrival of humans (chapter 7). Bear in mind throughout this chapter that knowledge of these other vertebrate faunas is less complete than knowledge of the corresponding avifaunas; new species are still being discovered, older specimens are being re-identified, the systematics of many groups are still poorly understood, and the superspecies concept has been little used.

Mammals

New Guinea today has about 189 native mammal species (table 4.1), of which two are monotremes (echidnas) and the rest are nearly equally divided among marsupials, bats, and murid rodents. At this level, New Guinea's mammal fauna is similar to that of Australia, with which New Guinea was joined as a single continent during Pleistocene times of low sea level. New Guinea's largest extant species is a kangaroo (*Macropus agilis*) weighing up to 26 kg. In the Pliocene and Pleistocene New Guinea supported larger species of now-extinct marsupials: the Tasmanian wolf, *Thylacinus cf. cynocephalus*, which survived in New Guinea until a few thousand years ago; at least five species of diprotodonts (rhinoceros-like herbivores) up to the size of a cow (ca. 400 kg); and about six species of kangaroos (*Dendrolagus, Protemnodon, Thylogale, Watutia*) weighing up to about 60 kg (Flannery et al. 1983, 1989, Flannery & Plane 1986, Flannery 1990, 1992, 1994). New Guinea's extinct megafauna is thus much less diverse than the rich extinct megafauna of Australia and may have coexisted with humans for longer (Flannery 1990, 1994).

Among the recent native mammals of Northern Melanesia (table 4.1), there are no monotremes. However, among bats, all six of New Guinea's families, 62% of the genera, and 46% of the species are represented (Bonaccorso 1998). Northern Melanesia also has 22 endemic bat species confined to three of the six families, plus two endemic genera (*Melonycteris* and *Anthops*). There are four additional nonendemic bat species absent from New Guinea itself and shared instead with the New Hebrides, Wallacea, or Southeast Papuan Islands. As a result, the total number of bat species in Northern Melanesia (ca. 58) is only modestly fewer than that on New Guinea (ca. 69): 29 species of flying foxes (family Pteropodidae), six sheath-tailed bats (Emballonuridae), two horseshoe bats (Rhinolophidae), nine leaf-nosed bats (Hipposideridae), 11 vespertilionid bats (Vespertilionidae), and one free-

Table 4.1. Comparison of recent native nonmarine mammal species of New Guinea and Northern Melanesia.

	Total species in New Guinea[a]		Total species in Northern Melanesia[b]		New Guinea species in Northern Melanesia[c]	
Mammal	Species	Genera	Species	Genera	Species	Genera
Monotremes	2	2	0	0	0	0
Marsupials	61	26	6	6	6	6
Bats	69	29	58	21	32	18
Rodents	57	23	20	6	6	5
Total	189	80	84	33	44	29

Sources: for Northern Melanesia, Flannery 1995 (see also Laurie & Hill 1954, Bonaccorso 1998) plus unpublished comments by T.J. Flannery, J.E. Hill, and J.D. Smith; for New Guinea, Flannery (1990) (see also Laurie & Hill 1954, Menzies & Dennis 1979, Ziegler 1982, Menzies 1991, Flannery et al. 1996, and Bonaccorso 1998).

[a]Total numbers of resident mammal species and genera on New Guinea (excluding species confined to offshore islands).

[b]Numbers resident in Northern Melanesia.

[c]Northern Melanesian species or genera among those occurring on New Guinea. (Some Northern Melanesian species are endemic or have arrived from elsewhere, and thus do not occur on New Guinea.) Species introduced to the New Guinea region in modern times are excluded.

tailed bat (Molossidae). One or more bat species occur on all Northern Melanesian islands; the largest two Bismarck islands (New Britain and New Ireland) have 30 species each, and the seven largest Solomon islands have 15–23 species each.

Marsupials present a different picture. All six of Northern Melanesia's marsupial species (the bandicoot *Echymipera kalubu*, the phalangers *Phalanger orientalis* and *Spilocuscus maculatus*, the glider *Petaurus breviceps*, and the kangaroos *Dendrolagus matschiei* and *Thylogale browni*) are shared with New Guinea; not a single one is endemic. Four of the six marsupials are confined to the Bismarcks, one (*Thylogale browni*) also reached the Solomon island of Buka, and one (*Phalanger orientalis*) is also widespread in the Solomons. New Guinea's marsupial fauna is far richer: Northern Melanesia shares only 23% of New Guinea's genera and only 10% of New Guinea's species.

Northern Melanesia's murid rodent fauna is not quite as poor: 20 species, compared to 57 on New Guinea. Six of those 20 species are shared with New Guinea, and the other 14 are endemic; five belong to an endemic genus confined to the Solomon Islands (*Solomys*), three to an endemic subgenus confined to the Solomon island of Guadalcanal (subgenus *Cyromus* of genus *Uromys*). Thus, endemism is highest in rodents (70%), intermediate in bats (38%), and lowest in marsupials (0%). The endemic rats include large species over 1 kg in weight. They are ecologically diverse and include aquatic, terrestrial, and forest canopy species.

It is entirely as expected that the ratio of Northern Melanesian to New Guinea species is highest for bats (0.84), intermediate for rodents (0.35), and lowest for marsupials (0.10). We expect bats to be most successful at dispersing over water (by flying), rodents to be less successful (by rafting), and marsupials to be least successful. Nevertheless, the identity of those few marsupials that occur in Northern Melanesia presents an obvious puzzle. Few Australian and New Guinea marsupials have been demonstrably successful at unassisted overwater colonization anywhere. If marsupials had reached Northern Melanesia unassisted, one would expect them to have done so long ago, only at long intervals, and to have differentiated into highly distinct endemic taxa as a result of colonizing an environment devoid of ecological equivalents. This reasoning proves correct for Celebes and the Moluccas (Flannery 1995), which have some distinctive endemic species and even genera of marsupials. Northern Melanesia's rodents present a similar puzzle, in that most are the expected distinct endemics but six are similar to their New Guinea conspecifics.

The likely explanation of these puzzles emerges from recent identifications of mammal bones at archaeological sites (Flannery et al. 1988, Flannery & Wickler 1990, Flannery & White 1991, Flannery 1995). Excavated levels on New Ireland predating the arrival of humans, who reached the island by around 33,000 BP (years before present), contain many species of bats and two rodents but no marsupials. The fauna was then transformed by a series of introductions of New Guinea species: the phalanger *Phalanger orientalis* by around 10,000 years ago, the wallaby *Thylogale browni* around 7,000 years ago, the rats *Rattus praetor* and *R. exulans* along with domestic pigs and dogs several thousand years ago, and the phalanger *Spilocuscus maculatus* in 1929. Excavations on other Northern Melanesian islands extend these findings. The wallaby *T. browni* was also introduced to Buka and Lihir but subsequently disappeared and no longer occurs on these islands; it persists on New Britain, New Ireland, and Umboi. *Rattus praetor*, *R. exulans*, or both were introduced to Buka, Nissan, and Tikopia a few thousand years ago, while the phalanger *Spilocuscus (maculatus) kraemeri* and the bandicoot *Echymipera kalubu* were introduced to Manus around 13,000 years ago (Spriggs 1997).

Conversely, the archaeological sites also reveal extinctions of native mammal populations that had arrived before humans and disappeared following human arrival. The endemic rat species *Rattus sanila* became extinct on New Ireland several thousand years ago, probably due to competition from the ecologically similar *R. praetor* that arrived then. At least two endemic rodent species disappeared on Buka, while other endemic rodents have undergone drastic range contractions on Bougainville, Guadalcanal, Malaita, and San Cristobal but survived until modern times. The two rodents that disappeared prehistorically on Buka and the two that disappeared within the last century on Guadalcanal were large and probably terrestrial, making them especially vulnerable to humans and their introduced mammalian commensals, while the rodents that have survived on Buka and Guadalcanal are arboreal and hence less vulnerable.

It thus seems likely that all of Northern Melanesia's marsupials and at least two of its six nonendemic rodents arrived with humans. Today, Melanesian people still often transport live phalangers and wallabies as pets or food. Even the name of the person who introduced *Phalanger orientalis* to Long Island in the nineteenth century is known: a man called Ailimai. The rats *R. praetor* and *R. exulans* probably arrived as stowaways in canoes; *R. exulans* dispersed in this manner to all Polynesian islands, including Hawaii. Originally, the Northern Melanesian mammal fauna may have consisted only of bats and rats. The diversity of native rats would have been much higher than now, with several sympatric species on Bougainville and other large islands of the Bukida group, Guadalcanal, and New Britain (where they survive) and with at least one extinct species each on Manus and New Ireland. Many other Northern Melanesian islands, including all small or outlying islands, may have had no native rats. As on so many other oceanic islands throughout the world, human colonization has been eliminating the native rodents by overhunting and by effects of introduced mammals (pigs, dogs, and *Rattus praetor* and *R. exulans*) as predators, competitors, and disease vectors. The recent arrival of cats and the rats *R. rattus* and *R. norvegicus* with Europeans threatens the survival of Northern Melanesia's native rodents that withstood the millennia of previous human impact.

Amphibia

Frogs are the sole amphibians native to the Australian region. Within the region, New Guinea has a rich frog fauna of about 200 species in 4 families, dominated by the families Microhylidae and Hylidae (more than 87 and 74 species, respectively; table 4.2). Remarkably, only six species in these two species-rich families occur in Northern Melanesia, two of them endemic and four shared with New Guinea. A third family, Myobatrachidae (or Leptodactylidae), with more than 100 species in Australia and six in New Guinea, is absent from Northern Melanesia. Instead, Northern Melanesia is rich in Ranidae, of which New Guinea has only about 14 species in three genera but Northern Melanesia has 37 species in 6 genera. Of those 37 species, 24 occur in the Solomons, 18 in the Bismarcks, and all except three are endemic to Northern Melanesia. Finally, one toad (*Bufo marinus*) was successfully introduced to New Guinea, the Bismarcks, and Solomons in 1937 (Pippet 1975).

It may be an artifact of incomplete exploration that the Bismarcks appear to have fewer native frog species than the Solomons (23 vs. 27), despite the Bismarcks' greater area and closer proximity to New Guinea. Thanks to the efforts of one energetic collector, Fred Parker, who discovered nine new species on the Solomon island of Bougainville between 1961 and 1966, that island appears to have more frog species (24) than the whole of the Bis-

Table 4.2. Comparison of native frogs of New Guinea and Northern Melanesia.

Family	Total species in New Guinea		Total species in Northern Melanesia		New Guinea species in Northern Melanesia	
	Species	Genera	Species	Genera	Species	Genera
Myobatrachidae	6	3	0	0	0	0
Hylidae	>74	2	4	1	3	1
Ranidae	~14	3	37	6	3	2
Microhylidae	>87	11	2	2	1	2
Total	>181	19	43	9	4	4

Sources: for Northern Melanesia, unpublished tabulations by R. Zweifel, W. Brown, and E.E. Williams and from Loveridge (1948), Brown & Myers (1949), Brown (1952, 1997), Zweifel (1960, 1969), Brown & Tyler (1968), Brown & Parker (1970), Menzies (1976, 1982), Brown & Menzies (1978), Zweifel & Tyler (1982), and Allison (1996); for New Guinea, unpublished tabulations by R. Zweifel and from Loveridge (1948), Zweifel (1956, 1958, 1962, 1969, 1972), Allison (1996), Menzies (1976), Zweifel & Tyler (1982), and Brown (1997). Numbers are only approximate because of incomplete exploration.

marcks. However, further exploration is unlikely to change several qualitative conclusions.

First, the Northern Melanesian frog fauna of 43 species is a mixture of three elements. An Asian element consists of three species of genus *Rana* (family Ranidae), a genus much richer in Asia than in New Guinea. Of the three Northern Melanesian *Rana* species, two are shared with New Guinea, and only one is endemic. A New Guinea element includes two species (an endemic *Oreophryne* and a *Sphenophryne* shared with New Guinea) in the subfamily Sphenophryninae of family Microhylidae (the distributions of the subfamily and of both of those genera are centered on New Guinea), plus four species of genus *Litoria* (one endemic, three shared with New Guinea) in the family Hylidae (the genus is centered on New Guinea). The numerically largest component of the Northern Melanesian frog fauna consists of 34 species in two radiations in the family Ranidae, discussed below.

Second, many species-rich genera of New Guinea frogs have failed completely or almost completely to reach Northern Melanesia. Absent are the genera *Nyctimystes*, *Cophixalus*, *Phrynomantis*, *Xenobatrachus*, and *Barygenys*, with about 21, 17, 10, 9, and 7 New Guinea species, respectively. Greatly underrepresented in Northern Melanesia are the genera *Litoria*, *Sphenophryne*, and *Oreophryne*, with, respectively, 53, 14, and 13 New Guinea species but only 4, 1, and 1 Northern Melanesian species.

Third, most of Northern Melanesia's species are endemic (37 out of 43).

Fourth, there is much endemism within Northern Melanesia; 18 of the 37 endemic species are confined to a single island. Seven other endemics are confined to modern fragments of the larger Pleistocene island Greater Bukida.

Finally, as mentioned above, the Northern Melanesian frog fauna is dominated (34 species) by two radiations in the family Ranidae. The smaller radiation comprises eight species in the genus *Batrachylodes*. The whole genus is confined to the Solomons, and seven of the eight species are confined to Bougainville or to that island plus other fragments of Pleistocene Greater Bukida, but the genus apparently belongs to the widely distributed subfamily Raninae. The larger and more distinctive radiation is of the ranid subfamily Platymantinae, which is centered in Northern Melanesia (Brown & Myers 1949, Zweifel 1960, 1969, Brown & Tyler 1968, Brown & Parker 1970, Brown & Menzies 1978, Menzies 1982). Of its four genera, two (*Ceratobatrachus* and *Palmatorappia*) are monotypic and confined to the Solomons, and one (*Discodeles*) consists of five species confined to the Bismarcks, Solomons, or both. The sole platymantine genus to range beyond Northern Melanesia is *Platymantis*, but it is still centered there with 19 species, 18 of them endemic. The sole *Platymantis* species outside Northern Melanesia are two Fijian species (Brown & Myers 1949), one species on Palau, three species endemic to North New Guinea or the West Papuan Islands, *P. papuensis* shared between North New Guinea and the Bismarcks, and about 15 Philippine species. These distributions

Table 4.3. Comparison of native lizards of New Guinea and Northern Melanesia.

	Total species in New Guinea		Total species in Northern Melanesia		New Guinea species in Northern Melanesia	
Family	Species	Genera	Species	Genera	Species	G enera
Agamidae	13	6	2	1	2	1
Pygopodidae	1	1	0	0	0	0
Gekkonidae	27	6	14	7	8	7
Varanidae	6	1	1	1	1	1
Scincidae	104	14	48	11	18	9
Dibamidae	1	1	0	0	0	0
Total	152	29	65	20	29	18

Sources: unpublished tabulations by W. Brown, plus Loveridge (1948), Brown & Parker (1977), McCoy (1980), Zweifel (1980), Allison (1982, 1996), Brown (1991), and Greer (1982). Numbers are only approximate because of incomplete exploration.

suggest that the subfamily Platymantinae, comprising half of all Northern Melanesian frogs, arose and radiated there but sent a few colonists east to Fiji and west to New Guinea and beyond.

Why should platymantines be much more successful colonists than other groups of frogs of New Guinea and Northern Melanesia? As noted by Gibbons (1985), a distinctive feature of platymantines is that they lay a terrestrial egg, from which they develop directly and forgo a tadpole stage, while *Rana* and almost all other Papuan hylids lay eggs in water and pass through an aquatic stage. These same attributes distinguish successful from unsuccessful frog colonists on Caribbean and Philippine islands as well. Freshwater eggs and tadpoles would obviously have no chance of crossing salt water, whereas terrestrial eggs might survive being rafted inside floating vegetation.

Evidently, as one would expect from the modest ocean-crossing abilities of frogs elsewhere in the world (Darlington 1957), frogs have found it difficult to reach Northern Melanesia. If all the platymantines are descended from a single ancestor, then Northern Melanesia's 43 known frog species may be derived from only 11 original colonists. Those colonists that arrived faced few competitors and were therefore able to radiate and to survive until they differentiated to high levels of endemism. Since frogs tend to live at much higher population densities than birds (many more individuals per hectare), this fact would also tend to yield low extinction rates and high levels of endemism. Thus, we interpret the high endemism of Northern Melanesian frogs as resulting from the combination of their poor ocean-crossing ability with their high population densities. This same explanation applies to those groups of Northern Melanesian birds with high endemism (chapter 11).

Reptiles

Tables 4.3 and 4.4 compare the known, possibly native, terrestrial and freshwater lizards and snakes of Northern Melanesia with those of New Guinea. Some of Northern Melanesia's nonendemic reptiles are likely to have been introduced by humans, either recently in the case of species characteristic of areas of human habitation and for which the sole Northern Melanesian records are recent (e.g., the geckoes *Gehyra mutilata*, *Hemidactylus frenatus*, and *Carlia fusca* and the snake *Ramphotyphalops braminus*) or else prehistorically. In addition, Northern Melanesia supports about a dozen species of marine reptiles (sea turtles and sea snakes) and one partly-marine crocodile, all of them widespread outside Northern Melanesia. Two of the world's three populations of sea snakes living in lakes are in Rennell's Lake Tegano.

New Guinea reptiles or reptile groups absent from Northern Melanesia include a freshwater crocodile, several species of freshwater turtles, and New Guinea's single species of legless fossorial lizard (family Dibamidae) confined to the western tip of New Guinea. Otherwise, all New Guinea reptile families are shared with Northern Melanesia,

Table 4.4. Comparison of native non-marine snakes of New Guinea and Northern Melanesia.

Family	Total species in New Guinea		Total species in Northern Melanesia		New Guinea species in Northern Melanesia	
	Species	Genera	Species	Genera	Species	Genera
Typhlopidae	11	2	6	1	3	1
Boidae	9	6	6	4	4	3
Acrochordidae	2	1	1	1	1	1
Colubridae	27	10	9	4	5	3
Elapidae	23	11	6	5	2	2
Total	72	30	28	15	15	10

Sources: unpublished tabulations by S. McDowell, plus Loveridge (1948), McDowell (1970, 1974, 1975, 1979, 1984), McCoy (1980), Malinote & Underwood (1988), Allison (1996), and O'Shea (1996). Numbers are only approximate because of incomplete exploration.

but many New Guinea genera are not shared, such as five of the six genera of dragon lizards (Agamidae). The New Guinea reptile families best represented in Northern Melanesia are the lizard families Gekkonidae and Scincidae (geckoes and skinks) and the snake families Typhlopidae and Boidae (blindsnakes, boas and pythons), each with about half as many species in Northern Melanesia as in New Guinea. Both lizard families are well adapted to overwater colonization (Gibbons 1985): the geckoes by their waterproof skins and eggs, fat storage, and ability to survive for months without food; the skinks by their frequent clutches (hence high frequency of gravid females capable of founding a new population from a single colonist) and their habit of getting inside wood (hence their likelihood of being rafted).

Reptilian endemism in Northern Melanesia is highest in elapid snakes; three of the four elapid species belong to monotypic endemic genera confined to the Solomons. The sole other endemic genera are the monotypic skink genus *Corucia* of the Solomons, and the skink genus *Geomyersia* with one species each in the Solomons and Bismarcks. The endemic species are also concentrated in the Solomons: endemics constitute 81% (29 out of 36) of reptile species confined in Northern Melanesia to the Solomons, but only 35% (9 out of 26) of species confined in Northern Melanesia to the Bismarcks, and only 10% (3 out of 31) of species occurring both in the Solomons and Bismarcks. About half of Northern Melanesian species are endemic among skinks, geckoes, and colubrid snakes, but few or none among agamid and varanid lizards and among typhlopid and boid snakes. All six Northern Melanesian species of *Tribolonotus* skinks are endemic; the genus may have arisen in Northern Melanesia, since there are only one or two species elsewhere (on New Guinea; Zweifel 1966, Cogger 1972). Three endemic species of *Sphenomorphus* skinks (*S. fragosus*, *S. taylori*, *S. transversus*) are confined to the mountains of Bougainville.

Among Northern Melanesian reptiles are many large and widespread species likely to prey on birds and/or their eggs: the lizard *Varanus indicus*, the six boid snakes (boas and pythons), and the colubrid snake *Boiga irregularis*. These include terrestrial species as well as arboreal ones that climb trees efficiently. The brown tree snake *Boiga irregularis* is notorious for having exterminated almost all forest bird species of Guam after being introduced there in recent decades (Savidge 1987). The coexistence of Northern Melanesian birds with this snake illustrates the point that they evolved in the presence of snakes and other vertebrate predators and developed behavioral defenses, unlike Guam birds, which evolved in the absence of snakes and proved defenseless.

Comparison of Vertebrate Classes

Similarities as well as differences emerge from comparing the distributions of the various terrestrial vertebrate classes with each other and with the bird distributions that we discuss in the remainder of this book. The conclusions that we now draw anticipate those drawn from our much larger avian database.

First, in all terrestrial vertebrate classes the endemic genera are concentrated in or confined to the Solomons. These include the rodent genus *Solomys* and subgenus *Cyromys*, the bat genus *Anthops*, the frog genera *Batrachylodes*, *Ceratobatrachus*, *Palmatorappia*, and *Discodeles*, the lizard genus *Corucia*, the snake genera *Loveridgelops*, *Parapistocalamus*, and *Salomonelops*, and all five endemic bird genera of Northern Melanesia (*Microgoura*, *Nesasio*, *Stresemannia*, *Guadalcanaria*, and *Meliarchus*). Likely reasons are that the Solomons are more remote from New Guinea than are the Bismarcks, may have included continuously emergent land for a longer time, and include Northern Melanesia's largest recent land mass (Pleistocene Greater Bukida).

Second, all classes include endemic species now confined to fragments of Pleistocene Greater Bukida. These stranded relicts evidently reached Greater Bukida over water, then underwent an evolutionary loss of overwater colonization ability. The relicts are drawn disproportionately from Northern Melanesia's endemic genera and most distinct endemic species. They include the rat *Melomys bougainville* and *Solomys* and *Uromys* rats, the bats *Anthops ornatus* and *Pterops mahaganus* and the *Pteralopex atrata-P. anceps* superspecies, seven frog species (*Litoria lutea*, *Batrachylodes trossulus*, *Palmatorappia solomonis*, and others), six lizard species (*Geomyersia glabra*, *Sphenomorphus cranei*, *Tribolonotus blanchardi*, and others), and a dozen bird species (*Accipiter imitator*, *Nesasio solomonensis*, and others: cf. species 24, 70, 97C, and 111B on maps 3, 14, 22, and 23).

Third, overwater colonizing ability differs greatly within each class and even within genera. Thus, while bats and birds tend to be more successful colonists than frogs for the obvious reason that bats and birds can fly, there are innumerable exceptions. Among *Pteropus* fruit bats, *P. tonganus* is distributed for 5,000 km from Karkar Island off New Guinea to Samoa, while its congener *P. gilliardorum* is confined to New Britain. Among birds, the hawk *Accipiter novaehollandiae* lives from Wallacea to the Solomons, while its congener *A. imitator* is confined to Greater Bukida. Among frogs, platymantines have dispersed over the whole of Northern Melanesia and North New Guinea, while most New Guinea species of hylids and microhylids have failed to reach Northern Melanesia. Lizards include geckoes and skinks that have spread over the whole tropical Pacific, along with three species of *Sphenomorphus* skinks confined to the mountains of Bougainville. Snakes include three monotypic elapid genera confined to the Solomons, along with a colubrid (*Boiga irregularis*) spread from Australia and New Guinea throughout Northern Melanesia. For the frogs and lizards we have

Table 4.5. Relation between colonizing ability and proneness to form endemics among Northern Melanesian terrestrial and freshwater vertebrate groups.

Group	Northern Melanesian species, as % of New Guinea species	% of Northern Melanesian species that are endemic[a]
Bats	84	38
Birds	43	51
(Lizards)[b]	43	(49)
(Snakes)	39	(32)
Rodents[c]	35	70
Frogs	24	84

Sources: data for all taxa except birds are from the data used in compiling tables 4.1–4.4; for birds, from Appendix 5 of this book, plus the Beehler et al. (1986) value of New Guinea's number of resident bird species (578). See text for discussion.

[a]Percentage of Northern Melanesian species endemic at the allospecies level or higher.

[b]For lizards and snakes the true percentage of endemics is probably considerably higher than the nominal values, shown in parentheses, because archaeological information is lacking for those taxa (cf. note c below for rodents).

[c]Values for rodents exclude the nonendemic rats *Rattus exulans* and *R. praetor*, because they were shown by archaeological studies to be prehistoric human introductions in Northern Melanesia.

discussed on pp. 20 and 21 physiological, reproductive, and behavioral features distinguishing successful from unsuccessful colonists. For birds and bats, the most important distinctions are probably behavioral: whether or not the species chooses to fly out across water gaps without land visible in the distance.

Those are examples of similarities among the vertebrate classes. There are also differences, especially in propensity for speciation and for forming endemic taxa. Radiations of birds and mammals to form many sympatric species from a single ancestor are conspicuous on the continents and the large islands of New Guinea and Madagascar, and (for birds) on remote archipelagoes such as the Galapagos and Hawaii. There have been no such radiations among Northern Melanesian birds, and few among the mammals and snakes (Bukida's *Solomys* rats are an exception), but striking ones among platymantine frogs and *Sphenomorphus* and *Tribolonotus* skinks. Corresponding differences exist among classes in their proportion of endemic species (table 4.5): 84% of Northern Melanesian frogs, but only 38% of the bats. How can these differences be explained?

Chapter 11 shows that propensity of birds to form endemics is related to dispersal ability and abundance: taxa living at high population densities but of only modest dispersal ability are overrepresented among endemics. To survive long enough to become a distinct endemic, an isolated population must avoid extinction, but risk of extinction for isolates is determined overwhelmingly by low population density, so only abundant species survive long enough to become higher endemics. Modest dispersal ability is characteristic of endemics because gene flow inhibits differentiation.

Table 4.5 compares, for terrestrial vertebrate groups, their tendency to form endemics (as reflected in the percentage of Northern Melanesian species that are endemic at the allospecies level or higher) with their overwater colonizing ability (as reflected in the ratio of Northern Melanesian to New Guinea species). The rodent analysis excludes two species now known from archaeological studies to be introductions by prehistoric humans (*Rattus exulans* and *R. praetor*). Among bats, birds, rodents, and frogs there is a monotonic relation between our two indices: as the colonization index decreases from 84% to 24% (from bats to birds to rodents to frogs), the endemism index rises from 38% to 84%. The much higher population densities of frogs than of birds and bats must also contribute, along with frogs' lower vagility, to frogs' long persistence times and frequency of endemics. Lizards and snakes do not fit neatly into this pattern, but the data are inadequate: they are even less well known distributionally and taxonomically than Northern Melanesian frogs, and many of Northern Melanesia's nonendemic reptiles must be suspected (by analogy with findings for archaeologically explored Pacific islands) to be introductions by prehistoric humans.

Summary

Northern Melanesia's mammals consist of a few nonendemic marsupials, all probably introduced prehistorically by humans; about 20 rodent species, mostly endemic; and nearly as many bat species as on New Guinea. Most species-rich groups of New Guinea frogs are absent from Northern Melanesia; instead, Northern Melanesia has experienced a radiation of the frog subfamily Platymantinae, whose reproductive biology may adapt its members to overwater colonization. Northern Melanesian lizards and snakes include numerous species that prey on birds and their eggs. All these vertebrate classes include endemic genera, mostly confined to the Solomons, and species confined to modern fragments of Pleistocene Greater Bukida. Within each class, overwater colonizing ability varies greatly among species, even within the same genus. Vertebrate groups that have had the greatest difficulty colonizing Northern Melanesia from New Guinea have the highest proportion of endemics among their Northern Melanesian species.

PART II

HUMAN HISTORY AND IMPACTS

5

Human History

Northern Melanesia's modern avifauna has been molded not only by the physical and biological environment in which it evolved, but also by a long history of human impacts. Part II of this book describes the history of human settlement (this chapter) and the history of ornithological exploration (chapter 6) in Northern Melanesia, in order to be able to assess the degree to which bird distributions as we know them may have been altered by human influences (chapter 7).

Modern Peoples and Languages

In 1990 the population of the Bismarcks was censused as 435,493; in 1993, the population of the Solomons was slightly larger (128,794 in 1980 on Bougainville, Buka, and nearby atolls, which belong to the nation of Papua New Guinea and which have not been subsequently censused; ca. 333,000 in the remainder of the Solomons, which belong to the nation of the Solomon Islands). The population is growing rapidly, at a rate in excess of 3% per year. The most densely populated areas are New Britain's Gazelle Peninsula, Buka, the Kieta-Panguna area of Bougainville, Malaita, and the north coast of Guadalcanal, where population densities approach or exceed 50 per km^2. Almost all the rest of Northern Melanesia's land area has population densities below 20 per km^2, mostly below 5 per km^2.

Most of the population is classified as Melanesian, related to people of New Guinea and more distantly related to the aboriginal people of Australia (Cavalli-Sforza et al. 1994). The peoples of certain outlying Solomon islands (Rennell, Bellona, Sikaiana, Ontong Java, Nukumanu, Tauu, and Fead) are classified as Polynesians. However, there is great diversity within these groupings and even within the same island (Chowning 1982). For example, within Northern Melanesia about 156 aboriginal languages are spoken (excluding English and Pidgin English), of which 136 are classified as Austronesian (a family of languages spoken from Madagascar and Southeast Asia east to Polynesia), 20 as East Papuan (a subgroup of a diverse family consisting mostly of languages spoken on New Guinea; Greenberg 1971, Wurm 1982, Foley 1986, Ruhlen 1987, Grimes 1988, Comrie et al. 1996). Papuan languages are generally considered the older substratum, while Austronesian languages are thought to have arrived from Southeast Asia via Indonesia several thousand years ago (e.g., Wurm 1983, Bellwood 1985, Bellwood et al. 1995, Spriggs 1997, but see Terrell 1986). Today, Papuan languages are confined to New Britain's Gazelle Peninsula, a few small groups elsewhere on New

Britain and New Ireland and Umboi, central and southern Bougainville, Savo, and parts of Vella Lavella, New Georgia, and Rendova.

Prehistory before European Exploration

When did the first human settlers of Northern Melanesia arrive? By 40 KYBP (40,000 years before the present) and perhaps as early as 60 KYBP, humans originating from Southeast Asia had crossed the last water barriers of eastern Indonesia to colonize New Guinea and Australia, joined in a single continent at Pleistocene times of low sea level (White & O'Connell 1982, Groube et al. 1986, Roberts et al. 1994, Kirch 2000). Until recently, there was no firm evidence that the water barriers separating the Solomons and Bismarcks from each other and from New Guinea had been crossed before a few thousand years ago. However, human occupation sites have now been found that date to 33 KYBP on New Ireland, 35 KYBP on New Britain, 28 KYBP on Buka, and 13 KYBP on Manus (Spriggs 1997, Kirch 2000). Naturally, these sites are unlikely to be the very first occupied; all one can say is that some islands of Northern Melanesia have been occupied for at least 33 millenia (Goodenough 1996, Spriggs 1997).

The water gaps separating the main Solomon islands from each other are relatively narrow, so it is likely that the whole main chain of the Solomons from Bougainville to San Cristobal was occupied soon after the colonization of Bougainville from the Bismarcks. Similarly, within the Bismarcks, New Britain, New Ireland, and New Hanover must all have been occupied within a short time. Even Manus, the most remote of the large Bismarck islands, has been occupied for at least 13,000 years. However, it may have taken longer for humans to discover remote Solomon outliers such as Rennell, Ontong Java, and Sikaiana.

There were undoubtedly subsequent further human colonizations of New Guinea from Indonesia. The next secure proof of fresh human colonization of Northern Melanesia after the first wave is the appearance of the cultural horizon known as Lapita, which spread rapidly from the Bismarcks (probably ultimately from Island Southeast Asia) to Samoa and Tonga around 3600 BP and became ancestral to Polynesians (Kirch 1988, 1997, 2000, Kirch & Hunt 1988, Terrell 1986, Bellwood 1987, Bellwood et al. 1995, Spriggs 1997). Among Northern Melanesian islands from which Lapita sites are known are Manus, New Britain, New Ireland, Mussau, Vuatom, Feni, Nissan, and Buka. Evidence of human colonization at the sites includes distinctive pottery, plant remains derived from tree-crop agriculture, and indirect evidence of root-crop agriculture, plus bones of many hunted marine and terrestrial animals, of domestic animals (pigs, dogs, chickens), and of the commensal rat *Rattus exulans* (Kirch 1988). Obsidian chemically identified as coming from quarries at Talasea on New Britain and at Lou Island in the Admiralties has been found at Lapita and pre-Lapita sites on other islands, indicating long-distance overwater trade. Talasea obsidian has been found as far east as Fiji and as far west as Borneo (Spriggs 1997). The Lapita people presumably spoke an Austronesian language, judging from the language of their descendants such as the Polynesians. They may have been the first people to bring pottery, agriculture, domestic animals, and Austronesian languages into Northern Melanesia. By the time that literate observers began to reach Northern Melanesia in the sixteenth century, all Northern Melanesian islands were occupied by neolithic Melanesians and Polynesians subsisting by root- and tree-crop agriculture, domestic animals, fishing, and hunting.

European Exploration and Modern History

Beginning in the sixteenth century, Northern Melanesian islands were visited by European explorers, followed by navy and merchant ships, then whalers, beachcombers, traders, labor recruiters, and missionaries, and finally colonial government officers and planters (Dumont d'Urville 1841–46, Guppy 1887a, Woodford 1890, Amherst & Thomson 1901, Wichmann 1909–12, Dunmore 1965, Beaglehole 1966, Coates 1970, Oliver 1973, Gash & Whittaker 1975, Laracy 1976, Hempenstall 1978, Wiltgen 1979, Bennett 1987). From 1545 (or perhaps 1528) until 1799, 24 European ships or groups of ships visited Northern Melanesia: 12 voyages of exploration before the British colonization of Australia in 1788, and 12 (possibly more) explorers and navy and merchant ships thereafter, when the Australian colony began to attract much more ship traffic (see table 5.1 for summary). In 13 of these cases there was no landing; the ship either sailed on, merely noting locations of islands seen and having no contact with natives, or else having

Table 5.1. European sightings of Northern Melanesian islands before 1800.

Year	Person or ship	Islands seen or visited: Bismarcks	Islands seen or visited: Solomons	Reference
1528	Alvaro de Saavedra	Man??		Wichmann (1909)
1545	Ynigo Ortiz de Retes	Nin, Her, Wuv		Wichmann (1909)
1568	Alvaro de Mendaña		OJ, Ys (87), Fla (5), Savo, Guad (36), Mal (1), Ul (1), 3S, Ugi (1), SCr (41), SCat (1), SAn (2)	Guppy (1887), Amherst & Thomson (1901), Beaglehole (1966)
1595	Mendaña's ship *Santa Ysabel*		SCr	Allen & Green (1972), Spriggs (1997)
1616	Le Maire and Schouten	Feni, NI, Lih, Tab, NH, SMt, Man, Ramb	Tauu, Nis	Wichmann (1909), Beaglehole (1966)
1643	Tasman	Feni, Tang, Lih, NI, Tab, NH, Ting, NB, Unea, Umb, Long, Cr	OJ, Tauu, Nis	Wichmann (1909), Beaglehole (1966)
1700	Dampier	SMt, Em, NH, Tab, Lih, Tang, Feni, NI, NB (7), Umb, Long, Cr		Wichmann (1909)
1722	Roggeveen	Tab, NI (1), Admiralties		Wichmann (1909)
1767	Carteret	Feni, NI (11), NB, DY, Dyl, NH, Ting, Man, Wuv	Gow, Mal, OJ, Nis, Buka	Wichmann (1909), Beaglehole (1966)
1768	Bougainville	NI (18), Feni, Tang, Lih, Tab, NH, SMt, Em, Anch, Nin	Gan, Bag, VL, Mono, Ch, Boug, Buka	Wichmann (1909), Dunmore (1965), Beaglehole (1966)
1769	Surville		Ch, Ys (8), Gow, Mal, Ul, 3S, Ugi, SCr, SAn, SCat	Dunmore (1965)
1781	Maurelle	Nin, Her, Man, SMig, SMt, Tench, NH, NI, Tab, Tang, Feni	Nis, OJ	Wichmann (1909)
1788	Shortland		SCr, Guad, Russell, NGa, Simb, Mono, Shtl, Boug, Ch, OJ	Dumont d'Urville (1841–46), Guppy (1887), Wichmann (1909)
1791	Hunter	NI, NB, DY (4), Dyl, NH, Ting, Ramb, Naun, SMig, Man, Anch	Sik, OJ, Nis, Buka	Wichmann (1909)
1791	Bowen		NGa	Guppy (1887)
1792	Manning		SCr, Guad, Russell, NGa, Ys, Ch	Guppy (1887)
1792–3	d'Entrecasteaux	NI (6), NB, Vu, Dyl, NH, Ting, Ramb, Naun, Man, Her, Nin, Wuv, Umb, NB, Unea, Witu, Lol, Dyl, Naun, Anch	NGa, Simb, Mono, Boug, Buka, Nis, SAn, SCat, SCr, Ul, Guad, NGa	Wichmann (1909), Dunmore (1965)
1973	Boyd		Ren, Bel	Wolff (1970)
1794	Ship *Indispensable*		Guad, Mal	Dumont d'Urville (1841-6)
1795	Raven	NB, NI (7)	Ch, Boug	Wichmann (1909)
1796	Hogan	NI, NB, DY, Dyl		Wichmann (1909)
1798	Ship *Ann and Hope*		SCr	Bennett (1987)
1799	Ship *Resource*		Fla, Guad, Mal, Scr	Bennett (1987)

This table lists all instances known to us in which Europeans sighted Northern Melanesian islands before 1800 (Wichmann 1909, Spriggs 1997). The details may be relevant to interpreting possible early European impacts on bird distributions (chapter 7). We list identifiable islands sighted (see appendix 1 for abbreviations). Islands on which a landing was made are underlined, and the number in parentheses gives the number of days that the landing lasted or on which landings were made.

Table 5.2. Distribution of nineteenth-century European impact on the Solomon Islands.

	Island												
	Shtl	Ch	Simb/ Gizo	NGa	Ys	Russells	Fla	Guad	Mal	SCr	Ren	OJ	Sik
Whalers	14	3	7	1	6	1	2	5	14	39	5	7	15
Beachcombers	—	—	2	—	—	—	—	1	2	11	—	—	1
Resident traders	5	3	11	38	5	1	30	45	—	44	—	4	—
Year of first resident trader	1885	1897	1869	1869	1877	1900	1869	1876	—	1868	—	1889	—

"Whalers" = number of whaling ships that had some communication with islanders in the years 1799–1887 (from appendix 1 of Bennett 1987); "beachcombers" = number of records of castaways, deserters, and escaped convicts in the years 1820–70 (from appendix 2 of Bennett 1987); "resident traders" = number of records of resident traders in the years 1859–1900, first noted in the year named in the last row (from appendix 3 of Bennett 1987). Shtl, Shortlands, includes Mono; Simb/Gizo, Simbo, Gizo, includes Ganonga and Vella Lavella; NGa, New Georgia, includes neighboring islands; Fla, Florida, includes Savo; SCr, San Cristobal, includes Ugi, Santa Ana, and Santa Catalina; and Ren, Rennell, includes Bellona. Ys, Ysabel; Guad, Guadalcanal; Mal, Malaita; OJ, Ontong Java; Sik, Sikaiana. Note how European impact varied among the islands: e.g., the San Cristobal group was contacted often and early, Choiseul (Ch) rarely and late, by whalers, beachcombers, and resident traders. The table considers only those islands within the modern nation of the Solomon Islands (i.e., Bougainville, Buka, and nearby atolls are not tabulated).

had some contact with natives in canoes who came out to meet the ship. Nine further ships sent landing parties for 1–18 days to gather water, and relations with natives were confined to brief skirmishes. The only long sojourns ashore were by Mendaña's 1568 expedition in the Solomons and by the lost ship *Santa Ysabel* from Mendaña's 1595 expedition.

The first known visit of a whaling ship (from America, Britain, Australia, or New Zealand) to Northern Melanesian waters was in 1799. A few more whalers appeared before 1820, after which visits increased, peaked in the 1840s and 1850s, declined after 1860, and ceased in 1887. Whalers occasionally landed, but more often traded with islanders in canoes that came out to the ships. The whalers sought fresh food, water, sex, tortoise shell, and curios, in return for which they gave iron tools (axes and machetes), nails, cloth, and glass bottles. From the 1830s some Solomon Islanders joined the whaling ships as crew members. For the Solomons south of Bougainville, the top row of table 5.2, summarizing visits by island, shows that the islands preferred by whalers were Simbo, Mono, and the Shortlands, Santa Ana and Santa Catalina (near San Cristobal), San Cristobal, and Sikaiana (Bennett 1987). Bougainville and Buka (Oliver 1973) and the Bismarcks (Wichmann 1909–12) were also visited by whalers.

Beachcombers—deserters, castaways, and runaway convicts from European ships—began to appear in the Solomons from the 1820s and more often after 1850 (Bennett 1987). By far the largest number was at Makira Harbor on San Cristobal (second row of table 5.2). As far as we know, there were no beachcombers on Bougainville, the Bismarcks (except for a short-lived group on New Ireland in 1839), or New Guinea before resident traders and missionaries became established in the late nineteenth century, for the simple reason that it would have meant prompt murder for a beachcomber to land on a Northern Melanesian island that had not already grown accustomed to Europeans through visits of whalers.

Traders from Sydney began to visit the Solomons regularly after 1850, seeking tortoise shell, pearl shell, beche de mer, and coconut oil in exchange for guns, metal axes, cloth, tobacco, and glass (Guppy 1887, Gash & Whittaker 1975, Bennett 1987). Around 1869 a few traders began to take up residence in the Solomons, but the bottom two rows of table 5.2 show that some islands received resident traders much earlier and in greater numbers than others. Preferred islands were initially the San Cristobal group (including Ugi and Santa Ana), New Georgia, Savo, Florida, and Simbo, then Guadalcanal, and later Ysabel and Fauro (Bennett 1987). At the other extreme, Choiseul had no resident trader until 1897, and Malaita none throughout the nineteenth century. In the Bismarcks traders from German firms such as Godeffroy and Hernsheim begain to take up residence in the early 1870s, at first on New Britain's Gazelle Peninsula and Duke of York, then on New Ireland and later on the islands east of New Ireland (Wichmann 1909–12, Gash & Whittaker 1975, Hempenstall 1978). With Godeffroy's development of the method of preserving coconuts in the form of dried copra, the copra trade became a big business from the late 1870s.

Labor recruiters began to visit the Solomons around 1870 to obtain term laborers for the sugar, coconut, or cotton plantations of Fiji, Queensland, Samoa, and New Caledonia (Coates 1970, Oliver 1973, Bennett 1987). When colonial governments and plantations were set up within Northern Melanesia, laborers were also recruited from islands with otherwise little direct access to trade goods (such as Malaita and Bougainville) to work on islands with many plantations (such as the Russells, Guadalcanal, and New Britain's Gazelle Peninsula). Initially, kidnapping or purchase of slaves were common methods of "recruitment," eventually yielding to more-or-less voluntary contracts.

The first European missionaries in Northern Melanesia were Catholics (French and Italian Marists), who resided briefly on Umboi (1848–49, 1852–55) and San Cristobal (1845–47) before succumbing to disease, starvation, and hostile natives (Laracy 1976, Wiltgen 1979). In the 1850s Anglicans began bringing Solomon Islanders to New Zealand for training and returned them to the Solomons as native missionaries. Not until around 1881 did the first European missionaries take up residence in the Solomons, on Florida and then Ysabel. The first European missionary in the Bismarcks after the brief attempt on Umboi was the Methodist George Brown, who came to Duke of York in 1875, set up stations on New Britain and New Ireland, and was followed by the Catholic Mission of the Sacred Heart on New Britain in 1882 (Wichmann 1909–12, Parkinson 1926, Gash & Whittaker 1975). Not until 1902 was the first mission station opened on Bougainville, and only in the 1930s was Rennell effectively converted to Christianity. Several missionaries, especially George Brown on Duke of York, Otto Meyer on Vuatom, and H. Welchman on Ysabel, collected birds and made important contributions to the ornithological exploration of Northern Melanesia.

The copra trade initially involved European traders buying copra from island natives. Around 1882 Europeans began buying land on New Britain's Gazelle Peninsula to develop as coconut plantations, with natives now in the role of laborers rather than owners. Following German annexation, the plantation system quickly spread to Duke of York and New Ireland (Hempenstall 1978), and then to Buka and Bougainville in 1905 (Oliver 1973), about the same time that the establishment of European-owned plantations began in the British Solomon Islands (Bennett 1987).

In the 1870s the commercial interests of traders and labor recruitment led British and German warships to begin to visit Northern Melanesia regularly. These visits culminated in Germany's official annexation of the Bismarcks and Northeast New Guinea in 1884; the 1886 Anglo-German Demarcation that declared Buka, Bougainville, the Shortlands, Choiseul, Ysabel, and Ontong Java to be in Germany's unofficial sphere of influence, the remainder of the Solomons to be in the British sphere; the official declaration of a British protectorate over the British sphere in 1893; and the Anglo-German Samoa Convention of 1899, by which Germany relinquished her Solomon holdings except for Buka and Bougainville to Britain. Rabaul immediately became the center of German administration, but it was not until 1896 that the British based a resident commissioner for the Solomons (C. M. Woodford, important earlier as a collector of Solomon birds, Woodford 1890) at Tulagi.

In World War I the German possessions were captured by Australia, which then administered them together with Papua (Southeast New Guinea). In 1942 Japan captured the Bismarcks and Solomons (except for the eastern islands of Malaita and the San Cristobal group) and lost them after fierce fighting, which involved the establishment of large military bases (and consequently the accidental introduction of some exotic animal species) on numerous islands. In 1975 Papua New Guinea, consisting of eastern New Guinea, the Bismarcks, and northern Solomons (Buka, Bougainville, Nissan, Nuguria, Kilinailau, Tauu, and Nukumanu) became an independent nation, while the British Solomon Islands achieved independence under the name of Solomon Islands in 1978.

European Impacts Relevant to Bird Distributions

Europeans carried animal species to Northern Melanesia that may have influenced modern bird distributions—especially cats, also the rats *Rattus rattus* and *R. norvegicus*, the toad *Bufo marinus*, and several lizard species. Europeans also changed the numbers and distribution of native people. Population numbers were initially reduced by introduced diseases, as is apparent by comparing accounts of the dense populations on Ysabel, North Guadalcanal, and San Cristobal seen by Mendaña's 1568 expedition with accounts of populations there in the twentieth century. Native warfare, which had previously been chronic but local, became more deadly as islanders acquired steel axes and guns. In particular, nineteenth-century raids by headhunters from the New Georgia group depopulated much of Choiseul and Ysabel and some areas of the New Georgia group itself. These population reductions were succeeded after the mid-twentieth century by population growth due to modern medicine and a declining death rate.

In recent decades there have been population shifts for economic reasons, notably immigration to Bougainville's copper mine before its closure by civil war, West New Britain's oil-palm plantations, and North Guadalcanal's commercial center. In addition, on all the larger islands there were formerly both coastal peoples, enjoying access to marine foods and trade, and interior peoples, enjoying protection from raiders from the sea. The end of interisland raiding and the rise of demand for manufactured goods and services arriving by sea have led throughout Northern Melanesia to population shifts from the interiors toward the coasts. Especially on West New Britain, New Ireland, Manus, New Georgia, Ysabel, Guadalcanal, San Cristobal, and parts of Bougainville, the interiors already are or else are becoming depopulated.

Finally, we stress that Europeans, and presumably their associated vermin, reached different areas of Northern Melanesia at different times. These differences relate to varying proximity to sailing routes to Australia, risks from reefs, safe anchorages, protection against attack afforded by small offshore islets, and availability of desired trade items. Areas impacted heavily by the end of the nineteenth century included the Gazelle Peninsula, Duke of York, North New Ireland, and the San Cristobal group (including Ugi, Santa Ana, and Santa Catalina). Choiseul and St. Matthias had little direct European influence until later; much of Bougainville and the main body of New Britain remained almost uncontacted until after World War I; and Malaita, Rennell, and Bellona were especially resistant to outside influence until the 1930s.

These details may be important for interpreting apparent modern bird distributions. For example, the ground dove *Gallicolumba salamonis*, known only from one specimen collected on San Cristobal in 1882 and another on Ramos in 1927, is one of only two Northern Melanesian bird species believed to have become extinct since ornithological exploration began. Its rapid disappearance may be connected with the fact that its known range was confined to the eastern Solomons, the area most visited by Europeans in the mid-nineteenth century (and hence likely first to acquire feral cats preying on ground birds). As another example, the now-extinct ground pigeon *Microgoura meeki* (plate 8) was collected only in 1904, on Choiseul, an island that has no other endemic species because it was joined to Bougainville and Ysabel at Pleistocene times of low sea level. Of those three islands, Choiseul was the one least visited by Europeans in the nineteenth century. We suggest that the pigeon originally occurred on all three islands but was eliminated by the arrival of cats on Bougainville and Ysabel before the first extensive collections of birds on all three islands in 1900–04.

Summary

The human population of Northern Melanesia currently numbers around 1 million and speaks 156 native languages. People reached some Northern Melanesian islands from New Guinea by at least 33,000 years ago. Around 1600 B.C., the Lapita cultural horizon spread from the Bismarcks out into the Pacific to become ancestral to the Polynesians. European explorers arrived by A.D. 1545, whalers around 1799, beachcombers in the 1820s, traders after 1850, and missionaries in 1875, culminating in Germany's annexation of the Bismarcks in 1884. European impact on Northern Melanesian birds has included effects of introduced species, especially cats and rats. Europeans and their associated vermin reached different islands at different times. This fact is significant in interpreting modern bird distributions, such as the apparent confinement of Northern Melanesia's most distinctive bird species (the pigeon *Microgoura meeki*) to Choiseul.

6

Ornithological Exploration of Northern Melanesia

Collections and observations of Northern Melanesian birds began in 1823 and are listed chronologically in appendix 4.[1] As reflected in figure 6.1, which plots descriptions of new taxa by decade, the peak periods of discovery were 1875–94, 1900–25, and 1927–35. The first specimens were obtained by French expeditions on New Ireland in 1823 and 1827 and on Ysabel in 1838. A trickle of further specimens continued until the first peak at 1875–1894. During that peak G. Brown, E.L.C. Layard and G.E. Richards, and others obtained the first large collections from the Bismarcks (New Britain, Duke of York, and New Ireland) in 1875–81; the Challenger Expedition of 1875 made the first collection on Manus; many collectors operated in the eastern Solomons in 1878–82; C. M. Woodford made the first significant collections on the Shortlands and New Georgia in 1886–87; and H. Welchman collected on Ysabel in the early 1890s. F. Dahl lived on New Britain in 1896–97 and wrote what are still the most detailed published field observations of New Britain birds (Dahl 1889).

1. The principal published syntheses have been, for both the Bismarcks and the Solomons, Gray (1859) and Salvadori (1881); for the Bismarcks, Reichenow (1899) and Meyer (1936); for the Solomons, Mayr (1945a); for the eastern Solomons, Cain and Galbraith (1956) and Galbraith and Galbraith (1962); for Rennell and Bellona, Mayr (1931b), Bradley and Wolff (1956), and Diamond (1984a); for Bougainville, Hadden (1981); and for the Bismarcks and Bougainville, Coates (1985, 1990).

The second peak, in 1900–25, largely represents collections made for the Tring Museum by Albert Meek (1900–13) and Albert Eichhorn (1923–25). In the Bismarcks they made the first collections on Emirau, Umboi, Feni, Witu, and Unea and also made large collections on all the other main islands (New Britain, New Ireland, New Hanover, Manus, St. Matthias). In three expeditions to the Solomons in 1900–08 Meek made the first collections on Bougainville, Choiseul, Mono, Vella Lavella, Kulambangra, Gizo, and Rendova and additional collections on Ysabel, Florida, Guadalcanal, New Georgia, and San Cristobal. The residence of Otto Meyer on the Bismarck island of Vuatom from 1902 to 1937 resulted not only in detailed published field observations of Bismarck birds, especially on their breeding biology, but also in the first collection on Lihir.

The third and last peak reflects collections by the Whitney South Sea Expedition in the Solomons from 1927 to 1930 (with Beck, Coultas, Drowne, Eyerdam, Hamlin, Mayr, and Richards as collectors), and by the Whitney collector William Coultas in the Bismarcks from 1932 to 1935. In the Solomons the Whitney Expedition operated on all ornithologically significant islands and made the first collections of note on Malaita, Rennell, Bellona, Ganonga, Buka, Tetipari, Simbo, Vangunu, and Gatukai and also in the mountains of Bougainville, Guadalcanal, Kulambangra, San Cristobal, and Ysabel, where previous collectors

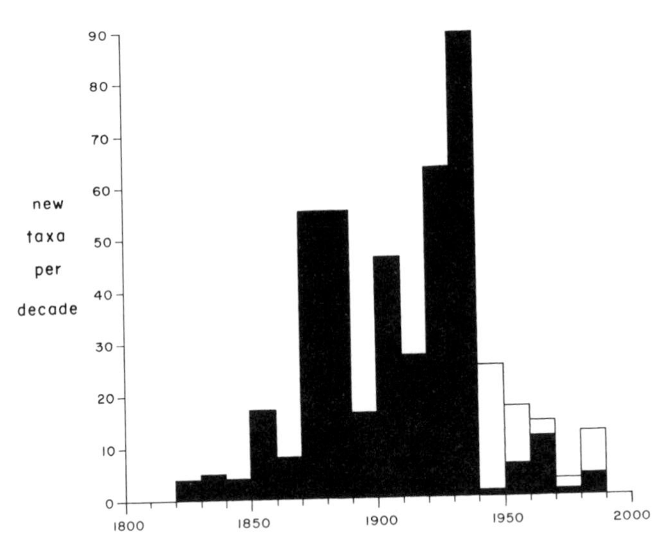

Fig. 6.1. The accumulation of knowledge of Northern Melanesian birds. The ordinate gives, by decade, the number of new taxa (species or subspecies) endemic or nearly endemic to Northern Melanesia that were described within that decade. Because of lags between collection of specimens and publication of descriptions, some descriptions were based on taxa collected in a previous decade. From the decade 1940–49 to the present, solid bars indicate the number of descriptions based on newly discovered populations; open bars indicate the number of descriptions based on study of populations discovered before 1940. Note the three peaks in ornithological exploration corresponding to collections in 1875–94, 1900–25, and 1927–35.

had been confined to the lowlands. Coultas made the first collections on Tabar, Tanga, and Long, plus large collections on New Britain and the Northwest Bismarcks. Ernst Mayr described most novelties resulting from the Whitney collections in a series of papers from 1931 to 1955 and in a book (Mayr 1945a) summarizing the avifaunas of Southwest Pacific archipelagoes, including the Solomons but not the Bismarcks.

Since World War II there have been significant further collections in the mountains of six large islands: New Britain (by Gilliard, Salomonsen, and Diamond), New Ireland (by Salomonsen and Beehler), Umboi (Diamond), Bougainville (by Schodde and Hadden), and Guadalcanal and San Cristobal (by Cain and Galbraith). There were also collections on four islands whose birds had not previously been studied: Dyaul (Salomonsen) and Tolokiwa, Crown, and Sakar (Diamond). Of the 71 new taxa described since 1940, most have been based on study of specimens collected before 1940, and only 23 have resulted from discoveries of new populations: one species, one allospecies, and one subspecies from the mountains of New Britain; one allospecies and one subspecies from the mountains of Bougainville; five, three, and one subspecies from the mountains of Guadalcanal, New Ireland, and Tolokiwa, respectively; one allospecies from the lowlands of the New Georgia group; six subspecies from Dyaul; and one subspecies each from Pavuvu and Nissan. The sole new full species of bird discovered in Northern Melanesia since 1937 is the honey-eater *Melidectes whitemanensis* (plate 9), collected in the mountains of New Britain in 1958; (see p. 124 of chapter 16 for the distinction between allospecies and full species). The new allospecies are the thicket-warbler *Cichlornis [whitneyi] grosvenori* (mountains of New Britain, 1958: plate 9), the rail *Gallirallus [philippensis] rovianae* (lowlands of New Georgia group, 1976), and the thicket-warbler *Cichlornis [whitneyi] llaneae* (mountains of Bougainville, 1979).

Ornithologists in Northern Melanesia since 1972 have shifted their efforts from collecting specimens and discovering new taxa to field observations of known taxa. Notable publications of observations have been by Cain and Galbraith (1956) for Guadalcanal, San Cristobal, Ugi, and Ulawa; by Diamond (1975b), Schodde (1977), and Hadden (1981) for Bougainville; by Diamond (1984a) and Filardi et al. (1999) for Rennell; by Blaber (1990) for New Georgia; by Webb (1992) and Kratter et al. (2000) for Ysabel; by Coates (1985, 1990) for many Bismarck islands; by Buckingham et al. (1996) for Manus and eight Solomon islands; and by Diamond (2001a) and Dutson (2001) for many islands. In 1974 and 1976 Diamond resurveyed most central Solomon islands and tabulated which of the widespread species (whose detailed distributions earlier collectors tended to ignore) actually occur on each island.

While most Northern Melanesian bird taxa and their island-by-island distributions have thus surely been discovered by now, a few unnamed taxa may await collection and description. Sight or voice records or native accounts that could refer to undescribed taxa include a ground bird in the mountains of Malaita that natives considered taboo and refused to collect for the Whitney Expedition, a *Microeca*-like flycatcher reported on New Ireland and New Britain, and the bird responsible for a song that ornithologists have regularly heard in the mountains of Bougainville and that local people know as the *kopipi* or *odedi* (names in two different local languages) (Diamond 1975b, Hadden 1981, Beehler 1983). Habitats in which we reckon with the possibility of further new taxa or records of distributional significance are the marshes of San Cristobal and the mountains of New Ireland, Malaita, Vella Lavella, New Georgia, and San Cristobal.

Summary

Collections of Northern Melanesian bird specimens began in 1823 and peaked in the years 1875–94, 1900–25, and 1927–30. No new full species has been discovered since 1958 and no new allospecies since 1979.

7

Exterminations of Bird Populations

Many bird species of oceanic islands have become extinct in modern times after Europeans reached the islands. These extinctions have been caused by direct and indirect effects of humans, such as habitat destruction, introductions of mammalian predators and herbivores, introductions of nonnative bird species, and overhunting. Recent discoveries of subfossil birds at archaeological sites have revealed even more extinctions that occurred after oceanic islands had first been settled by preindustrial peoples but before modern literate observers or ornithologists arrived. Still other island species, while still extant today, have suffered drastic range contractions through extinctions of their populations on some islands. Extinctions of island birds are best known for Hawaii, New Zealand, and the Mascarenes, but they have now been documented for all paleontologically explored Pacific islands east of the Solomons, as well as for the West Indies and South Atlantic islands (e.g., Olson & James 1982, Steadman & Olson 1985, Steadman 1989, 1993, 1995, James & Olson 1991, Steadman & Kirch 1998, Steadman et al. 1999).

The result of these human-caused extinctions is that modern bird distributions on oceanic islands depict only a sample of the original island avifaunas. Knowledge of island bird distributions before human impact alters biogeographic and evolutionary conclusions drawn from modern distributions alone. Thus, it is essential for the purposes of this book to consider to what extent Northern Melanesian bird distributions as we know them may have been altered by human impact. We summarize four types of evidence: Northern Melanesian species not recorded in the past several decades, hence possibly now extinct; comparison of the post-1850 avifauna with that noted by the first European explorers; bird fossils; and susceptibility of Northern Melanesian birds to known mechanisms of human impact.

What Bird Species May Now Be Extinct in the Bismarcks or Solomons?

On islands such as Hawaii and New Zealand, many modern bird species are now considered extinct because they have not been observed in recent decades. Standard reference books on extinct and endangered birds (e.g., King 1981, Collar & Andrew 1988, Collar et al. 1991) are somewhat out of date for Northern Melanesia. Hence we reassessed the current status of all resident bird species previously recorded for Northern Melanesia.

The 1953–54 Oxford University Expedition to the Solomons and Gilliard's 1958–59 expedition to New Britain serve as convenient time points to define the start of the "current era." Since those expeditions there has been much further field work,

Table 7.1. Resident Bismarck and Solomon bird species not definitely recorded since 1959 and 1954, respectively.

Species	Level of endemism	No. of Bismarck islands	Last specimen collected	Last reliable report
Bismarck Islands				
Turnix maculosa[a]	Subspecies	2	1896	ca. 1905
Gymnocrex plumbeiventris	—	1	ca. 1869	ca. 1869
Rallina tricolor	Subspecies	3	1944	1944
Tyto novaehollandiae	Subspecies	1	1934	1934
Cichlornis grosvenori	Allospecies	1	1958	1958
Rhipidura rufifrons[a,b]	Subspecies	2	1934 (1945)	1934 (1997)
Solomon Islands				
Anas gibberifrons	Subspecies	1	1928	1985(?)
Gallicolumba jobiensis[c]	Subspecies	3	1927	1927
Gallicolumba salamonis	Species	2	1927	1927
Microgoura meeki	Genus	1	1904	1904
Zoothera dauma[c]	Subspecies	1	1904	1904

"Level of endemism" denotes whether a taxon is endemic to Northern Melanesia at the subspecies, allospecies, species, or genus level or else is not endemic (i.e., even the Northern Melanesian subspecies occurs extralimitally). "No. of islands" gives the number of islands on which the species had been recorded earlier. The analysis for the Solomons updates Diamond (1987).

[a]Recent records for Solomons but not for Bismarcks.

[b]*Rhipidura rufifrons* was last collected and observed on Manus in 1934 but still survives on outlying islets of the Manus group (collected 1945, repeatedly observed from 1975 to 1997).

[c]Recent records for Bismarcks but not for Solomons.

including visits to virtually all ornithologically significant islands of Northern Melanesia. Most (96–97%) of the 169 and 164 species plus allospecies known for the Bismarcks and Solomons, respectively, have been reencountered repeatedly in this recent field work. Table 7.1 lists those few Bismarck and Solomon bird populations for which there have been no definite records since 1959 and 1954, respectively. Which of these species are probably extinct or rare in Northern Melanesia, and which are probably extant, unthreatened, and just hard to detect?

At the one extreme, as recently as 1934, the distinct fantail subspecies *Rhipidura rufifrons semirubra* was described as common to very common everywhere on Manus, where Albert Meek's collectors obtained 10 specimens in 1913. Manus has been revisited by at least 10 collectors or observers or groups since 1944, but none has recorded this fantail. The sole known surviving populations are on the offshore islets Tong, San Miguel, and Sivisa (Silva 1975, Tolhurst 1993, Eastwood 1995). Other subspecies of *R. rufifrons* remain common throughout the Solomons, but *R. r. semirubra* is now evidently extinct or very rare on Manus.

Another example of a species that must be extinct is the ground pigeon *Microgoura meeki*, the most spectacular endemic bird of Northern Melanesia (plate 8), which was the object of three intensive but unsuccessful searches (by the Whitney Expedition, Shane Parker, and Jared Diamond in 1929, 1968, and 1974, respectively) at the site on Choiseul where it was discovered in 1904. Local villagers with detailed knowledge of birds told Whitney Expedition members and Parker (1972) that the ground pigeon had been exterminated by introduced cats. Rennell islanders told Diamond (1984a) that the duck *Anas gibberifrons* disappeared after the fish *Tilapia mossambica* was introduced into Rennell's lake. Hence it seems probable that the Bismarck or Solomon populations of these three taxa (*Rhipidura rufifrons semirubra* on Manus itself, *Microgoura meeki*, and *Anas gibberifrons*) are extinct.

At the opposite extreme, few ornithologists have reached elevations above 1500 m on New Britain since Gilliard discovered the cryptic warbler *Cichlornis grosvenori* (plate 9) there in 1958. In 11 days spent on the summit of New Britain's Whiteman Mountains, Gilliard never saw this bird, which remains known only from two specimens collected by New Guinean hunters working with him. It is therefore unsurprising that the few other ornithologists to visit the habitat of this skulking species also failed to observe it, and we see no reason to doubt its continued existence.

Between these two extremes lie the cases of four species in the Bismarcks (the bustard quail *Turnix maculosa*, the rail *Gymnocrex plumbeiventris* [if it really did occur on New Ireland], the rail *Rallina tricolor*, and the barn owl *Tyto novaehollandiae*) and three species in the Solomons (the ground pigeons *Gallicolumba jobiensis* and *G. salamonis* and the ground thrush *Zoothera dauma*). All were cryptic and uncommon or rare to begin with and were known from few islands (only two on the average) and few specimens. It is difficult to evaluate whether lack of recent records for them indicates any change in status.

Thus, 2–5 of the Bismarcks' 169 known species and 2–5 of the Solomons' 164 known species might now be extinct or very uncommon. Of the 11 possible Northern Melanesian extinctions, nine (all except *Tyto novaehollandiae* and *Rhipidura rufifrons*) involve ground-dwelling or ground-nesting birds, suggesting that introduced cats may have been the culprits, as Choiseul islanders reported for the ground pigeon *Microgoura meeki*.

In short, Northern Melanesia has lost a few historically attested bird populations, but its losses to date (1–3% of the known avifauna) are trivial compared to those of Hawaii, New Zealand, and other remote Pacific islands. For example, 70% of Hawaii's historically attested land and freshwater bird species are now extinct or endangered.

What Bird Species Were Reported by the First European Explorers?

When Captain Cook and other early explorers reached Hawaii, Tahiti, and other Polynesian islands, they collected or observed bird species, some of which were never found again by subsequent ornithologists. Extinctions of Polynesian birds followed so rapidly upon European contact that the first thorough collectors of birds many decades later must already have been studying a partly depleted avifauna. To assess whether this might also be true of Northern Melanesia, we summarize the first six sets of bird records, from 1568 to 1838.

The Mendaña Expedition of 1568 constituted European discovery of the Solomons and also the first European landings in the Solomons. No bird specimens were collected, but expedition journals (Amherst & Thomson 1901, pp. 50, 88, 132, 213, 275) give five descriptions of three birds in sufficient detail to permit identification to species: the cockatoo *Cacatua ducorpsi* ("parrots entirely white, with a crest of feathers on their head"; plate 8) and the pigeons *Ducula pistrinaria* ("grey pigeons"; plate 8) and *D. rubricera* ("doves, larger than the largest wild pigeons of Spain, and some have a fleshy protuberance above the nostrils, like half a pomegranate, very red. Their plumage resembles that of the peacock's tail"; plate 8). All three are still among the most conspicuous coastal birds of the Solomons. The expedition journals also mention eagles, red parrots, green parrots, and other vague descriptions that could well correspond to other conspicuous extant species of the coasts, such as the eagle *Haliaeetus sanfordi*, the red parrot *Chalcopsitta cardinalis*, and the green parrot *Trichoglossus haematodus*.

William Dampier in 1700 was the fourth or fifth European to sail through the Bismarcks, and the first European to land there (on New Britain). He refers evidently to the cockatoo *Cacatua galerita* ("Cockadores") and the crow *Corvus orru* ("crows like those in England"), both still common coastal birds (Dampier 1939, p. 214).

In 1768 Bougainville (1771, p. 277) became the fourth European to land in the Bismarcks (on New Ireland) and observed what was evidently the pigeon *Ducula rubricera* ("gros pigeons de la plus grande beauté. Leur plumage est ver-doré. Ils ont le col et le ventre gris-blanc et une petite crête sur la tête.")

In 1823 and 1827 the Coquille Expedition and then the Astrolabe Expedition landed for 10 and 14 days, respectively, on New Ireland and made the first collections of Bismarck birds. In addition, R.-P. Lesson, the zoologist of the Coquille Expedition, wrote detailed descriptions of many species that he observed but did not collect. A complication in evaluating bird records of these two expeditions is that both, like other eighteenth and early nineteenth-century expeditions, failed to appreciate

Table 7.2. Species recorded for New Ireland in 1823 by the Coquille Expedition.

Vernacular name applied	Scientific name applied	Modern scientific name
		Fregata ariel or *F. minor*
Aigle oceanique		*Haliaeetus leucogaster*
Chevalier gris		*Tringa incana* or *T. brevipes*
	Columba nicobarica	*Caloenas nicobarica*
Colombe pinon, Colombe muscadivore	*Columba pinon*	*Ducula rubricera*
Perroquets à vestiture écarlate	*Psittacus lori*	*Lorius hypoinochrous*
Perroquet vert à plumes lustrées de Moluques	*Psittacus sinensis*	*Eclectus roratus*
Gros lori papou	*Psittacus grandis*	*Eclectus roratus*
Cuculus, uniform green		*Chrysococcyx lucidus*
Coucal atralbin	*Centropus ateralbus*	*Centropus ateralbus*
	Alcedo albicilla	*Halcyon saurophaga*
	Alcedo ispida var. *moluccana*	*Alcedo atthis*
	Halcyon cinnamominus	*Halcyon sancta*
Échenilleur karou	*Ceblepirus karu*	*Lalage leucomela*
Hirondelle		*Hirundo tahitica*
Gobe-mouches à tête d'acier	*Muscicapa chalybeocephalus*	*Myiagra alecto*
Gobe-mouches ornoir	*Muscicapa chrysomela*	*Monarcha chrysomela*
Traquet turdoide	*Saxicola merula*	*Pachycephala pectoralis*
Soui-manga à gorge bronzée		*Nectarinia jugularis*
Stourne metallise	*Lamprotornis metallicus*	*Aplonis metallica*
Drongo		*Dicrurus megarhynchus*
Corbeau à duvet blanc		*Corvus orru*

The first two columns give the vernacular and scientific names applied by Lesson & Garnot (1826–28) in their report on zoological results of the expedition. They also describe a metallic black, pale-eyed flycatcher ("le tenourikine") of which they lost the specimen; this might be *Myiagra alecto* or *M. hebetior*, if their recollection of eye color were incorrect.

the importance of recording exact collecting localities and mislabeled or failed to label some specimens throughout the whole voyage. On New Ireland the Coquille Expedition reported two species ("*Columba puella*" = *Ptilinopus magnificus* and "gobe-mouches télescophthalme" = *Arses telescophthalmus*), and the Astrolabe Expedition also reported two species ("lori papou" = *Charmosyna papou* and "columbi-galline Goura" = *Goura cristata*), that all subsequent collectors have found confined to New Guinea. However, both expeditions also collected on the Vogelkop Peninsula at the western end of New Guinea, also recorded these species there and the "New Ireland specimens" belong to the Vogelkop races. Similarly, the Coquille Expedition claimed an Australian parrot ("perruche de Latham" = *Lathamus discolor*) both for New Ireland and for Australia. These five records are far more likely to involve mislabeled specimens than vanished populations that by astonishing coincidence were identical to the populations at another expedition collecting site 3000–5000 km away. Tables 7.2 and 7.3 list the remaining 22 species collected or observed on New Ireland by the Coquille Expedition and the remaining 12 species collected by the Astrolabe Expedition. All are still common on New Ireland today.

Finally, Dumont d'Urville's Pôle Sud expedition made the first collection of Solomon birds in 1838 on San Jorge Island (virtually part of Ysabel). One specimen attributed to San Jorge ("*Myzomela solitaria*") is of the Fijian honey-eater *M. jugularis*, surely another case of mislabeling. The other nine species are common on Ysabel today: *Chalcophaps stephani*, "*Lorius cardinalis*" = *Chalcopsitta cardinalis*, "*Pionus heteroclitus*" and "*P. cyaniceps*" = male and female *Geoffroyus heteroclitus*, *Cacatua ducorpsi*, "*Athene taeniata*" = *Ninox jacquinoti*, "*Pachycephala orioloides*" = *P. pectoralis*, *Dicaeum aeneum*, *Myzomela lafargei*, and "*Lamprotornis fulvipennis*" = *Aplonis grandis*.

In summary, early explorers of Northern Melanesia between 1568 and 1838, including the first Europeans to land in the Bismarcks and

Table 7.3. Species recorded for New Ireland in 1827 by the Astrolabe Expedition.

Vernacular name applied	Scientific name applied	Modern name applied
Aigle de Pondichery		*Haliastur indus*
Colombe muscadivore, tubercule du bec rouge		*Ducula rubricera*
Colombe de Carteret		*Macropygia amboinensis*
Pigeon de Nicobar		*Caloenas nicobarica*
Coucal violet	*Centropus violaceus*	*Centropus violaceus*
Checéche bariolée	*Noctua variegata*	*Ninox variegata*
Martin-pécheur sacré		*Halcyon sancta*
Moucherolle à longues soies	*Muscipeta setosa*	*Rhipidura rufiventris*
Moucherolle noir et blanc	*Muscipeta melaleuca*	*Rhipidura leucophrys*
Gobe-mouches ornoir		*Monarcha chrysomela*
Gobe-mouches à tête d'acier		*Myiagra alecto*
Stourne		*Aplonis metallica* or *A. cantoroides*

The first two columns give the vernacular and scientific names applied by Quoy & Gaimard (1830) in their report on zoological results of the expedition.

Solomons and also the first to collect in both archipelagoes, provided a total of 49 species/island records by specimens or adequate descriptions. All involve species that are still extant and common today. In contrast, some species that Captain Cook collected on various Polynesian islands were never found again. One reason for this difference is clear. European colonization and impact in Polynesia followed soon after Cook's discoveries. In contrast, European colonization of Northern Melanesia did not begin until the 1870s. In the 1870s and 1880s, before Europeans could have had much impact on Northern Melanesian birds, some of the first European settlers were amateur naturalists who amassed large collections of birds (e.g., Brown, Cockerell, Finsch, Hübner, Layard, Richards, and Woodford). In addition, all known Northern Melanesian bird species except one have been collected or observed in Northern Melanesia in the twentieth century. (The sole exception is the cryptic rail *Gymnocrex plumbeiventris*, questionably collected once on New Ireland in the nineteenth century.) All except three known Northern Melanesian species survived at least until 1927. It therefore seems unlikely that many Northern Melanesian bird populations extant upon European discovery disappeared before they could be collected. The only hints of this happening are the distribution of *Microgoura meeki* (collected only on Choiseul, but we might expect former presence on Bougainville or Ysabel as well) and of *Gallicolumba salamonis* (known from two specimens from the eastern Solomons, where we hypothesize a formerly wider distribution).

Bird Fossils

We are indebted to David Steadman for accounts of the sole bird fossils examined to date from Northern Melanesia, in addition to his published accounts of the fossils. They come from the Bismarck islands of St. Matthias (Steadman & Kirch 1998) and New Ireland (Steadman et al. 1999).

The St. Matthias bones were excavated at three Holocene archaeological sites (Talepakemalai, Etakosara, and Epakapaka) of Lapita people, with radiocarbon dates of 1600–200 B.C. There is no known occupation of the St. Matthias group predating those Lapita sites. Most of the 58 identifiable bird bones belong to seabirds, migrant shorebirds, and the introduced chicken, but 13 bones are of land birds identified or tentatively referred to eight species. Six of these species—the osprey *Pandion haliaetus*, the starling *Aplonis metallica*, and the pigeons *Ptilinopus insolitus*, *P. solomonensis*, *Ducula pistrinaria*, and *Caloenas nicobarica*—still occur on St. Matthias. The other two land bird taxa do not still occur on St. Matthias but may be identifiable with taxa still occurring on other Bismarck islands: a large parrot, possibly related to the cockatoo *Cacatua galerita ophthalmica* of New Britain; and a large barn owl, possibly related to *Tyto novaehollandiae manusi* of Manus. None of the identified land bird bones from St. Matthias provide firm evidence of distinctive extinct species. If the Lapita sites really do represent the first human colonization of the St. Matthias group, then there could not have been earlier human-related extinctions in the group.

The New Ireland bones were excavated at five Late Pleistocene and Holocene pre-Lapita archaeological sites (Balof, Matenkupkum, Matenbek, Panakiwuk, and Buang Merabak), with radiocarbon dates of about 35,000–6,000 BP. In contrast to the Lapita sites on St. Matthias, at which bones of seabirds and migrant shorebirds and introduced chickens predominate, these pre-Lapita sites on New Ireland yielded no shorebird or chicken bones and only two petrel bones. The remaining 239 bones belong to about 49 species, mostly (43 species) nonpasserines. Of the 49 species, about 37 are or may be identified with species still present on New Ireland. Three taxa are surely extinct: a very large megapode, a flightless *Gallirallus* rail, and a very large flightless *Porphyrio* rail. Three other taxa are extant elsewhere, though unrecorded from New Ireland today: the swamphen *Porphyrio porphyrio* (recorded from seven Bismarck islands, including two of the New Ireland group), the rail *Porzana tabuensis* (recorded from two Bismarck islands), and the quail *Coturnix ypsilophorus* (known from New Guinea and Australia). The remaining six taxa are also unrecorded from New Ireland today, but it is uncertain whether they are conspecific with extant species or are extinct taxa similar to those extant species: two large goshawks (possibly including *Accipiter meyerianus*, widespread in Northern Melanesia), a cockatoo (possibly *Cacatua galerita* of New Britain), a small barn owl (possibly *Tyto alba* of Tanga in the New Ireland group or *T. aurantia* of New Britain), a large barn owl (possibly *Tyto novaehollandiae* of Manus), and a crow (possibly *Corvus tristis* of New Guinea). Thus, the 49 species consist of about 37 species still extant on New Ireland, between 2 and 7 species extant elsewhere in the Bismarcks, 1–2 species extant on New Guinea, and between 3 and 9 extinct species.

In short, on both St. Matthias and New Ireland, about 75% of the bird taxa identified from bones at archaeological sites are or may be identifiable with taxa still extant on the island. The remaining bones belong to taxa now absent from St. Matthias or New Ireland but possibly still extant on nearby islands, plus (on New Ireland) at least three extinct species. In contrast, on Hawaii and other Polynesian islands, the relative proportions of extant and extinct populations represented among fossil bird bones are reversed: 75% of the fossil taxa are now globally or locally extinct. Thus, these preliminary identifications suggest that the Northern Melanesian avifauna has suffered bird extinctions since human arrival, but many fewer extinctions than the avifaunas of more remote Pacific islands.

Susceptibility of Northern Melanesian Birds to Known Mechanisms of Extermination

To interpret the apparently modest impact of Europeans in Northern Melanesia and to guess the impact of human colonists before Europeans, we review the mechanisms by which humans have exterminated oceanic island birds around the world. Four mechanisms stand out: habitat destruction, introduced predators, introduced competing birds, and overhunting, with disease as an additional factor in some cases (Diamond 1984b).

Habitat destruction has caused bird exterminations especially on dry islands susceptible to fire (e.g., parts of Hawaii and New Caledonia), on islands where large populations of introduced domestic herbivores such as goats and sheep and cattle developed (Madagascar and Australia), and on islands with large agricultural human populations (many Polynesian islands). These factors have all been slight for European colonization of Northern Melanesia, where European settlers were few, occupied little land, introduced no feral herbivores except deer in some parts of New Britain, and did little pasturing. Regarding prehistoric Melanesian settlers, the only dry parts of Northern Melanesia susceptible to burning are small areas of Guadalcanal and New Britain. Prehistorically introduced herbivores were pigs, a phalanger, a wallaby, and a tree kangaroo. While human populations were evidently denser several centuries ago than today, none of the large islands became massively deforested, although some small outliers including Ontong Java, Sikaiana, and Bellona did. These three islands lack endemic taxa of modern birds, suggesting that they might have suffered prehistoric extinctions, but their area might also have been too small originally for evolution of endemics.

Introduced predators such as rats, cats, pigs, dogs, and snakes have caused bird exterminations on islands with few or no native mammalian and reptilian predators. Notable cases include extinctions of almost all forest bird species on the island of Guam (which lacks native snakes) following the introduction of the brown tree snake from the Solomons (!) (Savidge 1987) and extinctions due to arrival of rats on Big South Cape (a New Zealand island), Lord Howe, and Hawaii (Atkinson 1985). Islands with native rodents (e.g., Galapagos) or native land crabs (e.g., Fiji and many other tropical Pacific islands) suffered few or no losses of birds to introduced rats, presumably because birds of those islands had coevolved with and learned to cope with predators.

Most large central Northern Melanesian islands have native rodents, and as a result the commensal rats of Europeans have barely established a toehold in Northern Melanesia. Most islands, including even such outliers as Rennell and Bellona, have large native lizards (*Varanus*) and colubrid and boid snakes that prey on birds, plus numerous species of large land crabs (the coconut crab *Birgus latro* and others). New Britain and Guadalcanal have native *Uromys* rats, which are by far the most important predators on bird nests in Queensland (Laurance et al. 1993, Laurance & Grant 1994). Thus, one would expect Northern Melanesian birds not to be very susceptible to introduced predators, since they had evolved in the presence of a varied suite of native predators. The only evidence for predator-caused extinctions subsequent to European arrival are the already-discussed cases of possible cat predation on ground birds. The sole possible avian predators that Europeans introduced besides cats are local populations of *Rattus rattus* and *R. norvegicus*. Predators brought by earlier Melanesian colonists were *Rattus exulans*, *R. praetor*, phalangers, pigs, and dogs. Some of these surely contributed to the declines of native rodents already discussed in chapter 4.

Established introduced nonnative bird species are a third potential cause of declines of native island bird species, mediated either by competition or by transmission of avian diseases. It is certainly true that those oceanic islands on which many introduced bird species have established themselves are also the islands that have lost many native species: notably, New Zealand and Hawaii, each with dozens of introduced exotics and dozens of lost native species. Less certain is the chain of cause and effect: did the introductions cause the declines of natives, or did declines of natives for other reasons permit successful introductions (Diamond & Veitch 1981, Case 1996)? Whichever way the chain operates, effects of introduced bird species must be negligible in Northern Melanesia, where only a single nonnative bird species has established itself (the Indian myna *Acridotheres tristis* in nonforest habitats of four Solomon islands; see Appendix 3).

The remaining major mechanism of island bird exterminations is overhunting, which is especially likely to victimize large birds (e.g., *Ducula* pigeons in Polynesia; Steadman & Olson 1985), flightless birds (e.g., moas and elephant birds), and colonially nesting birds (e.g., seabirds). In Northern Melanesia the sole modern declines attributable to overhunting are recent ones of colonial pigeons (*Caloenas nicobarica*, *Ducula pistrinaria*, and *D. spilorrhoa*). The sole flightless extant birds known from Northern Melanesia are the rails *Gallirallus insignis* (New Britain; plate 8), *Nesoclopeus woodfordi* (Bougainville, Choiseul, Ysabel, Guadalcanal), and *Pareudiastes sylvestris* (San Cristobal; plate 8), plus the possibly flightless *Gallirallus rovianae* (New Georgia group). As already mentioned, two undescribed flightless, extinct rail species in the genera *Porphyrio* and *Gallirallus* formerly occurred on New Ireland (Steadman et al. 1999). Since fossils of extinct flightless rails have been found on almost all paleontologically explored oceanic islands, we assume that other Northern Melanesian islands besides New Ireland supported flightless rails that succumbed to prehistoric human colonists. However, because Northern Melanesia has large native rodents, we doubt that it supported as diverse a radiation of flightless taxa as did Hawaii and New Zealand. We expect that Northern Melanesia, like other oceanic islands, also formerly supported nesting colonies of shearwaters and petrels, of which only Heinroth's shearwater (*Puffinus heinrothi*) on Bougainville survives today.

Summary

Known post-European extinctions of birds in Northern Melanesia have been far fewer to date than in Polynesia or on other remote oceanic islands, though the current rapid spread of commercial logging is likely to increase the pace of extinctions. Only one nonnative introduced species has established itself. The relative intactness of Northern Melanesia's historical avifauna is as one would expect from the low European population, the varied fauna of native predators, the few and very restricted populations of introduced European mammals except for cats, and the paucity of avian targets to reward hunters with firearms except for colonial pigeons. The extent of prehistoric extinctions is poorly known. We would guess that they included colonial shearwaters and petrels, flightless rails, and populations of any species on small outlying islands cleared for agriculture. We would also guess, and the limited available information about bird fossils suggests, that prehistoric extinctions have been less severe on large or central Northern Melanesian islands than in Polynesia. Thus, we assume that the Northern Melanesian avifauna as known from modern distributions is worth analyzing.

PART III

THE NORTHERN MELANESIAN AVIFAUNA

8

Family Composition

Part III completes the presentation of background material by introducing Northern Melanesia's resident native avifauna, whose colonization routes and speciation are analyzed in the remainder of the book. This chapter describes the avifauna's family composition and compares it with that of the principal source avifauna, the avifauna of New Guinea. Chapters 9–12 go on to discuss variation in total number of resident bird species among Northern Melanesian islands and variation in ecologically important properties (level of endemism, habitat preference, abundance, and dispersal ability, plus number, area, and species richness of islands occupied) among Northern Melanesian bird species.

Most Northern Melanesian birds were derived from New Guinea. As the source pool of colonists, we take the birds of that portion of New Guinea facing Northern Melanesia—i.e., the northern watershed of eastern New Guinea, from the east tip of New Guinea west to Humboldt Bay and the Cyclops Mountains just west of the international border between Papua New Guinea and Indonesian New Guinea (141°E longitude). As the unit of analysis, here and elsewhere in this book, we take the zoogeographic species—i.e., superspecies, or else individual species that are not members of superspecies. By this definition, the New Guinea source pool contains 432 resident land and freshwater species, and Northern Melanesia contains 195.

Table 8.1 compares the family composition of the Northern Melanesian avifauna with that of the New Guinea source avifauna, using a conventional family-level classification similar to the one adopted with minor variations in most of the recent ornithological literature (e.g., the 16-volume *Peters Check-list of Birds of the World*: Peters 1931–87). Table 8.1 also shows how many of the New Guinea pool species in each family occur in Northern Melanesia. For many families this number is lower than the total number of species in Northern Melanesia because Northern Melanesia contains some endemic species not belonging to superspecies represented on New Guinea, and also contains some species derived from sources other than New Guinea, especially from Australia and the New Hebrides.

Disproportions Between the Northern Melanesian and New Guinean Source Avifaunas

To a first approximation, the source avifauna and the Northern Melanesian avifauna are fairly similar in composition. On closer examination, North-

Table 8.1. Family composition of the Northern Melanesian avifauna and of the New Guinean source avifauna.

Family	No. of species in NE New Guinea	No. of species in N Melanesia	No. of NE New Guinea species in N Melanesia
Nonpasserines			
Casuariidae (cassowaries)	2	1	1
Podicipedidae (grebes)	1	1	1
Pelecanidae (pelicans)	0	1	—
Phalacrocoracidae (cormorants)	2	1	1
Anhingidae (anhingas)	1	0	0
Ardeidae (herons)	11	7	6
Threskiornithidae (ibises and spoonbills)	1	2	1
Anatidae (ducks)	8	4	4
Accipitridae (hawks)	20	13	11
Falconidae (falcons)	3	3	2
Megapodiidae (megapodes)	3	1	1
Phasianidae (quail and pheasants)	2	1	1
Turnicidae (bustard-quail)	1	1	1
Rallidae (rails)	14	10	7
Jacanidae (jacanas)	1	1	1
Charadriidae (plovers)	1	1	1
Scolopacidae (sandpipers)	1	0	0
Recurvirostridae (avocets and stilts)	1	1	1
Burhinidae (thick-knees)	1	1	1
Columbidae (pigeons)	33	25	20
Psittacidae (parrots)	31	13	9
Cuculidae (cuckoos)	14	6	3
Tytonidae (barn owls)	3	2	1
Strigidae (owls)	4	2	1
Podargidae (frogmouths)	2	1	1
Aegothelidae (owlet-nightjars)	5	0	0
Caprimulgidae (goatsuckers)	3	2	1
Apodidae (swifts)	5	4	4
Hemiprocnidae (tree swifts)	1	1	1
Alcedinidae (kingfishers)	16	10	8
Meropidae (bee-eaters)	1	1	1
Coraciidae (rollers)	1	1	1
Bucerotidae (hornbills)	1	1	1
Passerines (songbirds)			
Pittidae (pittas)	2	3	2
Alaudidae (larks)	1	0	0
Hirundinidae (swallows)	1	1	1
Motacillidae (pipits and wagtails)	2	0	0
Campephagidae (cuckoo-shrikes)	12	7	4
Laniidae (shrikes)	1	0	0
Turdidae (thrushes)	5	3	3
Orthonychidae (log-runners)	8	0	0
Maluridae (wren-warblers)	5	0	0
Acanthizidae (Australasian warblers)	16	1	0
Sylviidae (Old World warblers)	4	8	4
Rhipiduridae (fantails)	10	6	4
Monarchidae (monarch flycatchers)	10	8	6
Eopsaltriidae (Australasian robins)	21	2	1
Pachycephalidae (whistlers)	22	3	2
Neosittidae (Australian nuthatches)	2	0	0
Climacteridae (Australian creepers)	1	0	0

Table 8.1. *Continued.*

Family	No. of species in NE New Guinea	No. of species in N Melanesia	No. of NE New Guinea species in N Melanesia
Dicaeidae (flower-peckers)	8	1	1
Nectariniidae (sunbirds)	2	2	2
Zosteropidae (white-eyes)	4	7	2
Meliphagidae (honey-eaters)	44	11	3
Estrildidae (estrildid finches)	9	3	3
Sturnidae (starlings)	4	6	3
Oriolidae (orioles)	1	0	0
Dicruridae (drongos)	2	1	1
Grallinidae (magpie-larks)	1	0	0
Artamidae (wood-swallows)	2	1	1
Cracticidae (butcherbirds)	4	0	0
Ptilonorhynchidae (bowerbirds)	8	0	0
Paradisaeidae (birds of paradise)	24	0	0
Corvidae (crows)	2	2	1
Total	432	195	137

The left-most column of numbers gives, for each bird family, the number of resident land and fresh-water bird species in Northeast New Guinea (as defined in the text), which served as the main source for Northern Melanesian birds. The right-most column gives the number of those New Guinea source species that also occur in Northern Melanesia. The middle column of numbers gives the number of species in Northern Melanesia, regardless of whether they belong to the New Guinea source pool. Species are assigned to bird families according to current standard classifications, such as Peters' *Check-list of Birds of the World* (Peters 1931–87).

ern Melanesia is enriched in some groups and deficient in others, as noted previously by Mayr (1965c). On the one hand, Northern Melanesia is enriched in Old World warblers, white-eyes, and starlings: it contains even more species of each of those families (including numerous endemic species) than does New Guinea. Northern Melanesia is also enriched in pigeons and swifts, of which it contains considerably more than half of the New Guinean source species, because of the superior overwater dispersal abilities of these families. A trivial difference is that Northern Melanesia has a pelican absent in the New Guinea source pool, but that pelican occurs in South New Guinea outside the Northeast New Guinean source region as we have defined it. On the other hand, Northern Melanesia contains no representative of several species-rich New Guinea families (owlet-nightjars, log-runners, wren-warblers, birds of paradise, and bowerbirds), only one representative of another species-rich New Guinea family (Australasian warblers), and disproportionately few representatives of the Australasian robins, whistlers, honey-eaters, and flower-peckers. Trivial deficiencies are that Northern Melanesia lacks representatives of nine families represented in the New Guinean source pool by only one or two species (anhingas, sandpipers, larks, pipits, shrikes, orioles, magpie-larks, Australian nuthatches, and Australian creepers).

Of the families in which Northern Melanesia is strikingly underrepresented, all except the owlet-nightjars belong to the order Passeriformes (songbirds), generally referred to as passerines. Table 8.2 shows that 48% of New Guinea's nonpasserines, but only 18% of its passerines, occur in Northern Melanesia. Alternatively (table 8.3), Northern Melanesia has 61% as many nonpasserines as New Guinea, but only 32% as many passerines. This finding corresponds to Faaborg's (1977) finding that, if one compares the family-level composition of the West Indian avifauna and that of the neotropical mainland colonist pool, the West Indies are relatively enriched in nonpasserines and relatively poor in passerines.

New Guinean source species can also be classified as to whether they are confined to the mountains of New Guinea or whether they also occur in the New Guinea lowlands. Table 8.2 shows that far fewer New Guinean montane than lowland species have reached Northern Melanesia: only 12%, compared to 44% of the lowland species.

Table 8.2. Comparison of the New Guinean source avifauna with that fraction of it that reached Northern Melanesia.

	Lowland species	Montane species	All species
Nonpasserines	149/82 (55%)	45/11 (24%)	194/93 (48%)
Passerines	112/34 (30%)	126/10 (8%)	238/44 (18%)
Sedentary Corvida	63/4 (6%)	94/3 (3%)	157/7 (4%)
Vagile Corvida	26/14 (54%)	13/2 (15%)	39/16 (41%)
Passerida and pittas	23/16 (70%)	19/5 (26%)	42/21 (50%)
Total	261/116 (44%)	171/21 (12%)	432/137 (32%)

Species in the Northeast New Guinea source pool are classified as lowland (present in the lowlands) or montane (confined to the mountains). They are also classified taxonomically as passerines (songbirds) or non-passerines (other birds). The passerines are subdivided according to Sibley & Ahlquist (1990), as belonging to the parvorder Corvida (believed to have originated in the Australian region) or to the parvorder Passerida or family Pittidae (believed to have originated in Eurasia or Africa). The Corvida are further subdivided into "vagile" or "sedentary" tribes and families, according to whether the tribe or family has many or else few or no species that reached the Asian continental shelf (see text for details). Each entry consists of the number of species in the New Guinean source pool, followed by the number of those species that also occur in Northern Melanesia, expressed in parentheses as a percentage. Note that lowland species are much better colonists of Northern Melanesia than are montane species, in every taxonomic category; the sedentary Corvida are much poorer colonists of Northern Melanesia than other taxonomic groups, both among lowland and montane species; and the other two categories of passerines (vagile Corvida, Passerida and pittas) differ little in ability to colonize Northern Melanesia from each other or from nonpasserines.

Disproportions Based on a Revised Classification

Further insight into the causes of these disproportions between the Northern Melanesian and New Guinea source avifaunas comes from taxonomic studies by Sibley and Ahlquist (1990; Sibley & Monroe 1990), who used DNA/DNA hybridization to reassess higher-level relationships of bird species. They showed that almost all New Guinea and Australian passerine species belong to two separate parvorders: the Passerida, which evidently radiated in Eurasia and Africa, but of which many colonists have reached New Guinea and Australia; and the Corvida, which evidently radiated in New Guinea and Australia but of which many colonists later reached Eurasia and Africa and further radiated there. The Passerida include the larks, swallows,

Table 8.3. Comparison of the New Guinean source avifauna with the Northern Melanesian avifauna.

	NE New Guinea	N Melanesia	N Melanesia/ NE New Guinea
Non-passerines	194	119	61%
Passerines	238	76	32%
Sedentary Corvida	157	18	11%
Vagile Corvida	39	24	62%
Passerida and pittas	42	34	81%
Total	432	195	45%

Species are classified taxonomically into nonpasserines and three groups of passerines, as explained in the legend of Table 8.2 and in the text. Within each taxonomic category, the left-most column of numbers gives the number of species in the Northeast New Guinea source pool; the middle column, the number of species in Northern Melanesia (regardless of whether those species also occur in the Northeast New Guinea source pool); and the right-most column gives the middle column as a percentage of the left-most column. Note that sedentary Corvida are underrepresented in Northern Melanesia compared to other taxonomic groups.

pipits and wagtails, thrushes, Old World warblers, starlings, sunbirds, flower-peckers, white-eyes, and various groups of finches. The Corvida include all the endemic or near-endemic traditionally recognized New Guinea and Australian families, such as the lyrebirds, log-runners, wren-warblers, Australasian warblers, Australasian robins, whistlers, birds of paradise, and bowerbirds. Groups of the Corvida that reached Eurasia and Africa and radiated there, and of which members of predominantly Eurasian and African genera subsequently recolonized Australia, include the crows, drongos, orioles, and shrikes. The sole Northern Melanesian and New Guinean passerines belonging to neither the Passerida nor Corvida are the pittas, which like the Passerida evidently radiated in Eurasia and then colonized New Guinea and Australia.

All of the species-rich New Guinea passerine families that are absent or underrepresented in Northern Melanesia belong to the Corvida. We divided the Corvida into two sets differing in dispersal ability. First, we tabulated those families, subfamilies, or tribes of the Corvida (as recognized by Sibley & Ahlquist 1990) that occur in New Guinea but of which few or no representatives have reached the Sunda Shelf or Asian mainland. In effect, we use colonization westward from New Guinea as a reflection of dispersal ability. Using Sibley & Ahlquist's (1990) terminology, those groups are the Climacteridae (Australian creepers), Ptilonorhynchidae (bowerbirds), Maluridae (wren-warblers), Meliphagidae (honey-eaters), Acanthizinae (Australasian warblers), Eopsaltriidae (Australasian robins), Orthonychidae (log-runners), Pomatostomidae (Australian pseudo-babblers), Cinclosomatinae (quail-thrushes), Pachycephalinae (whistlers and Australian creepers), Artamini (butcherbirds and wood-swallows), and Paradisaeini (birds of paradise). Of these 11 groups, six have no species extending west of Wallace's line, three have one species each, and Artamini and Pachycephalinae have three and four species, respectively (12% and 7% of the species in those respective groups). We refer to these groups as the "sedentary Corvida" because relatively few or no members have reached the Asian continental shelf. (It may at first seem disconcerting to include the honey-eaters and whistlers under this designation, since some species in each family have spread far over the Pacific. However, those whistler colonists belong to only five superspecies of the 57 species of Pachycephalinae, most of which are indeed poor overwater colonists as judged by failure to spread both east into the Pacific and west to Asia. Among the Meliphagidae, though, 34 species, or about 16% of the family, occur on Pacific islands east of New Guinea, suggesting that some factor other than problems in overwater dispersal, such as competition from Asian sunbirds, accounts for the near-failure of honey-eaters to colonize Asia.)

Thus, we split off a set of Corvida that we term "sedentary Corvida." We then separately tabulated those families, subfamilies, and tribes of Corvida that occur in New Guinea but of which numerous representatives have reached the Asian continental shelf. Those groups, which we term the "vagile Corvida," consist in Sibley & Ahlquist's (1990) terminology of the Rhipidurini (fantails), Dicrurini (drongos), Monarchini (monarchs and magpie-larks), Corvini (crows), Oriolini (orioles and cuckoo-shrikes), and Laniidae (shrikes). In four of those six groups, a great majority of the species occurs in Eurasia or Africa, while this is true of 24% or 26% of the other two groups, the Rhipidurini or Monarchini, respectively—still much higher than in any of the groups that we consider sedentary Corvida.

Tables 8.2 and 8.3 compare the Northern Melanesian and New Guinean avifaunas by this revised taxonomy, which we believe corresponds more closely to actual relationships and evolutionary history than does the conventional taxonomy. Disproportionately few New Guinean species of sedentary Corvida have reached Northern Melanesia: only 4%, compared to 41% for the vagile Corvida, 48% for the nonpasserines, and 50% for the Passerida and pittas (table 8.2). (One might wonder whether this trend is merely due to a disproportionately large fraction of sedentary Corvida being montane species and due to montane species being poorer colonists than lowland species. However, table 8.2 also shows that both the lowland and the montane species among the sedentary Corvida are underrepresented in Northern Melanesia compared to the other taxonomic groups.) Even when one considers all species of Northern Melanesia, regardless of whether they also occur in New Guinea (table 8.3), the sedentary Corvida are still underrepresented: Northern Melanesia has only 11% of New Guinea's species total of sedentary Corvida, compared to 62% for the vagile Corvida, 61% for the nonpasserines, and 81% for the Passerida and pittas.

The explanation for Northern Melanesia's underrepresentation of the group that we define as sedentary Corvida has nothing to do with their taxonomic status as Corvida, since other Corvida (those that we define as vagile Corvida) are well represented in Northern Melanesia. Instead, the explanation has to do with overwater dispersal ability. Within both the Corvida and the Passerida, there is much variation in dispersal ability. However, because the Passerida arose in Eurasia and Africa, only those vagile Passerida groups with good overwater dispersal ability could reach New Guinea, while the more sedentary Passerida groups (such as babblers) were unable to disperse from the Asian continental shelf to New Guinea. The Corvida also are a mixture of sedentary and vagile groups, but both the sedentary and vagile groups occur in New Guinea/Australia, since that was where the Corvida evolved. While we have identified poor overwater dispersers among the Corvida on the basis of failure to reach the Asian continental shelf, essentially the same definitions follow from ability to cross water to reach oceanic islands of the New Guinea region (islands near New Guinea but lying in deep water off the continental shelf and lacking Pleistocene land bridges to New Guinea). On those islands only 16% of the species in New Guinea's groups of sedentary Corvida (defined by having reached Asia) occur, but 45% of the vagile Corvida, 43% of the Passerida and pittas, and 51% of the nonpasserines are present.

An analysis of subgroups among the nonpasserines and Passerida reinforces these conclusions drawn from the Corvida. For example, among the nonpasserines, the family Aegothelidae (owlet-nightjars), which resembles the Corvida in having evolved in New Guinea/Australia, is underrepresented on oceanic islands of the New Guinea region (reached by only one of five aegothelid species in the Northeast New Guinean pool) and is absent from Northern Melanesia. Among the Passerida, the group that Sibley and Ahlquist (1990) term the Melanocharitini (Papuan sunbirds and berry-peckers) has secondarily lost its dispersal ability after reaching New Guinea, so that its nine species are now all endemic to New Guinea and its land-bridge islands; none occurs in Northern Melanesia or on an oceanic island of the New Guinea region. We shall discuss many similar examples of such secondary loss of dispersal ability in the Northern Melanesian avifauna.

Table 8.2 also shows that within each taxonomic group, the montane species are greatly underrepresented in Northern Melanesia compared to the lowland species. This is probably because New Guinea's montane species are especially poor at overwater dispersal, and also because Northern Melanesia has lower mountains and relatively less montane habitat than does New Guinea.

Summary

Most groups of birds in the Northeast New Guinean source pool have colonized Northern Melanesia. The disproportions between that source avifauna and the Northern Melanesian avifauna mainly involve groups underrepresented in Northern Melanesia because of poor overwater colonizing ability. Most of these groups belong to the songbird parvorder Corvida, which radiated in New Guinea/Australia, and many of whose members have never left the New Guinea/Australia continental shelf. In addition, the underrepresentation of New Guinea's montane species in Northern Melanesia may further reflect Northern Melanesia's deficit of montane habitats, as well as the poor dispersal ability of many montane species.

9

Determinants of Island Species Number

The number of bird species on Northern Melanesian islands varies from 3 species on some tiny islets up to 127 for the largest island, New Britain. What factors account for that variation?

Tables 9.1 and 9.2 summarize for each ornithologically significant island its number of resident species and also its area (A), height (H), and distance (D) from the nearest major source island. We list separately for each island the number of species normally occurring in the lowlands (S_{low}) and the number normally confined to the mountains on that island (S_{mt}). Here we discuss the dependence of S_{low} on A, the dependence of S_{low} on D, effects of volcanic defaunation, determinants of S_{mt}, and, finally, the effects of Pleistocene land bridges. Our discussion is based on (updated from) Diamond (1972a, 1974, 1982a, 1983), Diamond and Mayr (1976), Diamond et al. (1976), Gilpin and Diamond (1976, 1981), Mayr and Diamond (1976), and Diamond and Gilpin (1980).

Species/Area Relation

Bismarck Islands

Figure 9.1 is a double-logarithmic plot of S_{low} against island area for the 41 Bismarck Islands of table 9.1. For those 22 Central Bismarck Islands (coded by symbol ● in Fig. 9.1) that have not undergone recent volcanic defaunation and that were not connected by Pleistocene land bridges to much larger islands, area explains 94% of the variation in S_{low} according to a power function of the form $S = S_o A^z$:

$$S_{low} = 15.7\ A^{0.215} \tag{9.1}$$

This may be rewritten as $\log S_{low} = \log 15.7 + 0.215 \log A$, resulting in an approximately linear double-logarithmic graph (fig. 9.1). In the island biogeographic literature, one finds that some species/area relations are better fitted by a power function, while others are better fitted by a logarithmic function of the form $S = a + b \log A$. In the case of Bismarck birds, the best-fit logarithmic function ($S_{low} = 25.4 + 13.2 \log A$) yields lower explained variation than the best-fit power function (87% vs. 94%).

Four groups of islands are coded by different symbols in figure 9.1 and require separate discussion: defaunated volcanoes (Δ), land-bridge islands (O), the isolated St. Matthias group (x), and the even more isolated Northwest Bismarck group (+). First, the defaunated volcanoes include the most deviant island in figure 9.1, Ritter (point Δ at $A = 0.36$ km^2), whose S_{low} of 4 species is far below (Z^2 value of 17.4) the value of 13 species predicted for its area. The explanation is that Ritter's steep slopes are still largely bare of vegetation after the island's volcanic defaunation in 1888 (Diamond 1974). The three other volcanically defaunated Bismarck is-

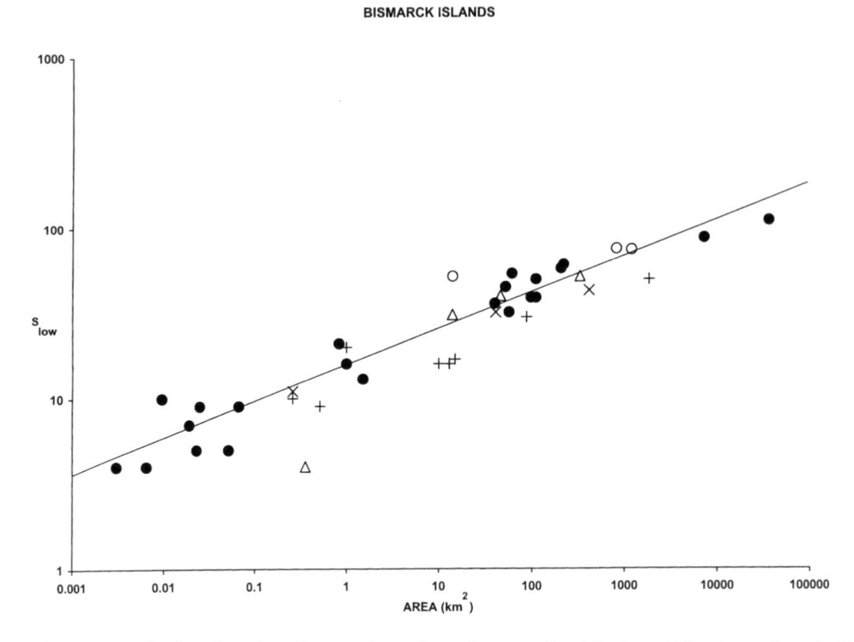

Fig. 9.1. Species/area relation for the Bismarcks, plotted on a double-logarithmic scale. Ordinate: S_{low}, the resident number of lowland species. Data are from table 9.1. ● = the Central Bismarck Islands New Britain, New Ireland, and their satellites (islands 1, 7, 9–17, 19, 20, and 22–30 in table 19.1, omitting the incompletely surveyed island 18). X = the isolated St. Matthias group (islands 31–33). + = the very isolated Northwest Bismarcks (islands 34–41). Δ = islands 3, 4, 6, and 8, recently defaunated by volcanic explosions. ○ = islands 2, 21, and 24, formerly joined by Pleistocene land bridges to much larger islands. The solid line is the best-fit line through the points ●; $S_{low} = 15.7\ A^{0.215}$. See text for discussion.

lands depicted in figure 9.1 (Long, Tolokiwa, and Crown, symbols Δ at A = 14–329 km²) have S_{low} values as expected for normal islands because their vegetation has already regenerated to tall forest after defaunation in the seventeenth century. This also proves to be true of recent volcanic islands in the Solomons, such as Savo, Simbo, Kulambangra, and Bougainville.

Second, the three islands formerly joined by Pleistocene land bridges to larger islands (symbols ○) all have S_{low} values slightly higher than predicted for islands lacking land bridges, but only for one of those three islands (Vuatom, at A = 14 km²) is the excess statistically significant ($Z^2 = 5.33$). We discuss this "supersaturation" at the end of this chapter.

Third, the three islands of the isolated St. Matthias group (sympols X in Fig. 9.1) have S_{low} values somewhat below those predicted for central islands of the same area, but these species deficits are neither individually nor collectively statistically significant. Finally, seven of the eight islands of the even more isolated Northwest Bismarck group (symbols + in Fig. 9.1) have S_{low} values even farther below those predicted, and these species deficits are collectively significant ($p = .03$).

Thus, species numbers on Bismarck islands show to varying degrees the effects of defaunation, supersaturation, and isolation familiar in island biogeography (Diamond 1972a, 1974).

Solomon Islands

Figure 9.2 is a double-logarithmic plot of S_{low} against island area, similar to figure 9.1 for the Bismarck Islands, for the 51 Solomon Islands of table 9.2. Whereas a power function gave a better fit than a logarithmic function to the data for the Central Bismarck Islands, the reverse is true for central, nonisolated Solomon islands (see Fig. 9.2 legend for definition). All Solomon islands with areas smaller than 0.4 km² have S_{low} values significantly

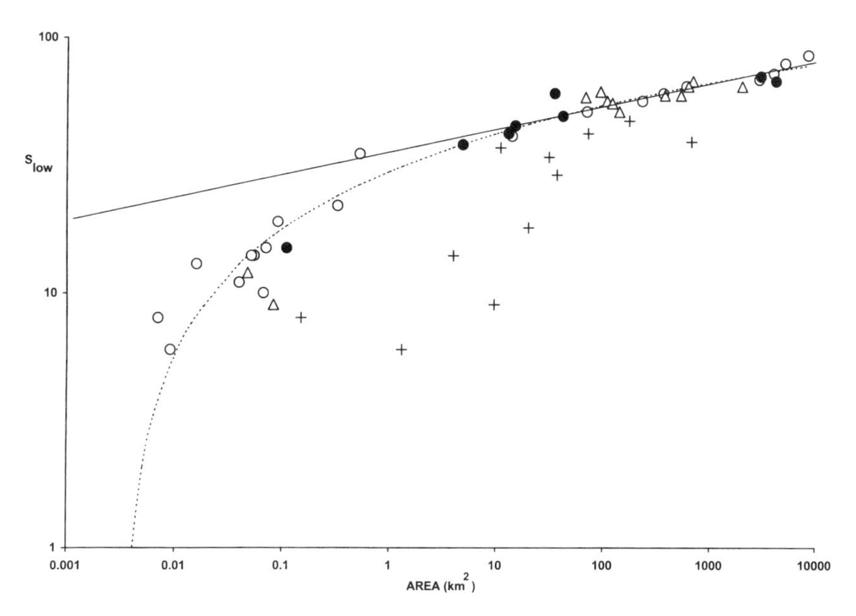

Fig. 9.2. Species/area relation for the Solomons, plotted on a double-logarithmic scale. Ordinate: S_{low}, the resident number of lowland species. Data are from table 9.2. ○, Δ, ● = "central" islands: i.e., islands with $S_{total} > 50$ species and islands within 10 km of such an island. ○ = islands derived from the expanded Pleistocene island of Greater Bukida (islands 1–11 and 13–21 of table 9.2). Δ = islands similarly derived from Pleistocene Greater Gatumbangra, Greater Vellonga, and Greater Rendipari (islands 22–33). ● = islands with no recent connections (islands 34, 35, 38, 40, 42, 43, 45, 47). + = isolated islands, defined as islands that support fewer than 50 species and that lie more than 10 km from the nearest island with more than 50 species (islands 12, 36, 37, 39, 41, 44, 46, and 48–51). The solid line is the best-fit line $\log S_{low} = \log 36.1 + 0.086 \log A$ fitted through the points for all central islands whose area exceeds 0.4 km^2. Note that the relation actually defined by the points is markedly curved. The dashed curve is the regression equation $S_{low} = 29.8 + 12.1 \log A$ depicted in figure 9.3. Modified from Diamond and Mayr (1976).

below those predicted by the best-fit power function ($S_{low} = 36.1\ A^{0.086}$; explained variation 86%) fitted to data for islands larger than 0.4 km^2 (fig. 9.2). Instead, the data for Central Solomon Islands over the whole range of areas are best fit by a logarithmic function (fig. 9.3; explained variation 97%):

$$S_{low} = 29.8 + 12.1 \log A \qquad (9.2)$$

Every isolated Solomon island (points + in Figs. 9.2 and 9.3) has an S_{low} value significantly lower than the value predicted by either the best-fit power function or the best-fit logarithmic function to data for central islands. The next section demonstrates that this deficit increases with island isolation.

Most of the Solomon islands of table 9.2 were connected to other islands by Pleistocene land bridges, making it difficult to test for supersaturation effects (i.e., S values higher than for never-connected islands of the same area). However, the slightly high values (above the best-fit straight line) for the two largest islands (right-most points) of figure 9.2, Bougainville and Guadalcanal, may be due to supersaturation, as discussed in the last section of this chapter.

Power Function Exponent

The steepness of the double-logarithmic form of the species/area relation is described by its slope, z, which is also the exponent of area if the relation is

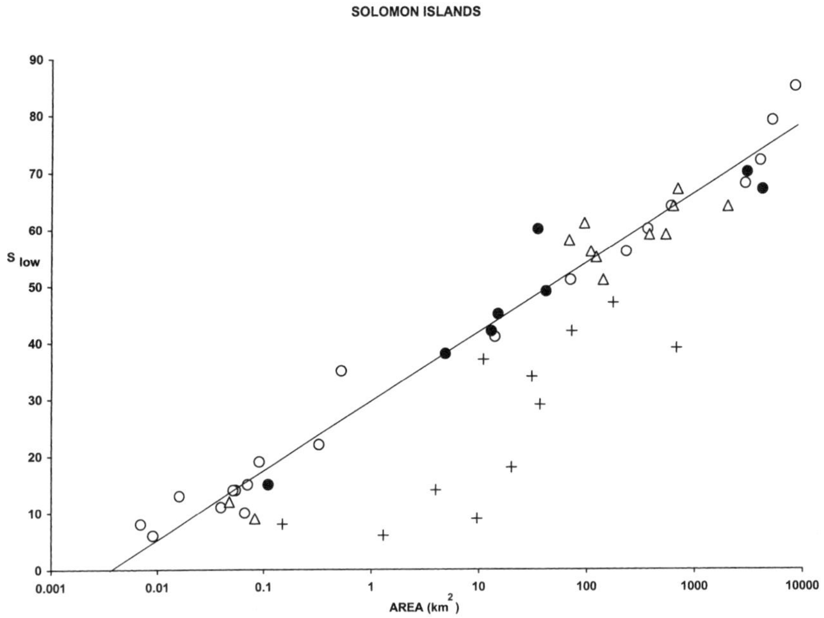

Fig. 9.3. Species/area relation for the Solomons: as Fig. 9.2, except that the ordinate is S_{low} instead of log S_{low}. The solid line is the best-fit line $S_{low} = 29.8 + 12.1 \log A$ fitted through the points for all central islands (i.e., all points except those coded +). Note the excellent fit of this line (explained variation 97%) over a 10^6-fold range in island area. Points + for isolated islands fall below this relation for central islands, as analyzed further in figure 9.4. Modified from Diamond and Mayr (1976).

reexpressed as a power function ($S = S_0A^z$). For well-surveyed islands exceeding about 1 km^2 in area, the value of z is 0.22 for the satellite islands of New Guinea; 0.22 for the Central Bismarcks (eq. 9.1, fig. 9.1); 0.086 for the Central Solomons (fig. 9.2); and 0.05 for the New Hebrides. Thus, z decreases for archipelagoes at increasing distances from New Guinea. The explanation for this (Diamond & Mayr 1976) is that increasing distances from the major source of colonists (New Guinea) select for an increasingly vagile subset of the source colonist pool, and that the species/area relation is flatter for a more vagile set of species (because vagile species can maintain themselves on small as well as on large islands).

Causes of the Species/Area Relation

There has been much discussion in the biogeographic literature as to why large islands generally have more species than small islands. Detailed examination of bird populations on Northern Melanesian islands reveals that three major factors contribute. First, habitat heterogeneity increases with island area, and some species are restricted to particular habitats present only on larger islands. For instance, only large islands can have rivers and swamps and hence support bird species confined to rivers and swamps. Second, even if appropriate habitat is present on an island, the island can support some particular species characteristic of that habitat only if the available area of habitat is at least equal to the territory size for one male and female of that species (or for one family group or flock, in the case of social species). Finally, a population consisting of a single pair is unlikely to exist for long. Instead, populations may occasionally go extinct and may eventually be refounded by new immigrants. Population size increases, and hence risk of extinction decreases, with area. Since, in addition, immigration rates increase with island area, the number of species on an island should fluctu-

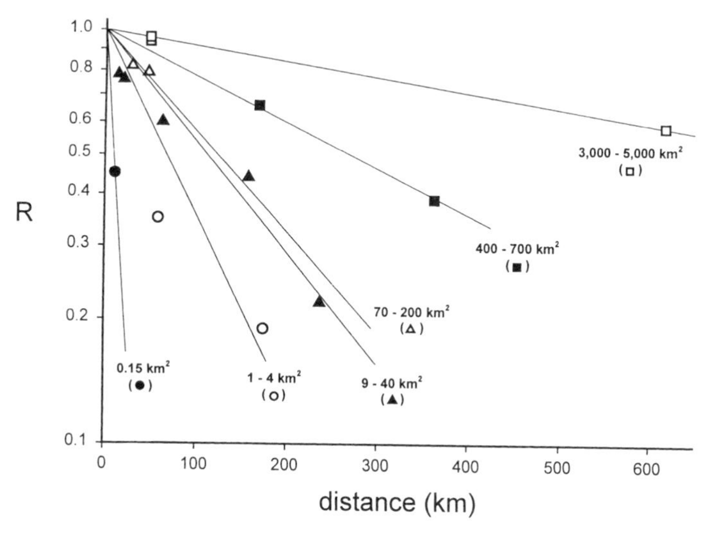

Fig. 9.4. Effect of isolation on bird species number of Solomon islands. Ordinate: the ratio R, defined as the actual S_{low} of an isolated island divided by the value expected for a central island of the same area. See legend of Fig. 9.2 for definition of "central" islands. Abscissa: distance from nearest island with $S_{low} > 50$. The points are based on islands 12, 34–37, 39, 41, 44, 46, and 48–51 of table 9.2, plus Ndeni of the Santa Cruz group and Espiritu Santo of the New Hebrides (see table 1 of Diamond et al. 1976 for data on those two islands). Different symbols are used for islands of six different size classes, as coded on the graph. Note that species number decreases with isolation, and that this decrease is steeper for smaller islands. The lines are exponentials ($R = \exp[-D/D_0]$) fitted through the points for islands of each size class. Modified from Diamond et al. (1976).

ate around or tend to approach a dynamic equilibrial value that increases with area (Mayr 1965a, MacArthur & Wilson 1967).

Effect of Isolation

Examining the species/area relation for the Bismarcks (fig. 9.1), we see that the outlying islands of the St. Matthias group, and especially of the even more remote Northwest Bismarcks, have S_{low} values below those expected for Central Bismarck Islands of the same area. A similar conclusion holds for the Solomons and is depicted quantitatively in figure 9.4, which shows that ratios of species number on isolated islands to species number on central islands of the same area decrease exponentially with distance for islands in a given size class. The decrease is steepest for the smallest islands. If one describes the exponential decrease of the ratio by a space constant, D_o (i.e., ratio $R = \exp[-D/D_0]$ for an island at distance D from the source), then the space constant increases from 15 km for an islet of 0.15 km^2 to 1200 km for large islands of 3000–5000 km^2. This finding follows straightforwardly from considerations of equilibria between immigration and extinction when species differ in their probabilities of immigrating and going extinct (Diamond et al. 1976).

Montane Species

Our analysis of species number has so far considered only those of an island's species that normally occur at sea level. In addition, islands higher than about 800–1000 m also have some populations that are confined to the mountains and that do not normally occur at sea level. As summarized in tables 9.1 and 9.2, the number of such montane populations (S_{mt}) ranges from only one species on islands barely reaching 800–1000 m in elevation, such as New Georgia and Vella Lavella, to 15–23 species on New Britain, New Ireland, Bougainville, and

Guadalcanal, Northern Melanesia's highest islands (2400–2600 m). S_{mt} increases with island height, H (in meters), such that, on average, S_{mt} equals 4.6 species per 1000 m of height. Alternatively, since one might expect S_{mt} to be related to island area as well as to height, and since S_{low} increases with area, one can calculate the ratio S_{mt}/S_{low}, which equals, on average, 0.07 per 1000 m of height. That is, each 1000 m of height adds a number of montane populations equal to 7% of the species number at sea level.

However, there is also much variation in S_{mt} not explained by island height or area alone. Examples are Gatukai (H = 888 m, A = 109 km^2), which has three montane populations compared to only one on the slightly higher and much larger New Georgia (H = 1006 m, A = 2044 km^2); Malaita (H = 1280 m, A = 4307 km^2) and Ysabel (H = 1250 m, A = 4095 km^2) have nearly the same heights and areas, but Malaita has seven montane populations compared to only three for Ysabel; and Ganonga (H = 854 m, A = 142 km^2) has three montane populations compared to only one on the much larger and nearly as high Vella Lavella (H = 793 m, A = 640 km^2).

An explanation for these paradoxes becomes apparent from contour maps (Mayr & Diamond 1976). Some islands, such as Kulambangra and Tolokiwa, are steep stratovolcanoes, for which the area of montane habitat is a significant fraction of the basal area. Other islands, such as New Georgia and Vella Lavella, are mostly flat and low, with just a few peaks. Not surprisingly, the number of montane populations reflects not only island height but also the available area of montane habitat. In each of the comparisons of the preceding paragraph, the island with more montane populations has proportionately much more montane habitat (area above 600 m) than the island with fewer montane populations: 2% versus 1% of lowland area in the comparison of Gatukai versus New Georgia, 10% versus 3% for Malaita versus Ysabel, and 3% versus 0.8% for Ganonga versus Vella Lavella. An equation in which island areas above and below 600 m and island elevation are the dependent variables explains 93% of interisland variation in S_{mt} in the Solomons (Mayr & Diamond 1976). The principal remaining discrepancy is the low S_{mt} value of the Bismarck island of Long, related to defaunation in a volcanic eruption several centuries ago. Long's montane avifauna has recovered much more slowly than its lowland avifauna because of the lower dispersal ability of montane bird species (Diamond 1974).

Supersaturation

In the New Guinea region, islands that were connected by land bridges to New Guinea at Pleistocene times of low sea level have today up to nearly three times more species than islands that have similar areas and distances from New Guinea but that lacked Pleistocene land bridges (Diamond 1972a). During the Pleistocene the land-bridge islands (or "continental islands" in the sense defined by Mayr 1941d) must have filled up with the number of species now found in an equivalent area of the New Guinea mainland. Ever since the land bridges were severed around 10,000 years ago, the avifaunas of the land-bridge islands have been relaxing back toward the species number expected for oceanic islands of their area, through selective extinctions of their now-isolated populations—especially through losses of species unable to disperse across water and hence unable to reverse what would otherwise be temporary extinctions. However, the largest New Guinea land-bridge islands are still supersaturated in species number today. Similar supersaturation effects are also conspicuous for large neotropical islands formerly connected to the American mainland (Terborgh 1974). Have the vanished Pleistocene land bridges of Northern Melanesia left similar legacies?

We show in chapter 33 that 13 bird species have distributions confined to the modern fragments of Greater Bukida, the largest Pleistocene island of the Solomons (fig. 1.3). One species is confined to the modern fragments of Greater Gatumbangra and Greater Rendipari (possibly joined to each other). However, figures 9.2 and 9.3 show no gross differences in species number between Northern Melanesian islands that were or were not part of larger Pleistocene islands. At most, the two largest Greater Bukida fragments, Bougainville and Guadalcanal, have, respectively, eight and four more species than predicted for their area by equation 9.2 (the deviation for Bougainville is almost significant statistically; $Z^2 = 3.5$). The three Bismarck land-bridge islands, New Hanover, Umboi, and Vuatom, also have species numbers somewhat above predicted values, as legacies of their Pleis-

Table 9.1. Island species number and its determinants in the Bismarcks.

Island	S_{low}	S_{mt}	A (km²)	H (m)	D (km)
1. New Britain	111	16	35,742	2,439	88 (NG)
2. Umboi	76	8	816	1,655	25 (1)
3. Long	52	2	329	1,304	65 (2)
4. Tolokiwa	40	4	46	1,396	22 (2)
5. Sakar	36	0	40	998	15 (2)
6. Crown	31	2	14	566	10 (3)
7. Malai	21	0	0.83	~0	9 (2)
8. Ritter	4	0	0.36	137	10 (2)
9. Tambiu	9	0	0.067	~0	8 (2)
10. Hein	5	0	0.052	~0	5 (2)
11. Midi	9	0	0.025	~0	24 (2)
12. Tamum	5	0	0.023	~0	9 (2)
13. Nup	7	0	0.019	~0	7 (2)
14. Noru	10	0	0.0096	~0	6 (2)
15. Matenai	4	0	0.0065	~0	8 (2)
16. Araltamu	4	0	0.0031	~0	7 (2)
17. Witu	32	0	57	351	63 (1)
18. Unea	20	0	31	591	60 (1)
19. Lolobau	54	0	61	932	5 (1)
20. Duke of York	45	0	52	~0	13 (1)
21. Vuatom	52	0	14	~350	8 (1)
22. Credner	16	0	1.0	~0	8 (1)
23. New Ireland	88	15	7,174	2,399	30 (1)
24. New Hanover	75	0	1,186	960	30 (23)
25. Dyaul	50	0	110	180	14 (23)
26. Tingwon	13	0	1.5	~0	29 (24)
27. Tabar group	61	0	218	621	26 (23)
28. Lihir	58	0	205	854	47 (23)
29. Feni group	39	0	110	521	55 (23)
30. Tanga group	39	0	98	500	59 (23)
31. St. Matthias	43	0	414	651	96 (24)
32. Emirau	32	0	41	~0	17 (31)
33. Tench	11	0	0.26	~0	74 (32)
34. Manus	50	0	1,834	718	274 (NG)
35. Rambutyo	30	0	88	308	43 (34)
36. Nauna	20	0	1.0	129	88 (34)
37. San Miguel	10	0	0.26	~0	26 (34)
38. Ninigo group	16	0	13	~0	235 (NG)
39. Hermit group	16	0	10	244	170 (34)
40. Wuvulu	17	0	15	37	162 (NG)
41. Anchorite group	9	0	0.52	~0	174 (34)

S_{low} is the number of resident land and freshwater bird species normally occurring near sea level; S_{mt} is the number of species confined to mountains on that island; A is area; H is height; and D is distance from the major source island identified by the island number in parentheses (NG = New Guinea). For example, the nearest major source island to island 2 is island 1, 25 km distant. S_{low} values omit species 1,3, 92, and 112 of appendix 1, because species 1 was probably introduced prehistorically to a single island and because species 3, 92, and 112 are winter visitors that breed only occasionally in small numbers (see chapter 17). Islands 2–16 are the Vitiaz-Dampier islands; 17 and 18, the French islands; 2–22, the satellites of New Britain; 24–30, the satellites of New Ireland; 31–33, the St. Matthias group; 34–37, the Admiralties; and 34–41, the Northwest Bismarcks. Islands 23 and 24 were joined during late Pleistocene periods of low sea level into a single expanded island. Ornithological surveys of islands 19, 25, 27, 30, and especially 18 were less intensive than those of other islands, so that actual S values for those islands are probably slightly higher than the tabulated values.

Table 9.2. Island species number and its determinants in the Solomons.

Island	S_{low}	S_{mt}	A (km^2)	H (m)	D (km)
1. Bougainville	85	18	8,591	2,591	1 (5)
2. Guadalcanal	79	23	5,281	2,448	24 (6)
3. Ysabel	72	3	4,095	1,250	36 (6)
4. Choiseul	68	3	2,966	970	41 (8)
5. Buka	64	0	611	402	1 (1)
6. Florida	60	0	368	400	24 (2)
7. Shortland	56	0	232	206	8 (1)
8. Fauro	51	0	71	587	11 (1)
9. Buena Vista	41	0	14	302	13 (6)
10. Nusave	35	0	0.53	70	12 (7)
11. Bagora	22	0	0.33	61	9 (8)
12. Nugu	8	0	0.15	~0	11 (6)
13. Samarai	19	0	0.091	~0	9 (8)
14. New	15	0	0.071	~0	5 (8)
15. Dalakalonga	10	0	0.067	39	16 (6)
16. Elo	14	0	0.055	15	11 (1)
17. Kosha	14	0	0.052	9	5 (2)
18. Kukuvulu	11	0	0.040	30	9 (1)
19. Tapanu	13	0	0.016	~0	8 (8)
20. Kanasata	6	0	0.0091	15	2 (8)
21. Nameless	8	0	0.0070	~0	7 (8)
22. New Georgia	64	1	2,044	1,006	0.1 (26)
23. Kulambangra	67	15	704	1,768	2 (26)
24. Vangunu	59	5	544	1,124	0.6 (22)
25. Gatukai	56	3	109	888	8 (24)
26. Kohinggo	61	0	95	124	0.1 (22)
27. Wana Wana	58	0	69	80	2 (26)
28. Nusalavata	9	0	0.083	~0	4 (22)
29. Kundu Kundu	12	0	0.048	~0	4 (22)
30. Vella Lavella	64	1	640	793	9 (31)
31. Ganonga	51	3	142	854	9 (30)
32. Rendova	59	4	381	1,063	4 (33)
33. Tetipari	55	0	122	405	4 (32)
34. Malaita	67	7	4,307	1,280	47 (35)
35. San Cristobal	70	6	3,090	1,040	47 (34)
36. Pavuvu and Banika (Russells)	47	0	176	543	46 (2)
37. Mono	42	0	73	355	29 (7)
38. Ugi	49	0	42	204	8 (35)
39. Nissan	29	0	37	34	63 (5)
40. Gizo	60	0	35	199	12 (30)
41. Savo	34	0	31	484	13 (2)
42. Santa Anna	45	0	15	159	8 (35)
43. Simbo	42	0	13	335	7 (31)
44. Three Sisters	37	0	11	76	20 (35)
45. Santa Catalina	38	0	4.9	98	9 (35)
46. Borokua	14	0	4.0	360	59 (25)
47. Kicha	15	0	0.11	~0	9 (25)
48. Rennell	39	0	684	154	168 (35)
49. Bellona	18	0	20	76	157 (2)
50. Ontong Java	9	0	9.6	~0	237 (3)
51. Sikaiana	6	0	1.3	~0	175 (34)

S_{low} is the number of resident land and freshwater bird species normally occurring near sea level; S_{mt} is the number of species confined to mountains on that island; A is area; H is height; D is distance from the nearest island with S_{total} ($= S_{low} + S_{mt}$) > 50, identified by the island number in parentheses. For example, the nearest island to island 2 is island 6, 24 km distant. Islands 1–21 were joined during late-Pleistocene periods of low sea level into a single expanded island, Greater Bukida. Islands 22–29, 30–31, and 32–33 were similarly joined into other Pleistocene islands, termed Greater Gatumbangra, Greater Vellonga, and Greater Rendipari, respectively. Islands 34–51 have had no recent connections. Small islands whose species are not listed in Appendix 1 are islands 10, 11, 13, 14, 16, and 18–21 in the Shortland group, 12 and 15 in the Florida group, 17 near Guadalcanal, and 28 and 29 near New Georgia. From Diamond and Mayr (1976), with revisions.

tocene bridges to the larger New Ireland or New Britain, but only for Vuatom is the deviation statistically significant (fig. 9.1).

This slightness of avifaunal supersaturation in Northern Melanesia is consistent with the fact that only 11% (13 out of 116) of the species of Greater Bukida do not disperse across water, whereas the comparable figure for Pleistocene New Guinea is about 70%. We suggest three reasons for this difference (Diamond 1983): species had to be superior overwater colonists to reach the Solomons or Bismarcks from New Guinea in the first place; Greater Bukida had a much smaller area and hence higher extinction rates than Pleistocene New Guinea, so that loss of water-crossing ability was much more likely to lead to extinction for species of Greater Bukida than of New Guinea; and Greater Bukida's smaller area relative to that of its major fragments also means that the initial supersaturation ratio for species number (the initial species number divided by that expected for an island of the same area at equilibrium) was much closer to 1 for Greater Bukida's fragments than for the New Guinea land-bridge islands. Greater Bukida's area of 45,870 km^2 was only 5.3 times that of its largest modern fragment, Bougainville, while Pleistocene New Guinea was more than 100 times larger than Aru, its largest modern fragment other than New Guinea. Similarly, Greater Gatumbangra (5,535 km^2) was only 2.7 times larger than its largest modern fragment New Georgia, Greater Vellonga (2,186 km^2) only 3.4 times the area of its fragment Vella Lavella, and Greater Rendipari (855 km^2) only 2.2 times the area of its fragment Rendova.

Summary

Island species number ranges from 3 to 127 species. Lowland species number, S_{low}, increases logarithmically or double-logarithmically with island area, A, according to a double-logarithmic area exponent of 0.18 for central Bismarck islands and 0.09 for central Solomon islands. Major factors responsible for this increase in S_{low} with A are habitat heterogeneity, species territory sizes, and immigration/extinction equilibria. Remote islands have fewer species than do similar-sized central islands, the deficit being greater for smaller islands, as expected from considerations of immigration/extinction equilibria for pools of species varying in their immigration and extinction probabilities. Vegetation and bird species numbers on most volcanically defaunated islands have recovered quickly. The total number of montane populations absent at sea level increases with island elevation and also with the available area of montane habitat. While now-severed Pleistocene land bridges have left some legacies in modern distributions of individual species, they have had little effect on modern island species numbers. Three factors contribute to this slightness of supersaturation effects for Northern Melanesian birds.

10

Level of Endemism, Habitat Preference, and Abundance of Each Species

One of the main problems in understanding speciation of Northern Melanesian birds is to explain why different species differ greatly in their proneness to speciate. Much of this book is devoted to interpreting those differing pronenesses to speciate in terms of species differences in six ecological, evolutionary, and distributional characteristics, tabulated for each species in appendix 5. In this chapter we introduce three of these ecological and evolutionary characteristics: level of endemism, habitat preference, and abundance. Chapter 11 introduces the remaining ecological characteristic, overwater dispersal ability, and chapter 12 interprets the two tabulated distributional characteristics (number of islands occupied and incidence category) in terms of the four ecological and evolutionary characteristics.

Level of Endemism

The first evolutionary characteristic we introduce in this chapter is level of endemism. For each of the 195 species making up the Northern Melanesian resident land and freshwater avifauna (appendix 1), the column "level of endemism" in appendix 5 lists whether the Northern Melanesian populations are endemic to Northern Melanesia at the level of genus (meaning that the populations belong to a genus not occurring outside Northern Melanesia) or only at the level of species, allospecies, or subspecies, or whether the populations are not endemic at all (meaning that they belong to a subspecies as well as allospecies, species, and genus occurring outside Northern Melanesia). In most cases the most closely related population occurs in New Guinea or occasionally in Australia, the Santa Cruz group, New Hebrides, or Fiji.

Of the 195 species, the Northern Melanesian populations of only 38 (19%) are not endemic, while 71, 51, 30, and 5 are endemic at the subspecies, allospecies, species, and genus level, respectively (36, 26, 15, and 3%, respectively). Northern Melanesia has no endemic bird family; in fact, all resident land and freshwater birds of all Pacific islands east of New Guinea and west of the Galapagos belong to families represented on New Guinea, Australia, or both, except for one New Caledonian species and many Hawaiian and New Zealand species. The five endemic genera of Northern Melanesia are all monotypic and each confined to one or three of the largest Solomon islands.[1]

1. The five monotypic endemic genera (plates 8 and 9) are the distinctive pigeon *Microgoura meeki* of Choiseul; the distinctive owl *Nesasio solomonensis* of Choiseul, Bougainville, and Ysabel; and three somewhat undistinctive honey-eaters currently placed in monotypic genera merely because of uncertainty about their affinities (*Meliarchus sclateri* of San Cristobal, *Stresemannia bougainvillei* of Bougainville, and *Guadalcanaria inexpectata* of Guadalcanal).

Only seven other Northern Melanesian species belong to genera absent from New Guinea, though they occur on some other island (generally another Melanesian or Polynesian island) outside Northern Melanesia.[2] Undoubtedly, paleontologists will discover Northern Melanesian bird fossils belonging to endemic genera exterminated by the first human colonists. However, it seems likely that, although most Northern Melanesian bird populations had begun to differentiate, differentiation did not proceed as far as on New Zealand and Hawaii, where many or most bird species belong to endemic genera and families. This is as one would expect, given that Northern Melanesia is much less isolated from the colonization sources of New Guinea and Australia than are Hawaii and New Zealand: the average level of endemism increases among Pacific island avifaunas with increasing isolation because immigration rates decrease with distance (see figure 4 of Diamond 1980).

Habitat Preference

The next ecological and evolutionary characteristic is preferred habitat, coded for each Northern Melanesian species in the column "habitat" of Appendix 5. Most species (about 55%) occur in lowland forest, though many of these species also occur in other habitats. The next largest categories are species confined to the mountains (16%; species of montane forest except for one swift and one river flycatcher) and species confined to open habitats (11%; species of grassland, secondary growth, forest edge, and gardens). The remaining categories are lowland aerial species (6%), lowland species confined to fresh water and its vicinity (6%), species of the sea coast (3%), species of swamps and marshes (2%), and mangrove species (1%; two species). The predominance of lowland and montane forest species reflects the fact that most of Northern Melanesia is forested.

Of Northern Melanesia's lowland and montane forest species, 57 also occur in the New Guinea region as conspecifics, and a further 24 occur in the New Guinea region as different allospecies. In most cases the New Guinea populations also live in forest, but 14 of the conspecifics and 4 of the allospecies do not, being confined either to nonforest habitats on the New Guinea mainland (e.g., the cuckoo *Cacomantis variolosus*, the kingfishers *Halcyon* [*diops*] and *H. chloris*, the cuckoo-shrike *Coracina tenuirostris*, and the crow *Corvus orru*) or else to small islands around New Guinea (the pigeons *Ptilinopus solomonensis*, *Ducula* [*myristicivora*], and *Macropygia mackinlayi* and the white-eye *Zosterops* [*griseotinctus*]). That is, these 18 species excluded from New Guinea mainland forest were able to shift into forest on the large islands of Northern Melanesia.

Northern Melanesian habitats differ greatly in the level of endemism of their avifaunas. All of the 5 endemic genera and 30 endemic species without exception, and 47 of the 51 endemic allospecies, live in lowland forest or montane habitats. (The sole exceptions are the river kingfisher *Alcedo websteri*, the grass finches *Lonchura* [*spectabilis*] and *L. melaena*, and the wood-swallow *Artamus insignis*; plate 9.) If one counts an endemic genus, species, allospecies, or subspecies as 4, 3, 2, or 1, respectively, and a nonendemic population as 0, then the level of endemism averaged over all montane species is 2.1, approximately the level of an endemic allospecies. Of the 31 montane species, all except *Columba vitiensis* and possibly *Rhipidura fuliginosa* have differentiated to at least the subspecies level, mostly to the allospecies level or higher. Lowland forest species are only slightly less distinct (average endemism level 1.7). At the opposite extreme, all marsh and swamp species and most coastal species have not differentiated even subspecifically (average endemism levels 0–0.2), suggesting that they are vagile and their populations well mixed. Aerial feeders and species of fresh water, open country, and mangrove are intermediate in their average level of endemism (0.5–1.0).

2. These seven species whose congeners are absent from New Guinea are the rail *Nesoclopeus woodfordi* of Greater Bukida, congeneric with and belonging to the same superspecies as *N. poecilopterus* of Fiji; the rail *Pareudiastes sylvestris* (plate 8) of San Cristobal, congeneric with *P. pacificus* of Samoa; the warbler *Ortygocichla rubiginosa* (plate 9) of New Britain, congeneric with and possibly belonging to the same superspecies as *O. rufa* of Fiji; the warbler superspecies *Cichlornis grosvenori* (plate 9) of New Britain, *C. llaneae* of Bougainville, and *C. whitneyi* in the New Hebrides; the warbler *Cettia parens* of San Cristobal, possibly congeneric with *C. ruficapilla* of Fiji and three other warbler species of Palau, Tenimbar, and far eastern Asia; the monarch flycatcher *Clytorhynchus hamlini* (plate 9) of Rennell, belonging to the same superspecies as *C. nigrogularis* of Fiji and Santa Cruz and congeneric with the superspecies consisting of *C. pachycephaloides* (New Caledonia and New Hebrides) plus *C. vitiensis* (Fiji and Samoa); and the white-eye *Woodfordia superciliosa* (plate 1) of Rennell, belonging to the same superspecies as *W. lacertosa* (Santa Cruz).

Abundance

The remaining ecological and evolutionary characteristic introduced in this chapter is abundance, crudely categorized in appendix 5 by assigning each species to one of five classes of differing abundance, based on our field experience and on published estimates by other field workers. By abundance, we do not mean population density in the most preferred habitat; we instead mean total population size for an entire Northern Melanesian community, consisting of both a large island and its fringing islets with their supertramps. Hence abundance or population size is a product of population density in preferred habitat times the total area of that habitat. In other words, a species may be rare either because it lives at low density in a widespread habitat or at high density in a very restricted habitat, while the most abundant species necessarily live at high density in the most widespread habitats. Here we merely introduce the five abundance classes, which we use to interpret other species characteristics in subsequent chapters. The five classes are described below.

Class 1 consists of the 25 rarest species, with estimated population sizes of fewer than 1000 pairs on a 10,000-km^2 island. These species turn out mostly to be ones that are not specialized in their habitat preference but live at very low population densities. All are nonpasserines, and most are big-bodied species with large territories or home ranges in forest: the eagle *Haliaeetus* [*leucogaster*] and nine other sparsely distributed hawks and falcons, five large pigeons and three large owls, the hornbill *Rhyticeros plicatus*, the largest Northern Melanesian coucal (*Centropus violaceus*) and kingfisher (*Halcyon bougainvillei*; plate 8), and four other rare species.

Class 2 consists of the 27 next rarest species, estimated at 1,000–10,000 pairs on a 10,000–km^2 island. These prove to consist mostly of species living at modest population densities and confined to the most restricted habitats that account for a tiny fraction of Northern Melanesia's area. They are species of the sea coast (the heron *Egretta sacra*, osprey *Pandion haliaetus*, thick-knee *Esacus magnirostris*, nightjar *Eurostopodus mystacalis*, and kingfishers *Alcedo atthis* and *Halcyon saurophaga*), lakes and large rivers (the grebe *Tachybaptus* [*ruficollis*], cormorant *Phalacrocorax melanoleucos*, spoonbill *Platalea regia*, three species of ducks, the jacana *Irediparra gallinacea*, plover *Charadrius dubius*, stilt *Himantopus leucocephalus*, and torrent flycatcher *Monachella muelleriana*), and marshes and swamps (three species of herons and four species of rails), plus four species living locally or at low population density in grassland or low, open habitats (the quail *Coturnix chinensis*, bustard quail *Turnix maculosa*, chat *Saxicola caprata*, and finch *Lonchura melaena*).

Class 3 consists of 74 species of intermediate abundance, estimated at 10,000–100,000 pairs on a 10,000-km^2 island. These are species of the major Northern Melanesian habitats. Most of them are the less common forest species, though not the rarities of class 1. However, others are the commonest species restricted to mangrove (the heron *Butorides striatus* and kingfisher *Alcedo pusilla*), islet forest (the supertramp passerines *Monarcha cinerascens*, *Pachycephala melanura*, *Myzomela sclateri*, and *Aplonis feadensis*), and secondary growth and open country (the nightjar *Caprimulgus macrurus*, bee-eater *Merops philippinus*, roller *Eurystomus orientalis*, swallow *Hirundo tahitica*, fantail *Rhipidura leucophrys*, three species of grass warblers, and the grass-finch *Lonchura* [*spectabilis*]). Despite the abundance of these species in their habitats, the somewhat restricted area of their habitats means that their population sizes are modest, though not as modest as the habitat specialists of class 2, confined to the most restricted habitats.

Class 4 consists of 43 species of moderately common forest birds, with estimated population densities of 10–100 pairs/km^2. About half of these are passerines, while the other half are the most abundant species of small-bodied nonpasserines (the commonest pigeons, parrots, and kingfishers).

Class 5 consists of the 22 most abundant Northern Melanesian forest birds, with estimated population densities of $\geq$100 pairs/km^2. All of these are small-bodied passerines: warblers, flycatchers, honey-eaters, white-eyes, the smallest campephagids, and a small whistler, flower-pecker, and thrush.

Summary

Most Northern Melanesian bird populations are endemic at some level: mostly at the subspecies or allospecies level, but there are also 30 endemic full species and five endemic genera. This modest level of endemism compared to the New Zealand and

Hawaiian avifaunas is as one would expect from Northern Melanesia's more modest isolation from its colonization sources.

The habitat occupied by the greatest number of Northern Melanesian species is forest, reflecting the fact that most of Northern Melanesia is forested. Most of those forest species also occur in the New Guinea region, and most of those New Guinea populations in turn also live in forest. However, 18 of those New Guinea populations are confined to nonforest habitats of the New Guinea mainland or to small islands around New Guinea, and they were able to shift into forest on colonizing large Northern Melanesian islands. The average level of endemism is highest among forest species and lowest among marsh and swamp and coastal species.

We tabulate Northern Melanesian species in five categories of abundance (total population size for a Northern Melanesian island and its fringing islets). The rarest species are mostly large-bodied, forest nonpasserines with large home ranges; the next rarest are mostly species of specialized habitats accounting for a tiny fraction of Northern Melanesia's area; and the most abundant species are small-bodied forest species, especially passerines, living at high population densities.

11

Overwater Dispersal Ability of Each Species

Species differences in dispersal are one of the most important explanatory factors underlying species differences in tendency to speciate, in geographic variability (level of endemism), and in breadth of distribution. Species differ enormously in their dispersal ability, ranging from species that can be seen flying over water any day in Northern Melanesia, to species that have never been seen crossing water and for which there is not even any evidence of overwater dispersal within their evolutionary lifetime as a species.

In this chapter we describe the evidence available for assessing dispersal ability, and we tabulate that evidence for all species and allospecies in the Bismarcks and Solomons. Comparing this evidence between the two archipelagoes makes it clear that dispersal ability evolves and is subject to natural selection. We then examine how dispersal ability is related to the three ecological and evolutionary characteristics of each species introduced in chapter 10: habitat preference, level of endemism, and abundance.

Indicators of Dispersal

We can draw on six types of evidence, described in more detail in appendix 6, to assess the overwater dispersal ability of each Northern Melanesian species or allospecies. Appendix 6 tabulates all these types of evidence explicitly for each species or allospecies in the Bismarcks or Solomons. The types of evidence are described below.

1. *Witnessed flights over water.* While at sea in the Solomons and Bismarcks, J.D. obtained hundreds of observations of birds flying over water between islands. Reports by other observers provide further instances.

2. *Presence on tiny islets.* Presence of a bird on a tiny islet guarantees that the species must have arrived there recently across water, because there are so few individuals in each bird population on an islet that populations must go extinct frequently. Pimm et al. (1988) showed that almost no island bird population averaging less than 10 pairs survives more than about 20 years and that survival times for populations of this size range from 1 to 20 years. In addition, small islets are generally low islets, whose bird populations are likely to be wiped out simultaneously every few decades by a tidal wave or cyclone (cf. Spiller & Schoener's [1998] observations of periodic eradication of lizard and spider populations on low Caribbean islets by hurricanes). We have taken 25 ha (0.1 square miles) as an arbitrary cutoff and reasoned that any bird species recorded from an island of that size or smaller must fly across water frequently.

3. *Records of a species appearing as a stray on an island on which it is not resident.* If a survey detects a species on an island some time after an earlier survey that failed to record the species and that was of adequate intensity to detect the species had

it been present, then the species must have arrived across water in the intervening time. Naturally, the validity of this evidence depends on the adequacy of the evidence that the species was indeed absent at the earlier survey. In practice, the evidence that we accept is of three sorts: surveys of tiny islets that J.D. visited repeatedly and that were small enough that every bird individual could be counted on each visit; records from the 30-year residence on Vuatom Island of Otto Meyer, a careful observer who was intimately familiar with Vuatom's entire avifauna and paid special attention to strays; and records by Bismarck or Solomon islanders equally familiar with their island's avifauna.

4. *Historical colonizations.* These are further records in which Meyer or Bismarck or Solomon islanders observed "new" species arriving at an island, the difference being that the arrivals were observed to colonize permanently rather than soon disappear.

5. *Recently defaunated volcanoes.* Several Northern Melanesian islands are known to have been defaunated by dated volcanic explosions: Ritter by an explosion on March 13, 1888; Long, Crown, and Tolokiwa by an explosion of Long late in the seventeenth century; and Vuatom by an explosion of the Rabaul volcano around A.D. 600. Several other Northern Melanesian islands (Savo, Simbo, and Borokua in the Solomons, and Sakar, Unea, and Garowe in the Bismarcks) have well-preserved cones or calderas or continuing volcanic activity and are judged by geologists to be young, surely Holocene, volcanoes, although their most recent major eruptions have not been dated. Hence any species recorded from these islands must have crossed water within the Holocene.

6. *Presence on islands separated by permanent water barriers.* While some islands are known to have been connected to each other by land bridges at Pleistocene times of low sea level, other islands are believed never to have had land connections with each other. For example, geologists conclude that New Britain is moving toward New Guinea, that the channel between New Britain and New Guinea is now the narrowest that it has ever been, and that there has never been a land connection between New Britain and New Guinea. Similarly, Manus and St. Matthias in the Bismarcks are believed never to have had connections with each other or with any other Bismarck island, while San Cristobal, Malaita, and the New Georgia group in the Solomons are believed not to have had land connections with each other or with the chain of Solomon Islands from Buka to Guadalcanal. Hence any species present both in New Guinea and the Bismarcks, or both in the Bismarcks and Solomons, or on pairs of Bismarck or Solomon islands separated by such permanent water barriers, must have crossed water at least within the evolutionary lifetime of that species.

Formulation of Dispersal Categories in the Bismarcks and Solomons

We now use all of the above evidence to formulate four classes of dispersal ability for the Bismarcks and Solomons separately, with class 1 consisting of the most vagile species and class 4 consisting of the most sedentary species.

The first four of the above six types of evidence demonstrate overwater dispersal within the twentieth century: witnessed flights over water, presence on tiny islets, twentieth-century records of strays, and twentieth-century records of colonization. Our class 1 of dispersal consists of species with at least two such records of twentieth-century overwater dispersal. We limit this class to species with at least two such records, so as not to include mistakenly, in our category of the most vagile species, a few species for which we have only a single anomalous record of twentieth-century dispersal. Species of class 1 have evidently dispersed repeatedly during the twentieth century.

Class 2 consists of species for which there are not at least two records of twentieth-century overwater dispersal but which nevertheless occur on Holocene volcanic islands, or for which there is a single record of twentieth-century dispersal. Species of class 2 have evidently dispersed during the Holocene.

Class 3 consists of species for which we have no evidence of either twentieth-century or Holocene dispersal, but whose presence on islands separated by permanent water barriers demonstrates overwater colonization at least within the evolutionary lifetime of the species.

Class 4 consists of species that are endemic to single islands or to sets of islands connected to each other at Pleistocene times of low sea level, and for which there are no records of presence on tiny islets or Holocene volcanic islands, straying to other islands, or observed flights across water. For these species we have no evidence of their ability to fly across water within the evolutionary lifetime of the species.

We have classified species in this way separately for the Bismarcks and Solomons to detect possible differences in dispersal ability between these archipelagos. We have carried out the analysis separately for each allospecies of a superspecies to detect possible differences among allospecies in their dispersal ability.

Comparison of the Types of Evidence

Many or most species recorded from tiny islets have also been seen flying across water, and vice versa. That is, there is fair agreement between these two types of evidence for twentieth-century overwater dispersal. For example, within the Solomons 39 species have both been recorded on tiny islets and observed flying over water, while only an additional 5 species have been observed flying over water but not recorded from islets, and 16 species have been recorded from islets but not observed flying over water. Within the Bismarcks, where such records are much scantier, eight species have both been recorded from islets and observed flying over water, while seven species have been observed flying over water but not recorded from islets, and 30 species have been recorded from islets but not seen flying over water.

The species seen flying over water but not recorded from islets consist mainly of species unlikely even to alight on islets because islets are ecologically unsuitable, such as the duck *Anas superciliosa*, which requires swamps or rivers present only on larger islands. Similarly, a low islet would provide no suitable habitat for a montane species, undoubtedly explaining why the montane pigeon *Ducula melanochroa* was recorded as a straggler to Vuatom but has never been recorded from any islet. Conversely, the species recorded from islets or as vagrants or colonists on large islands, but not seen flying over water, consist of species that disperse at night or only on one or a few occasions in their lifetimes.

The species seen flying over water during the day consist especially of species that commute daily between islet roosts and larger islands where they feed (e.g., the heron *Egretta sacra*, the colonial pigeons *Ducula pistrinaria* [plate 8] and *D. spilorrhoa*); species with large territories comprising many islets between which an individual commutes during the day (the eagle *Haliaeetus sanfordi*, the kingfisher *Halcyon saurophaga* [plate 6]); and nectarivores and frugivores that commute daily or seasonally in search of flowering and fruiting trees (lories, starlings, and the hornbill *Rhyticeros plicatus*). However, for some species whose presence on many islets shows that they must frequently cross water, J.D. never observed them doing so, and Solomon and Bismarck islanders told him that the flights of some of these species are at night (e.g., the megapode *Megapodius freycinet*, the heron *Nycticorax caledonicus*, and the rail *Porphyrio porphyrio*). Other species that occur frequently on islets but that J.D. never saw flying across water are solitary, territorial, sedentary species such as the kingfisher *Halcyon chloris* (plate 6) and the flycatchers *Monarcha cinerascens* (plate 5) and *Pachycephala melanura* (plate 2). Individuals of these species probably disperse just once in their lives, as immatures in search of a territory, and remain resident on their territory thereafter. Even J.D.'s most experienced Solomon Island informant, Teu Zinghite of Kulambangra, had only once in his life seen *Halcyon chloris* fly over water (a pair that he saw flying from Kulambangra to Gizo).

Most species whose twentieth-century water-crossing ability is attested by records of vagrants or historical colonizations are also attested by witnessed overwater flights or presence on tiny islets. This is true for 19 out of the 34 species recorded as vagrants or colonists in the Solomons; for 13 out of 22 recorded repeatedly as vagrants, or for three of the five recorded as colonists, on Vuatom; and for five of the eight recorded as vagrants on other Bismarck islands. However, for none of the four species recorded on Vuatom only once as a vagrant during Meyer's three decades of residence there (the heron *Ixobrychus sinensis*, the pigeon *Ducula finschi* [plate 8], the kingfisher *Ceyx websteri*, and the wood-swallow *Artamus insignis* [plate 9]) is there any other form of evidence attesting twentieth-century water-crossing ability. This supports the interpretation that species Meyer observed as vagrants only once on Vuatom cross water only infrequently.

Our classifications of species' overwater dispersal ability include some that we feel intuitively to be misclassifications. These are species we are forced to place in dispersal class 3 because we have no twentieth-century or Holocene evidence of overwater dispersal within Northern Melanesia, although other evidence such as banding studies in Australia prove that they disperse long distances across water. These cases in which we have surely underestimated the dispersal ability of actually vagile species include the heron *Ardea intermedia* and the rail *Poliolimnas cinereus* in the Bismarcks and Solomons, the heron *Butorides striatus* and the stilt *Himantopus himantopus* in the Bismarcks, and the falcon *Falco peregrinus* and the rail *Porzana tabuensis* in the Solomons. All of these species are very uncommon, locally distributed, and confined to specialized habitats in the Bismarcks or Solomons. Hence few individuals reside in Northern Melanesia. This paradox illustrates that frequency of evidence for overwater dispersal ability is actually a product of two factors: likelihood of dispersal per individual, multiplied by abundance. For a given dispersal ability, evidence of overwater dispersal will necessarily be biased toward the more abundant species.

Comparison of the Bismarcks and Solomons

Is there any average difference in overwater dispersal ability between Bismarck and Solomon bird populations? As summarized in table 11.1, far more Bismarck than Solomon species fall into dispersal class 2, while somewhat higher proportions of Solomon than Bismarck species fall into dispersal classes 1, 2, and 4.

These differences must at least partly be artifacts of differences in the evidence available to us. J.D. made special efforts to survey tiny islets in the Solomons and thereby accumulated species lists for 81 such Solomon islets, but for only 20 Bismarck islets. Similarly, J.D. spent much more time on a boat among Solomon than among Bismarck islands and thereby accumulated hundreds of observations of Solomon birds flying over water but only a handful of such observations for the Bismarcks. Those differences undoubtedly explain the excess of Solomon over Bismarck species in class 1. Conversely, six ornithologically surveyed, recently defaunated or young volcanic islands or sets of islands are available in the Bismarcks, but only three in the Solomons. That discrepancy, plus the fact that more of the Solomon species present on young Solomon volcanoes had already been allocated to dispersal class 1 because of the abundant evidence for twentieth-century overwater dispersal in the

Table 11.1. Comparison of the Solomon and Bismarck avifaunas with respect to dispersal classes.

Dispersal class	Number of species	
	Solomons	Bismarcks
1	59	43
2	13	59
3	60	38
4	38	26
Total	170	166

The dispersal class of each species (or allospecies) in each archipelago is given in appendix 6. Class 1 consists of the most vagile species; class 4 consists of the most sedentary species. See text for discussion.

Solomons, has resulted in fewer Solomon than Bismarck species being allocated to dispersal class 2 (Holocene evidence but no twentieth-century evidence for overwater dispersal). Hence we shall not attempt to draw conclusions about the average vagility of the Solomon avifauna compared to the Bismarck avifauna.

If one instead compares dispersal classifications for the same species in the Solomons and Bismarcks, discrepancies may be due to the same artifacts of available information mentioned in the preceding paragraph. However, detailed examination of the evidence for particular species reveals six cases in which a species is classified into dispersal class 1 in one archipelago, class 3 in the other, and in which the evidence is so overwhelming that the difference is likely to be real. In all six cases it is the Solomon population that belongs to the more vagile class 1 and the Bismarck population to the more sedentary class 3. The species involved are the heron *Butorides striatus*, the hawk *Aviceda subcristata*, the kingfisher *Alcedo pusilla*, the hornbill *Rhyticeros plicatus*, the cuckoo-shrike *Coracina lineata*, and the starling *Mino dumontii*. For these six species collectively, there are 78 records on islets, 15 observations of flying over water, 12 records of arrival as vagrants, 7 records on young volcanic islands, and one large modern wave of colonization for the Solomons, but not a single such record for the Bismarcks. Evidently, the Solomon populations of these species really are more vagile than the Bismarck populations. This interpretation is reinforced by the broader distribution of these species, extending to more numerous and smaller islands, in the Solomons than in the Bismarcks. For example, *Rhyticeros plicatus* and *Coracina lineata* are both confined in the Bismarcks to the two largest islands and one other nearby island, but in the Solomons they occupy 25 and 22 islands, respectively, including small and/or remote ones.

One correlate of vagility is breadth of distribution over Northern Melanesia. One might expect that the most vagile species would be more likely than more sedentary species to be shared between the two archipelagos, while the more sedentary species would be more likely to be confined to either the Bismarcks or Solomons. This expectation is confirmed: of the 14 species with the most records of twentieth-century-dispersal in the Bismarcks, all either occur in the Solomons or (two cases) are represented there by another allospecies. Similarly, of the 21 species with the most records of twentieth-century dispersal in the Solomons, all but four extend widely through the Bismarcks. Conversely, 13 of the 38 species classified in the more sedentary class 3 for the Bismarcks, and 16 of the 60 species similarly classified for the Solomons, are neither present nor represented by another allospecies in the other archipelago.

Evolution of Dispersal Ability

The above examples of six species whose Solomon populations are much more vagile than their Bismarck populations indicate that dispersal ability evolves, just as do morphology and clutch size. We now cite six other types of evidence to illustrate such evolutionary changes in dispersal ability.

First, all the species that we place in the sedentary dispersal class 4 (38 Solomon and 26 Bismarck endemic species or allospecies, confined to single islands or to fragments of a single larger Pleistocene island) show no evidence for having crossed water within their evolutionary history as a species or allospecies. Nevertheless, their ancestors must have arrived in the Solomons or Bismarcks over water, because Northern Melanesia was never connected to a source region, and hence its whole biota must have arrived over water. That is, subsequent to their arrival in Northern Melanesia, these now-sedentary species lost the dispersal ability that brought them to Northern Melanesia in the first place.

Second, there are 11 cases in which the Solomons or Bismarcks are occupied by a single endemic subspecies of a species that has crossed water gaps within its evolutionary history as a species (as demonstrated by twentieth-century dispersal, presence on young volcanic islands, or presence on islands separated by permanent water gaps), but the endemic subspecies shows no such evidence of overwater crossing ability and is confined to a single island or to fragments of a single Pleistocene island. That is, other races of the same species within its evolutionary history (or even within the twentieth century) have crossed water, but the endemic subspecies no longer does so and shows no evidence of having done so since the Pleistocene period. In these cases, the Bismarck or Solomon endemic subspecies must have lost the water-crossing ability that brought its ancestors to the Bismarcks or Solomons in the first place and that other races of the same species still retain.

For example, the frogmouth *Podargus ocellatus* of New Guinea and Australia is represented in Northern Melanesia only in the Solomons, where it occurs as an endemic subspecies confined to three large fragments of Pleistocene Greater Bukida. The frogmouth must have reached the Solomons over water, but its Solomon population no longer disperses over water. Similarly, New Britain is occupied by very distinctive endemic subspecies of the cockatoo *Cacatua galerita* and the flycatcher *Monachella muelleriana*, which must have reached New Britain over water from New Guinea; the Australian race of the cockatoo has reached islands off Australia as a vagrant within the twentieth century. However, the New Britain populations have never been recorded off New Britain in modern times, not even on the large, nearby, seemingly ecologically suitable island of New Ireland.

A third type of evidence involves differences in dispersal ability among allospecies of the same superspecies. There are numerous superspecies of which one or more allospecies evince no evidence for water-crossing ability (i.e., belong to dispersal class 4), but of which other allospecies do evince ability to cross water. For instance, the parrot *Lorius albidinucha* has never been recorded off of New Ireland, but two other allospecies of the same superspecies, *L. hypoinochrous* of the Bismarcks and *L. chlorocercus* (plate 6) of the Solomons, occur on many islands separated by permanent water gaps and have both crossed water within the twentieth century. Within the white-eye superspecies *Zosterops [griseotinctus]* (plate 1), three allospecies are confined as endemics to single islands or to two fragments of a Pleistocene island in the Solomons, but another allospecies, *Z. griseotinctus*, has colonized Long Island and its neighbors since the explosion of Long in the seventeenth century. Still another allospecies, *Z. rendovae*, occasionally flies over water to islets near New Georgia and Kulambangra.

A fourth example involves *Zosterops rendovae* again (plate 1). This species is famous in evolutionary biology because two quite distinct subspecies are, respectively, confined to Rendova and Tetipari islands, separated by a water gap of only 3.4 km, yet neither subspecies has ever been recorded from the other island or from any nearby islet. However, the remaining race of the same allospecies, *Z. r. kulambangrae*, is the one that wanders to small islets near New Georgia and Kulambangra. This example thus illustrates marked differences in dispersal ability among nearby subspecies of the same species.

Our fifth example again illustrates subspecific differences in dispersal ability, in this case for the kingfisher *Halcyon chloris* (plate 6). Two of its Northern Melanesian subspecies are quite vagile: *H. c. alberti*, which is widespread on Solomon islands of the Bukida chain and New Georgia group and has been recorded there from 35 tiny islets; and *H. c. stresemanni* of islands near New Britain (but not New Britain itself), which has colonized Long and its neighbors since Long's seventeenth-century eruption. However, the New Britain race *H. c. tristrami* is so sedentary that it extends beyond New Britain only to two nearby medium-sized islands, is absent from the nearby medium-sized Sakar and Duke of York, and was found on only 1 of 10 islets surveyed off the northeast coast of New Britain. Other similar examples in the Bismarcks are that the glossy swiftlet population of the small Nusa Island off New Ireland belongs to the vagile race *Collocalia esculenta stresemanni* otherwise confined to the Northwest Bismarcks 300 km distant, not to the less vagile race *C. e. kalili* of the large island of New Ireland only 1 km from Nusa; similarly, the rainbow lorikeet of New Hanover is the race *Trichoglossus haematodus flavicans* of the distant outlying Northwest Bismarcks and St. Matthias group, not the race *T. h. massena* of the Central Bismarcks.

As these examples show, small or remote islands favor the evolution of vagile taxa that tend to be excluded from nearby large islands: not only the many supertramp species to be discussed in chapter 12, but also supertramp subspecies. This phenomenon probably explains why the avifaunas of the two principal groups of remote islands in the Bismarcks, the St. Matthias group and the Northwest Bismarcks, are more closely related to each other subspecifically than either is related to the avifauna of the central New Hanover/New Ireland group, even though New Hanover is much closer than the Northwest Bismarcks to the St. Matthias group (chapter 30). Similarly, on the islands of Geelvink Bay off Northwest New Guinea, subspecies of the megapode *Megapodius freycinet geelvinkianus*, fruit dove *Ptilinopus rivoli prasinorrhous*, and sunbird *Nectarinia sericea nigriscapularis* live on small islands near large islands occupied by the subspecies *M. f. affinis, P. r. miquelii*, and *N. s. mysorensis*, respectively (Mayr 1941c). Salmonsen (1976) refers to the "insularization" of these small island subspecies and cites examples from other parts of the world.

Our final example of subspecific differences in dispersal ability involves the fruit dove *Ptilinopus solomonensis* (plate 6). The Bismarck races *P. s. meyeri* and *P. s. johannis* are vagile supertramps that colonized both Long and Ritter after modern volcanic eruptions, occur on three other Holocene volcanic islands, and are known from small, low islets. In contrast, two of the Solomon races, *P. s. bistictus* and *P. s. ocularis*, constitute a distinctive megasubspecies and are confined to high elevations on the two largest and highest Solomon islands, Bougainville and Guadalcanal. But the Solomon races *P. s. neumanni, P. s. vulcanorum*, and *P. s. solomonensis* are morphologically similar to the Bismarck supertramp races, which they also resemble ecologically in ranging to small, low islands. One individual of those Solomon small-island races was collected on the large island of Guadalcanal—presumably a vagrant individual of a vagile subspecies, straying to an island occupied by a sedentary subspecies (Galbraith & Galbraith 1962, pp. 24–25).

In short, the Northern Melanesian avifauna illustrates evolutionary loss of dispersal ability by vagile ancestors of now-sedentary endemic species and subspecies. It also illustrates marked differences in dispersal ability between different allospecies of the same superspecies, different subspecies of the same species, and even different populations of the same subspecies (most notably in the case of the starling *Mino dumontii kreffti*, of which the Solomon population crosses water very successfully today and the Bismarck population does not cross water at all).

Biological Correlates of Dispersal Abilities

We now examine how dispersal ability relates to level of endemism, habitat preference, and abundance. We begin with some theoretical predictions based on evolutionary reasoning.

Predictions

Dispersal brings potential benefits and also potential costs or risks. These benefits and costs vary among species, depending on their other biological attributes. Consideration of those attributes should permit predictions about species differences in dispersal ability, based on evolutionary cost/benefit reasoning.

Compared to a stationary individual, a dispersing individual is likely to incur an increased risk of death from any of several causes, including predation, physical exhaustion, unfavorable climatic conditions, and difficulties of foraging in unfamiliar surroundings encountered en route. In addition, the individual is likely to end up at a site less suitable than the natal site from which it dispersed, because the natal site was necessarily one suitable for its parents to survive and reproduce. As a result of these two costs, dispersal means death for most individuals of most species. If there were no further considerations operating, one would therefore predict that all species would be sedentary, which is, of course, incorrect.

The flaw in this prediction is that these two costs are offset by a potential benefit of dispersal, plus a potential cost of staying home. The potential benefit is that the dispersing individual may find an unoccupied block of habitat suitable for its species, may settle there, and may found a large new population inheriting its genes. The potential cost of staying at home is that the natal site may already be saturated with established conspecifics, in which case an individual that stayed at home would face low chances of winning a territory and reproducing.

The magnitudes of these costs and benefits, and hence the dispersal propensity that represents their evolutionary trade-off point, are expected to vary among species. The availability of suitable habitat patches currently unoccupied by the species is (1) higher for species inhabiting small habitat patches (with small population sizes and hence frequent population extinctions even at saturating population densities) than for species inhabiting large patches; (2) higher for rare than abundant species (again because of small population sizes and frequent extinctions); and (3) higher for species of unstable habitats or rapidly changing successional habitats (leading to frequent extinctions) than for species of stable habitats.

These considerations lead to the predictions that, all other things being equal, dispersal ability should evolve to be higher for rare species than abundant species, higher for species whose habitat occurs as small, scattered patches than for species of large, coherent blocks, and higher for species of unstable than stable habitats. In addition, species that fly long distances in the course of their normal daily foraging or seasonal movements (e.g., aerial feeders, species with large territories, and nectarivores and frugivores wandering in search of ephemeral food supplies) already require strong powers of flight, so dispersal costs them nothing additional anatomically. One expects lower dispersal ability in species that otherwise fly little or not at all (e.g., species with small territories or ground-dwelling species that forage by walking); for such species, the costs of strong powers of flight are budgeted solely against the ledger of dispersal.

A Dispersal Index

Appendix 6 gives a dispersal index for each Northern Melanesian allospecies, tabulated separately for Bismarck and Solomon populations. Detailed analysis was required to detect dispersal differences between allospecies of the same superspecies and between Bismarck and Solomon populations of the same allospecies. For the following analysis, however, we have combined the finely split indices of appendix 6 into a single index for each superspecies, representing a combined value for all of its Northern Melanesian allospecies (if there is more than one) and for both its Bismarck and Solomon populations. The column "dispersal" of appendix 5 lists values of the new combined index, and the legend to the appendix explains the derivation of

Table 11.2. Relation between dispersal and habitat.

	Number of Species							
Dispersal class	Coast	Aerial	Open country	Fresh water	Mangrove	Swamp	Lowland forest	Montane forest
A (vagile)	4.8	6	8	4	—	—	32.3	1
B	—	6	13	8	2	4	54.4	19.6
C (sedentary)	—	—	—	—	—	—	17.5	10.5
Average dispersal index	3.0	2.5	2.4	2.3	2.0	2.0	2.1	1.7

Entries in the upper three rows are numbers of Northern Melanesian species in each dispersal class and habitat, from the tabulations of Appendix 5. Noninteger values arise because some species are allocated to more than one dispersal or habitat class as a result of interisland variation. Note that species of nonforested habitats are confined to the two more vagile classes, while many montane forest species belong to the most sedentary class and only one belongs to the most vagile class. The association between dispersal and habitat is highly significant by a chi-square test (χ^2 41.62, df 14, $p < .0002$); the most significant individual values (highest Z^2 scores) are the big deficit of class A species and big excess of class C species in montane forest and the excess of class A species on the coast.

Entries in the bottom row are an average numerical dispersal index for species of each habitat; a higher number means higher average vagility. (The number is calculated by assigning values of 1, 2, and 3 to dispersal classes C, B, and A respectively, and averaging over all species). Note again that nonforested habitats (especially coast, aerial, open country, and fresh water) have relatively vagile species (average index 2.3–3.0), while montane forest has relatively sedentary species (index 1.7).

the index. The new combined index distinguishes just three groups of species: A, the 56 most vagile species; B, 107 species of intermediate vagility; and C, the 28 most sedentary species.

Dispersal and Habitat

Table 11.2 calculates an average numerical dispersal index for species of each habitat. The table also sets out explicitly how many species of each of the three classes of dispersal ability are represented in each habitat.

This procedure illustrates a type of analysis that we perform frequently throughout this book. We do not confine ourselves to searching for average trends by examining pairwise correlations between variables, or regressions of a variable on one or more other variables. That procedure would not explore the numerous exceptions to any average trends detected. The heterogeneous Northern Melanesian avifauna includes many species clusters that are ecologically and distributionally distinct from each other, and in some of which some particular correlation is opposite to the correlation in other clusters. To describe merely the average trend for the whole pooled avifauna overlooks the most important message of this book: that different species often behave differently.

To mention just a few examples drawn from the following paragraphs, nonforest species are, on average, somewhat more vagile than forest species. But closer inspection reveals important differences even among species of the same habitat. While all other species of montane forest are relatively sedentary, one (the pigeon *Ducula melanochroa*; plate 8) is a social frugivore that breeds in montane forest, regularly descends after breeding to the lowlands, and occasionally disperses over water. Lowland forest harbors sedentary species that are higher-level endemics, rare, and confined to the largest islands (e.g., the owl *Nesasio solomonensis*; plate 8), but lowland forest also harbors vagile species that are barely differentiated, abundant, and present on small as well as on large islands (e.g., the starling *Aplonis metallica*). Similar examples of heterogeneity within the Northern Melanesian avifauna abound throughout the chapters of this book.

In analyzing the relation between dispersal and habitat, let us begin by examining average trends. The bottom row of Table 11.2 presents an average numerical value for the dispersal ability of the species of each habitat by assigning values of 1, 2, and 3 to species of dispersal classes C, B, and A, respectively. The habitat whose species are on average least vagile is montane forest, with an average value of only 1.7. Species of lowland forest are on average nearly as sedentary, with an average value of 2.1. The habitats whose species are on average most vagile are nonforested habitats: fresh water, open country, and secondary growth, aer-

Table 11.3. Relation between dispersal and level of endemism.

	Number of Species				
Dispersal class	Not endemic	Endemic subspecies	Endemic allospecies	Endemic full species	Endemic genus
A (vagile)	16.5	29	7.5	3	—
B	24.5	44.3	26.2	12	—
C (sedentary)	—	—	9	14	5
Average dispersal index	2.4	2.4	2.0	1.6	1.0

Entries in the upper three rows are numbers of Northern Melanesian species in each dispersal class and at each level of endemism, from the tabulations of appendix 5. Noninteger values arise because some species are allocated to more than one dispersal or habitat class as a result of interisland variation. Note that the most vagile species are concentrated among the lowest levels of endemism, while the most sedentary species are concentrated among the highest levels of endemism. The association between dispersal and level of endemism is highly significant by a chi-square test (χ^2 80.97, df 8, $p < .0001$); the most significant individual values (highest Z^2 scores) are among class C species, in their deficits at the two lowest levels and their excesses at the two highest levels of endemism. There is a significant monotonic decrease in dispersal with level of endemism (Spearman correlation −.45), consisting of a linear trend (χ^2 41.03, df 1, $p < .0001$) and also a quadratic trend (χ^2 8.87, df 1, $p = .0029$).

Entries in the bottom row are an average numerical dispersal index for species of each level of endemism (1.0 = most sedentary . . . 3.0 = most vagile; see table 11.2 for explanation). Note that vagility decreases with increasing level of endemism.

ial, and the sea coast, with average values of 2.3, 2.4, 2.5, and 3.0, respectively. While these average values are higher than the average values of 1.7 and 2.1 for montane and lowland forest, the differences are modest. The two other nonforested habitats, swamp and mangrove, both yield average values of 2.0, but each of these habitats has very few species (four and two, respectively), reducing the confidence that can be attached to these average values.

Further insights can be gleaned from table 11.2 by searching for heterogeneity among the species of each habitat. First, without exception, all species of the six nonforested habitats are of high or medium vagility (dispersal class A or B). This conclusion conforms to the predictions made on p. 72. All of these habitats except aerial occur (unlike forest) as isolated, small, or narrow patches, whose populations face high risk of extinction and require high vagility to maintain themselves by immigration following inevitable frequent extinctions. In addition, open country and secondary growth habitats are unstable with time in Northern Melanesia: they are continually being created by disturbance and destroyed by forest succession, so that their populations, too, must be vagile to maintain themselves. The high vagility of aerial species reflects, as predicted, the fact that they fly long distances in the course of their normal daily foraging, so that the added costs associated with dispersal are modest for them.

A second insight gained by searching table 11.2 for species heterogeneity within a habitat is that almost all (30 out of 31) species of montane forest are of low or medium vagility (dispersal class C or B). The sole exception is the vagile, social, frugivorous pigeon *Ducula melanonochroa*, described earlier. Montane forest is Northern Melanesia's wettest and most stable habitat, where one expects low vagility.

Thus, for most habitats the search for heterogeneity reinforces conclusions drawn from the modest average trends, and the dispersal characteristics of the habitat's species agree with the predictions made earlier in this section. However, species of lowland forest are evidently a much more heterogeneous group, as already illustrated by the contrast between the sedentary owl *Nesasio solomonensis* and the vagile starling *Aplonis metallica*. Lowland forest species are distributed over all three dispersal classes (A, B, and C), and their heterogeneity is analyzed further below and in the next chapter.

Dispersal and Endemism

Table 11.3 is similar to table 11.2 analyzing the relation between dispersal and habitat, but instead

Table 11.4. Relation between dispersal and abundance.

Dispersal class	Abundance class				
	1 (rare)	2	3	4	5 (abundant)
	Number of species				
A (vagile)	1	9	30	14	2
B	15	18	35	24	15
C (sedentary)	9	—	9	5	5
Average dispersal index	1.7	2.3	2.3	2.2	1.9

Like tables 11.2 and 11.3, but for the relation between dispersal and abundance. Entries in the upper three rows are numbers of Northern Melanesian species in each dispersal class and each abundance class, from the tabulations of appendix 5. Entries in the bottom row are an average numerical dispersal index for species in each abundance class (1.0 = most sedentary, and 3.0 = most vagile). The association between dispersal and abundance is highly significant by a chi-square test (χ^2 27.51, df 8, p = .0008), but the relation is not monotonic (Spearman correlation is .03), and a quadratic trend is significant (p = .00001), but a linear trend is not significant (p = .54) by a chi-square test. Instead, dispersal peaks at intermediate abundance. See text for discussion.

analyses the relation between dispersal and endemism. Table 11.3 summarizes the average trend, which is simple: average dispersal ability decreases monotonically with increasing level of endemism, from 2.4 on a scale of 3 (very vagile) for nonendemic species down to 1.0 (completely sedentary) for members of endemic genera. An examination of table 11.3 for species heterogeneity yields the following additional conclusions.

First, the most sedentary species (class C) are all endemic to Northern Melanesia at the allospecies level or higher. That fact is an automatic consequence of our definition of class C, because any species with conspecific populations outside Northern Melanesia is ranked as having dispersed over water within its evolutionary lifetime as a species and excluded from class C.

Second, and conversely, the higher level endemics (the endemic genera and full species) are disproportionately sedentary: all of the endemic genera, and half of the endemic full species, belong to the most sedentary class (C). But that fact is not an automatic consequence of our definitions. It would be possible for higher endemics to be highly vagile within Northern Melanesia, to fly across water in the course of their daily foraging movements, and thus to belong to the most vagile dispersal class (A), but still to constitute higher endemics through lacking conspecifics or congeners outside Northern Melanesia. In fact, three full species endemic to Northern Melanesia (the itinerant flocking lory *Chalcopsitta cardinalis* (plate 8), which can be seen flying over water between Solomon Islands on any day, and the supertramp passerines *Aplonis [feadensis]* and *Myzomela sclateri* (plate 4)) are highly vagile within Northern Melanesia. Thus, the complete sedentariness (class C) of most other (19 out of 34) higher level endemics is striking.

Finally, the most vagile species (class A) are drawn disproportionately (81% of them) from species at the lowest levels of endemism (endemic subspecies or nonendemics). But even that most vagile class of species is heterogeneous and includes endemic allospecies, as well as the three endemic full species listed above.

In short, little-differentiated recent colonists tend to be vagile, while older colonists (higher endemics) tend to be sedentary. This suggests a tendency toward an evolutionary transformation of vagile invaders into sedentary endemics, as discussed by Wilson (1961) for Melanesian ants. This trend is called the "taxon cycle" and has occasioned much debate, in part because of the many exceptions to it. Our analysis of Northern Melanesian birds reveals both the trend and the exceptions, discussed further in chapter 35.

Dispersal and Abundance

Table 11.4 is similar to tables 11.2 and 11.3 analyzing the relation between dispersal and habitat or endemism, but instead analyzes the relation between dispersal and abundance. The bottom row of table 11.4 shows that a monotonic trend does not exist: average dispersal ability falls between 1.7 and 2.3 (on a scale of 1–3) for species of all five abundance classes, with an apparent slight peak at intermediate abundance. These results do not fit our prediction that (all other things being equal) vagility should decrease with increasing abundance.

Recall, however, the systematic bias in the data of table 11.4. As mentioned earlier in this chapter, frequency of evidence for overwater dispersal ability is actually a product of two factors: likelihood of dispersal per individual multiplied by abundance. Table 11.4 is therefore biased toward attributing high vagility to abundant species. Bearing in mind this bias, and examining table 11.4 as we examined tables 11.2 and 11.3 for species heterogeneity, what we can conclude?

It is striking that the 22 most abundant species (class 5) include only two of the most vagile species (class A). All of these most abundant species are small passerines inhabiting the largest blocks of forest, and most of them are territorial. Almost none of them has been observed crossing water: among our thousands of direct observations of Northern Melanesian species flying across water, only two are of these abundant forest passerines. That failure is noteworthy because the high abundance of these species would yield more observations of overwater dispersal for a given per-capita probability of dispersal; hence their actual dispersal ability must be even lower than indicated in table 11.4. Evidently, these most abundant species really are relatively sedentary, as predicted.

As for the 25 least common species (abundance class 1), they too include only one of the most vagile species (class A: the eagle *Haliaeetus* [*leucogaster*]). Is that apparent conclusion an artifact of their rareness, yielding few opportunities to observe dispersal? That sole very vagile species in the group, the eagle, is conspicuous because it is the largest Northern Melanesian bird, hunts by soaring high, and is thus very easy to detect, thereby yielding many observations of overwater dispersal. However, the hawks *Henicopernis infuscata* and *Accipiter imitator* of abundance class 1 also hunt by soaring, yet they have never been observed flying over water, so their apparent sedentariness is real. Two species of abundance class 1, the cuckoo *Centropus violaceus* and the pigeon *Reinwardtoena* [*reinwardtii*], often give loud and unmistakable calls that carry for several kilometers, yet neither the cuckoo nor the Solomon allospecies of the pigeon has ever yielded evidence of overwater dispersal by being recorded on a small island, a Holocene volcanic island, or as a vagrant. It therefore seems likely that at least some very uncommon species really are rather sedentary, contrary to the prediction that (all other things being equal) they should be vagile. Evidently, all other things are not equal. As we discuss in the next section, one important confounding factor proves to be island area.

The Most Sedentary Species

As we have seen, all of the 28 most sedentary species are higher-level endemics (endemic allospecies, full species, or genus; table 11.3), and all are species of lowland forest or mountain forest (table 11.2). In addition, they are virtually confined to the largest Northern Melanesian islands: the sole populations of the 28 species on islands smaller than 200 km^2 are the population of the pigeon *Henicophaps foersteri* on Lolobau and two populations each of the white-eye *Zosterops metcalfii* and the crow *Corvus woodfordi* on smaller fragments of Greater Bukida. With these few exceptions, 8 of the 28 sedentary species are confined to the largest modern Northern Melanesian island, New Britain (plus its large neighbor Umboi in three cases); 11 to the largest modern fragments of the largest Pleistocene island, Greater Bukida; five to the three next-largest islands (New Ireland, Malaita, and San Cristobal); and two each to Kulambangra and Rennell. Among the 28 sedentary species, the rarest (the nine belonging to abundance class 1) are virtually confined to the very largest islands, exceeding 2900 km^2 in area.

These distributions result from the inability of the most sedentary species to cross water and recolonize islands on which they have become extinct. Hence they survive only on islands large enough to support large populations, which can persist in isolation for a long time. The rarest of these sedentary species can sustain themselves only on the largest islands. These considerations explain the paradox mentioned in the preceding section: that the rarest species unexpectedly include highly sedentary species.

The Most Vagile Species

The 56 most vagile species are a heterogeneous group. Half of them live in lowland forest, where most are of intermediate abundance (class 3), and only two (the flycatchers *Rhipidura rufifrons* and *Monarcha* [*melanopsis*]) are very abundant; most are just endemic subspecies or nonendemic, but six are endemic allospecies. Nine are supertramps, confined as breeders or residents to small islands. One is an itinerant flocking frugivore of montane forest, the pigeon *Ducula melanochroa*. Finally, 23 are nonforest species of the sea coast, open country, fresh water, or air.

Summary

We used six indicators to assess the overwater dispersal ability of each Northern Melanesian species and allospecies in the Bismarcks and Solomons. Species differ greatly in their propensity for overwater dispersal. There are also marked differences between different allospecies of the same superspecies, different subspecies of the same allospecies, and even between Bismarck and Solomon populations of the same subspecies. Yet the ultimate ancestors of all Northern Melanesian taxa arrived over water from outside Northern Melanesia. Evidently, dispersal ability changes with evolutionary time. This is not just a matter of volant ancestors evolving into flightless insular endemics, because Northern Melanesia's extant avifauna includes only four flightless taxa. Hence there must be a large behavioral component to variation in dispersal ability: some taxa are more reluctant to disperse over water than other taxa.

Dispersal ability is related to a taxon's habitat preference, level of endemism, abundance, and area of island occupied. Thus, dispersal ability is subject to natural selection. An evolutionary cost/benefit analysis suggests that the costs of dispersal—the increased risk of death while dispersing and the risk of ending up at a site less suitable than the natal site—are more likely to be offset by the potential benefits of dispersal for uncommon species of unstable, small, or scattered habitat patches (i.e., most nonforest habitats in Northern Melanesia) than for abundant species of stable, large, coherent habitat blocks (especially montane forest); and more likely for species that fly a lot in the course of their normal foraging activities (e.g., aerial feeders, nomadic nectarivores and frugivores) than for species that fly little in the course of normal foraging. Dispersal ability tends to decrease at higher levels of endemism, but there are many exceptions.

12

Distributional Ecology

In this chapter we analyze two distributional characteristics of each Northern Melanesian bird species in terms of the four ecological and evolutionary characteristics introduced in the two preceding chapters. The distributional characteristics analyzed are the total number of Northern Melanesian islands that each species occupies, and the range of areas or species richnesses of the islands that each species occupies. Both characteristics vary greatly among species.

Number of Islands Occupied by Each Species

The number of tabulated Northern Melanesian islands (i.e., those tabulated in appendix 1) that each species occupies, which we abbreviate as i, ranges from 1 to 69 islands. The value of i is 1 for the 26 species confined in Northern Melanesia to a single island. i is 64–69 for the most widely distributed species, the pigeon *Caloenas nicobarica*, the heron *Egretta sacra*, and the mound-builder *Megapodius freycinet*, occurring in Northern Melanesia on 64, 67, and 69 islands, respectively. How can we account for this great variation in i (listed for each species in appendix 5)? This variation is related especially to variation in habitat preference and vagility, and also to variation in level of endemism. Let us begin with the relation between i and habitat.

Island Number and Habitat

Habitat availability affects species distributions within Northern Melanesia. This becomes apparent from table 12.1, which summarizes, for each of the eight habitat types recognized in appendix 5, the mean value and range of i for its species.

At the one extreme, every island necessarily has a sea coast. All five species of the sea coast are geographically widespread in Northern Melanesia and reside on virtually all islands except (in some cases) a few remote ones. These widespread coastal species are the eastern reef egret *Egretta sacra*, the sea eagle *Haliaeetus* [*leucogaster*], the osprey *Pandion haliaetus*, the beach stone-curlew *Esacus magnirostris*, and the beach kingfisher *Halcyon saurophaga* (plate 6), occurring on 67, 53, 60, 42, and 61 islands, respectively.

At the opposite extreme, lakes and rivers are confined to a few large islands, hence nine of the 12 freshwater species are confined to these few islands. (The three species with wider distributions—the cormorant *Phalacrocorax melanoleucos*, the duck *Anas superciliosa* and the common kingfisher *Alcedo atthis*, on 14, 38, and 38 islands, respectively—can use the sea coast and small streams as well as lakes and rivers.) Six freshwater species (the ducks *Dendrocygna arcuata*, *D. guttata*, and *Anas gibberifrons*, the spoonbill *Platalea regia*, the jacana *Irediparra gallinacea*, and the stilt *Himanto-*

Table 12.1. Relation between species habitat preference and number of islands occupied.

Habitat	No. of species in the habitat	Mean i value ±SD	Range of i values
Coast	5	57 ± 10	42–67
Mangrove	2	28 ± 5	24, 31
Aerial	12	24 ± 19	2–49
Open country	21	20 ± 21	1–58
Lowland forest	104	19 ± 18	1–69
Fresh water	12	10 ± 14	1–14, 38, 38
Swamp and marsh	4	7 ± 4	2–12
Montane	31	4 ± 3	1–12

The variable i is the number of Northern Melanesian islands that a given species occupies. The third column gives i's average value ± standard deviation for all species of the indicated habitat (the number of species given in the second column). Data are from appendix 5. Notice that coastal species are widely distributed with no exceptions; that freshwater, swamp and marsh, and montane species are narrowly distributed with only two exceptions (two fresh water species occupying 38 islands each); and that aerial, open country, and lowland forest species are intermediate in their average values but very heterogeneous (wide ranges of their i values). The association between i and habitat is highly significant by one-way ANOVA ($p < .0001$), and post hoc pairwise comparisons show i to be significantly higher (p .0001–.037) for the coast than for any other habitat.

pus leucocephalus) are confined to one or more of the four islands with the largest and most productive lakes (New Britain, Long, Rennell, and Umboi). Similarly, mountains and swamps and marshes are confined to a few large islands; hence all species of those habitats without exception are confined to a modest number of islands (i = 2–12 for the four species of marshes and swamps, i = 1–12 for the 31 montane species).

However, it is also clear that, while habitat availability is one factor limiting distributions, it is not the sole factor. Lowland forest and aerial habitats occur on all Northern Melanesian islands, and some open country occurs on most islands. Yet species of these three habitats occupy a range of island numbers: 1–69, 2–49, and 1–58 islands, respectively, with mean values (averaged over all the species of that habitat) of 19, 24, and 20 species, respectively. For example, lowland forest harbors not only the warbler *Ortygocichla rubiginosa* (plate 9), confined to a single island, but also the moundbuilder *Megapodius freycinet* on 69 islands; aerial foragers include the wood-swallow *Artamus insignis* (plate 9), confined to two islands, as well as the swallow *Hirundo tahitica* on 46 islands; and open-country species include the grass finch *Lonchura melaena* on two islands and the sunbird *Nectarinia jugularis* on 57 islands. Why are so many species of these virtually ubiquitous habitats restricted in their island distributions? Why do species vary so greatly in their i value, even within the same habitat? We discuss contributing factors throughout the remainder of this chapter.

Island Number and Endemism

Recall that Northern Melanesian habitats differ greatly in the average level of endemism of their species (chapter 10). Nevertheless, if one reanalyzes i values, comparing species at the same level of endemism in different habitats, one obtains essentially the same sequence of i values with habitat as in table 12.1. Hence the conclusions of table 12.1 concerning which habitats have the most widely or narrowly distributed species are not qualitatively altered by taking into account the differing average levels of endemism among habitats.

However, level of endemism does have some effect independent of habitat: it contributes to some of the variation in i values among species sharing the same habitat. Figure 12.1 plots i against level of endemism for the two habitats offering the largest number of species for analysis: lowland forest with 104 species and montane habitats with 31 species. Among the species of each habitat, the average level of endemism declines with i. That is, higher endemics (endemic genera and species) are overrepresented among species restricted to few is-

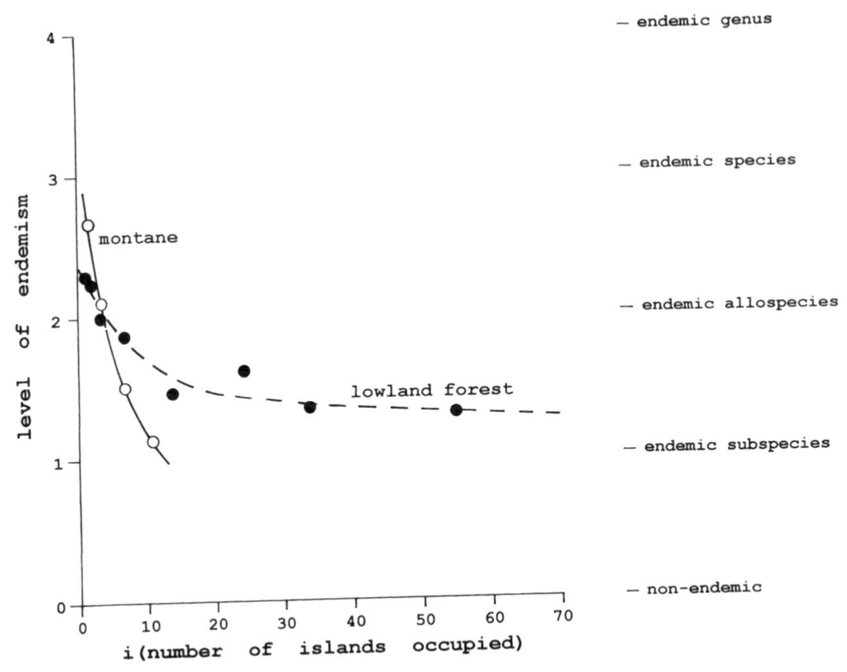

Fig. 12.1. Relation between the number of Northern Melanesian Islands that a species occupies (i, abscissa) and the species' level of endemism (ordinate), for species of two different habitats: lowland forest (●) and montane habitats (○). Level of endemism is graded as shown on the right, from 0 for a species whose Northern Melanesian populations are not endemic to 4 for a species belonging to an endemic genus. The species of each habitat are divided into classes according to their i value (e.g., 10–18 islands, 20–29 islands), and the mean level of endemism for a class's species is plotted against the class's mean i value. Data are from appendix 5. Note that, in each habitat, species occupying the fewest islands have, on average, the highest level of endemism and that no montane species occupies more than 12 islands but that lowland forest species occupy up to 69 islands.

lands, while species with undifferentiated populations tend to occupy the largest number of islands. For instance, among species of lowland forest, the three members of endemic genera occupy, respectively, only one, one, and three islands, while the 43 species represented only by endemic subspecies occupy, on average, 26 islands. Similarly, among aerial species the sole higher endemic (the woodswallow *Artamus insignis*, an endemic allospecies) is confined to two islands, while all other aerial species belong to only endemic subspecies or undifferentiated populations and occupy on the average 22 islands.

Island Number, Dispersal, and Abundance

Straightforward reasoning based on equilibria between immigration rates and extinction rates leads one to predict that i should increase with increasing vagility (because of higher immigration rates) and should also increase with increasing abundance (because of higher immigration rates and lower extinction rates).

Figure 12.2 depicts the average number of islands occupied by species of each dispersal class as a function of abundance class. At any level of abundance, i increases steeply with increasing vagility, just as predicted. The 28 most sedentary species occupy, on average, only two islands (never more than six islands), while the 56 most vagile species occupy, on average, 36 islands. The trend is an overwhelming one: 51 of the 56 most vagile species occupy more islands than any of the 28 most sedentary species.

While that predicted effect of dispersal on i is obvious from figure 12.2, the predicted effect of abundance is not obvious from the figure. Never-

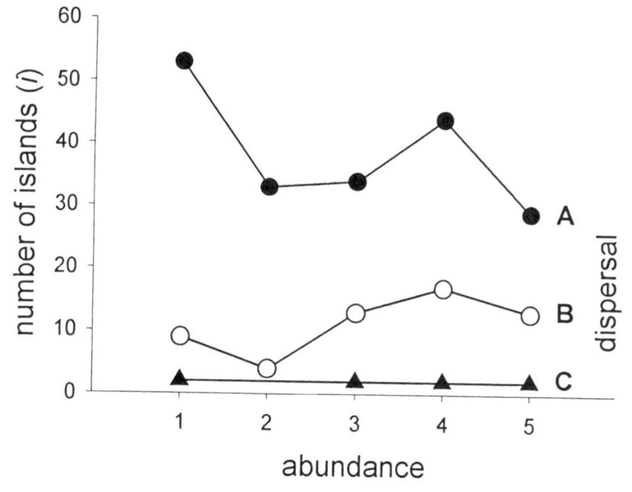

Fig. 12.2. Ordinate: mean value of i (number of islands occupied) for all the species of a given dispersal class (C = most sedentary, B = intermediate, A = most vagile) and abundance class (1 = rarest . . . 5 = most abundant). Abscissa: abundance class. The three sets of points are for the three dispersal classes. Sample size = 9–35 species per point, except n = 1, 2, 5, and 5 for classes A1, A5, C4, and C5, respectively. Data are from appendix 5. Note that i increases steeply with increasing vagility. Statistical analysis yields the following conclusions: there is significant variation in i (p = .0001 by two-way ANOVA); i varies significantly with vagility (p = .0001) and with abundance (p = .0095); vagility and abundance account for 51% of the variation in ln(i); and, in that model when sample sizes are taken into account, ln(i) increases significantly with abundance for dispersal classes A and B but not C. (The decrease in i with abundance that the graph suggests for dispersal class A is an artifact of the very low sample sizes for abundance classes A1 and A5.) In other words, vagile abundant species tend to occupy more islands than sedentary rare species, and the effect of vagility is greater than the effect of abundance.

theless, statistical analysis (see fig. 12.2 legend) reveals that i does increase with abundance as predicted, but the effect is much weaker than the effect of dispersal.

Species Incidences

Up to this point, we have analyzed only the total number of islands, i, that each Northern Melanesian species occupies. Let us now analyze in addition the characteristics of those islands—specifically, their species richnesses or areas—that each species occupies.

Calculation of Incidence Graphs

Northern Melanesian species differ greatly in the characteristics of the islands that they occupy. Some species occur only on large, central, mountainous, well-watered, species-rich islands. Others occur on large islands and also on small, species-poor islands down to some island area or species richness limiting for that particular bird species. Still other species are entirely absent from species-rich islands and confined to species-poor, small or remote islands.

These patterns can be displayed graphically as follows. Consider for each island its total number of resident land and freshwater species, S. Then group islands into classes with similar S values (e.g., 1–10 species, 11–20 species, 21–30 species, etc.). For a given species, calculate for each island class the fraction of Bismarck or Solomon islands in that class occupied by that species, and call that fraction the species' incidence, J. For example, if a certain species occurs on half of the islands whose species numbers fall in the range 51–60 species, then J is 0.5 for that species and that island size class. Finally, plot J against S for the given species.

Diamond (1975a) previously constructed such incidence graphs for the Bismarcks. We have constructed similar graphs for the Solomons. Note that we analyze incidences separately for the Bismarck

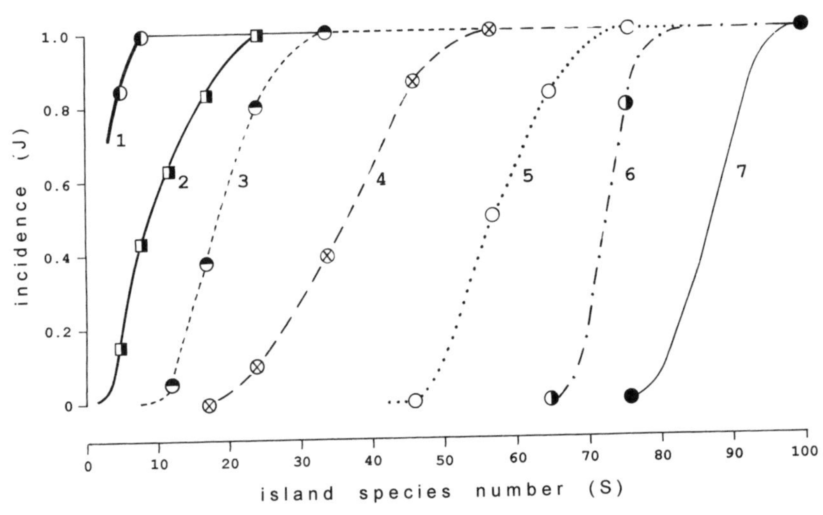

Fig. 12.3. Incidence as a function of island species number for eight bird species of the Solomons. Central Solomon islands were divided into 11 sets on the basis of their species number, S (2–6, 7–9, 10–14, 15–19, 20–29, 30–38, 41–51, 54–59, 60–65, 71–82, and 102–103 species). The number of islands per set was 10–19 for the five sets with lowest S, 5–7 for the next five sets, and 2 for the set with $S = 102$, 103. The ordinate J is the incidence of the given species (i.e., the fraction of the islands in the set on which the species occurs). The abscissa is the average value of S for the islands of the set. Thus, $J = 1.0$ or $J = 0$ means that the species occurs on all islands or on no island, respectively, that has approximately the indicated species number. For clarity, all points at $J = 1.0$ except the left-most one for each species, and all points at $J = 0$ except the right-most one for each species, are omitted. Islands with S values below 30 are 72 islets that Diamond surveyed in 1974 and 1976 (Diamond 1984d) and whose species lists are not included in appendix 1. The species are: 1 = the sunbird *Nectarinia jugularis*, 2 = the megapode *Megapodius freycinet*, 3 = the pigeon *Chalcophaps stephani*, 4 = the hawk *Aviceda subcristata* (map 2), 5 = the kingfisher *Ceyx lepidus*, 6 = the warbler *Phylloscopus* [*trivirgatus*] (map 36), 7 = the kingfisher *Halcyon bougainvillei* or the whistler *Pachycephala implicata*. By the criteria of table 12.3, we classify these species in the Solomons as, respectively, a D tramp, D tramp, D tramp, D tramp, C tramp, high S species, and high S species. In effect, each species is limited to islands above some characteristic species richness or area.

Archipelago and the Solomon Archipelago. We do not calculate incidences for both archipelagoes combined (i.e., for all Northern Melanesian islands) because the most important distributional barrier within Northern Melanesia lies between the Bismarcks and Solomons (chapter 29). We discuss later in this chapter the extent to which incidence graphs for a given species differ between the Bismarcks and Solomons.

Northern Melanesian islands vary in bird species number (S) from 1 to 127. If all species had equal probabilities of occurring on any island, then all species would yield identical incidence graphs, with J increasing linearly with S. In fact, as illustrated in figures 12.3 and 12.4, incidence graphs are rarely linear and vary greatly among species. As illustrated in figures 12.5 and 12.6, the same conclusions and graphs of similar form are obtained if islands are classed by area (A) and if J is plotted against log A rather than against S, since S varies closely and almost linearly with log A (fig. 9.3).

As discussed previously for the Bismarcks (Diamond 1975a), Northern Melanesian species can be grouped into six categories on the basis of their incidence graphs. These species differences arise from differences in species biology—especially in habitat preference, area requirements, dispersal ability, and competitive ability.

Supertramps

One extreme of the six categories is exemplified by figures 12.4 and 12.6: species absent as breeding

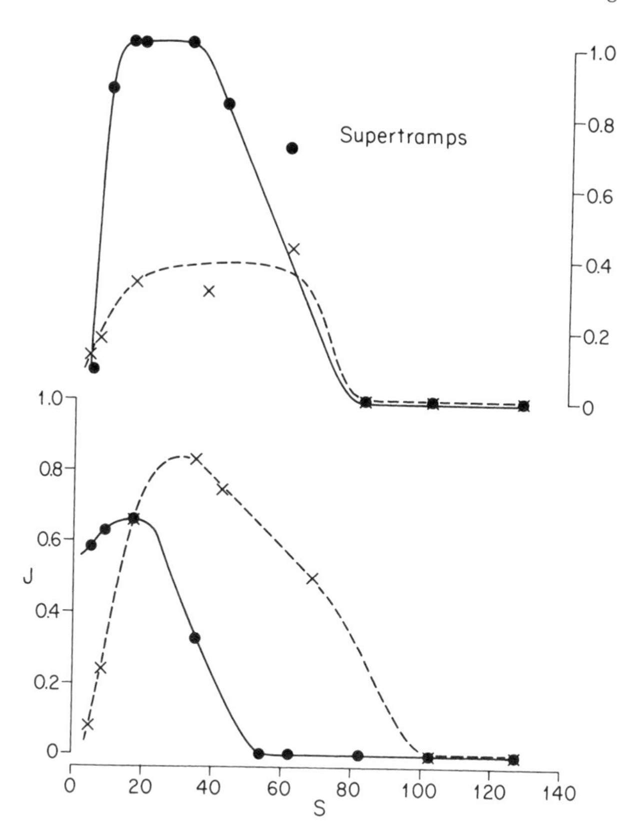

Fig. 12.4. Incidence (J) as a function of island species number (S) for four supertramp species of the Bismarcks. The curves were constructed analogously to the curves of figure 12.3 for seven species of the Solomons. (Top graph) the flycatcher *Monarcha cinerascens* (●; map 40) and the honey-eater *Myzomela sclateri* (X). (Bottom graph) the honey-eater *Myzomela pammelaena* (●; map 47) and the pigeon *Ptilinopus solomonensis* (X; map 7). From Diamond (1975a).

residents from species-rich or large islands and confined to species-poor or small islands. Such species are called "supertramps" because they represent an extreme development of species specialized for overwater colonization (tramp species), but we define species as supertramps purely on the basis of their island distributional pattern (absence from species-rich or large islands).

Supertramps tend to have high reproductive potential and broad habitat tolerance but low competitive ability (Diamond 1974, 1975a). Because of their overwater colonizing ability, they are among the first species to reoccupy an island that has been defaunated by a volcanic explosion and that is becoming revegetated. Eventually, species of more modest dispersal ability but greater competitive ability arrive at the revegetated island and crowd out the supertramps. Supertramps are chronically overrepresented on small islands, where extinctions and recolonizations occur often, and on remote islands, which other species have difficulty reaching.

An example of a supertramp distribution is that of the honey-eater *Myzomela sclateri* (plate 4), which is confined to islands of the Bismarck Volcanic Arc (fig. 3.2). Currently, most of the population of this species lives on the large (329 km^2) island of Long and on its neighbors Crown and Tolokiwa, which were defaunated by a massive eruption of Long in the seventeenth century (Diamond et al. 1989). On those islands *M. sclateri* is one of the commonest bird species at all elevations from sea level to the summits and occupies virtually all vegetated habitats from coconut plantations and savanna to closed forest and montane scrub. On Karkar, a similar-sized volcanic island which has had several modern eruptions but has not been recently defaunated, *M. sclateri* is confined to high elevations in and around the caldera. On Umboi, a similar-sized dormant volcanic island that has had no modern eruptions, there is only a single, uncertain sight record of *M. sclateri* (Eastwood 1995b) despite its abundance nearby on Long. *M. sclateri* has never been recorded on the much larger island of New Britain, despite living on islets within 1 km of New Britain. Its other occurrences are on small dormant or extinct volcanic islands (e.g., Vuatom, Witu, Unea) and very small coral islets (e.g., Credner) within the Bismarck Volcanic Arc.

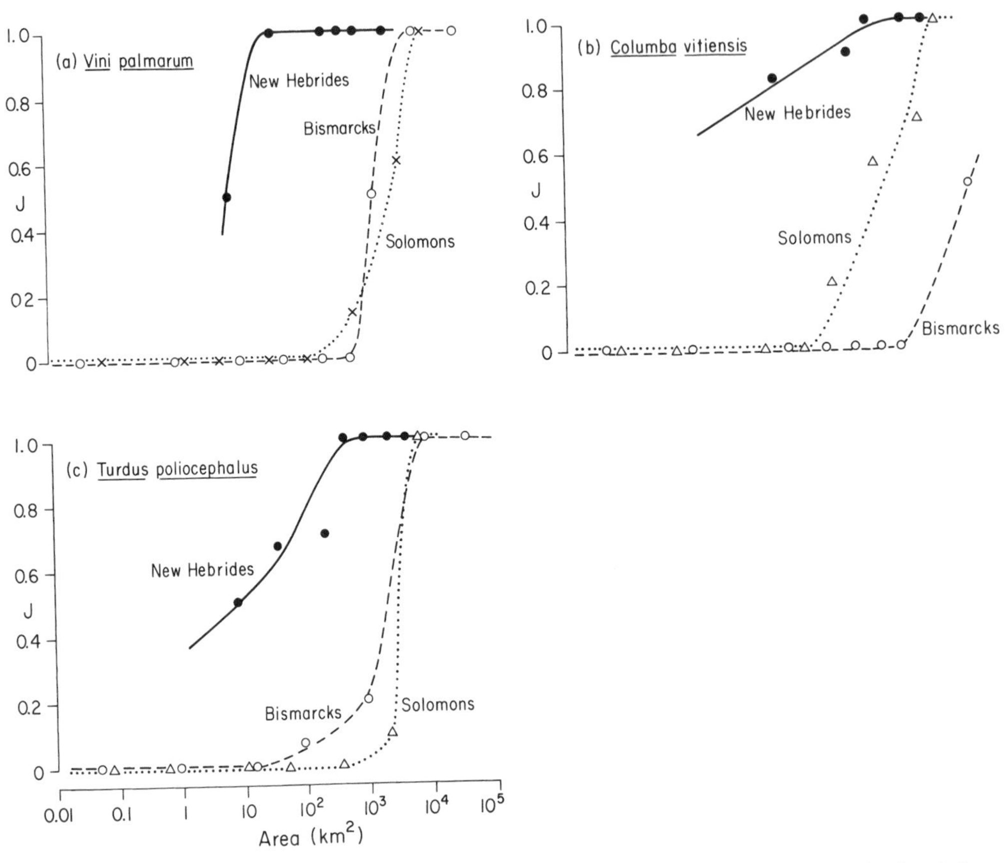

Fig. 12.5. Examples of how the incidence of a given species shared among the New Hebrides, Solomons, and Bismarcks tends to decrease from the New Hebrides to the Solomons to the Bismarcks, perhaps because the number of co-occurring, competing species increases from the New Hebrides to the Solomons to the Bismarcks. Incidence in this figure is plotted as a function of island area, rather than of island species number as in figures 12.3 and 12.4. Islands of an archipelago are grouped into area classes (*A*, in km^2). The fraction of a class's islands occupied by a particular species (i.e., the species' incidence, J) is plotted against log *A*. (a) Incidence of the lory *Charmosyna* (*Vini*) [*palmarum*]. It occupies all islands larger than about 100 km^2 and many smaller islands in the New Hebrides, which it shares with only one lory species. In the Bismarcks and Solomons, which it shares with five other lory species, it is confined to islands larger than 500–1000 km^2 (map 17). (b) The pigeon *Columba vitiensis* occupies all large and many small islands in the New Hebrides, which it shares with no similar pigeon, but is confined to the largest islands of the Bismarcks and Solomons (map 11), which it shares with two similar pigeon species. (c) The thrush *Turdus poliocephalus* (map 33) is one of 23 arboreal omnivores in the Solomons (including one other thrush), 14 in the Bismarcks (including one other thrush), and 9 in the New Hebrides (including no other thrush). As the number of competing species decreases, the incidence of this thrush increases to include more islands and small islands. From Diamond (1982a).

Table 12.2 lists all species with supertramp distributions in the Bismarcks and Solomons (see maps 7, 12, 40, 43, 45, and 47). It is apparent that supertramps are diverse ecologically and taxonomically, ranging from large frugivorous pigeons to a small insectivorous flycatcher, omnivorous white-eye, and nectarivorous honeyeaters. All are relatively vagile (dispersal categories A or B, mostly A) and relatively abundant (abundance categories 3, 4, or 5).

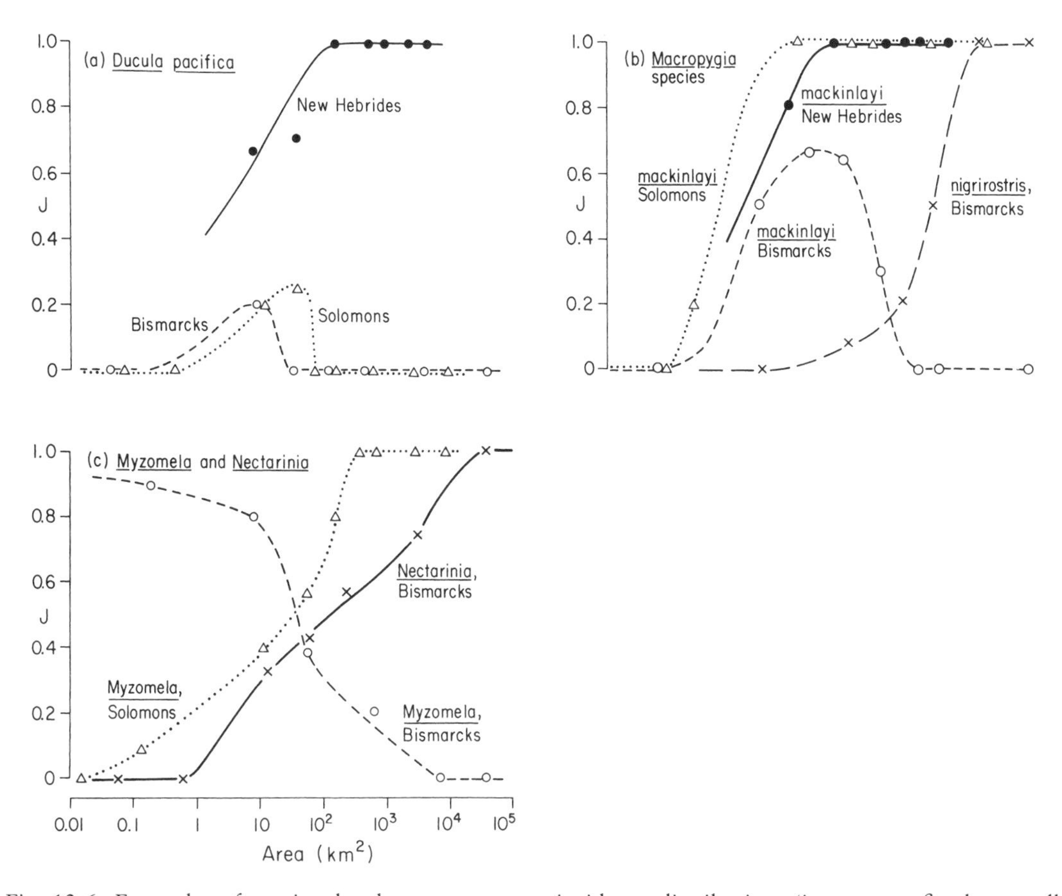

Fig. 12.6. Examples of species that have supertramp incidence distributions (i.e., are confined to small islands) in the Bismarcks or Solomons but that achieve tramp incidence distributions (i.e., occur on large as well as small islands) in the Solomons or New Hebrides, respectively, because of the reduced number of competing species in the latter archipelagoes. Incidence is plotted as a function of log (island area), as in figure 12.5. (a) The fruit pigeon *Ducula pacifica* is a supertramp confined to a few islands smaller than 100 km^2 in the Bismarcks and Solomons, which it shares with five and three congeners, respectively, including other supertramps. It occupies all islands larger than 100 km^2 plus many small islands in the New Hebrides, which it shares with only one congener. (b) The cuckoo-doves *Macropygia mackinlayi* and *M. nigrirostris* (map 12) are of the same size, similar in diet, and taxonomically the most closely related pair of species in their genus (allospecies of the same superspecies). *M. nigrirostris* occurs only in the Bismarcks; *M. mackinlayi* occurs in all three archipelagoes. In the Solomons and New Hebrides, from which *M. nigrirostris* is absent, *M. mackinlayi* occurs on every island larger than about 10 km^2 as well as on smaller islands. In the Bismarcks *M. mackinlayi* is a supertramp confined to small islands and absent from larger ones, which are occupied instead by *M. nigrirostris*. The Bismarck distributions define strictly non-overlapping checkerboards: *M. mackinlayi* on 15 islands, *M. nigrirostris* on 5, no islands shared despite vagrants of each species having been recorded on islands where the other is established. (c) Both the honey-eater *Myzomela* [*lafargei*] (map 47) and the ecologically and morphologically similar sunbird *Nectarinia sericea* occur in the Bismarcks. Only *M.* [*lafargei*] is in the Solomons, where it occupies all islands larger than 230 km^2 as well as some small islands. In the Bismarcks it is a supertramp confined to islands up to 414 km^2, whereas larger islands are occupied instead by *N. sericea*. The Bismarck distributions define strictly non-overlapping checkerboards: *M.* [*lafargei*] on 23 islands, *N. sericea* on 18 islands, no islands shared; see Diamond and Gilpin (1981) for further discussion. From Diamond (1982a).

Table 12.2. Northern Melanesian supertramps.

Species	Group	Plate no.	Distribution	
			Bismarcks	Solomons
Ptilinopus [*purpuratus*]	Fruit pigeon	—	—	Supertramp
Ptilinopus solomonensis	Fruit pigeon	6	Supertramp	Tramp
Ducula pacifica	Fruit pigeon	—	Supertramp	Supertramp
Ducula pistrinaria	Fruit pigeon	8	Supertramp	Tramp
Ducula spilorrhoa	Fruit pigeon	—	Supertramp	—
Macropygia mackinlayi	Pigeon	—	Supertramp	Tramp
Caloenas nicobarica	Pigeon	—	Supertramp	Supertramp
Monarcha cinerascens	Monarch flycatcher	5	Supertramp	Supertramp
Pachycephala melanura	Whistler	2	Supertramp	Supertramp
Zosterops [*griseotinctus*]	White-eye	1	Supertramp	Tramp
Myzomela sclateri	Honey-eater	4	Supertramp	—
Myzomela lafargei	Honey-eater	4	Supertramp	Tramp
Aplonis feadensis	Starling	—	Supertramp	Supertramp

This table lists all species having supertramp distributions (defined as species absent as regular breeding residents from large species-rich islands but present on smaller more species-poor islands) in either the Bismarcks or Solomons. "Plate no." refers to the number of the color plate, if any, on which the species is depicted. The right-most two columns indicate for each archipelago whether the species does not occur (—), occurs with a supertramp distribution, or occurs with a tramp distribution (present on large species-rich islands as well as small species-poor ones). Note that five species with supertramp distributions in the Bismarcks have tramp distributions in the Solomons (see fig. 12.6 and text discussion). *Ducula spilorrhoa* and *Caloenas nicobarica* breed on small islands but often feed in coastal forest of large islands; *Ducula pistrinaria* does so in the Bismarcks but actually breeds on large Solomon islands. *Macropygia mackinlayi*, *Monarcha cinerascens*, *Pachycephala melanura*, and possibly *Ptilinopus solomonensis* have occasionally been recorded as nonbreeding vagrants in coastal habitats of New Britain and some other large Bismarck islands on which they do not reside.

Incidence Categories Other Than Supertramps

Species other than supertramps occur on large, species-rich islands and on varying numbers of small, species-poor islands. Table 12.3 divides these species into five arbitrary categories, based on the number of islands occupied (i values). High S species are defined as species confined to the largest, most species-rich islands. A tramps, B tramps, C tramps, and D tramps are defined as occurring on species-poor as well as species-rich islands and are arbitrarily delineated by total number of islands of occurrence (increasing from A tramps to D tramps). With these definitions, as one proceeds from A tramps to D tramps, species distributions extend to increasingly small and species-poor islands. For instance, almost all Bismarck A tramps are confined to islands with more than 32 species, while half of the C tramps reach islands with 15–31 species, and most D tramps reach islands with under 10 species. Appendix 5 lists the incidence category of each Northern Melanesian species separately for the Bismarcks and for the Solomons.

To examine the relation between incidence category and habitat preference, we divided the eight habitat types used in appendix 5 into two groups: "ubiquitous habitats" present on all islands (lowland forest, lowland aerial, and sea coast), and "specialized habitats" present only on certain islands (mountains, freshwater, marshes and swamps, mangrove, and open habitats). Most (79%) B, C, and D tramps occur in ubiquitous habitats, as one would expect, since species of specialized habitats present only on certain islands are unlikely to occupy enough islands to fulfill the definition of these tramp categories. All supertramps also occur in ubiquitous habitats. However, among A tramps and high S species, only 50% occur in ubiquitous habitats, while 50% are confined to specialized habitats. For example, almost all montane species are high S species or A tramps, confined to few and species-rich islands, for the obvious reason that only large, species-rich islands have high mountains. Thus, half of the species largely confined to a few big islands might be restricted to such islands because of their specialized habitat preference.

Table 12.3. Definitions of incidence categories in Northern Melanesia.

	No. of islands occupied			
	A tramp	B tramp	C tramp	D tramp
Bismarcks	1–6	7–12	13–18	19–25
Solomons	1–7	8–15	16–23	24–31

Incidence categories are defined separately for the Bismarcks and Solomons, based on species distributions on the well-surveyed islands larger than 0.5 km^2, and excluding the Polynesian outlier islands of the Solomons. That leaves 25 Bismarck and 31 Solomon islands for analysis. Supertramps are defined by their absence as residents on all the most species-rich islands. High *S* species are defined as being confined to islands with >66 (Bismarck) or >71 (Solomon) species, plus not more than one island with 43–66 (Bismarck) or 38–65 (Solomon) species. Tramps are defined as occurring on the most species-rich islands and on sufficiently many or sufficiently poor species-poor islands to violate the above definition of high *S* species. Categories of tramps are delineated by the number of islands occupied, as listed in the table. Definitions of Bismarck categories are nearly the same as those used by Diamond (1975a). Solomon categories are defined so as to be as nearly similar as possible to the Bismarck categories, taking into account the fact that there are more islands with more than 65 or 66 species in the Solomons than in the Bismarcks.

Why are the other 50% of A tramps and high *S* species restricted to few islands, despite living in apparently ubiquitous habitat types? This is essentially the same question as posed earlier in this chapter (why are so many species of these virtually ubiquitous habitats restricted in their island distributions?). It is also one of the major questions arising from the familiar species/area relation of island biogeography: greater habitat diversity on larger islands provides only part of the explanation for the increase in species number with island area, so what is the other part of the explanation? Detailed examination of distributions of individual Northern Melanesian species suggests at least four answers: subtle habitat requirements, large territory requirements, large seasonal movements, and low dispersal.

First, as an example of subtle habitat requirements, the peregrine falcon (*Falco peregrinus*) requires suitable cliffs for breeding (available only on some islands), although its habitat (lowland air) in our crude classification is apparently ubiquitous. Again, species categorized as living in lowland forest may actually require a particular forest type present only on certain islands.

Second, it is not enough that a species' habitat merely be represented on an island: there must be enough of the habitat to provide a territory for at least one breeding pair (or colony in the case of social species). This fact alone would bar forest species with large territories or sparse distributions (such as the large goshawk *Accipiter meyerianus* and large owl *Nesasio solomonensis* [plate 8]) from small forested islands.

Third, species dependent on spatially and temporally patchy food supplies, such as many itinerant frugivores (e.g., pigeons), nectarivores (e.g., lories), and insectivores specialized on large-bodied, swarming insects (the bee-eater *Merops philippinus* and wood swallow *Artamus insignis* [plate 9]), can maintain a year-round resident population only on an island large enough that seasonal movements will permit an adequate food supply to be located in any month.

Finally, there remain many abundant species that occur throughout the forest and are not known to undertake large-scale seasonal movements, so that these above-mentioned three factors fail to account for their restriction to a few large islands. We suspect that the relation between immigration and extinction rates accounts for most of these cases. In the steady state, the incidence, J, is given by $I/(I + E)$, where I is immigration rate and E is extinction rate (Diamond & Marshall 1977b, Gilpin & Diamond 1981). If E is high, I must also be high for J to be appreciably greater than zero. Extinction rates increase with decreasing island size. Thus, if immigration rates are low, the species becomes restricted to (i.e., J becomes appreciable on) islands of large area, large population size, and hence low extinction probability. In practice, species that can often be seen flying across water are the species found on tiny islands holding only a single pair of the species (e.g., the flycatcher *Rhipidura leucophrys* and the sunbird *Nectarinia jugularis*).

We are struck by the distributions of 25 lowland forest species that are restricted to a few large islands (confined to eight or fewer Northern Melanesian islands; incidence category high *S* or A tramp), that are of modest or low dispersal ability (dispersal category B or C), and for which we can discern no reason why they could not survive on the medium-sized islands and on some other large islands from which they are nevertheless absent.[1] Two-thirds of these species are relatively abundant (abundance categories 3, 4, or 5), so that they exist only on islands large enough to support populations numbering thousands of pairs. The likely explanation is that their dispersal ability is sufficiently low that immigrations of colonists to an island only large enough to hold few hundred pairs are even rarer than the rare extinctions of such populations, with the result that J (= $I/[I + E]$) is low, and the island is usually unoccupied. In addition to these 25 species that are sedentary and restricted to large islands throughout Northern Melanesia, five other lowland forest species are vagile and widely distributed in the Solomons but more sedentary and restricted to large islands in the Bismarcks (see also notes 3 and 4).[2] The Northern Melanesian populations of all 30 of these species are endemic, mostly at higher levels (7 endemic subspecies, 14 endemic allospecies, 8 endemic full species, 1 endemic genus). Hence these 30 species conform to the picture of forest-dwelling endemics of restricted distribution, as often discussed in the literature on taxon cycles.

Consideration of species' habitat preferences, levels of endemism, and incidence categories simultaneously shows that all endemic genera and species are confined to mountains or to lowland forest. Not only the montane endemics but also most lowland forest endemics are confined to few and species-rich islands, probably due to their low dispersal ability, as discussed in the preceding paragraphs. At the opposite extreme, most species with wide distributions (C and D tramps) are little differentiated, vagile species of ubiquitous habitats (lowland forest, sea coast, and the air) and open habitats.

Differences Between the Bismarcks and Solomons in Species Distributions

Most species occurring in both the Bismarcks and Solomons have incidences that differ little between the two archipelagoes. The exceptions are of five types and are worth detailing. The ultimate cause of most of the exceptions is probably that the Bismarck avifauna, being closer to New Guinea, is somewhat richer than the Solomon avifauna: about 142 versus 122 widespread species, respectively (excluding geographically peripheral species in each archipelago). Because the Bismarcks have more competing species, the distribution of any given species is likely to be somewhat more compressed in the Bismarcks than in the Solomons. Solomon distributions, in turn, are compressed in comparison with those in the New Hebrides (fig. 12.5), which lie still farther from New Guinea and whose species pool consists of only 56 species (Diamond & Marshall 1977a, 1977b, Diamond 1982a).

The first type of exception (table 12.2) is that five species with tramp distributions in the Solomons (i.e., present on species-rich as well as species-poor islands) have supertramp distributions in the Bismarcks (i.e., confined to species-poor islands). In each case it is clear which Bismarck competing species absent in the Solomons are responsible for the supertramp's restricted distribution in the Bismarcks: *Ptilinopus* [*rivoli*] *rivoli*, *Macropygia* [*ruficeps*] *nigrirostris*, *Zosterops atrifrons*, several *Ducula* species, and *Nectarinia sericea* restrict the ranges of the Bismarck supertramps *P.* [*rivoli*] *solomonensis*, *M.* [*ruficeps*] *mackinlayi* (map 12), *Z.* [*griseotinctus*] (map 45), *D. pistrinaria*, and *Myzomela* [*lafargei*] (map 47), respectively. For example, the pigeon *Ptilinopus solomonensis* is a supertramp confined to species-poor small or remote islands in the Bismarcks, where the closely related *P. rivoli* occupies most of the larger and more species-rich islands, but *P. solomonensis* occupies large as well as small islands in the Solomons,

1. *Henicopernis infuscata, Accipiter imitator, A. luteoschistaceus, A. brachyurus, Nesoclopeus woodfordi, Ducula finschii, D. brenchleyi, Henicophaps foersteri, Cacatua galerita, Loriculus tener, Centropus violaceus, Nesasio solomonensis, Podargus ocellatus, Halcyon* [*diops*], *H. bougainvillei, Pitta anerythra, Ortygocichla rubiginosa, Monarcha chrysomela, Myiagra hebetior, Zosterops atrifrons, Z. metcalfii, Myzomela cineracea, Philemon* [*moluccensis*], *Dicrurus* [*hottentottus*], and *Corvus woodfordi*.

2. *Aviceda subcristata, Centropus* [*ateralbus*], *Coracina lineata, Monarcha manadensis*, and *Mino dumontii*.

Table 12.4. Comparisons of species incidences in three archipelagoes.

Measure	Bismarcks	Solomons	New Hebrides
Total species pool (no. of species)	142	122	56
High *S* species and A tramps (% of pool)	57	40	21[a]
C tramps and D tramps (% of pool)	23	45	55[a]
Average incidence (%)[b]	33	46	56

The first row is the total number of species in the archipelago, excluding geographically peripheral species. Row 2 gives the percentage of those species that are restricted to large, species-rich islands (high-S species and A-tramps); row 3, the percentage with wide distributions on small as well as large islands (C-tramps and D-tramps). Row 4 gives the average incidence: the percentage of the archipelago's islands (using island sets with similar distributions of areas in the three archipelagoes: see Diamond 1982a) occupied by a given species, averaged over all species in that archipelago's pool. By all these measures, species tend to be most narrowly distributed in the Bismarcks, which have the richest species pool, and most widely distributed in the New Hebrides, which have the poorest species pool.

[a]From Diamond and Marshall (1977b).

[b]From Diamond (1982a).

where *P. rivoli* is absent (map 7, plate 6). Similarly, there are two species with supertramp distributions in the Solomons (*Ptilinopus* [*purpuratus*] and *Ducula pacifica*) that have tramp distributions in the New Hebrides. Figure 12.6 illustrates three examples of these shifts between supertramp and tramp status. In contrast, no species with a supertramp distribution in the Solomons shifts to a tramp distribution in the Bismarcks, and no species has a supertramp distribution in the New Hebrides.

The second type of exception involves species present on species-rich islands in both the Solomons and Bismarcks. Recall that high *S* species, A tramps, B tramps, C tramps, and D tramps are defined as occurring on species-rich islands plus a number of species-poor islands that increases as one shifts along the spectrum of incidence categories from high-S species to D-tramps. Along this spectrum only eight out of 182 species shift by three or four categories between the two archipelagoes,[3] and only 13 species shift by two categories.[4] In 18 of these 21 cases (all except *Charmosyna placentis*, *Ixobrychus flavicollis*, and *Gallicolumba jobiensis*), it is the Solomon population that has the wider or "trampier" distribution. For example, in the Bismarcks the starling *Mino dumontii* is confined to six large islands and is classified as an A tramp, while in the Solomons, exclusive of outliers and the San Cristobal group, it occurs on every ornithologically explored island. Similarly, of the eight species that shift by two or more categories between the Solomons and the New Hebrides, the New Hebridean population has the wider distribution in six cases (Diamond & Marshall 1977b; see fig. 12.5 for examples).

The third type of exception is that the Bismarcks have relatively fewer species with wide distributions (C tramps and D tramps) and relatively more species with narrow distributions (high *S* species and A tramps) than do the Solomons. As rows 2 and 3 of table 12.4 show, the New Hebrides are at the opposite pole from the Bismarcks in both respects.

The fourth exception is another expression of the third. One can calculate an average incidence for all of an archipelago's species by calculating for each species the percentage of the archipelago's islands that the species occupies and then averaging over all species. Row 4 of Table 12.4 compares average incidences for the Bismarcks, Solomons, and New Hebrides, based on sets of islands chosen to have similar distributions of areas in the three archipelagoes (see Diamond 1982a for details). Aver-

3. *Butorides striatus*, *Aviceda subcristata*, *Porphyrio porphyrio*, *Charmosyna placentis*, *Alcedo pusilla*, *Rhyticeros plicatus*, *Coracina lineata*, and *Mino dumontii*.

4. *Ixobrychus flavicollis*, *Anas superciliosa*, *Amaurornis olivaceus*, *Ducula rubricera*, *Columba vitiensis*, *Gallicolumba jobiensis*, *Geoffroyus heteroclitus*, *Eudynamys scolopacea*, *Centropus* [*ateralbus*], *Ninox* [*novaeseelandiae*], *Ceyx lepidus*, *Coracina papuensis*, and *Monarcha* [*manadensis*].

age incidence is lowest for the Bismarcks (33%), intermediate for the Solomons (46%), and highest for the New Hebrides (56%). Part of the explanation is competitive release: when a given species occurs in more than one of the three archipelagoes, it usually has a lower incidence in the more species-rich archipelago (see tables 2 and 3 of Diamond 1982a for examples, and our figs. 12.5 and 12.6 for incidence graphs). Another part of the explanation may be that, since the New Hebrides lie farthest from the major colonization source of New Guinea, the New Hebrides' isolation tends to select for colonizing species with superior overwater colonizing ability, which are hence likely to achieve wide distributions within the archipelago.

A final exception concerns species present in the Bismarcks but entirely absent from the Solomons, and vice versa. Most such species (38 present only in the Bismarcks, 20 present only in the Solomons) are confined to a few species-rich islands (high S species or A tramps) in the archipelago where they do occur. Only 14 species that are widely distributed in one archipelago (C or D tramps: seven D tramps and three C tramps of the Solomons, four C tramps of the Bismarcks) are absent or confined to the geographic periphery of the other archipelago. For example, the monarch flycatchers *Monarcha [melanopsis]* (map 40, plate 5) and *Myiagra [rubecula]* (map 42, plate 5) are widespread in the Solomons ($i = 30$ and 27 islands, respectively) but absent in the Bismarcks; the lory *Chalcopsitta cardinalis* (map 15, plate 8) occupies virtually all central Solomon islands but only four Bismarck outliers nearest to the Solomons; and the pigeons *Ptilinopus insolitus* (map 6, plate 9) and *Macropygia amboinensis* are widespread in the Bismarcks ($i = 17$ and 16 islands, respectively) but absent in the Solomons. In a few of these cases one can suggest ecologically similar species in one archipelago that may contribute to excluding the other archipelago's tramp species, but most of these absences lack simple explanations.

Thus, species incidences are fairly similar between the Bismarcks and Solomons, but shifted in the direction of somewhat wider distributions in the more remote and species-poor Solomons, and even more so in the New Hebrides.

Summary

The number of Northern Melanesian islands (denoted i) occupied by a given species ranges from 1 for the 26 single-island endemics to 69 for the most widely distributed species. Some of this variation is related to habitat preference: the more widespread a species' preferred habitat, the more islands that species tends to occupy. However, lowland forest and aerial habitats occur on all Northern Melanesian islands, yet many species of these habitats have restricted island distributions, so some explanations other than availability of habitat must apply in these cases. For species of a given habitat, i decreases with increasing level of endemism within Northern Melanesia: Northern Melanesia's higher endemics have more restricted distributions than do species with undifferentiated populations. The value of i increases greatly with vagility and increases weakly with abundance, as expected from considerations of immigration/extinction equilibria.

Northern Melanesian species also differ greatly in their incidence graphs, which depict the fraction of islands occupied within a given area class or species-richness class, as a function of island area or species richness. At one extreme are Northern Melanesia's 13 supertramp species, vagile species confined to species-poor islands and excluded from species-rich islands by more sedentary competitors. At the opposite extreme are species confined to the largest, most species-rich islands. Four reasons contribute to explaining the hundreds of cases of species absent from islands offering apparently suitable habitat. In many cases of abundant but sedentary lowland forest species at higher levels of endemism, the most likely explanation is low dispersal ability combined with immigration/extinction equilibria: below a certain island size, population extinction rates become higher than immigration rates.

PART IV

COLONIZATION ROUTES

13

Proximate Origins of Northern Melanesian Populations

What are the main routes by which the ancestors of Northern Melanesia's bird populations arrived? Can one identify not only the proximate geographic source of the colonists, but also a more ultimate geographic source from which those colonists originated? These and related questions about colonization routes form the subject of part IV of our book. To illustrate the types of evidence available for answering these questions and to understand the distinction between proximate and ultimate origins, let us begin with an example: the origins of the Northern Melanesian population of Blyth's hornbill, *Rhyticeros plicatus*.

Blyth's hornbill, whose distribution, taxonomy, and relationships are detailed on p. 387 of appendix 1, is the sole species of its family occurring in New Guinea and Northern Melanesia. The hornbill family as a whole (Bucerotidae) consists of 14 genera with about 45 species, of which 22 species live in Africa, 19 in the Indomalayan region, four (including *R. plicatus*) in Wallacea, and only one (*R. plicatus*) in the Australian region (Sanft 1960, Forshaw & Cooper 1983–94, Kemp 1995). The four species of Wallacea and the Australasian region belong to two genera, both of which have their other populations concentrated in the Indomalayan region.

The Northern Melanesian species, *R. plicatus*, consists of seven subspecies: a very distinctive one (also treated as a separate allospecies) in Southeast Asia and the Sunda Shelf, three subspecies in the Moluccas and New Guinea, one subspecies in the Bismarcks, and two in the Solomons (map 25). *R. plicatus* has two closely related congeners in the Oriental region and one on Sumba Island in Wallacea. The Northern Melanesian subspecies are fairly similar to each other, and their closest extralimital relatives are the two New Guinea subspecies. Within Northern Melanesia, hornbills can often be seen flying over water for short distances in response to local availability of food. Until 1940, the only hornbills in the New Georgia group of the Solomons were confined to a few small islands; almost all other islands of the New Georgia group were occupied after 1940 (Diamond 2001a).

We draw the following inferences from these distributional and taxonomic facts. The Northern Melanesian populations of *R. plicatus* probably had their proximate origins in New Guinea. That is, New Guinea is probably the site where the first individuals to colonize Northern Melanesia hatched and whence they dispersed to reach Northern Melanesia. The colonization of Northern Melanesia was probably fairly recent on a geological time scale because the Northern Melanesian subspecies are still not very different from the New Guinea subspecies, and *R. plicatus* is still expanding its range in the part of Northern Melanesia farthest from New Guinea, the Solomons. *R. plicatus* has not yet reached the two most remote major Bismarck islands, Manus and St. Matthias, or the most remote major Solomon islands, the San Cristobal group. While the proximate origins of the Northern Melanesian population of *R. plicatus* were thus evidently in New Guinea, their ultimate origins lay in Asia. That is to say, the New Guinea populations were derived at some stage from Asia, certainly at the level of the genus, and possibly at the level of the species group or even of the species. In other words, the genus, and possibly the species group, to which *R. plicatus* belongs arose in Asia; even the species itself may have arisen there. There are also further questions about still more ultimate origins (i.e., whether the hornbill family arose in Asia or in Africa), but the answer to that question is not obvious and lies beyond the scope of this book.

This example of *R. plicatus* illustrates two points. First, one may attempt to deduce geographic origins on the basis of taxonomic affinities and distributions of the Northern Melanesian populations and their relatives. Second, since any extant population has a long history going back to the origins of life on Earth, questions about origins can be answered at different levels or time scales. For all Northern Melanesian populations this chapter will attempt to identify proximate origins, defined as the geographic source from which the colonizing individuals dispersed to reach Northern Melanesia. In the next chapter we use these conclusions about proximate origins to reassess the phenomenon of "faunal dominance" (the asymmetry of faunal exchanges between biogeographic provinces differing in species richness or area) as illustrated by Northern Melanesian birds. Finally, in chapter 15 we attempt to identify ultimate origins of Northern Melanesian populations at some higher taxonomic level, variously ranging from the subspecies to the parvorder. Appendix 7 summarizes, for each Northern Melanesian species or allospecies, the taxonomic and distributional data on which we base our conclusions about origins.

In this chapter we consider first, and in greatest detail, the proximate origins of those Northern Melanesian populations that are endemic to Northern Melanesia only at the subspecies level or else are not endemic at all—that is, populations of species or allospecies also occurring outside Northern Melanesia. We then consider allospecies endemic to Northern Melanesia, and finally endemic full species or genera. The lower the level of endemism, the more confidently one can draw conclusions about colonization routes, for three reasons: greater certainty with which the most closely related extralimital populations can be identified; less time available in which inferences of colonization routes from modern geographic distributions could have become confounded by extinction of some former potential source populations (see below); and less geological time available for the configuration of land to have become altered.

Endemic Subspecies and Nonendemic Populations

We base our conclusions about the proximate origins of endemic subspecies and nonendemic populations on the distribution of the Northern Melanesian subspecies, of the most closely-related subspecies, and of other conspecific populations (see appendix 7.1 for data). We assume that those subspecies most similar to each other phenotypically are the ones most likely to be related by common ancestry (although convergence is an alternative explanation, particularly when subspecific characters only involve size or color). We also assume that colonization is more likely to have proceeded from an area harboring many taxa of a group to an area harboring only a few taxa of that same group than vice versa (although the reverse may be true if there has been a secondary radiation in the species-rich area, or if there has been much recent extinction in the currently species-poor area). The following examples illustrate our reasoning and some of the types of complexities involved.

Examples

In the case of *Rhyticeros plicatus* discussed above, only New Guinea needs be considered as a proximate source of colonists for Northern Melanesia, since the species is absent in Australia and in Pacific archipelagos from the New Hebrides eastward, and since the New Guinea subspecies are more similar to the Northern Melanesian subspecies than are the subspecies west of New Guinea. Since all other members of the hornbill family live to the west of New Guinea, and since most live on the Asian and African mainland, it is far more likely that *Rhyticeros plicatus* spread from Asia to New Guinea and then to Northern Melanesia than that the hornbill family arose in Northern Melanesia, that *R. plicatus* as the sole Northern Melanesian survivor of those ancient origins spread from Northern Melanesia to New Guinea to Asia, and that hornbills subsequently radiated in Asia and Africa.

The goshawk *Accipiter fasciatus* occurs within Northern Melanesia only on Rennell and the neighboring island of Bellona, while it occurs extralimitally in Australia, New Guinea, some islands of Wallacea, the New Hebrides, and New Caledonia. The Rennell population is indistinguishable from the South Australian subspecies, and it is distinct from the subspecies of New Guinea, New Caledonia, and the New Hebrides. Hence we assume that the Rennell population was derived from Australia.

The barn owl *Tyto alba* is virtually confined in Northern Melanesia to the Solomons plus one adjacent Bismarck island (Tanga). Like *Accipiter fasciatus*, *T. alba* occurs in Australia, New Guinea, the New Hebrides, and New Caledonia. Unlike *Accipiter fasciatus*, the Solomon population of *T. alba* belongs to an endemic subspecies, but that subspecies is similar to the Australian subspecies and distinct from the subspecies of New Guinea, the New Hebrides, and New Caledonia (Mayr 1935). Thus, we assume that the Solomon population was derived from Australia. In addition, on Long Island in the western Bismarcks near New Guinea there is a *Tyto alba* population that has not been identified subspecifically and that might have been independently derived from New Guinea.

The pitta *Pitta erythrogaster* (map 26) is confined in Northern Melanesia to the Bismarcks, where it occurs as four endemic subspecies. Other subspecies occur in Australia, New Guinea, and islands from New Guinea west to the Philippines. The Australian population is confined to the tip of the Cape York Peninsula, but the species is distributed over all of New Guinea. The endemic Bismarck race with the widest distribution, *P. e. gazellae*, is most similar to the race *P. e. habenichti* of the adjacent part of northern New Guinea (Mayr 1955). Thus, we assume that *P. erythrogaster* colonized Northern Melanesia from New Guinea rather than from Australia. The sole other three pittas of Northern Melanesia are allospecies of widely ranging superspecies represented in New Guinea or Australia. All other pittas live on islands west of New Guinea and in Asia and Africa. Hence we assume that pittas spread ultimately from Asia to New Guinea and Northern Melanesia rather than vice versa.

The duck *Anas gibberifrons* occurs in Northern Melanesia only as an endemic subspecies on Rennell. Another subspecies is distributed over the land masses surrounding the Solomons (i.e., Australia, New Guinea, New Zealand, the New Hebrides, and New Caledonia). (Two other subspecies live on islands west of New Guinea.) Thus, subspecific characters cannot identify which of the various land masses surrounding the Solomons were the source of the Rennell population because the populations of all of those land masses belong to the same subspecies. However, the populations of New Guinea, New Zealand, the New Hebrides, and New Caledonia are very local or uncommon or founded only in modern times, probably transient, and maintained by occasional irruptions from the large, nomadic population of Australia. Thus, we assume that the Rennell population as well was derived from Australia.

Obviously, our deductions about colonization routes could be in error if other potential source populations existed in the recent past and became extinct. For example, perhaps the hornbill *R. plicatus* was formerly widespread in Australia, reached Northern Melanesia from there rather than from New Guinea, and subsequently became extinct in Australia. This interpretation would involve a purely hypothetical assumption based on no evidence whatsoever and would be a more complicated explanation than derivation from the extant similar population of New Guinea. However, in this and other cases fossils could prove the more complex explanation to be the valid one. For example, many extinct subfossil taxa are being discovered on Pacific islands east of the Solomons (Steadman 1989), and there are numerous cases in which discoveries of fossils have altered biogeographic conclusions drawn from studies only of living taxa. Therefore, since so few avian fossils from Northern Melanesia have been collected or studied, our interpretations should be considered as provisional ones that await testing on the basis of fossil evidence.

Colonization from Australia

We begin our analysis of proximate origins of endemic subspecies and nonendemic populations by considering colonists of Northern Melanesia from Australia. Although unequivocal Australian colonists make only a small contribution (10 species) to the Northern Melanesian avifauna, they are the group whose origins are most clearly documented by population movements continuing today. Appendix 7.1a lists these 10 species and their relevant data. As elsewhere in this book, our analysis considers only Northern Melanesia's breeding avifauna and omits species occurring solely as nonbreeding winter visitors.

For one of these species, the grebe *Tachybaptus novaehollandiae* (map 1), it is uncertain whether the proximate origins were in Australia or in the New Hebrides. The sole Northern Melanesian population is an endemic subspecies on Rennell Island, while the New Hebrides population belongs to a different subspecies from the Australian population, but it is not clear whether the Rennell subspecies is more similar to the New Hebrides or to the Australian subspecies. However, even if the most proximate origins were from the New Hebrides (as is clearly true for the case of the cuckoo *Chrysococcyx lucidus*, discussed below on p. 100), origins at the next higher level were still from Australia, since the New Hebrides populations themselves were derived from Australia.

All 10 of these Australian colonists carry out large-scale seasonal movements within or beyond Australia. The duck *Anas gibberifrons* is a nomad that lives in the dry interior of Australia and moves long distances in response to rainfall. The cuckoo *Scythrops novaehollandiae*, the nightjar *Eurostopodus mystacalis*, and the kingfisher *Halcyon sancta* (plate 6) perform regular north–south migrations each year to winter farther north in Australia or on islands north of Australia; *Scythrops novaehollandiae* and *Halcyon sancta* winter regularly in Northern Melanesia. The grebe *Tachybaptus novaehollandiae*, the pelican *Pelecanus conspicillatus*, the ibis *Threskiornis moluccus*, and the spoonbill *Platalea regia* sporadically irrupt within or outside Australia. The goshawk *Accipiter fasciatus* and the barn owl *Tyto alba* have smaller-scale local irruptions within Australia. Thus, massive, on-going dispersal is continuing to generate large numbers of colonists of these species today.

Four of the 10 species are not well established in Northern Melanesia but instead breed only occasionally (*Pelecanus conspicillatus*, *Scythrops novaehollandiae*, and *Halcyon sancta*), or it is uncertain whether they breed in Northern Melanesia at all (*Platalea regia*). The tiny Northern Melanesian breeding populations of these species are undoubtedly maintained or periodically refounded by migrants and vagrants from Australia. Irruptions from Australia to Northern Melanesia are well documented for *Pelecanus conspicillatus* (Bradley & Wolff 1956, Cain & Galbraith 1956, Galbraith & Galbraith 1962, Filewood 1969, 1972) and for the cormorant *Phalacrocorax melanoleucos* (Diamond 2001).

Of the six Australian colonists that are well established in Northern Melanesia, five belong to endemic subspecies; only one (*Accipiter fasciatus*) does not.

Within Northern Melanesia the Australian colonists are heavily concentrated in the Solomons. Of the 10, five are confined to Rennell in the eastern Solomons, one occurs more widely in the Solomons but not in the Bismarcks (*Eurostopodus mysticalis*), three occur mainly in the Solomons and either reach only one of the nearest Bismarck islands (the barn owl *Tyto alba*) or else visit but do not breed in the Bismarcks (*Pelecanus conspicillatus* and *Halcyon sancta*), and only one (*Scythrops novaehollandiae*) occurs mainly in the Bismarcks. A map (e.g., fig. 0.2, p. xiv) makes clear why these Australian colonists are concentrated in the Solomons rather than in the Bismarcks: there is no land lying between the Solomons and Australia, but the Bismarcks are shielded from Australia by eastern New Guinea and the Southeast Papuan Islands.

Colonization of Solomons Directly from New Guinea

We consider three species, listed in appendix 7.1b, to have reached the Solomons directly from New Guinea. All three are widespread in New Guinea and on the Southeast Papuan Islands and may have reached the Solomons via the Southeast Papuan Islands. The pigeon *Ptilinopus viridis* (map 8) is absent in Australia and the New Hebrides; the frogmouth *Podargus ocellatus* occurs in Australia, but the Solomon subspecies resembles that of the Southeast Papuan Islands rather than that of Australia; and the pigeon *Columba vitiensis* (map 11) occurs in the New Hebrides, but the superspecies is centered on New Guinea, Australia, and islands to the west, and the ranges of all but one of its 47 congeners are to the west, so that it is much more likely to have spread from New Guinea eastward to the Solomons to the New Hebrides than vice versa. *Podargus ocellatus* is absent in the Bismarcks, while the other two species are confined in the Bismarcks to single islands adjacent to the Solomons (*Ptilinopus viridis* on Lihir, *Columba vitiensis* on New Ireland), which they may have reached by a secondary spread from the Solomons to the Bismarcks. The Northern Melanesian populations of two of the three species (*Ptilinopus viridis* and *Podargus ocellatus*) belong to endemic subspecies.

Colonization of Bismarcks from New Guinea

We infer that 30 species (appendix 7.1c) have colonized the Bismarcks from New Guinea, but have not reached the Solomons. None leaps over the Solomons to reach the New Hebrides or any other eastern archipelago.

Of these 30 species represented by the same or similar subspecies on New Guinea, 17 are absent from Australia. The other 13 occur in Australia but are inferred to have reached Northern Melanesia from New Guinea instead because in eight cases the Bismarck and New Guinea populations belong to the same subspecies, but the Australian population is a different subspecies, or else the Bismarck population is an endemic subspecies that is more similar to the New Guinea than to the Australian subspecies; three are confined to tropical Australia but are widespread in New Guinea; and two (the falcon *Falco berigora* and the barn owl *Tyto novaehollandiae*) are confined in the Bismarcks to single islands north of New Guinea and shielded from Australia by the whole width of New Guinea.

The Bismarck populations of 11 of these species still belong to the same subspecies as the New Guinean source population, while the other 19 species have differentiated subspecifically in the Bismarcks.

Taken together, the Bismarck distributions of these 30 species suggest a series of snapshots of different stages in the spread of New Guinea colonists through the Bismarcks (fig. 13.1; stages 1–7). Two species (the cassowary *Casuarius bennetti* and the robin *Monachella muelleriana*) are confined to New Britain and two (*Falco berigora* and the pitta *Pitta sordida*) to the Long group, the Bismarck islands nearest New Guinea. Two others (the duck *Dendrocygna guttata* and the bee-eater *Merops philippinus*) are confined to New Britain and neighboring smaller islands plus the Long group. Five occur not only on New Britain plus in three cases the Long group, but also on New Ireland, the next large Bismarck island beyond New Britain: the plover *Charadrius dubius*, the pigeon *Gymnophaps albertisii* (map 10), the thrush *Saxicola caprata* (map 31), and the warblers *Cisticola exilis* and *Megalurus timoriensis*. One (*Tyto novaehollandiae*; map 21) is confined to the large island of Manus north of New Guinea. The other 18 species have variously spread more widely in the Bismarcks: to smaller islands west and east of New Ireland, to the St. Matthias group, and to Manus.

Colonization of Bismarcks and Western Solomons from New Guinea

Five species (appendix 7.1d) provide distributional snapshots in the next stage of the spread of species eastward from New Guinea (Fig. 13.1; stages 8 and 9). These five species occur not only in New Guinea and the Bismarcks but also on the westernmost Solomon islands. Two of them (the grebe *Tachybaptus ruficollis* [map 1] and the heron *Ixobrychus sinensis*) are confined in the Solomons to the westernmost large island, Bougainville; one species, the eagle *Haliaeetus leucogaster*, is confined in the Solomons to a small island even farther west, Nissan. The parrot *Charmosyna placentis* (map 17) occurs in the Solomons not only on Bougainville but also on three smaller Solomon islands west or north or east of Bougainville: Buka, Fead, and Nissan. The whistler *Pachycephala melanura* (map 43) is confined in the Solomons to small islets near Bougainville and near the adjacent islands of Buka and the Shortland group. All five species are absent not only from the Solomons east of Bougainville and the Shortlands but also from the New Hebrides and other archipelagos to the east.

The Northern Melanesian populations of all five of these species belong to the same subspecies as that of the New Guinea region. Three of the five are absent in Australia, while one (*Pachycephala melanura*) occurs in Australia as a different subspecies. The fifth (*Haliaeetus leucogaster*) is monotypic and occurs both in Australia and New Guinea, so it cannot be determined whether the Bismarcks were colonized from New Guinea or from Australia.

Colonization of Bismarcks and Solomons from New Guinea

Forty-three New Guinea species (appendix 7.1e) are widely distributed in both the Bismarcks and the Solomons. Twenty-seven of these species also occur in Australia, and 13 occur in the New Hebrides and/or other archipelagos east of the Solomons. However, the likelihood of New Guinea origins for Northern Melanesian populations of these species is suggested by their subspecific affinities or by their confinement in Australia to the tropical north. Northern Melanesian populations of 10 of these species belong to the same subspecies as the New Guinea population, while 25 have differentiated as endemic subspecies, and a further eight are represented both by endemic subspecies (on

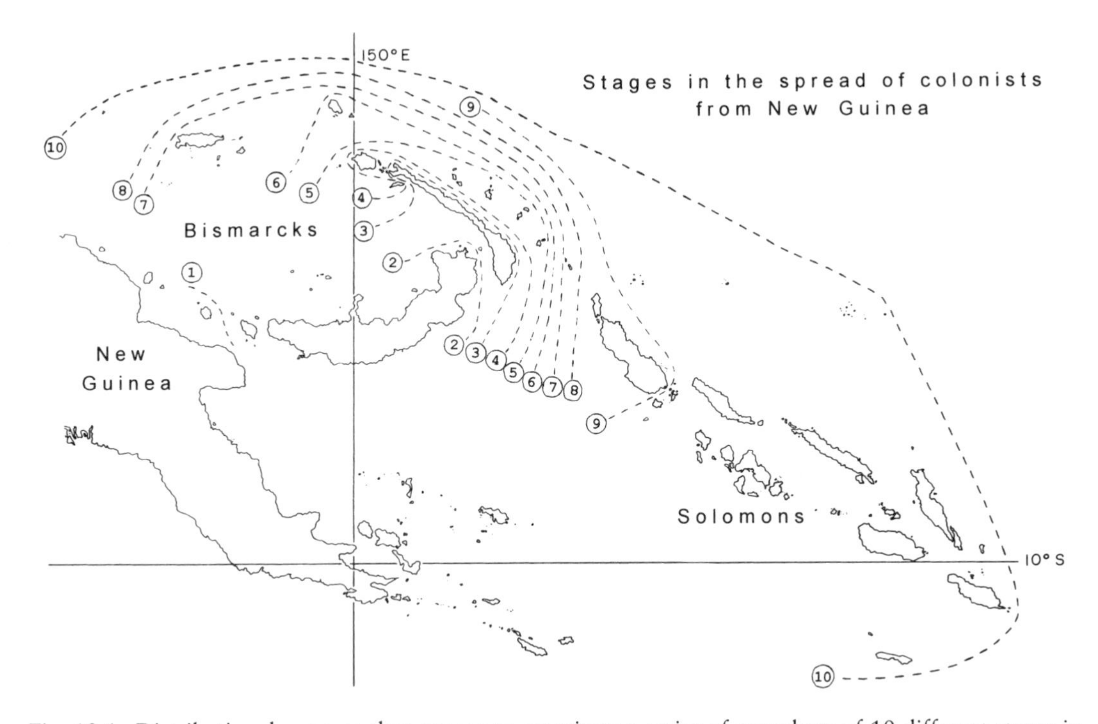

Fig. 13.1. Distributional patterns that appear to constitute a series of snapshots of 10 different stages in the progressive spread of New Guinean colonists through the Bismarcks and Solomons. All species occupy New Guinea, as well as the depicted range in the Bismarcks. 1. The pitta *Pitta sordida*: confined in the Bismarcks to the westernmost Bismarck islands nearest New Guinea (Long, Crown, Tolokiwa). 2. The bee-eater *Merops philippinus*: confined in the Bismarcks to Long, Umboi, and New Britain. 3. The thrush *Saxicola caprata* (map 31): Long and New Britain, plus New Ireland, the next large Bismarck island beyond New Britain. 4. The warbler *Megalurus timoriensis*: Tolokiwa in the Long group, New Britain, and New Ireland, plus New Hanover, the next large Bismarck island beyond New Ireland. 5. The sunbird *Nectarinia sericea*: Umboi, New Britain, New Ireland, New Hanover, and the islands west (Dyaul) and east (Tabar, Lihir, Feni) of New Ireland (i.e., virtually the whole Bismarcks except for the outlying St. Matthias group and Manus group). 6. The triller *Lalage leucomela* (map 30): Umboi, New Britain, New Ireland, New Hanover, islands west and east of New Ireland, and St. Matthias (i.e., virtually the whole Bismarcks except for the most remote outlier, the Manus group). 7. The fantail *Rhipidura rufiventris* (map 37): New Britain, New Ireland, New Hanover, islands west and east of New Ireland, St. Matthias and Manus (i.e., essentially the whole of the Bismarcks). 8. The eagle *Haliaeetus leucogaster*: the whole of the Bismarcks, plus Nissan, the westernmost Solomon island. 9. The parrot *Charmosyna placentis* (map 17): the Bismarcks, Nissan, and Buka and Bougainville, which are the next two Solomon islands east of Nissan. 10. The pigeon *Caloenas nicobarica*: the whole of the Bismarcks and Solomons.

some Northern Melanesian islands) and (on other Northern Melanesian islands) by populations belonging to a New Guinea subspecies. Some of these species may represent still further distributional snapshots of the spread of a colonizing species from New Guinea to the Bismarcks, then into the western Solomons and finally down the chain of the Solomons to its southeastern end at San Cristobal (fig. 13.1; stage 10). For example, distributions of the goshawk *Accipiter meyerianus* and the warbler *Acrocephalus stentoreus* are consistent with such a linear spread, having reached Guadalcanal in the Solomons but no farther.

However, many other species of appendix 7.1e have clearly had more complex distributional histories involving multiple colonizations of Northern Melanesia from New Guinea, or else a spread on a broad front from New Guinea over the Bismarcks and Solomons. For instance, Northern Melanesian populations of the cuckoo-shrike *Coracina tenuirostris* (plate 7, map 28) appear to represent at least five independent colonization waves from New Guinea (Mayr 1955). One reached Long and Sakar in the western Bismarcks near New Guinea following a volcanic eruption that defaunated Long 300 years ago (Diamond et al. 1989). The Long and Sakar populations still belong to the widespread New Guinea race *C. t. muellerii*. A second wave gave rise to the group of *C. tenuirostris* subspecies *admiralitatis, rooki, heinrothi*, and *matthiae* occupying New Britain and neighboring islands, the St. Matthias group, and Manus of the Bismarcks. A third colonization produced another group of subspecies that now occupies most of the Solomon islands plus New Ireland and neighboring islands, the Bismarck islands nearest the Solomons (subspecies *remota, ultima, saturatior*, and *erythropygia*). A fourth wave is represented by the subspecies *nisoria* of the Russell group of the Solomons, which is most similar to the populations of New Guinea or the Southeast Papuan Islands but shows some admixture with the other Solomon races. Finally, the population of San Cristobal, *C. salomonis*, is so distinctive that it is considered a separate allospecies and probably represents the oldest colonization wave.

As another example of multiple colonizations, the drongos (*Dicrurus hottentottus*) of Northern Melanesia appear to represent two independent colonizations from New Guinea, one giving rise to the Bismarck populations on the New Britain group and New Ireland (the latter ranked as a separate allospecies, *D. megarhynchus*: plate 8), the other giving rise to the two subspecies of *D. hottentottus* on Guadalcanal and San Cristobal of the Solomons (Vaurie 1949, Mayr 1955; map 52). Similarly, the Bismarck and Solomon populations may represent independent derivatives of the New Guinean source population in the cases of the cuckoo-shrikes *Coracina lineata* and *C. papuensis* (Mayr 1955).

At least four waves of the rail *Gallirallus* [*philippensis*] have reached Northern Melanesia. One of these is on Long in the western Bismarcks near New Guinea, belonging to the same subspecies of *G. philippensis* that occurs on the opposite coast of Northeast New Guinea. A second gave rise to several endemic subspecies elsewhere in the Bismarcks (*anachoretae, praedo, admiralitatis*, and *lesouefi*). A third wave evidently originating from the east gave rise to the endemic subspecies *christophori* of the eastern Solomons, which is most similar to the subspecies of the New Hebrides and Fiji. Finally, the population of the New Georgia group in the Solomons is a distinctive endemic and possibly flightless allospecies, *G. rovianae*, presumably derived from an earlier colonization.

Colonization of the Bismarcks from New Guinea or Australia

Five species (appendix 7.1f) are present both in New Guinea and Australia and are confined in Northern Melanesia to the Bismarcks: New Britain alone in the case of the parrot *Cacatua galerita* and the jacana *Irediparra gallinacea*, New Britain plus Umboi in the case of the duck *Dendrocygna arcuata* and the kingfisher *Tanysiptera sylvia*, and New Britain plus Umboi and Long in the case of the stilt *Himantopus leucocephalus*. Three of the five are represented in the Bismarcks by endemic subspecies, in the case of *Cacatua galerita* so distinct that it has also been considered as a separate allospecies. The Australian and New Guinean populations of *Dendrocygna arcuata, Irediparra gallinacea*, and *Himantopus leucocephalus* belong to the same subspecies, thus making it impossible to differentiate between Australia and New Guinea as the source of the Bismarck population. In the case of the other two species, the Australian race differs slightly from the New Guinean race, but it is not clear to which population the Bismarck subspecies is most closely related. Thus, all that we can infer is that these species reached the Bismarcks either from New Guinea or from Australia.

Colonization of the Bismarcks and Solomons from New Guinea and/or Australia

A further 13 species listed in appendix 7.1g occur both in the Bismarcks and Solomons, and both in New Guinea and Australia. Six of these species also reach Pacific archipelagos east of the Solomons, but in each case the species' center of distribution lies in New Guinea or Australia or islands to the west, so that these species probably spread from west to east rather than vice versa. In three cases the Northern Melanesian populations constitute endemic subspecies; in six cases they belong to the same subspecies as extralimital populations; and in four cases some of the Northern Melanesian populations belong to endemic subspecies while others are not endemic.

Subspecific characters do not permit us to decide in these 13 cases whether the Northern

Melanesian populations arose from New Guinea or from Australia. In five of the species (the herons *Egretta sacra*, *Ardea intermedia*, and *A. alba*, the thick-knee *Esacus magnirostris*, and the pigeon *Ptilinopus superbus*; map 5), the New Guinean, Australian, and Northern Melanesian populations all belong to the same subspecies, so that there are no subspecific differences to help determine origins. In one case (Northern Melanesian populations of the kingfisher *Halcyon chloris* except those of the eastern Solomons; map 24) the New Guinean and Australian populations belong to the same subspecies, the Northern Melanesian populations to endemic subspecies. In four other cases (the cormorant *Phalacrocorax melanoleucos*, the herons *Nycticorax caledonicus* and *Ixobrychus flavicollis*, and the hawk *Haliastur indus*), the New Guinean and Australian populations again belong to the same subspecies, while some of the Northern Melanesian populations belong to that same subspecies, and others are endemic. In the remaining three cases (the hawk *Aviceda subcristata*, the rail *Porphyrio porphyrio*, and the thrush *Zoothera heinei*; map 32), the Australian and New Guinean populations belong to different subspecies, while the Northern Melanesian populations are endemic, but it is not clear whether they are closer to the Australian or New Guinea subspecies.

Thus, we cannot discriminate between New Guinean and Australian origins for these species. Some of them may have colonized Northern Melanesia on a broad front from Australia and New Guinea—for example when New Guinea was joined to Australia at low-sea-level times of the Pleistocene. We note, however, that many wind-blown vagrants of *Phalacrocorax melanoleucos* reached the New Georgia group of the Solomons from Australia after the cyclone of 1952, and that *Ardea intermedia* and *A. alba* have not been proven to breed within Northern Melanesia but may instead be represented by vagrant individuals from Australia. In addition, racial characteristics indicate that one endemic subspecies of *Aviceda subcristata*, that on the Bismarck island of Manus north of New Guinea, was derived from the Northeast New Guinea population (Mayr 1945a).

Colonization from New Hebrides

For 13 Northern Melanesian populations (appendix 7.1h) the closest relatives are in the New Hebrides, New Caledonia, and other eastern archipelagos. Three cases (the rail *Gallirallus philippensis*, the kingfisher *Halcyon chloris* [map 24], and the whistler *Pachycephala pectoralis* [map 43]) involve species widespread in the Bismarcks and Solomons, represented there by numerous endemic subspecies, and also widespread in Australia and the New Guinea region, from whence most of the endemic races were probably derived. However, the endemic races of these three species in the easternmost Solomons are strikingly distinct from other Northern Melanesian races, and closely related instead to the New Hebrides subspecies, which we presume represent the source for these eastern Solomon populations.

All of the other 10 species are absent from New Guinea. Only three (the cuckoo *Chrysococcyx lucidus*, the fantail *Rhipidura fuliginosa*, and the robin *Petroica multicolor*) occur in Australia, but the endemic Northern Melanesian subspecies of those species are much more similar to the New Hebrides race than to the Australian race. Ten of the populations belong to endemic subspecies; one (the pigeon *Ptilinopus greyi*) is monotypic, and one (the pigeon *Ducula pacifica*) is represented by the New Hebrides subspecies in the Solomons and by an endemic subspecies in the Bismarcks.

Seven of the 10 species absent from New Guinea have their centers of distribution overwhelmingly to the east and thus probably reached Northern Melanesia from the east rather than vice versa. All occur in the New Hebrides and probably originated there proximately or nearly-proximately. (Three also occur in the Santa Cruz group, which is smaller than the New Hebrides but nearer the Solomons: hence the species could have arrived from the New Hebrides via the Santa Cruz group.) While subspecific characters relate the endemic race of *Chrysococcyx lucidus* proximately to the New Hebrides (as mentioned in the preceding paragraph), the species also occurs in Australia, and all other congeners live in Australia and New Guinea and areas to the west, where the ultimate origins of the Solomon population presumably lie. The cuckoo-shrike *Coracina caledonica* (map 27) is widespread in the Solomons and otherwise occurs only in the New Hebrides and New Caledonia, but the superspecies to which it belongs occurs otherwise in Australia and to the west, so that its ultimate origins also are probably from Australia.

The confined distributions of these eastern invaders within Northern Melanesia are striking. The three species (*Gallirallus philippensis*, *Halcyon chloris*, *Pachycephala pectoralis*) with races of other origins occurring elsewhere in Northern Melanesia have their eastern-derived races confined to the easternmost large Solomon island of San Cristobal and some nearby islands. Of the other 10 species, only one (*Ducula pacifica*) reaches the Bismarcks at all, where it is a supertramp confined to four small islands. Within the Solomons, three of the 10 are confined to the southeasternmost Solomon island, Rennell (plus the adjacent Bellona in the case of *Chrysococcyx lucidus*); one (*Ptilinopus greyi*) is confined to Gower, a small island in the eastern Solomons; three (the cuckoo-shrike *Lalage leucopyga* [map 30], the fantail *Rhipidura fuliginosa*, and the honey-eater *Myzomela cardinalis* [map 46, plate 4]) are confined to the San Cristobal group (plus Rennell in the case of *Myzomela cardinalis*); and *Ducula pacifica* occurs in the Solomons on Rennell and various smaller islands. Only *Coracina caledonica*, which may have arisen in the Solomons rather than in the east, is widespread in the Solomons. Thus, no proven eastern invader has been able to spread beyond the eastern Solomons to the large islands of the main Solomons or Bismarcks.

Interpretation Complicated or Difficult

Nine other species (appendix 7.1i) warrant individual discussion. The swallow *Hirundo tahitica* is an allospecies of Asian origin. One group of subspecies (including *frontalis*) occurs from Asia to New Guinea, plus the Long group and possibly Manus of the western Bismarcks. The other group of subspecies (*subfusca* and *tahitica*) occupies the Solomons, most of the Bismarcks, and islands east to Fiji and the Societies. The New Britain race *H. t. ambiens* is a hybrid between these two subspecies groups. The Australian population *neoxena* is very distinct and is sometimes separated as another allospecies. Thus, the ultimate origins are clearly from Asia via New Guinea, and western and eastern groups of populations now meet in Northern Melanesia. The population of most of Northern Melanesia (*subfusca*) could either have colonized from the east or else (less likely) differentiated in Northern Melanesia and spread eastward (Mayr 1934a).

Populations of five further species may have arisen at the allospecies or species level within the Bismarcks. The monarch flycatcher *Monarcha cinerascens* (map 40, plate 5) is a small-island species distributed from Wallacea to islands off North and East New Guinea, the Bismarcks, and western Solomons. It is closely related to the *Monarcha* [*melanopsis*] superspecies of New Guinea, Australia, and the Solomons, and probably arose as the Bismarck allospecies of that superspecies and subsequently spread westward and eastward.

Regarding the parrot *Lorius hypoinochrous* (map 16, plate 6), one subspecies occurs in the Bismarcks, Southeast Papuan Islands, and Southeast New Guinea, and the other two subspecies are endemic to the Southeast Papuan Islands. Other allospecies occupy New Guinea and the Moluccas, New Ireland, and the Solomons. *Lorius hypoinochrous* may either have originated as the allospecies of the Southeast Papuan Islands, from whence it colonized the Bismarcks and Southeast New Guinea, or it may have arisen as the New Britain allospecies and spread to the Southeast Papuan Islands and New Guinea.

The finch *Lonchura spectabilis* consists of one subspecies in the Bismarcks and three other subspecies in Northeast New Guinea. Two other allospecies occur in the Bismarcks, and there are many related species in New Guinea. The ultimate origins of the species group are clearly New Guinea, but it is uncertain whether the allospecies *Lonchura spectabilis* spread from North New Guinea to the Bismarcks or vice versa.

The Northern Melanesian population of the white-eye *Zosterops griseotinctus* is confined to five species-poor Bismarck islands (Nissan in the westernmost Solomons, Nauna in the Northwest Bismarcks, and the three islands of the Long group), while the four other subspecies live in the Southeast Papuan Islands, but the other five allospecies of the same superspecies are in the Solomons (see plate 1, map 45). This species may have colonized the Southeast Papuan Islands from Northern Melanesia, or vice versa.

Finally, the kingfisher *Halcyon saurophaga* (map 24, plate 6) consists of two endemic subspecies in the Northwest Bismarcks, plus one subspecies distributed over the rest of the Bismarcks, Solomons, North New Guinea, Moluccas, and Southeast Papuan Islands. It may have arisen in North New Guinea and/or the Northwest Bismarcks as an allospecies of its widely distributed, close relative *H. chloris* (Red Sea to Samoa), curiously absent from North New Guinea and the Northwest Bismarcks.

Thus, among Northern Melanesian birds, these five species are the most plausible candidates to have colonized the New Guinea region extensively from the Bismarcks, but in all five cases that direction of spread remains uncertain, and only three of these (*Lorius hypoinochrous, Lonchura spectabilis,* and *Halcyon saurophaga*) occur on the New Guinea mainland. Ten other Northern Melanesian taxa (p. 109) have colonized the New Guinea region only marginally to occupy a few fringe islands.

The whistler *Pachycephala pectoralis* (map 43) is distributed from Wallacea, Australia, and the New Guinea region to Fiji, with six closely related species in New Guinea (Galbraith 1956). There are several distinct groups of subspecies and related species evidently representing multiple colonization waves (see plate 2). Most Solomon populations are a distinct subspecies group (Galbraith's group C), related to other populations from the Lesser Sundas and New Guinea to Fiji. The Bismarck populations belong to a different subspecies group (Galbraith's group H), of uncertain origin and now distributed from Wallacea to Fiji. The San Cristobal race *P. p. christophori* is a hybrid between the other Solomon populations and the New Hebrides populations (Galbraith's group G).

The flycatcher *Rhipidura rufifrons* (map 39) has a far-flung distribution from Wallacea east to the Santa Cruz group and from Australia north to Micronesia. Its subspecies and related species define several waves of expansion (Mayr & Moynihan 1946). The Solomon populations are most similar to the race *R. r. louisiadensis* of the Southeast Papuan Islands, while the Manus race *R. r. semirubra* is distinct and of uncertain affinities.

The swiftlet *Aerodramus spodiopygius* is distributed from Wallacea east to Samoa. The related New Guinean population *hirundinaceus* has been considered a different allospecies on the basis of its dark rump but has also been considered conspecific (Salomonsen 1983). It is unclear whether the Northern Melanesian population is most closely related to the populations of Wallacea and those to the east of Northern Melanesia or whether it is derived from New Guinea and just convergent in rump color on other forms.

Endemic Allospecies

Having discussed proximate origins of endemic subspecies and non-endemic populations, we now consider more briefly Northern Melanesia's endemic allospecies. We classify their colonization routes into groups similar to those that we used for the endemic subspecies and nonendemic populations.

Colonization Directly from Australia

The two allospecies listed in appendix 7.2a (the cockatoo *Cacatua ducorpsi* [plate 8] and the pitta *Pitta anerythra* [plate 9]) appear to be colonists from Australia. Both occur in Northern Melanesia only as endemic allospecies confined to the Solomons. Other allospecies are in Australia but not New Guinea, except that there is a very local population of the Australian cockatoo *Cacatua sanguinea* (in the same superspecies as *C. ducorpsi* of the Solomons) in South New Guinea. Thus, as in the case of Northern Melanesia's endemic subspecies and nonendemic populations, the endemic allospecies that arrived from Australia are concentrated in (confined to) the Solomons.

Colonization of Bismarcks from New Guinea

Fourteen endemic allospecies (appendix 7.2b) are confined to the Bismarcks and are presumed to be derived from New Guinea. The superspecies to which they belong are absent from Australia and from Pacific islands east of Northern Melanesia, except that two occur in tropical Australia and two have a more widespread Australian distribution but as populations taxonomically more distinct from the Northern Melanesian populations than are the New Guinea populations. Two of these superspecies are also represented in Northern Melanesia by populations of the New Guinea allospecies, one of which reaches the Solomons. The other 13 superspecies are absent from the Solomons.

Colonization of Bismarcks and Solomons from New Guinea

Twelve superspecies (appendix 7.2c) are present both in the Bismarcks and Solomons, are represented either in the Bismarcks or Solomons or both by endemic allospecies, and are believed to be derived from New Guinea. In six cases all the Northern Melanesian populations are endemic at the allospecies level and belong to between one

(*Geoffroyus heteroclitus*; map 20, plate 9) and six (*Monarcha* [*manadensis*]; map 41, plate 3) endemic allospecies. In the other six cases both the New Guinea allospecies and one or more endemic allospecies occur in Northern Melanesia. The *Gymnophaps* superspecies (map 10) is represented by an endemic allospecies in the Solomons, while the Bismarck populations are conspecific with the New Guinean allospecies. The warbler *Phylloscopus* [*trivirgatus*] has an endemic allospecies (*P. makirensis*) on San Cristobal of the eastern Solomons, while the other Solomon and Bismarck populations belong to the New Guinea allospecies *P. poliocephalus* (map 36). The pigeons *Ptilinopus solomonensis* (map 7, plate 6) and *Macropygia mackinlayi* (map 12) are near-endemic allospecies shared by the Bismarcks and Solomons, while the New Guinea allospecies of the same superspecies also occurs in the Bismarcks. The parrots *Lorius* [*lory*] (map 16, plate 6) and *Micropsitta* [*pusio*] (map 19) are represented by different endemic allospecies in the Bismarcks and Solomons, while the New Guinean allospecies of the same superspecies also occurs in the Bismarcks.

All 12 of these superspecies are widespread on New Guinea. Only two also occur in Australia, as a population confined to the tip of the Cape York Peninsula (*Geoffroyus geoffroyi*; map 20), or as a more distinct allospecies (*Ninox novaeseelandiae*; map 22, plate 7). Only one of these 12 superspecies, *Macropygia mackinlayi*, extends to the Santa Cruz group and New Hebrides east of Northern Melanesia, as a population conspecific with the Northern Melanesian allospecies and presumably derived from it (the other two allospecies of the same superspecies are distributed from the Bismarcks to southeast Asia [map 12]). For these reasons it seems likely that these 12 allospecies or groups of allospecies were derived from New Guinea, not from Australia or archipelagos to the east. In many of these cases multiple colonizations of Northern Melanesia from New Guinea may be involved.

Colonization of Northern Melanesia from New Guinea or Australia

For nine superspecies (appendices 7.2d, 7.2e, and 7.2f) represented in Northern Melanesia by endemic allospecies and also occurring in both New Guinea and Australia, we cannot discern taxonomic characters that make clear whether the derivation was from New Guinea or from Australia. Only one of these nine superspecies reaches islands east of Northern Melanesia: the kingfisher *Halcyon* [*diops*], represented by an endemic allospecies in the New Hebrides (map 23). Since the other members of this superspecies are two allospecies in the Bismarcks and four in New Guinea and Australia and the Moluccas, we assume spread eastward to the New Hebrides rather than vice versa.

As in some previously discussed examples, the Northern Melanesian populations of four of these superspecies include not only endemic allospecies but also populations conspecific with New Guinea or Australian populations. Thus, multiple colonizations of Northern Melanesia may be involved. In several cases (the owl *Ninox* [*novaeseelandiae*] [map 22, plate 7], the kingfisher *Halcyon* [*diops*] [map 23], the thrush *Zoothera* [*dauma*] [map 32], and the monarch flycatcher *Monarcha* [*melanopsis*] [map 40, plate 5]), the Northern Melanesian populations belong to two, three, or four distinct endemic allospecies. Of these endemic allospecies or groups of allospecies, two are confined to the largest Bismarck island of New Britain, one also reaches the Solomon island of Buka, two are widespread in the Solomons but absent from the Bismarcks, two are widespread in the Bismarcks and Solomons, and one is confined to San Cristobal in the eastern Solomons.

Eastern Origins or Affinities

Six superspecies (appendix 7.2g) are represented in Northern Melanesia by endemic allospecies but have centers of distribution lying to the east of Northern Melanesia. Five of these superspecies (all except the superspecies to which the pigeon *Ptilinopus richardsi* belongs) are absent from New Guinea and Australia. The superspecies to which the pigeon *Ducula brenchleyi* (plate 8) belongs has three allospecies in archipelagos east of Northern Melanesia, only one in Northern Melanesia itself, so that this superspecies probably colonized Northern Melanesia from the east rather than vice versa. Similarly, although the superspecies containing the monarch flycatcher *Clytorhynchus hamlini* (plate 9) of Rennell includes just one other allospecies, living in the Santa Cruz group, the two other species of the genus live in archipelagoes from the New Hebrides eastward, so that this species, too, probably arrived from the east. In the other four cases it is as plausible that the superspecies spread from Northern Melanesia eastward as vice versa.

Table 13.1. Summary of proximate origins of Northern Melanesian bird populations.

Level of endemism	Proximate origins	Distribution within N Melanesia: Bismarcks only	Solomons only	Both	Total no. of species
Nonendemics and endemic subspecies	New Guinea	30	3	48	81
	Australia	1	9	—	10
	New Guinea or Australia	5	—	13	18
	East	—	12	1	13
	Complex or uncertain	3	—	6	9
	Total	39	24	68	131
Endemic allospecies	New Guinea	14	—	12	26
	Australia	—	2	—	2
	New Guinea or Australia	2	1	6	9
	East	—	6	—	6
	Complex or uncertain	1	1	7	9
	Total	17	10	25	52
Endemic species and genera	New Guinea or west	3	5	2	10
	East	1	1	—	2
	Uncertain	4	16	3	23
	Total	8	22	5	35

The first (left-most) column identifies the level of endemism of each species' Northern Melanesian populations: i.e., whether it belongs to an endemic genus, species (or superspecies), allospecies, or subspecies confined to Northern Melanesia, or whether it is non-endemic (belonging to the same subspecies as an extralimital population). The second column identifies the proximate origin of the Northern Melanesian population. The remaining columns denote whether the distribution within Northern Melanesia is virtually confined to the Bismarcks, or to the Solomons, or extends to both the Bismarcks and Solomons. Assignments for each species are given in appendices 7.1–7.3. The proportions of species among the various assignments (omitting the rows "complex or uncertain") do not differ significantly (χ^2 test, df 9, $p = .60$) if one compares endemic allospecies with nonendemics and endemic subspecies. That is, gross distributions within Northern Melanesia and proximate extralimital origins do not vary with level of endemism, at least up to the level of endemic allospecies.

None of these six superspecies occurs in the Bismarcks, two are confined in the Solomons to Rennell, and two others (*Ptilinopus richardsi* and *Ducula brenchleyi*) are confined to the eastern Solomons. Thus, as in the case of endemic subspecies or nonendemic populations with eastern affinities, the endemic allospecies with eastern affinities are concentrated in the eastern Solomons.

Interpretation Complicated or Difficult

For nine superspecies with endemic allospecies in Northern Melanesia (appendix 7.2h), the interpretation is complicated or difficult. Five belong to superspecies with other allospecies both in New Guinea (or New Guinea and Australia) and in archipelagos east of Northern Melanesia, so it is unclear whether these superspecies reached Northern Melanesia from the east or west. Three of these superspecies (those including the pigeons *Ducula rubricera* [map 9, plate 8] and *D. pistrinaria* [plate 8] and the cuckoo *Centropus ateralbus*) are absent from archipelagos east of Northern Melanesia, from Australia, and from the New Guinea mainland but include allospecies on small islands in the western part of the New Guinea region (Western Papuan Islands and the islands of Geelvink Bay), the Moluccas, or islands of western Indonesia. These species may have island-hopped from the west to Northern Melanesia by way of islands north of New Guinea, bypassing New Guinea. Finally, the white-eye *Zosterops* [*griseotinctus*] consists of five endemic allospecies in the Solomons plus one allospecies shared between the Bismarcks and the Southeast Papuan Islands (map 45, plate 1). The relationships of this superspecies, and hence its origins, are unclear.

Endemic Full Species and Genera

Thirty-five Northern Melanesia taxa (appendix 7.3) are endemic at the level of full species or genus.

COLOR PLATES

Zosterops metcalfii
Zosterops atrifrons
Zosterops griseotinctus
Zosterops vellalavella
Zosterops splendidus
Zosterops r. kulambangrae
Zosterops r. tetiparius
Zosterops luteirostris
Zosterops r. rendovae
Zosterops rennellianus
Zosterops murphyi
Zosterops stresemanni
Woodfordia superciliosa
Zosterops ugiensis
H. Douglas Pratt 1997

Plate 1 All Northern Melanesian species of the white-eye family Zosteropidae. The nine birds facing right are all eight Northern Melanesian taxa of the superspecies *Zosterops* [*griseotinctus*] (species 167), a striking example of geographic variation discussed in many textbooks of biology, plus the possibly closely related *Z. murphyi* (species 168). The members of the superspecies occupy different islands but, especially in the New Georgia group (*Z. vellalavella, Z. splendidus, Z. luteirostris*, and the three megasubspecies of *Z. rendovae*), they exhibit marked geographic variation across narrow water gaps. For instance, the distinct megasubspecies *Z. r. tetiparius* and *Z. r. rendovae*, of Tetipari and Rendova, respectively, occur on islands separated by a gap of 3.5 km; the distinct allospecies *Z. vellalavella* and *Z. splendidus* are separated by a gap of 8 km; and the other taxa of the New Georgia Group are separated by gaps of 12–24 km. The songs of these New Georgia group white-eyes are equally divergent (Diamond 1998).

Facing left are all other Northern Melanesian species of Zosteropidae (species 165, 166, and 169–171). *Z. stresemanni* and *Woodfordia superciliosa* are each confined to a single island and do not vary geographically, while the depicted subspecies of *Z. atrifrons, Z. metcalfii*, and *Z. ugiensis* are *Z. a. ultimus, Z. m. metcalfii*, and *Z. u. ugiensis*, respectively. The sole cases of sympatry among the taxa on this plate are between *Z. murphyi* and *Z. r. kulambangrae* (segregating by altitude on Kulambangra), *Z. ugiensis* and *Z. metcalfii* (segregating by altitude on Bougainville), and *Z. rennellianus* and *Woodfordia superciliosa* (segregating by ecological differences related to size on Rennell). The coexistence of *Z. murphyi* and *Z. r. kulambangrae* may result from a double invasion of Kulambangra by the *Z.* [*griseotinctus*] superspecies. All depicted species and allospecies are endemic to Northern Melanesia, except that the allospecies *Z. atrifrons* and *Z. griseotinctus* extend to New Guinea and the Moluccas and to the Southeast Papuan Islands, respectively. See p. 158 of chapter 21 and p. 177 of chapter 22 for discussion, and map 45 for ranges.

Pachycephala melanura dahli
♀
♂
Pachycephala p. ottomeyeri
♂
♀
♂
♀
Pachycephala p. melanoptera
♀
Pachycephala implicata richardsi
♂
♂
♀
Pachycephala p. melanota
♀
♂
Pachycephala i. implicata
♀
Pachycephala p. orioloides
♂
♂
♀
Pachycephala p. sanfordi
♂
♀
♂ = ♀
Pachycephala p. christophori
Pachycephala p. feminina
Pachycephala pectoralis
H. Douglas Pratt © 1997

Plate 2 Males and females of selected taxa of whistlers (family Pachycephalidae): seven of the 16 Northern Melanesian subspecies of *Pachycephala pectoralis* (species 159) facing left; *P. melanura dahli* (species 160), facing right in the upper left corner; and both subspecies of *P. implicata* (species 161), facing right along the middle and lower left sides. This plate illustrates geographic variation and also speciation.

Over its wide range from Java east to Samoa, *P. pectoralis* falls into 66 subspecies plus five other allospecies of the same superspecies. All populations are sexually dimorphic, except for the hen-feathered *P. p. feminina* of Rennell (lower left corner) and *P. p. xanthoprocta* of Norfolk (extralimital, not depicted) and the cock-feathered allospecies *P. flavifrons* of Samoa (extralimital, not depicted). Note the great geographic variation in plumage within Northern Melanesia, both among males and among females. There is also great geographic variation in song, altitudinal range, and foraging behavior. Despite this great variability, the 66 populations grouped under *P. pectoralis* are considered conspecific because of hybridization and intergradation between apparently very distinct subspecies. For instance, *P. pectoralis christophori* of San Cristobal (bottom middle and right) is a stabilized hybrid population between two distinct subspecies.

P. implicata is confined to the mountains of Guadalcanal (*P. i. implicata*) and Bougainville (*P. i. richardsi*), where it lives at altitudes above *P. pectoralis* without any hybridization. *P. melanura* is broadly sympatric with *P. pectoralis* from North Queensland through the Bismarcks to the Northwest Solomons, but the two species segregate by habitat: *P. melanura* in coastal mangrove (in Queensland) and on small or volcanically disturbed islands (in Northern Melanesia), but *P. pectoralis* in forests on larger islands. Vagrants of *P. melanura* often reach the breeding range of *P. pectoralis,* without any evidence of hybridization except for a variable hybrid population at the periphery of *P. melanura's* range in the Northwest Solomons. Nevertheless, *P. melanura* is very similar in plumage to some sympatric populations of *P. pectoralis* (compare males of *P. m. dahli* at upper left and of *P. p. ottomeyeri* at upper right); these constitute sibling species. See chapters 20, 22, and 23 for further discussion, and map 43 for ranges.

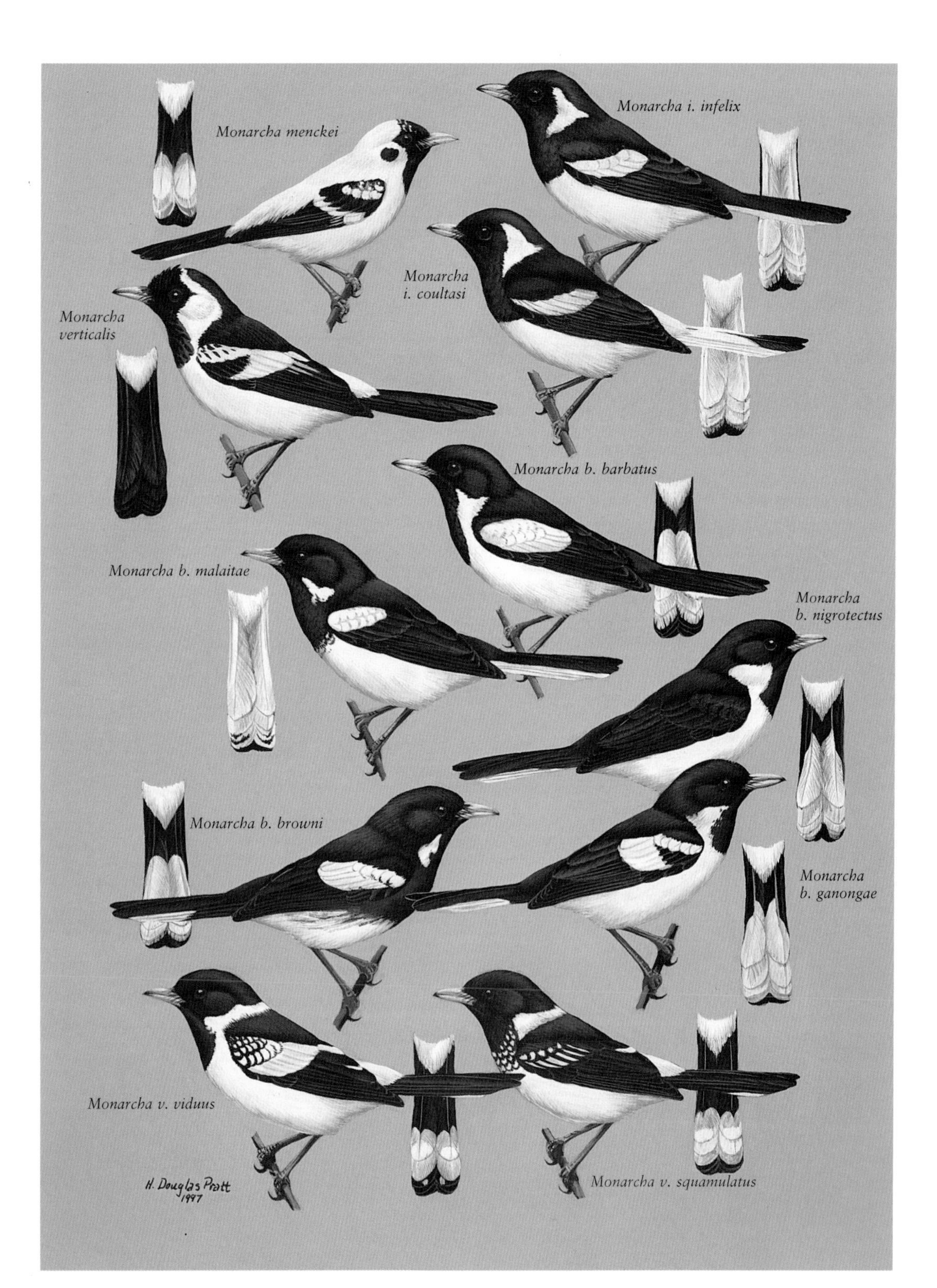
Monarcha menckei
Monarcha i. infelix
Monarcha i. coultasi
Monarcha verticalis
Monarcha b. barbatus
Monarcha b. malaitae
Monarcha b. nigrotectus
Monarcha b. browni
Monarcha b. ganongae
Monarcha v. viduus
Monarcha v. squamulatus
H. Douglas Pratt 1997

Plate 3 Geographic variation in the *Monarcha* [*manadensis*] superspecies (species 152) of pied monarchs. The plate depicts all six allospecies occurring in Northern Melanesia (*M. infelix, M. menckei, M. verticalis, M. barbatus, M. browni,* and *M. viduus*), as well as all separate megasubspecies within these allospecies except for the two megasubspecies of *M. verticalis*. Sympatric species occur among the pied monarchs of New Guinea and some other islands outside Northern Melanesia. However, all of these pied monarch populations of Northern Melanesia are allopatric (living on different islands) and do not come into contact, so there is no direct information about whether they would hybridize if given the opportunity. By comparison with the sympatric species pairs outside Northern Melanesia, many of the Northern Melanesian forms are on the borderline between megasubspecies and allospecies, and it is often difficult to decide how to rank them. See p. 150 of chapter 20 for further discussion, and map 41 for ranges.

Myzomela sclateri
Myzomela melanocephala
Myzomela eichhorni
Myzomela lafargei
Myzomela pammelaena
Myzomela tristrami
Myzomela malaitae
Myzomela erythromelas
Myzomela cineracea
Myzomela cardinalis
Myzomela pulchella
Myzomela cruentata
H. Douglas Pratt · 1997

Plate 4 All 12 Northern Melanesian species and allospecies of *Myzomela* honey-eaters (species 172–177). Six of the taxa belong to the *M.* [*lafargei*] superspecies, which is endemic to Northern Melanesia: from northwest to southeast (map 47), the allospecies *M. pammelaena, M. lafargei, M. eichhorni, M. melanocephala, M. malaitae,* and *M. tristrami.* Interestingly, *M. pammelaena* and *M. tristrami*, the two allospecies geographically most remote from each other (at opposite ends of Northern Melanesia), are superficially most similar in their all-black plumage. Two other taxa belong to the *M.* [*cardinalis*] superspecies that ranges both west and east of Northern Melanesia (map 46): the endemic allospecies *M. erythromelas* (confined to New Britain), plus endemic subspecies of *M. cardinalis* (Southeast Solomons), an allospecies also ranging beyond the Solomons. *M. cineracea* is an allospecies endemic to New Britain and its neighbor Umboi; other allospecies of the same superspecies occur in the New Guinea region and Moluccas. *M. pulchella* is a species endemic to New Ireland, but *M. cruentata* is widespread in the Bismarcks as well as in New Guinea. Two of these 12 honey-eaters, *M. pammelaena* and *M. sclateri,* are endemic supertramps, confined to species-poor (small, remote, or disturbed) islands and absent from nearby species-rich islands.

The five cases of sympatry among Northern Melanesian myzomelids all present interesting features. The larger *M. pammelaena* and the smaller *M. sclateri* are sympatric only on Long Island and its two closest neighbors, apparently as a result of recolonizing those islands after they were defaunated by a volcanic explosion three centuries ago. Compared to allopatric populations of each species, the sympatric populations are even more divergent in size (*M. pammelaena* even larger, *M. sclateri* even smaller), suggesting rapid evolution of increased character divergence due to competition in sympatry (Diamond et al. 1989). *M. cardinalis* is a modern invader of San Cristobal, where it is still extending its range and where it encountered the established *M. tristrami.* Hybrids between the two species have greatly decreased in frequency since the 1880's, suggesting either selection against hybridization, or else increased availability of conspecific mates for the initially rare invader *M. cardinalis.* New Britain and New Ireland both support a pair of small species, which segregate altitudinally and one of which is in each case *M. cruentata.* Interestingly, the altitudinal relations are reversed between the two islands: *M. cruentata* lives at altitudes above *M. erythromelas* on New Britain but at altitudes below *M. pulchella* on New Ireland. New Britain supports an additional, larger, endemic allospecies, *M. cineracea*, which co-exists with both *M. erythromelas* and *M. cruentata* by ecological adaptations related to size. See p. 108 of chapter 14 and p. 180 of chapter 23 for discussion.

Myiagra f. feminina
♀
♀
Myiagra
f. ferrocyanea
Monarcha c. ugiensis
♀
Myiagra
cervinicauda
Myiagra ferrocyanea
♂
Monarcha c. megarhynchus
♀
Myiagra
c. occidentalis
Monarcha erythrostictus
Monarcha richardsii
adult
Monarcha cinerascens
juvenile
♀
Myiagra
hebetior
hebetior
♀
Myiagra alecto
♂
♀
♂
Myiagra hebetior eichhorni
H. Douglas Pratt
1998

Plate 5 Speciation and geographic variation in three groups of monarch flycatchers. The *Myiagra* [*rubecula*] superspecies (species 156, five birds facing right at upper left) comprises 11 allospecies ranging from the Moluccas to Micronesia and Samoa, with sympatric sibling species in New Guinea, Australia, and Fiji as products of completed speciations. Northern Melanesian populations are all allopatric, confined to the Solomons, and divided into three allospecies (*M. cervinicauda, M. caledonica,* and *M. ferrocyanea*), of which *M. ferrocyanea* is further divided into three megasubspecies. Depicted are females of all three allospecies and of two of the three megasubspecies, which differ markedly. Males are similar to each other, so only one male (*M. f. ferrocyanea*) is depicted.

The *Monarcha* [*melanopsis*] superspecies (species 151, five birds facing left at upper right) comprises five allospecies, of which two are confined to New Guinea and Australia and the other three (*M. erythrostictus, M. castaneiventris*, and *M. richardsii*, depicted here) are endemic to the Solomons. *M. castaneiventris* includes two very distinct megasubspecies, of which the all-black *M. c. ugiensis* (upper right corner) is sometimes considered a distinct allospecies, but all-black individuals are also encountered (albeit rarely) in the other megasubspecies (depicted is the race *M. c. megarhynchus*). The four individuals depicted are adults, along with a juvenile of *M. richardsii*, to which juveniles of the other allospecies are similar. Adult *M. cinerascens* (species 150, left margin at middle), which may have arisen as the Bismarck allospecies of *M.* [*melanopsis*], is quite similar to juvenile *M. richardsii* but is a supertramp coarsely sympatric with all three Solomon allospecies of *M.* [*melanopsis*], occupying small islands of the Solomons and Bismarcks plus some remote or volcanically disturbed larger Bismarck islands. See p. 176 of chapter 22 for discussion and map 40 for ranges.

Myiagra alecto (species 154, male and female facing right in left bottom corner) and *M. hebetior* (one male and two females facing right in right bottom corner) provide a good example of speciation (see p. 148 of chapter 20 and p. 176 of chapter 22 for discussion and map 42 for ranges). *M. hebetior* consists of three megasubspecies distinguished by female plumage (two of the three females are depicted). Males of the three megasubspecies are identical to each other and very similar to male *M. alecto* in plumage. *M. hebetior* is an old endemic of the Bismarcks, while *M. alecto* is a recent invader from New Guinea, now occurring sympatrically with *M. hebetior* on Bismarck islands where the two subspecies segregate ecologically by habitat.

Halcyon saurophaga anachoreta
Halcyon chloris alberti
Halcyon chloris nusae
Halcyon sancta
Halcyon chloris amoena
Halcyon chloris matthiae
♂
♀
Ptilinopus solomonensis johannis ♂
♀
Ptilinopus solomonensis ocularis
♂
Ptilinopus rivoli
Lorius albidinucha
Lorius chlorocercus
Lorius hypoinochrous
H. Douglas Pratt
© 1997

Plate 6 Speciation and geographic variation in three groups of birds. Top: three species of *Halcyon* kingfishers (species 112–114, map 24). Most populations of *H. saurophaga* are white-headed, but head color is polymorphic in the population *H. s. anachoreta* (upper left corner), with some individuals being white headed and others blue headed. Most populations of the species *H. chloris* (four populations illustrated at upper right) are blue-headed, but occasional white-headed individuals occur, and the population *H. c. matthiae* (depicted) is white-headed. These three kingfisher species occur sympatrically on most Northern Melanesian islands, preferring different habitats and also differing in size. See p. 101 of chapter 13 and p. 175 of chapter 22 for discussion.

Middle: the fruit doves *Ptilinopus [rivoli] rivoli* (male and female at left) and *P. [r.] solomonensis* (two males and a female at right) constitute a superspecies with extensive interdigitation but little actual sympatry, representing an early stage in speciation (species 54, map 7). *P. rivoli* occurs on New Guinea and the larger Bismarck islands. *P. solomonensis* occurs on large as well as small Solomon islands, but in the Bismarcks it is virtually confined as a supertramp to small or remote or volcanically disturbed islands. Coexistence of the two allospecies is known from only three islands. Two megasubspecies of *P. solomonensis* occur in Northern Melanesia: the montane races *ocularis* (depicted) and *bistictus* of the mountains of Guadalcanal and Bougainville, and the lowland races (e.g., the depicted *johannis*) in the lowlands of other Northern Melanesian islands. See p. 148 of chapter 20, and p. 175 of chapter 22 for discussion.

Bottom: the parrot superspecies *Lorius [lory]* (species 78, map 16) illustrates completed speciation in Northern Melanesia. The genus *Lorius* consists of six or seven species that are distributed from the Moluccas to the Solomons and are largely allopatric. However, the range of *L. albidinucha* (left bottom corner), an endemic of New Ireland, is completely shared with the more wide-ranging *L. hypoinochrous* (right bottom corner). The two species segregate altitudinally on New Ireland, *albidinucha* in the mountains and *hypoinochrous* in the lowlands, with some altitudinal overlap made possible by the considerably smaller size and smaller bill of *albidinucha*. The superficial resemblance of *L. chlorocercus* (middle bottom) to the sympatric smaller parrot *Charmosyna margarethae* (see Plate 9) is striking. See p. 101 of chapter 13, p. 148 of chapter 20, and p. 165 of chapter 22.

Philemon albitorques
C. salamonis
C. t. nisoria
Philemon cockerelli
C. t. heinrothi
Philemon eichhorni
C. t. remota
Reinwardtoena crassirostris
Reinwardtoena browni
Ninox meeki
Ninox variegata
Ninox j. granti
H. Douglas Pratt
1998
Ninox odiosa
Ninox j. jacquinoti
Ninox j. roseoaxillaris

Plate 7 Geographic variation within four superspecies. Upper left: the three Northern Melanesian allospecies of the honey-eater superspecies *Philemon* [*moluccensis*] (friarbirds: species 178, map 48). They occur on the three largest Bismarck islands: *P. albitorques* (top) on Manus, *P. cockerelli* (middle) on New Britain (depicted is the race *P. c. umboi* of nearby Umboi Island), and *P. eichhorni* of New Ireland. Six other allospecies are distributed from New Guinea and Australia to the Moluccas.

Upper right: females representing the four main Northern Melanesian stocks of the *Coracina* [*tenuirostris*] superspecies (species 127, map 28), which ranges from Wallacea to the Solomons. (Only females are depicted because geographic variation in males is less striking.) The most distinctive population of this superspecies is the short-winged, heavy-bodied, ventrally rufous, peripheral isolate *C. salamonis* (upper right corner), ranked as a separate allospecies and confined to the easternmost Solomon island (San Cristobal). All other populations of the superspecies are sufficiently similar to be assigned to one allospecies, *C. tenuirostris*. Its Northern Melanesian populations fall into three megasubspecies: the ventrally barred *C. t. nisoria* confined to the Russell group of the Solomons (second from upper right corner); the ventrally barred *muelleri* group (depicted is the race *heinrothi* of New Britain, third from upper right corner), occupying New Guinea plus the Bismarcks except for the New Ireland group; and the ventrally unbarred *remota* group (depicted is the race *remota* of New Ireland), occupying the New Ireland group plus all of the Solomons except for the Russells and San Cristobal. The barred *nisoria* may have arisen as a hybrid population between the barred *muelleri* group and unbarred *remota* group (see p. 180 of chapter 23 for discussion).

Middle: giant cuckoo-doves of the *Reinwardtoena* [*reinwardtii*] superspecies (species 68, map 13), distributed from the Moluccas to the Solomons. Bismarck populations belong to one endemic allospecies (*R. browni*, right, nominate race depicted), Solomon populations to another endemic allospecies (*R. crassirostris*, left). The most distinctive population of the superspecies is its easternmost representative, the peripheral isolate *R. crassirostris*, unique in its crest and heavy bill.

Bottom: the owl superspecies *Ninox* [*novaeseelandiae*] (species 97, map 22), distributed as nine allospecies from Wallacea to New Zealand. Depicted are one representative each of the three Bismarck allospecies in different postures (*N. variegata superior, N. odiosa, N. meeki*), plus representatives of all three megasubspecies of the Solomon allospecies in identical postures looking to the right (*N. j. jacquinoti, N. j. granti,* and *N. j. roseoaxillaris*). The Northern Melanesia taxa vary more than two-fold in body mass (from the smallest, *roseoaxillaris*, at lower right, to the largest, *meeki*, at upper center).

Ducula pistrinaria
Ducula brenchleyi
Chalcopsitta cardinalis
Cacatua ducorpsi
Ducula rubricera
Ducula melanochroa
Halcyon bougainvillei
Ducula finschii
Dicrurus megarhynchus
Pareudiastes sylvestris
Nesasio solomonensis
Gallirallus insignis
Columba pallidiceps
Microgoura meeki
H Douglas Pratt 1998

Plate 8 Fourteen large-bodied endemics of Northern Melanesia. Northern Melanesia's two most distinctive endemics are two monotypic endemic genera: the large owl *Nesasio solomonensis* (lower left; species 98), confined to the three largest post-Pleistocene island fragments of Northern Melanesia's largest Pleistocene island, Greater Bukida; and the crowned ground pigeon *Microgoura meeki* of Choiseul (at bottom center; species 74, Map 4), exterminated by cats in the twentieth century.

Five endemic full species are depicted. The red parrot *Chalcopsitta cardinalis* (top center; species 76, map 15), recorded from almost every Solomon island, is one of the three most vagile higher endemics of Northern Melanesia. The kingfisher *Halcyon bougainvillei* (center of left margin; species 115, map 4; depicted is a male of the Bougainville race *H. b. bougainvillei*), the flightless forest gallinule *Pareudiastes sylvestris* (just below center of the plate; species 45, map 4), and the flightless forest rail *Gallirallus insignis* (lower right corner; species 38, map 4) are confined to Bougainville and Guadalcanal, San Cristobal, and New Britain, respectively. The ground-feeding forest pigeon *Columba pallidiceps* (lower left corner; species 65, Map 11) is scattered over eight large and two small Solomon and Bismarck islands. Note that five of these seven endemic species and genera are very sedentary and are confined to the largest Northern Melanesian islands.

Of the seven endemic allospecies depicted, six are vagile water crossers and occur on small as well as large islands. The exception is the drongo *Dicrurus megarhynchus* (lower right corner; species 192B, map 52), confined to New Ireland. With its bizarrely twisted tail, this peripheral isolate near the eastern extremity of the range of the widely distributed superspecies *D.* [*hottentottus*] is the most distinctive member of the superspecies. The cockatoo *Cacatua ducorpsi* (species 85A) is widespread in the Solomons and frequently observed flying over water. Five vagile species of imperial fruit pigeons of genus *Ducula* are depicted in a diagonal row from the upper left corner: left-most, the widespread supertramp *D. pistrinaria* (nominate race depicted; species 59A); second from left, *D. brenchleyi* (species 60A), an invader from the east that still occurs only in the eastern Solomons; third from left, the widespread *D. rubricera* (race *rufigula* depicted; species 57A, map 9), with a fleshy red bill-knob; and fourth from left the melanistic *D. melanochroa* (species 61A), and right-most, *D. finschii* (species 58A), both of which breed on a few large Bismarck islands but wander over water to smaller islands in search of fruit.

Meliarchus sclateri
Ptilinopus insolitus
Artamus insignis
Charmosyna margarethae
Geoffroyus heteroclitus
Melidectes whitemanensis
Guadalcanaria inexpectata
Stresemannia bougainvillei
Clytorhynchus hamlini
Cichlornis grosvenori
Pitta superba
Pitta anerythra
Ortygocichla rubiginosa
H. Douglas Pratt 1998

Plate 9 Thirteen small-bodied endemics of Northern Melanesia. Northern Melanesia has five endemic monotypic genera, two of which (*Nesasio* and *Microgoura*) were depicted in plate 8. The other three are honey-eaters confined to single large islands: *Meliarchus sclateri* of San Cristobal (upper left corner; species 181, map 49), *Guadalcanaria inexpectata* of the mountains of Guadalcanal (middle of left margin, with yellow tufts; species 182, map 49), and *Stresemannia bougainvillei* (middle of right margin; species 180, map 49) of the mountains of Bougainville. A fourth honey-eater confined to a large island, *Melidectes whitemanensis* (second from top at right margin, with pale-eye-ring; species 179, Map 49) of the mountains of New Britain, is considered an endemic full species rather than an endemic genus because it is believed to belong to *Melidectes*, a genus otherwise consisting of nine honey-eaters in the mountains of New Guinea.

The red parrot *Charmosyna margarethae* (second from top at left; species 81) is an endemic full species that is vagile, widely distributed in the Solomons, and remarkably similar in appearance to the larger sympatric parrot *Lorius chlorocercus* (plate 6). The endemic full species *Ortygocichla rubiginosa* (lower left corner; species 135, map 4) and the endemic allospecies *Cichlornis grosvenori* (second from bottom at left) are skulking, little-known thicket warblers confined to New Britain.

All of the remaining birds depicted are endemic allospecies. The fruit dove *Ptilinopus insolitus* (top center; species 53A, map 6; nominate race depicted) is widespread in the Bismarcks and differs from the New Guinea allospecies *P. iozonus* in its fleshy, red bill-knob. The wood-swallow *Artamus insignis* (upper right corner; species 193A) is confined to the two largest Bismarck islands. The yellow-headed parrot *Geoffroyus heteroclitus* (just above the center; species 87A, Map 20; nominate race depicted) is widely distributed over Northern Melanesia. The monarch flycatcher *Clytorhynchus hamlini* (just below the center; species 149A) is one of the five allospecies endemic to Rennell Island. The large black pitta *Pitta superba* (middle of bottom row; species 120A) is one the four allospecies endemic to Manus. The small pitta *Pitta anerythra* (lower right corner; species 121A) has exactly the same distribution as the owl *Nesasio solomonensis* (plate 8) and the hawk *Accipiter imitator* (not depicted); all three are non water-crossing species, confined to the three largest post-Pleistocene fragments of Northern Melanesia's largest Pleistocene island, Greater Bukida (see chapter 33 for discussion).

Fig. 13.2. Summary of inferred number of bird colonizations proximately connecting Northern Melanesia and the New Hebrides to each other, New Guinea, and Australia. Data are from this chapter, except that colonizations to the New Hebrides are from Diamond and Marshall (1976). Note that New Guinea has provided the most colonists of Northern Melanesia and that Australia and the New Hebrides have provided fewer colonists.

Thirty-three of these are isolated species, while two are superspecies with three allospecies (the warbler *Cichlornis* [*whitneyi*], which is nearly endemic to Northern Melanesia but has an additional population in the New Hebrides; map 35) or six allospecies (the honey-eater *Myzomela* [*lafargei*]: map 47, plate 4). Six of these 35 taxa are endemic or nearly so at the level of a monotypic genus; the other 29 are endemic at the full species (or superspecies) level. These old endemics are concentrated in the Solomons: 21 are confined to the Solomons, eight to the Bismarcks, and six occur both in the Solomons and Bismarcks.

It is difficult to assess the origins of many of these old endemics, since for only 10 is the identity of the most closely related species unambiguous. For those taxa whose nearest relatives can be assessed, five, or possibly seven, are derived from New Guinea; three are derived from New Guinea or islands to the west of New Guinea; and two may be derived from islands east of Northern Melanesia.

Summary

Our conclusions about proximate origins are summarized in table 13.1, in which species are grouped separately according to level of endemism (nonendemic populations and endemic subspecies, endemic allospecies, and endemic species and genera). We reiterate that, the lower the level of endemism, the greater the confidence with which proximate origins can be identified, for the three reasons given earlier.

Nevertheless, nonendemic populations, endemic subspecies, and endemic allospecies yield fairly similar conclusions, which do not vary with statistical significance between levels of endemism (see table 13.1 legend). The largest group of colonists (approximately half of the total) is derived from New Guinea; the next largest group is derived from New Guinea or Australia or both (fig. 13.2). A smaller group of colonists is derived from Australia, and another small group from the New Hebrides or other archipelagos to the east (fig. 13.2). Similarly, for those archipelagoes to the east, New Guinea has also contributed many more avian colonists than has Australia (Mayr 1941d). Origins are complex or uncertain for only 7% of the nonendemics and the endemic subspecies, but for 17% of the endemic allospecies.

Colonists from different sources tend to concentrate in different parts of Northern Melanesia. New Guinea immigrants occur either in the Bismarcks or else in the Bismarcks plus the Solomons. Only three New Guinean colonists have colonized the Solomons directly without having also colonized the Bismarcks. However, the Australian immigrants and the eastern immigrants (or the endemic allospecies with eastern affinities) occur almost exclusively in the Solomons, especially in the eastern Solomons.

On the average, taxa at higher levels of endemism probably colonized Northern Melanesia earlier than did nonendemic populations. Since proximate origins are similar for nonendemic populations, endemic subspecies, and endemic allospecies, this suggests that colonization routes have not varied much with time from the present back at least as far into the past as the time of origin of the endemic allospecies. For those of the endemic species and genera whose proximate origins can be inferred, 10 came from New Guinea or the west and two from the east—a pattern similar to that for the more recent colonists. However, some of the endemic species and genera of uncertain origins may date to earlier periods in the Tertiary, when land configurations of New Guinea and Northern Melanesia were quite different from the present ones.

14

Upstream Colonization and Faunal Dominance

As we have seen in chapter 13, the main direction of avian colonization of Northern Melanesia has been from west to east: from Asia through New Guinea to the Bismarcks, and from the Bismarcks to the Solomons. Similarly, for tropical Southwest Pacific archipelagoes east of the Solomons, the main direction of colonization has been from west to east: from the Solomons to the New Hebrides to Fiji to Samoa (Diamond & Marshall 1976). This pattern is to be expected because land masses decrease in area and species richness from west to east, from giant, species-rich Asia to New Guinea to the Bismarcks to the Solomons to the New Hebrides.

However, as discussed by Diamond and Marshall (1976), the asymmetry of faunal exchanges between neighboring tropical Southwest Pacific archipelagoes is quantitatively greater than expected from the ratio of their areas or species numbers. This phenomenon appears to be essentially the same as the phenomenon of faunal dominance often discussed in the biogeographic literature—for instance, in connection with Asian/Australian faunal exchanges in Wallacea, Old World/New World exchanges across the Bering Strait, and North American/South American exchanges across the Isthmus of Panama. In addition, those eastern colonists that do succeed in spreading "upstream" from east to west in the Pacific fit a distinctive pattern of ecological distributions in the colonized archipelagoes to the west. The eastern invaders are concentrated in species-poor western communities, while western invaders establish themselves in species-rich as well as in species-poor eastern communities. We now discuss these asymmetrical faunal exchanges between the New Hebrides and Northern Melanesia, between the Solomons and Bismarcks, and between Northern Melanesia and New Guinea. We then suggest a probable reason for these asymmetries.

Exchanges Between the New Hebrides and Northern Melanesia

About 17 species have spread from New Guinea via the Bismarcks and Solomons to the New Hebrides (fig. 4 of Diamond & Marshall 1976). All of these species are widespread in the New Hebrides, even in species-rich lowland forest communities of the largest islands.

At first glance, colonization in the reverse direction, from the New Hebrides to Northern Melanesia, appears approximately equal: appendices 7.1h and 7.2g list a total of about 19 possible colonists. However, distributions of the colonists from the New Hebrides are strikingly stratified within Northern Melanesia. They are concentrated in the eastern Solomons, and elsewhere within Northern Melanesia they are strictly confined to species-poor communities on small islands, on outlying islands, or in the mountains of large islands (e.g., see distributions of *Coracina caledonica*, *Lalage leucopyga*, *Gerygone flavolateralis*, *Rhipidura* [*spilodera*], and *Myzomela cardinalis* illustrated in maps 27, 30, 34, 38, and 46, respectively). There are nine New Hebridean invaders on San Cristobal, the easternmost large Solomon island, 11 on the large eastern outlier of Rennell, and one confined to the small outlier of Gower in the eastern Solomons. Five species reach the second largest Solomon island of Guadalcanal, and three reach the largest island of Bougainville at the northwest end of the Solomon chain, but seven out of those eight populations are confined to species-poor communities in the mountains of those islands, and the eighth (*Gallirallus philippensis*) is a recent invader of species-poor secondary growth communities on Guadalcanal. One New Hebridean invader, the pigeon *Ducula pacifica*, is a supertramp confined to small and outlying islands of the Solomons and Bismarcks. No other New Hebridean invader has reached any Bismarck island, and none lives in the species-rich lowland forest communities of the principal Solomon islands west of San Cristobal.

Thus, the apparent symmetry of exchanges between the New Hebrides and Northern Melanesia becomes grossly asymmetrical when one examines which species from each archipelago have actually established themselves in species-rich communities of the other archipelago.

Exchanges Between the Bismarcks and Solomons

In chapter 13 (see fig. 13.1) we discussed distributions that appeared to imply colonization of the Bismarcks overwhelmingly from New Guinea, followed by eastward spread from the Bismarcks to the western Solomons and then southeast down the chain of the Solomons. In addition, there are three types of distributional patterns that suggest a minority of westward upstream colonizations of the Bismarcks from the Solomons.

One such pattern involves species that are widespread in the Solomons and that occur in the Bismarcks only on eastern outlier Bismarck islands nearest the Solomons (mainly on small, species-poor islands). Examples are the hawk *Accipiter albogularis* (map 3), confined in the Bismarcks to Feni; the pigeon *Ptilinopus viridis* (map 8), confined in the Bismarcks to Lihir until its recent colonization of the northwestern outlier Manus around 1989; the barn owl *Tyto alba*, virtually confined in the Bismarcks to Tanga (the Solomon subspecies *crassirostris*); and the parrot *Chalcopsitta cardinalis* (map 15), confined in the Bismarcks to Feni, Tanga, Tabar, and Lihir, plus once recorded as a vagrant from New Hanover. An extension of this pattern is the pigeon *Columba vitiensis* (map 11), confined in the Bismarcks to New Ireland (and once recorded as a vagrant on New Britain).

The second pattern is defined by three species that have wide distributions on large as well as on small islands in the Solomons and that are also widespread throughout the Bismarcks but confined there to species-poor, small or outlying islands (the pigeons *Ptilinopus solomonensis* [map 7] and *Macropygia mackinlayi* [map 12], and the honeyeater *Myzomela* [*lafargei*] [map 47]). That is, these species were forced to become supertramps as they invaded the Bismarcks from the Solomons, just as *Ducula pacifica* was forced to become a supertramp on invading the Solomons and Bismarcks from the New Hebrides.

The remaining pattern suggesting colonization of the Bismarcks from the Solomons involves species of which different subspecies occur in the Bismarcks and Solomons, but one or a few small outlier Bismarck islands nearest the Solomons are occupied by the widespread Solomon subspecies rather than by the widespread Bismarck subspecies. Clear examples are that the Solomon races of the hawk *Haliastur indus flavirostris*, the tree swift *Hemiprocne mystacea woodfordiana*, and the roller *Eurystomus orientalis solomonensis* also occur on the nearest Bismarck island (Feni), and that the Solomon race of the pigeon *Ducula pistrinaria pistrinaria* also occurs on Feni plus the nearby Tanga and Tabar.

Exchanges Between Northern Melanesia and the New Guinea Region

About 15 species from Northern Melanesia appear to have invaded the New Guinea region (fig. 14.1). Ten of these species occur on islands along New Guinea's north coast; five occur on the Southeast Papuan Islands; but no more than three of these species occur on parts of the New Guinea mainland adjacent to Northern Melanesia. Thus, of mainland New Guinea's approximately 550 species, about 130 have established in the Bismarcks, but

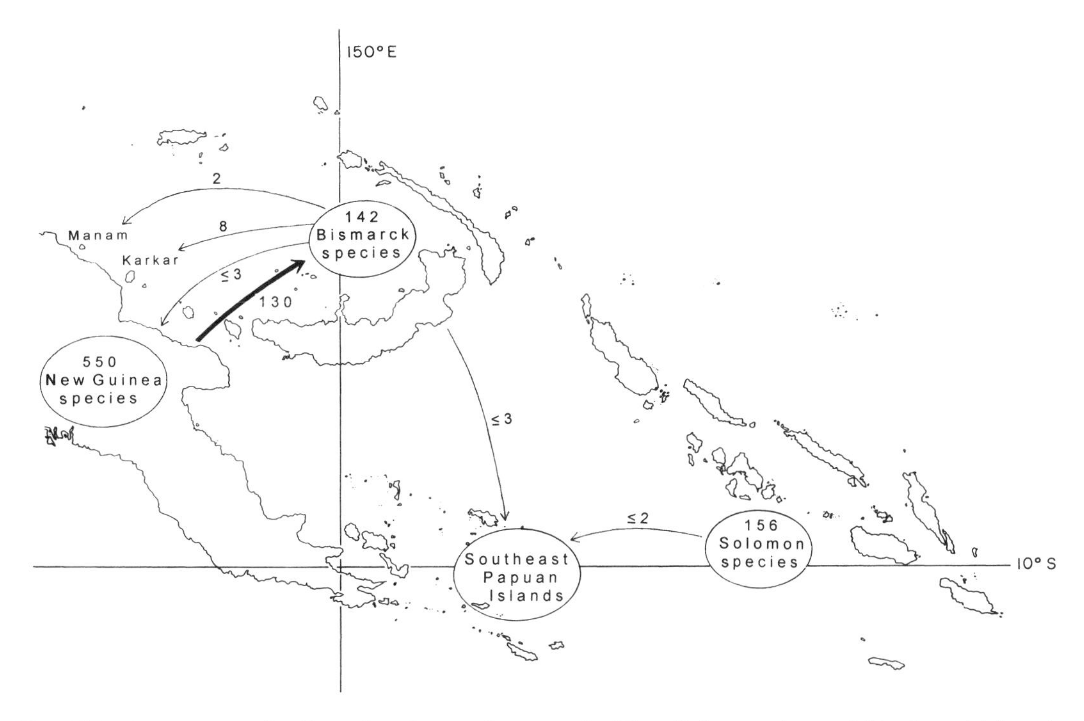

Fig. 14.1. Asymmetry of avifaunal exchanges between the New Guinea region and Northern Melanesia. Although most of the Bismarcks' 142 species and the Solomons' 156 species arose from New Guinea, few species from the Bismarcks and even fewer from the Solomons have conversely succeeded in penetrating the New Guinea region—most have colonized only the region's fringing islands (e.g., Karkar, Manam, Southeast Papuan Islands), and not more than three Bismarck species to the New Guinea mainland.

only 15 of the Bismarcks' approximately 142 species have established on islands fringing New Guinea, and only three or fewer have colonized mainland New Guinea.

The details are as follows. Karkar, the easternmost large island along New Guinea's north coast, supports eight Bismarck invaders: the megapode *Megapodius freycinet eremita*, the supertramp pigeons *Ducula pistrinaria rhodinolaema* and *Macropygia mackinlayi arossi* (map 12) and *Gallicolumba beccarii johannae,* the parrots *Trichoglossus haematodus massena* and *Charmosyna rubrigularis* (map 17), the supertramp monarch flycatcher *Monarcha cinerascens* (map 40), and the supertramp honey-eater *Myzomela sclateri* (Diamond & LeCroy 1979). On Manam, the next large island along New Guinea's north coast as one proceeds west, two of the same Bismarck supertramps have been recorded, *Ducula pistrinaria* and *Monarcha cinerascens*. On the islands of Tarawai and Seleo, farther west, the supertramp pigeon *Ducula pacifica* has been recorded along with *Monarcha cinerascens*. Still farther west, the supertramp pigeon *Ptilinopus solomonensis* occurs on islands of Geelvink Bay. *Monarcha cinerascens* occurs on most islands along New Guinea's north coast, from Karkar west through Geelvink Bay to the Western Papuan Islands at the western extremity of New Guinea and on to the Moluccas and other small islands of Wallacea.

The Southeast Papuan Islands support three species that may have come from the Bismarcks (the parrot *Lorius hypoinochrous* [map 16], *Monarcha cinerascens* [map 40], and the white-eye *Zosterops griseotinctus* [map 45]) and two species that may have come from the Solomons (the pigeon *Ducula p. pistrinaria* and the fantail *Rhipidura rufifrons louisiadensis*), but only for two of these species (*Ducula pistrinaria* and *Monarcha cinerascens*) is it highly probable that spread was from Northern Melanesia to the Southeast Papuan Islands rather than vice versa.

Table 14.1. Asymmetries of faunal exchanges.

Island or archipelago		Area (km^2)		Species no.		Colonists		West/east ratios		
West	East	West	East	West	East	W → E	E → W	Area	Species no.	Colonists
New Guinea	Bismarcks	808,300	48,800	550	142	130	≤3	16.6	3.9	≥43
Bismarcks[a]	New Hebrides	48,800	12,000	142	56	16	1	4.1	2.5	16
New Hebrides[a]	Fiji	12,000	18,300	56	56	26	5	0.66	1.0	5.2

For three pairs of archipelagoes the table gives for each archipelago its land area, number of resident land and fresh-water bird species, and number of species inferred to have colonized the other archipelago. W → E is the number of species that colonized the eastern archipelago from the western one; E → W is the number of species that colonized the western archipelago from the eastern one.

[a]Data in the lowest two rows are modified from Diamond and Marshall (1976). The right-most three columns give the ratio of areas, species numbers, and colonists. Note that, for each pair of archipelagoes, the ratio of colonists (5.2 to ≥43) is much greater than the ratio of areas (0.66–16.6) or of species numbers (1.0–3.9). That is, faunal exchange is more asymmetrical than expected from the disproportion in source areas or in numbers of potential colonizing species alone.

Finally, as discussed on pp. 101–102 of chapter 13, the distributions of three species suggest the possibility that they may have invaded the New Guinea mainland from the Bismarcks: *Lorius hypoinochrous* (map 16) in Southeast New Guinea, the kingfisher *Halcyon saurophaga* (map 24) in North New Guinea, and the finch *Lonchura spectabilis* in Northeast New Guinea. In all three cases, however, the direction of spread is not certain.

Asymmetry Ratios

For three adjacent pairs of Pacific islands or archipelagoes from New Guinea to Fiji, the asymmetry of their faunal exchanges (i.e., the ratio of their exchanges) is much greater than expected from the asymmetry of areas or of bird species numbers (table 14.1). That is, it is not simply the case that a western archipelago with five times more area or species than an eastern archipelago sends five times more successful colonists eastward than spread from east to west. Instead, the ratio of colonists is even greater than the ratio of areas or species numbers. As an extreme example, the New Hebrides archipelago has even less area and only the same number of bird species compared to the Fijian Archipelago to the east, yet New Hebridean invaders of Fiji outnumber Fijian invaders of the New Hebrides by 5 to 1.

Biogeographers from Darwin to Darlington have often noted such asymmetry of faunal exchange, and have often discussed it in terms such as "faunal dominance." The most familiar example from the Pacific is the greater penetration of Wallacea by Indomalayan birds than by Papuan/Australian birds (Mayr 1944c). Equally well-known examples elsewhere in the world are the dominance of the Old World fauna over the New World fauna (Chapter 9 of Darlington 1957) and the disproportionate extinction of South American mammalian families after North America and South America were rejoined in the late Pliocene (Simpson 1940). Thus, species from a richer fauna on a larger land mass tend to be more successful at colonizing a smaller land mass with fewer species than vice versa—a principle clearly stated by Darwin. For Pacific archipelagoes from Samoa westward, the eastern invaders tend to be concentrated in species-poor communities of those western archipelagoes that they do succeed in colonizing.

The explanation usually given for faunal dominance is the effect of increased interspecific competition in a richer fauna and increased intraspecific competition within larger populations on competitive ability. When a species that has already won or survived 500 fights comes up against a species that has survived 5 fights, the former is more likely to prevail. This interpretation is reinforced by our observation that, among Southwest Pacific archipelagoes, eastern invaders of western species-rich archipelagoes are concentrated in species-poor communities.

Summary

Avian colonization in the Southwest Pacific is predominantly downstream from west to east. Whereas Northern Melanesian invaders eastward into the New Hebrides are widespread in species-rich New Hebridean communities (such as lowland forest on the largest islands), New Hebridean invaders westward ("upstream") into Northern Melanesia are confined to the Eastern Solomons or to species-poor communities of more western islands (such as small or outlying islands, secondary growth, and mountains). The same pattern applies to Solomon invaders westward into the Bismarcks and to Bismarck invaders westward into the New Guinea region. For adjacent pairs of these archipelagoes, the asymmetry of exchanges is much greater than expected from the asymmetry of island areas or island species numbers. This phenomenon resembles asymmetrical exchanges and faunal dominance elsewhere in the world. We attribute faunal dominance to the effect of increased competition in a rich fauna on competitive ability.

15

Ultimate Origins of Northern Melanesian Populations

In chapter 13 we analyzed the proximate geographic origins of each Northern Melanesian bird population—the geographic source area from which colonizing individuals dispersed to reach Northern Melanesia. Not surprisingly, all three of the proximate source areas identified (New Guinea, Australia, and the New Hebrides) were regions near Northern Melanesia. However, we also noted the distinction between proximate origins and ultimate origins: while the colonizing individuals may have dispersed from some nearby area (e.g., New Guinea), that proximate source population may in turn have been ultimately derived, at some earlier geological time and higher taxonomic level, from a different area more distant from Northern Melanesia. In the present chapter we ask how many of the proximate source populations can be traced back ultimately to Asia, how many instead have had a long evolutionary history in the Australian region, and at what taxonomic level their ultimate origins can thus be traced to Asia or the Australian region. Mayr (1944a) previously sought to discriminate between proximate and ultimate origins of faunal elements in analyzing the avifauna of Timor and Wallacea.

Asian Immigrants

Although the proximate origins of Northern Melanesia's avifauna lie in New Guinea, Australia, and archipelagos to the east, many of those colonists of Northern Melanesia clearly reached New Guinea or Australia in turn from Asia. We illustrated our reasoning for the hornbill *Rhyticeros plicatus* on page 93. In all, there are at least 53 Northern Melanesian species whose origins can be traced at some level to Asia. That is, in the case of those 53 species the colonists of New Guinea and Australia arrived from Asia, regardless of whether the Asian ancestors can in turn be traced back even farther to Europe or Africa. (For example, the warbler genus *Cisticola* consists of 45 species, of which 43 are confined to Africa or adjacent islands. However, one of the two extra-African species, *C. exilis*, extends from India to New Guinea, Australia, and the Bismarcks. Thus, while the species *C. exilis* probably arose in Asia, the genus arose in Africa.)

We now provide six examples of how we assess the taxonomic level at which these ultimately Asian colonists trace their Asian ancestry. Our six examples illustrate ultimate Asian origins at six taxonomic levels: subspecies, allospecies, superspecies, species group, genus, and family.

The falcon *Falco peregrinus* has most of its subspecies and several related species distributed throughout the world west of Wallace's Line. Only four subspecies occur east of Wallace's Line. One of these, *F. p. ernesti*, extends from the Asian continental shelf (Greater Sunda Islands) to the Solomons. Hence the species *F. peregrinus* surely, and the subspecies *F. p. ernesti* probably, arose outside the Australian region and arrived from Asia. Subspecific affinities place the proximate origins of the Northern Melanesian population in New Guinea, not in Australia. Thus, we conclude that *F. peregrinus* reached Northern Melanesia proximately from New Guinea but ultimately (at the subspecies level) from Asia.

The warbler genus *Acrocephalus* consists of 17 species in Eurasia and Africa, plus a superspecies of nine allospecies distributed from Europe and North Africa to the Pacific islands. One of the nine allospecies, *A. stentoreus*, is distributed from Northeast Africa through Asia to New Guinea, Australia, and Northern Melanesia. The Northern Melanesian subspecies is that of North New Guinea and North Queensland, not that of the rest of Australia. Thus, we conclude that the Northern Melanesian population of this warbler was derived proximately from New Guinea, but ultimately (at the allospecies level) from Asia.

The duck *Anas superciliosa* belongs to a superspecies whose other six allospecies are distributed throughout Eurasia, Africa, and North America. The allospecies *A. superciliosa* is distributed from the Greater Sunda Islands to Polynesia, with its center of distribution in the Australian region. The Northern Melanesian population belongs to the North New Guinea subspecies, not to the Australian subspecies. Thus, while *Anas superciliosa* reached Northern Melanesia proximately from New Guinea, and while the allospecies might have differentiated in the Australian region, this duck is derived ultimately (at the superspecies level) from Asia.

The thrush genus *Turdus* is distributed throughout Eurasia, Africa, and the New World. The species occurring in Northern Melanesia, *T. poliocephalus*, cannot be assigned to a superspecies but does belong to a species group consisting of many species in the Old World and New World. *T. poliocephalus,* the sole species of *Turdus* in the Australian region, is distributed from the Greater Sundas, Philippines, and Taiwan to Samoa, with its center of distribution in Wallacea and the Australian region. The species is absent in Australia, and some of the Northern Melanesian subspecies are morphologically close to New Guinea subspecies. Thus, while this thrush reached Northern Melanesia proximately from New Guinea, and while the species may have arisen in Wallacea and the Australian region, the species group did come ultimately from Asia.

We already discussed in chapter 13 how the hornbill *Rhyticeros plicatus* belongs to a genus derived ultimately from Asia, although it is uncertain whether the species itself or its species group also arose in Asia or else in Wallacea and the Australian region. Thus, the Northern Melanesian population owes its proximate origins to New Guinea, but its ultimate origins (at the genus level) to Asia.

Finally, the warbler *Megalurus timoriensis* belongs to the family Sylviidae, most of whose species live in Eurasia and Africa. The genus *Megalurus* consists of six species distributed from India to New Zealand, but the center of distribution of the genus is not clearly in Asia. Similarly, the species *M. timoriensis* is distributed from the Philippines to New Guinea, Australia, and the Bismarcks, so that it could not be claimed that the species arose in Asia. Species relationships within the Sylviidae are too unclear to assign *Megalurus* to a group of genera, tribe, or subfamily. Thus, all that can be said is that *M. timoriensis* reached Northern Melanesia ultimately (at the family level) from Asia. The proximate origins of the endemic Bismarck subspecies are in New Guinea, since the Bismarck race resembles the New Guinea race rather than the Australian race.

The right-most column of appendix 7 notes each Northern Melanesian species that we identify as being ultimately of Asian origin, and also notes the taxonomic level at which the ancestral population arose ultimately from Asia.

Old Australian Taxa

Sibley and Ahlquist (1985, 1990) showed that most songbirds (order Passeriformes) of the Australian region belong to a parvorder Corvi, a few of whose tribes (such as the drongos and crows) emigrated to the Old World, radiated there, and then reinvaded the Australian region. Thus, Northern Melanesian species belonging to the Corvi trace their ultimate origins back for a long time in the Australian region. The ancestors of the Northern Melanesian species of predominantly Australian groups of Corvi such as honey-eaters, Australian robins (such as *Petroica* and *Monachella*), and whistlers (*Pachycephala*) have presumably lived in the Australian region continuously since the ancestral Corvi reached the region, although one or two modern species of honey-eaters and *Pachycephala* have spread secondarily to the Indomalayan region.

Hence these Northern Melanesian taxa have not only their proximate origins, but also their ultimate origins back to the parvorder (suborder) level, in the Australian region.

In the case of those Northern Melanesian species belonging to groups of Corvi that did radiate in Asia, we cannot automatically claim that the ancestors of the Northern Melanesian species have resided continuously in the Australian region since ancestral Corvi arrived. Perhaps, instead, they returned from the Australian region to Asia, reradiated in Asia, and then reinvaded the Australian region. However, the predominantly Australian fantails (tribe Rhipidurini) clearly arose in the Australian region at least at the tribe level, although it is possible that the subfamily Monarchinae to which they belong (consisting of the fantails plus the drongos and monarch flycatchers, the latter two tribes with many Asian and African species) differentiated in Asia and Africa after escaping to there from Australia, and that the Rhipidurini then reinvaded the Australian region. Similarly, the group of predominantly Australian genera including the cuckoo-shrikes *Lalage* and *Coracina*, and the group including the flycatchers *Monarcha*, *Myiagra*, and *Clytorhynchus*, arose in the Australian region, even though the tribes to which they belong (Oriolini and Monarchini, respectively) may have differentiated earlier in Asia and Africa after emigrating to there from Australia. Hence these Northern Melanesian taxa belong ultimately to the Australian region at least at the level of the tribe (fantails) or the group of genera (*Lalage* and *Coracina*, *Monarcha* and *Myiagra* and *Clytorhynchus*).

Among nonpasserines as well, many have had an old history in the Australian region. For example, it is uncertain whether the parrot order (Psittaciformes) arose in the Australian region, but several subfamilies of parrots, including the cockatoos (*Cacatua*, etc.), lories (*Charmosyna*, *Lorius*, *Trichoglossus*, *Chalcopsitta*, etc.), and pygmy parrots (*Micropsitta*), certainly did. The groups of genera that comprise the fruit pigeons *Ptilinopus* and *Ducula*, the bronzewing pigeons *Chalcophaps* and *Henicophaps*, the parrots *Geoffroyus* and *Eclectus*, and the warblers *Ortygocichla* and *Cichlornis* differentiated within the Australian region, whatever their origins at higher levels. The genera *Podargus* (frogmouths), *Gallicolumba* (ground doves), *Ninox* (hawk owls), and *Tanysiptera* (paradise kingfishers) differentiated within the Australian region, as did the families Casuariidae (cassowaries) and Megapodiidae (megapodes).

The right-most column of Appendix 7 notes each Northern Melanesian species that we identify as having had a long history within the Australian region. The appendix 7 tables also note the level (e.g., genus, tribe, or parvorder level) to which that history within the Australian region can be traced back.

Other Taxa

Many other taxa evidently arose in the Australian region at the level of the species group or superspecies, but we cannot decide whether their origins at the level of the genus or higher lie within the Australian region or outside it (e.g., Asia). Examples include Northern Melanesian species of hawks (*Accipiter*), cuckoos (*Chrysococcyx*, *Eudynamys*, *Centropus*, *Scythrops*), barn owls (*Tyto*), white-eyes (*Zosterops*), and grassfinches (*Lonchura*).

There are other Northern Melanesian species whose ultimate origins we have not attempted to classify at all. Some of these are wide-ranging species belonging to cosmopolitan families (herons, ibises, and spoonbills) whose histories are especially difficult to deduce on distributional evidence alone and will require fossil evidence. Others belong to groups (cuckoos [*Cacomantis*], swiftlets [*Aerodramus* and *Collocalia*], tree swifts [Hemiprocnidae], dwarf kingfishers [*Ceyx*]) with numerous representatives both in the Australian region and the Indomalayan region, so we hesitate to say in which region the group arose.

Total Numbers of Traceable Species

As summarized in table 15.1, 53 Northern Melanesian species can be traced ultimately to Asia. In contrast, 134 Northern Melanesian species have a traceable history to at least the superspecies level in the Australian region and cannot be clearly traced to Asia. Of these 134, 81 are "old Australians" that have been in the Australian region from the time of differentiation of their group of genera, or even longer.

Recall that Northern Melanesian populations differ in their level of endemism, from endemic genera to taxa that still belong to the same subspecies

Table 15.1. Summary of ultimate origins of Northern Melanesian bird populations.

Level of endemism within Northern Melanesia	Proximate origin	Ultimate origin: Asia 0	1	2	3	4	5	6	8	Total	Australia 1	2	3	4	5	6	7	8	9	Total
Endemic subspecies or nonendemic	NG	1	12	4	4	5	1	3	1	31	2	8	2	8	13	2	5	2	3	45
	Au	—	—	1	—	1	—	—	—	2	—	—	3	1	—	—	—	—	—	4
	NG or Au	—	4	2	1	—	—	—	—	7	—	—	1	—	1	—	—	—	4	2
	East	—	—	—	—	—	—	—	—	—	—	—	3	—	5	1	—	—	4	13
	Complex	—	1	—	—	—	—	—	—	1	—	2	2	—	1	1	1	—	1	8
	Total	1	17	7	5	6	1	3	1	41	2	10	11	9	20	4	6	2	8	72
Endemic allospecies	NG	—	—	—	1	2	—	—	—	3	1	2	1	3	9	1	1	—	3	21
	Au	—	—	—	—	—	—	—	—	—	—	—	—	—	—	—	1	—		
	—	1																		
	NG or Au	—	—	1	—	1	—	—	—	2	—	3	1	—	2	1	—	—	—	7
	East	—	—	—	—	—	—	—	—	—	—	—	—	2	3	1	—	—	—	6
	Complex	—	—	—	—	—	—	—	—	—	—	2	2	—	3	—	1	—	1	9
	Total	—	—	1	1	3	—	—	—	5	1	7	4	5	17	3	3	—	4	44
Endemic species or genus	NG or west	—	—	—	1	—	—	2	—	3	—	—	1	—	1	—	2	—	1	5
	East	—	—	—	—	1	—	—	—	1	—	—	—	—	1	—	—	—	—	1
	Uncertain	—	—	—	—	2	—	1	—	3	—	—	2	1	2	—	—	—	7	12
	Total	—	—	—	1	3	—	3	—	7	—	—	3	1	4	—	2	—	8	18
Grand total		1	17	8	7	12	1	6	1	53	3	17	18	15	41	7	11	2	20	134

The first column identifies the level of endemism of the Northern Melanesian population of a species: i.e., whether it belongs to an endemic subspecies, allospecies, species (or superspecies), or genus confined to Northern Melanesia, or whether it is non-endemic (belonging to the same subspecies as an extralimital population). The second column identifies the proximate origin of the Northern Melanesian population (from New Guinea = NG, Australia = Au, etc.). The remaining columns, denoting ultimate origin, indicate whether the ancestry of the Northern Melanesian population can be traced back ultimately to Asia (at the level of the subspecies itself = 0; allospecies = 1; superspecies = 2; group of species = 3; genus = 4; group of genera = 5; tribe = 6; subfamily = 7; family = 8; parvorder = 9), or whether the ancestry can be traced back within the Australian region up to a certain taxonomic level (same codes 1 . . . 9) and cannot be traced back further outside the Australian region. Assignments of each species are given in Appendix 7. See text for discussion.

as extralimital populations. At every level of endemism within Northern Melanesia, species of "ultimately Australian origin" (i.e., the 134 species defined above) predominate over Asian invaders. The predominance is most extreme among Northern Melanesia's endemic species and genera (18 Australians to 7 Asians) and among the endemic allospecies (44 Australians to 5 Asians), and is less marked among Northern Melanesia's endemic subspecies and nonendemics (72 Australians to 41 Asians). If we narrow our categories of ultimate origin to recent Asian invaders (at the level of the superspecies or younger) and old Australians (at the level of the group of genera or higher), the recent Asian invaders contribute negligibly to Northern Melanesia's endemics: they contribute none of the endemic species and genera, among which are 14 old Australians, and they contribute only one of the endemic allospecies, among which are 27 Old Australians.

In one respect, species of ultimate Asian and Australian origins differ in their directions of approximate entry into Northern Melanesia. All proximate invaders from the east whose ultimate origins can be traced are ultimately Australian; none is Asian. However, the proportion of ultimately Australian to ultimately Asian species is approximately the same among proximate invaders from New Guinea (66:34) as among proximate invaders from Australia (4:2).

Asian and Australian invaders also differ somewhat in their distributions within Northern Melanesia (i.e., whether they are confined to the Bismarcks, confined to the Solomons, or present in both the Bismarcks and the Solomons). Many old Australians (24 out of 81) are confined to the Solomons, but almost no recent Asians are (1 out of 26). Correspondingly, relatively more recent Asians (18 out of 26) than old Australians (31 out of 81) occur both in the Bismarcks and Solomons. Thus, the recent Asian invaders have tended to spread over both the Bismarcks and Solomons and have not had much time to differentiate within Northern Melanesia. In contrast, many of the old Australians have contracted their distributions onto the Solomons alone and have differentiated to higher levels of endemism (endemic allospecies, species, or genera) within Northern Melanesia.

Summary

Whereas in chapter 13 we analyzed proximate geographic origins of Northern Melanesia's avifauna (i.e., the probable birth places of the colonizing individuals themselves), in this chapter we instead analyzed more ultimate origins of the colonizing populations, at an earlier geological time and higher taxonomic level. We have examined how far into the evolutionary past the ancestors of those colonizing individuals can be traced back to Asia or within the Australian region.

For 53 Northern Melanesian species termed "recent Asians," we can trace their origins to Asia at some higher taxonomic level—mainly at the level of the superspecies, group of species, genus, or group of genera. Conversely, 134 Northern Melanesian species termed "old Australians" have a long evolutionary history in the Australian region, in most cases going back at least to the time of differentiation of their group of genera.

At every level of endemism within Northern Melanesia, but especially at higher levels, colonists of old Australian origins predominate over colonists of recent Asian origins. All proximate invaders of Northern Melanesia from the east with traceable ultimate origins are old Australians. The Northern Melanesian populations of recent Asian ancestry tend to be widespread in Northern Melanesia and tend to be at a lower level of endemism, while many of the old Australians have contracted their distributions onto the Solomons and have differentiated to higher levels of endemism.

PART V

TAXONOMIC ANALYSIS: DIFFERENCES AMONG SPECIES

16

The Problem of Speciation

One of the most unyielding problems of evolutionary biology has been the problem of speciation. How does one species give rise to several daughter species? Darwin was confronted by this problem in 1837, when he began to study the collections that he had accumulated during the voyage of the Beagle. The philosopher John Herschel referred to speciation as the "mystery of mysteries." Sadly, Darwin failed to arrive at the correct answer to this question in his book *On the Origin of Species*, whose title suggests that it is devoted to the question's solution. In reality, Darwin's book is instead devoted to establishing the fact of evolutionary change with time (see fig. 0.1 for the distinction) and to elucidating the mechanism of natural selection that underlies evolutionary change.

A thorough understanding of the problem of speciation was not achieved until the time of the so-called evolutionary synthesis in the 1930s and 1940s (Mayr & Provine 1980). Perhaps the major reason for the long persistence of this uncertainty was a failure to clarify what the terms "species" and "speciation" mean. Below we attempt such a clarification.

What Are Species?

Among the ancients, particularly among the followers of Plato, a species was a class of objects that differed substantially from other classes of objects. There are different "species" of inanimate objects, like crystals, gems, or rocks, but the term "species" with the same definition was also applied to plants and animals. Such a species definition, based entirely on degree of similarity (or difference), is referred to as a "typological" species concept. This was the species concept adopted by all zoologists and botanists until long into the nineteenth century.

Eventually, naturalists became dissatisfied with a species concept based entirely on similarity or differences because they realized that there is something more to biological species than just the arbitrary similarity-based distinctions among "species" of rocks. Linnaeus encountered this problem as early as 1758, when striking differences in plumage led him to describe the male and female of the mallard (*Anas platyrhynchos*) as two different species, and also to describe the immature and the adult of the goshawk (*Accipiter gentilis*) as two different species. When Linnaeus learned that these different plumage types were only different sexes or different life-history stages of the same biological entity, he considered them to be plumage types of the same species, and he discarded the later proposed name as a synonym of the earlier proposed name.

In time, many other such cases of different-appearing kinds of organisms (not just different sexes or age classes), coexisting within what any biologist would intuitively consider to be the same species, were discovered within many species of animals and plants. For example, in the North American Arctic live two geese that look very different (the all-white snow goose and the dark-bodied blue goose), but that are identical in size and habits and that associate in mixed flocks. Both plumage types encompass adult males as well as females, and they co-occur in the same interbreeding population. Eventually, it became universally agreed that these geese are just color phases of the same species. In certain plants the aquatic and the terrestrial forms of the same species are even more drastically different than are the snow goose and the blue goose.

In addition, proponents of the typological species concept encountered still another difficulty that was exactly the opposite of the first one: the occurrence in nature of exceedingly similar (or virtually indistinguishable) populations at the same place, but without any interbreeding between those populations. In spite of their virtual identity to the human eye, the populations behaved like different species. Such co-occurring, non-interbreeding, similar populations are now called "cryptic" or "sibling" species.

These two types of observations—the regular interbreeding of very dissimilar-appearing animals or plants, and the non-interbreeding of very similar-appearing animals or plants—forced biologists to adopt a new concept of species in sexually reproducing organisms, a concept based not on similarity but based instead on reproductive behavior. This actually represented an entirely different species concept, the "biological species concept" (BSC), defined as follows: Species are groups of interbreeding natural populations that are reproductively isolated from other such groups. This reproductive isolation has a genetic basis, consisting of characteristics ("isolating mechanisms") that make it difficult or impossible for individuals of one species to interbreed with individuals of other species. Every species is separated from all other species by such isolating mechanisms. We emphasize that the isolating mechanisms are often behavioral: an individual wild animal that encounters individuals of its own species as well as of other species *chooses*, overwhelmingly or solely, to mate with individuals of its own species. And yet, that animal, when caged with an individual of the other sex of another species, may interbreed freely and produce fertile hybrids. Thus, interfertility or hybridization of caged animals is not relevant to the definition of species. There are numerous examples of species that rarely or never hybridize with each other in the wild, but that do hybridize and produce fertile offspring if a male of one species and a female of the other species are put together in a cage where no potential mate of their own species is available to them.

These observations lead us to the question why the enormous diversity of animals or plants in nature does not constitute a continuum of individuals, each of which is slightly different from other individuals, but is instead separated into discrete packages that do not interbreed with each other. Why are different gene pools packaged into reproductively isolated species? The inferiority of most hybrids provides the answer to this question. The populations of each species possess a well-balanced genotype that had been selected for this internal harmony by thousands of generations of selection. Hence hybridization with a different species of a very different genotype is likely to produce in the hybrids a deleterious recombination. It is the role of species-specific isolating mechanisms to prevent such hybridization, or to reduce greatly its frequency.

Even though the meaning of the biological species concept is now clear, certain practical difficulties may arise for human observers (although not for the individual animals involved) in delimiting species taxa. When two populations coexist in the same area, it is easy to decide whether they belong to the same species: if they interbreed, they are conspecific; if not, they are different species. However, this criterion of interbreeding cannot be applied if the populations are geographically isolated. For instance, there are many cases, particularly in island regions, of two somewhat similar populations geographically isolated by water barriers separating islands. Do the two populations possess reproductive isolating mechanisms that would prevent them from interbreeding if they came into contact, or do they lack such reproductive isolating mechanisms and would they interbreed if they came into contact? To answer this question requires analyzing the process of speciation.

Speciation

The term "speciation" means "multiplication of species": the derivation of several daughter species from a single parental species. This process is entirely different from evolutionary change with time, called "phyletic evolution," in which a phyletic lineage evolves over time from one species as defined by paleontologists into a single later daughter species defined by paleontologists (see fig. 0.1). Because in that case the parent "species" and the daughter "species" exist at different geological times, they cannot come into contact with each other, and it is meaningless to discuss whether they constitute separate species according to the biological species concept that can be applied only to simultaneously existing populations. Phyletic evolution of one such ancestral "species" (a "chronospecies") into a daughter chronospecies does not constitute genuine speciation (see Bock 1986). The chronospecies are just arbitrarily defined divisions of a temporal continuum.

Speciation can occur through different processes in different organisms. These processes include increasing genetic divergence in asexually reproducing organisms ("agamospecies"); instantaneous speciation by chromosomal mutations; speciation through hybridization; sympatric speciation (incipient daughter populations achieving reproduction isolation while remaining within cruising range of each other) through mate selection and differing niche preferences (e.g., Bush 1975, Schliewen et al. 1994, Howard & Berlocher 1998, Tregenza & Butlin 1999); and geographic speciation (allopatric speciation), as discussed in detail below. The available evidence indicates that in mammals and birds (and probably in many or most other kinds of organisms) only allopatric speciation occurs. As we shall see, our analysis of Northern Melanesian birds confirms that, without even a single exception, speciation proceeds through allopatric speciation.

The process of allopatric speciation unfolds as follows. When a population of a species becomes geographically isolated from other conspecific populations, its genetic similarity to the other populations is no longer maintained by gene flow. Instead, the isolate is subject to a number of genetic processes inducing genetic divergence: new mutations, loss of alleles by accidents of sampling, the production of novel genotypes by recombination, and (most important) differential survival of different genes under novel selection pressures, because the isolated population lives in a (even though perhaps only slightly) different physical and biotic environment.

In the course of this divergence, the isolate first acquires differences that the taxonomist rates as being only of subspecific value (i.e., differences that would probably not prevent interbreeding if the isolate reencountered the parental population). Such an isolate is then considered a different subspecies. As divergence continues, the isolate may eventually become a distinct species, incapable of breeding with the parental species, which has thus split into two non-interbreeding daughter species. Familiar examples are the anthropoid apes. Humans, common chimpanzees, pygmy chimpanzees, and gorillas today constitute at least four distinct species that have never been observed to hybridize in nature, or even in cages. Yet these four species or sets of species are believed to have shared a common ancestor until about 7 million years ago, and the common chimpanzee and pygmy chimpanzee shared a common ancestor even more recently.

There are two ways in which geographic separation of the new incipient species from the parental species can arise. In the process called "dichopatric speciation" (Cracraft 1984), the previously continuous range of a species is split into two parts by a new barrier, such as a mountain range, water gap, or a belt of vegetation. Examples of such new barriers include the isolation of Britain from continental Europe by the English Channel, or of Alaska from Siberia by the Bering Strait (when sea level rose at the end of the Pleistocene), or the postulated Pleistocene fragmentation of the Amazonian rainforest into wet forest refugia separated by open, dry habitats (Haffer 1974). Such dichopatric speciation is characteristic for speciation in continental areas.

The other process of allopatric speciation is called "peripatric speciation" (Mayr 1982a,b). In this case a new incipient species is initiated as the result of a founder population becoming established beyond the periphery of the parental population by a process of primary isolation. Such an isolated population either eventually becomes extinct, or it merges again with the parental population, or it diverges to such an extent that it eventually becomes a new species. For example, when Darwin discovered three different species of mockingbirds on three different islands of the Galapagos, he realized that they had descended by speciation from colonists of a single mockingbird species on the Pacific coast of mainland South America. For about 15 years Darwin considered allopatric speciation the normal process of speciation. Unfortunately, he later became confused due to equivocal use of the term "variety," and he turned to an invalid ecological theory of sympatric speciation without geographic isolation (Mayr 1992). Sympatric speciation then became the more popular theory among evolutionary biologists for the next 80 years, although Moritz Wagner and others of Darwin's contemporaries and followers continued to adhere to the theory of allopatric speciation.

Thus, a parental population can become split into geographically isolated daughter populations. However, there is no proof for the completion of speciation until the isolates subsequently develop geographic overlap. Suppose that the daughter populations come into contact again, either because the geographic barrier becomes bridged as a result of geological changes or because colonists from one daughter population manage to cross the barrier and reach the other daughter population. If, during the period of geographic isolation, few differences in mate selection or in reproductive biology have developed, and if the now-reunited daughter populations interbreed, hybridization will prevent any incipient reproductive isolation from developing further. If, at the opposite extreme, the daughter populations have already evolved genetic differences such that they do not hybridize or such that they produce only sterile or nonvariable hybrids, then the reunited daughter populations already constitute distinct species. But reproductive isolation between two reunited daughter populations does not guarantee ecological differences as well, since different sets of characters are involved. There are many examples of populations whose ranges abut geographically and which rarely or never interbreed at the zone of contact, yet the two populations are still ecologically similar to each other in their respective areas. Evidently the daughter populations in these cases diverged more markedly in mate selection criteria than in ecology. The two populations are not yet able to reinvade each others' geographic ranges and live in sympatry: speciation is not yet completed. For the daughter populations to live in sympatry, they must develop not only reproductive differences but also ecological differences permitting them to occupy different niches.

The attainment of geographic overlap may arise in any of three ways, whose relative frequencies differ between continents and islands. The main way in which isolates achieve overlap on continents is by expanding their ranges to meet on a broad geographic front. When that happens, the outcome depends on whether the geographic isolates have acquired reproductive isolation and differ in their ecological requirements. If the former geographic isolates are not reproductively isolated, a zone of hybridization develops at the contact border. If the isolates are reproductively isolated but not ecologically distinct, they establish "parapatric contact" (i.e., abutting but mutually exclusive geographic ranges) with minimal interbreeding at the contact border. If the isolates are reproductively isolated and ecologically distinct, the outcome is completion of speciation and achievement of broad geographic overlap with minimal interbreeding and with ecological segregation.

The main way in which isolates achieve overlap on islands differs from that on continents. Since islands are separated by water gaps, it is impossible for populations isolated on separate islands to expand their geographic ranges until they meet on a broad front, as on continents. Instead, colonists fly from one island to another island occupied by another isolate. In some cases, several colonist individuals will arrive simultaneously, will find each other as mates, will be reproductively isolated and ecologically distinct from the colonized island's original resident population, and will try to establish themselves beside (but separate from) the original resident population. Far more often, though, the colonizing individuals find no conspecifics and hence either die without reproducing or else reproduce (hybridize) with a member of the resident population. The colonists' genotype then disappears into the vast gene pool of the resident species. Natural selection will probably eventually eliminate most introgressed genes of the colonists, but some of their genes may make a beneficial contribution and will be permanently retained. Thus, the large resident gene pool acts as a sink into which the genes of occasional colonists from other islands disappear. This is why, among Northern Melanesian birds, there are hundreds of cases in which neighboring islands are occupied by distinct allospecies, but only a few cases (chapter 23) in which hybridization between those allospecies is detectable.

The last of the three possible outcomes when two former isolates meet is especially likely when the meeting takes place on islands without a resident population of either isolate, either because the island lacks that particular group of species or because the island initially lacks any biota (because it had recently been defaunated by a volcanic eruption or tidal wave or had recently arisen from the sea). Such an empty island may be colonized simultaneously by individuals from two isolates in two different source areas (two different source islands). The Northern Melanesian avifauna provides several examples of hybrid taxa that arose when colonists from two sources met on islands defaunated by volcanic explosions (chapter 23).

Speciation in nature occurs very slowly on the time scale of a scientist's life. Therefore, biologists attempting to understand the course of speciation have had to resort to the "snapshot" method that we described on pages xii–xiii. Such evidence has led to the conclusion that geographic speciation is the prevailing mode of speciation in birds. This conclusion is based on numerous examples of obvious intermediate stages in the process of geographic speciation, as described for Northern Melanesian birds in chapters 17–24.

But there are still many unanswered questions about geographic speciation, and the answers may differ among taxa and among regions. For example, at any given time, are most taxa in nature representative of only one particular stage in speciation, and are some stages rarely seen because they are passed through so quickly? Does geographic isolation for, say, 10,000 years usually lead to geographic variation and speciation, or is that the exception? Are ecological differences usually developed more rapidly or more slowly than reproductive isolation, hence is it rare or common for reunited daughter populations to remain "stuck" for a long time with abutting geographic ranges and without sufficient ecological differences to permit sympatry? To what extent do ecological differences develop in allopatry, as a result of the daughter populations living then in different physical environments and exposed to other species? To what extent does ecological segregation instead develop when the daughter populations come into contact again, as a result of competition (selection for individuals of each species that are ecologically unlike the other species and that therefore survive better; Diamond 1986a)? Is there any particular preferred sequence to the development of ecological differences between species? That is, do daughter species that have just achieved sympatry tend to segregate by habitat and only later evolve differences in foraging technique, or vice versa?

A major reason that our understanding of these and other questions about geographic speciation has remained incomplete is that most studies have been based on selected examples. Among the few quantitative studies of aspects of geographic variation in a whole fauna are the analyses for Australian birds by Keast (1961) and for North American birds by Mayr and Short (1970). We are not aware of a case in which all stages of geographic speciation in a whole fauna have been studied with the degree of completeness that we now attempt for the Northern Melanesian avifauna, in an effort to answer such questions.

Taxonomic Terms

We conclude this chapter by defining some taxonomic terms. "Subspecies" are local populations that are recognizably different from each other but that are nevertheless considered to belong to the same species, because they are observed to interbreed in nature or because it is inferred that they are likely to interbreed. The only circumstance under which one can decide by actual observation whether two populations are reproductively isolated is if individuals of the two populations actually do come in contact in nature, because the populations occupy different interconnected portions of a geographical or altitudinal gradient on the same land mass, or because the two populations live on different islands but individuals from each island occasionally appear on the other island. However, there is a problem that commonly arises in deciding whether two populations have achieved species status: if they live on different islands, and if individuals from one island are not observed to arrive at the other island, then one has no opportunity to observe directly whether individuals of the two populations interbreed. Thus, one cannot be certain whether they constitute subspecies or species. How can one attempt to make an inference?

One may attempt to infer reproductive isolation, or the lack thereof, by examining how much the two geographically isolated populations differ in mating behavior, song, or morphology, and then comparing these differences with the degree of differences between populations that do come into contact and are thus known to interbreed or not to interbreed (Mayr & Ashlock 1991, pp. 103–105). Obviously, the resulting inference is likely to be uncertain because one may have information about morphological differences but not about differences in mating behavior, whose correlation with morphological differences is imperfect. Also, it is essential to use as a standard the differences between populations of known reproductive status that belong to the same taxonomic group as the pair of populations whose reproductive status is uncertain. The reason is that much greater morphological differences are compatible with interbreeding in some taxa than in others. For example, among whistlers of the *Pachycephala pectoralis* complex, there are several areas where morphologically different subspecies groups hybridize freely, so that morphologically distinct populations on different islands cannot be safely assumed to be reproductively isolated (Mayr 1932d, Galbraith 1956). On the other hand, sympatric species of honey-eaters of the *Meliphaga analoga* group in New Guinea are extremely similar to each other morphologically, so that morphological differences between apparently representative *Meliphaga* populations would be more likely to suggest species status.

The next stage after the subspecies is the "allospecies": two or more populations that are derived from a common ancestor and that are still allopatric in distribution but that are inferred to be reproductively isolated. If the ranges of the allospecies actually abut geographically, the ranges are said to be "parapatric." One can then observe whether individuals of the two populations interbreed at the boundary, and one can thus test reproductive isolation directly. If the ranges do not abut but are separated by a geographic barrier, one must attempt to assess subspecies versus allospecies status by the indirect criteria explained above. A group of allospecies is referred to as a "superspecies."

The next stage is one of partial sympatry, in which the geographic ranges of the daughter populations now overlap in part. Since individuals of the daughter populations coexist with each other, one can infer reproductive isolation with confidence. When geographic overlap (sympatry) is extensive, one refers to speciation as having been completed, and the two daughter populations are considered to constitute full species rather than allospecies. As time passes, and as these sympatric species evolve and diverge further, they may eventually become sufficiently unlike each other to warrant being classified in separate genera. For example, although humans, the one or more species of gorillas, and the two or more species of chimpanzees are believed to have shared a common ancestor 7–10 million years ago, they are now sufficiently unlike each other that they are usually classified in three genera (*Homo, Pan,* and *Gorilla*).

Because speciation is a continuous process, the delineation of stages is inevitably arbitrary. Especially arbitrary is the taxonomic status of a group of related populations of which all are allopatric except that two of the populations occur sympatrically. If it were not for the two sympatric populations, all the populations would be grouped into a superspecies. But the two sympatric populations violate a strict definition of a superspecies, as a group of reproductively isolated and recently diverged but still allopatric populations. Which one of the two sympatric populations should be considered to belong to the superspecies, and which one should be ranked separately as a full species? The decision would be totally arbitrary if the two sympatric populations were roughly similar to the allopatric populations. Should one instead dismember the whole superspecies, just because two populations have achieved sympatry?

A convenient alternative is to retain the concept of the superspecies but to regard it as being represented in one area by a "doublet" (two sympatric taxa; see van Bemmel 1948, Cain 1954). Several such cases occur within Northern Melanesia, such as the sympatry of the pigeons *Macropygia mackinlayi* and *M. nigrirostris* on three Bismarck islands plus Karkar (map 12), or the sympatry of the parrots *Lorius albidinucha* and *L. hypoinochrous* on New Ireland (map 16, plate 6). There are also examples in which a superspecies whose Northern Melanesian representatives are strictly allopatric is represented by a doublet outside Northern Melanesia, as in the case of the pigeon superspecies *Ptilinopus* [*purpuratus*] and *Ducula* [*rufigaster*] on New Guinea. In appendix 1 we have thus extended the concept of a superspecies to comprise not only a set of strictly allopatric taxa, but also a set of taxa that are allopatric except for a doublet. In this way we avoid dismembering a superspecies merely because speciation has proceeded to the next stage in only one part of the superspecies' geographic range.

Summary

Species of sexually reproducing organisms are not just arbitrary divisions of a continuum but possess a reality, documented by reproductive isolation from each other. In the process of speciation (not to be confused with phyletic evolution; i.e., evolutionary change with time), an ancestral population constituting one reproductive community becomes two or more non-interbreeding daughter populations. How can this split develop? In birds, mammals, and probably in most higher animals and plants, incipient daughter populations achieve reproductive isolation while isolated geographically. Those geographic barriers required for allopatric speciation depend either on the establishment of an isolated new population by colonizing individuals crossing an existing geographic barrier, or else on the development of a new barrier transecting the formerly continuous geographic range of a species. Even though allopatric speciation seems to be the rule in sexually reproducing animals generally, a host of questions about it remain unsolved—questions for whose solution the Northern Melanesian avifauna is especially suitable.

17

Stages of Geographic Speciation Among the Birds of Northern Melanesia

If speciation is an on-going process that is not closely synchronized among a region's species, then it should be possible to find different stages of speciation represented among a region's species. This expectation is confirmed for Northern Melanesian birds, as shown in tables 17.1–17.3, based on 191 out of the 195 breeding land and freshwater bird species of Northern Melanesia. Our unit is the "zoogeographic species," or "species" for short (i.e., superspecies, plus species not belonging to superspecies). Thus, when a superspecies breaks up into several allospecies, we consider the superspecies (not each allospecies) as a single unit, comparable to an isolated species that does not belong to a superspecies.

Our analysis omits seabirds, nonbreeding visitors to Northern Melanesia (appendix 2), and the single surviving introduced species (appendix 3). Our analysis also omits four of the 195 breeding native land and freshwater species: the cassowary *Casuarius bennetti*, because its presence on a single Northern Melanesian island (New Britain) appears to be due to prehistoric introduction by humans, and three species that are common in Northern Melanesia as winter visitors from Australia, but that breed only occasionally in small numbers (the pelican *Pelecanus conspicillatus*, the cuckoo *Scythrops novaehollandiae*, and the kingfisher *Halcyon sancta*).

A simple classification of stages of geographic speciation recognizes the following five stages:

Stage 1: The species exhibits no geographic variation (within the region under study).

Stage 2: Geographic variation at the subspecies level.

Stage 3: Geographic variation at the allospecies level—i.e., formation of a superspecies whose component populations are still allopatric, though inferred to be reproductively isolated.

Stage 4: Recently completed speciation—i.e., achievement of partial or complete sympatry between taxa that are still closely related to each other.

Stage 5: Ancient speciation, as evidenced by distinct endemic taxa (possibly, but not necessarily, with a relict distribution as a result of extinction from some former parts of the geographic range).[1]

Not all Northern Melanesian species can be neatly placed into one of these five stages. This in-

1. In defining what constitutes a taxon endemic to Northern Melanesia, we have broadened the definition slightly to include not only taxa that are strictly confined to Northern Melanesia, but also five taxa that clearly originated in Northern Melanesia and whose ranges now include one or two nearby extralimital island groups: the hawk *Accipiter albogularis* (on the Santa Cruz group east of the Solomons; map 3), the pigeon *Ptilinopus solomonensis* (on islands of Geelvink Bay off Northwest New Guinea; map 7), the pigeon *Ducula pistrinaria* (on the Southeast Papuan Islands, and on Karkar and Manam Islands off Northeast New Guinea), the parrot *Charmosyna rubrigularis* (on Karkar; map 17), and the finch *Lonchura hunsteini* (introduced to Ponape north of the Solomons).

Table 17.1. Geographic variation in Northern Melanesian bird species.

I. Species not varying geographically within Northern Melanesia: 91
  A. Not occurring outside Northern Melanesia; endemic monotypic species, not belonging to a superspecies: 22
  B. Occurring outside Northern Melanesia: 69
    1. Nonendemic species not belonging to a superspecies: 30
      a. Monotypic species: 1
      b. Polytypic species with a single, nonendemic, subspecies: 17
      c. Polytypic species with a single, endemic, subspecies: 12
    2. Superspecies represented in Northern Melanesia by a single allospecies: 39
      a. Endemic monotypic allospecies: 16
      b. Nonendemic allospecies with a single, non-endemic, subspecies: 14
      c. Nonendemic allospecies with a single, endemic, subspecies: 9
II. Species varying geographically within Northern Melanesia: 100
  A. Species not belonging to a superspecies but represented by more than one subspecies: 34
  B. Superspecies represented by a single allospecies with more than one subspecies in Northern Melanesia: 31
  C. Superspecies represented by several (on average, 2.7) allospecies: 35
III. Not analyzed: 4
  A. Introduced species: 1
  B. Occasionally breeding winter visitors: 3
Total considered: 191 analyzed species, 4 nonanalyzed species

convenience is to be expected because speciation is a continuous process rather than a march of abruptly delineated steps. For example, the allospecies of several superspecies show, in turn, subspeciation. Some species are at different stages of speciation in different parts of their geographic range (see discussion of *Coracina* [*tenuirostris*], *Rhipidura rufifrons*, and *Pachycephala pectoralis* on pp. 99, 102, and 150). However, one can cope with some of this heterogeneity by appropriate subdivision of the five major stages (table 17.1).

We infer 69 species (group IB of table 17.1) to be at stage 1 because they occur outside Northern Melanesia but show no geographic variation within Northern Melanesia. Most of these appear to be taxa in the process of invading Northern Melanesia from the outside (mainly from New Guinea), but some appear to be taxa that differentiated to the species level within Northern Melanesia and are now invading extralimital regions (mainly archipelagoes to the east). We infer 65 species (groups IIA and IIB of table 17.1) to be at stage 2 because they have subspeciated within Northern Melanesia but still belong to a single allospecies (group IIB) or to a species not a member of a superspecies (group IIA). We infer 35 species (group IIC of table 17.1) to be at Stage 3 because they are superspecies that have formed two or more allospecies within Northern Melanesia. As discussed in chapter 22, we recognize 19 examples of stage 4 (closely related species with partial or complete sympatry). Finally, stage 5 is exemplified by 22 endemic species (group IA) that are monotypic (no subspecies recognized) and do not belong to a superspecies.

While Northern Melanesian birds are famous for geographic variation, many (91 out of 191 analyzed, or 48%) show no geographic variation within Northern Melanesia (table 17.1). However, out of the 69 of these 91 that occur outside Northern Melanesia, all but one (the duck *Dendrocygna*

Table 17.2. Kinds of species in Northern Melanesia.

I. Superspecies: 105
  A. Superspecies with a single Northern Melanesia allospecies: 70
  B. Superspecies with more than one Northern Melanesia allospecies: 35
II. Isolated species not belonging to superspecies: 86
  A. Without geographic variation (represented by a single subspecies) in Northern Melanesia: 52
  B. With geographic variation (represented by more than one subspecies) in Northern Melanesia: 34
Total considered: 191 analyzed species

Table 17.3. Allospecies in Northern Melanesia.

I. Allospecies: 165
  A. Belonging to superspecies with a single Northern Melanesia allospecies: 70
    1. Endemic allospecies: 23
    2. Nonendemic allospecies: 47
  B. Belonging to superspecies with several Northern Melanesia allospecies: 95
    1. Endemic allospecies: 72
    2. Nonendemic allospecies: 23
II. Isolated species not belonging to a superspecies: 86
    1. Endemic species: 30
    2. Nonendemic species: 56
Total allospecies, plus isolated species not belonging to a superspecies, considered: 251 (127 endemic, 124 nonendemic)

guttata) vary geographically outside Northern Melanesia.

More than half (105 out of 191) of Northern Melanesian species are superspecies (table 17.2). Of the 86 that are not, many (34) still vary subspecifically within Northern Melanesia.

Finally, for those readers accustomed to considering allospecies rather than zoogeographic species as the unit of analysis, we resummarize the Northern Melanesian avifauna in table 17.3. Much of the literature of biology does not distinguish between allospecies and isolated species not belonging to superspecies and merely refers to both as "species." In contrast, as already mentioned in the first paragraph of this chapter, we use "species" to mean either a superspecies or an isolated species, but not an allospecies. Since 35 of Northern Melanesia's 191 analyzed zoogeographic species break up into more than one allospecies within Northern Melanesia (on the average, 2.7 allospecies, for a total of 95 such allospecies), the total number of allospecies plus species not belonging to superspecies is 251 (= 95 + [191 − 35]).

Summary

Of Northern Melanesia's 195 breeding land and freshwater bird species, we analyze 191 and exclude four (one prehistorically introduced species, plus three occasionally breeding winter visitors). Our unit is the "zoogeographic species"—i.e., superspecies, plus isolated species not belonging to a superspecies. (More than half of our zoogeographic species are superspecies.) We tabulate how these species fall into five stages of geographic variation, starting with species exhibiting no geographic variation in Northern Melanesia, and ending with distinct endemic taxa resulting from ancient speciation.

18

Absence of Geographic Variation

Factors Predisposing to Geographic Variation

One might expect that each species would be represented by a separate subspecies or allospecies on each island where the species occurs, if all of the following four conditions were met: 1) if the population of each island had been established by a single founding event; 2) if that event had occurred in the distant past, so that there would have been ample time since then for evolution to follow a different trajectory on each island; 3) if the populations of all islands had remained genetically isolated from each other since the founding event, without any subsequent gene flow between islands; and 4) if the phenotype of the species were sufficiently plastic. In addition, the likelihood of a distinct population on each island would be greater if each island differed considerably from all other islands in its physical environment (habitats, elevation, and area) and/or in its biological environment (competing species and predators), so that each population would have been exposed to different pressures.

This expectation is confronted with reality in figure 18.1, which depicts, for each Northern Melanesian species, the number of Northern Melanesian islands that it occupies and its number of distinguishable Northern Melanesian populations (subspecies plus monotypic allospecies). The number of islands occupied by each species ranges from only 1 island (26 cases) up to 69 (the mound-builder *Megapodius freycinet*). The number of distinguishable populations into which each species falls in Northern Melanesia ranges from only one (91 cases) up to 16 (the whistler *Pachycephala pectoralis*; plate 2, map 43). For species occupying two to four islands, there are only eight cases in which each island is occupied by a distinct subspecies, and no such case for species occupying five or more islands. Many species, including Northern Melanesia's three most widely distributed species (*Megapodius freycinet*, the heron *Egretta sacra*, and the pigeon *Caloenas nicobaria*, on 69, 67, and 64 islands, respectively), occupy dozens of islands without showing any geographic variation.

Thus, the theoretical extreme of geographic variation—every island population different from every other—is rarely achieved by Northern Melanesian birds. Instead, in nearly every species that occurs on more than a single island, some groups of islands are occupied by indistinguishable populations. Furthermore, as documented in the preceding chapter (table 17.1), there are great differences among Northern Melanesian species in their degree of geographic variation: about half of Northern Melanesian species show no geographic variation within Northern Melanesia, while others break up into many subspecies or allospecies. What factors account for these great differences among species?

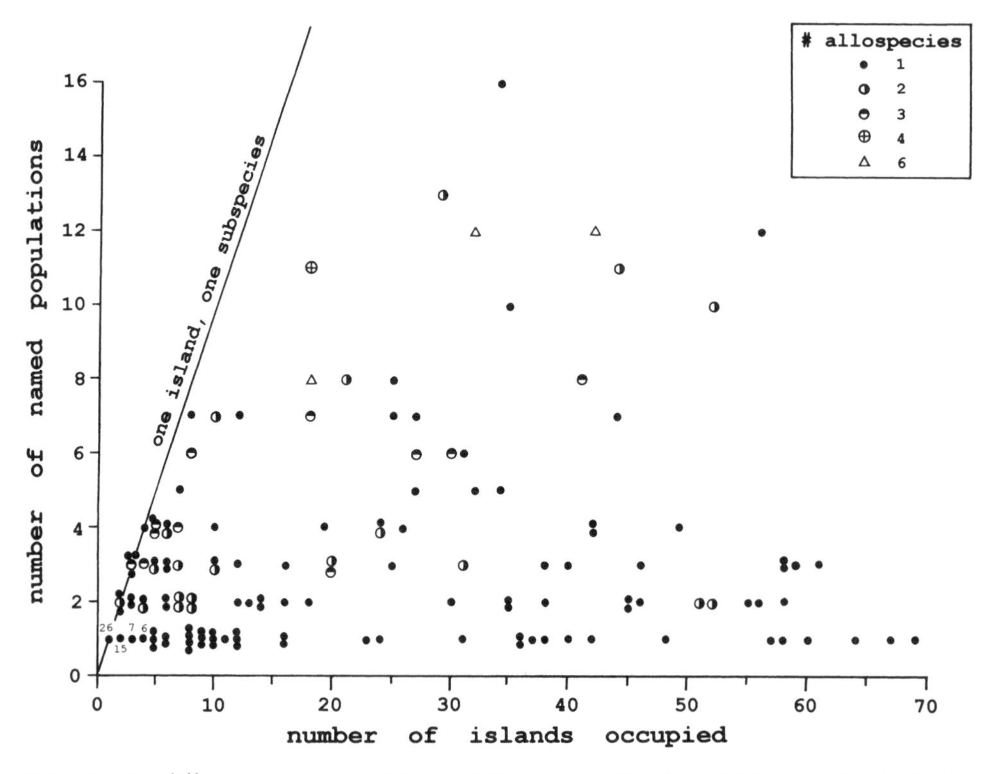

Fig. 18.1. Species differences in geographic variability among Northern Melanesian birds. Each of the 191 analyzed zoogeographic species is represented by a point whose abscissa value is the number of well-surveyed Northern Melanesian islands that it occupies, and whose ordinate value is the number of named, distinguishable populations (subspecies plus monotypic allospecies) that we recognize in Northern Melanesia. Points are coded by the number of allospecies into which the species' Northern Melanesian populations fall. The points at an ordinate value of a single subspecies, abscissa values of 1, 2, 3, and 4 islands, correspond, respectively, to 26, 15, 7, and 6 species. At other points corresponding to multiple species, symbols are slightly displaced to permit symbols to be displayed individually. The line of identity ("one island, one subspecies") is the line on which points would fall if every island population were subspecifically distinct; this is the case for a few species occupying up to four islands, but for no species occupying five or more islands. Note that many species exhibit no geographic variation within Northern Melanesia (i.e., consist of only a single subspecies), despite occupying up to 69 islands.

The reasons why some species show no geographic variation are likely to be the mirror image of the reasons why other species show great geographic variation. Species whose populations were founded only recently, or whose populations are still connected by much gene flow between islands, or which occur on only a few islands, or which are phenotypically very stable, are likely to show little or no geographic variation. Conversely, species whose populations were founded in the distant past, and whose populations today are not connected by gene flow, and which occupy many islands, and whose phenotypes are plastic, are likely to show geographic variation.

In the next chapters we consider species that do show marked geographic variation. The present chapter considers those 91 species that do not show geographic variation within Northern Melanesia. One of us (E.M.) has long been interested in the problem of the absence of geographic variation (Mayr 1942:46). Since it is obvious that not all examples of this phenomenon are due to the same, single cause, the explanation of the phenomenon requires analysis of an entire fauna. The present chapter begins by exploring in turn each of the above-mentioned factors that might contribute to lack of geographic variation. We then examine the application of these factors to all of the 91 nonvariable species grouped into taxonomic categories. In this way we can provide explanations for the absence of geographic variation in most of these species.

Factors Predisposing to Lack of Geographic Variation

We consider in turn four factors whose contributions to lack of geographic variation in Northern Melanesian birds we can discern: recency of isolation, small geographic range, high gene flow, and phenotypic stability.

Recency of Colonization

There are four circumstances leading us to infer, in the cases of 21 species, that their lack of geographic variation within Northern Melanesia is due to populations on different islands having become isolated only recently.

First, there are five species that have wide geographic ranges in the Australian region but that occur within Northern Melanesia just on one island or group of islands geographically close to sources of colonists from outside Northern Melanesia, and whose populations on these islands are not endemic (i.e., they belong to the same subspecies as an extralimital population). We infer that these species may have colonized Northern Melanesia only recently, may only have established a first "bridgehead" within Northern Melanesia, and may not have had time to differentiate from the source population. The species are the duck *Dendrocygna arcuata* and jacana *Irediparra gallinacea* on New Britain or the New Britain group, the fantail *Rhipidura fuliginosa* on San Cristobal, and the hawk *Falco berigora* and pitta *Pitta sordida* on the Long group. In the case of the latter two species, we can be especially confident that recency of colonization is the correct explanation for lack of geographic variation, because all existing plant and animal populations of the Long group are derived from colonists that arrived after the group was defaunated by a volcanic explosion in the seventeenth century (Diamond et al. 1989).

Second, there are nine other species that also have a large range outside Northern Melanesia and whose Northern Melanesian populations are not endemic and show no geographic variation, but whose distributions within Northern Melanesia are wider than those of the five species cited in the previous paragraph. Nevertheless, only one of these nine species is widespread both in the Bismarcks and Solomons; most are confined to either the Bismarcks or the Solomons or to scattered islands. Five of these species (*Ixobrychus sinensis*, *Caprimulgus macrurus*, *Merops philippinus*, *Saxicola caprata*, and *Acrocephalus stentoreus*) are widely distributed in Asia as well as in the Australian region, while four others (*Columba vitiensis* [map 11], *Rhipidura leucophrys*, *Myiagra alecto* [map 42], and *Pachycephala melanura* [map 43]) are widely distributed in the Australian region. We infer that these nine species have colonized Northern Melanesia recently from the outside and that their distributions have extended beyond an initial bridgehead, but that the colonization has still been too recent for differentiation or for expansion throughout the whole archipelago. In the case of the heron *Ixobrychus sinensis*, it was discovered breeding in Northern Melanesia only in 1979.

Third, there are four taxa that are endemic to Northern Melanesia and that occur on three to eight islands but nevertheless show no geographic variation. However, all of the occupied islands belong to the Greater Bukida group of the Solomons, which was connected together at Pleistocene times of low sea level until about 10,000 years ago (fig. 3.3). Until then, gene flow within Greater Bukida may have prevented differentiation, and isolation of the fragments of Greater Bukida by post-Pleistocene water gaps may have been too recent for appreciable differentiation. These four examples are the endemic frogmouth subspecies *Podargus ocellatus inexpectatus*, the endemic kingfisher allospecies *Halcyon leucopygia* (map 23), the endemic hawk species *Accipiter imitator* (map 3), and the endemic owl genus *Nesasio solomonensis*.

Finally, three species that have huge ranges reaching Asia or even Europe occur in Northern Melanesia as a single endemic subspecies that is widespread either in the Bismarcks or in the Solomons but not in both. These species are the quail *Coturnix chinensis*, barn owl *Tyto alba*, and warbler *Cisticola exilis*. Their colonization may have been relatively recent, though less recent than that of the 14 nonendemic colonizers mentioned above. That is, their colonization may have occurred just long enough ago to have permitted differentiation of an endemic Northern Melanesian subspecies, but not sufficiently long enough ago to have permitted further differentiation within Northern Melanesia.

Small Geographic Range

It is impossible for a species confined to a single Northern Melanesian island to exhibit inter-island geographic variation within Northern Melanesia. All other things being equal, the potential for geographic variation increases with the number of islands that a species occupies. In fact, 32 species that exhibit no geographic variation within Northern Melanesia have restricted geographic ranges: either a single island, or a single large island with one or two smaller satellite islands. Naturally, it is possible that some of these species occupied more islands in the past, that they exhibited geographic variation then, and that the extinction of geographically distinctive populations left them in their current status of lacking geographic variation within Northern Melanesia. The 32 species in this category consist of 15 endemic full species, 7 endemic allospecies, 6 endemic subspecies, and 4 nonendemics.[1] This list omits species discussed in the section "Recency of Colonization."

High Gene Flow (Highly Vagile Species)

Species of which individuals frequently disperse between islands are unlikely to differentiate geographically because populations on different islands are joined by gene flow. We can test this prediction by examining geographic variation in the 56 Northern Melanesian species that disperse most frequently over water, according to the evidence summarized in chapter 11. (These are the 56 species assigned in that chapter to dispersal category A.) Of these 56 species, the Northern Melanesian populations of most (40 species) fall into only one or two subspecies. Ten additional vagile species are represented by three or four subspecies, but all of these 10 species occur on at least 26 islands, so that each subspecies is still widespread. Only six of these 56 most vagile species show marked geographic variation (5–12 subspecies).

An alternative test is to ask how large a contribution high vagility makes to explaining the absence of geographic variation in 91 Northern Melanesian species. We have already seen that many of these 91 cases can be explained by recency of colonization or small geographic range, leaving 39 cases unexplained. Of those 39 cases, 21 belong to the set of the 56 most vagile species. In particular, of the 17 geographically invariant species with wide distributions (on 12 or more islands), all except one (the starling *Mino dumontii*) are among the 56 most vagile species.

Thus, vagility is an important factor in explaining lack of geographic variation, especially among species that do not vary despite wide geographic distributions.

Phenotypic Stability

Some species groups are notorious for consisting of species that are exceedingly similar to each other morphologically (i.e., sibling species), but that are quite different ecologically. Among birds, the *Aerodramus* and *Collocalia* swiftlets of the Indomalayan and Australian regions, the *Meliphaga* honey-eaters of New Guinea, and the *Empidonax* flycatchers of North America are well-known examples of such phenotypic stability. For whatever reason, the morphology of these species is far more conservative than other aspects of their biology.

While these examples refer to interspecific conservatism within a genus, there are also examples of intraspecific stability within a species—that is, species that exhibit little morphological variation, although one might expect variation on other grounds. One suspects phenotypic stability in the case of a species distributed over a large area that

1. Endemic full species: *Accipiter luteoschistaceus, Gallirallus insignis, Pareudiastes sylvestris, Gallicolumba salamonis, Microgoura meeki, Ortygocichla rubiginosa, Cettia parens, Phylloscopus amoenus, Zosterops murphyi, Z. stresemanni, Myzomela pulchella, Melidectes whitemanensis, Stresemannia bougainvillei, Meliarchus sclateri*, and *Guadalcanaria inexpectata*. Endemic allospecies: *Henicopernis infuscata, Accipiter princeps, A. brachyurus, Henicophaps foersteri, Clytorhynchus hamlini, Woodfordia superciliosa*, and *Myzomela cineracea*. Endemic subspecies: *Threskiornis moluccus, Anas gibberifrons, Chrysococcyx lucidus, Lalage leucopyga, Gerygone flavolateralis*, and *Monachella muelleriana*. Nonendemic species: *Platalea regia, Dendrocygna guttata, Accipiter fasciatus*, and *Gymnocrex plumbeiventris*.

has apparently been occupied for a long time, especially if the species is not highly vagile. For example, phenotypic stability might help explain why the New Britain and New Ireland populations of the cuckoo *Centropus violaceus* are not different, even though this cuckoo is a very distinctive endemic that has presumably been in the Bismarcks for a long time, and although its behavior suggests that it is very poor at dispersing across water gaps.

Local Absence of Geographic Variation

Our examples so far have been drawn from species that exhibit no geographic variation anywhere within Northern Melanesia. A related but less extreme phenomenon involves species that show no geographic variation through a large part of Northern Melanesia, though they have differentiated elsewhere within Northern Melanesia. One set of examples includes superspecies represented by two or more allospecies, one of which occupies a wide range within Northern Melanesia with no geographic variation. The examples are the pigeon allospecies *Gymnophaps solomonensis* (map 10) and the parrot allospecies *Charmosyna meeki* (map 17) on all high mountainous Solomon islands, the pigeon allospecies *Reinwardtoena crassirostris* (map 13, plate 7) in all the Solomons, the parrot allospecies *Lorius chlorocercus* (map 16, plate 6) in the eastern Solomons, and the flycatcher subspecies *Monarcha v. verticalis* (map 41, plate 3) in the Bismarcks from New Britain to New Hanover. Other allospecies and subspecies of these five superspecies occupy much of the rest of Northern Melanesia. Similarly, the parrot *Geoffroyus h. heteroclitus* (map 20, plate 9) occupies all of Northern Melanesia with no geographic variation except for an endemic subspecies on Rennell; and the parrot *Trichoglossus haematodus massena* occupies all of the New Hebrides, Solomons, and most of the Bismarcks without geographic variation except for two endemic subspecies in the Northwest Bismarcks. The reasons for local lack of geographic variation in these cases probably include three of the reasons we already discussed for lack of geographic variation throughout all of Northern Melanesia: recency of colonization of the area involved, high vagility, or localized phenotypic stability.

Classification of Geographically Nonvariable Species by Level of Endemism

The 91 species that do not vary geographically within Northern Melanesia can be classified according to their level of endemism within Northern Melanesia. Thirty-two are species with nonendemic populations, 21 are endemic at the subspecies level, 16 are endemic at the allospecies level, and 22 are endemic at the level of the full species or genus (table 18.1). We now consider each of these four groups of species separately.

Species with Nonendemic Populations

Of the 32 species that do not vary geographically within Northern Melanesia and whose Northern Melanesian populations are not endemic even at the subspecies level (table 18.1, entries 1a - 1c), only one, the duck *Dendrocygna guttata*, is monotypic (lacks geographic variation) throughout its entire geographic range (from the Philippines through New Guinea to Northern Melanesia). The other 31 do vary geographically outside of Northern Melanesia, and 14 of the 31 exhibit such marked variation that they are considered to be superspecies with two or more allospecies. Thus, it is evident that these 31 species are capable of geographic variation, and the question arises why they failed even to subspeciate within Northern Melanesia, an area where so many species do vary geographically. Table 18.2 summarizes, for each of these species, whether they are characterized within Northern Melanesia by recent colonization or isolation, small geographic range, or high vagility. We have not tabulated the fourth possible explanatory factor for lack of geographic variation, phenotypic stability, because we do not know how to assess it for each species.

Twelve of these species (table 18.2, first row, left column) appear to be recent colonists of Northern Melanesia, as judged by their widespread extralimital ranges contrasting with Northern Melanesian ranges confined to bridgeheads. Four more species have small geographic ranges confined to one or just a few neighboring islands, reducing the opportunities for geographic variation. Eleven other species are highly vagile and thus unlikely to develop interisland differences. That leaves five

Table 18.1. Species that do not vary geographically within Northern Melanesia.

	No. of species	
	Subtotal	Total
1. Not endemic to N Melanesia (not even at the subspecies level)		
a. Monotypic species, no geographic variation within or outside N Melanesia: #15	1	
b. Polytypic species not belonging to a superspecies; all N Melanesian populations belong to a single non-endemic subspecies: #6, 7, 8, 23, 30, 33, 46, 47, 48, 51, 75, 101, 131, 143, 146, 154, 160	17	
c. Superspecies with a single non-endemic allospecies in N Melanesia (monotypic allospecies, or else polytypic allospecies with a single non-endemic subspecies in N Melanesia): #10, 13, 16, 21, 31, 32, 41, 42, 49, 50, 64, 117, 138, 186	14	
		32
2. Endemic to N Melanesia at the subspecies level (non-endemic allospecies or species, all N Melanesian populations belong to a single endemic subspecies)		
a. Polytypic species not belonging to a superspecies: #5, 80, 90, 99, 100, 130, 134, 139, 140, 157, 191, 194	12	
b. Superspecies with a single non-endemic allospecies in N Melanesia, represented in N Melanesia by a single endemic subspecies: #12, 14, 17, 34, 35, 62, 84, 95, 164	9	
		21
3. Endemic to N Melanesia at the allospecies level (superspecies represented in N Melanesia by a single, endemic, monotypic allospecies): #19, 27, 28, 58 60, 61, 70, 85, 88, 108, 135, 149, 171, 172, 185, 193		16
4. Endemic to N Melanesia at the level of full species or genus (endemic monotypic species not a member of a superspecies): #24, 26, 38, 45, 65, 72, 74, 76, 81, 94, 98, 137, 142, 168, 170, 174, 176, 179, 180, 181, 182, 190		22
Grand total		91

Numbers at the end of each description correspond to species codes from appendix 1 for all species fulfilling the given definition.

species (table 18.2, first row right column), within this group of 32 species, for which we have not succeeded in identifying a correlate of geographic invariance.

Species Endemic at the Subspecies Level

In the cases of 21 species that do not vary geographically within Northern Melanesia (table 18.1, entries 2a and 2b), all of their Northern Melanesian populations belong to a single endemic subspecies. Nine of these 21 species exhibit sufficiently marked geographic variation outside Northern Melanesia that they are classified as superspecies, although the allospecies to which their Northern Melanesian population belongs is not endemic to Northern Melanesia. Thus, since these 21 species are clearly capable of geographic variation, the question again arises why they do not vary geographically within Northern Melanesia.

At least five of these species (table 18.2, second row left column) may be relatively recent colonists of Northern Melanesia, as suggested by their wide extralimital ranges extending to the Philippines (the duck *Dendrocygna arcuata*), India (the quail *Coturnix chinensis* and warbler *Cisticola exilis*), or even the New World (the barn owl *Tyto alba*). The frogmouth *Podargus ocellatus* is confined to three post-Pleistocene fragments of Greater Bukida and may not have had time to differentiate further since the time when rising sea levels carved up its formerly continuous range. Six of these species have small geographic ranges within Northern Melanesia, being confined to Rennell, the San Cristobal group, or the New Britain group. Finally, seven of

Table 18.2. Reasons for absence of geographic variation in Northern Melanesia.

Level of endemism	Recent colonization or isolation	Small geographic range	Highly vagile	Unexplained
Not endemic	#10, 33, 47, 64, 101, 117, 131, 138, 143, 146, 154, 160	#13, 15, 23, 41	#6, 8, 16, 21, 30, 32, 46, 50, 51, 75, 186	#7, 31, 42, 48, 49
Single endemic subspecies	#14, 35, 95, 99, 139	#12, 17, 90, 130, 134, 157	#5, 34, 62, 80, 100, 164, 194	#84, 140, 191
Single endemic allospecies	—	#19, 27, 28, 70, 135, 149, 171, 172	#61, 85	#58, 60, 88, 108, 185, 193
Monotypic endemic species	#24, 98	#26, 38, 45, 72, 74, 137, 142, 168, 170, 174, 179, 180, 181, 182	#76, 176	#65, 81, 94, 190
Total	19	32	22	18

Numbers following # signs correspond to species codes from appendix 1. For each species exhibiting no geographic variation within Northern Melanesia, this table identifies which of three potential contributing factors seems most important. The right-most column ("unexplained") lists species for which none of the three potential contributing factors applies, and whose absence of geographic variation therefore remains unexplained. "Total" adds up the number of species for which that factor is considered most important.

these species are highly vagile, leaving three unexplained species to which none of these explanatory factors applies.

Species Represented by a Single Endemic Allospecies

Sixteen superspecies that range outside Northern Melanesia are represented within Northern Melanesia by a single endemic allospecies that does not vary geographically (table 18.1, entry 3). Most (10) of these 16 allospecies are nonpasserines. In contrast, most widespread endemic allospecies of passerines in Northern Melanesia show at least some geographic variation.

Eight of these 16 endemic allospecies have little or no opportunity for geographic variation within Northern Melanesia because their ranges are confined to a single island or a couple of nearby islands (New Britain or Rennell; table 18.2, third row second column). The other eight allospecies have somewhat wider ranges, but none is really widespread; for example, none occurs both in the Bismarcks and Solomons, and none of the five that occurs in the Bismarcks reaches the Admiralties or St. Matthias group. Two of these allospecies are highly vagile: for example the cockatoo *Cacatua ducorpsi* (plate 8) can be seen flying over water almost any day in the Solomons. That leaves six endemic allospecies whose absence of geographic variation remains unexplained.

Endemic Full Species or Genera

Twenty-two species that exhibit no geographic variation within Northern Melanesia are endemic to Northern Melanesia at the level of full species or (in five cases) genus (table 18.1, entry 4). Most (16) of these species are very distinctive and isolated taxa with no close relatives, while six do have close relatives and are presumably the products of a relatively recent speciation or colonization (the hawk *Accipiter luteoschistaceus*, the pigeons *Columba pallidiceps* [map 11, plate 8] and *Gallicolumba salamonis*, the warbler *Phylloscopus amoenus* [map 36], the white-eye *Zosterops murphyi* [map 45, plate 1], and the honey-eater *Melidectes whitemanensis* [map 49, plate 9]).

For 14 of these 22 species, the reason they exhibit no geographic variation is obvious: they are confined to a single island or a pair of nearby islands (table 18.2, row 4 column 2). Two other species (the hawk *Accipiter imitator* [map 3] and the owl *Nesasio solomonensis*) are confined to post-Pleistocene fragments of Greater Bukida, so that their populations would have been joined in a single land mass until 10,000 years ago. Finally, two more species (the lory *Chalcopsitta cardinalis* [map

15] and the supertramp honeyeater *Myzomela sclateri*) are highly vagile and lack geographic variation for that reason. *Myzomela sclateri* is one of the vagile, small-island specialists that colonized Long Island after its eruption three centuries ago. *Chalcopsitta cardinalis* is distributed over the entire Solomon chain, is often seen flying from island to island, and colonized Ontong Java in 1972 across a 170-km water gap (Bayliss-Smith 1973). Absence of geographic variation remains unexplained for four monotypic endemic species.

Summary

Our conclusions about the identifiable reasons underlying the absence of geographic variation within Northern Melanesia for 91 species are summarized in table 18.2. Among species whose Northern Melanesian populations are not endemic at all, high vagility and recency of colonization appear to be the two most important factors. These two factors decrease in importance with increasing level of endemism for Northern Melanesia, until one reaches species endemic to Northern Melanesia at the level of the full species or genus. For such species the main reason behind lack of geographic variation is a restricted geographic range, confined to one or a few islands.

That leaves 18 species (right-most column of Table 18.2) for whose lack of geographic variation we cannot demonstrate a reason. Most of these species are ones that we would predict to disperse readily across water, such as the heron *Ardea intermedia*, the falcon *Falco peregrinus*, the rail *Porzana tabuensis*, and the stilt *Himantopus leucocephalus*. We merely do not happen to have proof of their water-crossing ability within Northern Melanesia. That is, among the 91 species lacking geographic variation, high vagility within Northern Melanesia is known to characterize 22, and probably also characterizes up to 18 others.

19

Geographic Variation: Subspecies

The Frequency of Geographic Variation

Let us consider the frequency of geographic variation among Northern Melanesian species, at first without inquiring into the actual number of distinct populations per variable species. We can evaluate this frequency in two ways.

First, take as the unit the zoogeographic species (superspecies, plus full species not belonging to superspecies). As explained in the first two paragraphs of chapter 17, Northern Melanesia has 191 such zoogeographic species that we analyze. How many of these analyzed species exhibit geographic variation within Northern Melanesia, regardless of whether it is at the level of the subspecies or allospecies? Of the 191 analyzed species, 100 do exhibit geographic variation (table 17.1): 65 of them are represented by multiple subspecies but not multiple allospecies within Northern Melanesia; 10 are represented by multiple allospecies, but none of those allospecies is divided further into multiple subspecies; and 25 are divided into multiple allospecies, some of which are further divided into multiple subspecies (appendix 5, columns "allo" and "(s + A)").

Alternatively, instead of considering these zoogeographic species as the unit, one could take as the unit the "species" as usually construed: that is, isolated species not members of superspecies, plus each individual allospecies of a superspecies. Of the 251 analyzed Northern Melanesia species by this definition, 102 vary geographically: 34 out of Northern Melanesia's 86 isolated species not belonging to superspecies, plus 68 of the 165 allospecies.

The Variable Degree of Geographic Variation

The preceding paragraphs consider only the presence or absence of geographic variation, not its degree. As illustrated by figure 18.1, the actual fineness of geographic variability varies greatly among Northern Melanesian birds. At the one extreme are invariant species like the mound-builder *Megapodius freycinet*, recorded from 69 Northern Melanesian islands but with all those populations belonging to the same subspecies. At the opposite extreme are some species in which every island is occupied by a distinct subspecies or allospecies, and other species with up to 16 subspecies of which each occupies one or several islands. How can we account for these enormous species differences in geographic variability, in terms of species differences in biological properties?

As a simple index of the geographic variability of each species, we take a ratio: the number of geographically distinct populations (distinct subspecies and allospecies) into which the species (in the sense of zoogeographic species) falls within

Northern Melanesia, divided by the number of Northern Melanesian islands that the species occupies. (Many more such geographically distinct populations are subspecies than allospecies.) We define this ratio (tabulated in the right-most column of appendix 5) as the $(s + A)/i$ ratio, where s stands for number of subspecies within Northern Melanesia, A stands for number of monotypic allospecies, and i stands for the number of islands occupied. This ratio would be 1.0 if every island were occupied by a distinct population.

In the preceding chapter we plotted the numerator of this ratio ($[s + A]$, the number of distinguishable populations), against i, for all 166 of the 191 analyzed species occupying two or more islands (fig. 18.1). Points would fall on the line of identity ($[s + A]/i = 1.0$) if each island that a species occupied harbored a distinct population. In fact, it is obvious from the trend of figure 18.1 that the vast majority of points falls below the line of identity and that the ratio's value must tend to decrease with i. For instance, all eight species falling on the line of identity occupy only two to four islands. Only two species occupying more than 50 islands have an $(s + A)/i$ ratio exceeding 0.05.

One might at first consider this finding trivial: the ratio's denominator is i, so of course the ratio should decrease with i. But the numerator $(s + A)$ also increases with i, for the simple reason that more islands occupied means more opportunities for distinguishable populations. The question is, why does $(s + A)$ increase at a decreasing rate with i? In other words, why, for species occupying many islands, is a lower proportion of those island populations distinguishable than for species occupying few islands? And what variables account for the enormous interspecific variation in the ratio $(s + A)/i$ at a given i value? From the following analysis we shall see that vagility and abundance are the most important explanatory variables and that level of endemism is not an important factor.

Effects of Vagility and Abundance on Geographic Variability

Are vagile or sedentary species, or abundant or rare species, more prone to geographic variation? To answer these two questions, we divided species into the three classes of vagility discussed in chapter 11, where class A refers to the most vagile species, class B to less vagile species, and class C to the most sedentary species of Northern Melanesia. We also divided species into the five classes of abundance defined in chapter 10. Recall that our abundance measure is of population size, taking account both of population density and of extent of habitat available to the species. Abundance class 1 consists of the rarest species with the smallest populations, all of which prove to be nonpasserines, most of them large-bodied ones. Abundance classes 2, 3, and 4 consist of progressively more abundant species, until abundance class 5 consists of the 22 most abundant species of Northern Melanesia, all of which prove to be small-sized passerines widely distributed in forest.

As a measure of geographic variability, table 19.1 compares $(s + A)/i$ values for species of the five abundance classes and three vagility classes, controlling for number of islands occupied. For a given vagility class and number of islands occupied, geographic variability (i.e., $[s + A]/i$ values) increases with abundance, being several times greater for the most abundant species than for the least abundant species occupying a similar number of islands. For example, among species of intermediate vagility (class B) occupying 7 to 29 islands, the number of distinct subspecies and allospecies per island averages only 0.10 for the rarest species (abundance class 1), increases to 0.28 for species of intermediate abundance (abundance class 3), and to 0.59 for the most abundant species (abundance class 5). For a given abundance class and number of islands occupied, geographic variability ($(s + A)/i$ values) decreases with vagility: the least vagile class-C species are several times more variable than the most vagile class-A species. For example, among species of intermediate abundance (class 3) occupying 2–6 islands, the number of distinct subspecies and allospecies per island averages 0.42 for the most vagile species (class A), 0.45 for less vagile species (class B), and 0.69 for the least vagile species (class C). Table 19.1 also illustrates, as mentioned above, that $(s + A)/i$ decreases steeply with number of islands occupied (i), if comparisons are made within the same abundance and vagility classes.

These conclusions from inspection of Table 19.1 are confirmed by statistical analysis. The best-fit model that we obtain to the data is

$$(s + A)/i = (0.0398)(1.178^{a})\ (4.45^{B})(8.89^{C}) \tag{19.1}$$

where the meaning of the exponents a, B, and C is as follows: a is the abundance class (1, 2, 3, 4, or

Table 19.1. Dependence of geographic variability on abundance and dispersal.

Islands occupied	Vagility[a]	Abundance class 1	2	3	4	5
2–6	A	—	0.25 (1)	0.42 ± 0.12 (2)	0.50 (1)	—
7–29	A	—	0.11 ± 0.04 (2)	0.12 ± 0.08 (9)	0.13 (1)	0.26 (1)
30–69	A	0.04 (1)	0.03 ± 0.02 (5)	0.07 ± 0.07 (19)	0.07 ± 0.06 (12)	0.20 (1)
2–6	B	0.49 ± 0.34 (8)	0.44 ± 0.17 (11)	0.45 ± 0.24 (11)	0.70 ± 0.19 (8)	0.72 ± 0.26 (5)
7–29	B	0.10 ± 0.03 (5)	0.18 ± 0.06 (3)	0.28 ± 0.19 (18)	0.22 ± 0.12 (11)	0.59 ± 0.18 (7)
30–69	B	0.10 (1)	—	0.13 ± 0.10 (4)	0.20 ± 0.12 (5)	0.38 ± 0.13 (2)
2–6	C	0.50 ± 0.26 (6)	—	0.69 ± 0.34 (3)	0.75 ± 0.35 (2)	0.33 (1)
7–29	C	—	—	—	—	—
30–69	C	—	—	—	—	—

Entries are average values ± standard deviations (not standard errors of the mean), and (in parentheses) number of species averaged for $(s + A)/i$, the number of geographically distinct populations (subspecies or allospecies) per island that a species occupies. For example, if a species occupies 12 islands and breaks up into four subspecies, then $(s + A)/i = 4/12 = 0.33$; if it breaks up into two allospecies, one of which breaks up into two subspecies, then $(s + A)/i = 3/12 = 0.25$. Species are grouped by number of islands occupied (2–6, 7–29, or 30–69); by abundance class (1 = least abundant, 5 = most abundant); and by dispersal class (A = the most frequent water crossers, B = less frequent water crossers, C = the most sedentary species). $(s + A)/i$ values, listed in appendix 5 along with values of abundance, dispersal, and i, were calculated for all of Northern Melanesia's analyzed breeding land and freshwater bird species except for those confined to a single island. Dashes indicate cells that contain no species. Note that $(s + A)/i$, an index of geographic variability, increases with abundance and decreases with vagility (i.e., index lowest for the most vagile species, class A). The index also decreases with number of islands occupied, though the absolute number of geographically distinct populations $(s + A)$ increases with number of islands occupied (fig. 18.1). See text for discussion and statistical analysis.

5), $B = 1$ for dispersal class B but 0 for classes A and C, and $C = 1$ for dispersal class C but 0 for classes A and B. In other words,

$$(s + A)/i = (0.0398)(1.178^{a}) \text{ for dispersal class A,}$$
$$(s + A)/i = (0.0398)(4.45)(1.178^{a}) \text{ for dispersal class B,}$$
$$(s + A)/i = (0.0398)(8.89)(1.178^{a}) \text{ for dispersal class C.}$$

That is, $(s + A)/i$, our measure of geographic variability, increases 4.45-fold from dispersal class A to class B, increases a further $8.89/4.45 = 2$-fold from class B to class C, and increases 1.178^{5} (i.e., 1.178 to the 5th power) = 2.27-fold from abundance class 1 to class 5. The effect of dispersal on geographic variability is thus larger than the effect of abundance. However, the most sedentary species (class C), while they have the highest $(s + A)/i$ values, tend to have modest absolute numbers of distinct populations $(s + A)$ as well as low i values, because they can reach so few islands. The highest $(s + A)$ values are for species of intermediate vagility (class B), as we discuss later in the section "The Great Speciators").

Equation 19.1 accounts for 48% of the variation in $(s + A)/i$. All terms in the equation are statistically significant, with p values of .0038 for the abundance term and <.0001 for B and C. It follows from the definition of $(s + A)/i$, with i in its denominator, that an equation incorporating a term with i should explain more variation than does equation 19.1 lacking a term with i. In fact, an equation incorporating a term for i does yield a higher percentage of the variation, 73%, but with the effects of abundance and dispersal class still significant ($p < .0004$). The best-fit equation obtained by statistical analysis proves to be:

$$(s + A)/i = (0.240)\ (1.337^{a})\ (i^{-0.651})\ (1.96^{B})\ (2.20^{C}) \qquad (19.2)$$

where the exponents B and C are as defined earlier.

We can now return to the question posed earlier: why, for species occupying many islands, is a lower proportion of those island populations distinguishable than for species occupying few islands? Why does $(s + A)$ increase at a decreasing rate as i increases, with the consequence that the ratio $(s + A)/i$ decreases with i but that the exponent of i in equation 19.2 is not -1.0 (as one might at first have expected from the definition of $[s + A)/i]$ but has the best-fit value of -0.651? The explanation is that i, the number of islands that each species occupies, increases with vagility (of course, more vagile species occupy more islands), but that $(s + A)/i$ decreases with vagility and hence with i (of course, more vagile species are less likely to subspeciate). In addition, i increases with both abundance and vagility (fig. 12.2), while $(s + A)/i$ also increases with abundance but decreases with vagility; the effect of vagility is stronger than the effect of abundance in both relations.

Interpretation of the Effects of Vagility and Abundance

It is obvious why geographic variability decreases with vagility: because high rates of gene flow between islands inhibit interisland genetic differentiation. All other things being equal, less vagile species, with lower interisland gene flow, should be more variable.

Why should geographic variability increase with abundance? We suggest that part of the explanation involves the steep inverse dependence of risk of extinction on population size. The risk that a population will go extinct due to demographic accidents rises steeply with decreasing population size (Mayr 1965a, MacArthur & Wilson 1967, Diamond 1984f, Pimm et al. 1988). But a population must escape extinction for a significant time if it is to differentiate to the level of a distinct subspecies or allospecies. Hence populations of rare species are less likely to survive long enough to differentiate geographically than are populations of abundant species, if one confines comparisons to species of similar vagility. In addition, more mutations are introduced into large populations than into small populations, so that large populations are expected to have more genetic variation and hence higher rates of evolution.

(Lack of) Relation between Geographic Variability and Degree of Endemism

Genetic divergence between formerly identical populations takes time. One might initially expect that, the longer a species has been present in Northern

Table 19.2. Relation between geographic variability and degree of endemism.

	Islands occupied		
Level of endemism	2–6	7–29	30–69
Nonendemic	0.32 ± 0.14 (12)	0.10 ± 0.02 (6)	0.02 ± 0.01 (9)
Endemic subspecies	0.57 ± 0.19 (14)	0.24 ± 0.20 (26)	0.10 ± 0.10 (28)
Endemic allospecies	0.61 ± 0.27 (19)	0.34 ± 0.21 (18)	0.17 ± 0.12 (11)
Endemic species or genus	0.59 ± 0.29 (13)	0.22 ± 0.13 (7)	0.12 ± 0.14 (3)

As Table 19.1, except that species are grouped by number of islands occupied (2–6, 7–29, 30–69) and by level of endemism: "nonendemic" = Northern Melanesian populations are not endemic but belong to the same subspecies as extralimital populations; "endemic subspecies" = non-endemic species, but with one or more subspecies endemic to Northern Melanesia; "endemic allospecies," "endemic species or genus" = allospecies, full species, or genus endemic to Northern Melanesia. Entries are average values ± standard deviations, and (in parentheses) number of species averaged for $(s + A)/i$, an index of geographic variability. Note that values of $(s + A)/i$ for species with only nonendemic populations are low (because all Northern Melanesian populations of the species belong to just one or at most two extralimital subspecies). The remaining species have differentiated within Northern Melanesia into 1–16 endemic taxa, but note that $(s + A)/i$ values do not vary with level of endemism between endemic subspecies, allospecies, full species, and genera. $(s + A)/i$ values do decrease with number of islands occupied. See text for discussion and statistical analysis.

Melanesia, the greater the number of geographically distinct populations it will have had the opportunity to form. One would also expect that, all other things being equal, time elapsed since first colonization of Northern Melanesia increases with level of endemism: endemic genera, species, and allospecies have probably survived in Northern Melanesia for longer, on average, than species represented only by endemic subspecies, and the latter in turn are presumably on the average older than extralimital species whose Northern Melanesian populations have not even differentiated to the level of an endemic subspecies. One might therefore expect that geographic variability within Northern Melanesia would increase with level of endemism.

In fact, this is not true, as table 19.2 shows. When one controls for number of islands, while nonendemics do have lower $(s + A)/i$ values, there are no differences between species endemic at the subspecies, allospecies, species, and genus level with respect to $(s + A)/i$ values (i.e., with respect to degree of geographic variation). Statistical analysis shows that, when one controls for abundance and dispersal, there is no significant effect of level of endemism, from the level of endemic subspecies to endemic genus, on $(s + A)/i$. That is, a full three-way analysis of variance model including abundance class, dispersal class, and level of endemism plus all possible two-way and three-way interactions does not explain significantly more variation in $(s + A)/i$ than does a model including only abundance class and dispersal class ($F = 1.174$, df = 13,102, $p = .28$). Why is there no evidence for the expected relation between geographic variability and level of endemism?

We suggest that the expectation ignored two considerations. First, subspeciation may develop rapidly. For example, among the populations founded on Long Island since its defaunation by a volcanic explosion in the seventeenth century, four species (the cuckoo *Eudynamys scolopacea*, swallow *Hirundo tahitica*, and honey-eaters *Myzomela pammelaena* [map 47] and *M. sclateri*) have already formed morphologically distinct populations (Diamond et al. 1989). Only 33 of Northern Melanesia's 191 analyzed species have failed to differentiate at least to the level of an endemic subspecies.

Second, once geographic differentiation is thus underway, a steady state may be reached between the rate of formation and the rate of extinction of distinct populations. As a result, the $(s + A)/i$ ratio for a species may attain a steady-state value determined by that species' vagility and abundance. With further time after attaining that steady-state value, the ratio would not change, although some of the distinct populations would eventually survive long enough to attain higher levels of endemism. That is, a particular species, provided that its populations were sufficiently large to escape quick extinction, might subspeciate to attain an $(s + A)/i$ ratio of 0.33, meaning that its populations now fall into a number of endemic subspecies, each occupying, on average, three nearby islands. With further time, some of those subspecies would differentiate to the level of endemic allospecies, while others would go extinct, leaving the latter islands

Table 19.3. The great speciators of Northern Melanesia.

Species	Abundance	Vagility	i	s	A	$(s + A)/i$	Map	Plate
Accipiter novaehollandiae	3	A	35	10	(1)	0.29	—	—
Gallirallus [*philippensis*]	3	B	21	6	2	0.38	—	—
Ptilinopus [*rivoli*]	4	B	52	8	2	0.19	7	6
Micropsitta [*pusio*]	4	B	41	6	3	0.22	19	—
Ninox [*novaeseelandiae*]	3	B	18	7	4	0.61	22	7
Collocalia esculenta	4	A	44	7	(1)	0.16	—	—
Ceyx lepidus	4	B	25	8	(1)	0.32	—	—
Halcyon chloris	4	A	56	12	(1)	0.21	24	6
Pitta erythrogaster	4	B	10	4	(1)	0.40	26	—
Coracina caledonica	3	B	6	4	(1)	0.67	27	—
Coracina lineata	3	B	25	7	(1)	0.28	—	—
Coracina [*tenuirostris*]	3	B	44	9	2	0.25	28	7
Lalage leucomela	5	B	12	7	(1)	0.58	30	—
Zoothera [*dauma*]	3	B	8	3	3	0.75	32	—
Turdus poliocephalus	5	B	8	7	(1)	0.88	33	—
Phylloscopus [*trivirgatus*]	5	B	10	5	2	0.70	36	—
Rhipidura [*rufiventris*]	5	B	29	11	2	0.45	37	—
Rhipidura [*spilodera*]	4	B	4	1	3	1.00	38	—
Rhipidura rufifrons	5	A	27	8	(1)	0.30	39	—
Rhipidura [*rufidorsa*]	4	B	5	1	3	0.80	39	—
Monarcha [*manadensis*]	4	B	32	7	6	0.41	41	3
Monarcha chrysomela	4	B	5	4	(1)	0.80	—	—
Petroica multicolor	4	B	4	4	(1)	1.00	—	—
Pachycephala pectoralis	5	B	34	16	(1)	0.47	43	2
Dicaeum [*erythrothorax*]	5	B	18	4	3	0.39	44	—
Zosterops [*griseotinctus*]	5	B	18	2	6	0.44	45	1
Myzomela cruentata	5	B	7	5	(1)	0.71	—	—
Myzomela [*lafargei*]	5	B	42	6	6	0.29	47	4
Philemon [*moluccensis*]	4	B	5	1	3	0.80	48	7
Lonchura [*spectabilis*]	3	B	7	1	3	0.57	—	—
Dicrurus [*hottentottus*]	4	B	6	2	2	0.67	52	8

Great speciators are defined as species which either are represented by seven or more geographically distinct populations (subspecies and/or allospecies) within Northern Melanesia, regardless of their $(s + A)/i$ ratio, or are represented by four to six distinct populations and have an $(s + A)/i$ ratio of 0.33 or greater. A = number of allospecies in Northern Melanesia, s = number of subspecies, $(s + A)$ = number of geographically distinct populations. [If there is only a single allospecies, we give A as (1), s as the total number of subspecies, and the number of geographically distinct populations is simply s. If there are two or more allospecies, we give A as that number of allospecies, s as the number of subspecies beyond 1 in each allospecies summed over all allospecies, so that the number of geographically distinct populations is $(s + A)$]. Columns 2, 3, and 4 respectively give the abundance (1 = least abundant, 5 = most abundant), vagility (A = most vagile, B = less vagile), and number of islands occupied (i). The column $(s + A)/i$ gives the number of geographically distinct populations divided by i. Data are from appendix 5. Note that all great speciators are relatively abundant (abundance class at least 3, mostly 4 or 5), none is sedentary (vagility class C), few are highly vagile (vagility class A), and most are of intermediate vagility (class B). The right-most two columns give the number of the map and plate, if any, depicting the species' distribution and the species itself, respectively.

to be repopulated by individuals of surviving subspecies. The result would be $(s + A)/i$ ratios independent of level of endemism after the first stages of differentiation, as observed in table 19.2.

The Great Speciators

The feature that makes Northern Melanesian birds famous in texts of evolutionary biology is their nu-

Table 19.4. Abundance and vagility of the great speciators.

	Abundance				
Vagility	1	2	3	4	5
A	0/1 (0%)	0/9 (0%)	1/30 (3%)	2/14 (14%)	1/2 (50%)
B	0/15 (0%)	0/18 (0%)	7/35 (20%)	11/24 (46%)	9/15 (60%)
C	0/9 (0%)	0/0 (0%)	0/9 (0%)	0/5 (0%)	0/5 (0%)

Great speciators are the 31 Northern Melanesian species with the greatest geographic variability, as listed in table 19.3. This table classifies them simultaneously by vagility (A = the most vagile species, B = less vagile, C = most sedentary) and abundance (1 = least abundant, 5 = most abundant). Each entry gives the number of great speciators, followed after a stroke by the total number of Northern Melanesian species, with that combination of vagility and abundance, followed in parentheses by the percentage of all species with that vagility/abundance combination that are great speciators. For example, nine great speciators are very abundant (class 5) and only moderately vagile (class B), accounting for 60% of the 15 Northern Melanesian species with that abundance and vagility. Note that great speciators are drawn disproportionately from abundant species of intermediate vagility, and include no rare species (abundance classes 1 or 2), no sedentary species (vagility class C), and few very vagile species (vagility class A). Statistical analysis yields an excellent fit ($p = .89$ by a chi-square test) of the data to the logistic regression model, $\ln [P/(1 - P)] = -7.29 + 2.1B + 1.2a$, where P is the proportion of species in a given abundance × dispersal class that are great speciators, $P/(1 - P)$ is the odds ratio for being a great speciator, B equals 1.0 for vagility class B and equals 0 for vagility classes A or C, and a is the abundance class (1, 2, 3, 4, or 5). That is, the odds of being a great speciator are 8.0 times higher for species of intermediate vagility (class B) than for species of high or low vagility, and the odds increase 3.3-fold with each increase in abundance from class 1 to class 5. The increase in P with abundance, and the peak in P for vagility class B, are both highly significant ($p < .0001$ and $p < .001$, respectively).

merous instances of marked geographic variation and speciation. For example, the Northern Melanesian populations of *Pachycephala pectoralis* (plate 2, map 43) fall into 16 subspecies, while the populations of *Monarcha* [*manadensis*] (plate 3, map 41), *Zosterops* [*griseotinctus*] (plate 1, map 45), and *Myzomela* [*lafargei*] (plate 4, map 47) fall into six allospecies plus additional subspecies. Table 19.3 lists Northern Melanesia's 31 "great speciators," defined either as zoogeographic species represented in Northern Melanesia by at least 7 subspecies and/or allospecies, or as zoogeographic species represented by at least 4 subspecies and with an $(s + A)/i$ ratio of at least 0.33. Are these great speciators characterized by the same properties of high abundance and intermediate vagility that accounts for the general patterns of geographic variability (specifically, of $[s + A]$ rather than of $[s + A]/i$)?

The answer is yes. Of the 31 great speciators, none is rare (abundance classes 1 or 2); they account for, respectively, 11%, 30%, and 45% of the species in abundance classes 3, 4, and 5. Similarly, no great speciator is very sedentary (vagility class C), because such sedentary species cannot reach enough islands to form many distinct populations (i.e., modest $[s + A]$ values despite high $[s + A]/i$); and only 7% are species of the highest vagility (vagility class A), whose high gene flow inhibits differentiation. Instead, great speciators account for 25% of species of intermediate vagility (vagility class B). Thus, great speciators are drawn disproportionately from abundant species of intermediate vagility.

This point is further illustrated by table 19.4, which indicates how great speciators are distributed over classes defined by combinations of abundance and vagility. Fully 60% of the most abundant (class 5) species of intermediate vagility (class B) are great speciators. In addition, as we have discussed elsewhere (Diamond et al. 1976), the vagility of the great speciators is of a particular type: they are short-distance colonists. That is, they are good at crossing short water gaps but poor at crossing wide water gaps, with the result that they tend to be absent from the more remote islands of Northern Melanesia.

Geographic Variability of Montane Species

As an example of how the preceding considerations can be used to interpret differences in geographic variability between and within sets of Northern Melanesia birds, consider Northern Melanesia's 23 species whose populations are confined to the mountains of some or all Northern Melanesian islands that they occupy, and which occur on more than one island. (There are also other montane species that are confined to a single island and thus cannot vary geographically.) Those montane species provide some of the most spectacular examples of geographic variation in the Northern Melanesian avifauna, with 10 species falling into a distinct subspecies or allospecies on every island occupied. For example, each of the three Northern Melanesian populations of the montane warbler *Cichlornis* [*whitneyi*] (map 35), which occurs in the mountains of Guadalcanal, Bougainville, and New Britain, belongs to a distinct allospecies, while each of the four populations of the montane robin *Petroica multicolor* belongs to a distinct subspecies. The high variability of the montane species, and also the exceptions to their variability, can be explained in terms of abundance and vagility. Twenty of the 23 species belong to the class of intermediate vagility, class B, the vagility class most prone to geographic variation. All nine montane species belonging to abundance classes 4 and 5 (the abundance classes most prone to geographic variation) vary geographically, with five of those nine species having distinct subspecies or allospecies on every island occupied. Among the less common species (abundance classes 1–3), half of the species show either no or little geographic variation. In addition, four of the six montane species that exhibit little or no geographic variation (the pigeons *Ducula melanochroa* and *Gymnophaps* [*albertisii*], the parrot *Charmosyna* [*palmarum*], and the finch *Erythrura trichroa*) are nomadic species that feed on transient blooms of fruit, flowers, or bamboo seeds and that wander in flocks, so that gene flow between islands must be especially effective in reducing geographic variability. Thus, it is possible to understand why most montane species are so prone to geographic variation, and also why a few of them are not.

Summary

Northern Melanesian species vary greatly in geographic variability, ranging from invariant species (of which a single subspecies occupies 69 islands) to highly variable species (a different subspecies or allospecies on each island occupied). As a measure of geographic variability, we analyze the ratio $(s + A)/i$: the number of geographically distinct populations (subspecies or allospecies) per island that a species occupies. Because this ratio tends to decrease with the number of islands occupied (i), our analysis controls for variation in i. The ratio decreases with vagility because gene flow inhibits differentiation. The ratio increases with abundance: the risk of extinction decreases steeply with increasing population size, so populations of abundant species are more likely than are populations of rare species to survive long enough to differentiate geographically. The ratio bears no relation to degree of endemism; this suggests that, once geographic differentiation is underway, a steady state is reached between the rates of formation and of extinction of distinct populations. Northern Melanesia's 31 famous "great speciators," with many distinct subspecies and allospecies, fit these patterns; they tend to be abundant species of intermediate vagility, which can reach many islands and survive on them but can also differentiate on many of those islands. Northern Melanesia's 21 montane species also fit these patterns; many of them show spectacular geographic variation, befitting their high abundance and intermediate vagility.

20

Geographic Variation

Megasubspecies

Like almost any classificatory system that classifies anything, the Linnaean system of taxonomic classification makes arbitrary divisions between items, some of which, in fact, grade continuously into each other. Dividing the continuous processes of speciation and geographic variation into hierarchical taxonomic levels is an obvious example. The formal alternatives available to us for expressing the relationship between two populations are to consider them as members of the same subspecies; as members of different subspecies, but of the same species or allospecies; as members of different allospecies, but of the same superspecies; as members of different superspecies or species, but of the same genus; and so on.

In fact, any taxonomist knows how often such taxonomic decisions are difficult or arbitrary. This is as expected from our understanding of speciation: divergence between populations arises and proceeds, for the most part, gradually, not in sharp steps. (We acknowledge the concept of punctuated evolution—the possibility of relatively abrupt evolutionary change—but it does not conflict with what we have just said. Changes that appear abruptly on a geological time scale are, on the time scale of an individual animal's lifetime, unusually large gradual changes taking place over "just" some dozens or thousands of generations.) The two most distinct steps in this gradual divergence are the two culminating steps of speciation: the transition from subspecies to allospecies status, when two formerly interbreeding populations become non-interbreeding; and the transition from allospecies to full species status, when the two populations that were formerly too similar to coexist finally develop niche differences sufficient to permit sympatry.

Even recognition of the first of these two distinct transitions involves some arbitrariness, for several reasons. First, when two geographically isolated and formerly conspecific populations newly regain geographic contact, their frequency of hybridization may not immediately be zero but may instead be significant and then decline over time. Second, hybridization may continue to occur, albeit rarely, for a long time (especially at locations where one of the two species is at the periphery of its geographical range). Third, when one is making taxonomic judgments about two geographically isolated populations that are not in contact, one cannot observe directly whether they regularly interbreed, and therefore one can only attempt to infer whether they would interbreed if given the opportunity (and hence whether they should be considered conspecific subspecies or else distinct allospecies). Beyond this stage of completed speciation, it becomes especially arbitrary how to group species at higher levels into genera, tribes, families, and so on.

Taxonomy would be much simpler if we could express differences by means of a continuously varying quantitative parameter, instead of by discrete arbitrary levels. For classification above the full species level, the percent difference in whole-genome, single-copy DNA has been suggested as such a parameter (Sibley & Ahlquist 1990). Unfortunately, DNA differences have yet to be determined for Northern Melanesian species pairs, though they have been determined for many New Guinean and Australian species pairs. In addition, even if such information were available, the borders between subspecies and allospecies, and between allospecies and full species, do not depend on those differences between whole-genome DNA but instead on differences between just those portions of the genome responsible for reproductive isolation and for ecological segregation, respectively. Hence, to analyze speciation, we are forced to use a system of taxonomic classification based on discrete levels, despite its imperfections.

Nevertheless, we wish in this chapter to give readers a feeling for the range of geographic variation that we group into the three categories: "no geographic variation," "geographic variation at the subspecies level," and "geographic variation at the allospecies level." We want, in addition, to identify for readers all cases of subspecific geographic variation that seem especially marked and close to the allospecies level. Naturally, we do so by recognizing yet one more arbitrary taxonomic level between the levels of the subspecies and allospecies! We refer to this intermediate level as the "megasubspecies," defined as a subspecies or a group of subspecies whose distinctness is closer to the allospecies level than is the distinctness of subspecies not so identified (Amadon & Short 1992). Quite a few taxonomists today are more disposed than we are to recognize insular isolates as species rather than subspecies: we thereby make the task of such taxonomists easier by explicitly naming the most distinct populations that we still consider to be subspecies.

In practice, we denote 21% of Northern Melanesia's endemic subspecies as megasubspecies. Of course, recognizing yet one more arbitrary intermediate level in no way solves the basic problem of analyzing a continuous process by recognizing arbitrary categories. Instead, we hope to call explicit attention to the continuous nature of geographic variation and to identify those cases among Northern Melanesian birds of particular interest for further study.

In the first part of this chapter, we describe 10 examples to give a sense of the range of geographic variation lumped into the category "subspecies," and to indicate the practical problems encountered when one tries to decide whether insular populations already constitute allospecies or are still just subspecies. In the second part of this chapter we briefly summarize some numerical aspects of our decisions about megasubspecies.

Examples Illustrating the Range of Geographic Variation at the Subspecies Level

Minor Geographic Variation Not Worth Recognizing by Subspecies: Porphyrio porphyrio

The highly vagile swamphen *Porphyrio porphyrio* exhibits minor but complex geographic variation throughout its enormous range across the Pacific, from Northern Melanesia to Samoa. Within any given population, there is much individual variation in color and size (Mayr 1949a). If one compares average values among populations, there are some minor geographic trends; for example, the populations of Ysabel and the New Hebrides are, on average, relatively large, those of Manus and San Cristobal are relatively small, and the Manus population contains more greenish or bluish individuals than other populations. However, most of this geographic variation is irregular from island to island and defines no trends. There are no indications that this variation represents significant progress toward incipient speciation. The sole example of completed speciation in Pacific swamphens that we can use to calibrate our standard for the degree of difference required to achieve speciation by swamphens is on New Zealand, where *P. porphyrio* occurs sympatrically with its famous relative the takahe, *Porphyrio ("Notornis") mantelli*. The takahe is several times the size of *P. porphyrio*, flightless, and with much stouter legs and bill. By this standard, existing differences among Pacific populations of *P. porphyrio* represent insignificant progress toward speciation. Even if one wanted to recognize the population of every Pacific island as a distinct subspecies, most individuals could fit into many of the resulting subspecies, whereas the usual criterion for distinguishing subspecies is that 70% of individuals must be assignable to one subspecies or another without overlap. Hence we consider all swamphen populations from Northern Melanesia to Samoa to belong to a single subspecies, *Porphyrio porphyrio samoensis*.

A Single Weak Subspecies: Chalcophaps stephani

Populations of the ground dove *Chalcophaps stephani* from the Solomons average slightly larger than populations from the Bismarcks and New Guinea. The average length of the male's wing is 150 mm (range 147–154 mm) in the Solomons, as compared to 140 mm (range 135–148 mm) in the Bismarcks and New Guinea. We can detect no differences in song among the Solomon, Bismarck, and New Guinea populations. The sole congeneric species is *Chalcophaps indica*, with which *C. stephani* occurs sympatrically in New Guinea and Celebes, and from which it differs greatly in plumage and in voice. Hence we consider the Solomon population to constitute only a weak subspecies, *Chalcophaps stephani mortoni*, and consider the Bismarck population to belong to the nominate subspecies of New Guinea, *Chalcophaps stephani stephani.*

Several Weak Subspecies: Rhyticeros plicatus

The hornbill *Rhyticeros plicatus* (map 25), which flies readily over water and has colonized most of the New Georgia group within the twentieth century, exhibits minor variation in size and in darkness of coloration within Northern Melanesia and New Guinea. New Guinea birds are somewhat larger than those of the Bismarcks and northwestern Solomons, which are in turn larger than those of the remainder of the Solomons. New Guinea and Bismarck birds are somewhat darker than Solomon birds. Hence we recognize three weak endemic subspecies for Northern Melanesia: *Rhyticeros plicatus dampieri* for the medium-sized, dark birds of the Bismarcks; *R. p. harterti* for the medium-sized, paler birds of the Northwest Solomons; and *R. p. mendanae* for the small, paler birds of the rest of the Solomons.

Single Megasubspecies: Chrysococcyx lucidus

The small cuckoo *Chrysococcyx lucidus* breeds in Australia, New Zealand, New Caledonia, and the New Hebrides, and the Australian and New Zealand breeding populations are common winter visitors to Northern Melanesia, but the sole breeding population of Northern Melanesia occurs on neighboring Rennell and Bellona Islands in the southeastern Solomons. Males of the Rennell/Bellona population are similar to males of the subspecies *Chrysococcyx lucidus layardi* of the New Hebrides and New Caledonia, differing only in minor plumage characteristics. However, whereas all other populations of this cuckoo are nearly monomorphic sexually (males and females similar to each other), the Rennell/Bellona population is sexually dimorphic: the female's crown is purple, unlike the copper-colored crown of the male, and the female (but not the male) has a brown wash on the underparts and on the tail. These same characters differ among sympatric species of the genus *Chrysococcyx* in New Guinea and Australia, but the degree of difference between Rennell/Bellona females and *layardi* females is less than that between sympatric *Chrysococcyx* species. The songs of *C. lucidus* that we heard on Rennell and Bellona are similar to songs of the same species that we heard in New Guinea and Australia, and distinct from songs of other *Chrysococcyx* species that we heard in New Guinea. Hence the Rennell/Bellona population should surely be assigned to the species *Chrysococcyx lucidus* rather than to a distinct allospecies, but it does constitute a quite distinct subspecies (*C. lucidus harterti*), which we consider distinct enough to rate as a megasubspecies.

Population on Borderline Between Megasubspecies and Allospecies: Ptilinopus viridis

The fruit dove *Ptilinopus viridis* (map 8) is distributed from the Moluccas through New Guinea, is absent from most of the Bismarcks, and reappears in the Solomons. With one exception, all populations are fairly similar to each other in plumage, though there are sufficient differences to permit recognition of subspecies, and though there are also moderate differences in vocalizations between locations. The only strikingly distinct population is that of the southeastern Solomons (San Cristobal and its neighbors), where birds have a white head instead of a green head. No white-headed birds have been found elsewhere in the range, and no green-headed birds have been found in the southeastern Solomons. Throughout the range of the species, all populations have two distinct songs, both of them recognizably different from songs of the many other congeneric species of *Ptilinopus*; the southeastern Solomons population has those usual two songs. The population of the remainder of the Solomons has been described as a subspecies *P.*

viridis lewisi on the basis of minor plumage distinctions from New Guinea and Moluccan populations, while the southeastern Solomons population has been named *eugeniae*.

Most modern authors have considered *eugeniae* as a subspecies of *Ptilinopus viridis*, but Goodwin (1983) recognized it as a distinct allospecies, *Ptilinopus eugeniae*, though admitting that it may be conspecific with *Ptilinopus viridis*. This is a typical borderline case because *lewisi* and *eugeniae* do not actually come into contact, and thus we do not know whether they would interbreed. Our own inference, based on comparison with sympatric *Ptilinopus* species or with integrading *Ptilinopus* subspecies with whose voice differences and plumage differences we are familiar, is that *P. v. eugeniae* has not quite reached the level of reproductive isolation and is still conspecific with *P. v. lewisi*, though *P. v. eugeniae* unquestionably represents a distinctive megasubspecies.

Several Megasubspecies: Myiagra hebetior

The flycatcher *Myiagra alecto* (map 42, plate 5) is widespread from the Moluccas, New Guinea, and Australia through most of the Bismarcks. The closely related *Myiagra hebetior* is confined to six Bismarck islands, five of which it shares with *M. alecto*, while it occupies the most remote island of its Bismarck range, St. Matthias, by itself. Males of *M. hebetior* and *M. alecto* are both iridescent black, differing slightly in size and sheen, but are so similar to each other that the taxonomic distinctness of *hebetior* was not recognized until 1924. Females of *M. alecto* are very different from males, having a rufous back, wings, and tail and white underparts and being black only on the crown. Females of the isolated St. Matthias population of *M. hebetior* are confusingly similar to females of *M. alecto*, differing only by having the wings and tail black like the crown (plate 5). Females of the Dyaul population of *M. hebetior* are even more similar to sympatric *M. alecto* females, differing only in a light grey crown, while females of *M. hebetior* from the other four Bismarck islands are much more distinct from sympatric *M. alecto* females in the color of every part of the plumage.

Thus, differences between females of the three *M. hebetior* populations are greater than differences between sympatric *M. hebetior* and *M. alecto* females on Dyaul. However, males of the three *M. hebetior* populations are essentially identical to each other and consistently exhibit only the same modest differences vis à vis males of *M. alecto*. Clearly, the differences among *M. hebetior* females are approaching those separating sympatric species. Since, however, there is no geographic differentiation among the males, we assume that the three populations of *M. hebetior* have not yet reached the allospecies level, but we recognize them as three megasubspecies; some of our readers may disagree.

Another Case of Several Megasubspecies: Ptilinopus solomonensis

The two fruit doves *Ptilinopus solomonensis* (map 7, plate 6) and *P. rivoli* are clearly each other's closest relatives within the genus *Ptilinopus*, sharing many features of plumage as well as similar calls. The range of *P. rivoli* is centered to the west (Moluccas eastward to Bismarcks), the range of *P. solomonensis* to the east (Bismarcks east to the Solomons), but both species are widely distributed over the Bismarcks and over the islands of Geelvink Bay. In these areas of sympatry at a coarse level, their distributions are nearly allopatric at a fine level: *P. rivoli* occupies the larger islands, being confined to the mountains of the highest islands, while *P. solomonensis* occupies the smaller islands. However, vagrants of *P. solomonensis* reach the coastal lowlands of the large islands occupied by *P. rivoli*, and both species have been recorded on Umboi Island of the Bismarcks and on two islands of Geelvink Bay.

To the east of the range of *P. rivoli*, in the Solomons, *P. solomonensis* has montane populations on some islands, lowland populations on other islands, and occurs both in the lowlands and mountains of still other islands. Interestingly, the montane populations of Bougainville, Malaita, and Guadalcanal are most similar to each other in having red-violet areas of the head confined to two spots around the eye; red-violet areas occupy the entire forecrown of the Bismarck and lowland Solomon populations (plate 6). Among these three montane populations, the two most similar are *P. s. bistictus* of Bougainville and *P. s. ocularis* of Guadalcanal. We consider *P. s. ocularis* and *P. s. bistictus* together to constitute one megasubspecies, the lowland races of the Solomons and Bismarcks another megasubspecies, and the distinct race *P. s. speciosus* of Geelvink Bay a third megasubspecies, with *P. s. ambiguus* of Malaita somewhat intermediate between the first two of these megasubspecies.

What makes this case especially interesting is that the two megasubspecies of Northern Melanesia are separated not only by plumage differences but also by ecological differences, the populations of *P. s. bistictus* and *P. s. ocularis* being largely confined to elevations above 800 m. Lest one be tempted to consider this as a case of incipient sympatric speciation, we point out that the lowland and montane megasubspecies occupy different islands rather than sharing the same island. In this case as in other cases, what has prevented sympatric speciation is gene flow between montane and lowland populations on nearby islands. For example, Galbraith and Galbraith (1962) recorded, at a site in Guadalcanal montane forest where the montane race *P. s. ocularis* was common, an individual evidently of the lowland megasubspecies, presumably having arrived from low islands near Guadalcanal.

Multiple Stages of Speciation: Aplonis [grandis] Superspecies

The *Aplonis [grandis]* superspecies (map 51), confined to the Solomons, consists of four named taxa: *dichroa* of San Cristobal, *malaitae* of Malaita, *macrura* of Guadalcanal, and *grandis* of all the rest of the Solomons. Compared to *grandis* and *macrura*, *dichroa* is only half the size, lives in flocks rather than in pairs, and has a different voice. The race *malaitae* is more similar to *grandis* than to *dichroa* but is still distinct from *grandis* in being 15% smaller, having a white rather than a red eye, and having differences of plumage coloration. The races *macrura* and *grandis* are very similar.

We consider *dichroa* a distinct allospecies because it differs morphologically and in voice from *grandis* more than do some sympatric species pairs within the genus *Aplonis*, such as *A. cantoroides* and *A. insularis*, *A. metallica* and *A. mystacea*, and *A. metallica* and *A. brunneicapilla*. We consider *malaitae* sufficiently distinct to rank as a megasubspecies within the species *A. grandis*, while the two weak subspecies *grandis* and *macrura* collectively constitute the other megasubspecies of that species. Thus, this superspecies is interesting because it simultaneously exemplifies several different degrees of distinctiveness, corresponding to very weak subspecies, strong subspecies, and allospecies.

Very Complex Example: Pachycephala pectoralis

The whistler *Pachycephala pectoralis* (map 43, plate 2) provides one of the most famous examples of geographic variation among birds, and one of the most complex and richly instructive examples of speciation.

Briefly, the golden whistler group to which *P. pectoralis* belongs is centered in New Guinea, where there are seven sympatric species, some of them so similar to others that their distinctness was not recognized for many decades. Within Northern Melanesia occur many island populations whose range of variation spans that of New Guinea's sympatric species, so that initially many of them were described as distinct species. However, it eventually became apparent that all of these Northern Melanesian populations were allopatric, except for the coexistence of very distinct montane and lowland populations on Bougainville and Guadalcanal. Hence Mayr (1945a) assigned all Northern Melanesian populations to the single species *P. pectoralis* except for the high-altitude taxa of Bougainville and Guadalcanal, populations whose distinctness as a separate species (*P. implicata*) has never been in doubt.

Galbraith (1956, 1967) then recognized that in tropical Australia two taxa formerly considered subspecies of *P. pectoralis* actually breed together in very close proximity, one (*melanura*) breeding in mangrove and on nearby islets, the other (*pectoralis*) breeding in nearby rainforest. In the Bismarcks Diamond (1976) similarly found that the Bismarck representatives of *melanura* and *pectoralis* (*dahli* and *citreogaster*, respectively) breed in close proximity, the former breeding on islets and young volcanic islands, the latter on older and larger islands, but individuals of *dahli* frequently reach coastal forest of the larger islands occupied by *citreogaster*. Diamond took both species in the same mist net on Tolokiwa Island. Intergrades between *dahli* and *citreogaster* have not been observed, but intergrades between *dahli* and the *pectoralis* representative *bougainvillei* do occur in the population termed *whitneyi* in the Shortland group of the Solomons, at the eastern margin of *dahli*'s range. Thus, it is now clear that the golden whistler complex includes at least two sympatric species in tropical Australia and the Bismarcks, *P. pectoralis* (including *citreogaster*) and *P. melanura* (including *dahli*).

Compared to the minor plumage differences between *P. melanura* and *P. pectoralis*, or between the sympatric New Guinea species *P. schlegelii*, *P. lorentzi*, and *P. meyeri*, the famous geographic variation of allopatric populations of the golden whistler group in the Solomons is far more striking. For example, while all other Solomon populations show marked sexual dimorphism of plumage between male and female, the Rennell population *P. feminina* is hen-feathered; that is, males and females identically share a plumage similar to females of other populations. Males of most populations have olive-colored backs, but *melanonota* males of Vella Lavella have black backs. Males of the population *sanfordi* of Malaita, described as recently as 1931 as a separate species, have uniformly yellow underparts and lack the black breastband of most other populations. Accompanying this geographic variation in plumage is equally spectacular variation in song and ecology. The Rennell population *feminina* often feeds on the ground, while the other populations are arboreal. Some populations are virtually confined to the mountains, others live in the lowlands as well, and others live on flat low islands.

If one takes as a standard the modest morphological differences that permit sympatry between related whistlers in New Guinea, Australia, and the Bismarcks, one would feel compelled to divide these Solomon populations into numerous allospecies. However, several islands provide clear evidence of hybridization between such seemingly distinct whistler populations. The San Cristobal population, *christophori*, appears to have arisen as a hybrid between quite different whistler populations of the Solomons and the New Hebrides. Some Fijian populations are clearly hybrids between very distinct forms, and some Bismarck populations may be hybrids as well (Mayr 1932d, Galbraith 1956).

Thus, we have the paradox that, in the Bismarcks, Australia, and especially in New Guinea, some very similar whistlers occur sympatrically without any signs of hybridization, whereas in the Solomons and Fiji, far more distinct parapatric whistlers have hybridized. Presumably the difference in outcome has to do with the greater species richness of New Guinea and Australia, which favors the coexistence of several narrow-niched species, whereas a single broad-niched species is favored in the more species-poor environment of the Solomons and Fiji. In judging the taxonomic status of allopatric Solomon populations, we therefore take as our standard the hybridization between distinct populations in the Southeast Solomons (San Cristobal) and Fiji rather than the reproductive isolation observed in New Guinea, Australia, and the larger Bismarck Islands. Hence we consider the Solomon populations as conspecific but as falling into four megasubspecies or megasubspecies groups, with the Bismarck populations belonging to a fifth megasubspecies.

Many Borderline Cases: the Monarcha [manadensis] *Superspecies*

Northern Melanesia has representatives of a widespread group of black-and-white monarch flycatchers (map 41, plate 3), whose Northern Melanesian populations fall into 13 forms distinct enough to be named taxonomically. All the Northern Melanesian forms are allopatric; no island supports two forms. Thus, there is no example of completed speciation within Northern Melanesia; instead, the practical question concerns which of these forms to rank as subspecies and which to rank as allospecies. Related flycatchers occur in New Guinea, Australia, eastern Indonesia, and Micronesia, with most of the forms being allopatric, but New Guinea and three other islands support sympatric species pairs. The forms differ especially in size and in the pattern of the black-and-white plumage.

Among the Northern Melanesian forms, the most distinctive is *Monarcha menckei* of St. Matthias, whose tiny size and largely white plumage has always caused it to be regarded as a distinct allospecies. There has also been agreement that *M. infelix* of the Admiralty Islands is a separate allospecies from *M. verticalis* of the rest of the Bismarcks. Disagreement focuses on whether the form *ateralbus*, confined to Dyaul Island in the Bismarcks, should be ranked as a separate allospecies from *M. verticalis* of the rest of the Bismarcks (Salomonsen 1964) or merely as a megasubspecies. There is even more disagreement as to the status of the eight forms of the Solomon Islands. Four of them were initially described as separate species. Mayr (1945a) ranked all eight as subspecies of the same allospecies, *M. barbatus*. Later, Mayr (1955) suggested that the Solomon populations might be divided into three allospecies: *M. barbatus* of the Bukida Islands and Malaita, *M. browni* of the New Georgia group, and *M. viduus* of the southeastern

Solomons. Galbraith and Galbraith (1962) agreed with the separation of *M. viduus* but not of *M. browni*.

Our own conclusion is to divide these Northern Melanesian monarchs into six allospecies, within which we recognize eight megasubspecies and four weak subspecies. We follow Mayr (1955) in recognizing three allospecies for the Solomons. We rank *ateralbus* as a megasubspecies under *M. verticalis*. *M. browni* falls into four weak subspecies, and *M. infelix* and *M. viduus* each fall into two megasubspecies.

This example is instructive because, as in the case of *Aplonis* [*grandis*], within the same superspecies one encounters forms of a wide range of distinctness, ranging from weak subspecies through strong subspecies to undoubted allospecies. The case is also instructive because it includes many "toss-ups"—cases in which, even by using extralimital species pairs as standards, it remains difficult to infer whether allopatric Northern Melanesian forms have or have not reached the level of reproductive isolation. We emphasize that, while the former instructive feature (wide range of distinctness of forms within the same superspecies) corresponds to a truly continuously varying aspect of geographic variation, the latter instructive feature (many taxa about whose reproductive isolation we are uncertain) merely corresponds to ignorance on our part. If the birds from different islands actually came into contact with each other, they would quickly demonstrate whether they are reproductively isolated and deserve to be ranked as an allospecies. We, not the birds, are the ignorant ones, because in this case the birds have not performed the test for us by regaining geographic contact.

Numbers of Megasubspecies

In appendix 1 we have ranked all endemic subspecies in Northern Melanesia as either weak or strong, wherever possible as judged by the standard of differences among species within that same genus. The results are as follows.

Of Northern Melanesia's 191 zoogeographic species, 114 are represented in Northern Melanesia by endemic subspecies (the others have only nonendemic subspecies in Northern Melanesia or else are monotypic). Of those 114, we recognize 48 as containing endemic megasubspecies (that is, strong endemic subspecies), and 66 as containing only weak subspecies. By "megasubspecies," we mean either a single strong subspecies quite distinct from conspecific populations, or else a group of subspecies only weakly distinct from each other but collectively quite distinct from other conspecific populations or groups of conspecific populations (e.g., the two weak subspecies of the starling *Aplonis grandis*, *A.g. grandis* and *A.g. macrura*, discussed on p. 149).

Out of the 385 endemic subspecies in Northern Melanesia, we recognize 82 megasubspecies (21% of 385). This percentage is approximately the same for species from all three classes of vagility or all five classes of abundance. Most Northern Melanesian species with endemic megasubspecies have only one (26 species) or two (16 species) such megasubspecies. Two endemic species (the cuckoo-shrike *Coracina holopolia* [map 29] and the already-discussed flycatcher *Myiagra hebetior*) fall into three subspecies, all of which rate as megasubspecies. Four species that we rank as great speciators, by virtue of their large number of geographically distinct populations within Northern Melanesia (8–16 named populations in these cases), are represented by three, three, four, and eight megasubspecies (the owl *Ninox* [*novaeseelandiae*], the kingfisher *Ceyx lepidus*, the just-discussed whistler *Pachycephala pectoralis*, and the just-discussed flycatcher *Monarcha* [*manadensis*]), respectively.

To reiterate, the main conclusion that we draw from this artificial division of subspecies into strong and weak is that subspecies differ continuously in their distinctness because geographic differentiation proceeds gradually. Only when two distinct allopatric populations regain geographic contact does the moment of truth arrive, when it becomes clear whether the first of two discontinuities in the process of speciation has been attained. That discontinuity is reproductive isolation, corresponding to the transition from megasubspecies to allospecies.

Summary

The stages of allopatric speciation form a virtual continuum, from species exhibiting no geographic variation, through species with varying degrees of distinctness of allopatric populations, to cases of completed speciation. This virtual continuum is broken by only two discontinuities: the transition from subspecies to allospecies status (corresponding to the achievement of reproductive isolation), and the transition from allospecies to full species status (corresponding to the achievement of ecological niche differences and sympatry). Nevertheless, we arbitrarily divide the continuous spectrum of distinctness among Northern Melanesian subspecies into two sets: more distinct subspecies (megasubspecies) and less distinct subspecies. We do so to call attention to borderline cases of particular interest for further study by the student of speciation, and for reconsideration by those taxonomists who may prefer to rank some of our megasubspecies instead as allospecies. We have described in detail 10 sets of examples illustrating the wide range of geographic variation encompassed by Northern Melanesian subspecies, from trivial and/or inconsistent to striking and bordering on allospecies status. Among Northern Melanesia's 385 endemic subspecies, we recognize 82 as megasubspecies. Most Northern Melanesian species with endemic subspecies have only one or two such megasubspecies; the record is held by the *Monarcha* [*manadensis*] superspecies, with six endemic allospecies subdivided into eight megasubspecies and four weak subspecies.

21

Geographic Variation

Allospecies

Closely related allopatric populations are often so distinct that little doubt exists that they have passed from the subspecies to the species level. Such species are called "allospecies," and a set of allospecies is called a "superspecies." A superspecies is defined as a monophyletic group of closely related and largely or entirely allopatric species—that is, a group of taxa that have achieved reproductive isolation but little or no geographic overlap. Completely allopatric distributions suggest one of three possible interpretations: that colonists of one such taxon arrive too rarely at the geographic range of the other to become established; that the two taxa are still so similar ecologically that they cannot achieve sympatry by occupying distinct niches, and that arriving colonists are thus prevented by competitive exclusion from becoming established; or that arriving colonists hybridize with the resident population, and their genes thus become lost in the large gene pool of residents.

The decision of whether closely related allopatric taxa should be considered as allospecies, subspecies, or full species poses several practical difficulties. First, direct observation of whether two taxa are reproductively isolated and do not often hybridize with each other is possible only if the taxa are actually in contact (i.e., if their geographic ranges are abutting [parapatric] or actually overlap slightly [see below] or if their ranges are separated by a gap frequently bridged by dispersing individuals). In Northern Melanesia, however, most putative allospecies occupy different islands, and colonists of one taxon have not been observed arriving at the other taxon's island. Hence one can only attempt to infer, by comparison with the morphological differences between related species known to have achieved reproductive isolation, whether the taxa have achieved reproductive isolation (i.e., whether they should be ranked as subspecies or as allospecies). There can never be complete certainty about that inference, because morphological divergence and the acquisition of reproductive isolation are not necessarily strictly correlated with each other, as we have already illustrated by cases on the borderline between megasubspecies and allospecies, discussed in chapter 20.

A second practical difficulty is that, in cases involving sets of allopatric taxa distributed over much of the world, the uncertainty whether a Northern Melanesian taxon and a putative extralimital allospecies belong to the same superspecies inevitably increases with distance from Northern Melanesia and with the number of intervening taxa. For example, sets of ecologically and morphologically similar hawks, called goshawks (large species of genus *Accipiter*), sea eagles (genus *Haliaeetus*), and hobbies (small species of species *Falco*), are distributed around the world. Their Northern Melanesian representatives are *Accipiter meyerianus*, *Haliaeetus leucogaster* and *H. sanfordi*, and *Falco severus*, respectively. The geographically nearest representatives of these sets in Australia or Asia are clearly closely related to the Northern Melanesian representatives and belong in the same superspecies. In each case, though, there is some doubt whether the apparent representatives in the New World or Africa, which can obviously not be so closely related, also deserve to be grouped in the same superspecies or instead owe some of their similarity to convergent adaptations to a similar lifestyle.

Finally, in numerous cases involving a set of taxa all allopatric to each other except for some geographic overlap between two of the taxa, the question arises of how to treat such a set taxonomically (see chapter 16). Strictly speaking, the two taxa that have achieved partial sympatry violate the fundamental definition of allospecies and should not be grouped in the same superspecies. However, if one is forced to dismember a large superspecies merely because two of the taxa have achieved partial sympatry, it often becomes completely arbitrary which of the two sympatric taxa should be considered the local representative of the superspecies and which should be considered a separate full species. A useful procedure in such cases, introduced into ornithology by van Bemmel (1948) and followed by other authors such as Galbraith and Galbraith (1962) and Goodwin (1983), is to avoid dismembering the superspecies by considering it as being represented by a doublet (two taxa) in a small part of its range. It then becomes an arbitrary decision how much sympatry the taxonomist will tolerate before dismembering a superspecies containing a doublet. The two of us occasionally disagreed between ourselves about these decisions, one of us (E.M.) preferring not to combine into superspecies those taxa showing wide or (as in the case of doublets) complete overlap, while the other of us (J.D.) prefers to broaden somewhat the concept of superspecies and to express the obviously close relationship between the taxa by including them in the same superspecies (p. 124).

Allospeciation in Northern Melanesia

There is no other small area in the world in which geographic speciation has produced as many superspecies as Northern Melanesia. Many superspecies extend westward far beyond the Australian region, posing the above-mentioned problem of whether to group their distant apparent representatives in the same superspecies. Authors who in the course of comprehensive avifaunal studies have systematically grouped taxa or considered superspecies groupings at least within a single continent include Hall and Moreau (1970) for African birds, Harrison (1982) for Palaearctic birds, and Mayr and Short (1970) for North American birds. The list of birds of the world by Sibley and Monroe (1990) considers superspecies groupings on a worldwide basis. In our taxonomic revision of all Northern Melanesian species underlying our list in appendix 1, we have evaluated such worldwide superspecies groupings for all Northern Melanesian species. These groupings of allospecies of far-flung (sometimes even cosmopolitan) superspecies are in many cases tentative and will undoubtedly have to be modified in the future.

Among the 191 analyzed zoogeographic species that we recognize in Northern Melanesia, we consider 105 (55% of the total) to be superspecies and 86 to be isolated species not belonging to a superspecies. This figure of 55% is remarkably high in comparison with corresponding figures for other parts of the world. Exactly comparable tabulations are available only for North America (Mayr & Short 1970) and are summarized in table 21.1, which shows that proportionately twice as many Northern Melanesian species as North American species belong to a superspecies, either within Northern Melanesia or North America itself or else on a worldwide basis. This excess of allospeciation in Northern Melanesia compared to North America is all the more striking when one realizes that the land area and geographic distances within North America are far greater (area either 300 or 12 times greater, depending on whether one includes water area in the area of Northern Melanesia). Thus, the geographic variability for which Northern Melanesian birds are famous is expressed not only at the subspecies level but also at the allospecies level, presumably because of either or both of two factors: water barriers stop bird dispersal more effectively than do barriers on land, so that geographic variability is greater in insular than in continental environments; and bird dispersal is lower in the tropics than in the temperate zones.

Just as Northern Melanesian species vary greatly in their propensity to form subspecies, they also

Table 21.1. Comparison of superspecies in Northern Melanesia and North America.

	Northern Melanesia	North America
Land area (km^2)	85,000	24,300,000
Land + ocean area (km^2)	~2,100,000	24,300,000
Zoogeographic species	191	517
Superspecies		
On a worldwide basis	105 (55%)	127 (25%)
Within N Melanesia or N America	35 (18%)	44 (9%)

Entries are based on breeding land and freshwater bird species. "Superspecies on a worldwide basis" refers to that number, of the region's zoogeographic species, which belong to superspecies when taxa outside Northern Melanesia or North America are considered. "Superspecies within N Melanesia or N America" refers to that number of the region's superspecies represented by two or more allospecies within the region itself. Data for North America are from Mayr & Short (1970).

vary greatly in their propensity to form allospecies. As mentioned in the preceding paragraph, 86 of the 191 analyzed zoogeographic species have formed no allospecies at all (do not belong to a superspecies). The remaining 105 zoogeographic species that do belong to superspecies have a total of 422 allospecies worldwide, but the number of allospecies per superspecies varies greatly, from 2 in many cases, to 5–9 in 22 cases, 11 allospecies in the *Monarcha* [*manadensis*] and *Myiagra* [*rubecula*] superspecies, and the astonishing total of 16 allospecies in the *Ptilinopus* [*purpuratus*] superspecies. On the average, the number of allospecies per superspecies is slightly higher for passerines (4.5) than for nonpasserines (3.7), and higher for small-bodied nonpasserines (3.9; pigeons, parrots, kingfishers, etc.) than for large-bodied nonpasserines (3.1; waterbirds and hawks). Such differences are to be expected from the average higher abundance and lower vagility of passerines than nonpasserines, and of small, forest nonpasserines than large waterbirds and hawks. These correlations of high abundance and low vagility with allospeciation are similar to the correlations already discussed for subspeciation in the previous chapter.

The same correlations with abundance and vagility are evident for allospeciation within Northern Melanesia. Of the 35 superspecies represented by multiple allospecies within Northern Melanesia, all but four have only two or three allospecies, so that we cannot analyze variation in number of allospecies per zoogeographic species, as we previously analyzed variation in number of subspecies per zoogeographic species. However, we can still ask whether the presence or absence of multiple allospecies within Northern Melanesia is related to abundance and vagility. When we compare species in the five abundance classes previously defined, we find that the percentage of species in each class that are represented by multiple allospecies within Northern Melanesia increases from 13% for the rarest species (abundance class 1) to 33% for the most abundant species (abundance class 5). The correlation with vagility is even more striking: of the three vagility classes previously defined, 11% of Northern Melanesian species in the most vagile class (class A) are represented by multiple allospecies, 28% of species in the class of intermediate vagility (class B), and no species at all in the most sedentary class (class C). Thus, just as we previously noted for subspeciation within Northern Melanesia, and undoubtedly for the same reasons, allospeciation within Northern Melanesia is greatest for the most abundant species and for species of intermediate vagility. An analysis similar to that carried out previously for subspeciation (table 19.1) shows that these effects of abundance and vagility on allospeciation are independent of each other.

Of those 35 superspecies represented by two or more allospecies within Northern Melanesia, one, both, or all of the allospecies are endemic to Northern Melanesia in almost all of the superspecies. The only exceptions are two superspecies each represented by two nonendemic allospecies: the grebe

Tachybaptus [*ruficollis*], of which one allospecies invaded the Bismarcks from New Guinea, while another allospecies invaded the Solomons from Australia (map 1); and the pigeon *Macropygia* [*ruficeps*], which evidently reached Northern Melanesia by two successive invasions from New Guinea, of which the older invader (*M. mackinlayi*) now occurs mainly to the east in the Solomons and New Hebrides, while the more recent invader (*M. nigrirostris*) is confined within Northern Melanesia to the west in the Bismarcks and also occurs in New Guinea (map 12).

Of Northern Melanesia's 105 superspecies, 49 are represented within Northern Melanesia just by nonendemic allospecies (a single allospecies except in the two just-mentioned cases involving pairs of nonendemic allospecies). The other 56 superspecies are represented by at least one endemic allospecies, totaling 95 endemic allospecies, of which 38 occur just in the Bismarcks, 49 just in the Solomons, and 8 both in the Bismarcks and Solomons. Of those 95 endemic allospecies, 38 are nonpasserines, while 57 are passerines.

Of the 105 Northern Melanesian superspecies, most are confined to the Australian region, but 43 (51% of the nonpasserine superspecies, 25% of the passerine superspecies) extend west beyond Wallace's Line, and 21 even reach Europe, Africa, and/or the Americas. Of those 43 wide-ranging superspecies, only 9 are represented by an endemic allospecies within Northern Melanesia, and only 2 of those 9 belong to the superspecies that reach Europe, Africa, and/or the Americas. Thus, wide-ranging (hence presumably vagile) superspecies are unlikely to form endemic allospecies within Northern Melanesia.

In short, there are many expressions of the fact that geographic variation at the allospecies level, as well as at the subspecies level, is most marked for abundant species of intermediate vagility. Those expressions include the larger number of allospecies (on a worldwide basis) per superspecies for passerines than nonpasserines, and for small-bodied forest nonpasserines than for large waterbirds and hawks; the fact that the highest proportion of Northern Melanesia's zoogeographic species represented by multiple allospecies within Northern Melanesia are abundant species of intermediate vagility; the higher frequency of endemic allospecies in passerine than nonpasserine superspecies; the wider world ranges of nonpasserine than passerine superspecies; and the infrequency with which wide-ranging and presumably vagile superspecies form endemic allospecies within Northern Melanesia.

Snapshots of Colonization

We now divide Northern Melanesian superspecies into eight groups, each characterized by a particular geographic distribution within Northern Melanesia, breadth of distribution, and taxonomic differentiation. (We thus assign to groups only those 56 superspecies that have evolved at least one endemic allospecies in Northern Melanesia.) Most of these groups appear to correspond to successive snapshots in the colonization of Northern Melanesia. The eight groups are as follows: confined to the Bismarcks; confined to the Solomons; a single allospecies ranging throughout Northern Melanesia; separate wide-ranging allospecies in the Bismarcks and Solomons; widespread but highly differentiated superspecies; irregular distributional patterns, suggesting range retractions; superspecies composed of allospecies differing in their degree of differentiation within Northern Melanesia; and isolated relicts.

1. A Single Colonization of the Bismarcks

The single colonization category consists of 13 superspecies, most represented by a single allospecies (one represented by three allospecies) in the Bismarcks, a different allospecies in New Guinea, and absent from the Solomons. We infer that these species invaded the Bismarcks from New Guinea, differentiated there to the level of allospecies, and have so far failed to colonize the Solomons (unless a former Solomon population became extinct).

Five of these superspecies (the hawks *Henicopernis* [*longicauda*], *Accipiter* [*poliocephalus*] and *A.* [*cirrhocephalus*], the pigeon *Henicophaps* [*albifrons*] [map 14], and the honey-eater *Myzomela* [*eques*]) have an endemic allospecies confined to New Britain. Three more (the pigeons *Ducula* [*rufigaster*] and *D.* [*pinon*], and the wood-swallow *Artamus* [*maximus*]) are resident not only on New Britain but also on New Ireland, plus variously on Umboi and/or Dyaul. Two more (the parrot *Loriculus* [*aurantiifrons*] and the kingfisher *Alcedo* [*azureus*]) extend to New Hanover as well.

The finch *Lonchura* [*castaneothorax*] occurs on New Britain and also on the closest Solomon island, Buka. The Northern Melanesian populations of all 11 of these superspecies consist of one monotypic allospecies that has not differentiated taxonomically.

Finally, two further superspecies have still wider distributions within the Bismarcks and have differentiated taxonomically: the pigeon *Ptilinopus* [*hyogaster*] (map 6, plate 9), absent from the most outlying Bismarck group, the Admiralties, but otherwise widespread through the Bismarcks and with a distinct subspecies on the next most isolated Bismarck island, St. Matthias; and the honey-eater *Philemon* [*moluccensis*] (map 48, plate 7), with one endemic allospecies sharing New Britain and Umboi, another on New Ireland, and a third in the Admiralties.

The decreasing representation of these 13 superspecies as one proceeds from New Britain to Bismarck islands more remote from New Guinea may correspond to various stages in the spread of colonists from New Guinea, or else to the decreasing size of Bismarck islands from New Britain through New Ireland to New Hanover and St. Matthias, resulting in greater ecological opportunities on the larger nearer Bismarcks and in greater risk of extinction on the more remote smaller Bismarcks.

2. Colonization of the Solomons but Not of the Bismarcks

Twelve superspecies have developed endemic allospecies in the Solomons but do not occur in the Bismarcks. Six of these seem to have come from the east, as the most closely related allospecies occurs in the Santa Cruz group, New Hebrides, or Fiji: the rail *Nesoclopeus* [*poecilopterus*], the pigeons *Ptilinopus* [*purpuratus*] and *Ducula* [*latrans*] (plate 8), the flycatcher *Rhipidura* [*spilodera*] (map 38), the flycatcher *Clytorhynchus* [*nigrogularis*] (plate 9), and the white-eye *Woodfordia* [*superciliosa*] (plate 1) (the latter two confined in the Solomons to the outlying Rennell Island). Four others may have arrived directly from New Guinea or Australia, where the most closely related allospecies occurs: the parrot *Cacatua* [*sanguinea*] (plate 8), the pitta *Pitta* [*brachyura*], and the flycatchers *Monarcha* [*melanopsis*] (map 40, plate 5) and *Myiagra* [*rubecula*] (map 42, plate 5). The hawk *Accipiter* [*rufitorques*] (map 3) may have arrived either from New Guinea or the east, where related allospecies occur, while the starling *Aplonis* [*grandis*] (map 51) lacks close relatives outside Northern Melanesia and hence is of unknown derivation. Four of the superspecies have developed not just one but two or three endemic allospecies within the Solomons, and eight have differentiated into multiple subspecies within the Solomons.

3. Superspecies Represented by a Single Allospecies Ranging Throughout Northern Melanesia

The category of superspecies represented by a single allospecies consists of the pigeon *Ducula* [*myristicivora*] (map 9, plate 8) and parrot *Geoffroyus* [*heteroclitus*] (map 20, plate 9), absent only from the Admiralties and St. Matthias; and the pigeon *Ducula* [*rosacea*] (plate 8), which ranges throughout Northern Melanesia and also extends just outside Northern Melanesia to the Southeast Papuan Islands plus several islands off the northeast coast of New Guinea. None of these superspecies occurs east of the Solomons. All three are superior overwater colonists (vagility class A). The parrot and *Ducula* [*rosacea*] have formed only two weak subspecies within Northern Melanesia, while *Ducula* [*myristicivora*] has distinct megasubspecies in the Bismarcks and Solomons and thus forms a bridge to the category that we discuss next.

4. One Widespread Allospecies in the Bismarcks, Another in the Solomons

The fourth category consists of the pigeon *Reinwardtoena* [*reinwardtii*] (map 13, plate 7) and the parrot *Charmosyna* [*palmarum*] (map 17). The pigeon is widespread in the Bismarcks and occupies all of the larger Solomon islands, while the parrot occupies all mountainous Solomons islands exceeding 1,200 m in elevation and occupies the two largest Bismarck islands. These two superspecies may represent a later stage in differentiation of a widespread superspecies in Northern Melanesia, from the stage of a single, wide-ranging allospecies with only weak subspecies to a single allospecies divided into Bismarck and Solomon megasubspecies, as already illustrated in the preceding paragraph.

5. Widespread and Highly Differentiated Superspecies

The three superspecies that are widespread and highly differentiated (the owl *Ninox* [*novaeseelandiae*] [map 22, plate 7], flycatcher *Monarcha* [*manadensis*] [map 41, plate 3], and the honeyeater *Myzomela* [*lafargei*] [map 47, plate 4]) have differentiated still further to form four to six endemic allospecies in Northern Melanesia. Each of these superspecies has, in addition, formed several distinct megasubspecies on the borderline of allospecific status, plus further weak subspecies, within Northern Melanesia. These three superspecies represent the peak of geographic differentiation in Northern Melanesia. Members of the next group of superspecies to be discussed are also represented within Northern Melanesia by two or more endemic allospecies, but they differ in their more fragmented range within Northern Melanesia, suggesting that they have receded from their peak of distribution through extinction of some populations.

6. Highly Differentiated Superspecies with Fragmented Ranges

Fragmented ranges of well-differentiated, presumably old colonists suggest either failure to reach all parts of the archipelago in the first place, or else extinction of populations in some parts of the archipelago, whether due to subsequent arrival of a closely related, competing species or to other causes. Six superspecies, each represented within Northern Melanesia by at least two endemic allospecies, have such fragmented distributions.

The cuckoo *Centropus* [*ateralbus*] has a Bismarck allospecies confined to five of the larger islands, plus a Solomon allospecies with a restricted range (the New Georgia group, plus Guadalcanal and Florida, but absent from Malaita, the San Cristobal group, and Buka through Ysabel in the Bukida chain). The kingfisher superspecies *Halcyon* [*diops*] (map 23) ranges from the Moluccas through Australia and New Guinea to the New Hebrides, but its Bismarck allospecies is confined to New Britain, its Solomon allospecies to the Bukida chain. The great differentiation among the allospecies of this superspecies suggests that it expanded long ago and that its fragmented Northern Melanesian distribution reflects secondary extinction. The flycatcher allospecies *Rhipidura* [*rufidorsa*] (map 39) has one allospecies on the three large central Bismarck islands (New Britain, New Ireland, Umboi), another allospecies on the Bismarck outlier St. Matthias, and an extreme outlier in the Solomons in the mountains of Malaita. The flower-pecker *Dicaeum* [*erythrothorax*] (map 44) has differentiated into nine allospecies in its range from Celebes to the Solomons, with one allospecies widespread in the central Bismarcks (but absent from the Admiralties and St. Matthias), another allospecies in the Solomons on the Bukida chain and Malaita, the extremely aberrant *D. tristrami* on the southeasternmost Solomon island of San Cristobal, and no representative in the New Georgia group.

The white-eye superspecies *Zosterops* [*griseotinctus*] (map 45, plate 1) has a remarkable distribution: one allospecies on the outlier Rennell of the Solomons; another allospecies, *Z. griseotinctus*, confined in the Bismarcks to small islets or recently defaunated volcanoes and extending to some islets of the Southeast Papuan Islands; and, remarkably, four very distinct allospecies in the New Georgia group of the Solomons, one of them even confined to little Gizo Island which otherwise lacks even any endemic subspecies, and another allospecies (*Z. rendovae*) divided by a water gap of only 3.5 km between Rendova and Tetipari into two famous subspecies cited in many textbooks as examples of geographic differentiation (see plate 1). The absence of this superspecies from all large Bismarck islands and from all other Solomon islands may be due to the presence on those islands of many other ecologically similar but taxonomically not closely related species of white-eyes (*Zosterops* [*atriceps*], *Z. metcalfii*, *Z. ugiensis*, and *Z. stresemanni*).

Finally, the starling *Aplonis* [*feadensis*] (map 50) consists only of one allospecies on the Solomon outlier Rennell and another allospecies on nine remote islets scattered throughout the Bismarcks and Solomons. *Aplonis* [*feadensis*] is a derivative of the ubiquitous and successful *A. cantoroides*, sympatric with it on only three islands, and possibly eliminated elsewhere by competition with *A. cantoroides*.

7. Superspecies Represented by Multiple Allospecies of Unequal Levels of Endemism

Seventeen more superspecies occur as one or more allospecies endemic to Northern Melanesia, plus one nonendemic allospecies extending to New

Guinea or the New Hebrides. In some of these cases the most distinctive allospecies is the one most peripheral geographically, in keeping with the frequent pattern that peripheral subspecies are often also the most distinctive ones in many species. This distinctiveness of peripheral populations is to be expected from reduced gene flow and hence earlier achievement of reproductive isolation, plus low species diversity and hence greater opportunities for ecological divergence on peripheral islands. The most striking examples are the cuckoo-shrike *Coracina* [*tenuirostris*] (map 28, plate 7) and the warbler *Phylloscopus* [*trivirgatus*] (map 36), both with a distinctive endemic allospecies confined to San Cristobal (the most peripheral, large Solomon island) and with the New Guinea allospecies widely distributed throughout the remaining Bismarcks and Solomons. Nearly as striking is the pitta *Pitta* [*sordida*], with a remarkable endemic allospecies *P. superba* (plate 9) on Manus, the most remote, large Bismarck island, plus the non-endemic New Guinea allospecies *P. sordida* on the three Bismarck islands nearest New Guinea. Other examples include the eagle *Haliaeetus* [*leucogaster*], the pigeon *Gymnophaps* [*albertisii*] (map 10), and the fantail *Rhipidura* [*rufiventris*] (map 37), each with an endemic allospecies widespread in the Solomons and with the New Guinea allospecies widespread in the Bismarcks.

Differentiation has proceeded slightly further in the widespread parrot superspecies *Lorius* [*lory*] (map 16, plate 6), distributed in several allospecies from the Moluccas to the Solomons. The Southeast New Guinea allospecies *L. hypoinochrous* occupies the major Bismarck islands, while the endemic allospecies *L. chlorocercus* occupies the eastern Solomons, but in addition the endemic *L. albidinucha* is confined to the mountains of New Ireland, where *L. hypoinochrous* also occurs in the lowlands. Thus, this superspecies is represented by a doublet on New Ireland, presumably the result of *L. hypoinochrous* having invaded the range of the superspecies' representative *L. albidinucha* on New Ireland.

In most of these examples, the superspecies reached Northern Melanesia from New Guinea, and the Northern Melanesian islands nearest New Guinea still share the New Guinea allospecies, while an endemic species occupies the eastern part of Northern Melanesia most remote from New Guinea, the Solomons, or even the easternmost Solomon island of San Cristobal. In the warbler *Cichlornis* [*whitneyi*] (map 35, plate 9), the direction of colonization was reversed: it may have reached Northern Melanesia from the east (the New Hebrides), with the result that the nonendemic allospecies is on the eastern Solomon island of Guadalcanal, while the northwestern Solomon island of Bougainville and the Bismarck island of New Britain each have an endemic allospecies. That is, in this case the Bismarcks and western Solomons rank as peripheral, while in the previous cases the Solomons or eastern Solomons rank as peripheral.

Multiple waves of colonization may also be involved in these examples, wherein endemic and nonendemic allospecies divide up Northern Melanesia. For example, in the *Pitta* [*sordida*] superspecies, represented in Northern Melanesia by the endemic *P. superba* on the Bismarck outlier Manus plus the nonendemic *P. sordida* on Long, Tolokiwa, and Crown in the western Bismarcks, the latter three islands were colonized only within the last three centuries from New Guinea after the volcanic defaunation of Long.

Some other examples exemplifying multiple colonization waves are the following ones. The volant rail *Gallirallus* [*philippensis*] is widespread from the Philippines to Samoa but has formed old, flightless derivatives on numerous Pacific islands, including the endemic allospecies *G. rovianae* of the New Georgia group of the Solomons. The barn owl *Tyto* [*novaehollandiae*] (map 21) of Australia, South New Guinea, and the Moluccas has an undistinctive subspecies of the Australia/New Guinea allospecies on Manus, but a distinctive endemic allospecies (*T. aurantia*) on New Britain. Presumably the former is a much more recent colonist than the latter. Similarly, the fruit dove *Ptilinopus* [*purpuratus*] is represented in the Solomons by two allospecies: the New Hebrides allospecies *P. greyi* occurs without even subspecific differentiation on Gower Island in the Solomons, while the endemic allospecies *P. richardsii* occupies several islands in the southeastern Solomons. Presumably the latter is an older colonist, while the former is a recent invader that has barely established a beachhead on Gower.

The pigeon *Ptilinopus* [*rivoli*] (map 7, plate 6), thrush *Zoothera* [*dauma*] (map 32), and drongo *Dicrurus* [*hottentottus*] (map 52, plate 8) are represented on some Northern Melanesian islands by the New Guinea allospecies, on others by an endemic allospecies evidently derived from an older invasion wave. Especially instructive is the example of *Zoothera* [*dauma*], in which the New Guinea allospecies *Z. heinei* occupies Choiseul in the Central Solomons and the Bismarck outlier of St. Matthias as two weakly differentiated subspecies, while the distinctive allospecies *Z. margaretae* occupies two mountainous islands in the Southeast Solomons, and the distinctive *Z. talaseae* occupies three mountainous islands in the Bismarcks and Northwest Solomons. Thus, the ranges of the newer and older invaders *Z. heinei* and *Z. talaseae* are mutually exclusive but interdigitate in a geographically complex checkerboard. Further examples of endemic and nonendemic allospecies with complex distributions reflecting multiple colonization waves are the pigeon *Ptilinopus* [*rivoli*] (map 7), the parrot *Micropsitta* [*pusio*] (map 19), and the honeyeater *Myzomela* [*cardinalis*] (map 46).

Finally, the finch superspecies *Lonchura* [*spectabilis*] has two endemic allospecies in the Bismarcks, while the third allospecies (*L. spectabilis*) occurs on New Britain and four neighboring Bismarck islands plus adjacent areas of Northeast New Guinea. This is either the sole Northern Melanesian species, or else one of only two or three species (see chapter 13), for which the distributional evidence favors colonization of the New Guinea mainland from Northern Melanesia.

8. Relict Distributions

In a few cases the allospecies of a superspecies are separated from each other by wide geographical gaps and are absent from most intervening, ecologically apparently suitable islands. This situation may either arise from the chance element in long-distance colonization, or else from fragmentation of a formerly more continuous distribution by extinction. The latter interpretation is more plausible in the case of distinctive endemic taxa that have few or no close relatives, have presumably existed in Northern Melanesia for a long time and hence must have had ample opportunity to colonize other islands, and now appear to be completely sedentary by the various criteria described in chapter 11. The species discussed above under highly differentiated superspecies with fragmented ranges provide examples of such relict distributions. Two more extreme examples are worth mentioning here.

First, the flightless rail *Nesoclopeus* [*poecilopterus*] consists of two allospecies, one confined to the largest islands of the Bukida chain of the Solomons, the other confined to Fiji. The flightless rail genus *Pareudiastes* (plate 8) consists of two species (too different to be regarded as allospecies of the same superspecies) with an even more disjunct distribution: *P. sylvestris* on San Cristobal of the Solomons and *P. pacificus* of Samoa. These are surely relict distributions resulting from colonization of Pacific islands by a formerly volant ancestor, followed by evolution toward flightlessness and extinction of most of the colonist populations. This course of colonization, loss of flight, and extinction has been followed by several other Pacific rails and their flightless derivatives, including *Gallirallus* [*philippensis*], *Porzana tabuensis*, and *Porphyrio porphyrio* (Diamond 1991). Subfossil discoveries attest to the former existence of many other such flightless rails that disappeared following Polynesian or Melanesian settlement, resulting in the present relict distributions (Steadman 1995).

The second example is the warbler *Ortygocichla rubiginosa* (plate 9), confined to New Britain. Its nearest relative seems to be *O. rufa* of Fiji, which is sufficiently different that it is uncertain whether the two forms deserve to be grouped in the same superspecies. Other more distant relatives of this genus include the *Cichlornis* superspecies of the New Hebrides and Northern Melanesia, *Megalurulus mariei* and *M. bivittata* of New Caledonia and Timor, respectively, and possibly *Eremiornis carteri* of Australia. The distinctiveness of all these warblers from each other, the large distances between islands occupied by the most closely related species, and the absence of representative populations on most intervening islands exemplify a relict distribution.

Discussion

We draw two generalizations from the allospecies distributions that we have considered. First, there are many superspecies that have colonized all or most of Northern Melanesia and have since formed between one and six local allospecies. This pattern illustrates that dispersal is not necessarily a continuous, on-going process but often takes place in waves. Once such a wave has colonized an archipelago, the resulting populations may gradually lose the dispersal ability that spread them over the archipelago in the first instance. With loss of dispersal ability, they gradually form endemic allospecies. Patterns 3 through 5 described above illustrate stages in this progressive loss of dispersal ability, beginning with species that have occupied all of Northern Melanesia without even forming an endemic subspecies, leading eventually after subspeciation to a single, widespread endemic allospecies ranging through Northern Melanesia (pattern 3), then one widespread allospecies in the Bismarcks and another in the Solomons (pattern 4), and finally multiple endemic allospecies in each half of Northern Melanesia (pattern 5).

Many of the resulting endemic allospecies are now confined to single islands or to fragments of a single Pleistocene island, in some cases with nearby satellites. There are 31 such localized endemic allospecies in the Bismarcks and 32 in the Solomons. In the many cases where most suitable islands of Northern Melanesia are not occupied by any member of the superspecies, leaving many suitable but vacant islands, the localization of the endemic allospecies testifies to how sedentary they have become. For example, the large, mountainous, ecologically diverse Bismarck island of New Britain is separated by a narrow (30-km) channel from the large, mountainous, ecologically diverse island of New Ireland. Not surprisingly, New Britain shares many of its specialties with New Ireland, including the pigeons *Ducula finschii* and *D. melanochroa*, the parrot *Loriculus tener*, the cuckoos *Centropus ateralbus* and *C. violaceus*, the flycatcher *Rhipidura dahli*, and the wood-swallow *Artamus insignis*. Surprisingly, however, there are 10 other Bismarck endemic allospecies that are confined to New Britain (plus in two cases its neighbor Umboi, joined to New Britain in the Pleistocene) but that are absent from New Ireland, which was not joined to New Britain in the Pleistocene. All of these 10 allospecies are widespread in New Britain lowland or montane forest, and no ecological explanation is apparent for why they should not be equally able to occupy New Ireland. Whether or not they formerly occurred on New Ireland, their absence there now indicates their present inability to cross the narrow channel separating New Britain from New Ireland. These species are the hawks *Henicopernis infuscata* and *Accipiter princeps*, the rail *Gallirallus insignis* (map 4), the pigeon *Henicophaps foersteri*, the barn owl *Tyto aurantia*, the kingfisher *Halcyon albonotata*, the warblers *Ortygocichla rubiginosa* (map 4) and *Cichlornis grosvenori* (map 35), and the honeyeaters *Myzomela cineracea* and *Melidectes whitemanensis* (map 49). A reciprocal example is the parrot *Lorius albidinucha* (map 16), confined to New Ireland montane forest and absent from the more extensive areas of that habitat on New Britain.

These allospecies are virtually restricted to single islands, despite the nearby presence of ecologically suitable but vacant islands. Many other equally localized endemic allospecies, however, are replaced on neighboring islands by other endemic allospecies of the same superspecies. In these cases the localization of the taxa may be due not just to poor dispersal ability but also to competitive exclusion of any colonists by sister taxa on neighboring islands, or to absorption of arriving colonist individuals (by hybridization) into the large gene pool of the resident sister taxon. There are 32 superspecies represented by two or more allospecies within Northern Melanesia but without any sympatry between these allospecies. (This statement is not a circular result of the definition of allospecies, since we accept some sympatry between components of a superspecies before we would dismember a superspecies.) That is, the process of speciation often remains caught at the stage of reproductive isolation for a long time, without ecological differences sufficient for sympatry being achieved.

This outcome is especially likely in island archipelagoes, where dispersal involves occasional individuals or small flocks whose most likely fate on reaching an island already occupied by another allospecies is either death without mating or else loss of genetic distinctness through mating with the resident allospecies. The lower frequency of superspecies in North America than in Northern Melanesia (table 21.1) reflects the greater potential on continents for frequent dispersal setting up contact zones between former isolates. Such contact zones tend eventually to lead either to perfection of reproductive isolation, achievement of ecological segregation, and completion of speciation, or to breakdown of incipient reproductive isolation and termination of the speciation process. We cannot identify any clear examples of such contact zones in Northern Melanesia, in contrast to the many examples for the continents of North America (Mayr & Short 1970), South America (Haffer 1974), Australia (Keast 1961), Eurasia (Harrison 1982), and Africa (Hall & Moreau 1970). In the next chapter we discuss those cases in which members of a Northern Melanesian superspecies have nevertheless succeeded in achieving sympatry through ecological segregating mechanisms.

Summary

A superspecies is a monophyletic group of closely related, largely or entirely allopatric species termed "allospecies"—i.e., a group of taxa that have achieved reproductive isolation but little or no geographic overlap. Of Northern Melanesia's 191 analyzed zoogeographic species, 105 (i.e., 55%) are superspecies, a much higher percentage than for the North American avifauna. These 105 superspecies contain from 2 to a maximum of 16 allospecies. More than half of these 105 superspecies are represented in Northern Melanesia by at least one endemic allospecies. As true of subspeciation, allospeciation within Northern Melanesia is greatest for the most abundant species and for species of intermediate vagility. On the basis of geographic distribution and taxonomic differentiation, Northern Melanesian superspecies fall into eight groups that appear to correspond to successive snapshots in a history that includes the colonization of Northern Melanesia, achievement of widespread geographic distribution within Northern Melanesia, taxonomic differentiation within Northern Melanesia, and subsequent range retraction. Many superspecies have been involved in multiple waves of colonization. We cannot identify a single Northern Melanesian example of a phenomenon common on all the continents: contact zones between former isolates with limited hybridization.

22

Completed Speciation

The ultimate step in the process of geographic speciation consists of one allospecies invading the range of a sister allospecies. The result is geographical overlap (sympatry) of two species, both of which descended directly from the same ancestral species.

We can identify 19 Northern Melanesian species pairs that appear to exemplify stages in this process. Collectively, these and other species pairs constitute a series of snapshots permitting one to reconstruct the completion of speciation. The series begins with the 35 superspecies, already discussed in the preceding chapter, that are represented in Northern Melanesia by two or more morphologically similar allospecies with completely allopatric distributions. The next stage consists of pairs of two former allospecies whose ranges are still largely allopatric but which now overlap in a small zone of sympatry. In further stages, the zone of sympatry becomes extensive, although each species still retains some zone of allopatry. In the final stage, the geographic range of one species has been completely overrun by its sister species, so that the former no longer retains any zone of allopatry. This progression from allopatric to sympatric distributions is accompanied by evolution of progressively greater morphological differences between the sister species.

Completed speciation poses numerous questions that we address in the light of these 19 Northern Melanesian species pairs (cf. Lack 1944, 1945, 1947, 1949, 1971; Grant 1981, 1999; Bock 1986, 1995). How often does speciation involve an invasion from outside the archipelago, and how often does it instead involve range extensions of sister species both confined within the same archipelago, as has accounted for speciation of Darwin's finches in the Galapagos Archipelago? In the cases where speciation involves invasions from outside the archipelago, do any particular types of ecological niche relations tend to characterize older and younger invaders? What ecological niche differences permit the achievement of sympatry? As speciation proceeds, is there any particular sequence in which various types of niche differences tend to appear?

Our analysis to address these questions is similar to Diamond's (1986a) analysis addressing the same questions for completed speciation in montane birds of New Guinea. However, before addressing these general questions, we describe four of these 19 Northern Melanesian examples of speciation in progress, in order to give readers a feeling for the evidence provided by these snapshots.

Examples

Macropygia nigrirostris *and* M. mackinlayi

The pigeons *Macropygia nigrirostris* and *M. mackinlay* (map 12) belong to a superspecies whose only other allospecies occurs to the west of New Guinea. *M. nigrirostris* and *M. mackinlayi* are similar to each other in proporons and plumage and are of identical weight (average weight 87 g for each species), but they have different calls and are separated by constant plumage differences, and no hybrids are known. The range of *M. nigrirostris* is centered to the west; the range of *M. mackinlayi* is centered to the east. Specifically, *M. nigrirostris* occupies New Guinea and several large neighboring islands, with an additional subspecies on the major islands of the Bismarck Archipelago. *M. mackinlayi*'s center of distribution is to the east, in the Solomons and New Hebrides, where it occupies almost all large and many small islands, but it also occurs throughout the Bismarcks on outlying and smaller islands not occupied by *M. nigrirostris*. No island (except possibly Karkar) supports breeding populations of both species, but vagrants of *M. mackinlayi* frequently reach Bismarck islands with resident *M. nigrirostris* populations.

We infer that *M. nigrirostris* was originally the allospecies of New Guinea and the Bismarcks, that *M. mackinlayi* was the allospecies of the Solomons and New Hebrides, and that the present beginnings of sympatry were achieved by *M. mackinlayi*'s spreading westward from the Solomons into the Bismarcks, where it remains competitively excluded from islands occupied by the very similar *M. nigrirostris*.

Charmosyna [palmarum] *and* C. placentis

Related small lorikeets of genus *Charmosyna* (map 17) are distributed from the Moluccas across the Southwest Pacific to New Caledonia and Fiji. Among these lorikeets, allopatric populations that are especially similar in plumage and clearly belong to the same superspecies are *C. palmarum* of the New Hebrides, *C. diadema* of New Caledonia, *C. meeki* of the Solomons, *C. rubrigularis* of the Bismarcks, and *C. toxopei* of Buru in the Moluccas. Slightly less similar to those four taxa are two other species closely similar to each other, *C. rubronotata* of North New Guinea and *C. placentis*. *C. placentis* occurs from the Moluccas, through South New Guinea and parts of North New Guinea, to the Bismarcks and northwestern-most Solomon islands. This distribution brings it into sympatry with *C. rubronotata*, *C. rubrigularis*, *C. meeki*, and possibly *C. toxopei*.

We infer that *C. rubronotata* and *C. placentis* arose as the North and South New Guinea representatives, respectively, of the superspecies *C. [palmarum]*, and that *C. placentis* has now expanded to become sympatric with several other sister species. While *C. placentis* still occupies all of South New Guinea alone in allopatry, it has overrun most of the North New Guinea range of *C. rubronotata*, and it has also overrun most of the Bismarck range of *C. rubrigularis*, which now shares two of its three islands with *C. placentis* (*C. rubrigularis* still lives alone in allopatry on Karkar Island). However, most of the range of *C. meeki* in the Solomons has not yet been reached by *C. placentis*, which is sympatric with *C. meeki* only on the northwestern-most large Solomon island of Bougainville.

Where *C. placentis* occurs sympatrically with *C. rubronotata*, *C. rubrigularis*, or *C. meeki*, the two species segregate altitudinally, with *C. placentis* in the lowlands, the other species at high elevations. In the absence of its montane congeners (for example, on Umboi and Long Island), *C. placentis* occurs up to elevations of nearly 1000 m, considerably higher than in the presence of its congeners. Similarly, in the absence of *C. placentis*, *C. rubrigularis* and *C. rubronotata* extend down to sea level on Karkar Island and certain North New Guinea mountains, respectively, but both species are mainly confined to the mountains in the presence of *C. placentis*. Thus, the altitudinal niches of *C. placentis*, *C. rubrigularis*, and *C. rubronotata* are compressed competitively in the presence of a congener. However, *C. meeki*, which segregates ecologically from *C. placentis* on Bougainville by being confined to the mountains, remains confined to the mountains on other Solomon islands lacking *C. placentis*, so that *C. meeki* has become an obligate montane species that is unable to expand into the lowlands, even in the absence of its lowland competitor.

This case in the genus *Charmosyna* represents a stage of speciation slightly more advanced than the case of *Macropygia mackinlayi* and *M. nigrirostris*, in that the morphological differences between *C. placentis* and its sympatric congeners are slightly greater than those between the two *Macropygia* species, and in that *C. placentis* has now overrun most (but not all) of the ranges of *C. rubronotata* and *C. rubrigularis*.

Lorius albidinucha and *L. hypoinochrous*

The large lories *Lorius albidinucha* and *L. hypoinochrous* (map 16, plate 6) belong to a widespread superspecies distributed as seven allospecies from the Moluccas to the Solomons. The center of diversity of *L. hypoinochrous* is in the Southeast Papuan Islands and Southeast New Guinea, where it occurs as three subspecies, of which one extends over much of the Bismarck Archipelago. *L. albidinucha* is confined to the Bismarck island of New Ireland, which it shares with *L. hypoinochrous*. We infer that *L. hypoinochrous* recently expanded from Southeast New Guinea and the Southeast Papuan Islands over the Bismarcks, while *L. albidinucha* is an older resident of the Bismarcks (or at least of New Ireland).

On New Ireland the two species segregate altitudinally, with *L. albidinucha* at higher elevations and *L. hypoinochrous* at lower elevations (up to 750 m). In the absence of *L. albidinucha* (for example, on Umboi and New Britain), *L. hypoinochrous* occurs up to 1200 m. Thus, the altitudinal range of *L. hypoinochrous* is compressed in the presence of its montane competitor, but one cannot determine whether *L. albidinucha* retains its ability to occupy the lowlands and is similarly compressed by *L. hypoinochrous* because the sole island occupied by *L. albidinucha* is shared with *L. hypoinochrous*.

This case represents a more advanced stage of speciation than the previous two cases, in that the entire geographic range of *L. albidinucha* has now been overrun by its sister species.

Rhipidura tenebrosa and *R. fuliginosa*

The fantail-flycatcher *Rhipidura fuliginosa* is distributed from Australia through New Zealand to New Caledonia and the New Hebrides. The sole Northern Melanesian population is in the mountains of the Solomon island nearest the New Hebrides, San Cristobal, and may belong to the same subspecies as the New Hebridean population. A related superspecies, *R.* [*spilodera*], is distributed as five allospecies from Northern Melanesia through the New Hebrides and New Caledonia to Fiji and Samoa. Two Northern Melanesian allospecies live in the mountains of Bougainville and Guadalcanal, beyond the range of *R. fuliginosa*, while the third Northern Melanesian allospecies, *R. tenebrosa* (map 38), is confined to San Cristobal. Thus, *R. tenebrosa* shares its whole range (a single island) with *R. fuliginosa*, but that island constitutes only a tiny fraction of the range of *R. fuliginosa*, although the latter shares the New Hebrides and New Caledonia with another allospecies (*R. spilodera*) of the superspecies to which *R. tenebrosa* belongs. *R. fuliginosa* is presumably a recent invader of San Cristobal (since it may not even be subspecifically distinct), while *R. tenebrosa* is a much older derivative of an earlier expansion wave of the same stock.

R. tenebrosa is the most distinctive allospecies of the *R.* [*spilodera*] superspecies. Thus, the relationship between *R. tenebrosa* and *R. fuliginosa* is more distant than in the three cases of completed speciation discussed above, since *R. tenebrosa* and *R. fuliginosa* do not belong to the same superspecies, and *R. tenebrosa* is fairly distinct even within its own superspecies. In particular, *R. tenebrosa* is 2.4 times heavier than *R. fuliginosa*. Ecological segregation between *R. tenebrosa* and *R. fuliginosa* is partly spatial, since both species occur together at high elevations on San Cristobal but only *R. tenebrosa* occurs in the lowlands and lower hill slopes. In their zone of altitudinal overlap the two species segregate by foraging technique, *R. tenebrosa* being much more sluggish in its movements, in keeping with its considerably larger size.

Data on All Cases of Completed Speciation

For each species pair involved in the 19 cases of completed speciation in Northern Melanesia, we summarize in table 22.1 which of the two species is the older invader of Northern Melanesia (listed first in table 22.1), and which is the younger invader (listed second in table 22.1); their taxonomic closeness, graded from 1 (very close) to 3 (less close); their geographic overlap, graded from 2 (largely allopatric, only marginal sympatry) to 4 (one species occupies the entire geographic range of the second); their mean body weights, and the ratio of their weights, as an indicator of ecological segregation related to body size; the types of niche differences that permit the species to occur in sympatry; whether or not either of the two species shifts

Table 22.1. Cases of completed speciation in Northern Melanesian birds.

Species pair	Taxonomic closeness[a]	Geographic overlap[b]	Weight (g)[c]	Niche differences in sympatry[d]	Niche shifts[e]	Niche differences in allopatry[f]
Accipiter imitator/A. albogularis	3	4	♀ 212/393 = 0.54	S;F?D?	?/?	
Accipiter luteoschistaceus/A. novaehollandiae	3	4	259/253 = 1.02	H	?/—	
Ptilinopus solomonensis/P. rivoli	1	3	97/135 = 0.72	I	I/I	—
Macropygia mackinlayi/M. nigrirostris	1	2	87/87 = 1.00	I	I/—	I
Columba pallidiceps/C. vitiensis	3	3	459/466 = 0.98	V	?/?	
Gallicolumba salamonis/G. beccarii	3	4	(2.9)	S;D?	?/?	
Lorius albidinucha/L. hypoinochrous	1	4	♀ 132/211 = 0.63	A	?/A	
{ *Charmosyna rubrigularis/C. placentis*	2	3	36/31 = 1.16	A	A/A	A
{ *C. meeki/C. placentis*	2	2	25/31 = 0.81	A	—/?	A }
Halcyon saurophaga/H. chloris	2	3	119/76 = 1.57	H	-/?	}
Phylloscopus amoenus/P.poliocephala	2	4	(1.17)	pA;F	?/?	
Rhipidura tenebrosa/R. fuliginosa	3	4	17.2/7.3 = 2.4	pA,F,S	?/?	
Rhipidura malaitae/R. rufifrons	2	4	12.0/13.3 = 0.90	pA;F?	?/?	
Monarcha [melanopsis]/M. cinerascens	1	3	25/23 = 1.09	I	?/I	—
Myiagra hebetior/M. alecto	1	3	19/24 = 0.79	pA,H	?/A,H	
Pachycephala pectoralis/P.melanura	1	3	26/27 = 0.96	I	—/I	I
Pachycephala implicata/P/pectoralis	2	4	35/47 = 0.74	A	?/A	
Zosterops murphyi/Z. rendovae	1	4	(1.24)	A	?/?	
Aplonis [feadensis]/A. cantoroides	1	3	72/53 = 1.36	I,H	?/?	
Aplonis brunneicapilla/A. metallica	1	4	67/57 = 1.18	A	?/—	

This table characterizes all 19 species pairs that completed speciation within Northern Melanesia—i.e., pairs of closely related species that met in Northern Melanesia and achieved partly or wholly sympatric ranges there. The interactions of *Charmosyna placentis* with *C. rubrigularis* and *C. meeki*, which are, respectively, the Bismarck and Solomon allospecies of *C. palmarum*, are tabulated separately. Various attributes are coded according to abbreviations used by Diamond (1986a) in a similar analysis of New Guinea's montane avifauna. The first-listed species and second-listed species of each pair are inferred to be the older and younger invader, respectively, of Northern Melanesia (or of the zone of sympatry, in the case of *Zosterops murphyi/Z. rendovae*).

[a]Closeness of taxonomic relation, assessed by morphological criteria: 1 = divergence as for allospecies in that same genus, and little greater than for megasubspecies of those same species; 2 = less close, 3 = still less close.

[b]Degree of geographical overlap between the two species: 2 = marginal sympatry (zone of sympatry less than one-quarter of the whole geographical range of either species); 3 = extensive sympatry (zone of sympatry more than one-quarter of the whole geographical range for at least one of the species, but each species still has a zone of allopatry); 4 = one species shares its whole range with its sister species, but the latter still has a zone of allopatry.

[c]Average weight of the first species, followed by that of the second species, in the zone of sympatry, followed after an equal sign by the ratio of weights. Average male weight and average female weight were combined to obtain an average weight for each species ("♀" indicates that only weights of females were available for comparison). In a few cases where weights were unavailable, the value in parentheses is a weight ratio estimated as the cube of the ratio of wing lengths.

[d]Niche differences in the zone of sympatry: A = non-overlapping altitudinal ranges; pA = only partly overlapping altitudinal ranges (in the case of lories and white-eyes, which are socially itinerant frugivores and nectarivores, broader altitudinal overlap was required for a pA designation than in the case of sedentary nonflocking species); H = differing habitats (e.g., forest vs. secondary growth). V = differing vertical strata (e.g., canopy vs. ground); I = one species occupies small or remote islands, the other occupies large central islands; F = conspicuously differing foraging techniques for capturing prey; D = differing diets; S = ecologically significant size difference (body weight ratio >1.75 or <0.57).

[e]Whether the niche is compressed within the zone of sympatry, due to competition, for the first and second species, respectively. I, A, H = the species narrows its range of island types occupied, its altitudinal range, or its range of habitats, in the zone of sympatry compared to the zone of allopatry; — = the species does not narrow its niche; ? = it is not known whether the species narrows its niche. See text for examples.

[f]For species pairs at an early stage of speciation, such that each species still occurs partly in allopatry, do the two allopatric populations differ conspicuously in any of the niche parameters by which they differ in sympatry? A, I = yes, with respect to altitudinal range or to range of island types occupied; — = no; no entry = not known. See text for examples.

its niche in these respects between areas of sympatry and allopatry; and whether or not allopatric populations of the two species exhibit the same niche differences as do the sympatric populations. At the end of this chapter we provide narrative details for the remaining 15 species pairs, similar to the narrative details just provided for four of these species pairs. Below, we use this database to draw a series of conclusions about completed speciation.

Single or Multiple Achievements of Sympatry?

Of these 19 species pairs, four not only met to achieve sympatry within Northern Melanesia, but also did so independently outside Northern Melanesia. In three cases the meeting outside Northern Melanesia involved a different allospecies of the same superspecies: *Rhipidura rufifrons* meeting the allospecies *R.* [*rufidorsa*] *rufidorsa* in South New Guinea, as well as meeting *R.* [*rufidorsa*] *malaitae* on Malaita (map 39); *R. fuliginosa* meeting the allospecies *R.* [*spilodera*] *spilodera* in the New Hebrides and New Caledonia, as well as meeting *R.* [*spilodera*] *tenebrosa* on San Cristobal; and *Monarcha cinerascens* meeting the allospecies *M.* [*melanopsis*] *frater* in the New Guinea region, as well as the allospecies *M.* [*melanopsis*] *erythrostictus*, *M.* [*melanopsis*] *castaneiventris*, and *M.* [*melanopsis*] *richardsii* in the Solomons (map 40). In the remaining case, the same species or allospecies met outside Northern Melanesia as inside it: the whistlers *Pachycephala melanura* and *P.* [*pectoralis*] *pectoralis* met on the coast of Queensland and in the Southeast Papuan Islands as well as in the Bismarcks and northwestern Solomons (map 43).

Concordance Between Taxonomic Closeness and Geographic Overlap

If morphological divergence and achievement of sympatry proceeded at the same rate in all species involved in completed speciation, one would observe perfect concordance between our measures of taxonomic closeness and geographic overlap. That is, the species pairs most similar to each other morphologically would be largely allopatric, while more and more divergent species pairs would exhibit progressively greater degrees of sympatry.

Table 22.2. Relation between taxonomic similarity and degree of sympatry during speciation.

Taxonomic closeness	Degree of sympatry	No. of species pairs
1	2	1
	3	5
	4	3
2	2	$^1/_2$
	3	$1^1/_2$
3	2	0
	3	1
	4	4

In Table 22.1 all 19 Northern Melanesian species pairs involved in completed speciation were categorized according to their taxonomic closeness (1 = the two species are very similar, 2 = less close, 3 = still less close) and geographic overlap (2 = only marginal sympatry, 3 = extensive sympatry, 4 = whole range of one species is shared with its sister species). This table shows how the 19 species pairs are distributed over values of these two indices. The "$^1/_2$ species" values under "no. of species pairs" refer to the species pair including the superspecies *Charmosyna* [*palmarum*], whose two Northern Melanesian allospecies have different values of the index "degree of sympatry," and each of which is thus treated as a half-case. Statistical analysis shows that taxonomically less close species do tend to have more extensive sympatry, but the sample size is small, and the trend falls short of statistical significance (p = .15 and .16 for Pearson and Spearman correlations of .38 and .39, respectively). See text for discussion.

Table 22.2 illustrates that this expectation does apply, but only as a weak trend. Even among species pairs whose degree of morphological divergence we judge to be similar, there are differences in the degree of sympatry achieved. This should come as no surprise, since some clades are morphologically much more variable than are others, and since we do not have a good quantitative measure of degree of morphological divergence.

Older and Younger Invaders

In almost every case there is no difficulty in identifying which of the two sympatric species reached the present zone of sympatry more recently. The more recent arrival usually has a much wider range outside the zone of sympatry, and outside Northern Melanesia, than does the older resident. In addition, the Northern Melanesian population of the recent arrival is often not even subspecifically distinct, or else it is an endemic subspecies but one ex-

Table 22.3. Endemism level of newer and older invaders.

	Nonendemic	Endemic weak subspecies	Endemic megasub-species	Endemic allospecies	Endemic full species
Newer invader	6	7	4	1	0
Older invader	0	1	2	5	10

Of the 19 Northern Melanesian species pairs involved in completed speciation, 18 (all except *Zosterops murphyi/Z. rendovae*) are thought to have arisen by double invasion of Northern Melanesian islands from sources outside Northern Melanesia. Each species pair appears to consist of an older invader and a more recent invader, distinguished by criteria discussed in the text. This table summarizes the level of endemism of the 18 older and newer invaders. Note that the older invaders are mostly endemic to Northern Melanesia at the level of full species or allospecies, while the Northern Melanesian populations of most of the recent invaders are mostly only weakly distinct endemic subspecies or are not endemic at all (they belong to the same subspecies as extralimital populations). There is a high correlation between level of endemism and percentage of older invaders (Pearson and Spearman correlations .80 and .79, respectively; $p < .001$).

tending outside the area of sympatry (table 22.3). In contrast, most of the older invaders are endemic to Northern Melanesia at the level of allospecies or full species, and most of them now share their entire geographic range with the recent invader (table 22.3).

Mechanisms of Allopatric Speciation

In terrestrial organisms, three modes of allopatric speciation can be distinguished, depending on whether the isolating geographic barrier is within a single land mass ("continental speciation"), is a water gap between islands of the same archipelago ("intra-archipelagal speciation") or is a water gap between different archipelagos ("interarchipelagal speciation" or "double invasion"; see fig. 22.1; see Diamond 1977 for further details). The first of these mechanisms is the main mode of speciation on large land masses such as New Guinea and the continents. The second is the main mode on remote archipelagos, such as the Galapagos and Hawaiian archipelagoes. The third mode proves to be the one that has produced most of the sympatric bird species pairs in the Southwest Pacific from the Bismarcks to Samoa.

If continental speciation were significant for Northern Melanesian birds, we would find cases of subspeciation and allospeciation within the same large island, and cases of related species living allopatrically at opposite ends of the same islands but sympatrically in the middle of that island. In reality, there are only two possible examples among Northern Melanesian birds, both on the island of New Ireland, which is 340 km long. Two species are known to be differentiated between northern and southern New Ireland: the kingfisher weak subspecies *Halcyon chloris nusae* in the north versus *H. c. novaehiberniae* in the south, and the grassfinch allospecies *Lonchura [spectabilis] hunsteini* in the north versus *L. [spectabilis] forbesi* in the south. In each case, though, the North New Ireland population also occurs on nearby New Hanover, now separated from North New Ireland by a shallow water gap that arose at the end of the Pleistocene. Hence it is not certain that the observed differentiation actually arose within modern New Ireland; perhaps instead a primary isolate from a New Ireland source arose on New Hanover and then recolonized North New Ireland after a temporary extinction of the North New Ireland population.

The only Northern Melanesian island with longer intra-island distances than New Ireland is New Britain (470 km long). However, we do not yet know whether West and East New Britain bird populations have become differentiated, because bird specimens from the west end of New Britain have not yet been identified subspecifically. Other than those two New Ireland examples, there is no known instance—not even on the large Solomon islands of Bougainville and Guadalcanal—of different parts of the same island being occupied by different subspecies or different allospecies. The reason is surely that Northern Melanesian islands are all just too small, and birds too vagile, for geographic differentiation to take place within a sin-

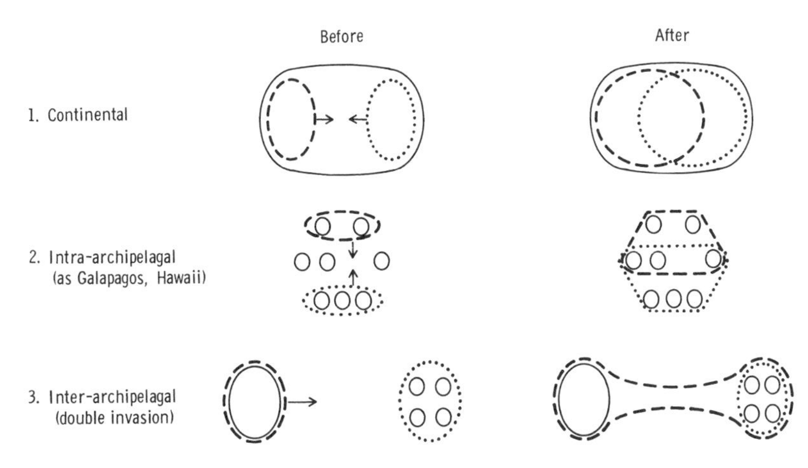

Fig. 22.1. Three alternative speciation mechanisms, which differ as to whether the barrier permitting divergence of two initially conspecific populations lies within a single land mass (case 1), or is a water gap between islands of the same archipelago (case 2), or is a water gap between two archipelagoes (case 3). In each case the sketch on the left illustrates distributions of two related (initially conspecific) taxa before an expansion of one or both taxa resulted in sympatry; the sketch on the right illustrates distributions after the achievement of sympatry. Island borders are shown by solid lines, the range of one taxon by dashes, the range of the other taxon by dots, and arrows indicate directions of expansion. In case 1, continental speciation, the sister populations were originally subspecies occupying different parts of the same land mass. In case 2, intra-archipelagal speciation, the sister populations originally occupied different islands of the same archipelago (e.g., the Galapagos), before expanding to share some islands. In case 3, interarchipelagal speciation or double invasion, the sister populations originally occupied different archipelagoes, until one population invaded the other's archipelago. From Diamond (1977).

gle land mass, as has happened so often on the much larger island of New Guinea and on the major continents.

Intra-archipelagal speciation is the mechanism for which birds of the Galapagos and Hawaiian archipelagoes are famous in evolutionary biology, and which is responsible for all cases of completed speciation among birds of these archipelagoes. In the tropical Southwest Pacific there are a few possible extant examples among species pairs of the more remote archipelagoes (the Marquesas, Societies, and Fiji group), plus other examples involving extinct taxa. We can detect only one possible example in Northern Melanesia, the case of the white-eyes *Zosterops murphyi* and *Z. rendovae* (map 45, plate 1). The former is confined to the mountains of Kulambangra, while the latter occupies the lowlands of Kulambangra plus seven other islands of the New Georgia group. The Kulambangra subspecies of *Z. rendovae* occurs on five of these seven islands. *Z. rendovae* in turn belongs to a superspecies with three other allospecies on other islands of the New Georgia group, one allospecies on Rennell, and the remaining allospecies (*Z. griseotinctus*) on some outlying Bismarck islands plus the Southeast Papuan Islands. *Z. murphyi* is closely related to this *Z.* [*griseotinctus*] superspecies and is apparently closest to *Z. griseotinctus* itself (Mees 1961). This suggests that *Z. murphyi* is a *Z. griseotinctus* derivative that was the older invader of the New Georgia group or of Kulambangra, while *Z. rendovae* is a more recent invader of the New Georgia group and has only recently extended its range to Kulambangra to become sympatric with *Z. murphyi*. This could easily have happened at Pleistocene times of low sea level, because Kulambangra was then connected to New Georgia and the other islands now occupied by the Kulambangra subspecies of *Z. rendovae*. On Kulambangra the two species segregate altitudinally, *Z. murphyi* occurring mainly from 1000 m up to the summit, *Z. rendovae* from 1000 m down to sea level.

All other cases of completed speciation in Northern Melanesian birds appear to involve interarchipelagal speciation (i.e., double invasions of Northern Melanesia by the same stock). We reach this conclusion because, in 17 of these sympatric species pairs, the inferred more recent invader has an extensive range outside of Northern Melanesia, with dozens of extralimital subspecies in some cases. In the 18th case, the inferred more recent invader, *Accipiter albogularis*, is an allospecies virtually endemic to the Solomons, but closely related allospecies occur outside Northern Melanesia in New Caledonia, Fiji, and New Guinea (map 3). In many of these species pairs, the derivation of the inferred older invader from outside Northern Melanesia is also clear, because the Northern Melanesian population of the inferred older invader is an endemic subspecies or allospecies of a wide-ranging extralimital allospecies or superspecies, respectively. Thus, we conclude that there has been no intra-archipelagal speciation within Northern Melanesia, except perhaps in the case of *Zosterops rendovae* and *Z. murphyi*.

The reason for the predominance of interarchipelagal over intra-archipelagal speciation among Northern Melanesian birds is surely that Northern Melanesia is accessible to the nearby rich avifaunas of New Guinea and Australia. Thus, available niches are generally preempted by colonists from outside the archipelago. Only the most remote Pacific archipelagos (Hawaii and the Galapagos) are so inaccessible to bird colonists from outside the archipelago that geographic isolation within the archipelago itself can make the major contribution to speciation.

Of the species inferred to be the recent invader in the 18 cases of interarchipelagal speciation, most (14) came from the New Guinea region, either from New Guinea itself or from the Southeast Papuan Islands. Clear examples are *Lorius hypoinochrous*, *Macropygia nigrirostris*, *Gallicolumba beccarii*, *Charmosyna placentis*, and *Phylloscopus poliocephala*. Only in one case, that of *Rhipidura fuliginosa*, did the recent invader clearly come from the east (the New Hebrides), and even in that case the New Hebrides population ultimately came from Australia. In the remaining three cases (*Accipiter albogularis*, *Halcyon chloris*, and *Pachycephala pectoralis*), either the immediate origin of the recent invader is equivocal, or else both New Guinea and the New Hebrides contributed colonists to different parts of Northern Melanesia.

Modes of Ecological Segregation

How did the 19 species pairs involved in completed speciation achieve ecological segregation in the zone of sympatry? The niche differences involved are tabulated in column 5 of table 22.1. In table 22.4 we use morphological criteria and degree of sympatry to categorize the species pairs according to stage of speciation, and we then tabulate the mode of ecological segregation for each of the three resulting stages of speciation.

In the earliest stage, involving sister species that are morphologically still very similar and/or still have largely allopatric distributions (11 cases), ecological segregation is strictly spatial. Four of these species pairs (*Ptilinopus solomonensis*/*P. rivoli*, *Macropygia mackinlayi*/*M. nigrirostris*, *Monarcha* [*melanopsis*]/*M. cinerascens*, and *Pachycephala pectoralis*/*P. melanura*) segregate by type of island, with one species occupying small or remote islands or ones recently defaunated by volcanic explosion, and the other species occupying large, central, old islands. In four additional cases (*Charmosyna* [*palmarum*]/*C. placentis*, *Lorius albidinucha*/*L. hypoinochrous*, *Zosterops murphyi*/*Z. rendovae*, and possibly *Aplonis brunneicapilla*/*A. metallica*) segregation is altitudinal, with one species living at high altitudes and the other at low altitudes on the same island. In one further case (*Aplonis* [*feadensis*]/*A. cantoroides*) there is partial segregation by type of island, combined with segregation by habitat within a few islands supporting both species. For the species pair *Myiagra hebetior*/*M. alecto*, spatial segregation is by a combination of altitude and habitat type, the latter species living in the coastal lowlands and open or second-growth habitats, the former species in the thickets and mature forest of the island interior from the lowlands extending up into the mountains. Finally, of the kingfishers *Halcyon saurophaga* and *H. chloris*, the former lives on the sea coast and the latter mainly behind the coast.

Four species pairs exemplify a middle stage of speciation (moderate taxonomic divergence but complete sympatry, or else greater taxonomic divergence but incomplete sympatry). Only one of these four pairs exhibits complete spatial segregation: the whistlers *Pachycephala implicata* and *P.*

Table 22.4. Relation between stage of speciation and mode of ecological segregation.

Stage of speciation	Mode of ecological segregation		
	Spatial	Spatial + nonspatial	Nonspatial
Early	4 × I 4 × A 1 × H 1 × I,H 1 × pA,H	—	—
Middle	1 × A	2 × pA,F?	1 × V
Late	1 × H	1 × pA,F,S	1 × S,F?,D? 1 × S,D?

The mode of ecological segregation in the zone of sympatry is tabulated for species pairs in the process of achieving completed speciation, as a function of the stage of speciation. Data are from table 22.1. Early stage = species pairs that are still very close taxonomically (closeness index 1 in column 2 of Table 22.1), or else fairly close (index 2) but still with incomplete sympatry (geographic overlap index 2 or 3 in column 3 of Table 22.1). Middle stage = fairly close taxonomically (index 2) but already completely sympatric (geographic overlap index 4), or else closeness and overlap indices 3. Late stage = more distant taxonomically (closeness index 3) and completely sympatric (geographic overlap index 4). Modes of ecological segregation are A = non-overlapping altitudinal ranges; pA = only partly overlapping altitudinal ranges; H = differing habitats; V = differing vertical strata; I = one species occupies small or remote islands, the other occupies large central islands; F = conspicuously differing foraging techniques; D = differing diets; S = ecologically significant size difference. Modes are classified as either spatial (the two species occupy different altitudes, habitats, or island types), nonspatial (the two species occupy the same horizontal space but use different foraging techniques, diets, vertical strata, or size-related traits), or a combination of spatial and nonspatial. Numbers are the number of species pairs segregating by each mode. (For example, at the earliest stage of speciation 10 species pairs segregate spatially: four of them by island type, four by altitude, and so on). Note that segregation is strictly spatial at the earliest stage, and that nonspatial segregation becomes more important and spatial segregation less important at later stages of speciation.

pectoralis, segregating altitudinally. Two other pairs (the warblers *Phylloscopus amoenus*/*P. poliocephala* and the flycatchers *Rhipidura malaitae*/*R. rufifrons*) segregate partly by altitude, partly by differences in foraging technique. The pigeons *Columba pallidiceps* and *C. vitiensis* overlap extensively in altitude and habitat preference, but the former is believed to forage extensively on the ground, the latter mainly in trees.

The last stage of speciation is represented by four species pairs that have not only diverged considerably taxonomically but also now exhibit complete sympatry. Only one of these pairs segregates spatially: the hawk *Accipiter luteoschistaceus* lives in forest, its sister species *A. novaehollandiae* lives mainly in open habitats and at the forest edge. Members of the remaining three species pairs (the hawks *Accipiter imitator* and *A. albogularis*, the ground doves *Gallicolumba salamonis* and *G. beccarii*, and the flycatchers *Rhipidura tenebrosa* and *R. fuliginosa*), in contrast to all other 16 species pairs discussed previously, differ greatly in body size: one member of the pair weighs twice or more the weight of the other member. In general, body size has diverse ecological consequences, such as determining size of prey that can be captured, strength of perch necessary for support, density of habitat compatible with maneuvering, concentration and nutritional quality of foods that can be harvested economically, metabolic requirements, home range size, antipredator strategy, and position in a dominance hierarchy (Terborgh 1983). Body-size differences make it easy for related species to coexist in the same microhabitat. Few details are available for the particular three species pairs involved, since they include some of the rarest or least known birds of Northern Melanesia, and since one of them (*Gallicolumba salamonis*) has never been observed in life and may now be extinct. However, the available information suffices to indicate that at least two of these three species pairs (the pair of hawks and the pair of flycatchers) do overlap extensively in habitat, and that the members of at least one of them (the pair of flycatchers) do exhibit differences in foraging technique related to difference in body size.

Thus, when closely related Northern Melanesian allospecies first begin to achieve sympatry, they usually achieve coexistence by occupying mutually exclusive spaces: either exclusive altitudinal zones, exclusive habitats, or exclusive sets of islands differing in area and remoteness. In effect, these newly speciated pairs of species become sympatric by "doing the same thing" in different habitats. The species thereby do not yet become syntopic (i.e., they do not yet coexist in the same microhabitat). They do, however, become sympatric (i.e., they either divide the altitudinal zones and habitats on the same island or else divide up a set of islands in a complex checkerboard). Only later do additional niche differences evolve that permit them to coexist within the same habitat. These differences include differing foraging techniques, preferences for different foraging heights above the ground, and ecological differences related to differing body size. That is, Northern Melanesian birds exhibit a trend from spatial toward nonspatial segregation as speciation proceeds. The same conclusion emerges from an analysis of the evolution of ecological segregation among montane birds of New Guinea (Diamond 1986a).

Niche Differences Between Older and Younger Invaders

In the preceding section we surveyed niche differences between newly sympatric species pairs. We noted previously that almost all such species pairs consist of a resident of Northern Melanesia derived from an older invasion, plus a related recent invader that has just arrived from outside Northern Melanesia. Do the older and more recent invaders tend to assume any characteristic niche relations?

In the case of segregation by altitude, the older invader almost always lives at higher altitudes in the mountains, and the more recent invader lives in the lowlands. This is true of all five species pairs that segregate strictly by altitude (*Charmosyna [palmarum]/C. placentis*, *Lorius albidinucha/L. hypoinochrous*, *Pachycephala implicata/P. pectoralis*, *Zosterops murphyi/Z. rendovae*, and possibly *Aplonis brunneicapilla/A. metallica*), and also in three of the four species pairs that exhibit partial altitudinal segregation (*Phylloscopus amoenus/P. poliocephala*, *Rhipidura malaitae/R. rufifrons*, and *Myiagra hebetior/M. alecto*). In each of these species pairs we have listed first the older invader that lives at high altitudes, then the more recent invader that lives at lower altitudes. The sole example in which the more recent invader lives at higher altitudes involves the fantail *Rhipidura fuliginosa*, a recent invader from the New Hebrides, living in the mountains of San Cristobal, and its relative *R. tenebrosa*, an old San Cristobal endemic derived from the same stock, living in the mountains at the same elevations as *R. fuliginosa* and also in the lowlands.

In both of the two species pairs that segregate through one species living in the forest, the other species living at the forest edge or in secondary growth or open habitats, the older invader (*Accipiter luteoschistaceus* and *Myiagra hebetior*) is in the forest, and the more recent invader (*A. novaehollandiae* and *M. alecto*) is in nonforest habitats. Finally, there is no clear pattern to the five cases in which species segregate by occupying islands of different sizes or degrees of isolation. In three cases (*Ptilinopus solomonensis/P. rivoli*, *Macropygia mackinlayi/M. nigrirostris*, and *Aplonis [feadensis]/A. cantoroides*) the older resident of Northern Melanesia lives on small or remote islets, while the more recent invader is on large, central islands. These relations are reversed in the remaining two cases (*Monarcha [melanopsis]/M. cinerascens* and *Pachycephala pectoralis/P. melanura*), where it is the more recent invader that occupies small or remote islets while the older resident is on large or central islands.

These relations are relevant to the question, discussed in chapter 35, of "taxon cycles"—that is, whether invaders of Northern Melanesia tend to evolve in characteristic ecological directions.

Does Ecological Segregation Evolve in Allopatry or Sympatry?

Let us now ask at what stage in speciation the niche differences permitting sympatry develop (Lack 1944, 1945, 1949, 1971, Diamond 1986a). Possible answers to this question fall between two extremes.

At one extreme, the differing physical or biotic environments to which the taxa were exposed while still in allopatry may have led to the development there of large niche differences that required no further amplification for the taxa to become sympatric. For example, of sister species segregating altitudinally in Northern Melanesia, the montane species and the lowland species may already have been confined to the mountains and lowlands, respectively, even before arriving in Northern Melanesia. At the opposite extreme, the two taxa might have occupied similar niches in allopatry, and the niche differences permitting coexistence might have developed (or been greatly amplified) through divergence in sympatry, perhaps because of competitive interactions between the taxa.

Among the 19 cases of completed speciation in Northern Melanesia, in five cases both members of the species pair exist in allopatry and have been studied in allopatry as well as in sympatry. In four of these cases the niche differences permitting sympatry in Northern Melanesia already existed between allopatric populations of the two species. Of the pigeons *Macropygia mackinlayi* and *M. nigrirostris*, which coexist in the Bismarcks as a result of the former occupying small islands and the latter occupying large islands, *M. nigrirostris* occupies only large islands even in the New Guinea region, where it lives alone without *M. mackinlayi*. Similarly, in the Solomons and New Hebrides, where *M. mackinlayi* lives in the absence of *M. nigrirostris*, it occupies small as well as large islands. Thus, *M. nigrirostris* already eschewed small islands and *M. mackinlayi* used them even before achieving sympatry.

An analagous case involves *Pachycephala pectoralis* and *P. melanura*, which coexist in the Bismarcks as a result of the former occupying large and central islands and the latter occupying small islets or remote islands. In South New Guinea, outside the range of *P. pectoralis*, *P. melanura* still occupies small islets, while in the Solomons, outside the range of *P. melanura*, *P. pectoralis* is virtually confined to the larger islands. Thus, even in allopatry, *P. pectoralis* eschews, and *P. melanura* uses, small islands. Similarly, in much of the Solomons the monarch flycatcher *Monarcha cinerascens* coexists with its close relative *M.* [*melanopsis*] by being confined to the smallest or most remote islands, but this is also true of *M. cinerascens* in the Bismarcks, where *M.* [*melanopsis*] is absent.

A further example is that the parrots *Charmosyna* [*palmarum*] and *C. placentis* segregate altitudinally in the Bismarcks and Northwest Solomons, the former living at higher altitudes than the latter. In allopatry *C. placentis* and the Bismarck allospecies *C.* [*palmarum*] *rubrigularis* both live in the lowlands, but the former extends in the mountains only up to 900 m, the latter up to twice as high (1800 m). Where the Solomon allospecies *C.* [*palmarum*] *meeki* lives in allopatry beyond the range of *C. placentis*, it is confined to the mountains and does not even descend to sea level. Thus, already in allopatry *C.* [*palmarum*] was better adapted to or confined to the mountains, while *C. placentis* was less adapted to the mountains.

In another case, however, allopatric populations of the two species do not exhibit the niche differences that permit them to coexist in sympatry. The pigeons *Ptilinopus solomonensis* and *P. rivoli* coexist in Northern Melanesia by occupying islands of different size, the former being confined to small or remote islets and the latter to larger islands in the zone of sympatry. However, both species occupy small as well as large islands in all or parts of their zones of allopatry. Thus, the niche relations in sympatry could not have been readily predicted by examining the allopatric populations.

In short, the niche differences permitting completed speciation in Northern Melanesia in some cases develop already in allopatry, and in other cases do not become apparent until the achievement of sympatry. For comparison, niche differences permitting completed speciation in lowland birds of New Guinea often develop in allopatry, while for montane birds of New Guinea they develop mainly in sympatry (Diamond 1986a). This difference is interpreted as follows. Most cases of speciation in New Guinean montane birds proceed via allopatric taxon pairs along New Guinea's Central Range, which is fairly homogeneous in habitat and biota from one end to the other. Strong forces that could produce divergence in allopatry are thus usually lacking. In contrast, speciation in New Guinean lowland birds often develops when sister taxa from wet New Guinea and arid Australia come into contact, having already evolved differing habitat preferences in their different lands of origin.

Northern Melanesia is not as drastically different from the main source region for Northern Melanesian birds (New Guinea) as New Guinea is from Australia, but it still differs significantly, particularly when one compares Northern Melanesian forest habitats with the secondary growth habitats preferred in New Guinea by many of the New Guinea species that colonized Northern Melanesia. Thus, it is not surprising that development of ecological segregation in allopatry has been more important in Northern Melanesia than in the mountains of New Guinea, but less important than for faunal exchanges between New Guinea and Australia. This conclusion is only preliminary and must be tempered by the small number of cases (five) available for analysis of this question in Northern Melanesia.

Summary

For each of 19 species pairs illustrating various stages in speciation from the beginning to the completion of sympatry, we assembled data to address many questions about speciation. The data consist of taxonomic closeness, degree of sympatry, ratio of body weights, sequence of invasion of Northern Melanesia, niche differences in sympatry, whether those differences also exist in allopatry, and whether those differences shift between sympatry and allopatry.

Four of the 19 species pairs achieved sympatry at two or more areas independently within their geographic range. Estimated taxonomic closeness tends to decrease, but only imperfectly so, with extent of sympatry. Among three possible mechanisms of allopatric speciation, interarchipelagal speciation (double invasion from outside the archipelago) accounts for 18 of the 19 species pairs (because Northern Melanesia is relatively accessible to nearby rich avifaunas). Intra-archipelagal speciation (as in the Galapagos) accounts for at most one species pair, and continental (intra-island) speciation does not account convincingly for any case (because Northern Melanesian islands are too small by the standards of bird dispersal distances).

At the earliest stage of incipient sympatry, ecological segregation between sympatric species involves mutual spatial exclusion, with the two species occupying different habitats, altitudes, or island types. Usually, the older invader is the species occupying higher elevations or the forest interior, while the younger invader lives in the lowlands or at the forest edge. As sympatry becomes more extensive, syntopy (occupation of the same space) becomes possible through the development of nonspatial niche differences—especially differences in foraging technique or height (including niche differences related to differing body size). Among Northern Melanesian birds, niche differences permitting sympatry developed in some cases already in allopatry, but in other cases they do not become apparent until the achievement of sympatry.

Appendix. Remaining Cases of Completed Speciation

In this chapter we described four examples of completed speciation among Northern Melanesian birds. We now describe the remaining 15 cases of species pairs involved in completed speciation.

Accipiter imitator *and* A. albogularis (*map 3*)

These two hawks appear to be divergent members of a species group (Mayr 1957, Wattel 1973). *A. imitator* is an old endemic of the Solomons, confined to the largest fragments of the Bukida chain, whereas *A. albogularis* belongs to a superspecies whose other allospecies occupy New Caledonia, Fiji, and New Guinea. *A. albogularis* may have been derived either from New Caledonia or New Guinea, now occupies the Solomons, and has spread from the Solomons eastward to the Santa Cruz group and westward to the easternmost Bismarck island (Feni). *A. albogularis* has a body weight double that of *A. imitator* and has much longer toes. Thus, the species are likely to differ in foraging technique and/or diet.

Accipiter luteoschistaceus (*map 3*) *and* A. novaehollandiae

These two hawks may be distant relatives belonging to the same group that includes *A. imitator* and *A. albogularis*. Within this group, *A. luteoschistaceus* is a Bismarck endemic confined to two of the largest Bismarck islands, where it coexists only with *A. novaehollandiae*. Hence *A. luteoschistaceus* is surely the older invader, while *A. novaehollandiae* has a large extralimital range to the west and must be a recent invader of Northern Melanesia from New Guinea. The two species are the same size but differ in habitat: *A. luteoschistaceus* lives in the forest interior, and *A novaehollandiae* lives at the forest edge and in the open.

Ptilinopus rivoli *and* P. solomonensis (*map 7, plate 6*)

These are closely related species of fruit doves with similar plumage and calls—so similar, in fact, that it was for a long time uncertain which species the race *speciosus* belonged to. Their centers of distribution are, respectively, to the west and to the east. *P. rivoli* extends from the Moluccas and New Guinea to the Southeast Papuan Islands and Bismarcks. While the *P. rivoli* subspecies of the Southeast Papuan Islands and Geelvink Bay are confined to small islets, the Bismarck and mainland New Guinea subspecies are confined to large islands. *P. solomonensis* occupies both large and small islands of the Solomons, but (as in the case of *Macropygia mackinlayi* discussed earlier in this chapter) it is confined to outlying or small islands of the Bismarcks. Both species have been recorded together only on two Bismarck islands and two Geelvink Bay islets, but in each case one species appears to be a resident, the other only a vagrant.

We infer that the history of this species pair is similar to that of *Macropygia nigrirostris* and *M. mackinlayi*. That is, the superspecies initially consisted of two entirely allopatric allospecies, *P. rivoli* in New Guinea and the Bismarcks and *P. solomonensis* in the Solomons, until the beginnings of sympatry were achieved by *P. solomonensis* spreading westward from the Solomons into the Bismarcks and then onward to the islands of Geelvink Bay, where the two species still largely exclude each other on an island-by-island basis. In the Bismarcks this exclusion takes the form of *P. solomonensis* being excluded from the big islands occupied by *P. rivoli*.

Columba pallidiceps *and* C. vitiensis (*map 11, plate 8*)

Of these related large pigeons, *C. pallidiceps* is endemic to Northern Melanesia, while the Northern Melanesian population of *C. vitiensis* belongs to a subspecies extending to New Guinea and the Moluccas, to an allospecies extending from the Philippines to Samoa, and to a superspecies extending to Japan and Australia. Thus, *C. vitiensis* must be a recent invader of Northern Melanesia from New Guinea, while *C. pallidiceps* represents an older invader and a modified former allospecies of the superspecies. Most islands occupied by *C. pallidiceps* are also occupied by *C. vitiensis*. The two species both occur in forest (on the larger islands mainly in montane forest). Ecological segregation may depend upon *C. pallidiceps* being mainly a ground feeder and *C. vitiensis* feeding arboreally.

Gallicolumba salamonis *and* G. beccarii

These two ground doves belong to a group that includes another representative in Northern Melanesia (*G. jobiensis*) plus numerous taxa on Pacific islands east of the Solomons. Relationships within the group are too poorly understood to reconstruct the details of speciation. However, it is clear that *G. salamonis* is the older invader, since it is confined to the eastern Solomons, while *G. beccarii* also occupies New Guinea and is presumably the more recent invader. Nothing is known about the ecology of *G. salamonis*, which is known from only two specimens and may be extinct. At least one of the two islands for which *G. salamonis* has been recorded (San Cristobal) is also inhabited by *G. beccarii*. Ecological segregation is likely to depend upon the much larger size of *G. salamonis* (nearly triple the weight of *G. beccarii*) and its thicker bill.

Halcyon saurophaga *and* H. chloris (*map 24, plate 6*)

Of these two kingfishers, *H. saurophaga* has two races confined to the Northwest Bismarcks, while its third race extends throughout the rest of Northern Melanesia to North New Guinea and the Moluccas. *Halcyon chloris* has an enormous range from the Red Sea through South New Guinea out to Samoa, with dozens of subspecies and with eight related allospecies on other Pacific islands. It is striking, however, that the otherwise widespread *H. chloris* is absent from the Northwest Bismarcks (where *H. saurophaga* has its endemic subspecies) and from North New Guinea. The two species are quite similar in appearance and call, and there may be some introgression in the Northwest Bismarcks.

A suggested scenario for speciation is that *H. saurophaga* arose in the Northwest Bismarcks and North New Guinea, while *H. chloris* was the representative in South and West New Guinea. Despite the subsequent great range expansion of *H. chloris*, it has still not occupied the original range of *H. saurophaga*. Ecological segregation is by habitat, *H. saurophaga* being entirely restricted to the seacoast, while *H. chloris* also occurs inland and in heavier vegetation. A third member of this group, *H. sancta*, breeds abundantly in Australia, winters commonly in Northern Melanesia, and occasionally breeds in the Solomons. It is ecologically most similar to *H. chloris*, from which it sorts by its smaller size, more open habitats, and lower perches.

Phylloscopus amoenus *and* P. poliocephala (*map 36*)

These two tree warblers belong to a group whose center of distribution is on the Sunda Shelf and in Wallacea. *P. poliocephala* is distributed in numerous subspecies from the Sunda Shelf through New Guinea to the Solomons and must be a recent invader of Northern Melanesia from New Guinea. *P. amoenus* differs from its relatives in plumage and heavier bill, is confined to the Solomon island of Kulambangra, and presumably represents an older invader of the same stock. Both species occur together on the summit of Kulambangra, but *P. poliocephala* extends to lower elevations than does *P. amoenus*. In addition to this partial spatial segregation, the differences in bill suggest differences in foraging behavior, confirmed by field observations: *P. amoenus* and *P. poliocephala* feed on mossy stems and in foliage, respectively (Buckingham et al. 1996).

Rhipidura malaitae *and* R. rufifrons (*map 39*)

These two fantail flycatchers belong to different but closely related superspecies that overlap in South New Guinea as well as on Malaita. While *R. malaitae* is confined to Malaita Island of the Solomons, the superspecies *R.* [*rufidorsa*] to which it belongs consists otherwise of two allospecies in the Bismarcks and one in New Guinea. *R. rufifrons* is distributed in many subspecies over a large range from Wallacea through New Guinea, Australia, and Northern Melanesia to Micronesia and the Santa Cruz Islands, with six related allospecies from Wallacea and Micronesia out to Fiji (Mayr & Moynihan 1946). Thus, *R. rufifrons* is presumably the more recent invader of Northern Melanesia (subspecific traits suggest derivation from the Southeast Papuan Islands), while *R.* [*rufidorsa*] *malaitae* is the older invader. As in the just-described case of the sister species of *Phylloscopus* warblers, these two fantails share the summit of Malaita, to which *R. malaitae* is confined, while *R. rufifrons* extends to lower elevations. These species are similar in size and probably segregate by foraging technique within their shared altitudinal band on the summit of Malaita.

Monarcha [melanopsis] *and* M. cinerascens (*map 40, plate 5*)

The flycatcher *Monarcha* [*melanopsis*] is a superspecies consisting of one allospecies in eastern Australia, another in the mountains of New Guinea and in the Cape York Peninsula, and three allospecies in the Solomons, but no representative in the Bismarcks. The related *M. cinerascens* is distributed as many subspecies on small islands from Wallacea through the New Guinea region and Bismarcks to the western Solomons. We infer that *M. cinerascens* reached the Solomons more recently than did the *M.* [*melanopsis*] superspecies and may have arisen as the representative of that superspecies in either the Bismarcks or Moluccas. It now encounters all three allospecies of the superspecies in the Solomons, plus the allospecies *M. frater* in the New Guinea region. If *M. cinerascens* did originate in the Bismarcks (as suggested by the absence of *M.* [*melanopsis*] there), then the relations of *M.* [*melanopsis*] and *M. cinerascens* would be analogous to the previously discussed relations of *Halcyon saurophaga* and *H. chloris*: one taxon of a pair (*M. cinerascens* or *H. saurophaga*) having originated in an area (the Bismarcks in the former case, the Northwest Bismarcks and North New Guinea in the latter case), from which the otherwise widespread other member of the pair (*M.* [*melanopsis*] or *H. chloris* respectively) is still absent.

M. cinerascens is confined to small, remote, or recently defaunated volcanic islands, while all three Solomon allospecies of *M.* [*melanopsis*] occur both on large and small Solomon islands. The two species are of similar body weight. Only three Solomon islets have been reported as supporting both species: Bates and Fara islets near Ysabel, with *M. cinerascens* and *M.* [*melanopsis*] *castaneiventris*; and Poharan islet near Buka, with *M. cinerascens* and *M.* [*melanopsis*] *erythrostictus*.

Myiagra hebetior *and* M. alecto (*map 42, plate 5*)

Males of these two monarch flycatchers are so similar that their distinctness as species was not even recognized until 1926. *M. hebetior* is confined to the Bismarcks, where it has clearly been resident for a long time, since the three Bismarck races are as distinct from each other in female plumage as each of the races is from female *M. alecto* (see chapter 20 for further details). *M. alecto* is distributed from the Moluccas through Australia and New Guinea to the Bismarcks and is a recent invader of the Bismarcks from New Guinea, since the Bismarck population is not even subspecifically differentiated from the New Guinea population. *M. hebetior* shares with *M. alecto* five of its six Bismarck islands, where the two species segregate through *M. alecto* being confined to secondary growth habitats of the coastal lowlands, *M. hebetior* living in inland forest of the lowlands and lower mountain slopes. *M. hebetior* lives in allopatry only on St. Matthias, the most remote Bismarck island that it occupies.

Pachycephala pectoralis *and* P. melanura (*map 43, plate 2*)

The whistler *Pachycephala pectoralis* breaks up into 65 subspecies plus five related allospecies in a huge range from Java to Fiji, colonized by a series of waves that have generally interbred with each other wherever they have met (Galbraith 1956). One of those waves became *P. melanura*, which completed speciation with *P. pectoralis* on islands of Torres Straits and along the coasts of North Australia and South New Guinea, from whence it spread to the Bismarcks and Northwest Solomons. Coexistence is achieved through *P. pectoralis* occupying larger islands, *P. melanura* small islets and recently defaunated volcanic islands. *P. pectoralis* and *P. melanura* seem to coexist without interbreeding in the Bismarcks and tropical Australia, but there are some signs of introgression in South New Guinea, the Louisiades (race *collaris*), possibly the Lesser Sunda island of Damar, and possibly the Bismarck island of Tabar (Galbraith 1956). The easternmost *P. melanura* population (*P. m. whitneyi*) occurs in the Shortland Group of the Northwest Solomons, where small islets support either individuals of pure *P. melanura* or else hybrids closer to *P. melanura*, while the largest Shortland island and Bougainville support pure *P. pectoralis*. No Northern Melanesian island supports breeding populations of both species, though vagrants of *P. melanura* are frequently observed on islands occupied by *P. pectoralis*. *P. melanura* is evidently a recent invader of the Solomons that has not yet spread beyond the Shortlands. In the Solomons beyond the range of *P. melanura*, *P. pectoralis* is virtually confined to large islands and does not occupy the small islets characteristic of *P. melanura*.

Pachycephala implicata (*plate 2*) *and* P. pectoralis

The whistler *P. implicata*, an old endemic species of the Solomons, is a member of the group to which *P. pectoralis* and *P. melanura* belong. *P. implicata* is confined to elevations above about 1,000 m in the mountains of Bougainville and Guadalcanal, with those two islands being occupied by two different subspecies. *P. pectoralis* is the more recent invader of the Solomons, as evidenced by its huge extralimital range with dozens of extralimital subspecies. On Bougainville and Guadalcanal *P. pectoralis* is strictly confined to elevations below those occupied by *P. implicata*, but on high Northern Melanesian islands lacking *P. implicata* (Kulambangra, New Britain, New Ireland, and Umboi) *P. pectoralis* extends up to elevations of at least 1,800 m.

Zosterops murphyi *and* Z. rendovae (*map 45, plate 1*)

These related white-eyes are interesting as the only Northern Melanesian species pair that appears to have completed speciation by intra-archipelagal speciation (i.e., by double invasion from within Northern Melanesia). The *Z.* [*griseotinctus*] superspecies consists of four allospecies in the New Georgia group of the Solomons, a fifth allospecies on Rennell, and a sixth (*Z. griseotinctus*) on islets of the Bismarcks and the Southeast Papuan Islands. Among these allospecies, the one most similar to *Z. murphyi*, endemic to the mountains of Kulambangra, may be *Z. griseotinctus*. Below the altitudinal range of *Z. murphyi*, in the lowlands of Kulambangra, lives the allospecies *Z.* [*griseotinctus*] *rendovae*, whose Kulambangra population belongs to the same subspecies as that of five other islands of the New Georgia group, all of which were connected to Kulambangra at Pleistocene times of low sea level. These facts suggest that the group to which *Z. murphyi* and *Z.* [*griseotinctus*] belong arose in Northern Melanesia, that *Z. murphyi* is a derivative of the first member of the group to reach Kulambangra, and that *Z. rendovae* is a later arrival on Kulambangra. Because both taxa may have colonized Kulambangra from sources within Northern Melanesia, this sympatric species pair represents an intra-archipelagal double invasion of Kulambangra, just as the sympatric sets of Darwin's finches on various Galapagos Islands represent double invasions of those islands from within the Galapagos.

Aplonis [feadensis] *and* A. cantoroides (*map 50*)

The starling *A. cantoroides* is a monotypic species distributed from the Bismarcks and Solomons throughout the New Guinea region, with a closely related allospecies on the Tenimbar Islands of Wallacea. Related to *A. cantoroides* is the *A.* [*feadensis*] superspecies, which is endemic to Northern Melanesia and consists of two allospecies with a fragmented distribution. Hence *A.* [*feadensis*] is the older resident, and *A. cantoroides* the more recent invader, of Northern Melanesia. *A.* [*feadensis*] is confined to remote and peripheral Northern Melanesian islands, most of them small. *A.* [*feadensis*] and *A. cantoroides* have been recorded together only on Wuvulu and the Hermit Islands of the northwestern-most Bismarcks and on Rennell, the southeastern-most Solomon island. The ecological segregation permitting this sympatry has been studied only for Rennell, where *A.* [*feadensis*] *insularis* is distributed over most of the island, while *A. cantoroides* is largely confined to the immediate vicinity of Rennell's lake. In addition, *A. insularis* has a more diverse diet than *A. cantoroides*, and there are differences in nest sites of the two species (Diamond 1984a).

Aplonis brunneicapilla *and* A. metallica

Of these two similar colonial starlings, *A. brunneicapilla* is endemic to the Solomons and confined to three islands, while *A. metallica* ranges from the Solomons and Bismarcks through the New Guinea region and Cape York Peninsula of Australia to eastern Wallacea. We infer that *A. brunneicapilla* is the older resident of the Solomons. Sympatry may depend upon altitudinal segregation, with *A. brunneicapilla* reported as nesting in the mountains above the altitudinal range of *A. metallica* (Cain & Galbraith 1956).

23

Hybridization

The preceding chapters may have given the impression that geographic isolation always leads to differentiation and speciation, proceeding inexorably through the stages of weak subspecies, megasubspecies, and allospecies to sympatric sister species. In fact, this progress is not inexorable. Many distinctive and presumably old but vagile species continue to exhibit no geographic variation over a huge range (e.g., the colonial pigeon *Caloenas nicobarica*) because high interisland gene flow renders populations of the species essentially panmictic. Many isolates become extinct before they can become morphologically distinct. Other incipient isolates are eventually reached by colonists from other isolates with which they hybridize, aborting further differentiation. Speciation often remains stalled at the stage of allospeciation, if reproductively isolated allospecies do not develop ecological segregating mechanisms and remain unable to invade each others' ranges (chapter 21). In this chapter, we discuss hybridization in Northern Melanesian birds.

On continents or large islands such as New Guinea, New Zealand, and Madagascar, distances are sufficient for geographic differentiation to proceed within a single land mass. When two such isolates that have formed within the same land mass meet before having achieved full reproductive isolation, they will hybridize in a zone of contact and form a hybrid belt. Such hybrid zones are extremely common on New Guinea and all the continents and have been described by numerous ornithologists and researchers of other animal and plant groups (see Mayr 1942, 1963, Keast 1961, Mayr & Short 1970, Haffer 1974, Gill 1998, and Rohwer & Wood 1998 for avian examples).

The establishment of an evident hybrid population on islands too small to permit geographic differentiation within the island's boundaries is much more difficult. On the continents, incipient isolates may regain contact on a broad front, but in archipelagoes expansion is effected by colonizing individuals crossing water gaps, and these individuals are usually absorbed by the host population. The morphological intermediacy of many insular subspecies between the morphological characteristics of subspecies on adjacent islands is surely due in part to such gene flow. Perhaps the greatest opportunity for development of a hybrid population in archipelagoes exists when a previously uninhabited island is simultaneously colonized by individuals of two different subspecies or allospecies.

Genuine hybrid populations (as opposed to homogeneous populations morphologically intermediate between neighboring populations) are characterized by greatly increased individual variability within the population, encompassing the characters of both of the parental taxa. Such hybrid populations, sometimes designated as subspecies when restricted to a well-defined area, have been noted in Northern Melanesia in about six genera, as discussed below.

1. The swallows *Hirundo tahitica* of eastern Northern Melanesia are dark-colored, dark-tailed birds that range from New Ireland and the Solomons to Polynesia, with little geographic variation in this huge area (subspecies *subfusca*). The swallows of New Guinea and the westernmost Northern Melanesian islands nearest New Guinea (Umboi and its neighbors) form a light-colored population with much white in the tail, recognized as the subspecies *H. t. frontalis*. The population of the Bismarck island of New Britain, between Umboi and New Ireland, is not only intermediate in coloration, size, and other characteristics distinguishing *H. t. subfusca* from *H. t. frontalis*, but also shows extreme variability in the amount of white in the tail. This suggests that the Bismarcks were invaded by swallows both from the west and the east, and that the two colonizing streams met on New Britain and there formed a hybrid population recognized as subspecies *H. t. ambiens* (Mayr 1934a, 1955). In turn, the swallow population of Long Island, lying between New Britain and New Guinea, is morphologically intermediate between *H. t. frontalis* and *H. t. ambiens*. This intermediate population of Long must have arisen by mixture of *H. t. ambiens* and *H. t. frontalis* colonists in the three centuries since Long was defaunated by a volcanic explosion (Mayr 1955).

2. The population of the cuckoo *Eudynamys scolopacea* in the Bismarcks (subspecies *salvadorii*) is a megasubspecies differing strikingly from the New Guinea population (subspecies *rufiventer*) in larger body size, much more whitish and less ochraceous underparts of the female, and broader black bands in the tail of the female. The population of Long Island and its neighbors is similar to the New Guinea population in body size, to the Bismarck population in broad tail bands, and is intermediate in color of the female's underparts (Diamond 2001a). Like the swallows of Long Island, this intermediate cuckoo population of Long Island must have arisen by mixture of colonists from New Guinea and New Britain in the centuries following Long Island's defaunation.

3. The scrub-fowl *Megapodius freycinet*, in its occupation of a large range from Indonesia to the New Hebrides, breaks up into some distinctive subspecies that have been considered full species by Schodde (1977) and White and Bruce (1986), and many earlier authors. For example, the Northern Melanesian race *eremita* differs from the North New Guinea race *affinis* in many characters including body size, length of crest, extent of feathering on the forehead, and coloration of the wings, flanks, and mantle. Schodde (1977) and White and Bruce (1986) both consider *affinis* and *eremita* as distinct allospecies. However, on Karkar, a volcanic island off the northeast coast of New Guinea, one finds a variable population containing some individuals almost indistinguishable from *eremita*, others almost indistinguishable from *affinis*, and still others showing a mosaic of characters of these two "parent" races. Specimens collected in 1969 are more similar to *eremita* than specimens collected in 1914, at which time females were much closer to *affinis* and males to *eremita*. Like Long Island, Karkar is an active volcano that has experienced major eruptions in the last several thousand years, but it is not known whether any of these eruptions defaunated the island. Thus, the Karkar population is a variable hybrid population between the source populations of New Guinea and New Britain, and the relative contributions of the two parental stocks appear to have shifted in favor of the Bismarck stock within the twentieth century (Mayr 1938a, Diamond & Lecroy 1979).

4. Females of the cuckoo-shrike *Coracina tenuirostris* on almost all Solomon islands have unbarred underparts, while females of the Bismarcks and New Guinea have barred underparts. The exception in the Solomons is the population of Pavuvu Island (race *nisoria*; Mayr 1950, 1955), whose barred underparts, small size, and rufous coloration are compatible with origin as a hybrid between New Guinea and Solomon colonists, though other interpretations are also possible (map 28, plate 7).

5. The honey-eater *Myzomela cardinalis* (map 46, plate 4) lives on some of the southeastern-most Solomon islands and on Pacific islands east of the Solomons, while *M. tristrami* is endemic to San Cristobal in the southeastern Solomons. It is uncertain how closely related these two species are within genus *Myzomela*. The plumage of *M. tristrami* is uniformly black, while the former is largely red. On San Cristobal *M. tristrami* occupies virtually all habitats and all of the island, while *M. cardinalis* is confined to the north coast. Residents of San Cristobal report that *M. cardinalis* is a recent invader from nearby smaller islands of the southeastern Solomons. Interestingly, some of the first known specimens of *M. tristrami*, collected on San Cristobal in 1908, have patches of red feathers suggestive of hybridization with *M. cardinalis*. A col-

lection from 1927 shows less hybridization (Mayr 1932a), while 1953 specimens included no hybrids. These facts suggest that the frequency of hybridization may have been decreasing with time, perhaps because the earliest colonists of *M. cardinalis* were unable to find conspecific mates, whereas *M. cardinalis* is now sufficiently well established that individuals have no difficulty finding conspecifics (Diamond 2001a). This reconstructed scenario is similar to many well-studied examples of hybridization on continents, where hybrids are most often observed at the periphery of a species' range or else early in an expansion cycle—conditions under which individuals have difficulty finding conspecific mates.

6. The richest material for studying hybridization in Northern Melanesia is provided by the closely related whistlers *Pachycephala pectoralis* and *P. melanura*. As discussed in chapter 22, these two taxa have evidently completed speciation and come recently into widespread sympatry. *P. melanura* (map 43) occupies small, remote islands, while *P. pectoralis* occupies large central islands. Both species are widespread in the Bismarcks, and *P. pectoralis* is widespread in the Solomons, but *P. melanura* penetrates only the northwestern-most Solomons as far as the Shortland group. The only obvious hybrid population is that peripheral population of the Shortlands, where whistlers of small islets (described as the race *P. pectoralis whitneyi*) range variably from individuals resembling the Northern Melanesian form *P. melanura dahli* to individuals resembling the Bougainville race *P. pectoralis bougainvillei* (Mayr 1932c). The *P. pectoralis* races of Manus and Tabar islands in the Bismarcks show subtler indications of hybridization with *P. melanura*. The San Cristobal population *P. pectoralis christophori* shows much plumage variation, suggesting hybridization between two distinct megasubspecies groups of *P. pectoralis*, those of the Solomons and of islands to the east (subspecies groups C and E of Galbraith 1956). Hybridization between megasubspecies of *P. pectoralis* is obvious in Fiji (Mayr 1932d) and is also suggested for several *P. pectoralis* populations elsewhere in the range of the species (Galbraith 1956). Thus, the megasubspecies of *P. pectoralis* exhibit numerous examples of failure to complete speciation, while the encounter of *P. melanura* and *P. pectoralis* is an instance of successful speciation but with breakdown of reproductive isolation at the periphery of *P. melanura*'s range.

Summary

It is by no means true that geographic isolation always leads to differentiation and completed speciation. Among the processes that can abort progress toward speciation is hybridization between isolates bridged by colonists. The morphological intermediacy of many Northern Melanesian insular subspecies between morphological characteristics of subspecies on adjacent islands testifies to the frequency with which colonizing individuals become genetically incorporated into an established host population. We have described in detail six hybrid populations encompassing the characters of both of the parental taxa and some of them characterized by greatly increased variability within the population. Most of these hybrid populations evidently arose when a previously uninhabited island was simultaneously colonized by individuals of two different subspecies or allospecies.

24

Endemic Species and Genera

The final stage in speciation is the development of an endemic full species—a species that does not belong to a superspecies ranging extralimitally. Of Northern Melanesia's 191 analyzed zoogeographic species, 35 fall into this category (table 24.1). They include four superspecies endemic to Northern Melanesia. These four superspecies are as distinct from extralimital populations as are the 31 endemic monotypic species; they differ only in that they vary geographically within Northern Melanesia, whereas endemic monotypic species do not vary geographically.

Naturally, it is in some cases uncertain whether to consider a Northern Melanesian endemic as an isolated full species, or else as a distinct allospecies belonging to a superspecies with extralimital representatives. Among the uncertain cases are *Pareudiastes sylvestris* (plate 8), *Gallicolumba salamonis*, *Chalcopsitta cardinalis* (plate 8), and *Ortygocichla rubiginosa* (plate 9), which we consider as isolated species whose nearest congeneric relatives are identifiable (see their species accounts in appendix 1) but in our view not sufficiently similar to place in the same superspecies. Lumpers might nevertheless rate these four taxa as allospecies and assign them to superspecies. Conversely, we consider *Cacatua ducorpsi* (plate 8), *Pitta anerythra* (plate 9), and *P. superba* (plate 9) as allospecies within superspecies ranging extralimitally, but splitters might consider each of these three taxa too distinct to include within a superspecies.

These borderline cases should not conceal the fact that Northern Melanesia does contain some very distinct endemic bird taxa, though not nearly as high a proportion as the avifaunas of Hawaii, New Zealand, or the Galapagos. For example, of the 35 endemic species, 5 belong to endemic monotypic genera. It is even unclear to which other genus each of these monotypic genera is most closely related. Two of these genera are very distinctive: the pigeon *Microgoura meeki* (plate 8) and the owl *Nesasio solomonensis* (plate 8). The other three monotypic genera (plate 9), "*Stresemannia*" *bougainvillei*, "*Guadalcanaria*" *inexpectata*, and "*Meliarchus*" *sclateri*, are undistinctive honey-eaters placed in monotypic genera by default: each has been variously assigned to alternative genera, and the closest relative is still uncertain. All five of these monotypic genera are confined to the Solomons, and none occurs in the Bismarcks, a point to whose significance we shall return below.

Among the 30 full species not belonging to monotypic genera, eight are closely related to a congener with which they now occur sympatrically as a result of recent completed speciation (chapter 22): the pigeons *Columba pallidiceps* (plate 8) and *Gallicolumba salamonis*, the warbler *Phylloscopus amoenus*, the flycatcher *Myiagra hebetior* (plate 5), the whistler *Pachycephala implicata* (plate 2), the white-eye *Zosterops murphyi* (plate 1), and the starlings *Aplonis* [*feadensis*] and *A. brunneicapilla*. Most of the other 22 full species not belonging to

Table 24.1. Endemic species of Northern Melanesia.

Species	Level of endemism[a]	Abundance, vagility[b]	Montane?[c]	Solomons (S) or Bismarcks (B)[d]	No. of islands[e]	Geographic variation[f]			Map[g]	Plate[h]
						s	M	A		
Accipiter imitator	S	1C	No	S	3(P)				3	
Accipiter luteoschistaceus	S	1C	No	B	2(P)				3	
Gallirallus insignis	S	3C	No	B	1				4	
Pareudiastes sylvestris	S	1C	Yes	S	1				4	8
Columba pallidiceps	S	1B	Yes	S,B	10				11	8
Gallicolumba salamonis	S	3B	No	S	2					
Microgoura meeki	G	1C	No	S	1				4	8
Chalcopsitta cardinalis	S	4A	No	S,B	36				15	8
Charmosyna margarethae	S	3B	No	S	9					9
Centropus violaceus	S	1B	No	B	2				4	
Nesasio solomonensis	G	1C	No	S	3(P)					8
Aerodramus orientalis	S	1B	Yes	S,B	3	3				
Halcyon bougainvillei	S	1C	Yes	S	2(P)	2			4	8
Coracina holopolia	S	3B	No	S	10		3		29	
Ortygocichla rubiginosa	S	3C	No	B	1				4	9
Cettia parens	S	3C	Yes	S	1				4	
Phylloscopus amoenus	S	3C	Yes	S	1				36	
Myiagra hebetior	S	4B	No	B	6		3		42	5
Pachycephala implicata	S	3C	Yes	S	2(P)		2			2
Zosterops metcalfii	S	5C	No	S	6(P)	2			45	1
Zosterops [griseotinctus]	SS	5B	No	S,B	18		3	6	45	1
Zosterops murphyi	S	5C	Yes	S	1				45	1
Zosterops ugiensis	S	5B	Yes	S	3	1	2		45	1
Zosterops stresemanni	S	5C	No	S	1				45	1
Myzomela pulchella	S	5C	Yes	B	1				4	4
Myzomela sclateri	S	3A	No	B	8					4
Myzomela [lafargei]	SS	5B	No	S,B	42	2	6	6	47	4
Melidectes whitemanensis	S	4C	Yes	B	1				49	9
"Stresemannia" bougainvillei	G	3C	Yes	S	1				49	9
Meliarchus sclateri	G	4C	No	S	1				49	9
"Guadalcanaria" inexpectata	G	3C	Yes	S	1				49	9
Aplonis [feadensis]	SS	3A	No	S,B	10	2		2	50	
Aplonis [grandis]	SS	4B	No	S	24	2	1	2	51	
Aplonis brunneicapilla	S	3B	Yes	S	3					
Corvus woodfordi	S	3C	No	S	6(P)		2			

[a]Level of endemism: S = endemic isolated full species, SS = endemic superspecies, G = endemic monotypic genus.
[b]Abundance and vagility, as classified in chapters 10 and 11: 1 = rarest . . . 5 = most abundant; A = most vagile . . . C = most sedentary.
[c]Yes = confined to the mountains on some or all Northern Melanesian islands occupied.
[d]S = occurs in Solomons, B = in Bismarcks, S,B = in both.
[e]Number of islands occupied; P indicates that all islands occupied are post-Pleistocene fragments of the same Pleistocene island.
[f]Geographic variation: no entry = none; s = number of weak subspecies; M = number of megasubspecies; A = number of allospecies.
[g,h]The number of the map and plate, if any, depicting the species' distribution and the species itself, respectively.

monotypic genera are taxonomically rather isolated; for many it is not even clear to which congener they are most closely related. Three of these 22 had even been placed in monotypic genera, which we no longer recognize: the rail *Pareudiastes "Edithornis" sylvestris*, the warbler *Phylloscopus "Mochthopoeus" amoenus*, and the honey-eater *Melidectes "Vosea" whitemanensis* (plate 9). Even some of the endemic allospecies that we discussed in the previous chapter have at one time been separated in monotypic genera, mainly on the basis of single exaggerated characters likely to function in sexual selection: the pigeon *Ptilinopus "Oedirhinus" [hyogaster] insolitus* (plate 9) because of the red, fleshy knob on its bill; the pigeon *Reinwardtoena "Coryphoenas" [reinwardtii] crassirostris* (plate 7) because of its crest and thick, curved bill; and the drongo *Dicrurus "Dicranostreptus" [hottentottus] megarhynchus* (plate 8) because of its twisted and elongated outer tail feathers.

Among the endemic full species, geographic origins can be clearly identified for the eight involved in recent completed speciation with close relatives. Most of them came from the New Guinea region, except that *Gallicolumba salamonis* may have come from the east. Two of the other endemic full species (*Pareudiastes sylvestris* and *Ortygocichla rubiginosa*) have their closest relative on islands to the east, while others (*Gallirallus insignis* [plate 8], *Chalcopsitta cardinalis*, *Charmosyna margarethae* [plate 9], *Halcyon bougainvillei* [plate 8], and *Melidectes whitemanensis* [plate 9]) have their closest relatives in the New Guinea region or to the west. Most of the remaining endemic full species are taxonomically isolated, meaning that almost by definition it is unclear where they came from, since the identity of their closest relatives is uncertain.

We now discuss a series of problems posed by the endemic full species: their localized geographic distributions; interpretation of those localized distributions in terms of their biological attributes of vagility and abundance; distribution between the Solomons and Bismarcks; the historical interpretation of their localized distributions; habitat preferences; and geographic variability.

Geographic Distributions of the Endemic Full Species

The most striking feature of the distributions of the endemic full species is how localized they are (columns 5 and 6 of table 24.1). Considering that all of Northern Melanesia is rather homogeneous in climate and in species distributions, it is remarkable that only six of the 35 endemic full species occur both in the Bismarcks and Solomons: the pigeon *Columba pallidiceps* (map 11), the parrot *Chalcopsitta cardinalis* (a Solomon species confined in the Bismarcks to the four or five Bismarck islands nearest the Solomons [map 15]), the swiftlet *Aerodramus orientalis*, the white-eye *Zosterops [griseotinctus]* (map 45), the honey-eater *Myzomela [lafargei]* (map 47), and the starling *Aplonis [feadensis]* (map 50). The latter three species also seem to be recent invaders of the Bismarcks from the Solomons, since the taxonomic diversity of each superspecies is much greater in the Solomons than in the Bismarcks. Of the remaining 29 species confined to either the Solomons or Bismarcks, only three are widespread within the Solomons (the parrot *Charmosyna margarethae*, the cuckoo-shrike *Coracina holopolia* [map 29], and the starling *Aplonis [grandis]* [map 51]), and only one is widespread in the Bismarcks (the flycatcher *Myiagra hebetior* [map 42].

Most of the endemic full species are either confined to a single island (13 species) or else to two or more post-Pleistocene fragments of a single Pleistocene island (seven species; column 5 of table 24.1). Two species occur on two islands not formerly joined in the same Pleistocene island, and three species occur on three islands of which two were not joined in the same Pleistocene island. Only 10 species occur on numerous islands (6–42 islands) of which many were not joined in the same Pleistocene island. Particularly widely distributed are the four endemic superspecies of Northern Melanesia: *Zosterops [griseotinctus]*, *Myzomela [lafargei]*, *Aplonis [feadensis]*, and *A. [grandis]*, on 18, 42, 10, and 24 islands, respectively.

The geographically localized, endemic full species are overwhelmingly confined to the largest Northern Melanesian islands. The 23 species confined to 1–3 islands consist of 36 island populations, of which all but four are on islands of area about 3,000 km^2 or greater. Three further populations are on islands of about 700 km^2 (two on Kulambangra, one on Umboi). Only one population is on a very small island: the Ramos population of *Gallicolumba salamonis*, otherwise known only from San Cristobal. No island smaller than 700 km^2 supports an endemic full species confined to that island. Let us now consider the biological interpre-

Table 24.2. Vagility of endemic full species.

	Vagility class		
	A	B	C
Number of endemic species	3	12	20
Total number of species	56	107	28
Endemics/total (%)	5	11	71

The first row gives the number of endemic full species in each of the three vagility classes defined in chapter 11: A = most vagile . . . C = most sedentary. The second row gives the total number of Northern Melanesian species (regardless of level of endemism) in each class. The last row gives the number in the first row, as a percentage of the number in the second row. Note that endemic full species are drawn disproportionately from the most sedentary species; they constitute 71% of the most sedentary species, but only 5% of the most vagile species ($p < .001$ by chi-square test, Pearson correlation .46, Spearman correlation .44).

tation of these localized distributions and virtual confinement to the largest islands.

Vagility and Abundance of Endemic Full Species

It is striking how disproportionately the endemic full species are drawn from the most sedentary of the three vagility classes that we defined previously. The endemics constitute 71% of those most sedentary species (class C), only 11% of the species of intermediate vagility (class B), and only 5% of the most vagile species (class A; table 24.2).

The low vagility of most of the endemic full species accounts for their distributional peculiarities discussed in the preceding section. They are mostly confined to one or a few islands because they rarely or never cross water to colonize other islands (or to recolonize islands that they formerly occupied and on which they are now extinct). They tend to be confined to the largest islands because those are the islands with the largest populations and hence lowest risk of extinction. Only the more vagile species are likely to be found on small islands, where populations frequently go extinct and where one is therefore likely to encounter only vagile species that recolonize frequently. The 20 full endemic species of the lowest vagility (class C) are all confined to a single island or to fragments of the same Pleistocene island. (This conclusion is not circular. Although presence of a species on islands not joined during the Pleistocene is one criterion by which we assigned species to vagility classes, we used several other criteria as well, and no species in vagility class C evinced any type of evidence for water-crossing ability.) All 10 of the endemic full species that occur on many islands (on seven or more), and all six that occur both in the Bismarcks and Solomons, belong to the more vagile species classes (class B or A).

As for abundance, the endemic full species are distributed over four of the five abundance classes defined previously and are absent only from class 2. That class consists of vagile species (vagility classes A and B) of specialized habitats including the seacoast, lakes and rivers, and grassland. The

Table 24.3. Abundance of endemic full species.

	Abundance class				
	1	2	3	4	5
Number of endemic species	9	0	14	5	7
Total number of species	25	27	74	43	22
Endemics/total (%)	36	0	19	12	32

The first row gives the number of endemic full species in each of the five abundance classes defined in chapter 10: 1 = rarest . . . 5 = most abundant. The second row gives the total number of Northern Melanesian species (regardless of abundance) in each class. The last row gives the number in the first row, as a percentage of the number in the second row. Note that endemic full species are drawn from all abundance classes except class 2 (see text for discussion). The relative proportions of endemic species do vary significantly with abundance class ($p = .004$ by chi-square test, class 1 significantly overrepresented and class 2 underrepresented), but there is no monotonic trend (Pearson and Spearman correlations close to zero).

lack of endemic full species in this abundance class reflects the sedentariness of the endemic full species discussed above (table 24.3).

Distribution Between the Solomons and Bismarcks

The endemic full species are concentrated in the Solomons rather than in the Bismarcks (column 4 of table 24.1). Only eight are confined to the Bismarcks, whereas 21 (including all five of the endemic genera) are confined to the Solomons. Six of the endemic full species occur both in the Solomons and Bismarcks. However, as discussed previously, four of those six are concentrated in the Solomons and probably colonized the Bismarcks only recently.

This concentration of endemic full species in the Solomons becomes more striking when one realizes that the total number of species of all levels of endemism combined is higher in the Bismarcks than in the Solomons. One contributing factor is surely that the Solomons are more peripheral from the main colonization source of New Guinea. This has two consequences: divergence is expected to be more rapid in the Solomons because there is less gene flow from extralimital conspecific populations during the early stages of differentiation; and faunal turnover rates should be lower in the Solomons because fewer competing invaders arrive, so that Solomon endemics are more likely to survive long enough to reach the level of distinctness of an endemic full species or endemic genus. An additional likely contributing factor is that more of the area of the Solomons than of the Bismarcks is geologically old, so that more species may have colonized the Solomons earlier.

How Did Many Endemics Become So Locally Distributed?

The very local distributions of so many of the old endemics present a paradox. The ancestral species must have had sufficient dispersal ability to reach Northern Melanesia from New Guinea or from another colonization source in the first place. As old endemics, they have had more time than other Northern Melanesian species to spread throughout Northern Melanesia. How did most of them (25 out of 35) nevertheless come to have such localized ranges, with 20 confined to a single island or fragments of a single Pleistocene island, and five more confined to only two or three islands? At least five alternative explanations suggest themselves:

Artifact of Human-caused Extinctions

Paleontological explorations of Pacific islands beyond Northern Melanesia have revealed, on every studied island, bones of now-extinct species that disappeared soon after the first prehistoric colonization by humans. These explorations have also revealed that some extant species that in modern times have been confined to one or a few islands, and hence have been considered as single-island endemics, were actually more widespread until prehistoric humans arrived. We know, by comparing twentieth-century distributions with records of eighteenth- and nineteenth-century European explorers, that still other species suffered range contractions and local extinctions soon after the arrival of Europeans with their rats and cats. Other species surely suffered similar range contractions after the arrival of Europeans but because the first European explorers did not collect specimens we are not aware of these contractions.

Could the restricted ranges of Northern Melanesia's old endemics similarly be an artifact of human-related extinctions? Of course, the question of the uncertain influence of human-related extinctions bedevils the interpretations of all Northern Melanesian species distributions. However, the question is particularly acute in interpreting the distributions of Northern Melanesia's old endemics, because on other Pacific islands (such as New Zealand and Hawaii) the old endemics have been much more susceptible to extermination by humans (and their rats and pigs and cats) than have recent colonists. This uncertainty regarding Northern Melanesia's old endemics will not be settled until Northern Melanesia becomes well-explored paleontologically. Nevertheless, some tentative guesses can be offered.

On the one hand, several facts concerning the locally distributed large ground pigeons *Microgoura meeki* (map 4) and *Gallicolumba salamonis* strongly suggest that the historic ranges of those two species are artifacts of post-European range contractions. *Microgoura meeki* is known only from seven specimens taken on Choiseul by Albert Meek in January 1904. Ornithologists who subsequently visited Choiseul have specifically searched for this bird, because it is the most distinctive avian endemic of Northern Melanesia, but have not encountered it. Residents of Choiseul say that it was exterminated by cats. Except for *Microgoura meeki* and the ground thrush *Zoothera dauma*, all other bird species of Choiseul also occur on other islands of the Bukida chain, to which Choiseul was connected during Pleistocene times of low sea level. The apparent restriction of *Microgoura meeki* to Choiseul is probably an artifact of extermination of other Bukida populations by cats before European ornithological exploration began. Choiseul would have been the last large Bukida island to receive cats because it was the last large Bukida island to be visited frequently by Europeans.

Similarly, *Gallicolumba salamonis* is known only from one specimen taken on the large southeastern Solomon island of San Cristobal in 1882, plus another taken on the tiny Solomon island of Ramos in 1927. The distance between San Cristobal and Ramos, and the tiny size and isolation of Ramos, suggest that *Gallicolumba salamonis* may formerly also have occurred on intervening islands such as Malaita, Guadalcanal, and Florida. Nothing specific is known about the cause of extinction of *G. salamonis*, but it too would have been susceptible to cats and rats, like other *Gallicolumba* species on other Pacific islands.

Thus, two of the isolated endemics became extinct in the twentieth century. Their twentieth-century distributions would have made no sense as pristine distributions, and their habits made them especially susceptible to human-introduced predators. In contrast, none of the 33 other isolated endemics is known to have suffered a contraction in its distribution or abundance since it was first discovered. All now coexist with introduced rats and cats. Twenty-one are small species of passerines unlikely to have been under hunting pressure from humans. Most of them are not rare species (they belong to abundance class 3 or higher), and 12 are common or among the most abundant species of Northern Melanesia (abundance classes 4 and 5). The present abundance of these species in the presence of humans and their commensals makes it unlikely that their distributions are relicts of human-related extinctions.

Obviously, we must suspend judgment on this issue until Northern Melanesia is better known paleontologically. For the present, we assume that some other distributions besides those of *Microgoura meeki* and *Gallicolumba salamonis* will prove to be recent artifacts but that many distributions will prove not to be artifacts.

Nonartifactual Explanations

At least four explanations can be suggested for how the old endemics may have achieved restricted distributions even before human arrival. Those explanations include an isolated chance colonization by a poorly vagile ancestor; an isolated colonization by a vagile ancestor prevented from occupying other islands because they were occupied by competitors; colonization resulting in former occupation of a wider range, followed by range contraction due to evolutionary loss of dispersal ability, with the result that occasional local extinctions of island populations could not be reversed by recolonization; and similar former occupation of a wider range, followed by range contraction due to subsequent colonization of Northern Melanesia by a competing species. Examples illustrating each of these interpretations are presented below.

The montane honey-eater *Melidectes whitemanensis* (plate 9, map 48) is confined to New Britain, the large Northern Melanesian island nearest New Guinea. All other *Melidectes* species are confined to the mountains of New Guinea. *Melidectes* species must be very poor overwater colonists, since there is no other *Melidectes* population on any island other than New Guinea or New Britain. Hence *M. whitemanensis* is likely to have arisen from a fortuitous colonization of the island nearest New Guinea by a poorly vagile ancestor.

In Northern Melanesia there is no instance of two species of white-eyes (*Zosterops*) being able to coexist in the same habitat of the same island. (The sole cases of sympatry involve segregation by altitude). Ten different species or allospecies of *Zosterops* occupy the lowlands of various Northern Melanesian islands, to the strict exclusion of each other. Hence the reason why the old Solomon endemics *Zosterops metcalfii*, *Z* [*griseotinctus*], and *Z. stresemanni* now occupy, respectively, the lowlands of the Bukida chain, New Georgia group, and Malaita to the exclusion of each other may well be that each ancestral colonist was excluded from other islands by established competing species of *Zosterops*. Similarly, within Northern Melanesia the similar-sized species of *Myzomela* honey-eaters tend to exclude each other by habitat or island, with 10 different species or allospecies occupying the lowlands of different islands to the exclusion of each other. The old endemics *Myzomela sclateri* and *Myzomela* [*lafargei*], both of which are vagile overwater colonists (dispersal class A), are probably restricted to their particular islands because of competitive exclusion by other similar-sized *Myzomela* honey-eaters on other islands.

Evolutionary loss of dispersal ability has surely befallen the rails *Gallirallus insignis* and *Pareudiastes sylvestris* (plate 8), whose ancestors must have reached Northern Melanesia over water but which are now flightless. Many of Northern Melanesia's other old endemics, which are volant but disperse poorly across water (dispersal class C), may analogously have become "behaviorally flightless" (i.e., reluctant to cross water).

Finally, five species are virtually confined to montane habitats on high islands, where they occupy altitudinal ranges mutually exclusive of the range of a lowland congener that represents a more recent invader of Northern Melanesia. These altitudinal pairs, citing in each case the old endemic first and the recent invader second, are the whistlers *Pachycephala implicata* and *P. pectoralis*, the white-eyes *Zosterops murphyi* and *Z.* [*griseotinctus*], the white-eyes *Z. ugiensis* and *Z. metcalfii*, the honey-eaters *Myzomela pulchella* and *M. cruentata*, and the starlings *Aplonis brunneicapilla* and *A. metallica*. In each case, the ancestor of the montane endemic may formerly have occupied a broader altitudinal range (including the lowlands) on many islands and may have become restricted to montane habitats on the highest islands by subsequent arrival of a lowland competitor.

Thus, there are examples illustrating each of four possible historical scenarios to account for the localized distribution of most of the old endemics within Northern Melanesia.

Habitat Preference

Of the 35 full endemics, all without exception are forest species. One (the swiftlet *Aerodramus orientalis*) is an aerial forager over forest, while the other 34 forage inside the forest. Most are virtually confined to forest and absent in open habitats; only the parrot *Chalcopsitta cardinalis* and occasionally the honey-eater *Meliarchus sclateri* extend out of the forest to occupy coconut plantations.

Fourteen of the 35 full endemics are confined to the mountains on some or all islands that they occupy (column 4 of table 24.1). The full endemics account for a disproportionate percentage of Northern Melanesia's montane species. As already discussed, some probably arose from lowland ancestors that were pushed up into the mountains by subsequently invading congeners. Others were probably already montane species at the time that they invaded Northern Melanesia, because their closest extralimital relatives are conspecific montane source populations (*Gymnophaps albertisii*, *Micropsitta bruijnii*, and others) or else related montane species (*Phylloscopus amoenus*, *Melidectes whitemanensis*, and possibly *Pareudiastes sylvestris* and *Pachycephala implicata*).

In chapter 35 we relate these facts to the hypothesis of taxon cycles, according to which colonists tend with time to undergo certain characteristic shifts in habitat preference.

Geographic Variation

Of the 35 full endemics, only 12 exhibit any geographic variation. This is largely because they have few opportunities to do so. Thirteen species cannot vary geographically because they are confined to a single island. Seven have few opportunities to vary because they are confined to fragments of a single Pleistocene island. Of those seven, three in fact do not vary geographically, two fall into weak subspecies, and two others fall into megasubspecies. Two more species that do not vary geographically are confined to two islands that were not joined in the same Pleistocene island. Three more species are confined to three islands; one of those species does not vary geographically, one falls into three weak subspecies, and one falls into three subspecies of which two are megasubspecies. Finally, 10 species are widespread, occupying 6–42 islands, including many not joined in the Pleistocene. Four of those species exhibit no geographic variation, while six do vary geographically, all of them strongly (differentiation into megasubspecies, allospecies, or both).

Summary

Northern Melanesia has 35 endemic full species (including five monotypic endemic genera), most of which lack identifiable close relatives. Most have quite localized geographic ranges, being confined to either the Bismarcks or Solomons but not both, and confined within either archipelago to just a single island or to post-Pleistocene fragments of a single Pleistocene island. They are overwhelmingly confined to the largest Northern Melanesian islands. Both of these distributional peculiarities result from their low vagility. The endemic species, including all five of the endemic genera, are concentrated in the Solomons rather than in the Bismarcks. Five historical factors contribute to these localized distributions, including recent anthropogenic extinctions, isolated chance colonization by a poorly vagile ancestor, exclusion by competing related species, and range contractions due to evolutionary loss of dispersal ability or due to subsequent colonization by a competitor. All 35 of the full endemics are forest species, and a disproportionate percentage is confined to the mountains. Few exhibit geographic variation, because their restricted distributions give them little opportunity to do so.

PART VI

GEOGRAPHIC ANALYSIS: DIFFERENCES AMONG ISLANDS

25

Endemism Index

Our analysis in parts III, IV, and V of this book has been at the level of individual species. We asked how the 191 analyzed species of Northern Melanesian birds differ from each other in their patterns of geographic variation and speciation, and how those species differences can be explained. Now, in part VI, we switch to an analysis at the level of islands. We ask how islands differ from each other in the patterns of geographic variation and speciation of their avifaunas, and how those island differences can be explained.

Northern Melanesian islands vary greatly as regards geography. They come in all sizes, from giant New Britain (35,742 km^2) to tiny islands of a fraction of a hectare. Four islands reach elevations of more than 2,000 m, while many others barely rise above the high-tide mark. Some are located centrally within Northern Melanesia, some peripherally. Some islands are isolated, while others are clustered into discrete groups, which lie, in turn, at various distances from each other. Some islands were connected to each other during the Pleistocene, while others rose recently from the sea by uplift or volcanic activity and have never been connected. As we explain in this part of our book, all of these geographic factors affect the formation of endemic taxa and the completion of speciation.

Much of part VI consists of discussions of individual islands or island groups, of the differentiation that has taken place on them, and of the effectiveness of the geographic barriers that surround them. We begin, however, with quantitative comparisons of all Northern Melanesian islands by means of four numerical indices that express somewhat different measures of the effectiveness of geographic barriers. We call these indices the "endemism index" (this chapter), the "pairwise differentiation index" (chapter 26), and two different "pairwise nonsharing indices" (chapter 27).

The endemism index expresses, for each island, the average degree to which the differentiation of its species populations has proceeded. This index, in effect, measures the combined effectiveness of all the geographic barriers surrounding an island. The second index, the pairwise differentiation index, expresses for each pair of islands the average degree of differentiation between the islands' populations of those species that share the two islands. This index measures the effectiveness of the geographic barrier separating only those two islands. Finally, the two pairwise nonsharing indices express, for each pair of islands, what proportions of the islands' avifaunas are confined to one or the other of the islands and are not shared between the islands. These two nonsharing indices are alternative expressions of the effectiveness of the geographic barriers separating that pair of islands.

Calculation of an Endemism Index

Northern Melanesian islands vary greatly in the degree of differentiation of their avifaunas. Some islands have not only numerous endemic subspecies, but also endemic allospecies, endemic isolated species, and even endemic genera. Other islands have no endemic taxa at all: all of their bird populations are shared even at the subspecies level with other islands.

To analyze these island differences, we have calculated a simple index of the average degree of differentiation of an island's bird populations. Increasing weight is given to higher degrees of differentiation. A weight of zero is assigned to populations that have not differentiated even subspecifically; a weight of 1 is assigned to populations endemic at the level of a weak subspecies; and weights of 2, 3, 4, and 5 are assigned to populations endemic at the level of a megasubspecies, allospecies, isolated full species, and genus, respectively. This measure of differentiation of each population is then averaged over all bird populations on the island (including those that have not differentiated even subspecifically) to obtain an average endemism index, *Y*, for the island's populations:

$$Y = \frac{(1)(s) + (2)(M) + (3)(A) + (4)(S) + (5)(G)}{(\text{total number of species})}, \tag{25.1}$$

where *s*, *M*, *A*, *S*, and *G* stand, respectively, for the numbers of populations endemic at the level of weak subspecies, megasubspecies, allospecies, full species, and genus. The analysis is confined to the 191 species of Northern Melanesian birds analyzed in part V of this book.

Two practical details arise in calculating this index. First, three Solomon islands with high mountains (Guadalcanal, Bougainville, and Kulambangra) are well isolated from each other and from other mountainous islands and consequently have numerous endemic taxa of montane birds. However, on all three islands the endemism of the lowland avifauna considered alone is much lower than that of the montane avifauna; in fact, Kulambangra has no lowland endemics at all. Hence for these three islands we calculate two alternative versions of the endemism index: one based on the island's entire avifauna (including the montane populations), and another based solely on the lowland avifauna.

Second, a complication arises with respect to large islands surrounded by smaller satellite islands. The avifaunas of the satellites are overwhelmingly shared at both the species and subspecies level with their larger neighbor. Conversely, while each satellite thus has only a few endemic taxa of its own, the big island has many endemic taxa, and in addition there are numerous taxa endemic to the big island plus one or more of its small satellites. A typical example involves the big, mountainous island of New Ireland, adjacent to which is the smaller, low island of New Hanover that was joined to New Ireland at Pleistocene times of low sea level, plus five much smaller islands or island groups (Dyaul, Tabar, Lihir, Tanga, and Feni) that had no Pleistocene connections to New Ireland. Eleven endemic taxa are confined to New Ireland; five are confined to New Ireland plus New Hanover; two to New Ireland plus Dyaul; two to New Ireland plus New Hanover and Dyaul; and seven to New Ireland plus one or more of the satellites Tabar, Lihir, Tanga, and Feni, plus in three cases Dyaul. Thus, the New Ireland group has 27 endemics, of which 11 are strictly confined to New Ireland and 16 others are shared with one or more of its satellites. Similar situations arise with respect to New Britain, Manus, and San Cristobal as well.

In these cases we have calculated an endemism index for the island group (the big island plus its satellites), and separate indices for individual satellites to reflect taxa strictly confined to that satellite. In this chapter we discuss how many of the taxa that we list as endemic to the larger island are strictly confined to that larger island, as opposed to taxa that the larger island shares with its satellites. Not surprisingly, the taxa strictly endemic to the larger island and not shared with its satellites are drawn disproportionately from the higher levels of endemism. For example, of the 27 endemic taxa of the New Ireland group as defined above, one full species, four allospecies, and only six weak subspecies are strictly confined to New Ireland, while no full species, only two allospecies, but 14 weak subspecies are shared with 1 or more of the satellites. The reason is obvious: those taxa that succeeded in colonizing satellites were also more likely to spread to other island groups as well and thus not to be counted as endemics of the island group under consideration.

Interpretation of the Endemism Index

For each ornithologically significant Northern Melanesian island, table 25.1 lists its total number

Table 25.1. Bird endemism on Northern Melanesian islands.

Island	Area (km^2)	Distance (km)	Total species	Endemic					Endemism index
				Subspecies	Mega-subspecies	Allospecies	Species	Genus	
1. Bismarcks	49,658	46 (NG)	146	41	14	30	7	—	1.28
2. New Britain	35,742	30 (11)	127	18	3	11	3	—	0.54
3. Duke of York	52	9 (2)	45	—	—	—	—	—	0
4. Umboi	816	25 (2)	84	3	—	—	—	—	0.04
5. Vuoatom	14	8 (2)	52	—	—	—	—	—	0
6. Lolobau	61	5 (2)	54	—	—	—	—	—	0
7. Witu	57	63 (2)	32	—	—	—	—	—	0
8. Long	329	50 (NG)	52	—	—	—	—	—	0
9. Tolokiwa	46	22 (4)	44	1	—	—	—	—	0.02
10. Crown	14	10 (8)	31	—	—	—	—	—	0
11. New Ireland	7,174	30 (2)	103	20	—	6	1	—	0.41
12. New Hanover	1,186	30 (11)	75	6	—	—	—	—	0.08
13. Dyaul	110	14 (11)	44	3	3	—	—	—	0.21
14. Tabar	218	26 (11)	61	4	1	—	—	—	0.10
15. Lihir	205	47 (11)	58	3	—	—	—	—	0.05
16. Tanga	98	59 (11)	39	1	—	—	—	—	0.03
17. Feni	110	55 (11)	39	2	—	—	—	—	0.05
18. Tingwon	1.5	29 (12)	13	—	—	—	—	—	0
19. St. Matthias	414	96 (12)	43	16	2	2	—	—	0.60
20. Emirau	41	17 (19)	32	—	—	—	—	—	0
21. Tench	0.26	74 (20)	11	—	—	—	—	—	0
22. Manus	1,834	241 (19)	50	16	4	4	—	—	0.72
23. Rambutyo	88	43 (22)	30	—	1	—	—	—	0.07
24. Ninigo	13	235 (NG)	16	1	—	—	—	—	0.06
25. Hermits	10	170 (22)	16	—	—	—	—	—	0
26. Anchorites	0.52	174 (22)	9	—	—	—	—	—	0
27. Wuvulu	15	162 (NG)	17	—	—	—	—	—	0
28. Solomons	36,000	171 (11)	141	30	21	30	17	5	1.81
29. Nissan	37	63 (31)	29	2	1	—	—	—	0.14
30. Bukida group	22,512	51 (62)	119	22	8	7	5	4	0.83
31. Buka	611	1 (32)	64	—	—	—	—	—	0
32. Bougainville	8,591	52 (36)	103	10	2	1	—	1	0.21
33. Bougainville lowlands	8,591	52 (36)	85	3	—	—	—	—	0.04
34. Shortland	232	8 (32)	56	—	—	—	—	—	0
35. Fauro	71	11 (32)	51	—	—	—	—	—	0
36. Choiseul	2,966	52 (32)	71	2	—	—	—	1	0.10
37. Ysabel	4,095	36 (38)	75	3	—	—	—	—	0.04
38. Florida	381	24 (39)	60	3	—	—	—	—	0.05
39. Guadalcanal	5,281	24 (38)	102	21	2	—	—	1	0.29
40. Guadalcanal lowlands	5,281	24 (38)	79	11	1	—	—	—	0.16

(continued)

Table 25.1. *Continued.*

Island	Area (km^2)	Distance (km)	Total species	Endemic: Subspecies	Endemic: Mega-subspecies	Endemic: Allospecies	Endemic: Species	Endemic: Genus	Endemism index
41. Mono	73	29 (34)	42	1	—	—	—	—	0.02
42. Savo	31	13 (39)	34	—	—	—	—	—	0
43. New Georgia group	4,924	53 (30)	85	13	4	5	2	—	0.52
44. Gatumbangra group	3,566	9 (58)	82	5	1	—	2	—	0.18
45. Gatukai	109	8 (46)	59	—	—	—	—	—	0
46. Vangunu	544	0.6 (47)	64	—	—	—	—	—	0
47. New Georgia	2,044	0.6 (46)	65	—	—	—	—	—	0
48. Kohinggo	95	0.1 (47)	61	—	—	—	—	—	0
49. Wana Wana	69	2 (48)	58	—	—	—	—	—	0
50. Kulambangra	704	2 (48)	82	3	—	—	2	—	0.13
51. Kulambangra lowlands	704	2 (48)	67	—	—	—	—	—	0
52. Vellonga group	805	24 (50)	65	2	2	1	—	—	0.14
53. Vella Lavella	640	9 (54)	65	1	1	1	—	—	0.09
54. Ganonga	142	9 (53)	54	1	1	1	—	—	0.11
55. Simbo	13	7 (54)	42	—	—	—	—	—	0
56. Gizo	35	12 (53)	60	—	—	1	—	—	0.05
57. Rendipari group	502	9 (47)	64	1	2	—	—	—	0.08
58. Rendova	381	4 (59)	63	—	1	—	—	—	0.03
59. Tetipari	122	4 (58)	55	—	1	—	—	—	0.04
60. Borokua	4.0	59 (45)	14	—	—	—	—	—	0
61. Russells	176	46 (39)	47	3	1	—	—	—	0.11
62. Malaita	4,307	47 (63)	74	10	5	2	1	—	0.41
63. San Cristobal	3,090	47 (62)	76	14	6	8	3	1	0.88
64. Ugi	42	8 (63)	49	—	2	—	—	—	0.08
65. Three Sisters	11	20 (63)	37	1	—	—	—	—	0.03
66. Santa Anna + Santa Catalina	20	8 (63)	48	1	—	—	—	—	0.02
67. Rennell	684	168 (63)	39	8	7	5	—	—	0.95
68. Bellona	20	23 (67)	18	—	—	—	—	—	0
69. Ontong Java	9.6	237 (37)	9	—	—	—	—	—	0
70. Sikaiana	1.3	175 (62)	6	—	—	—	—	—	0

The Bismarcks consist of islands 2–27; the Solomons, 28–70; the Bukida Group, 31–40; the New Georgia Group, 44–59; the Gatumbangra Group, 45–51; the Vellonga Group, 53–54; the Rendipari Group, 58–59; and the San Cristobal Group, 63–66.

The column "distance" gives island isolation, measured as distance from the nearest major colonization source identified by island number in parentheses (e.g., the major source nearest New Britain is island number 11, New Ireland, 30 km distant; NG = New Guinea). For a few islands the source island identified in this table differs from the one identified in table 9.1; this table identifies a more proximate source with more shared endemics.

The columns "endemic" give the number of taxa endemic to the island at the level of weak subspecies, megasubspecies, allospecies, full species, and genus, respectively. The column "endemism index" gives the average level of endemism of the island's species, on a scale ranging from 0 for non-endemic taxa to 5 for endemic genera (see p. 194 for formula). Data are tabulated for the isolated high Solomon islands Bougainville, Guadalcanal, and Kulambangra including or excluding their montane species (e.g., nos. 32 vs. 33, Bougainville vs. Bougainville lowlands). In cases of some larger islands that share some endemic taxa with one or more nearby smaller satellite islands, the shared taxa are counted among the endemics of the large island. Those cases consist of taxa shared between New Britain and islands 3–10; New Ireland and islands 12–17; St. Matthias and island 20; Manus and islands 23–27; Bougainville and islands 31, 34, 35, and 41; San Cristobal and islands 64–66; and Rennell and island 68.

At Pleistocene times of low sea level, islands 3, 5, 6, and possibly 4 were joined to New Britain (island no. 2); island 12, to New Ireland (island no. 11); and each island of the Bukida group, Gatumbangra group, Vellonga group, and Rendipari group, to the other members of its respective group.

Active volcanoes are islands nos. 6, 8, 42, and 55. Young volcanoes, or islands recently defaunated by volcanic explosions or by ash fallout from explosions of neighboring volcanic islands, are islands nos. 3, 5, 7, 9, and 10.

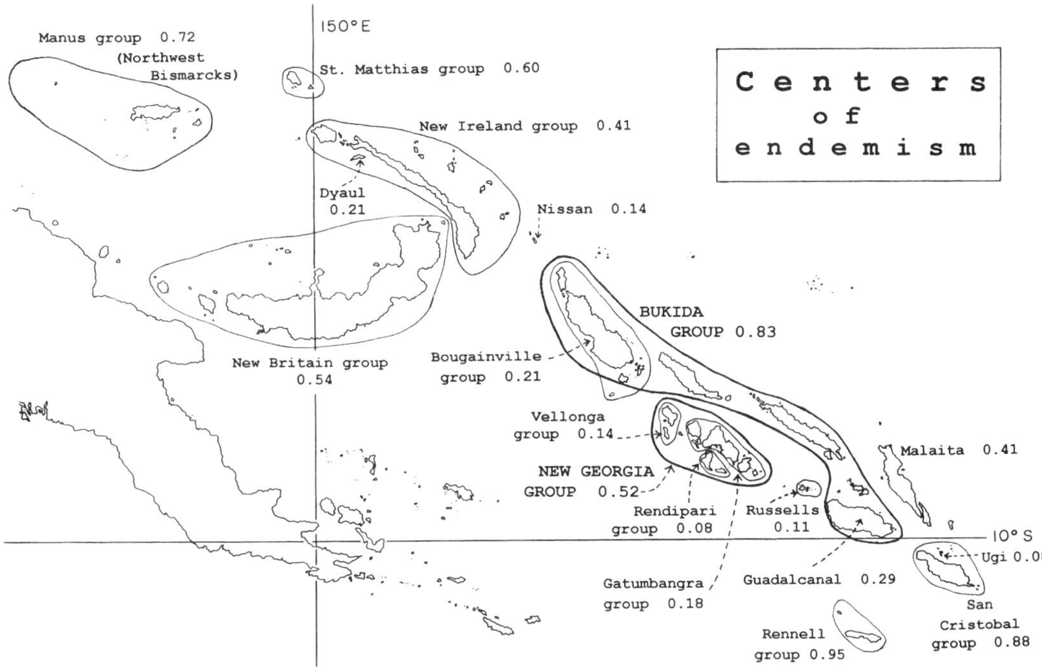

Fig. 25.1. Centers of endemism in Northern Melanesia. This map depicts values of the endemism index (defined in eq. 25.1 and tabulated in table 25.1) for the islands or island groups with the highest value of the index. The index expresses the level of endemism, averaged over the island's (or group's) entire avifauna, on a scale ranging from 0 (subspecifically identical to populations on other islands) and 1 (weak endemic subspecies) to 5 (endemic genus). Note, for example, that the Rennell, San Cristobal, Bukida, and Manus groups have the most distinct avifaunas (highest average level of endemism).

of species, its number of endemic taxa at each level of endemism, its average endemism index calculated by the above formula, its area, and its distance from its nearest larger neighbor or major colonization source, and figure 25.1 is a map of the major centers of endemism identified by this calculation. The highest values of the endemism index are for the Solomon Archipelago as a whole (1.81) and for the Bismarck Archipelago as a whole (1.28). That is to say, the avifaunas of these archipelagoes are well differentiated from each other and from those of the neighboring archipelagoes (New Guinea, Santa Cruz Archipelago, and New Hebrides Archipelago), with the Solomon avifauna being more distinct than the Bismarck avifauna. Within Northern Melanesia, the most distinct islands are Rennell (endemism index 0.95), San Cristobal (0.88), the Bukida group (0.83), and Manus (0.72), followed by St. Matthias (0.60), New Britain (0.54), the New Georgia group (0.52), and Malaita and New Ireland (0.41 each). The isolated, mountainous Solomon islands Guadalcanal, Bougainville, and Kulambangra have indices of 0.29, 0.21, and 0.13, respectively, for their total avifaunas, but these indices decline to 0.16, 0.04, and 0, respectively, for their lowland avifaunas alone. Only eight other Northern Melanesian islands have indices of 0.1 or higher (0.21–0.10, in descending sequence Dyaul, the Gatumbangra group, Nissan, the Vellonga group, the Russells, Ganonga, Tabar, and Choiseul). Many islands have no endemics whatsoever (indices of 0.0; e.g., Buka, Shortland, Bellona, and Simbo in the Solomons, and Vuatom, Duke of York, Witu, and Emirau in the Bismarcks).

To understand these island differences in level of endemism, let us initially postpone discussion of two groups of islands that have low endemism indices for obvious reasons: young volcanoes and neighboring islands recently defaunated by their explosions; and 24 islands recently joined to other islands by Pleistocene land bridges. This leaves for

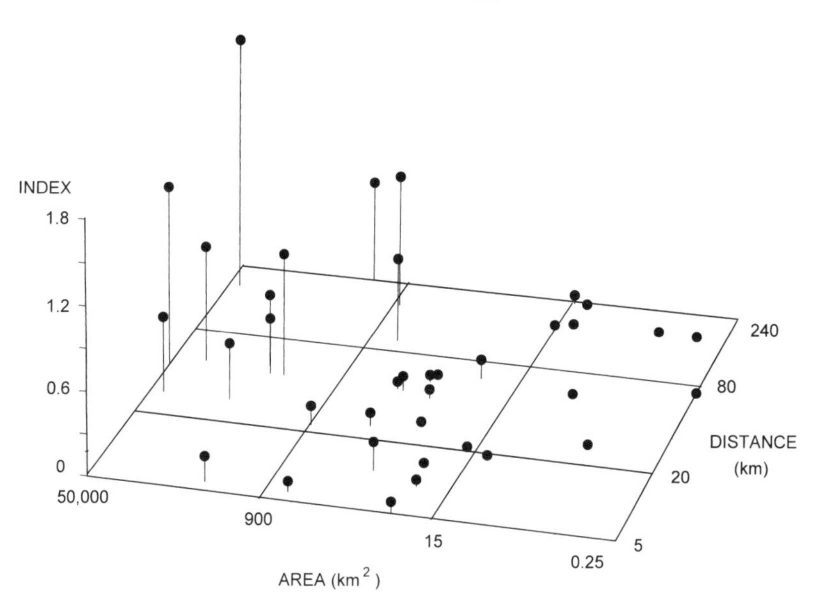

Fig. 25.2. Values of the endemism index for avifaunas of Nothern Melanesian islands or island groups, as a function of island area (km^2) and distance from the nearest major colonization source (km). Data are from table 25.1, for avifaunas of islands not joined by Pleistocene land bridges to larger islands and not recently defaunated by volcanism. The index (see eq. 25.1 for definition) measures the average level of endemism of all species populations in the island avifauna; a high value means a strongly differentiated avifauna. The area and distance scales are logarithmic. Note that index values increase with island area and isolation: only the largest and most isolated islands have high values. See text for discussion and statistical analysis.

discussion 37 islands (islands 1, 2, 11, 13–30, 41, 43, 44, 52, 56, 57, 60–65, and 67–70 of table 25.1) that have been isolated from each other and that have been able to support life continuously for longer times. As figure 25.2 illustrates, endemism scores for these islands define two clear patterns: endemism increases with island area, and it also increases with island isolation from the nearest colonization source. Only two islands with area ≤35 km^2 (Three Sisters and Ninigo, with one weak subspecies each), and only five islands with isolation <20 km (Dyaul, Gatumbangra, Gizo, Rendipari, and Ugi, with index values of only 0.08–0.21), have any endemics. All islands with index values >0.40 (named in the preceding paragraph) are both large (area > 400 km^2) and remote (isolation ≤30 km). Identical patterns apply when endemism indices are calculated for all the major archipelagoes of the tropical Southwest Pacific contrasted with each other (see fig. 4 of Diamond 1980, fig. 3 of Diamond 1984c).

Statistical analysis confirms these conclusions derived from visual inspection of figure 25.2. An equation with only two adjustable parameters gives an excellent fit to the data for the 37 islands and explains 86% of the variation in the value of the endemism index, Y:

$$Y = 0.00595\ (AD)^{0.36}, \qquad (25.2)$$

where A is island area and where D is distance from the nearest major colonization source (table 25.1). An equation with three adjustable parameters yields a slightly but significantly ($p = .024$ by the F-test) better fit and explains 88% of the variation in Y:

$$Y = 0.00445\ (A)^{0.33}\ (D)^{0.50}. \qquad (25.3)$$

Thus, the endemism index increases nonlinearly with both island area and isolation, and these two variables account remarkably well for the great variation in avifaunal endemism among Northern Melanesian islands.

The causes of these two patterns become clear on reflection. First, as for why level of endemism should increase with island area, consider that a population must survive uninterruptedly for some time if it is to differentiate. A population that lasts only a few years is unlikely to differentiate even subspecifically; a population that lasts thousands of years may have time to differentiate subspecifically but is unlikely to reach the level of an endemic full species; and populations that last millions of years may have sufficient time to differentiate to the level of endemic full species or genus. But overwhelmingly, the strongest determinant of population lifetime is population size, which (all other things being equal) is proportional to island area (Diamond 1984f, Pimm et al. 1988). On small islands, where populations are composed of few individuals, few populations survive long enough to differentiate to higher endemism levels. On large islands, where many species have large populations that survive for a long time, a higher fraction of populations can survive long enough to differentiate up to higher endemism levels. This pattern is one that Mayr (1965a) discerned previously, when he compared levels of endemism among island archipelagoes differing in area but at a similar distance from the nearest major colonization source.

Nevertheless, area alone is not the sole determinant of level of endemism. Elsewhere in the Pacific and other oceans, there are small islands with distinct endemics, such as Wake Island (7 km^2) and Pitcairn Island (5 km^2). In Northern Melanesia no island so small supports even a weak endemic subspecies. The explanation, of course, is that the small islands supporting endemic bird species are very remote ones in the world's oceans. Figure 25.2 shows that, for islands of a given area, the endemism index even within Northern Melanesia tends to increase with distance from the nearest colonization source.

There are two reasons for this effect of distance, both related to the decrease in immigration rates with increasing isolation. First, at early stages of differentiation, when a population is not yet reproductively isolated from conspecifics on other islands, immigrants from those other islands introduce their genes into the local gene pool and thereby inhibit the formation even of endemic subspecies. Second, once reproductive isolation has been achieved and the level of endemic allospecies has been reached, immigrants no longer inhibit differentiation by introducing their genes, but species number does increase with immigration rates and thus decreases with distance (MacArthur & Wilson 1967). But more coexisting species also mean more competition, and extinction rates rise steeply with species number (Gilpin & Diamond 1981). Thus, the explanation for why tiny, remote islands such as Wake and Pitcairn have endemic species, while islands of similar size within Northern Melanesia do not, is twofold: phyletic evolution (including subspeciation) proceeds more rapidly on the remote islands because fewer conspecifics arrive from elsewhere to dilute the differentiating gene pool with their genes; and in the steady state the remote islands harbor fewer species and hence less interspecific competition, so that populations that have already reached the level of reproductive isolation are less likely to go extinct before they have gone on to the level of endemic full species or genus.

The effects of area and distance evidenced for Northern Melanesian birds in figure 25.2 also operate on birds elsewhere in the world, and on other taxa. For example, the Pacific islands with the highest percentage of avian endemics consist of huge New Guinea, despite its proximity to Australia, big and isolated New Zealand, and the smaller but very remote Hawaiian Archipelago and the Galapagos. Within a given archipelago, endemism is higher in beetles than in birds, and higher in birds than in ferns. These taxonomic differences in endemism can be related to taxonomic differences in population density and immigration rates, and hence in population lifetimes and differentiation (Diamond 1984c): beetles tend to have higher population densities (hence lower extinction rates) than birds, but ferns have higher immigration rates than birds. We have already seen in part V of this book that the effect of population size on population lifetime is a major determinant of species differences among birds in their proneness to geographic differentiation.

An improved measure of island isolation would surely explain an even higher percentage of the variation in our endemism index than does equation 25.3. Our measure of isolation (distance from the single nearest major colonization source) is crude. Naturally, effective isolation depends on immigrants arriving from all sources, not just from the single nearest major source. Islands with just one nearest neighbor are effectively more isolated than islands surrounded by several neighbors at the same distance.

For example, San Cristobal, Malaita, and the New Georgia group are all located at similar distances from their nearest neighbor (about 50 km) and are of rather similar areas (3,000–5,000 km^2). However, the island with the highest endemism index is the smallest of these three islands (San Cristobal, index 0.88), rather than the largest islands as initially expected (New Georgia with 0.52, Malaita with 0.41). The reason is that San Cristobal is a peripheral island at the eastern end of the Solomons, with a narrow end separated by 50 km from the narrow ends of Malaita and Guadalcanal (fig. 25.1). In contrast, Malaita is aligned along its entire length with Guadalcanal and is also pointed at two other islands (San Cristobal and Ysabel), while the New Georgia group is aligned along its entire length with the Bukida group (fig. 25.1). That is, although San Cristobal, Malaita, and the New Georgia group all have the same crude isolation measure of about 50 km as measured by distance from nearest colonization source, San Cristobal is actually much less accessible to immigrants than are Malaita and the New Georgia group. Out of the 37 islands depicted in figure 25.2, San Cristobal and Rennell are the two most deviant (with unexpectedly high values of the endemism index) and are the only two islands that yield significant misfits to equation 25.3 (Z^2 scores corresponds $p < .002$ and $p < .001$, respectively). These deviations are related to their being the most peripheral Solomon islands at the southeast end of the Solomon chain, as discussed further in chapter 31. Thus, a future extension of our analysis will be to develop a more complex and predictive measure of isolation, which involves an integral of all source area at a given distance, weighted by some inverse measure of distance and integrated over distance.

Volcanically Defaunated Islands

The islands that we have been discussing so far have avifaunas believed to have been in continuous existence for at least tens of thousands of years. In addition, there are 10 other ornithologically explored islands (nine in table 25.1, plus Ritter, not included in the table) whose former avifaunas must have been destroyed by volcanic activity and whose existing avifaunas are presumably only a few centuries or millennia old: Crown, Duke of York, Lolobau, Long, Ritter, Savo, Simbo, Tolokiwa, Vuatom, and Witu. Among these 10 islands, Lolobau, Long, Savo, and Simbo are still volcanically active. Long and Ritter must have been defaunated by volcanic explosions that removed most of the volume of these two islands in the seventeenth and nineteenth centuries, respectively. Fallout beds of Long ash on Long's neighbor Crown, and the proximity and faunal composition of Tolokiwa, suggest that these two nearby islands were defaunated by the Long eruption. Ash beds on Duke of York and on Vuatom suggest that they were defaunated by the eruption of New Britain's nearby Rabaul Volcano 1,400 years ago. The form of Witu's caldera suggests that it is recent, and in addition Witu may have been damaged by the eruption of the nearby Dakataua Volcano on New Britain 1,150 years ago. Lolobau, in addition to being active itself, may have been similarly damaged by eruptions of Ulawun Volcano and other active nearby volcanoes on New Britain.

Except for Long and Tolokiwa, these volcanically disturbed islands lack endemic bird populations. Most of these islands are in any case sufficiently small and close to neighbors that equation 25.3 would lead one to expect low values of the endemism index even if the islands were undisturbed. The island for which equation 25.3 predicts the highest endemism index (0.22) is the largest and second-most remote of these volcanic islands, Long. Its slight endemism is surely because volcanic activity destroyed its former avifauna.

The main exception to this lack of endemism on volcanically disturbed islands is a weak subspecies of thrush, *Turdus poliocephalus tolokiwae*, on the summit of Tolokiwa (Diamond 1989). Populations of this thrush are extremely prone to differentiate: its 52 recognized subspecies are exceeded in number among Northern Melanesian birds only by the subspecies of the whistler *Pachycephala pectoralis*. The Tolokiwa thrush differs from known populations of the species on New Guinea and other Northern Melanesian islands in being somewhat grayer and less brown. Either this modest color difference arose within the few centuries available for vegetation on Tolokiwa to regenerate since the Long Island ash fallout of the seventeenth century; or Tolokiwa's thrush and some forest managed to survive the Long ash fallout; or the Tolokiwa thrush population is derived from neighboring New Britain's thrush population, which has been observed but not collected, so that it is unknown whether the Tolokiwa thrush differs from the New Britain thrush.

Land-bridge Islands

Twenty-four islands of table 25.1 (numbers 4, 12, 31–40, 45–51, 53, 54, 58, 59, and 66) were not volcanically defaunated but were joined at Pleistocene times of low sea level into much larger islands, or (in the case of Guadalcanal and Umboi) were nearly joined to other islands by such Pleistocene land emergence. For 21 of these 24 islands, the endemism index is below that predicted from equation 25.3 for islands of similar area and isolation that had no Pleistocene land connections. Collectively, the 24 land-bridge islands deviate significantly from equation 25.3 ($p = .018$ by a chi-square test). These deficits of endemism are greatest for the four largest such fragments of Pleistocene islands (the Bukida fragments Bougainville, Guadalcanal, Ysabel, and Choiseul) because their large area would otherwise have endowed them with many endemics. The reason, of course, is that Pleistocene land bridges connected the lowland avifaunas of what are now these islands to each other or to avifaunas of other now-separate islands, thereby tending to homogenize their lowland avifaunas.

Summary

Northern Melanesian islands vary greatly in geographic parameters such as area, distance, and location. This chapter and the following chapters examine how these geographic parameters cause islands to differ from each other in the patterns of geographic variation and speciation of their avifaunas. We calculate four numerical indices to express somewhat different measures of the effectiveness of the geographic barriers setting off each island's avifauna.

The first index calculates the average level of taxonomic differentiation (ranging from not even subspecifically distinct up to endemic genus) of an island's bird populations. By the criterion of this endemism index, the ornithologically most distinctive Northern Melanesian islands are Rennell, San Cristobal, the Bukida Group, and Manus. The numerical value of this endemism index increases with island area (because of the steep decline in extinction rates with area) and with island isolation (because of the decline in gene flow, and in species number and hence competition and extinction rates, with isolation). The equation $Y = 0.00445\ A^{0.33}\ D^{0.50}$, where A is island area (km^2) and D is isolation (km), accounts for 88% of the variation in the endemism index. A task for the future will be to calculate a more refined measure of isolation than merely nearest-neighbor distance, taking into account the areas, shapes, and orientations of all potential source islands at different distances. As expected, islands that have recently been defaunated by volcanic activity, or that were connected to other islands at Pleistocene times of low sea level, have values of the endemism index lower than for undisturbed and long-isolated islands with the same area and isolation.

26

Pairwise Differentiation Index

The endemism index discussed in chapter 25 analyzes the entire avifauna of a particular island. In effect, it compares that avifauna with the avifaunas of all other islands. The endemism level for a particular species population on the given island is taken as the lowest level of differentiation found when the population is compared to related populations on any other island. That is, if for a certain species on island A there is a population on some other island B that is differentiated only at the level of weak subspecies from the population on island A, the endemism level of that species on island A is taken as only that of a weak subspecies, even if the population on island A differs at the level of megasubspecies or allospecies from related populations on all other islands.

The pairwise differentiation index considered in this chapter differs in two respects from the endemism index considered in chapter 25. First, the analysis in the present chapter is confined to a particular pair of islands. If some species X is represented by different megasubspecies on two islands, then the differentiation level for that species is taken as megasubspecies, even though the populations of each island in turn may be subspecifically identical to the populations on some other islands. Second, the analysis is confined to those species (zoogeographic species, including superspecies) actually shared between the two islands and ignores species absent from one of the islands. For those shared species (not the entire avifauna), we calculated the average level of differentiation by the same formula used to calculate the endemism index (eq. 25.1).

Examples of the Pairwise Differentiation Index

Two examples illustrate the differences between the pairwise differentiation index and the endemism index. First, the endemism index of New Hanover is only 0.08, because it shares all except two of its 75 species with the much larger nearby island of New Ireland, to which it was joined at Pleistocene times of low sea level. The New Hanover populations of only eight of those shared species are differentiated at the level even of a weak subspecies from their New Ireland conspecifics, and the two New Hanover species not shared with New Ireland are subspecifically identical to populations on other islands. Of the eight populations differing as weak subspecies from their New Ireland conspecifics, only six constitute subspecies endemic to New Hanover; the other two are subspecifically identical to populations on other islands. Thus, the endemism index of New Hanover is only 0.08, largely because of sharing of species and subspecies with

New Ireland. That sharing is reflected in a pairwise differentiation index of only 0.11 for the shared avifaunas of New Ireland and New Hanover. However, the pairwise index for New Hanover and the next nearest large island, St. Matthias, is much higher, 0.64, because of the permanent wide water gap (now 96 km) between New Hanover and St. Matthias.

As a second example, New Britain, the largest Northern Melanesian island, has an endemism index of 0.57, largely because of its numerous endemic full species and allospecies. However, for those species that New Britain shares with its nearest large neighbor of New Ireland, the pairwise differentiation index is only 0.36; populations of most shared species (75 out of 96) belong to the same subspecies. For New Guinea, three times farther from New Britain than is New Ireland, the pairwise differentiation index for shared species is considerably higher, 1.19; populations of most of the 112 shared species belong to different subspecies (39), different megasubspecies (11), or different allospecies (24).

For selected pairs of Northern Melanesian islands, table 26.1 lists the level of differentiation of all shared species (including superspecies), along with the distance between the islands, the area of each island, and the mean area of the two islands. The island pairs selected consist of neighboring islands, plus some other island pairs of interest for comparison. We calculated differentiation indices for only a few pairs involving young volcanoes or recently defaunated islands because those indices were invariably zero or close to zero.

Geographic Barriers in Northern Melanesia

Northern Melanesia's major geographic barriers, as deduced from values of the pairwise differentiation index, are mapped in figure 26.1. The highest values of the index, and hence Northern Melanesia's most important geographic barriers to differentiation, are between the Bismarcks and New Guinea (index value 1.19 for New Guinea/New Britain, 1.15 for New Guinea/Manus); between the Bismarcks and the Solomons (Bougainville/New Britain 1.05, Bougainville/New Ireland 1.00); and between the Solomons and the Santa Cruz and New Hebrides Archipelagoes lying to the east (1.14–1.50 for the barrier separating San Cristobal, Rennell, or Guadalcanal from the New Hebrides or Santa Cruz). Within the Bismarcks, the major barriers are those separating the large, isolated islands of Manus and St. Matthias from their neighbors (Manus/New Hanover 0.79; Manus/St. Matthias 0.50; St. Matthias/New Hanover 0.64), followed by the barrier between the much closer but much larger Bismarck islands New Britain and New Ireland (0.36). Within the Solomons, the most important barriers are those surrounding the most remote, medium-sized island, Rennell (0.79 and 0.75 vs. Guadalcanal and San Cristobal, respectively) and the most peripheral, large island, San Cristobal (0.70 and 0.65 vs. Guadalcanal and Malaita, respectively). Other major barriers in the Solomons are those surrounding the next-most-peripheral large island, Malaita (0.65, 0.45, and 0.36 vs. San Cristobal, Ysabel, and Guadalcanal, respectively) and the barriers separating the New Georgia group from the Bukida group (0.42–0.45 for the barriers between New Georgia, Kulambangra, and Vella Lavella versus Choiseul or Ysabel). The lowest values of the pairwise differentiation index, and the weakest barriers, involve small and/or close islands, as well as islands joined by Pleistocene land bridges.

Dependence of the Index on Island Areas and Distance

For those 51 pairs of islands lacking Pleistocene land connections, values of the pairwise differentiation index increase with island areas and with distance between the islands (fig. 26.2). An equation with only three adjustable parameters gives a good fit to the data and explains 82% of the variation in the value of the differentiation index, W:

$$W = -1.33 + 0.12 \ln A + 0.23 \ln D, \tag{26.1}$$

where A is the mean area of the two islands, and D is the distance between them. More complex equations, with four or five adjustable parameters and incorporating terms in D^2 or in each island's separate area, explain 87% or 88% of the variation. As in the case of the endemism index of chap-

Table 26.1. Pairwise differences between island avifaunas.

Island 1/island 2	A_1/A_2 (km^2)	Mean A (km^2)	D (km)	Difference in species composition					Differentiation				
				S_1	S_2	S_J	Ochiai	$\frac{(S_2 - S_J)}{S_2}$	Same subspecies	Different subspecies	Different megaspecies	Different allospecies	Index
Bismarcks													
New Guinea/ New Britain	808,000/35,742	169,900	88	432	127	115	0.51	0.09	38	39	11	24	1.19
New Guinea/Manus	808,000/1,834	38,500	274	432	50	48	0.67	0.04	16	18	5	9	1.15
New Britain/ New Ireland	35,742/7,174	16,010	30	127	103	96	0.16	0.07	75	12	4	5	0.36
New Ireland/Dyaul	7,174/110	888	14	103	44	43	0.36	0.02	34	3	3	—	0.23
New Ireland/Tabar	7,174/218	1,251	26	103	61	58	0.27	0.05	48	8	2	—	0.21
New Ireland/Lihir	7,174/205	1,213	47	103	58	55	0.29	0.05	47	7	1	—	0.16
New Ireland/Tanga	7,174/98	838	59	103	39	35	0.45	0.10	30	5	—	—	0.14
New Ireland/Feni	7,174/110	888	55	103	39	35	0.45	0.10	29	6	—	—	0.17
New Hanover/ St. Matthias	1,186/414	701	96	75	43	33	0.42	0.23	19	9	3	2	0.64
St. Matthias/Emirau	414/41	130	17	43	32	32	0.14	0	32	—	—	—	0
Manus/St. Matthias	1,834/414	871	241	50	43	28	0.40	0.35	17	9	1	1	0.50
Manus/New Hanover	1,834/1,186	1,475	285	50	75	42	0.31	0.16	20	14	5	3	0.79
Manus/Rambutyo	1,834/88	402	43	50	30	28	0.28	0.07	27	—	1	—	0.07
Manus/Ninigo	1,834/13	154	255	50	16	11	0.61	0.31	9	2	—	—	0.18
Manus/Hermit	1,834/10	135	170	50	16	10	0.65	0.38	9	1	—	—	0.10
Manus/Anchorite	1,834/0.52	30.8	174	50	9	5	0.76	0.44	4	1	—	—	0.20
Solomons													
New Guinea/ Bougainville	808,000/8,591	83,400	636	432	103	77	0.63	0.25	20	28	9	20	1.38
New Hebrides/ San Cristobal	12,190/3,090	6,140	618	56	76	25	0.62	0.55	6	11	4	4	1.24
New Hebrides/ Guadalcanal	12,190/5,281	8,023	811	56	102	30	0.60	0.46	6	13	5	6	1.37
New Hebrides/ Rennell	12,190/684	2,888	657	56	39	22	0.53	0.44	5	9	5	3	1.27
Guadalcanal/ Santa Cruz group	5,281/829	2,092	541	102	33	21	0.64	0.36	7	7	4	3	1.14
San Cristobal/ Santa Cruz group	3,090/829	1,601	362	76	33	22	0.56	0.33	7	8	4	3	1.14
Santa Cruz group/ Rennell	829/684	753	574	33	39	20	0.44	0.39	5	5	5	5	1.50

Bougainville/Nissan	8,591/37	564	106	103	29	25	0.54	0.14	20	1	3	1	0.40
Buka/Nissan	611/37	150	63	64	29	22	0.49	0.24	19	1	1	1	0.27
Mono/Fauro	73/71	72	63	42	51	38	0.18	0.10	37	—	—	—	0
Shortland/Mono	232/73	130	29	56	42	39	0.20	0.07	38	1	—	—	0.03
Choiseul/New Georgia	2,966/2,044	2,462	54	71	65	57	0.16	0.12	41	10	3	3	0.44
Choiseul/Vella Lavella	2,966/640	1,378	53	71	65	58	0.15	0.11	42	9	3	3	0.42
Ysabel/New Georgia	4,095/2,044	2,893	88	75	65	60	0.14	0.08	42	12	3	3	0.45
Ysabel/Kulambangra	4,095/704	1,698	126	75	82	65	0.17	0.13	47	12	3	3	0.42
Kulambangra/ Vella Lavella	704/640	671	24	82	65	63	0.14	0.03	58	2	2	1	0.14
New Georgia/Rendova	2,044/381	882	8.5	65	63	61	0.05	0.03	58	1	2	—	0.08
Kulambangra/Gizo	704/35	157	14	82	58	58	0.16	0	57	—	—	1	0.05
Ganonga/Gizo	142/35	70	18	54	60	49	0.14	0.09	46	1	—	1	0.08
Russells/Gatukai	176/109	139	90	47	59	43	0.18	0.09	38	3	1	—	0.12
Guadalcanal/Russells	5,281/176	964	46	102	47	47	0.32	0	40	4	1	—	0.13
Malaita/Ysabel	4,307/4,095	4,200	78	74	75	65	0.13	0.12	46	10	8	1	0.45
Guadalcanal/Malaita	5,281/4,307	4,769	51	102	74	71	0.18	0.04	54	8	7	1	0.36
Malaita/San Cristobal	4,307/3,090	3,648	47	74	76	55	0.27	0.26	35	9	6	5	0.65
Guadalcanal/ San Cristobal	5,281/3,090	4,040	55	102	76	67	0.24	0.12	42	10	8	7	0.70
San Cristobal/Ugi	3,090/42	360	8.4	76	49	46	0.25	0.06	43	—	3	—	0.13
Russells/Ugi	176/42	86	294	47	49	35	0.27	0.29	27	4	3	—	0.29
Guadalcanal/Rennell	5,281/684	1,901	175	102	39	24	0.62	0.38	14	3	5	2	0.79
San Cristobal/Rennell	3,090/684	1,454	168	76	39	20	0.63	0.49	11	4	4	1	0.75
Rennell/Ugi	684/42	169	211	39	49	16	0.63	0.59	11	3	1	1	0.50
Rennell/Russells	684/176	347	270	39	47	12	0.72	0.69	7	3	2	—	0.58
Solomons/Bismarcks													
New Britain/ Bougainville	35,742/8,591	17,523	270	127	103	73	0.36	0.29	29	24	7	13	1.05
Bougainville/ New Ireland	8,591/7,174	7,851	172	103	103	66	0.36	0.36	26	24	6	10	1.00
New Ireland/Nissan	7,174/37	515	111	103	29	20	0.63	0.31	13	6	1	—	0.40
Feni/Nissan	110/37	64	58	39	29	22	0.35	0.24	17	4	1	—	0.27
Land-bridge pairs													
New Britain/Umboi	35,742/816	5,401	25	127	84	81	0.22	0.04	69	8	1	—	0.13
New Ireland/ New Hanover	7,174/1,186	2,917	30	103	75	72	0.18	0.04	64	8	—	—	0.11
Bougainville/Buka	8,591/611	2,291	1.0	103	64	63	0.22	0.02	63	—	—	—	0
Bougainville/Shortland	8,591/232	1,412	8.4	103	56	56	0.26	0	55	1	—	—	0.02
Bougainville/Choiseul	8,591/2,966	5,048	52	103	71	70	0.18	0.01	59	8	1	2	0.23

(continued)

Table 26.1. *Continued.*

Island 1/island 2	A_1/A_2 (km^2)	Mean A (km^2)	D (km)	Difference in species composition					Differentiation				
				S_1	S_2	S_J	Ochiai	$\frac{(S_2 - S_J)}{S_2}$	Same subspecies	Different subspecies	Different megaspecies	Different allospecies	Index
Choiseul/Shortland	2,966/232	830	73	71	56	55	0.13	0.02	48	5	1	1	0.18
Choiseul/Fauro	2,966/71	459	41	71	51	50	0.17	0.02	46	3	—	1	0.12
Shortland/Fauro	232/71	128	20	56	51	47	0.12	0.08	47	—	—	—	0
Ysabel/Choiseul	4,095/2,966	3,485	65	75	71	67	0.08	0.06	64	2	—	—	0.03
Ysabel/Florida	4,095/368	1,228	36	75	60	55	0.18	0.08	49	4	—	1	0.13
Guadalcanal/Florida	5,281/368	1,394	24	102	60	58	0.26	0.03	47	8	1	—	0.17
Guadalcanal/Ysabel	5,281/4,095	4,650	75	102	75	69	0.21	0.08	52	14	2	1	0.30
Vella Lavella/Ganonga	640/142	301	9.0	65	54	54	0.09	0	51	1	1	1	0.11
New Georgia/ Kulambangra	2,044/704	1,200	10	65	82	65	0.11	0	65	—	—	—	0
New Georgia/Vangunu	2,044/544	1,054	0.6	65	64	61	0.05	0.05	61	—	—	—	0
Vangunu/Gatukai	544/109	244	8	64	59	56	0.09	0.05	56	—	—	—	0
Rendova/Tetipari	381/122	216	3.5	63	55	54	0.08	0.02	53	—	1	—	0.04
Volcanic effects													
New Britain/ D. of York	35,742/52	1,363	13	127	45	45	0.40	0	45	—	—	—	0
New Britain/Vuatom	35,742/14	707	8	127	52	46	0.43	0.12	46	—	—	—	0
New Britain/ Witu Group	35,742/88	1,773	63	127	37	32	0.53	0.14	31	—	1	—	0.06
Ganonga/Simbo	142/13	43	7	54	42	40	0.16	0.05	40	—	—	—	0

The leftmost columns 1 and 2 name the two islands whose avifaunas are to be compared, with the larger island named first. The first 51 entries are for island pairs not connected to each other by Pleistocene land bridges, and not recently defaunated by volcanism nor risen from the sea. Among those 51 entries, the first two are New Guinea plus a Bismarck island; the next 14, a pair of Bismarck islands; the next seven, a Solomon island plus an archipelago (or an island in an archipelago) outside the Solomons; the next 24, a pair of Solomon islands; and the next four, a Solomon island plus a Bismarck island. The next 17 entries are island pairs joined or nearly joined to each other during the Pleistocene. The last four entries are island pairs of which the smaller has a young avifauna because of active volcanism, recent defaunation by ash fall-out, or recent volcanic origin.

Columns 3 and 4 are the areas A_1 and A_2 of islands 1 and 2, respectively; column 5, the mean area = $\sqrt{A_1 A_2}$); and column 6, interisland distance D.

Columns 7 and 8 are the number of species, S_1 and S_2, on the larger and smaller island, respectively; and column 9, the number of species, S_J, shared between the two islands. Column 10 ("Ochiai") gives the value of an index expressing the difference in species composition between the two islands, calculated as $1 - S_J/\sqrt{S_1 S_2}$, and derived from the Ochiai similarity index. Column 11 "$(S_2 - S_J)/S_2$," is another index expressing the difference in species composition where S_2 now refers to the more species-poor island (usually the smaller one); this index is 0 if all species of the more species-poor island also occur on the more species-rich island ($S_J = S_2$). Those two indices of columns 10 and 11 will be discussed in Chapter 27.

Columns 12–15 allocate the shared species (S_J) according to level of taxonomic differentiation between the populations of the two islands: same subspecies, different weak subspecies, different megaspecies, or different allospecies. For some island pairs the sum of columns 12–15 is less than S_J (column 9) because one or more populations of shared species have not been identified to subspecies. Column 16 ("Index") gives the average value (averaged over all S_J species) of this level of differentiation, assigning values of 0, 1, 2, or 3 to same subspecies, different weak subspecies, different megasubspecies, or different allospecies, respectively.

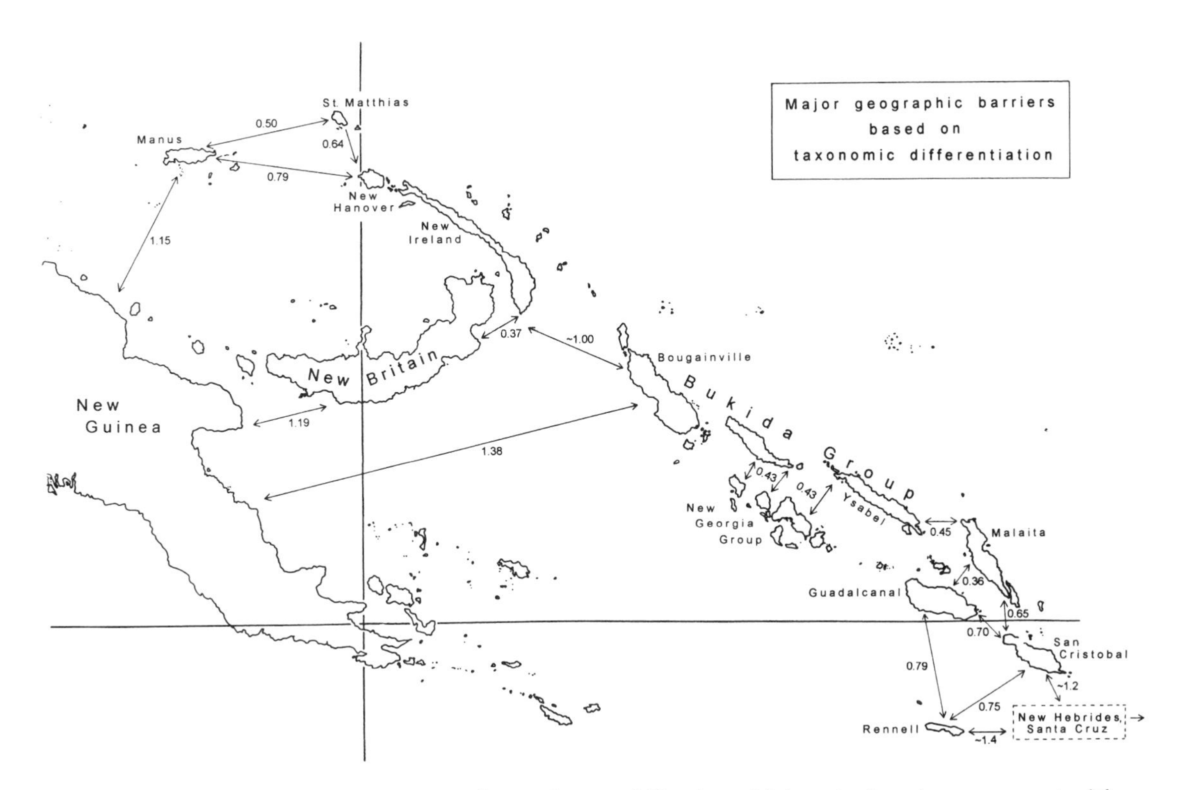

Fig. 26.1. Major geographic barriers within and around Northern Melanesia, based on taxonomic differentiation. Arrows connect island pairs with especially strongly differentiated avifaunas, indicating strong barriers to overwater dispersal. The number beside each arrow is the average level of differentiation (averaged over all shared species) between populations of species shared between the two islands, on a scale ranging from 0 (populations subspecifically identical) and 1.0 (different weak subspecies) to 3.0 (different allospecies). Data are from table 26.1. Note, for example, that the barrier between Bougainville and New Guinea (arrow marked 1.38) is among the strongest barriers.

ter 25, the most marked deviation from equation 26.1 involves Rennell: the Rennell/Santa Cruz island pair has a differentiation index of 1.50, higher than the predicted value of 0.88 (Z^2 score corresponding to $p < .001$), as will be discussed further in chapter 31.

The factors underlying those area and distance dependences of the differentiation index are similar to the factors discussed in the previous chapter in connection with the endemism index. The differentiation index between conspecific populations on pairs of islands increases with the distance between the islands, because of decreasing exchange of immigrants and decreasing gene flow between the islands. The differentiation index increases with island areas, because populations on larger islands survive for longer times and are thus more likely to last long enough to reach higher levels of differentiation.

Collectively, the 17 pairs of islands that were joined to each other by Pleistocene land bridges deviate significantly from equation 26.1 ($p = .0008$ by a chi-square test), mostly having index values lower than those for never-connected island pairs of similar distance and mean area, especially in the cases of the larger island pairs. The reason, of course, is that the avifaunas of those island pairs tended to become mixed and homogenized when the islands were joined during the Pleistocene. Island pairs that include an island recently defaunated by volcanism (last four entries of table 26.1) also have low values of the index because their avifaunas are too young for much differentiation to have occurred.

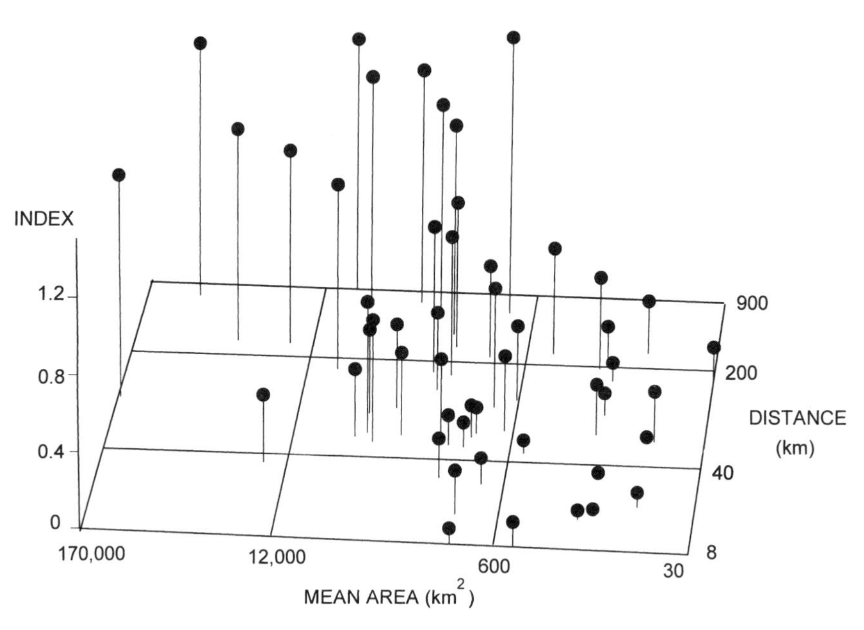

Fig. 26.2. Values of the pairwise differentiation index for avifaunas of pairs of Northern Melanesian islands, as a function of island mean area and the distance between the islands. Data are from table 26.1, for avifaunas of islands not joined by Pleistocene land bridges and not recently defaunated by volcanism. The index (see eq. 25.1 and second paragraph of this chapter for definition) measures the average level of taxonomic differentiation between conspecific populations on the two islands: high values mean strongly distinct avifaunas. The area and distance scales are logarithmic. Note that values tend to increase with island mean area and separation: only large pairs of islands, distant from each other, have high values. The most deviant point, the very high index value of 1.50 at the large distance of 574 km but modest mean area of 753 km^2, is for the Rennell/Santa Cruz island pair. See text for discussion and statistical analysis.

Summary

Our pairwise differentiation index indicates, for each of 51 pairs of nearby islands and for all species shared between those two islands, the average level of differentiation of the two islands' conspecific populations (i.e., whether they belong to the same subspecies or differ on the average at the level of weak subspecies, megasubspecies, or allospecies). High values of the index provide one means of defining the most important geographic barriers in Northern Melanesia. These prove to be the barriers between the Bismarcks and New Guinea, between the Bismarcks and Solomons, and between the Solomons and the archipelagoes to the east; the barriers around the large peripheral Bismarck islands of Manus and St. Matthias, and the large peripheral Solomon islands of Rennell and San Cristobal; and the barriers between New Britain and New Ireland, between the Bukida and New Georgia groups, and around Malaita. As expected, values of the pairwise differentiation index increase with distance between the two islands (because immigration rates decrease with distance) and with island areas (because extinction rates decrease with area). The equation $W = -1.33 + 0.12 \ln A + 0.23 \ln D$, where A is island mean area and D is the distance between the two islands, accounts for 82% of the variation in the index. Also as expected, islands that were connected by Pleistocene land bridges have lower values of the index than do never-connected islands of similar distance and areas.

27

Pairwise Nonsharing Indices

Differences in Island Species Compositions

In the previous two chapters we analyzed two indices of geographic barriers. This chapter introduces two more indices, closely related to each other, which evaluate barriers between island pairs, as did the pairwise differentiation index of chapter 26. However, whereas the pairwise differentiation index was based on taxonomic differentiation of populations of species shared between the two islands, our third and fourth indices are instead based on species not shared between the two islands—that is, the species of either island that are stopped completely by the water barrier between the islands. As discussed on the following pages, these two aspects of water barriers (differentiation of shared species vs. stopping of nonshared species) are related but not identical.

Causes of Pairwise Differences in Island Species Compositions

To understand the meaning of these indices, consider three types of cases. First, the Solomon island pair Malaita/Ysabel and the pair Malaita/San Cristobal both consist of islands similar in area and nearly identical in species number (74 vs. 75 species in the first case, 74 vs. 76 in the second case). However, the water barrier between each pair of islands constitutes a major barrier to species spread in either direction. For example, Malaita and Ysabel share only 65 species: nine of Malaita's species are absent on Ysabel, while 10 of Ysabel's species are absent on Malaita. In addition, many populations of the species shared between Malaita and Ysabel or between Malaita and San Cristobal are differentiated at the level of the subspecies or allospecies: the pairwise differentiation index is 0.45 for the former pair and 0.65 for the latter pair. Thus, the water barriers between these pairs of islands are strong barriers that stop the movement of species, as well as impede gene flow between conspecific populations.

Second and at the opposite extreme are pairs of nearby islands where all or almost all species of the smaller island occur on the larger island, but the larger island has many species absent on the smaller island. For example, all 32 species of the small Bismarck island Emirau occur on the nearby larger island of St. Matthias, whereas 11 of the 43 species of St. Matthias are absent on Emirau. Again, New Hanover shares 72 of its 75 species with the nearby larger island of New Ireland, but 31 of New Ireland's 103 species are absent on New Hanover. To use the terminology of Patterson and Atmar (1986), the avifauna of the smaller island constitutes a perfectly or near-perfectly "nested subset" of the avifauna of the larger island. Nesting arose in the former case because Emirau is a small satellite of St. Matthias and its avifauna is entirely derived over water from St. Matthias, while in the latter case New Ireland and New Hanover were joined at Pleistocene times of low sea level and the smaller

New Hanover has subsequently lost many more species populations of that Pleistocene shared avifauna by differential extinction than has the larger New Ireland. That is, in these cases of Emirau/St. Matthias and New Hanover/New Ireland, many species of the larger island are unable to maintain populations on the smaller island, for reasons related at least partly to small population sizes or to lack of suitable habitats on the smaller island; but all or nearly all species capable of maintaining populations on the smaller island are also able to maintain populations on the larger island.

Despite these substantial differences in species composition between the avifaunas of the smaller and larger island, the pairwise differentiation index for shared species is zero in the case of Emirau/St. Matthias and low (0.11) in the case of New Hanover/New Ireland, indicating that the intervening water barrier is only a weak barrier to gene flow between conspecific populations. These island examples approach the situation that would obtain within a single island if one compared species lists for nearby small and large census areas: many species recorded in the larger census area would be absent in the smaller area, but most or all species of the smaller area would be present in the larger area, and there would be no subspecific differentiation between populations of the two census areas.

Finally, still other pairs of islands, such as the New Britain/Vuatom or New Britain/Witu pairs, involve a large source island and a nearby small satellite island colonized mainly from that larger source, as in the case of Emirau/St. Matthias. Pairwise differentiation of populations of shared species is low or absent in the New Britain/Vuatom and New Britain/Witu cases, as in the Emirau/St. Matthias case. Also as in the Emirau/St. Matthias case, there are substantial differences in species composition between the avifaunas of the smaller and larger islands, despite the low pairwise differentiation of shared species. However, these two cases differ from the Emirau/St. Matthias case in that the small-island avifauna is not a nested subset of the large-island avifauna: not only does the big island have many species absent on the small island, but the smaller island also has five or six species absent on the larger island. Those latter absentees are supertramps as defined in chapter 12 and listed in table 12.2: vagile species confined as residents to small islands and competitively excluded as residents on large islands, despite frequent arrivals at large islands as vagrants. In effect, the supertramps are part of the large-island species pool but are confined to its fringing islets.

Thus, the water barrier between New Britain and Vuatom or Witu, just as the water barrier between Emirau and St. Matthias, constitutes only a weak barrier to gene flow and has produced little or no differentiation of shared species. Instead, the differences in island areas, population sizes, and habitats have produced some specialization of small-island and large-island avifaunas: small-island species absent on larger islands, as well as large-island species absent on smaller islands.

These three sets of examples illustrate that three distinct phenomena contribute to differences in species composition between neighboring islands: blockage of species movements in either direction by the water barrier, absences of many large-island species from small islands for ecological reasons, and absences of a few small-island supertramps from large islands. One expects that the first two of these phenomena will contribute more species differences than the third phenomena, that all three phenomena will be related to island distance, and that the second and third phenomena but not the first will be related to differences in area between the two islands.

Definitions of Two Indices

As for how to express these pair-wise differences in the species compositions of island avifaunas, several alternative indices are available to calculate faunal similarity based on presence/absence data (see Ludwig & Reynolds 1988, pp. 127-131 and 166-167, for discussion). A convenient starting point is the Ochiai similarity index, defined as $S_J/(S_1S_2)^{1/2}$, where S_1 is the total number of species on one island, S_2 is the total number of species on another island being compared with that first island, and S_J is the number of species jointly shared between the two islands. Hence $(S_1 - S_J)$ and $(S_2 - S_J)$ are the numbers of species of islands 1 and 2, respectively, that are absent from the other island. The Ochiai similarity index equals 1.0 for two identical faunas ($S_1 = S_2 = S_J$) and equals zero for totally dissimilar faunas ($S_J = 0$). We reverse this conventional similarity index by considering instead the expression (1 − Ochiai similarity index), as a measure of faunal dissimilarity. This expression, which we call the "Ochiai dissimilarity index," is 1.0 for totally dissimilar faunas and 0 for identical

Table 27.1. Comparison of four indices to assess geographic barriers.

	Endemism index	Pairwise differentiation index	Ochiai dissimilarity index	Non-nestedness index
Discussed in:	Chapter 25	Chapter 26	Chapter 27	Chapter 27
Islands compared	Target island vs. all other islands	A pair of islands	A pair of islands	A pair of islands
Measure of	Taxonomic differentiation of all species	Taxonomic differentiation of shared species	Species not shared	Species not shared
Influenced by	Water barriers to gene flow and dispersal	Water barriers to gene flow	Water barriers to dispersal; large-island species absent from small islands; small-island species absent from large islands.	Water barriers to dispersal; small-island species absent from large islands.
Predictors	A, D	$(A_1 A_2)^{1/2}$, D	A_1/A_2, D	D

A = island area; D = distance between islands. See text for discussion.

faunas. Columns 7–10 of Table 26.1 list respectively S_1, S_2, S_J, and the Ochiai dissimilarity index for selected pairs of islands. (We also calculated another often-used similarity index, the Dice index, defined as $2S_J/[S_1 + S_2]$; however, table 26.1 omits these Dice index values because they are closely correlated with the Ochiai values.)

For reasons previously discussed, Ochiai dissimilarity values may or may not be greater than zero for islands with identical species numbers (S_1 and S_2 values), depending on whether their species compositions are different or similar. However, Ochiai dissimilarity values are guaranteed to be greater than zero for islands with different S_1 and S_2 values, even if the species composition of the smaller island is a perfectly nested subset of the larger island's avifauna, as in the case of Emirau and St. Matthias. That is, the Ochiai dissimilarity index reflects all three of the above-mentioned phenomena (table 27.1). Not only does it reflect differences in species composition arising from water barriers to dispersal between island avifaunas, it may also reflect differences between faunas of very nearby islands differing more because of differences in island areas than because of the water barrier.

We want to use presence/absence data to calculate a barrier measure that might be more closely related to the barrier measure reflected in the pairwise differentiation index of chapter 26 (i.e., the effectiveness of a water barrier in inhibiting exchanges of conspecific individuals). Hence column 11 of Table 26.1 gives values of another index, $(S_2 - S_J)/S_2$, where S_2 is the number of species on the more species-poor island and $(S_2 - S_J)$ is the number of its species lacking on the richer island. This index may be called a "non-nestedness index," because it equals zero for perfectly nested faunas and 1 for totally unshared faunas. This index takes no account of large-island species absent on the smaller island. In effect, it ignores large-island species stopped by the water barrier, as the price for having an index uninfluenced by the ecological unsuitability of the smaller island for large-island species. Even this index is not a pure measure of the water barrier's effectiveness: supertramps of small islands absent on even close, large islands still cause this non-nestedness index to have values greater than zero. That is, the index reflects the first and third but not the second of the above-mentioned three phenomena (table 27.1).

Comparisons of the Three Pairwise Indices

We calculated the Pearson correlation coefficients among values of our three pairwise indices for the 51 pairs of islands listed in table 26.1, excluding land-bridge island pairs and volcanically disturbed islands. The closest correlation (0.74) is between

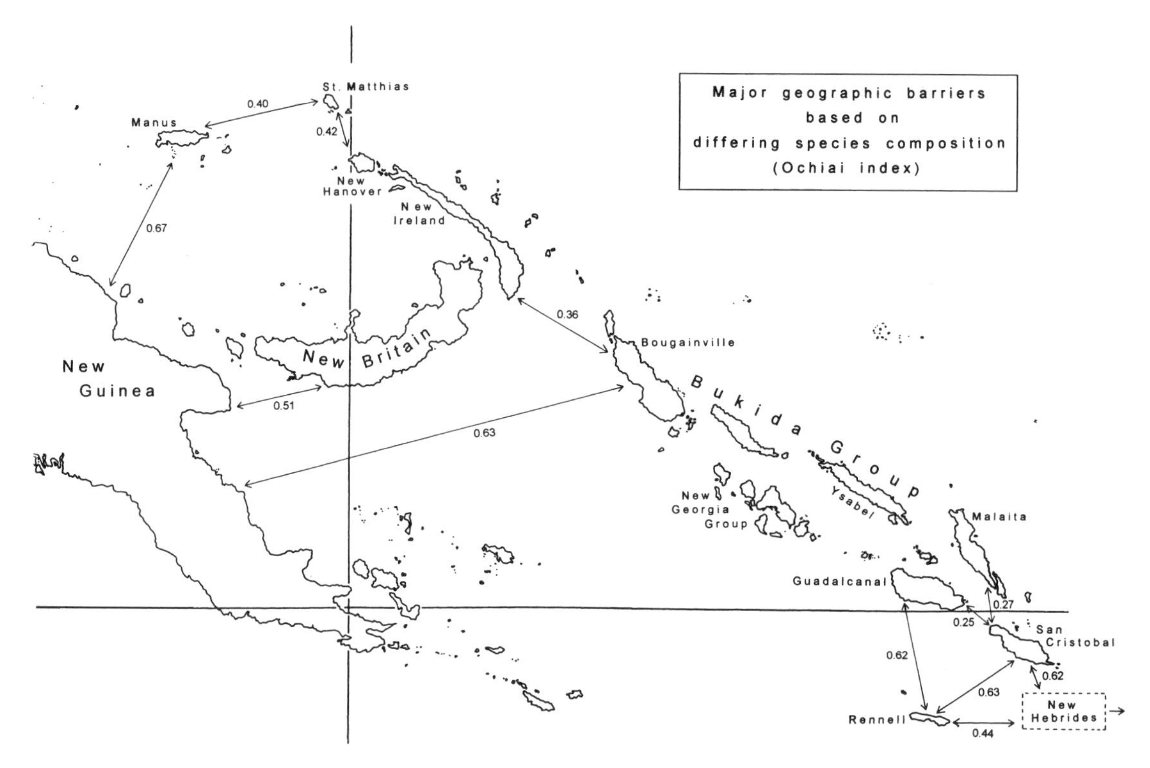

Fig. 27.1. Major geographic barriers within and around Northern Melanesia, based on differences in species composition between pairs of island avifaunas. Arrows connect island pairs with especially different avifaunas, indicating strong barriers to overwater dispersal. The number beside each arrow is the difference in species composition between the avifaunas of the two islands, as measured by the Ochiai dissimilarity index (see text for explanation). Data are from column 10 of Table 26.1. Note the high index values (strong barriers) separating the Bismarcks from the Solomons and from New Guinea, the eastern Solomons from the New Hebrides and Santa Cruz archipelagoes farther east, and Manus, St. Matthias, Rennell, and San Cristobal from their neighbors.

the Ochiai dissimilarity index and the non-nestedness index, because both of them measure aspects of nonshared species composition. The differentiation index of chapter 26 has a lower correlation with either the Ochiai index (0.50) or the non-nestedness index (0.52), because it measures a different phenomenon, taxonomic differentiation of shared species. All three indices are nevertheless correlated because all are influenced by water barriers between islands (table 27.1).

Locations of Major Barriers

Ochiai Dissimilarity Index

The major barriers reflected in the Ochiai dissimilarity index are mapped in figure 27.1. High values of this index express the barriers setting off the Solomons from the Santa Cruz and New Hebrides archipelagoes in the east (index values 0.44–0.62 for Rennell or San Cristobal vs. either of these two archipelagoes) and from the Bismarcks in the west (Bougainville/New Ireland 0.36). The index also reflects the barriers separating the Bismarcks from New Guinea (New Guinea/Manus 0.67, New Guinea/New Britain 0.51). Within the Solomons, the index reflects the isolation of Rennell (Rennell/San Cristobal 0.63, Rennell/Guadalcanal 0.62), of San Cristobal (San Cristobal/ Malaita 0.27, San Cristobal/Guadalcanal 0.25), and of Malaita (Malaita/San Cristobal 0.27). Within the Bismarcks, the index reflects the isolation of Manus (Manus/St. Matthias 0.40, Manus/New Hanover 0.31) and of St. Matthias (St. Matthias/Manus 0.40, St. Matthias/New Hanover 0.42). In addition,

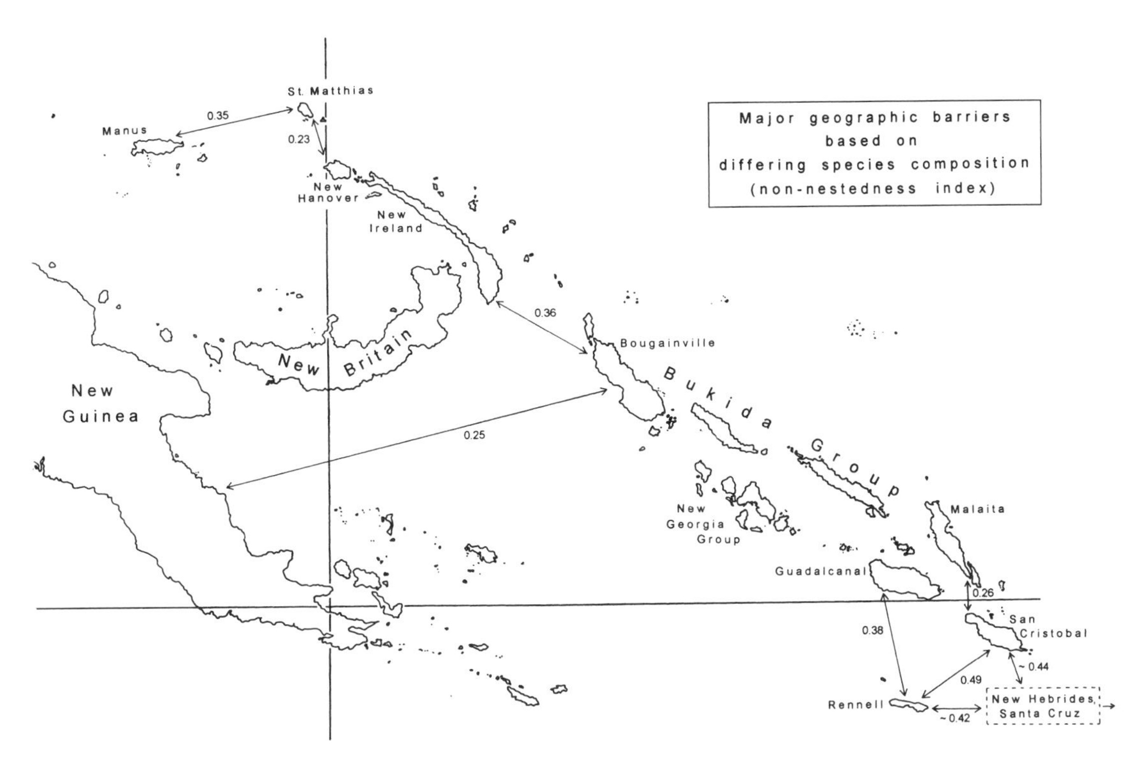

Fig. 27.2. As Fig. 27.1, except that pair-wise differences in island species compositions are measured by the non-nestedness index rather than by the Ochiai dissimilarity index. Note the high index values (strong barriers) at the same locations as in figure 27.1, except that the non-nestedness index fails to reflect the barrier separating New Guinea from the nearest Bismarck islands (Manus and New Britain).

the index also yields high values (0.27–0.45) for all five satellite islands of New Ireland (Dyaul, Tabar, Lihir, Tanga, and Feni) and for Manus compared to the Ninigo, Hermit, and Anchorite groups (0.61–0.76).

Non-nestedness Index

Selected values of the non-nestedness index are mapped in figure 27.2. As reflected by this index, as well as by the Ochiai index, high values separate the Solomons from the Santa Cruz and New Hebrides archipelagoes in the east (index values 0.33–0.55 for Rennell or San Cristobal vs. either of these two archipelagoes) and from the Bismarcks in the west (0.36 for Bougainville/New Ireland). However, in contrast to the Ochiai index, the barrier between the Bismarcks and New Guinea is poorly reflected in the non-nestedness index (only 0.09 for New Guinea/New Britain, 0.04 for New Guinea/Manus) because the overwhelming majority of Manus and New Britain species are shared with the much larger New Guinea. The absence of most New Guinea species from New Britain or Manus is not reflected in the non-nestedness index, which ignores large-island species absent from small islands.

Within the Solomons, the major barriers identified by the non-nestedness index are those setting off Rennell (e.g., index value 0.49 for Rennell/San Cristobal), San Cristobal, and Malaita (0.26 for Malaita/San Cristobal). The index also reflects the barrier between the Bukida group and the New Georgia group (Ysabel/Kulambangra 0.13, Choiseul/Vella Lavella 0.11). Within the Bismarcks, the index reflects the barriers setting off Manus (0.35 for Manus/St. Matthias, 0.16 for Manus/New Hanover) and St. Matthias (0.35 for St. Matthias/Manus, 0.23 for St. Matthias/New Hanover). These conclusions about major barriers within the Bismarcks and Solomons agree with the conclusions that we drew from the pairwise differentiation index (chapter 26). However, the non-nestedness index also yields high values (0.31–0.44) for pairwise comparisons of Manus with the small Ninigo, Hermit, and Anchorite groups to the west because these three groups have supertramp species lacking on Manus.

Dependence of the Nonsharing Indices on Island Distance and Areas

Ochiai Dissimilarity Index

An equation incorporating only island distance (D) explains 47% of the variation in the Ochiai dissimilarity index (H):

$$H = 0.0633\ D^{0.36} \qquad (27.1)$$

Adding a term for the ratio of island areas, A_1/A_2, increases the explained variation to 72%:

$$H = 0.0424\ D^{0.37}\ (A_1/A_2)^{0.16}. \qquad (27.2)$$

Adding a quadratic term and interaction term increases the explained variation only slightly further, to 78%.

It makes sense that the Ochiai index should increase with island distance: a water barrier causing differences in island species compositions even without any difference in island areas or species numbers creates positive values of the index. It also makes sense that the index should increase with the ratio of island areas: differences in island areas create differences in species numbers and hence also create positive values of the index.

The 17 pairs of land-bridge islands collectively deviate significantly ($p = .015$) from equation 27.2, with 13 of those 17 pairs having lower index values than predicted by the equation. This is as expected: Pleistocene land bridges connecting a pair of islands tended to erase differences in species composition and to make the islands share more species (higher S_J value) than would a never-connected island pair at the same distance and with the same area ratio.

Non-nestedness Index

An equation incorporating only island distance (D) explains 60% of the variation in the non-nestedness index (N), a higher percentage than D explains for the Ochiai index (only 47%):

$$N = 0.0092\ D^{0.61} \qquad (27.3)$$

Adding a term for island mean area (A) increases the explained variation only modestly, from 60% to 69%, and A enters only in an interaction term with D:

$$\ln N = -4.96 + 0.89 \ln D - 0.03 (\ln D)(\ln A), \qquad (27.4)$$

where $\ln N$, $\ln D$, and $\ln A$ are the natural logarithms (to the base $e = 2.718$) of N, D, and A respectively.

Thus, island areas have much less effect on the non-nestedness index than on the Ochiai index. This is as expected, because large-island species absent from a smaller island influence the Ochiai index but not the non-nestedness index (table 27.1).

The 17 pairs of land-bridge islands collectively deviate significantly ($p < 0.0001$) from equation 27.4 for the non-nestedness index, just as for the Ochiai index, and for the same reason. Of the 17 land-bridge island pairs, 16 have index values lower than predicted by equation 27.4.

Summary

We have calculated four indices to assess geographic barriers dividing Northern Melanesia and setting it off from the outside. The endemism index measures the distinctiveness of a given island's avifauna from the avifaunas of all other islands, while the three pairwise indices (the pairwise differentiation index and the two forms of a pairwise nonsharing index) compare islands pairwise in species composition or else in taxonomic differentiation of shared conspecific populations (table 27.1). These indices converge on the conclusions discussed below.

First, the avifaunas of the Bismarcks and Solomons are distinct from each other, as well as from the avifaunas of New Guinea in the west and of the Santa Cruz Archipelago and New Hebrides in the east.

Second, within the Bismarcks, the islands proportionately richest in endemics are New Britain, New Ireland, St. Matthias, and Manus, while Manus and St. Matthias have the most differentiated avifaunas. Within the Solomons, the islands proportionately richest in endemics are Rennell,

San Cristobal, the Bukida group, the New Georgia group, and Malaita, with the mountains of Bougainville and Guadalcanal also being proportionately rich, while Rennell, San Cristobal, and Malaita have the most differentiated avifaunas.

Third, the two indices analyzed in this chapter, the Ochiai dissimilarity index and the non-nestedness index, are based on those species confined to either one of a pair of islands and not shared between the two islands. Such nonsharing arises from a combination of three phenomena: one island's species were stopped by the intervening water barrier from establishing on the other island, even for two islands identical in area and in species number (but not identical in species composition); large-island species are absent from a nearby smaller island, due at least in part to the latter's ecological unsuitability or smaller population sizes, even if the small-island fauna is a perfectly nested subset of the large-island fauna; and small-island specialists (supertramps) are absent from nearby large islands. While the non-nestedness index is intentionally designed to be uninfluenced by the second of these three factors, the Ochiai dissimilarity index is influenced by all three. These two nonsharing indices are roughly correlated with each other and mostly identify the same major barriers, but with differences that are understandable from the differing definitions of these two indices.

Finally, the Ochiai dissimilarity index is influenced both by island distance and by the ratio of island areas. The non-nestedness index is influenced by island distance, but only to a minor degree by area. Avifaunas of pairs of land-bridge islands are much more similar to each other than avifaunas of pairs of islands of similar distance and area but never connected by land bridges. All of these findings are in accord with theoretical expectations.

28

The Establishment of Geographic Isolates

Genetic differentiation within an initially conspecific population can proceed in three different ways. Two of those ways were already mentioned in chapter 16 in connection with peripatric and dichopatric speciation.

The first is through primary isolation: colonizing individuals from a source population may succeed in crossing a major barrier to reach an area or island previously unoccupied by their species. There, they found a new population, which thus is immediately isolated from the source population by the barrier. The barrier may be a water gap between islands, or an ecological barrier within a single land mass (such as a mountain range separating lowland areas, a low-lying area separating mountains, or a dry savanna or desert separating forested areas). Already at its moment of founding, the new population may differ from its parent source population as a result of the founder effect; the gene frequencies of the founding individuals may happen to differ from the average gene frequencies in the source population. Once founded, the new population may or may not survive long enough for genetic differentiation to proceed further, depending on the colonized island's area and the population density of the species. In addition, the continued flow of conspecific individuals from the source population across the barrier may or may not be high enough to inhibit differentiation, depending especially on the width of the barrier and the vagility of the species.

In Northern Melanesia this process of primary isolation is clearly the main process involved in differentiation, as illustrated by the high average level of differentiation between conspecific populations on large islands separated by permanent water gaps. This process has operated even within modern times to produce a few distinct populations on volcanic islands defaunated within recent centuries (Long and Tolokiwa).

The second process leading to genetic differentiation is secondary isolation: a newly arising physical barrier may divide the geographic range of a species whose range was formerly uninterrupted. Causes of such newly arising barriers include the uplift of mountains, a decline in rainfall that makes habitats in part of a species' range unsuitable, and a rise in sea level that floods low-lying land. This process is illustrated in Northern Melanesia by post-Pleistocene fragmentation of former large Pleistocene islands such as Greater Bukida, Greater Gatumbangra, and New Ireland/New Hanover.

Northern Melanesia evinces a modest amount of differentiation between modern populations now separated by such post-Pleistocene water gaps, such as cases of subspecific differentiation between Bougainville and Choiseul or between New Ireland and New Hanover. In only three cases, however (the different allospecies of *Monarcha* [*melanopsis*] on Bougainville and Choiseul, of *Myzomela* [*lafargei*] on Ysabel and Florida, and of *Zosterops* [*griseotinctus*] on Vella Lavella and Ganonga;

chapter 32), are such water gaps associated with differentiation that has reached the allospecies level. In addition, we cannot be sure that such differentiation on post-Pleistocene island fragments really has arisen from secondary isolation; it may instead have arisen from primary isolation during post-Pleistocene or earlier Pleistocene times of high sea level and island fragmentation.

The remaining process leading to genetic differentiation is reduction of gene flow by distance within a species' continuous geographic range on a single land mass. This process, which may operate in the speciation mechanism known as "continental speciation" (chapter 22), tends to lead to clinal variation. The effect of distance must be gauged relative to the dispersal abilities of the particular species. For example, even small Pacific islands are famous for intra-island geographic variation within populations of land snails, which are poorly mobile. However, large islands, such as New Guinea, New Zealand, and Madagascar, are required for intra-island differentiation within populations of birds. As discussed in chapter 22 (p. 168), the only modern Northern Melanesian island known to provide possible examples of differentiation within an island is New Ireland, 340 km long, and even those examples may have a different explanation.

The remaining chapters of part VI analyze effects of modern water barriers within Northern Melanesia on the avifaunas that they separate. We begin by considering water barriers between Northern Melanesia and neighboring archipelagoes (chapter 29); then, the barrier between the Bismarcks and Solomons (chapter 29); next, other permanent water barriers within Northern Melanesia (chapters 30 and 31); and finally, Northern Melanesia's water barriers that have arisen since the Pleistocene period (chapters 32 and 33).

Summary

Among Northern Melanesian birds, most genetic differentiation within initially conspecific populations has proceeded through formation of primary isolates by colonists crossing permanent water barriers between islands. Some genetic differentiation may instead have involved secondary isolates formed when the rise of sea level at the end of the Pleistocene fragmented large Pleistocene islands into smaller modern islands. Because of the modest size of Northern Melanesian islands, there are few examples of genetic differentiation resulting from reduction of gene flow by distance within a species' continuous geographic range on a single land mass.

29

Interarchipelagal Barriers

Until now, we have assumed that Northern Melanesia is a zoogeographic region distinct from the New Guinea region to the west and from the Santa Cruz and New Hebrides archipelagoes to the east. We have also assumed that the Bismarck Archipelago and the Solomon Archipelago constitute distinct zoogeographic subregions of Northern Melanesia. In the present chapter we evaluate the barriers that divide these archipelagoes and that justify these zoogeographic definitions.

Barrier Between Northeast New Guinea and Northern Melanesia

We begin by comparing the Northern Melanesian avifauna as a whole with the avifauna of the adjacent part of New Guinea, Northeast New Guinea. We already compared the family compositions of these two avifaunas in chapter 8. The gap separating Northeast New Guinea from Northern Melanesia is about 48 km to the nearest Bismarck islands (Umboi and Long) and 636 km to the nearest Solomon island (Bougainville).

Northeast New Guinea Species Absent in Northern Melanesia

Northeast New Guinea harbors 432 breeding land and freshwater bird species, Northern Melanesia has 195, and 138 species are shared between Northeast New Guinea and Northern Melanesia. (As elsewhere in this book, "species" in these tabulations refers to superspecies plus isolated species.) Thus, 69% (294) of Northeast New Guinea's species are absent from Northern Melanesia. This is unsurprising because it exemplifies the most familiar fact of island biogeography: smaller islands lack many species of nearby larger islands. New Guinea has 9.4 times the combined area of all Northern Melanesian islands and hence harbors far more species.

The usual reasons advanced for this basic fact of island biogeography in comparisons of unequal-sized islands elsewhere in the world apply also to the comparison of Northeast New Guinea with Northern Melanesia. First, the larger New Guinea land mass has habitats lacking on Northern Melanesia's smaller and lower islands, such as alpine grassland and subalpine forest. Naturally, New Guinea's species of those habitats are absent from Northern Melanesia (except for the alpine grassland thrush *Turdus poliocephalus*, which has adapted to low-elevation habitats in Northern Melanesia).

Second, some habitats present in both New Guinea and Northern Melanesia are represented by not only absolutely but also proportionately far more area in New Guinea. Those include habitats such as mid-montane forest, large rivers, marshes, and grasslands. Naturally again, many New Guinea species of those habitats could not establish in

Northern Melanesia. The effect of these first two factors combined is that 150 New Guinea montane species, 88% of the total, are absent from Northern Melanesia.

Third, the larger New Guinea land mass has evolved many sedentary species that have never been recorded dispersing over water. As discussed in chapter 8, 145 species (96%) of a sedentary group of Northeast New Guinea species termed the "sedentary Corvi" are absent from Northern Melanesia. These are species so poor at crossing water that they have never been recorded, even as vagrants, from any Papuan island lacking a Pleistocene land-bridge connection to New Guinea—even Papuan islands just a few kilometers from New Guinea. Of course, the water barriers of 48–636 km between Northeast New Guinea and Northern Melanesia also stopped these species.

Finally, there are many bird genera of which New Guinea supports many more species than does any single Northern Melanesian island, even though each species individually would find suitable habitat in Northern Melanesia. Examples, listed as the number of species in Northeast New Guinea followed by the maximum number on any single Northern Melanesian island, are the parrot genus *Charmosyna* (6 vs. 2), the fantail-flycatcher genus *Rhipidura* (10 vs. 3), and the whistler genus *Pachycephala* (11 vs. 2). This exemplifies the denser species packing typical of larger islands, even in habitats shared with smaller islands.

All of these familiar factors contribute to Northern Melanesia's avifaunal poverty relative to New Guinea.

Northern Melanesian Species Absent in Northeast New Guinea

Also to be explained is why 57 of Northern Melanesia's 195 species (i.e., 29% of them) are absent from Northeast New Guinea. Three main reasons and some minor factors are apparent.

First, 16 of the absentees invaded Northern Melanesia from the east, and 13 of them are still confined to the Solomons, nine of them to the easternmost Solomons.[1] These species may not yet have had time to spread westward through the Bismarcks to New Guinea.

Second, at least eight of the absentees may be competitively excluded from New Guinea by ecologically equivalent congeners that are closely related, though not quite close enough to be classified within the same superspecies. Examples include the Northern Melanesian parrot *Charmosyna [palmarum]*, excluded by New Guinea's similar *Charmosyna rubronotata* (map 17), and the Solomon pitta *Pitta anerythra*, excluded by its New Guinea relative *P. sordida*. Numerous other Northern Melanesian species, especially white-eyes and honey-eaters, may also fall into this category.

Third, 20 of the absentees are sedentary old endemics of Northern Melanesia that have never been observed to cross a water barrier. Sixteen of the 20 are confined to the Solomons, and 16 are confined to single Northern Melanesian islands or to fragments of single Pleistocene islands. These species have evidently lost the colonizing ability required to reach New Guinea.

Further minor factors are that three of the Northern Melanesian species absent in Northeast New Guinea (the honey-eaters *Myzomela sclateri* and *M. [pammelaena]* and the starling *Aplonis [feadensis]*) are supertramps confined to small islands within Northern Melanesia; and that three others (the parrot *Charmosyna margarethae*, the nightjar *Eurostopodus mystacalis*, and the starling *Aplonis grandis*) occur in the Solomons but are absent from the Bismarcks.

These considerations still leave unexplained the absences in New Guinea of two Northern Melanesian species—the pigeon *Columba pallidiceps* (map 11) and the flycatcher *Myiagra hebetior* (map 42)—that are demonstrably good to fair at colonizing over water gaps within Northern Melanesia, that are old endemics of Northern Melanesia, and that do reach the Bismarcks, the western half of Northern Melanesia and the part closest to New Guinea. Both appear to be descended from a former Bismarck representative or allospecies of a widespread species or superspecies (*Columba [leucomela]* and *Myiagra alecto*, respectively). In each case the New

1. The nine of these eastern invaders confined to the easternmost Solomons are the rail *Pareudiastes sylvestris*, the pigeons *Ducula brenchleyi* and *Gallicolumba salamonis*, the cuckoo *Chrysococcyx lucidus*, the cuckoo-shrike *Lalage leucopygia*, the warbler *Gerygone flavolateralis* (map 34), the flycatchers *Rhipidura fuliginosa* and *Clytorhynchus hamlini*, and the white-eye *Woodfordia superciliosa*. Four of these eastern invaders extend over the full length of the Solomons: the rail *Nesoclopeus woodfordi*, cuckoo-shrike *Coracina caledonica* (map 27), flycatcher *Rhipidura [spilodera]* (map 38), and robin *Petroica multicolor*. Three of these eastern invaders reach the Bismarcks: the warblers *Ortygocichla rubiginosa* and *Cichlornis [whitneyi]* (map 35), and the honey-eater *Myzomela [cardinalis]* (map 46).

Guinea representative (*Columba vitiensis* and *Myiagra alecto*, respectively) now occurs as a recent invader in Northern Melanesia, sympatrically with the old Northern Melanesian endemic. Thus, these two species do have New Guinea representatives, although broad sympatry in Northern Melanesia prevents us from ranking the New Guinea representative as a member of the same superspecies.

Differentiation Between Northern Melanesian and Northeast New Guinea Populations of Shared Species

The preceding sections examined the effectiveness of the barrier between New Guinea and Northern Melanesia in entirely stopping species on one side of the barrier from colonizing the other side. Let us now turn to the 138 species shared between Northeast New Guinea and Northern Melanesia, and let us examine the degree to which the conspecific populations separated by the water barrier have become genetically differentiated.

Only 28 of the shared species (20% of the total) have subspecifically identical populations in Northern Melanesia and Northeast New Guinea and have thus failed to differentiate at all (e.g., the heron *Egretta sacra sacra*, the nightjar *Caprimulgus macrurus yorki*, and the fantail flycatcher *Rhipidura leucophrys melaleuca*). As discussed in chapter 26, this high degree of differentiation (involving 80% of the shared species) reflects the moderate width of the water barrier between New Guinea and Northern Melanesia, combined with the large area of New Guinea and the moderate areas of numerous Northern Melanesian islands. The water barrier reduces gene flow, while the very large or moderate areas of the islands on opposite sides of the barrier ensure that many pairs of conspecific populations separated by the barrier will survive for a long time (and thus have ample time to differentiate) before going extinct.

Details of the degree of differentiation are as follows. Twenty species shared between Northeast New Guinea and Northern Melanesia are represented in parts of Northern Melanesia by the same subspecies in Northeast New Guinea, but in other parts of Northern Melanesia by an endemic taxon (endemic weak subspecies in 10 cases, endemic megasubspecies in five cases, endemic allospecies in five cases). For example, the eagle *Haliaeetus* [*leucogaster*] occurs in New Guinea and Bismarcks as the undifferentiated allospecies *H. leucogaster*, but in the Solomons as the endemic allospecies *H. sanfordi*; and the finch *Erythrura trichroa* occurs in New Guinea and most of the Bismarcks as the subspecies *E. trichroa sigillifera*, on the outlying Bismarck island of St. Matthias as the endemic weak subspecies *E. t. eichhorni*, and in the Solomons as the endemic weak subspecies *E. t. woodfordi*. All Northern Melanesian populations belong to endemic weak subspecies in 34 cases, endemic megasubspecies in 18 cases, and endemic allospecies in 24 cases. For example, the swift *Collocalia esculenta* is represented in New Guinea by two weak endemic subspecies and in Northern Melanesia by seven weak endemic subspecies, while the owl *Ninox* [*novaeseelandiae*] occurs in New Guinea as the endemic allospecies *N. theomacha* and in Northern Melanesia as three allospecies endemic to the Bismarcks and one allospecies endemic to the Solomons (map 22, plate 7). Finally, 10 shared species are represented by endemic subspecies in some parts of Northern Melanesia and by endemic allospecies in other parts of Northern Melanesia. For example, the thrush *Zoothera* [*dauma*] occurs in New Guinea as the endemic subspecies *Z. heinei papuensis*, on three Northern Melanesian islands as two endemic subspecies of *Z. heinei*, and on five other Northern Melanesian islands as the endemic allospecies *Z. talaseae* and *Z. margaretae* (map 32).

Northeast New Guinea versus New Britain

The preceding sections have compared the avifaunas of Northeast New Guinea and of Northern Melanesia as a whole, separated by a water barrier varying in width from 48 to 636 km depending on the particular Northern Melanesian island. Let us now specifically compare Northeast New Guinea with New Britain, the large Northern Melanesian island nearest to New Guinea, separated from it by a water barrier of 88 km. With an area of 35,742 km^2, New Britain is by far the largest modern Northern Melanesian island, and with an elevation of 2439 m it is the third highest. Hence New Britain's 127 breeding species make it the most species-rich Northern Melanesian island.

Of New Britain's 127 species, 115 are shared with Northeast New Guinea. The 12 not shared consist of four New Britain endemics and eight Northern Melanesian endemics (endemic allospecies or full species). The Ochiai dissimilarity index for the New Guinea/New Britain island pair, an index which expresses these differences in species composition of the two avifaunas, is 0.51, a higher value than for all but a few comparisons of island pairs within Northern Melanesia (table 26.1).

Considering the 115 shared species, only 38 are represented by subspecifically identical populations on New Britain and Northeast New Guinea. In 39 cases the New Britain and Northeast New Guinea populations constitute different weak subspecies, in 11 cases different megasubspecies, and in 24 cases different allospecies (the remaining three cases cannot be analyzed). As a result, the differentiation index, expressing the average degree of differentiation between conspecific populations of the two islands, is 1.19 (table 26.1). That is, on average, the shared populations are differentiated at a level slightly greater than that of a weak subspecies. This value is higher than that for any adjacent pair of islands within Northern Melanesia.

This value of the differentiation index is as predicted from equation 26.1 for the mean area and distance of the New Guinea/New Britain species pair. Because New Guinea and New Britain are both so large, the mean area is the highest of any island pair compared. This fact, combined with the moderate water barrier of 88 km, accounts for the differentiation index of 1.19. The large areas combined with a moderate distance also explain the high value of the Ochiai index expressing differences in species composition of the two avifaunas.

Barriers to the East of Northern Melanesia

The nearest two archipelagoes lying to the east of Northern Melanesia are the smaller (829 km^2) but closer Santa Cruz Group (362 km east of San Cristobal, the eastern-most Northern Melanesian island), and the larger (12,190 km^2) but more distant New Hebrides Archipelago (618 km southeast of San Cristobal). Table 26.1 compares the avifaunas of these two eastern archipelagoes with those of San Cristobal, Guadalcanal, and Rennell in the eastern Solomons.

Solomon Species Absent to the East

The Ochiai dissimilarity index, expressing differences in species composition, is 0.44–0.64 for Santa Cruz and 0.53–0.62 for the New Hebrides in comparison with San Cristobal, Guadalcanal, and Rennell. These are among the highest Ochiai indices in table 26.1, as high as the value of 0.51 at the western border of Northern Melanesia (between New Guinea and New Britain) and higher than the value of 0.36 between the two halves of Northern Melanesia (the Bismarcks and Solomons).

This eastern barrier mainly reflects the fact that most Solomon species fail to extend east of the Solomons. Of 143 Solomon species, only 42 reach the New Hebrides and/or Santa Cruz group, while one other (the rail *Nesoclopeus* [*poecilopterus*]) skips over both Santa Cruz and the New Hebrides to reappear farther east in Fiji. In addition, five other Solomon species absent in Santa Cruz and the New Hebrides are present to the southeast in New Caledonia, but most of those five species probably reached New Caledonia from Australia rather than from the Solomons. In short, the sea barrier east of the Solomons stops about 70% of Solomon bird species from spreading eastward. The reasons are threefold: the width of the barrier; the modest areas of the Santa Cruz group and the New Hebrides, which constitute the targets beyond the barrier; and, as a minor factor, the somewhat restricted ecological opportunities in the smaller eastern archipelagoes compared with the larger Solomons. For example, the Solomons include two islands (Bougainville and Guadalcanal) considerably higher than the highest island of the New Hebrides (Espiritu Santo, 1,890 m), and the highest island of the Santa Cruz Group (Ndeni) is only 579 m, so that the Solomons have many more montane species.

Eastern Species Absent in the Solomons

While the high value of the Ochiai dissimilarity index stems especially from the fact that most Solomon species fail to cross the barrier eastward, a contributing factor is that a minority of Santa Cruz and New Hebrides species fail to cross the barrier westward to the Solomons. Of 56 New Hebrides species, 16 are absent from the Solomons: eight that colonized the New Hebrides from Australia, three that colonized from the east (Fiji and/or Samoa), and five old endemics of the New Hebrides. Similarly, of the 33 Santa Cruz species, six are absent in the Solomons: two that reached the Santa Cruz group from Australia, two that arrived from the east, and two old endemics.

Differentiation Between Solomon and Eastern Populations of Shared Species

The previous sections examined the effectiveness of the barrier between the Solomons and New Hebrides or Santa Cruz in entirely stopping species on one side of the barrier from colonizing the other side. Let us now turn to the species shared between the Solomons and eastern archipelagoes, and let us examine the degree to which the conspecific populations separated by the barrier have become genetically differentiated.

Only 20–33% of the shared species have subspecifically identical populations in the eastern Solomon Islands and in the New Hebrides or Santa Cruz group and have thus failed to differentiate at all. Six superspecies are represented by different allospecies on opposite sides of the barrier: *Ducula* [*latrans*], *Gallicolumba* [*canifrons*] or *G.* [*erythroptera*], *Charmosyna* [*palmarum*], *Halcyon* [*diops*], *Rhipidura* [*spilodera*], and *Myiagra* [*rubecula*]. The differentiation index, expressing the average degree of differentiation between conspecific populations of the western and eastern archipelagoes, is 1.14–1.50. These values are comparable to the differentiation index across the barrier at the western end of Northern Melanesia (i.e., between New Guinea and New Britain or Manus) and higher than indices within the Bismarcks or Solomons. These high values for the eastern barrier of Northern Melanesia fit the general pattern for the index's dependence on island areas and distance (eq. 26.1); the values arise from the width of the barrier, combined with the moderate areas of the islands on opposite sides of the barrier.

Barrier Between the Bismarcks and Solomons

We have now examined the barriers separating Northern Melanesia from the nearest archipelagoes to the west or east.

We next examine the major division within Northern Melanesia: the sea barrier between the Bismarcks and Solomons. This barrier is what warrants separating Northern Melanesia's islands into western and eastern groups termed the Bismarcks and Solomons, respectively. We specifically consider the barrier between the westernmost large Solomon island, Bougainville, and the easternmost two large Bismarck islands, New Britain and New Ireland.

Bismarck Species Absent on Bougainville

Bougainville lacks 54 of New Britain's 127 species and 37 of New Ireland's 103 species. Two-thirds of those Bismarck species missing from Bougainville are New Guinea species that reach their eastern range limit in the Bismarcks; they are absent not only from Bougainville but also from all other Solomon islands and from Pacific archipelagoes east of the Solomons. The other absentees fall into three categories: Bismarck species that reappear on other Solomon islands, though not on Bougainville; old endemics of the Bismarcks (or of New Britain and New Ireland), absent from New Guinea as well as from the Solomons; and two species (the duck *Dendrocygna arcuata* and cuckoo-shrike *Lalage* [*aurea*]) present on New Britain, absent in the Solomons, and reappearing on other archipelagoes east of the Solomons. The absence of so many Bismarck species on Bougainville is primarily because the Bismarcks are nearer the main colonization source of New Guinea, and secondarily because the aggregate area of the Bismarcks is somewhat greater than that of the Solomons.

Bougainville Species Absent on New Britain or New Ireland

Conversely, of Bougainville's 103 species, 30 are absent on New Britain and 37 on New Ireland. Whereas the straits between the Bismarcks and Solomons are a major barrier to the eastern spread of western species, they are only a minor barrier to the western spread of eastern species: just three eastern species that spread west to Bougainville failed to spread farther west and reach New Britain or New Ireland (the rail *Nesoclopeus* [*poecilopterus*], cuckoo-shrike *Coracina* [*caledonica*], and fantail flycatcher *Rhipidura* [*spilodera*]). (We explained in chapter 13 the type of evidence permitting us to infer the direction of spread across the Bismarck/Solomon barrier.) The reason for this asymmetry is that the Santa Cruz and New Hebrides archipelagoes east of the Solomons are small, remote, and species-poor and have therefore sent few emigrants to the Solomons; most of those emigrants have stopped in the eastern Solomons and not spread to Bougainville. Instead, the Bougainville species absent on New Britain or New Ireland are a mixture of Solomon species that reappear elsewhere in the Bismarcks, though not on New Britain or New Ireland; old endemics of the Solomons; and species that colonized the Solomons directly from the west (mostly from Australia, few from New Guinea), bypassing the Bismarcks.

The combined result of these absences of Bougainville species on New Britain or New Ireland, plus the large number of absences of New Britain or New Ireland species on Bougainville, is that the Ochiai dissimilarity index for the barrier between the Bismarcks and Solomons is 0.36 (Bougainville vs. either New Britain or New Ireland). This barrier is lower than for the barriers delineating Northern Melanesia as a whole from the archipelagoes to the west or east, but higher than the value for the barrier between any other adjacent pair of large, central Northern Melanesian islands. This is as expected from the large areas of Bougainville, New Britain, or New Ireland (the largest islands in their respective archipelagoes), plus the wide water gap between them (270 km to New Britain, 172 km to New Ireland).

Differentiation Between Bismarck and Bougainville Populations of Shared Species

Of the conspecific populations shared between Bougainville and New Britain or New Ireland, only 40% have subspecifically identical populations on Bougainville and on New Britain or New Ireland. Most of those species show no geographic variation anywhere in Northern Melanesia. Thus, if a species varies geographically in Northern Melanesia at all, it is likely to do so across the barrier between the Bismarcks and Solomons. Across the Bougainville/Bismarck barrier, 24 shared species differ at the level of weak subspecies, 6–7 at the level of megasubspecies, and 10–13 at the level of allospecies. The average value of the differentiation index for the whole shared avifauna is 1.05 (for New Britain vs. Bougainville) or 1.00 (New Ireland vs. Bougainville). This value is somewhat lower than the values for Northern Melanesia compared to either New Guinea or to Pacific archipelagoes to the east, but higher than the value for any other pair of adjacent islands within Northern Melanesia.

Comparison of the Whole Bismarck and Solomon Avifaunas

At the risk of providing several indigestible lists of species names, we now summarize the differences between the Bismarck and Solomon avifaunas as a whole.

There are 18 superspecies (18% of the 98 species or superspecies shared between the two archipelagoes) for which Bismarck and Solomon populations belong to distinct allospecies: the eagle *Haliaeetus* [*leucogaster*], the pigeons *Gymnophaps* [*albertisii*] and *Reinwardtoena* [*reinwardtii*] (plate 7), the parrots *Lorius* [*lory*] (plate 6) and *Charmosyna* [*palmarum*], the cuckoo *Centropus* [*ateralbus*], the owl *Ninox* [*novaeseelandiae*] (plate 7), the kingfisher *Halcyon* [*diops*], the warbler *Cichlornis* [*whitneyi*], the fantail flycatchers *Rhipidura* [*rufiventris*] and *R.* [*rufidorsa*], the monarch flycatcher *Monarcha* [*manadensis*] (plate 3), the flower-pecker *Dicaeum* [*erythrothorax*], the honey-eaters *Myzomela* [*cardinalis*] and *M.* [*lafargei*] (plate 4), and the white-eye *Zosterops* [*griseotinctus*] (plate 1), plus the pigeons *Ptilinopus* [*rivoli*] (plate 6) and *Macropygia* [*ruficeps*], of which the Solomon allospecies extends as a supertramp into the range of the Bismarck allospecies.

Each archipelago harbors numerous endemics (isolated species not belonging to a superspecies ranging outside the archipelago) confined to single islands. In addition, each archipelago harbors endemics occurring on more than one island of that archipelago but absent from the other archipelago. These latter endemics consist of four confined to the Bismarcks (the hawk *Accipiter luteoschistaceus*, the cuckoo *Centropus violaceus*, the monarch flycatcher *Myiagra hebetior* (plate 5), and the honey-eater *Myzomela sclateri* (plate 4)) and 13 confined to the Solomons (the hawk *Accipiter imitator*, the pigeon *Gallicolumba salamonis*, the parrots *Chalcopsitta cardinalis* (extending to two small islands of the eastern Bismarcks; plate 8, map 15) and *Charmosyna margarethae* (plate 9), the owl *Ne-*

sasio solomonensis (plate 8), the kingfisher *Halcyon bougainvillei* (plate 8), the cuckoo-shrike *Coracina holopolia*, the whistler *Pachycephala implicata* (plate 2), the white-eyes *Zosterops metcalfii* and *Z. ugiensis* (plate 1), the starlings *Aplonis grandis* and *A. brunneicapilla*, and the crow *Corvus woodfordi*).

There are 40 western species that reach a part of the Bismarcks but reach none of the Solomons nor any archipelago farther east: *Dendrocygna guttata*, *Henicopernis* [*longicauda*], *Accipiter* [*poliocephalus*] and *A.* [*cirrhocephalus*], *Falco berigora*, *Coturnix chinensis*, *Rallina tricolor*, *Gymnocrex plumbeiventris*, *Irediparra gallinacea*, *Charadrius dubius*, *Himantopus leucocephalus*, *Ptilinopus* [*hyogaster*] (map 6), *Ducula* [*rufigaster*] and *D.* [*pinon*] and *D.* [*spilorrhoa*], *Macropygia amboinensis*, *Henicophaps* [*albifrons*], *Cacatua galerita*, *Loriculus* [*aurantiifrons*], *Tyto* [*novaehollandiae*], *Caprimulgus macrurus*, *Alcedo* [*azurea*], *Tanysiptera sylvia*, *Merops philippinus*, *Pitta* [*sordida*] and *P. erythrogaster* (map 26), *Saxicola caprata* (map 31), *Cisticola exilis*, *Megalurus timoriensis*, *Monarcha chrysomela*, *Myiagra alecto* (map 42), *Monachella muelleriana*, *Nectarinia sericea*, *Zosterops atrifrons* (map 45), *Myzomela* [*eques*] and *M. cruentata*, *Philemon* [*moluccensis*] (map 48), *Lonchura* [*spectabilis*], *Artamus* [*maximus*], and *Corvus orru*.

Conversely, there are 11 eastern species that reach all or part of the Solomons but reach none of the Bismarcks or the New Guinea region farther west: *Nesoclopeus* [*poecilopterus*], *Pareudiastes sylvestris*, *Ptilinopus* [*purpuratus*], *Ducula* [*latrans*], *Gallicolumba salamonis*, *Coracina* [*caledonica*] (map 27), *Lalage leucopyga* (map 30), *Gerygone flavolateralis* (map 34), *Rhipidura* [*spilodera*] (map 38), *Clytorhynchus* [*nigrogularis*], and *Woodfordia superciliosa*.

Finally, there are 15 western species (mainly from Australia, with a few from New Guinea) that reach the Solomons and bypass the Bismarcks. (Three or four of these 15 species appear to have spread first from Australia to the New Hebrides or New Caledonia and thus staged their entry into the Solomons from the southeast.) Five of the 15 species are confined to Rennell in the southeastern Solomons: *Threskiornis moluccus*, *Platalea regia*, *Anas gibberifrons*, *Accipiter fasciatus*, and *Chrysococcyx lucidus*. One species, *Rhipidura fuliginosa*, is confined to San Cristobal, the easternmost large Solomon island. The remaining nine species are more widespread in the Solomons: *Cacatua* [*tenuirostris*], *Podargus ocellatus*, *Eurostopodus mystacalis*, *Pitta* [*brachyura*], *Monarcha* [*melanopsis*] (map 40), *Myiagra* [*rubecula*] (map 42), and *Petroica multicolor*, plus *Accipiter* [*rufitorques*] and *Ptilinopus viridis* (map 8), which extend slightly from the Solomons into the easternmost small islands of the Bismarcks (but *P. viridis* has colonized the northwestern Bismarck island of Manus in recent decades).

Definition of Ornithogeographic Regions

As mentioned in the Introduction to this chapter, we have been taking for granted the naturalness of Northern Melanesia, the Bismarcks, and the Solomons as ornithogeographic regions, without justifying their distinctness. We were able to do so because they are actually set off by such obvious water barriers and avifaunal differences that there has never been any question in the ornithological literature about their distinctness. The quantitative evaluations of sea barriers in this chapter and in chapters 26 and 27, which we now recapitulate, justify the delineation of these regions.

To the east, the easternmost Solomon islands (San Cristobal plus its satellites Santa Anna and Santa Catalina) are separated from the next Pacific islands to the east, the Santa Cruz Group, by a water gap of 362 km—much wider than (mostly, more than double the width of) the water gap between any adjacent pair of islands within Northern Melanesia. The pairwise differentiation index across this water gap (1.14; see table 26.1) is greater than for any adjacent pair of islands within Northern Melanesia, and the Ochiai index and nonnestedness index (table 26.1) are greater than for all but a few island pairs.

To the north, the northernmost Northern Melanesian islands (St. Matthias and the Northwest Bismarcks) are separated by 500 km from the nearest atolls, and by 1000 km from the nearest high islands, of avifaunally distinct Micronesia (Mayr 1945a). Not a single native bird species is endemic to (confined to) the combination of Micronesia and Northern Melanesia.

To the west and south, New Britain and its satellite Umboi are separated by 51 km from New Guinea and by 228 km from the Southeast Papuan Islands, which support a depauperate New Guinea avifauna. The pairwise differentiation index between New Britain and New Guinea (1.19) is higher than between any adjacent pair of Northern Melanesian islands; the Ochiai index (0.51) is also high; but the non-nestedness index (0.09) is low, for reasons discussed in chapter 27.

A glance at a map shows that Northern Melanesia must have a southwestern boundary somewhere along the Bismarck Volcanic Arc, the chain of volcanoes extending along New Britain's north coast and continuing west as a chain of islands just off the coast of Northeast New Guinea (fig. 1.2). Among these islands the percentage of Bismarck bird taxa decreases, and the percentage of New Guinea bird taxa increases, from east to west (p. 109 of chapter 14). The 80-km water barrier separating Long Island and its satellite Crown in the east from Karkar Island and its satellite Bagabag in the west constitutes Northern Melanesia's southwestern boundary (fig. 0.3). Among zoogeographically informative bird populations (i.e., excluding species that occur both in the Bismarcks and in the New Guinea region as the same subspecies), 72% of the taxa of Long and Crown belong to the Bismarck region, only 28% to the New Guinea region, but 84% of Karkar's and Bagabag's birds are New Guinean and only 16% are derived from the Bismarcks (Diamond & LeCroy 1979).

Wuvulu Island, west of the Ninigo and Hermit groups and Manus (fig. 0.3), constitutes Northern Melanesia's northwest limit. Among Wuvulu's six informative bird populations, all belong to the Bismarck avifauna (five Bismarck supertramp species and a race of the kingfisher *Halcyon chloris*); none belongs to the New Guinean avifauna. A 720-km water barrier separates Wuvulu from the next island to the west, Biak off Northwest New Guinea; only 4% of Biak's avifauna is traceable to the Bismarcks, and most of the rest is traceable to New Guinea (Mayr 1941c).

As for the division of Northern Melanesia into the Bismarcks and Solomons, the water barrier separating the westernmost major Solomon island of Bougainville from the easternmost major Bismarck islands of New Britain (270 km) and New Ireland (172 km) is not only the second widest water gap in Northern Melanesia (second only to the Manus/St. Matthias gap), but it is also the most marked ornithogeographic border (highest pairwise differentiation index: Bougainville/New Britain 1.05, Bougainville/New Ireland 1.00). Within this water gap lies the small island of Nissan, nearly equidistant from the New Ireland (Bismarck) satellite of Feni (58 km) and the Bougainville (Solomon) satellite of Buka (63 km), and with virtually the same pairwise differentiation indices and Ochiai indices vis-à-vis Bougainville and Buka as vis-a-vis New Ireland and Feni, but with slightly higher non-nestedness indices vis-a-vis the latter two Bismarck islands (on the average, 0.28) than vis-a-vis the former two Solomon islands (0.19). Hence Nissan is nearly intermediate geographically and zoogeographically; we have listed it among Solomon islands on the basis of its slightly closer non-nestedness index.

Summary

This chapter discusses effects of the sea barriers separating Northern Melanesia from archipelagoes to the west and east and dividing the two halves of Northern Melanesia.

Regarding the sea barrier west of Northern Melanesia, separating Northern Melanesia from Northeast New Guinea, Northern Melanesia lacks 69% of Northeast New Guinea's species—for reasons related to New Guinea's much larger area, greater habitat diversity, and richness in sedentary species unable to disperse over water. A smaller proportion (29%) of Northern Melanesia's species are absent from Northeast New Guinea for various reasons that include western range limits, competitive exclusion, and poor overwater dispersal. Of the species shared between Northern Melanesia and Northeast New Guinea, 80% have differentiated taxonomically across the water barrier, including many represented by different allospecies in Northern Melanesia and Northeast New Guinea. Similar conclusions emerge if one instead compares Northeast New Guinea with the nearest large Northern Melanesian island, New Britain.

The sea barriers east of Northern Melanesia separate the easternmost Solomon islands from the Santa Cruz and New Hebrides archipelagoes. Most Solomon species (71%) fail to reach those eastern archipelagoes because of their smaller areas and lower habitat diversity. Conversely, about one-quarter of the species in those eastern archipelagoes fail to reach the Solomons. Most of the shared species (about 75%) have differentiated taxonomically.

The sea barrier between the Bismarcks and Solomons is expressed specifically in the avifaunal differences between the westernmost large Solomon island (Bougainville) and the easternmost large Bismarck islands (New Britain and New Ireland). Bougainville lacks 40% of New Britain's or New Ireland's species; most of those absentees are New Guinean species reaching their eastern range limit in the Bismarcks. However, few eastern species reach their western range limit on Bougainville. Most of the shared species (60%) have differentiated taxonomically, a dozen at the allospecies level. Similar conclusions emerge if one instead compares the entire Bismarcks with the entire Solomons.

The quantitative strengths of these three sea barriers at the west, in the middle, and at the east of Northern Melanesia are well accounted for by the widths of the barriers and the areas of the archipelagoes that they separate. Considered quantitatively, these three barriers plus the sea barriers north, northwest, and southwest of the Bismarcks are what delineate Northern Melanesia as an ornithogeographic region and delineate the Bismarcks and Solomons as its two subregions.

30

Barriers Within the Bismarcks

Having discussed in the preceding chapter the barriers separating Northern Melanesia from the archipelagoes to the east and west, and the barrier dividing Northern Melanesia into the Bismarcks and Solomons, we now devote this chapter to barriers between Bismarck islands. In so doing, we initially treat as wholes those groups of islands that were connected to each other during Pleistocene drops of sea level, because the modern barriers between islands within each such group arose only within the last 10,000 years. For much of the Pleistocene, the barriers within the Bismarcks were those between such Pleistocene islands, not the barriers between modern islands. There were two major Pleistocene islands in the Bismarcks: the large island formed by the accretion of modern Umboi, Lolobau, and Duke of York islands to New Britain, and the large island formed by the accretion of modern New Hanover to New Ireland (fig. 3.3). In chapters 32 and 33 we discuss the post-Pleistocene barriers between the modern fragments of those former large islands.

Barrier Between New Britain and New Ireland

Modern New Britain is by far the largest Northern Melanesian island, while modern New Ireland is exceeded in area only by New Britain and also by Bougainville, the largest Solomon island. New Britain and New Ireland are now separated by only 30 km, and they were even closer (possibly as little as 8 km) when their fringing, shallow-water shelves were dry land at Pleistocene times of low sea level (fig. 1.3). The New Britain group consists of New Britain, the nearby smaller land-bridge islands of Umboi and Lolobau and Duke of York, and several nearby smaller volcanic islands that had no recent connection to New Britain (Vuatom, Witu, Unea, and Sakar). The New Ireland group similarly consists of New Ireland plus some adjacent smaller islands discussed on p. 237. However, all species occurring anywhere within the New Britain group occur on New Britain itself, except for a few supertramps confined to the smaller islands. The same finding is almost true of the New Ireland group: all but a few of its species occur on New Ireland itself. We discuss in turn four aspects of the barrier between New Britain and New Ireland: the differences in species composition that it produced, the differentiation that it produced between separated populations of the same species, the resulting endemism of the New Britain avifauna, and the endemism of the combined avifaunas of New Britain plus New Ireland.

Differences in Species Composition Between New Britain and New Ireland

New Britain has 127 species; New Ireland has 103 species. There are 96 species shared between New Britain and New Ireland, 31 New Britain species absent on New Ireland, and seven New Ireland species absent on New Britain. The Ochiai dissimilarity index, expressing the difference in species composition between the New Britain and New Ireland avifaunas, is only 0.16, the lowest value for any adjacent pair of big Northern Melanesian islands without a Pleistocene land connection. However, this value is as predicted by equation 27.2, which is our best fit to the Ochiai index as a function of island areas and distance. All other things being equal, big pairs of islands tend to have high values of the index, while close islands tend to have low values. In effect, while the large areas of New Britain and New Ireland by themselves would tend to lead to marked differences in species composition, this effect is reduced by the proximity of New Britain and New Ireland (even closer during the Pleistocene), which tends to facilitate dispersal between them and thus to equalize the composition of their avifaunas.

Qualitatively, one might have expected the avifaunas of New Britain and New Ireland to be similar, since the two islands are both big, approximately equal in elevation, and near each other. Four factors contribute to the absences of species present on one island from the other island.

First, the endemic small honey-eater of the New Britain lowlands, *Myzomela erythromelas*, is replaced ecologically by the similar-sized *M. cruentata* of the New Ireland lowlands, while the endemic small honey-eater *M. pulchella* of the New Ireland mountains is replaced ecologically by the similar-sized *M. cruentata* of the New Britain mountains (*M. cruentata* lives in the mountains of New Britain but in the lowlands of New Ireland). Thus, each of these two endemics, *M. erythromelas* and *M. pulchella*, is currently excluded from the other island by a similar competing species.

Second, 10 of these apparent absentees are highly vagile species (most of them living at low population densities) that have been recorded from one of the two islands but not the other and that seem likely to be found eventually on the island from which they have yet to be recorded. These vagile species are the rails *Rallina tricolor* and *Gymnocrex plumbeiventris* and the pigeon *Columba vitiensis*, present on New Ireland but not recorded from New Britain (except for one vagrant individual of the pigeon); and the cormorant *Phalacrocorax melanoleucos*, the herons *Ardea intermedia* and *A. alba*, the hawks *Accipiter meyerianus* and *Falco peregrinus* and *F. severus*, and the rail *Porphyrio porphyrio*, which have been recorded on New Britain but not on New Ireland. The three species apparently confined to New Ireland are cryptic and easily overlooked, so they could easily occur on New Britain. Similarly, *Porphyrio porphyrio*, although not yet recorded from New Ireland, has already been recorded from New Hanover, which is part of the New Ireland group and close to New Ireland itself. Thus, we do not attach significance to the apparent confinement of these vagile species to New Britain or New Ireland.

The third category consists of six species of water and marsh birds recorded only from New Britain and not from New Ireland: the heron *Ixobrychus sinensis*, the ducks *Dendrocygna guttata* and *D. arcuata*, the stilt *Hymantopus leucocephalus*, the jacana *Irediparra gallinacea*, and the river flycatcher *Monachella muelleriana*. All of these species are vagile except for the flycatcher. The fact that they have been recorded from New Britain but not from New Ireland is readily explainable by the fact that New Britain, but not New Ireland, has numerous ornithologically explored lakes, rivers, and marshes.

Finally, all these considerations still leave 20 species to which none of the above three explanations apply. They consist of three New Ireland species (*Lorius albidinucha*, *Aerodramus orientalis*, and *Monarcha chrysomela*) and 17 New Britain species (*Henicopernis longicauda*, *Accipiter luteoschistaceus*, *A. princeps*, *Turnix maculosa*, *Gallirallus insignis*, *Henicophaps foersteri*, *Cacatua galerita*, *Tyto aurantia*, *Tanysiptera sylvia*, *Halcyon albonotata*, *Merops philippinus*, *Zoothera talasaea*, *Ortygocichla rubiginosa*, *Cichlornis grosvenori*, *Myzomela cineracea*, *Melidectes whitemanensis*, and *Lonchura melaena*). Of these 20 absentees, five (*Aerodramus orientalis* of New Ireland, and the two *Accipiter* species, *Zoothera talasaea*, and *Cichlornis grosvenori*, of New Britain) are sufficiently cryptic and elusive to be found eventually on the other island, but all the other absences are probably real. All of these absentees except for *Merops philippinus* are highly sedentary species for which there are no records on any island of the New Britain or New Ireland group lacking a Pleistocene land connection to New Britain or New Ireland itself. Most are endemic to the New Britain or New Ireland groups, if not to New Britain or New Ireland itself: 15 of the 17 New Britain specialties (three endemic full species, eight allospecies, two megasubspecies, and two weak subspecies) and all three of the New Ireland specialties (one endemic allospecies and two weak subspecies). Thus, most of these specialties represent old endemics that have lost their colonizing ability and have become confined to a single large island or Pleistocene island.

There are only two apparent surprises among these absentees. It is initially surprising that the vagile bee-eater *Merops philippinus*, an Asian species that extends through New Guinea to New Britain, has reached the New Britain satellites Sakar and Umboi and has reached Long Island since its seventeenth-century explosion but is absent from New Ireland. However, this bee-eater is confined to dry, open habitats, of which there may be too few on New Ireland. The other surprise, a real one, is the monarch flycatcher *Monarcha chrysomela*, a New Guinea species absent from New Britain but represented by endemic subspecies not only on New Ireland but also on its satellites Lihir, Tabar, Dyaul, and New Hanover. The absence of this common, conspicuous forest species from New Britain is the most surprising distributional gap in the whole Bismarck avifauna. As a forest flycatcher of modest dispersal ability, it could hardly have colonized New Ireland directly from New Guinea; it must have arrived via New Britain and subsequently disappeared from New Britain, but there is no other forest flycatcher species present on New Britain and absent from New Ireland to explain its absence. The reason for its presumed disappearance on New Britain remains mysterious.

Differentiation Between the New Britain and New Ireland Avifaunas

For species shared between New Britain and New Ireland, the average differentiation index is only 0.36, on the scale where 0 represents no differentiation and 1.0 represents differentiation at the level of weak subspecies. That is, most (75 out of 96) of New Britain's and New Ireland's shared species are subspecifically identical on the two islands. The resulting differentiation index of 0.36 is the lowest for any adjacent pair of large Northern Melanesian islands without a Pleistocene land connection, except for the equally low value for Guadalcanal and Malaita. As discussed above, this modest degree of differentiation, despite the large areas of New Britain and New Ireland permitting populations to survive and differentiate for a long time on each island, is due to the narrow water gap between the two islands. The most marked differentiation, listing for each of the following pairs the New Britain taxon first and its New Ireland equivalent second, involves the allospecies *Micropsitta pusio* versus *M. finschii*, *Ninox odiosa* versus *N. variegata* (plate 7), *Philemon cockerelli* versus *P. eichhorni* (plate 7), *Lonchura spectabilis* versus *L. hunsteini*/*L. forbesi*, and *Dicrurus hottentottus* versus *D. megarhynchus* (map 52), and the megasubspecies *Halcyon chloris tristrami* versus *H. c. novaehiberniae*/*H. c. nusae*, *Hirundo tahitica ambiens* versus *H. t. subfusca*, *Coracina tenuirostris heinrothi* versus *C. t. remota* (plate 7), and *Myzomela cruentata coccinia* versus *M. c. erythrina*.

New Britain Endemics

Despite being such a large island, New Britain's endemism is modest. Out of its 127 species, the only ones endemic to New Britain or the New Britain group are 3 endemic full species, 11 allospecies, three megasubspecies, and 18 weak subspecies. Still, New Britain's endemism index of 0.54 is topped in the Bismarcks solely by the indices for the smaller but much more isolated islands Manus and St. Matthias. According to our best-fit equation 25.3, the value of 0.54 is close to the value expected from New Britain's area and distance: the effect of its large area is offset by its proximity to both New Guinea and New Ireland, with one or both of which it shares most of its taxa.

The most distinctive endemics of New Britain are the three endemic full species *Accipiter luteoschistaceus*, *Gallirallus insignis* (plate 8), and *Melidectes whitemanensis* (plate 9); the 11 endemic allospecies *Henicopernis longicauda*, *Accipiter princeps*, *Henicophaps foersteri*, *Tyto aurantia*, *Ninox odiosa* (plate 7), *Halcyon albonotata*, *Ortygocichla rubiginosa* (plate 9), *Cichlornis grosvenori* (plate 9), *Myzomela cineracea* (plate 4), *M. erythromelas* (plate 4), and *Philemon cockerelli* (plate 7); and the endemic megasubspecies *Cacatua galerita ophthalmica*, *Tanysiptera sylvia nigriceps*/*T. s. leucura*, and *Monachella muelleriana coultasi*. Of these 18 higher endemics, 10 are confined to New Britain; six are represented in addition by eight populations on the land-bridge satellites Umboi, Lolobau, or Duke of York; but only two are represented by populations on New Britain's satellite islands lacking a Pleistocene land connection (*Tanysiptera sylvia nigriceps* and *Ninox odiosa* on Vuatom). That is, all but two of the endemics of the New Britain group are now so poorly vagile and require such large areas that most former populations on the New Britain land-bridge satellites have disappeared, those satellites have not been recolonized over water since the Pleistocene, and the non–land-bridge satellites have not been colonized.

Joint Endemics of the New Britain and New Ireland Groups

We have just listed the endemics of the New Britain group, and we shall list below the endemics of the New Ireland group. In addition, since New Britain and New Ireland are so close, there are quite a few taxa shared by the New Britain and New Ireland groups, and endemics to these two groups considered jointly. The most notable such joint endemics are one endemic full species (the cuckoo *Centropus violaceus*); 11 endemic allospecies (*Accipiter brachyurus*, *Ducula finschii* [plate 8], *D. melanochroa* [plate 8], *Charmosyna rubrigularis*, *Loriculus tener*, *Centropus ateralbus*, *Alcedo websteri*, *Rhipidura dahli*, *Monarcha verticalis* (plate 3), *Dicaeum eximium*, and *Artamus insignis* [plate 9]); and six endemic megasubspecies (*Ducula rubricera rubricera*, *Reinwardtoena browni browni*, *Eudynamys scolopacea salvadorii*, the *Pitta erythrogaster novaehibernicae* group, *Cisticola exilis polionota*, and *Myiagra hebetior eichhorni* [plate 5]).

The New Ireland Group

The New Ireland group consists of New Ireland and New Hanover, joined to each other in the Pleistocene, plus the never-connected satellite islands Dyaul, Tabar, Lihir, Tanga, and Feni. The nearest large islands are New Britain (30 km distant, closer during the Pleistocene), St. Matthias (96 km from New Hanover), and the Solomon island of Bougainville (172 km). We already discussed the barriers between New Ireland and New Britain (p. 228) and between New Ireland and Bougainville (p. 222), and the barrier between the New Ireland group and St. Matthias group is discussed below (p. 232). Hence we discuss here only the resulting differentiation of the New Ireland avifauna (i.e., its combined distinction from all those adjacent avifaunas).

New Ireland's endemism index of 0.41 is as expected (eq. 25.3) from its large area but modest distance from its nearest neighbor, New Britain. Of its 103 species, the populations of 76 are sub-specifically identical to populations outside the New Ireland group, especially on New Britain and New Guinea. The 27 endemics consist of one full species, six allospecies, and 20 weak subspecies. Nearly half of these (11 of 27), including most of the higher endemics (the one full species and four of the allospecies), are confined to New Ireland itself, while two of the allospecies and 14 of the subspecies reached satellite islands of the New Ireland group (especially New Hanover, Dyaul, Tabar, and Lihir). New Ireland's most notable endemics are its sole endemic full species, the montane honey-eater *Myzomela pulchella* (plate 4), and its six endemic allospecies, the montane parrot *Lorius albidinucha* (plate 6), the owl *Ninox variegata* (plate 7), the honey-eater *Philemon eichhorni* (plate 7), the finches *Lonchura forbesi* and *L. hunsteini*, and the bizarre drongo *Dicrurus megarhynchus*, distinguished by its very long and spirally twisting tail (plate 8, map 52).

The St. Matthias Group

The St. Matthias group consists of the medium-sized St. Matthias (414 km^2) and the small Emirau (41 km^2) only 17 km away. All 32 species of Emirau are included in the richer avifauna of St Matthias (43 species). Lying 96 km northwest of New Hanover and 241 km east of Manus, St. Matthias is second only to Manus as the most isolated large Bismarck island, and after Manus and Rennell it is the third most isolated large island of Northern Melanesia.

Two positive features distinguish the species composition of the St. Matthias avifauna. (Absentees are listed in table 30.1 and discussed on p. 233.) First, because of its isolation, modest area, and resulting low species number, St. Matthias harbors five supertramp species, all shared with Emirau: the pigeons *Ptilinopus solomonensis* (map 7), *Ducula pistrinaria*, and *Macropygia mackinlayi* (map 12), the monarch flycatcher *Monarcha cinerascens* (map 40), and the honey-eater *Myzomela pammelaena* (map 47). In the Central Bismarcks (the New Britain and New Ireland groups), islands as large as St. Matthias are richer in species and hence harbor fewer supertramps, which tend to be competitively excluded.

The second noteworthy feature of the St. Matthias avifauna is that it includes sea-level populations of four species that are confined to the

Table 30.1. Widespread Bismarck species absent from the two largest Bismarck outliers.

Species	Absent from: St. Matthias Group	Absent from: Northwest Bismarcks	Species	Absent from: St. Matthias Group	Absent from: Northwest Bismarcks
Tachybaptus ruficollis	– (H)	– (H)	*Ninox [novaeseelandiae]*	–	
Butorides striatus	–	–	*Caprimulgus macrurus*	–	–
Nycticorax caledonicus	–		*Caprimulgus macrurus*	–	–
Anas superciliosa	–		*Collocalia esculenta*	–	
Aviceda subcristata	–		*Alcedo websteri*	– (H)	– (H)
Accipiter brachyurus	–	–	*Alcedo pusilla*	–	–
Coturnix chinensis	– (H)	– (H)	*Ceyx lepidus*	–	
Gallirallus philippensis	–		*Halcyon chloris*		–
Poliolimnas cinereus		–	*Eurystomus orientalis*		–
Amaurornis olivaceus	–	–	*Rhyticeros plicatus*	–	–
Porphyrio porphyrio	–		*Pitta erythrogaster*	–	– (C)
Charadrius dubius	– (H)	– (H)	*Hirundo tahitica*	–	
Ptilinopus superbus	–		*Lalage leucomela*		–
Ptilinopus insolitus		–	*Coracina papuensis*	–	
Ptilinopus rivoli	– (C)	– (C)	*Coracina lineata*	–	–
Ducula rubricera	–	–	*Turdus poliocephalus*		– (L)
Ducula melanochroa	– (L)	– (L)	*Saxicola caprata*	– (H)	– (H)
Ducula finschii	–	–	*Acrocephalus arundinaceus*	– (H)	– (H)
Ducula spilorrhoa	–	–	*Cisticola exilis*	– (H)	– (H)
Gymnophaps albertisii	– (L)	– (L)	*Megalurus timoriensis*	– (H)	– (H)
Columba pallidiceps	–	–	*Phylloscopus trivirgatus*		– (L)
Macropygia amboinensis	–		*Rhipidura leucophrys*		–
Macropygia nigrirostris	– (C)	– (C)	*Rhipidura [rufidorsa]*		– (C)
Reinwardtoena browni	–		*Myiagra hebetior*		–
Gallicolumba jobiensis	–	–	*Myiagra alecto*	–	
Lorius [hypoinochrous]	–	–	*Dicaeum eximium*	–	–
Charmosyna placentis	–		*Nectarinia sericea*	–	–
Charmosyna rubrigularis	– (L)	– (L)	*Zosterops atrifrons*	–	
Eclectus roratus	–		*Myzomela cruentata*	– (L)	– (L)
Micropsitta bruijnii	– (L)	– (L)	*Philemon [buceroides]*	–	
Geoffroyus heteroclitus	–	–	*Erythrura trichroa*		– (L)
Loriculus tener	–	–	*Lonchura [spectabilis]*	– (H)	– (H)
Cacomantis variolosus	–		*Artamus maximus*	–	–
Eudynamis scolopacea	–	–	*Mino dumontii*	–	–
Centropus violaceus	–	–	*Dicrurus [hottentottus]*	–	–
Centropus ateralbus	–	–	*Corvus orru*	–	–

This table lists all "widespread Bismarck species" (defined as species present both on the New Britain and New Ireland groups) that are absent from the St. Matthias group ("–" in the middle column), from the Northwest Bismarcks ("–" in the right column), or from both ("–" in middle and right columns). A minus sign indicates that the species is absent: C, excluded by a competitor (*Ptilinopus rivoli* by *P. solomonensis*, *Macropygia nigrirostris* by *M. mackinlayi*, *Rhipidura [rufidorsa]* and *Pitta erythrogaster* excluded from Manus by *R. rufifrons* and *P. superba* respectively); L, absent because the island group is too low for this montane species; H, absent because the island group lacks enough suitable habitat for this species of lakes, rivers, or grassland. Other absences (dash without C, L, or H) are not because of any of these specific reasons, but just because of the probabilistic decline in species incidences with island distance and decreasing area.

mountains elsewhere in the Bismarcks: the thrushes *Zoothera [dauma]* (map 32) and *Turdus poliocephalus* (map 33), the warbler *Phylloscopus trivirgatus* (map 36), and the fantail flycatcher *Rhipidura [rufidorsa] matthiae* (map 39). All except *Z. dauma* are absent from Emirau. It seems initially surprising that species restricted to elevations above 1,200 m or even 2,700 m on large islands such as New Guinea should occur in the very different habitat and climate of lowland St. Matthias. However, at least part of the explanation for the species being confined to the mountains of large, species-rich islands is that high-elevation habitats of those islands have many fewer species than at sea level. Comparison of the lower altitudinal limit of *Turdus poliocephalus* on different Northern Melanesian islands shows that on any given island it descends to that altitude at which it encounters a community of no more than about 30 other forest bird species (see fig. 42 of Diamond 1975a). Because of the modest area and isolated location of St. Matthias, *Turdus poliocephalus* encounters a community of only 30 species already at sea level on St. Matthias, but only at elevations above 2,700 m on New Guinea are communities with no more than 30 species encountered. In effect, these four species are somewhat akin to supertramps, restricted to communities with few competing species, except that islands with a high number of competing species entirely lack supertramps but harbor those sometimes montane species in high-elevation habitats of low species number.

Let us now consider in turn the overall level of endemism of the St. Matthias avifauna, its distinction from the avifauna of New Hanover, and its distinction from the avifauna of Manus.

St. Matthias Endemics

The endemism score for the St. Matthias group is 0.60, higher than that for any other Bismarck island except the even more remote Manus. As calculated from equation 25.3, this high endemism is as expected from the isolation and moderate area of St. Matthias. The endemics consist of 16 weak subspecies, two megasubspecies (the cuckoo-shrike *Lalage leucomela conjuncta* and the monarch flycatcher *Myiagra hebetior hebetior* [plate 5]), and two allospecies (the fantail flycatcher *Rhipidura [rufidorsa] matthiae* and the monarch flycatcher *Monarcha [manadensis] menckei* [plate 3]). All four of St. Matthias's sea-level populations of otherwise montane species are endemic. Of these 20 endemics of the St. Matthias group, Emirau shares with St. Matthias 10 of the endemic weak subspecies but none of the endemic megasubspecies or allospecies.

St. Matthias/New Hanover Barrier

New Hanover has 75 species, St. Matthias has 43, and 33 are shared. New Hanover is much richer in species because its area is nearly triple that of St. Matthias and because it was joined in the Pleistocene to the larger and avifaunally much richer New Ireland, to which New Hanover is still close. The absences of 42 New Hanover species from St. Matthias are readily explained by St. Matthias's isolation and smaller area. In table 30.1 we list and discuss the otherwise widespread Bismarck species absent from St. Matthias.

As for the 10 St. Matthias species absent from New Hanover, three of them are supertramps, and four are the sea-level populations of species montane elsewhere in the Bismarcks. Their absences from New Hanover are probably due to competition. The remaining three species recorded from St. Matthias but not from New Hanover are the eagle *Haliaeetus leucogaster*, the swiftlet *Aerodramus spodiopygius*, and the finch *Erythrura trichroa*, which are variously uncommon, difficult to collect, or cryptic, and hence possibly present but overlooked on New Hanover.

The average differentiation score for populations of species shared between St. Matthias and New Hanover is 0.64, the highest for any adjacent pair of large Northern Melanesian islands except for the Manus/New Hanover pair. This high score is as predicted from equation 26.1 for the moderate areas of both St. Matthias and New Hanover (reducing extinction rates) and the distance between them. The most marked differences involve three pairs of distinct megasubspecies and two of distinct allospecies. The St. Matthias populations of these five pairs are three of the endemic megasubspecies or allospecies of St. Matthias (*Lalage leucomela conjuncta*, *Monarcha [manadensis] menckei*, and *Myiagra hebetior hebetior*) plus two endemics shared between Manus and St. Matthias (*Micropsitta [pusio] meeki* and the *Coracina tenuirostris admiralitatis* megasubspecies group).

St. Matthias/Manus Barrier

Manus has 50 species, St. Matthias has 43, and 28 are shared. Most of the differences involve tramp species that could equally occur on either island but that have only a moderate probability of being on either because of both islands' isolation and moderate area. In addition, the endemic fantail *Rhipidura rufifrons semirubra* is closely related and ecologically equivalent to the endemic *Rhipidura matthiae* of St. Matthias. Finally, St. Matthias has two supertramps plus three sea-level populations of otherwise montane species that are absent from

Manus, while Manus has one old endemic (*Pitta superba*) absent from St. Matthias.

For the 28 species shared between St. Matthias and Manus, the average differentiation index for the pairs of conspecific populations is 0.50, close to the value predicted for their areas and distance from equation 26.1. That is, conspecific populations shared between St. Matthias and Manus are less differentiated than those shared between either St. Matthias and New Hanover (differentiation index 0.64) or between Manus and New Hanover (differentiation index 0.79). The explanation is that St. Matthias and Manus tend to share some vagile, wide-ranging subspecies, unlike the more sedentary populations of New Hanover (forming part of the much larger Pleistocene New Ireland). Most of the differences between St. Matthias and Manus populations are at the level of weak subspecies; there is only one pair of megasubspecies (of the kingfisher *Halcyon saurophaga*) and one pair of allospecies (of the *Monarcha* [*manadensis*] superspecies [plate 3]).

We call attention to the striking fact that conspecific populations on these two remote islands, St. Matthias and Manus, are more similar to each other than the populations of either are to populations on New Hanover, despite New Hanover being much closer to St. Matthias than is Manus, and despite New Hanover being nearly as close to Manus as is St. Matthias. This illustrates that proximity is not the only factor promoting gene flow between populations of two islands: selection for the vagility required to found conspecific populations on two remote islands is another factor.

The Northwest Bismarcks

The most remote islands of Northern Melanesia are the Northwest Bismarcks, consisting of the Admiralty group and its western outliers (the three small Ninigo, Hermit, and Anchorite archipelagoes plus a few isolated islands such as Wuvulu). By far the largest island is Manus (1,834 km^2) in the Admiralty group. It harbors 50 species, while smaller islands of the Northwest Bismarcks harbor an additional six species not on Manus (the rail *Gallirallus philippensis* and the parrot *Charmosyna placentis*, plus four supertramps [the monarch flycatcher *Monarcha cinerascens*, the white-eye *Zosterops griseotinctus*, the honey-eater *Myzomela pammelaena*, and the starling *Aplonis feadensis:* maps 40, 45, 47, and 50]). Below we discuss the Northwest Bismarck avifauna's missing species, endemism, differentiation, and derivation.

Missing Species

The main islands of the Bismarcks consist of two large, central islands close to each other (New Britain and New Ireland) plus two medium-sized, isolated island groups (the St. Matthias group and the Northwest Bismarcks). As summarized in table 30.1, the two isolated groups lack many species shared by the two central groups. Sixty species are absent from the St. Matthias group, 52 are absent from the Northwest Bismarcks, and 41 of those species are absent from both groups. Thus, there are a total of 60 + 52 = 112 absences of central species from the two isolated island groups.

Of those 112 absences, six can be attributed straightforwardly to competitive exclusion by an ecologically similar and in some cases closely related congener on the outlying islands: the replacements of *Ptilinopus rivoli* by *P. solomonensis* (map 7), *Macropygia nigrirostris* by *M. mackinlayi* (map 12), *Pitta erythrogaster* by *P. superba*, and *Rhipidura* [*rufidorsa*] by *R. rufifrons* (map 39). Another 13 cases involve species that are confined to the mountains of New Britain and New Ireland and are hence unlikely to occur in the St. Matthias Group or Northwest Bismarcks because of their low elevation: *Ducula melanochroa*, *Gymnophaps* [*albertisii*], *Charmosyna rubrigularis*, *Micropsitta bruijnii*, *Turdus poliocephalus*, *Phylloscopus trivirgatus*, *Myzomela cruentata*, and *Erythrura trichroa*. A further 18 cases involve species of lakes, rivers, or grassland, for which the two outlying island groups may offer insufficient habitat: *Tachybaptus ruficollis*, *Coturnix chinensis*, *Charadrius dubius*, *Alcedo websteri*, *Saxicola caprata*, *Acrocephalus arundinaceus*, *Cisticola exilis*, *Megalurus timoriensis*, and *Lonchura* [*spectabilis*].

However, those three factors still leave 75 cases of absences of central species on the outliers without any obvious ecological explanation (the residue of 112 absences after 6 + 13 + 18 cases have thus been explained). These 75 cases are typical examples of absences of large-island species from ecologically suitable smaller and more remote islands. These absences are reflected in the shapes of incidence functions (chapter 12): for each species, its probability of occurring on any particular Northern Melanesian island increases in a characteristic way with the island's area and proximity; the detailed form of the incidence function depends on the particular species' vagility, population density, and habitat requirements. The forms of incidence functions result from the trade-off between immigration, which decreases with an island's isolation, and risk of extinction, which decreases with an island's area. That is, the smaller and more isolated an island, the fewer the number of species from an available pool of ecologically suitable species that the island can support in the steady state.

Endemism

The Northwest Bismarck avifauna includes 34 endemic taxa. Five are weak subspecies confined to the small western outliers (*Gallirallus philippensis anachoretae*, *Trichoglossus haematodus nesophilus*, *Halcyon saurophaga anachoretae*, *Myzomela pammelaena ernstmayri*, and *Aplonis feadensis heureka*). The other 29 endemics occur in the Admiralties, where five are confined to smaller Admiralty islands and are absent from Manus (the megasubspecies *Monarcha infelix coultasi* of Rambutyo Island [plate 3, map 41], the second-largest island of the Northwest Bismarcks, plus the four weak subspecies *Gallirallus philippensis praedo* and *G. p. admiralitatis*, *Monarcha cinerascens fulviventris*, and *Myzomela pammelaena pammelaena*); 10 are endemics confined to Manus; and 14 are endemics shared by Manus and one or more smaller islands. The 24 endemics occurring on Manus consist of 16 weak subspecies, four megasubspecies (the pigeon *Reinwardtoena brownei solitaria* [map 13], the kingfishers *Ceyx lepidus dispar* and the *Halcyon saurophaga admiralitatis* group [map 24], and the fantail flycatcher *Rhipidura rufifrons semirubra* [map 39]), and four allospecies (the owl *Ninox meeki* [plate 7], the pitta *Pitta superba* [plate 9], the monarch flycatcher *Monarcha infelix* [plate 3], and the honey-eater *Philemon albitorques* [plate 7]).

The average endemism index for species occurring on Manus is 0.72, the highest for any Bismarck island, and exceeded in the Solomons only by San Cristobal and Rennell. This high value is as predicted for Manus's isolation and moderate area from equation 25.3.

Differentiation

The differentiation index for conspecific populations shared between Manus and other islands is 0.50 for Manus/St. Matthias, 0.79 for Manus/New Hanover, and 1.15 for Manus/New Guinea. These values, all of which are as predicted from equation 26.1 for areas and distance, make the Manus avifauna the most differentiated of any Bismarck island. The Bismarck island to whose avifauna the shared populations of Manus are most similar is (as discussed above) the isolated St. Matthias rather than the larger and only slightly more distant New Hanover.

With New Guinea, Manus shares 48 of its 50 species, of which only 16 have subspecifically identical populations on the two islands. The 48 pairs of conspecific populations differ at the level of weak subspecies in 18 cases, megasubspecies in five cases, and allospecies in nine cases (listing the Manus allospecies first in each case, *Ptilinopus solomonensis* vs. *P. rivoli*, *Macropygia mackinlayi* vs. *M. nigrirostris*, *Reinwardtoena browni* vs. *R. reinwardtii*, *Micropsitta meeki* vs. *M. pusio*, *Ninox meeki* vs. *N. theomacha*, *Aerodramus spodiopygius* vs. *A. hirundinaceus*, *Pitta superba* vs. *P. sordida*, *Monarcha infelix* vs. *M. manadensis*, and *Philemon albitorques* vs. *P. novaeguineae*).

Derivation of the Northwest Bismarck Avifauna

We have grouped islands off Northeast New Guinea into a zoogeographic unit, which we termed the Bismarck Archipelago. The justification for doing so is that the islands resemble each other in their avifaunas more than they resemble the adjacent two regions, the New Guinea region and the Solomon Archipelago. The simplest interpretation of this observation is that Bismarck islands have immediately derived more of their avifaunas from each other than from New Guinea or the Solomons.

However, table 30.2 shows that an additional factor contributes to the avifaunal similarities among Bismarck islands. This table identifies the immediate origins, insofar as possible, of all taxa resident in the Northwest Bismarcks. There are 43 taxa whose origins we can identify, while the origins of 20 other taxa cannot, for various reasons, be closely pinpointed. Among the 43 traceable taxa, 33 can be traced directly to the Central Bismarcks (the New Britain and New Ireland groups) and 10 to New Guinea. The grounds for tracing are either that the Northwest Bismarck species (or superspecies) occurs in only one of the two possible source avifaunas; or else that it occurs in both source avifaunas, but that its taxonomic affinities at the allospecies, megasubspecies, or weak sub-

Table 30.2. Sources of the Northwest Bismarck avifauna.

(Bis:4)	Species shared with Central Bismarcks (C Bis), absent from New Guinea (NG): *Ducula pistrinaria rhodinolaema, Myzomela pammelaena pammelaena/ M. p. ernstmayri, Aplonis feadensis heureka*
(NG:2)	Species shared with NG, absent from C Bis: *Tyto novaehollandiae manusi, Pitta [sordida] superba*
	Species shared both with C Bis and with NG:
(Bis:29)	Source in C Bis:
	(5) Belongs to same weak subspecies as C Bis, different from NG subspecies: *Porphyrio porphyrio samoensis, Ducula spilorrhoa subflavescens, Charmosyna placentis pallidior, Eclectus roratus goodsoni, Nectarinia jugularis flavigaster*
	(2) Belongs to same megasubspecies as C Bis, different from NG megasubspecies: *Megapodius freycinet eremita, Zosterops atrifrons admiralitatis*
	(5) Belongs to same allospecies as C Bis, different from NG allospecies: *Ptilinopus solomonensis johannis, Macropygia mackinlayi arossi, Reinwardtoena browni solitaria, Aerodramus spodiopygius delichon, Pachycephala pectoralis goodsoni*
	(11) Endemic weak subspecies more similar to C Bis than to NG subspecies: *Accipiter novaehollandiae manusi, Gallicolumba beccarii admiralitatis, Cacomantis variolosus blandis, Aerodramus vanikorensis coultasi, Collocalia esculenta stresemanni, Hemiprocne mystacea macrura, Coracina papuensis ingens, Coracina tenuirostris admiralitatis, Rhipidura rufiventris niveiventris, Monarcha cinerascens fulviventris, Aplonis metallica purpureiceps*
	(1) Endemic megasubspecies more similar to C Bis than to NG population: *Ceyx lepidus dispar*
	(5) Endemic allospecies more similar to C Bis than to NG allospecies: *Micropsitta meeki, Ninox meeki, Monarcha infelix/M. I. coultasi, Philemon albitorques*
(NG:8)	Source in NG:
	(3) Belongs to same weak subspecies as NG, different from C Bis subspecies: *Nycticorax caledonicus hilli, Hirundo tahitica frontalis* (?), *Aplonis metallica metallica*
	(5) Endemic weak subspecies more similar to NG than to C Bis subspecies: *Gallirallus philippensis praedo, Aviceda subcristata coultasi, Macropygia amboinensis admiralitatis, Trichoglossus haematodus nesophilus/T. h. flavicans*
	Uninformative:
(17)	(15) Vagile species; same subspecies in NW Bis, C Bis, and NG: *Egretta sacra sacra, Ardea intermedia intermedia, Ixobrychus flavicollis australis, Anas superciliosa pelewensis, Haliastur indus girrenera, Haliaeetus leucogaster, Pandion haliaetus melvillensis, Esacus magnirostris, Ptilinopus superbus superbus, Chalcophaps stephani stephani, Caloenas nicobarica nicobarica, Alcedo atthis hispidoides, Myiagra alecto chalybeocephala, Zosterops griseotinctus eichhorni, Aplonis cantoroides*
	(2) Endemic megasubspecies; separate megasubspecies shared between C Bis and NG: *Halcyon saurophaga admiralitatis/ H. s. anachoretae*
(3)	NW Bis endemics whose affinities require further study:
	(2) Endemic weak subspecies: *Gallirallus philippensis anachoretae/G. p. admiralitatis*
	(1) Endemic megasubspecies: *Rhipidura rufifrons semirubra*

This table analyses the origins of all taxa resident in the Northwest Bismarcks. Two alternative source regions are considered: the Central Bismarcks (New Britain and New Ireland groups) or New Guinea. Taxa are considered as derived from the Central Bismarcks rather than New Guinea if the species is present in the former but not the latter region, or if the Northwest Bismarck population belongs to the same lower-level taxon as the Central Bismarck population but not the New Guinean population, or if the Northwest Bismarck taxon is a lower-level endemic with affinities in the Central Bismarcks. Taxa are considered as derived from New Guinea rather than from the Central Bismarcks if the opposite findings apply. Some taxa are uninformative for either of two indicated reasons, while the affinities of some Northwest Bismarck endemics require further study to identify their source. Numbers in the left column are the number of taxa in each category.

species level lie with the conspecific population in one of the two sources.

Naturally, most of the 33 Central Bismarck populations that served as proximate sources stem ultimately in turn from New Guinea. However, it is noteworthy that so many Northwest Bismarck populations (10 out of 43, or 23% of the analysable total) arose not just ultimately but also proximately from New Guinea. Furthermore, eight of those 10 New Guinea colonists have founded conspecific populations in the Central Bismarcks. It was only by analysis of affinities below the species (or below the superspecies) level that we could recognize that their Northwest Bismarck populations had been derived directly from New Guinea, not from their Central Bismarck conspecifics.

In effect, then, two distinct factors contribute to the avifaunal similarities among Bismarck islands. One is interisland colonization between Bismarck islands, because for every Bismarck island the nearest neighbor is another Bismarck island. The second factor is independent colonization of many target islands by the same species from the New Guinean source pool. New Guinean species differ greatly in their colonizing abilities. Target islands lying at a given range of distances from New Guinea are therefore likely to be colonized again and again by the same set of superior colonists from New Guinea. That is, the avifaunal cohesion of the Bismarck Archipelago stems partly from convergent assembly (parallel but independent derivations) and partly from interisland colonization within the Bismarcks. Mayr (1941d, p. 214) reached the same conclusion for the cohesion of avifaunas (as well as of spider faunas and mollusc faunas) among Polynesian islands: "The open sea is like a sieve which lets pass only certain specially adapted elements, and that, in my opinion, is the reason for the relative uniformity of the animal life of Polynesia."

The same two factors (convergent assembly and interisland colonization) undoubtedly contribute to the avifaunal similarities among other Bismarck islands, and among other islands sharing the same archipelago elsewhere in the world. However, we can distinguish the separate contributions of these two factors especially clearly in the case of the Northwest Bismarcks because such a large percentage of their avifauna (29%) can be traced to independent colonization. An even clearer example of avifaunal convergence through independent colonization is the convergence between the avifaunas of the Bismarcks and of Biak Island (Diamond 2001b). Biak lies off Northwest New Guinea, whose avifauna is largely shared with Northeast New Guinea at the species level but is largely distinct at the subspecies level. Hence it is easy to establish the independent origins of much of the Biak and Bismarck avifaunas through their affinities to subspecies of Northwest and Northeast New Guinea, respectively. But the avifaunas of Biak and the Bismarcks are remarkably similar to each other at the species level because of convergent community assembly: a few dozen of the most vagile species of the New Guinea lowlands independently colonized one island after another.

The Smaller Bismarck Islands

Of the smaller Bismarck islands, the satellites of New Britain will not be considered below because all of them were either recently defaunated by volcanic activity or were connected to New Britain by Pleistocene land bridges. Those facts alone would ensure low endemism in their avifaunas, in addition to the fact that most of them have modest areas and isolations. Similarly, Emirau, which is small and close to St. Matthias, has no endemics, just as one would predict from its area and isolation (eq. 25.3). All the species of Emirau are also represented on St. Matthias as subspecifically identical populations. The remaining smaller Bismarck islands fall into two groups: the Northwest Bismarcks and the satellites of New Ireland.

Small Northwest Bismarck Islands

By far the largest Northwest Bismarck island is Manus, which has numerous species and endemics absent from the smaller islands, and which supports most species occurring on the smaller islands as subspecifically identical populations. Table 30.3 lists the 10 endemics of the smaller Northwest Bismarcks (those other than Manus). All are weak subspecies, except for the megasubspecies *Monarcha infelix coultasi* of Rambutyo (map 41, plate 3), the next largest Northwest Bismarck island after Manus. The sole other single-island endemics are the weak subspecies *Trichoglossus haematodus nesophilus* of Ninigo and *Gallirallus philippensis praedo* of Skoki Island (probably present on other islands near Manus). The other seven endemics are shared among two or more of the smaller Northwest Bismarcks.

As a result, all of the smaller Northwest Bismarcks have endemism scores of zero except for Rambutyo (0.07 as expected from its moderate area) and Ninigo (0.06 as expected from its area of 13 km^2, tied for the next largest after Rambutyo, combined with its isolation of 235 km, making it the most remote Northern Melanesian island after Manus). The other small Northwest Bismarck islands are too small, too close to Manus or New Guinea, or both too small and too little isolated to have developed endemics. These values of the endemism index fit the pattern of equation 25.3 for the islands' areas and distances.

Table 30.3. Endemics of the smaller Northwest Bismarcks.

Level of endemism[a]	Taxon	Islands
s	*Gallirallus philippensis anachoretae*	Anch, Her, Nin; Wuv (subsp?)
s	*Gallirallus philippensis praedo*	Skoki
s	*Gallirallus philippensis admiralitatis*	Papenbush, Pityili
s	*Trichoglossus haematodus nesophilus*	Nin
s	*Halcyon saurophaga anachoretae*	Anch, Her, Nin; Wuv (subsp?)
s	*Monarcha cinerascens fulviventris*	Ramb, Nau, SMig, Anch, Her, Nin; Wuv (subsp?)
M	*Monarcha infelix coultasi*	Ramb
s	*Myzomela pammelaena ernstmayri*	Anch, Her, Nin, Manu; Wuv (subsp?)
s	*Myzomela pammelaena pammelaena*	Ramb, Nau, SMig
s	*Aplonis feadensis heureka*	Her, Nin, Wuv; Tench (subsp?)

This table lists the 10 taxa endemic or nearly endemic to Northwest Bismarck islands other than Manus, the largest island. Island names are abbreviated as in Appendix 1.
[a]Level of endemism: s = weak subspecies; M = megasubspecies.

Except for these endemics, all species shared between the smaller islands and Manus occur as subspecifically identical populations. The smaller islands support populations of only a few species that are absent from Manus, and most of those are supertramps (*Monarcha cinerascens*, *Myzomela pammelaena*, and *Aplonis feadensis*).

Satellites of New Ireland

New Ireland satellites consist of tiny Tingwon (1.5 km^2, too small to support any endemics) plus the five larger islands or island groups Dyaul (west of New Ireland), Tabar, Lihir, Tanga, and Feni (east of New Ireland). Table 30.4 lists the 21 endemics of these New Ireland satellites. Most of them are weak subspecies, but Dyaul supports three striking megasubspecies (the monarch flycatchers *Monarcha verticalis ateralbus* [map 41], *Monarcha chrysomela pulcherrimus*, and *Myiagra hebetior cervinicolor*), of which *M. v. alteralbus* is so distinct that it was originally described as a distinct allospecies. In addition, there is one endemic megasubspecies on Tabar (*Pitta erythrogaster splendida* [map 26]) and one megasubspecies shared between Lihir and Tabar (the *Lalage leucomela ottomeyeri/tabarensis* megasubspecies group [map 30]). Of these 21 endemics, six are confined to Dyaul, five to Tabar, three to Lihir, two to Feni, and one to Tanga, while four are shared between Lihir and 1–3 of the other satellites east of New Ireland. Endemism indices for the New Ireland satellites are 0.03–0.21, approximately in accord with values predicted from equation 25.3 for their areas (98–218 km^2) and distances from New Ireland (14–59 km).

By far the highest endemism index among the New Ireland satellites is for Dyaul (0.21), while the other four satellites have values of 0.03–0.10. This is initially unexpected because Dyaul has only half the area of Tabar and Lihir and virtually the same area as Feni and Tanga, and Dyaul is the closest to New Ireland of the five satellites (14 km, compared to 26–59 km for the others). Why is the avifauna of Dyaul so distinct?

The probable explanation is the different geological histories of these five satellites. Volcanism ended on Dyaul by the Miocene, but the other four satellites all had Late Cenozoic volcanism. Evidently, there has been insufficient time since the most recent defaunations of those four younger islands for them to develop as many megasubspecies as has Dyaul.

Among the species shared between New Ireland and these five satellites, most populations on the satellites are subspecifically identical to the New Ireland populations. The sole exceptions are the above-mentioned endemics, plus the satellites' populations of *Ducula pistrinaria*, *Collocalia esculenta*, and *Hemiprocne mystacea*. The populations of

Table 30.4. Endemics of the New Ireland satellites.

Level of endemism[a]	Taxon	Islands
s	*Accipiter novaehollandiae lihirensis*	Lih, Tang
s	*Accipiter albogularis eichhorni*	Feni
s	*Aerodramus vanikorensis lihirensis*	Lih, Feni, Tab, Fead
M	*Pitta erythrogaster splendida*	Tab
s	*Coracina tenuirostris ultima*	Lih, Tab, Tang
s	*Lalage leucomela sumunae*	Dyl
M { s	*Lalage leucomela ottomeyeri*	Lih
M { s	*Lalage leucomela tabarensis*	Tab
s	*Rhipidura rufiventris gigantea*	Lih, Tab
s	*Rhipidura rufiventris tangensis*	Tang
M	*Monarcha verticalis ateralbus*	Dyl
M	*Monarcha chrysomela pulcherrimus*	Dyl
s	*Monarcha chrysomela whitneyorum*	Lih
s	*Monarcha chrysomela tabarensis*	Tab
M	*Myiagra hebetior cervinicolor*	Dyl
s	*Pachycephala pectoralis tabarensis*	Tab
s	*Pachycephala pectoralis ottomeyeri*	Lih
s	*Dicaeum eximium phaeopygium*	Dyl
s	*Nectarinia sericea eichhorni*	Feni
s	*Myzomela cruentata vinacea*	Dyl
s	*Myzomela cruentata cantans*	Tab

This table lists the 21 taxa endemic or nearly endemic to the satellite islands of New Ireland.
[a]Level of endemism: s = weak subspecies; M = megasubspecies.

those three species on the four satellites east of New Ireland are subspecifically identical to populations of the Solomons to the east, rather than to populations of New Ireland to the west. Evidently, virtually all populations of Dyaul and most populations of the four eastern satellites were derived from New Ireland, except for the few populations on the eastern satellites derived from the Solomons.

Similarly, most species occurring on the satellites also occur on New Ireland; the satellites support only 15 populations of species absent on New Ireland. Of those 15 populations, five are of the supertramp flycatcher *Monarcha cinerascens*, present on all five satellites (map 40); six are populations of Solomon species that spread west to the satellites east of New Ireland but not to New Ireland itself (*Accipiter albogularis* [map 3], *Ptilinopus viridis* [map 8], *Chalcopsitta cardinalis* [map 15], and *Tyto alba*); and four are populations of vagile species likely to be present but overlooked on New Ireland (*Falco peregrinus*, *Falco severus*, and *Porphyrio porphyrio*).

Summary

The two largest Bismarck islands, New Britain and New Ireland, which together with their satellites constitute the Central Bismarcks, share most species (including numerous joint endemics) because they are close to each other and are similar in elevation. Differentiation between New Britain and New Ireland populations of shared species is modest (notably, five pairs of allospecies and four pairs of megasubspecies). Most differences in species composition between the New Britain and New Ireland avifaunas are readily understandable in terms of competition, habitat differences, sedentariness, and possibly overlooked populations.

The two most isolated Bismarck islands, which also have the highest endemism in the Bismarcks (including endemic allospecies), are Manus in the Northwest Bismarcks and St. Matthias. Remarkably, conspecific populations of species shared between these two remote outliers are taxonomically more similar to each other than to conspecific populations on the closer but much more species-rich New Hanover—probably because of shared selection for the vagility required to found populations on remote islands. Both the Northwest Bismarcks and St. Matthias lack several dozen species shared between the central New Britain and New Ireland groups. Minor reasons include competitive exclusion and lack of montane habitats, but the major reason is the decrease in steady-state incidence (probability of occurrence on an island) with increasing distance (because of decreasing immigration rates). Two distinctive features of St. Matthias's avifauna, related to its species poverty caused by its isolation, are five supertramp species absent from the large, species-rich Central Bismarck islands and sea-level populations of four species confined elsewhere in the Bismarcks to mountains.

Detailed analysis of taxonomic affinities shows that 10 Northwest Bismarck populations originated directly from New Guinea rather than from Central Bismarck islands, even though eight of those 10 New Guinea colonists separately founded populations in the Central Bismarcks. That is, while the main reason for the species composition of the Northwest Bismarck's avifauna being like that of other Bismarck islands is colonization from the Central Bismarcks, a significant minor reason is community convergence: independent colonization of the Northwest Bismarcks and the Central Bismarcks by the same vagile species of the New Guinean source pool.

Smaller Bismarck islands with some endemics are the smaller Northwest Bismarcks and the satellites of New Ireland. The New Ireland satellite with by far the highest endemism is Dyaul, despite its being the least isolated and having one of the smallest areas. The likely explanation is that volcanism ended on Dyaul by the Miocene but continued into the Late Cenozoic on the four other satellites, so that much more time has been available for differentiation on Dyaul than on the other satellites.

31

Barriers Within the Solomons

The preceding chapter having discussed barriers between Bismarck islands, we devote this chapter to barriers between Solomon islands. At Pleistocene times of low sea level, the Solomons consisted of four main islands or island groups, plus several, much smaller, isolated islands (fig. 1.3). The four main groups were 1) the Bukida group, consisting of the chain from Buka to Florida and Guadalcanal, which are the modern fragments of the Pleistocene island Greater Bukida; 2) the New Georgia group (from Vella Lavella to Gatukai), consisting of modern fragments of the three nearby Pleistocene islands Greater Vellonga, Greater Gatumbangra, and Greater Rendipari, plus Gizo and Simbo; 3) Malaita; and 4) San Cristobal. While all four of these groups or islands share most species and many widespread Solomon endemics, they differ from each other by local endemism and differences in species composition. This is illustrated by table 31.1, which shows how geographic borders between allospecies, megasubspecies, and distributional gaps for seven species that are geographically highly variable coincide with the division of the Solomons into these four major island groups.

We now discuss in turn each of these four island groups and the barriers between them. The chapter then concludes with a discussion of the smaller isolated islands, especially the one with the most distinctive avifauna, Rennell. Two volcanic Solomon islands (Simbo and Savo) with young avifaunas are discussed in chapter 32, and in chapter 33 we discuss differences between modern islands derived by fragmentation of a single larger Pleistocene island.

The Bukida Group

Within the Bukida group reside 119 of the Solomons' 141 species. The two largest Bukida islands, Bougainville and Guadalcanal, have 103 and 102 species, respectively. Thus, the Bukida group is not only the largest (22,512 km^2) and highest (2,591 and 2,448 m on Bougainville and Guadalcanal, respectively) but also avifaunally the richest group of the Solomons, somewhat exceeded in both respects by New Britain, the largest Bismarck island (35,742 km^2, 127 species). Geologically, it is unknown whether Guadalcanal was joined during the Pleistocene to the rest of Greater Bukida, because modern sea contours indicate a channel (Sealark Channel) of up to 2 km wide and exceeding 200 m in depth between the shallow shelves fringing Guadalcanal and Florida. We nevertheless include Guadalcanal in discussing the Bukida group, because in this tectonically active region it is undertain whether the channel existed throughout the Pleistocene and because Guadalcanal's avi-

Table 31.1. Superspecies structure in the Solomons.

	Representatives			
Superspecies	Bukida group	New Georgia group	Malaita	San Cristobal
Coracina holopolia	*h. holopolia* (M)	*h. pygmaea* (M)	*h. tricolor* (M)	—
Rhipidura [*spilodera*]	*drownei*	—	—	*tenebrosa*
Monarcha [*manadensis*]	*barbatus barbatus* (M)	*browni*	*barbatus malaitae* (M)	*viduus*
Myiagra [*ferrocyanea*]	*f. ferrocyanea/cinerea* (M)	*f. feminina* (M)	*f. malaitae* (M)	*cervinicauda*
Dicaeum [*erythrothorax*]	*aeneum*	—	*aeneum*	*tristrami*
Myzomela [*lafargei*]	*lafargei, melanocephala*	*eichhorni*	*malaitae*	*tristrami*
Aplonis [*grandis*]	*g. grandis* (M)	*g. grandis* (M)	*g. malaitae* (M)	*dichroa*

This table illustrates how the geological division of the Solomon Archipelago by Pleistocene water barriers into four main islands or island groups coincides with the taxonomic divisions of many superspecies into allospecies and megasubspecies. For the superspecies named at the left, and for each of the four main islands or groups, the table names the representative of that superspecies. A single word refers to an allospecies (e.g., "*drownei*" = *Rhipidura* [*spilodera*] *drownei*), while two words followed by (M) refer to a megasubspecies.

fauna is broadly similar to that of other Bukida islands.

The most distinctive endemics of the Bukida group (plates 1–5 and 7–9), with montane species marked by asterisks, consist of four monotypic genera (the pigeon *Microgoura*, the owl *Nesasio*, and the two montane honey-eaters *Stresemannia** of Bougainville and *Guadalcanaria** of Guadalcanal); five full species (the hawk *Accipiter imitator*, kingfisher *Halcyon bougainvillei**, whistler *Pachycephala implicata**, white-eye *Zosterops metcalfii*, and crow *Corvus woodfordi*); eight allospecies belonging to seven superspecies (the rail *Nesoclopeus woodfordi*, kingfisher *Halcyon leucopygia*, pitta *Pitta anerythra*, thicket-warbler *Cichlornis llaneae**, monarch flycatcher *Monarcha erythrostictus*, fantail flycatcher *Rhipidura drownei**, and honey-eaters *Myzomela lafargei* and *M. melanocephala*); and nine megasubspecies belonging to eight species (the *Ptilinopus solomonensis bistictus-ocularis* group, *Centropus m. milo*, *Ninox jacquinoti granti*, the *Ninox jacquinoti jacquinoti* group, *Ceyx lepidus meeki*, *Coracina h. holopolia*, *Monarcha b. barbatus*, the *Myiagra ferrocyanea cinerea-ferrocyanea* group, and *Zosterops ugiensis hamlini**). Because Bougainville and Guadalcanal have the highest mountains of Northern Melanesia, seven of the 24 endemics (marked with asterisks above) are montane. The endemism index of the whole Bukida group is 0.83, exceeded in the Solomons only by the much smaller San Cristobal and Rennell. This index value arises because, although the Bukida group is large, it is also close to both New Georgia (53 km) and Malaita (51 km) so that Bukida shares many taxa with those two nearby islands. The endemism index of the Bukida group is as expected for its area and isolation (eq. 25.3).

In addition to these Bukida endemics, there are seven undifferentiated or weakly differentiated species that range extralimitally to New Guinea or even farther west but that occur in the Solomons only in the Bukida group: the heron *Ixobrychus sinensis*, the button-quail *Turnix maculosa*, the parrot *Charmosyna placentis* (map 17), the frogmouth *Podargus ocellatus*, the swiftlet *Aerodramus orientalis*, the thicket-warbler *Cichlornis* [*whitneyi*] (map 35), and the whistler *Pachycephala melanura* (map 43). There are no flagrant absentees for the Bukida group as a whole. That is, no species that reaches at least two of the other three main Solomon islands or island groups fails to reach Bukida, although, of course, local endemics confined to one of those three other groups are absent from Bukida.

The New Georgia Group

Species Composition

Whereas the Bukida group consisted of a single Pleistocene island (possibly with Guadalcanal slightly detached), the New Georgia group consisted of three Pleistocene islands (plus Gizo and Simbo, which remained separate): Greater Vellonga (now fragmented into Vella Lavella, Bagga, and Ganonga), Greater Gatumbangra (now Kulambangra, Wana Wana, Kohinggo, New Georgia, Vangunu, and Gatukai), and Greater Rendipari (now Rendova and Tetipari). However, these three Pleistocene islands were very close to each other, and the modern islands derived from them are still close. As a result, only five species vary geographically within the group, and it is useful to discuss the group as a unit.

The whole New Georgia group has 85 species, of which 82 occur on its richest island, Kulambangra. The endemics (plates 1–5) include two full species (the warbler *Phylloscopus amoenus* and white-eye *Zosterops murphyi*), both of them confined to the mountains of Kulambangra above the altitudinal ranges of more wide-ranging relatives. The other endemics of the New Georgia group consist of eight allospecies belonging to 5 superspecies (the rail *Gallirallus rovianae*, monarch flycatchers *Monarcha richardsii* and *M. browni*, honey-eater *Myzomela eichhorni*, and four allospecies in the white-eye *Zosterops* [*griseotinctus*] superspecies), five megasubspecies belonging to four species (*Centropus milo albidiventris*, *Coracina holopolia pygmaea*, *Myiagra ferrocyanea feminina*, and *Pachycephala pectoralis melanoptera* and *P. p. melanonota*), and 13 weak subspecies, three of them confined to the mountains of Kulambangra (subspecies of *Turdus poliocephalus*, *Phylloscopus poliocephala*, and *Petroica multicolor*). The other 61 species of the New Georgia group are subspecifically identical to populations elsewhere in the Solomons, especially on Bukida. The resulting endemism index for the New Georgia group is 0.52, much lower than for San Cristobal, Rennell, and

Bukida, but higher than that for Malaita. As already mentioned, the modest endemism of the New Georgia group despite its large area is due to its proximity to Bukida.

The sole species of the New Georgia group absent on Bukida are the two montane endemics of Kulambangra, plus the white-eye superspecies *Zosterops* [*griseotinctus*], occurring elsewhere in the Solomons only on Nissan and Rennell (map 45). In contrast, the New Georgia group lacks many species of Bukida, some of them occurring elsewhere in the Solomons as well. These absentees include the four endemic monotypic genera and five endemic full species of Bukida. They also include three Bukida endemic allospecies of superspecies that range outside the Solomons but do not occur beyond Bukida in the Solomons (*Nesoclopeus woodfordi*, *Halcyon leucopygia* [map 23], and *Pitta anerythra*). The New Georgia group also lacks the seven species, named in our discussion of the Bukida group (*Ixobrychus sinensis*, etc.), ranging extralimitally beyond the Solomons but confined in the Solomons to Bukida and only weakly differentiated there.

In addition, the New Georgia group lacks 14 species that occur not only on the Bukida group but also elsewhere in the Solomons, such as San Cristobal (10 of these 14 species), Malaita (5 species), Rennell (3), and smaller or more remote islands. These 14 absentees are the grebe *Tachybaptus* [*ruficollis*], the heron *Ardea intermedia*, the rail *Poliolimnas cinerea*, the pigeons *Ducula pacifica* and *D. brenchleyi*, the parrot *Lorius chlorocercus*, the owl *Ninox jacquinoti*, the thrush *Zoothera* [*dauma*], the warbler *Acrocephalus stentoreus*, the flycatchers *Rhipidura* [*spilodera*] and *Monarcha cinerascens*, the flower-pecker *Dicaeum* [*erythrothorax*], the white-eye *Zosterops ugiensis*, and the drongo *Dicrurus hottentottus*. The most striking of these absentees are *Ninox jacquinoti* (map 22) and *Dicaeum* [*erythrothorax*] (map 44), because they are widespread in the Solomons (occurring on Malaita and San Cristobal as well as on the Bukida Group) and occur on many Bukida islands, and because *Dicaeum* [*erythrothorax*] is common. However, Malaita and San Cristobal each lack an even larger number of species otherwise widespread in the Solomons. That is, these absentees imply nothing peculiar about the New Georgia group. Instead, among the four major groups of the Solomons, the largest, the Bukida group, lacks the fewest widespread species; the next largest, the New Georgia group, lacks 14; and the smaller and more isolated Malaita and San Cristobal lack even more.

Intragroup Differences in Species Composition

On all the main islands of the New Georgia group (Vella Lavella, Ganonga, Gizo, Kulambangra, Wana Wana, Kohinggo, New Georgia, Vangunu, Gatukai, Rendova, and Tetipari), most species are ones occurring on most other islands and on all three subgroups of islands (the Vellonga, Gatumbangra, and Rendipari subgroups). However, each subgroup lacks some otherwise widespread species.

How can one account for the distributions of 21 patchily distributed species, summarized in table 31.2? Of them, 11 are patchily distributed because they are montane species: all 11 are confined in the New Georgia group to one or more of the four islands exceeding 1,000 m in elevation (Kulambangra, New Georgia, Vangunu, and Rendova), and eight of them are confined to Kulambangra, by far the highest island (1768 m). Two of those eight species, *Phylloscopus amoenus* and *Zosterops murphyi*, are endemic to the summit of Kulambangra, and six of the other nine montane species are confined in the New Georgia group to Kulambangra, although they also occur in the Solomons beyond the New Georgia group. The other 10 patchily distributed species exhibit the expected effect of island area: the largest subgroup, the Gatumbangra group with 6–7 times the area of the other two groups, has the fewest absentees. Of these 21 patchily distributed species, the sole two unreported for Gatumbangra, the ground pigeons *Columba pallidiceps* and *Gallicolumba jobiensis*, are very cryptic and uncommon and could have been overlooked.

Patchy distributions on individual islands, as opposed to subgroups, are also readily understood in terms of montane habitats and areas of lowland habitats. The most striking distributional gaps are that Ganonga and Simbo, the most outlying members of the New Georgia group, lack three conspicuous, vagile, and otherwise widespread species of the Solomons: the parrots *Geoffroyus heteroclitus* (map 20) and *Cacatua ducorpsi*, and formerly the hornbill *Rhyticeros plicatus* (but the hornbill colonized Ganonga between 1974 and 1998).

Table 31.2. Patchy distributions on Pleistocene islands of the New Georgia group.

Species	Pleistocene Island		
	Greater Vellonga	Greater Gatumbangra	Greater Rendipari
Accipiter meyerianus (mt.)	–	+ Kul	–
Falco severus	–	+	–
Gallirallus rovianae	–	+	+
Gymnophaps solomonensis (mt.)	–	+ Kul, Vang	+ Rnd
Columba vitiensis	+	+	–
Columba pallidiceps	+	–	–
Gallicolumba beccarii	–	+	–
Gallicolumba jobiensis	+	–	–
Charmosyna meeki (mt.)	–	+ Kul, (?) Vang	–
Charmosyna margarethae	–	+	–
Micropsitta bruijnii (mt.)	–	+ Kul	–
Cacomantis variolosus	–	+	–
Aerodramus spodiopygius	–	+	–
Coracina holopolia	–	+	–
Coracina caledonica (mt.)	–	+ Kul, NGa, Vang	–
Turdus poliocephalus (mt.)	–	+ Kul	–
Phylloscopus poliocephalus (mt.)	–	+ Kul	–
Phylloscopus amoenus (mt.)	–	+ Kul	–
Petroica multicolor (mt.)	–	+ Kul	–
Zosterops murphyi (mt.)	–	+ Kul	–
Erythrura trichroa (mt.)	–	+ Kul	–

Of the 85 species of the New Georgia group, these 21 are the sole ones that do not occur on modern islands derived from all three of the New Georgia group's Pleistocene islands. For these 21 patchily distributed species, the table shows on which Pleistocene islands the species now occupies at least one modern island. Predominantly montane species are denoted "(mt.)", and their islands of occurrence are named (Kul = Kulambangra, NGa = New Georgia, Vang = Vangunu, Rnd = Rendova).

Geographic Variation Within the New Georgia Group

Most species have not subspeciated within the New Georgia group because its islands are so close to each other. Even the group's most isolated island, Gizo, lies only 12 km from its nearest neighbor (Vella Lavella), and in the Pleistocene no New Georgia island was separated by more than a few kilometers from its nearest neighbor. The endemism indices of the four main islands or Pleistocene islands are all only 0.05–0.18, and all fall in the sequence of island areas: Greater Gatumbangra (the largest) 0.18, Greater Vellonga 0.14, Greater Rendipari 0.08, and Gizo (the smallest, whose sole endemic is its famous allospecies *Zosterops luteirostris*) 0.05. Although Greater Vellonga had less than one-quarter the area of Greater Gatumbangra, its endemism index of 0.14 is only slightly lower than Greater Gatumbangra's (0.18) because Greater Vellonga is more than three times more isolated. These values are in accord with the pattern expected for the island areas and distances (eq. 25.3).

These low endemism indices arise from the fact that, of the 85 species of the New Georgia group, only five show any geographic variation within the group. However, among those five is the most famous example of geographic variation in Northern Melanesia (map 45, plate 1): the four allospecies of the *Zosterops* [*griseotinctus*] superspecies, all lying within a radius of 20 km (including two megasubspecies separated by the 3.5-km strait between Rendova and Tetipari) and differing strikingly not only in plumage (Mees 1961) but also in voice (Diamond 1998). Table 31.3 summarizes all five examples of geographic variation: one (*Zosterops* [*griseotinctus*]) at the allospecies level, two (*Monarcha browni* [plate 3] and *Pachycephala pectoralis* [plate 2]) at the megasubspecies level, and two

Table 31.3. Geographic variation within the New Georgia group.

Rhipidura cockerelli
R. c. lavellae: GV (VL, Gan)
R. c. albina: GG (Kul, Koh, NGa, Vang), GR (Rnd, Tet)

Monarcha browni
(M) *M. b. nigrotectus*: GV (VL, Bag)
(M) *M. b. ganongae*: GV (Gan)
(M) { *M. b. browni*: GG (Kul, Wana, Koh, NGa, Vang)
M. b. meeki: GR (Rnd, Tet)

Pachycephala pectoralis
(M) *P. p. melanonota*: GV (VL, Bag, Gan)
(M) *P. p. centralis*: GG (Kul, Koh, NGa, Vang, Gat)
(M) *P. p. melanoptera*: GR (Rnd, Tet)

Zosterops [griseotinctus]
Z. [g.] *vellalavellae*: GV (VL, Bag)
Z. [g.] *splendidus*: GV (Gan)
Z. [g.] *luteirostris*: Gizo
Z. [g.] *rendovae*: GG, GR
(M) *Z.* [g.] *r. rendovae*: GR (Rnd)
(M) *Z.* [g.] *r. tetiparius*: GR (Tet)
(M) *Z.* [g.] *r. kulambangrae*: GG (Kul, Wana, Koh, NGa, Vang, Gat)

Myzomela eichhorni
M. e. atrata: GV (VL, Bag)
M. e. ganongae: GV (Gan)
M. e. eichhorni: Gizo, GG (Kul, Wana, Koh, NGa, Vang), GR (Rnd, Tet)

This table lists all taxa of the sole five species that vary geographically within the New Georgia group. Island abbreviations are as in appendix 1; Pleistocene island abbreviations are GV = Greater Vellonga, GG = Greater Gatumbangra, GR = Greater Rendipari. *Z.* [g.] *vellalavellae*, etc., refers to the four allospecies of the *Zosterops* [*griseotinctus*] superspecies in the New Georgia group. (M) refers to a megasubspecies; weak subspecies (e.g., the taxa of *Rhipidura cockerelli* and *Myzomela eichhorni*) have no special designation.

(*Rhipidura cockerelli* and [plate 4] *Myzomela eichhorni*) at the level of weak subspecies. In chapter 32 we discuss geographic variation within each Pleistocene island. Here we comment only on geographic variation between the three Pleistocene islands, abbreviated as GG, GV, and GR (= Greater Gatumbangra, Greater Vellonga, and Greater Rendipari, respectively).

It is apparent from table 31.3 that the GG and GR populations belong together taxonomically in four of the five cases; only in the case of *Pachycephala pectoralis* are they as distinct as the distinction between the GV and GG populations of the same superspecies. In contrast, the GV populations are distinct from the GG and GR populations in all five cases. The explanation is surely that, even at Pleistocene times of low sea level, the gap between GV and GG was nearly the same as its present value of 24 km, while GR is now only 9 km from GG, and the Pleistocene gap between GR and GG contracted to less than 1 km or may even have been obliterated (fig. 1.3).

The Barrier Between the Bukida and New Georgia Groups

To assess the differences between the avifaunas of the two largest groups of Solomon islands, we compared the avifaunas of the two Bukida islands nearest the New Georgia group, Choiseul and Ysabel, with the avifaunas of the nearest and physically most similar New Georgia islands, Vella Lavella and Kulambangra and New Georgia itself (table 26.1). As reflected in our two indices to measure differences in species composition (the Ochiai index and the non-nestedness index), the barrier between the Bukida group and the New Georgia group is overshadowed by the other major barriers within the Solomons (those between Bukida,

Malaita, San Cristobal, and Rennell; figs. 27.1 and 27.2). In addition, the differentiation index for populations of species shared between the Bukida and New Georgia groups is 0.42–0.45, comparable to the index for the barrier between Bukida and Malaita but weaker than index values for the barriers setting off San Cristobal and Rennell (fig. 26.1).

The explanation for this modest effect of the barrier between the Bukida and New Georgia groups becomes apparent from the maps in figures 26.1, 27.1, and 27.2. While both the Bukida group and the New Georgia group are large, and while the closest distance between them is comparable to the distances setting off Malaita and San Cristobal, the Bukida and New Georgia groups are aligned parallel to each other for 230 km, whereas the narrow end of San Cristobal points at the narrow ends of Malaita and Guadalcanal. That is, the Bukida and New Georgia groups present to each other a broad target of aligned islands, resulting in much intergroup dispersal and hence only modest differentiation.

The differences in species composition between the two groups consist of the group endemics, specialties, and absentees already discussed under the Bukida and New Georgia groups individually. The most marked differentiation between populations shared between the two groups is in four pairs of allospecies (*Gallirallus philippensis* vs. *G. rovianae*, *Monarcha castaneiventris* and *M. erythrostictus* vs. *M. richardsii* [plate 5], *Monarcha barbatus* vs. *M. browni* [plate 3], and *Myzomela lafargei* and *M. melanocephala* vs. *M. eichhorni* [plate 4]) and five pairs of megasubspecies (*Centropus milo milo* vs. *C. m. albidiventris, Ceyx lepidus meeki* vs. *C. l. collectoris*, *Coracina h. holopolia* vs. *C. h. pygmaea*, *Myiagra ferrocyanea ferrocyanea* vs. *M. f. feminina* [plate 5], and *Pachycephala pectoralis orioloides* vs. *P. p. melanonota* [plate 2]), listing the Bukida population first in each case. In addition, the Bukida and New Georgia groups differ in 9–12 pairs of weak subspecies. However, 41–47 species shared between the two groups are represented by subspecifically identical populations in both groups.

Malaita

With an area of 4,307 km^2 and an elevation of 1280 m, Malaita is the third largest and fourth highest Solomon island. With 74 species, it virtually ties with Ysabel and San Cristobal as the fourth richest, after the much larger and higher Guadalcanal and Bougainville and the higher though smaller Kulambangra.

Species Composition

Malaita endemics (plates 1–4) consist of one full species (the white-eye *Zosterops stresemanni*), two allospecies (the fantail flycatcher *Rhipidura malaitae* and honey-eater *Myzomela malaitae*), five megasubspecies (*Coracina holopolia tricolor*, *Monarcha barbatus malaitae*, *Myiagra ferrocyanea malaitae*, the very distinctive whistler *Pachycephala pectoralis sanfordi*, and *Aplonis grandis malaitae*), and 10 weak subspecies. Malaita's endemism index of 0.41 is exceeded in the Solomons by Bukida, the New Georgia Group, San Cristobal, and Rennell. Malaita's value is as expected for its area and distance (eq. 25.3).

Of Malaita's 74 species, 71 are shared with Guadalcanal, 66 with Ysabel, and 55 with San Cristobal. Among these three islands, Malaita's lowest value of the pairwise non-nestedness index (0.04) is with Guadalcanal, because the Malaita avifauna is almost perfectly nested within the richer Guadalcanal avifauna. Malaita's lowest value of the pairwise Ochiai index (0.13) is with Ysabel because each island has only a few specialties not shared with the other island. For reasons discussed below, San Cristobal is the one of these three major islands that has the highest values of both the non-nestedness index and the Ochiai index with Malaita: that is to say, the species composition of the San Cristobal avifauna is more distinct from that of Malaita than are the avifaunas of Guadalcanal and Ysabel.

Malaita's three species not shared with Guadalcanal are the supertramp pigeon *Ducula pacifica* (once recorded on Malaita, possibly only as a vagrant) and the Malaita endemics *Zosterops stresemanni* and *Rhipidura malaitae*. Malaita's eight species not shared with Ysabel are the same three, plus *Ixobrychus flavicollis*, *Ptilinopus solomonensis*, *Ducula brenchleyi*, *Gymnophaps solomonensis*, *Lorius chlorocercus*, and *Tyto alba*. Of those six, *Ducula brenchleyi* and *Lorius chlorocercus* are eastern species reaching their western range limit on Malaita and Guadalcanal; *Gymnophaps solomonensis* and possibly *Ptilinopus solomonensis* are montane species for which Ysabel may offer insufficient area at high elevations (Ysabel and Malaita have virtually the same elevation and total area, but proportionally over three times more of Malaita's area than of Ysabel's area lies above an elevation of 600 m; Mayr & Diamond 1976); and *Tyto alba* is an inconspicuous species that might be present but overlooked on Ysabel. Conversely, the 10 Ysabel species absent from Malaita consist of eight Bukida specialties plus two species absent from Malaita probably merely by chance. The much larger number of species on the larger and much higher Guadalcanal absent on Malaita (31 absentees) similarly consists of five Bukida specialties and 10 chance absentees, plus 14 species of Guadalcanal's high mountains (for which Malaita is too low) and two Guadalcanal lowland endemics.

Differentiation

Among those species that Malaita shares with those three large neighboring islands, the pairwise differentiation index for conspecific populations is 0.36 for Malaita/Guadalcanal, 0.45 for Malaita/Ysabel, and 0.65 for Malaita/San Cristobal. One might initially expect the lowest differentiation to apply to the Malaita/San Cristobal island pair because San Cristobal is the neighbor with the smallest area and closest distance to Malaita. In reality, the San Cristobal avifauna is much more differentiated from the Malaita avifauna than are the Guadalcanal or Ysabel avifaunas. This is yet another example of the distinctness of the San Cristobal avifauna, discussed in the following section. (Briefly, Malaita and San Cristobal are long, slender islands with their narrow western and eastern ends, respectively, pointed towards each other, but Malaita and Guadalcanal are aligned nearly parallel to each other over much of their length, providing a much broader target for dispersing colonists.) The Guadalcanal and Ysabel avifaunas are nearly equally differentiated from the Malaita avifauna because Guadalcanal and Ysabel were formerly connected in Greater Bukida and still share similar avifaunas. Hence the conspecific populations differentiated between Malaita and Guadalcanal are virtually the same as those differentiated between Malaita and Ysabel: one pair of differing allospecies (in the *Myzomela* [*lafargei*] superspecies), seven pairs of differing megasubspecies (in the species *Ninox jacquinoti*, *Coracina holopolia*, *Rhipidura cockerelli*, *Monarcha barbatus*, *Myiagra ferrocyanea*, *Pachycephala pectoralis*, and *Aplonis grandis*), plus megasubspecies of *Ceyx lepidus* (differing only between Ysabel and Malaita) and 8–10 pairs of weak subspecies.

San Cristobal

Among Northern Melanesian islands, the avifauna of San Cristobal is exceeded in distinctiveness only by the avifauna of the much more isolated Rennell. This distinctiveness of the San Cristobal avifauna is expressed in four ways: a high degree of endemism; absences of numerous species widespread over the rest of Northern Melanesia, the presence of species that immigrated from archipelagoes east of Northern Melanesia and that did not spread west beyond San Cristobal, and, among those species shared between San Cristobal and the rest of the Solomons, a high degree of differentiation of populations of those shared species. We summarize each of these features before discussing explanations for San Cristobal's distinctiveness.

Endemism

As summarized in table 31.4, San Cristobal (or San Cristobal plus other islands of its group) has 32 endemics or near endemics, including an endemic genus of honey-eater (*Meliarchus sclateri* [plate 9]), three endemic full species (the rail *Pareudiastes sylvestris* [plate 8], the pigeon *Gallicolumba salamonis*, and the warbler *Cettia parens*), and eight endemic allospecies (plates 3, 4, 5, and 7), plus six megasubspecies (plates 2 and 7) and 14 weak subspecies. There are additional endemics confined to other islands of the San Cristobal group. Ugi has an endemic megasubspecies (*Monarcha castaneiventris ugiensis* [plate 5]) of a species present but more weakly differentiated on San Cristobal, plus two megasubspecies (*Rhipidura rufifrons ugiensis* and *Monarcha viduus squamulatus* [plate 3]) of species equally strongly differentiated on San Cristobal. The weak subspecies *Halcyon chloris sororum* is present on Three Sisters, and *Rhipidura rufifrons kuperi* is on Santa Anna and Santa Catalina. The supertramp species *Ptilinopus richardsii* is absent from San Cristobal but is present as a weak subspecies on all other islands of the group. San Cristobal's endemism index, 0.88, is the second highest of any Northern Melanesian island, barely exceeded by the value for the isolated Rennell (0.95), and double the value predicted by equation 25.3 for an island of its area and isolation.

Table 31.4. Endemics of the San Cristobal Group.

Level of endemism[a]		Taxon	Islands of occurrence
s		*Accipiter albogularis albogularis*	SCr, Ugi, SAn, SCat
s		*Amaurornis olivaceus ultimus*	SCr, Ugi, SAn, SCat, Gow
S		*Pareudiastes sylvestris*	SCr
s		*Ptilinopus richardsii richardsii*	Ugi, 3S, SAn, SCat
s		*Ptilinopus solomonensis solomonensis*	SCr, Ugi, 3S
M		*Ptilinopus viridis eugeniae*	SCr, Ugi, 3S, SAn, SCat
S		*Gallicolumba salamonis*	SCr, Rmos
s		*Micropsitta finschii finschii*	SCr, Ugi, SAn, Ren
s		*Ninox jacquinoti roseoaxillaris*	SCr, Ugi, SCat
s		*Collocalia esculenta makirensis*	SCr, Ugi, 3S
s		*Hemiprocne mystacea carbonaria*	SCr, SAn, SCat
M		*Ceyx lepidus gentianus*	SCr
s		*Halcyon chloris solomonis*	SCr, Ugi, SAn, SCat
s		*Halcyon chloris sororum*	3S
s		*Coracina lineata makirae*	SCr
A		*Coracina salamonis*	SCr
s		*Lalage leucopyga affinis*	SCr, Ugi
s		*Zoothera margarethae margarethae*	SCr
S		*Cettia parens*	SCr
A		*Phylloscopus makirensis*	SCr
A		*Rhipidura tenebrosa*	SCr
M	s	*Rhipidura rufifrons russata*	SCr
	s	*Rhipidura rufifrons kuperi*	SAn, SCat
M		*Rhipidura rufifrons ugiensis*	Ugi
s		*Monarcha [melanopsis] castaneiventris megarhynchus*	SCr
A		*Monarcha [melanopsis] ugiensis*	Ugi, 3S, SAn, SCat
A	M	*Monarcha viduus viduus*	SCr, SAn
	M	*Monarcha viduus squamulatus*	Ugi
A		*Myiagra cervinicauda*	SCr, Ugi, SAn
M		*Petroica multicolor polymorpha*	SCr
s		*Pachycephala pectoralis christophori*	SCr, SAn
A		*Dicaeum tristrami*	SCr
s		*Zosterops ugiensis ugiensis*	SCr
M		*Myzomela cardinalis pulcherrima*	SCr, Ugi, 3S
A		*Myzomela tristami*	Scr, SAn, SCat
G		*Meliarchus sclateri*	SCr
A		*Aplonis dichroa*	SCr
M		*Dicrurus hottentottus longirostris*	SCr

All endemics or near-endemics of the San Cristobal group (consisting of San Cristobal = SCr, Ugi, Three Sisters = 3S, Santa Anna = SAn, and Santa Catalina = SCat) are listed together with their islands of occurrence. Three taxa extending to Gower = Gow, Ramos = Rmos, and Rennell = Ren rate as near-endemics.

[a]Level of endemism: endemic genus (G), full species (S), allospecies (A), megasubspecies (M), or weak subspecies (s).

Absentees

The San Cristobal group strikingly lacks nine conspicuous, common, widespread lowland species that are surely absent on San Cristobal and could not be merely overlooked there: the goshawk *Accipiter novaehollandiae*, pigeon *Ptilinopus superbus* (map 5), cockatoo *Cacatua ducorpsi* (plate 8), kingfisher *Alcedo pusilla*, hornbill *Rhyticeros plicatus* (map 25), cuckoo-shrikes *Coracina holopolia* (map 29) and *C. papuensis*, fantail flycatcher *Rhipidura cockerelli* (map 37), and starling *Mino dumontii*. (The heron *Ixobrychus flavicollis* is also unrecorded in the San Cristobal group, but it is inconspicuous and somewhat patchily distributed elsewhere in Northern Melanesia.) Seven of these nine absentees (all except *Cacatua ducorpsi* and *Coracina holopolia*) are widespread in the Bismarcks as well as in the Solomons. Three of the absentees (*Ptilinopus superbus*, *Coracina papuensis*, and *Mino dumontii*) are present on every ornithologically explored Solomon island exceeding about 13 km^2 in area outside the San Cristobal group, except for absences on the volcanically active, small islands Savo and Simbo. Three others (*Accipiter novaehollandiae*, *Cacatua ducorpsi*, and *Rhyticeros plicatus*) are present on every well-explored Solomon island larger than 25 km^2 outside the San Cristobal group, except for one absence each on islands in the range 25–180 km^2. Four of these absentees (*Cacatua ducorpsi*, *Rhyticeros plicatus*, *Coracina papuensis*, and *Mino dumontii*) are among the noisiest and most conspicuous, common, and widespread species of Northern Melanesia. Thus, their absences from the San Cristobal group are noteworthy. For none of these nine species can one point to a San Cristobal endemic or eastern invader that ecologically fills the absentee's niche.

Eastern Specialties

The San Cristobal group supports populations of four species that invaded from the New Hebrides archipelago to the east and spread no farther along the main chain of the Solomons: the pigeon *Ptilinopus richardsii*, cuckoo-shrike *Lalage leucopyga* (map 30), fantail flycatcher *Rhipidura fuliginosa*, and honey-eater *Myzomela cardinalis* (map 46). Three other species that do occur with wide distributions west along the main chain of the Solomons and in the Bismarcks are represented on San Cristobal not by a widespread Solomon subspecies or megasubspecies but by endemic subspecies related to populations of the New Hebrides: the rail *Gallirallus philippensis*, kingfisher *Halcyon chloris*, and whistler *Pachycephala pectoralis*. Two of San Cristobal's endemic full species (*Pareudiastes sylvestris* and *Gallicolumba salomonis*) may have their nearest relatives in archipelagoes to the east and may thus also have been derived from the east. That is, the avifauna of San Cristobal, the easternmost Northern Melanesian island, includes a considerable component of specialties derived from the east.

Differentiation

The large Solomon islands nearest San Cristobal to the west are Malaita and Guadalcanal. Values of the pairwise differentiation index for populations of species that San Cristobal shares with these neighbors are 0.65 for San Cristobal/Malaita and 0.70 for San Cristobal/Guadalcanal. These values are higher than for any neighboring pair of large Northern Melanesian islands except the Rennell/San Cristobal pair (0.75) and the Manus/New Hanover pair (0.79). However, Manus and Rennell are much more isolated than is San Cristobal: Rennell lies 168 km from San Cristobal, and Manus lies 285 km from New Hanover, while San Cristobal is only 47 km from Malaita and 55 km from Guadalcanal.

These two pairwise differentiation values for San Cristobal are 30% higher than the values expected from equation 26.1 for San Cristobal's area and isolation. For example, Malaita is even larger than San Cristobal and lies at nearly the same distance from Guadalcanal as does San Cristobal, but the differentiation index for Malaita/Guadalcanal is only 0.36, half the value for San Cristobal/Guadalcanal (0.70). Ysabel also is larger than San Cristobal and lies 65% farther from Malaita than does San Cristobal, yet the differentiation index for Ysabel/Malaita is only 0.45, compared to 0.65 for San Cristobal/Malaita.

Interpretation of San Cristobal's Distinctness

Why is the San Cristobal avifauna so distinctive in its endemism, absentees, eastern specialties, and differentiation? We can point to two contributing obvious factors and one subtle factor, but some mystery remains, and we do not claim to have a complete answer.

First, San Cristobal shares with Manus and Rennell the distinction of being the most peripheral island in Northern Melanesia. Colonization sources close to San Cristobal lie only to the west, whereas most other Northern Melanesian islands are targets for colonists originating nearby from several directions.

Second, San Cristobal's position at the eastern periphery of Northern Melanesia makes it the first island to receive colonists from the New Hebrides, New Caledonia, and Fiji to the east. Naturally, since these eastern archipelagoes lie considerably farther from San Cristobal than does the main chain of the Solomons, the avifauna of San Cristobal is still predominantly a Solomon avifauna, but it nevertheless includes a significant eastern component. San Cristobal's avifauna is not impoverished for its area by Solomon Islands standards: for instance, San Cristobal holds even a few more species than the larger Malaita. The accumulation of eastern species on San Cristobal thus implies some reduction in the number of established Solomon Islands species from the west, though there is not a one-to-one replacement of western species by ecologically equivalent eastern species. Instead, diffuse competition may be involved in some of the replacements.

Finally, a more subtle point is that the narrow western end of San Cristobal points at the narrow eastern ends of Malaita and Guadalcanal, whereas the long axis of Malaita was aligned parallel to the long axis of Pleistocene Greater Bukida and is still aligned nearly parallel to the axis of modern Guadalcanal. Thus, Malaita receives immigrants over its entire length from Guadalcanal (and formerly from Greater Bukida), whereas San Cristobal presents a much smaller target to emigrants from Guadalcanal and Malaita, which in turn present only narrow "launching pads."

Rennell

Rennell has the most distinctive avifauna of any Northern Melanesian island. Its avifauna is also interesting because it provides a classical example of peripheral isolation and biotic origins by overwater colonization. Rennell is an isolated, uplifted coral atoll surrounded by water several thousand meters deep. The atoll's former central lagoon is now the largest lake of Northern Melanesia. Hence Rennell's biota originated by overwater colonization and evolved *in situ*. Today, Rennell provides examples of every stage in insular evolution, from species such as the pelican *Pelecanus conspicillatus*, which arrived as vagrant groups in recent decades but has not (or not yet) established itself as a breeder, through the nonendemic cormorant *Phalacrocorax carbo* and pigeon *Macropygia mackinlayi*, which arrived within this decade and within this century, respectively, to very distinctive endemic megasubspecies and allospecies that must have reached Rennell in the distant past.

Endemism

Table 31.5 summarizes Rennell's avifauna and that of its nearby smaller satellite Bellona, all of whose 18 species also occur on Rennell. Of Rennell's avifauna of 39 species, half are endemic. These endemics consist of eight weak subspecies (plate 5), seven very distinctive megasubspecies (plates 2, 4, and 6), and five endemic allospecies (the fantail flycatcher *Rhipidura rennelliana*, the monarch flycatcher *Clytorhynchus hamlini* [plate 9], the white-eyes *Zosterops rennelliana* and *Woodfordia superciliosa* [plate 1], and the starling *Aplonis insularis*). As an example of the distinctiveness of Rennell's megasubspecies, its whistler *Pachycephala pectoralis feminina* (plate 2) is, as the subspecies name implies, almost unique among the dozens of *Pachycephala pectoralis* races in that the male is dull-plumaged and resembles the female. The Rennell whistler is also ecologically and vocally distinctive, being a thrushlike ground feeder, unlike its arboreal conspecifics elsewhere, and having a thrushlike song. Of these 20 endemics of the Rennell/Bellona group, the higher endemics are disproportionately concentrated on Rennell; Bellona shares none of the endemic allospecies, and only three of the seven megasubspecies and three of the eight weak subspecies.

Rennell's endemism index is 0.95, the highest of any Northern Melanesian island. This is double the value predicted by equation 25.3, even for an island as large (684 km^2) and remote (168 km from San Cristobal) as Rennell. The most distinctive Bismarck island, Manus, has a lower index of 0.72, even though Manus is nearly three times Rennell's area and 44% more distant from its nearest neighbor. The explanation may involve Rennell's peripheral location: Manus receives immigrants from three nearly equidistant sources to the south, northeast, and northwest (New Guinea, St. Matthias, and New Hanover respectively), but Rennell's nearest sources lie only to the northeast (San Cristobal and Guadalcanal), and other sources of the Rennell avifauna lying in other directions (Australia, New Guinea, and Santa Cruz) are several times more distant.

Stages of Speciation

As mentioned above, the Rennell avifauna illustrates a wide spectrum of stages in speciation (table 31.5). The species of the first stage, "nonstarters" (denoted 0a and 0b in the table), are 15 species that show no or only slight geographic variation in Northern Melanesia, and in which the Rennell population belongs to the widespread Northern Melanesian or Solomon race. In nine of these

Table 31.5. Resident Avifaunas of Rennell (Ren) and Bellona (Bel) islands.

Level of endemism[a]	Taxon	Islands	Stage of speciation[b]
s	*Tachybaptus novaehollandiae rennellianus*	Ren	1
M	*Phalacrocorax melanoleucos brevicauda*	Ren	2
O	*Egretta sacra sacra*	Ren, Bel	Oa
O	*Ardea alba modesta*	Ren (vagrant?), Bel (vagrant)	Oa
O	*Ixobrychus flavicollis woodfordi*	Ren	Ob
M	*Threskiornis moluccus pygmaeus*	Ren, Bel	2
O	*Platalea regia*	Ren, Bel	Oa
O	*Anas superciliosa pelewensis*	Ren	Oa
s	*Anas gibberifrons*	Ren	1
O	*Accipiter fasciatus fasciatus*	Ren, Bel	Oc
O	*Pandion haliaetus melvillensis*	Ren, Bel	Oa
O	*Porzana tabuensis tabuensis*	Ren	Oa
O	*Porphyrio porphyrio samoensis*	Ren, Bel	Oa
s	*Ptilinopus richardsii cyanopterus*	Ren, Bel	Od
O	*Ducula pacifica pacifica*	Ren, Bel	Oa
O	*Macropygia mackinlayi arossi*	Ren, Bel	Oa
O	*Gallicolumba beccarii solomonensis*	Ren, Bel	Oc
O	*Caloenas nicobarica nicobarica*	Ren, Bel	Oa
O	*Lorius chlorocercus*	Ren, Bel (vagrant)	Oc
O	*Micropsitta finschii finschii*	Ren	Oc
M	*Geoffroyus heteroclitus hyacinthus*	Ren	2
M	*Chrysococcyx lucidus harterti*	Ren, Bel	2
O	*Tyto alba crassirostris*	Ren, Bel	Oa
O	*Aerodramus vanikorensis lugubris*	Ren, Bel	Ob
s	*Collocalia esculenta desiderata*	Ren, Bel	Od
O	*Hemiprocne mystacea woodfordiana*	Ren	Ob
s	*Halcyon chloris amoena*	Ren, Bel	Od
M	*Coracina lineata gracilis*	Ren, Bel	3
s	*Turdus poliocephalus rennellianus*	Ren	Od
M	*Gerygone flavolateralis citrina*	Ren	3
A	*Rhipidura rennelliana*	Ren	3
A	*Clytorhynchus hamlini*	Ren	2
s	*Myiagra caledonica occidentalis*	Ren	Od
M	*Pachycephala pectoralis feminina*	Ren	3
A	*Zosterops rennellianus*	Ren	3
A	*Woodfordia superciliosa*	Ren	3
M	*Myzomela cardinalis sanfordi*	Ren	3
O	*Aplonis cantoroides*	Ren, Bel	Oa
A	*Aplonis insularis*	Ren	2

[a]Level of endemism of the Rennell/Bellona population: A = endemic allospecies, M = endemic megasubspecies, s = weak subspecies, O = not endemic.

[b]Stage of speciation on Rennell/Bellona: Oa = widespread vagile species of which all Northern Melanesian populations (including that of Rennell) belong to the same subspecies; Ob = widespread vagile species with only slight geographic variation in Northern Melanesia, and the Rennell population belonging to the widespread Solomon race; Oc, Od = species with much geographic variation elsewhere, but the Rennell population surprisingly not endemic (Oc) or else only a weak endemic subspecies (Od); 1 = species with only slight geographic variation elsewhere and a weak endemic subspecies on Rennell; 2, 3 = species with no or only modest geographic variation elsewhere (2), or else marked variation elsewhere (3), but a very distinct endemic megasubspecies or allospecies on Rennell, often the most distinctive population of that species or superspecies. We have omitted *Phalacrocarax carbo*, recorded breeding on Rennell only within this decade.

species the populations of all Northern Melanesian islands belong to the same wide-ranging subspecies that also ranges extralimitally beyond Northern Melanesia; in two others, all Northern Melanesian populations belong to the same endemic weak subspecies (the supertramp pigeon *Macropygia mackinlayi* [map 12] and the barn owl *Tyto alba*); and four other species have developed one or more weak endemic subspecies in Northern Melanesia but no endemic population on Rennell (the heron *Ixobrychus flavicollis*, pigeon *Ducula pacifica*, swiftlet *Aerodramus vanikorensis*, and tree swift *Hemiprocne mystacea*). Thirteen of these 15 species fall into the most vagile of our three dispersal categories, category A (chapter 11). These species are too vagile for differentiation to have occurred even on Rennell.

A second stage (stage 1 in table 31.5) consists of two species of vagile water birds (the grebe *Tachybaptus novaehollandiae* and the duck *Anas gibberifrons*) that exhibit only weak geographic variation elsewhere in their range. However, these two species have managed to evolve weak endemic subspecies in Rennell's lake.

The third stage (stage 2 in table 31.5) consists of six species with little or no geographic variation elsewhere in their range, but with endemic megasubspecies (four cases) or allospecies (two cases) on Rennell. The details are of interest. The cormorant *Phalacrocorax melanoleucos* is otherwise distributed as a single uniform subspecies from Indonesia to New Caledonia (plus another subspecies in New Zealand), but a very distinctive, short-tailed, small-bodied megasubspecies has evolved in Rennell's lake. The ibis *Threskiornis moluccus* has differentiated into only two similar subspecies over its range through Australia, New Guinea, and the Moluccas, but the Rennell/Bellona population is a megasubspecies distinguished by smaller size (weight 25% less than its conspecifics) and short bill (36% less than its conspecifics). The parrot *Geoffroyus heteroclitus* (map 20) occupies the rest of Northern Melanesia without geographic variation; the sole differentiated form is the distinctively colored Rennell megasubspecies. The cuckoo *Chrysococcyx lucidus* is uniform throughout its Southern Melanesian range from New Caledonia to the Santa Cruz Islands; the Rennell/Bellona megasubspecies is unique in its marked sexual dimorphism. The populations of the supertramp starling *Aplonis [feadensis]* scattered on atolls for 2000 km along the northern fringe of Northern Melanesia have differentiated into only two weak subspecies, but the short-tailed Rennell population is an endemic allospecies (map 50). Finally, the monarch flycatcher *Clytorhynchus [nigrogularis]* populations of Fiji and Santa Cruz are only subspecifically distinct from each other, but the Rennell population is a long-billed, distinctively colored allospecies, *Clytorhynchus [nigrogularis] hamlini* (plate 9).

The next seven species show much geographical variation elsewhere in their range and in addition have formed distinctive populations (four endemic megasubspecies, three endemic allospecies) on Rennell (stage 3 in table 31.5). We mentioned above the distinctively hen-feathered race of *Pachycephala pectoralis* on Rennell. The Rennell megasubspecies of the cuckoo-shrike *Coracina lineata* and warbler *Gerygone flavolateralis* (map 34) are the most distinctive races of those species, the megasubspecies of the honey-eater *Myzomela cardinalis* (map 46) nearly so, while Rennell's fantail flycatcher (*Rhipidura rennelliana*), small white-eye (*Zosterops rennellianus*), and large white-eye (*Woodfordia superciliosa*) are so distinctive that they rate as allospecies. These populations are distinctive not only morphologically but also vocally and in their feeding habits. For example, as mentioned above, the Rennell whistler is specialized as a ground feeder, unlike its arboreal relatives, and converges vocally on the Rennell ground-feeding thrush (*Turdus poliocephalus*), while the Rennell small white-eye is a nuthatch-like insectivore (unlike its leaf-gleaning omnivorous relatives) and converges vocally on the Rennell large white-eye (Diamond 1984a).

Lest we leave the impression that Rennell populations are always the most distinctive ones of their species, we mention, finally, nine species where, for whatever reason, this is not the case (table 31.5). These nine species exhibit marked geographic variation in Northern Melanesia or elsewhere in the range, but no (four cases) or only weak (five cases) endemism on Rennell (stage 0c and 0d respectively in Table 31.5).

Absentees and Specialties

Because of Rennell's only moderate area and its isolated and peripheral location, it lacks 42 lowland species that are widespread in the Solomons and reach the San Cristobal Group. The list includes many of the most conspicuous, widespread, abundant species of San Cristobal and the Solomons. The 42 species are: *Butorides striatus*, *Nycticorax*

Table 31.6. Comparison of Rennell avifauna with other avifaunas.

Other avifauna compared	No. of species in other avifauna	No. of species shared with Rennell	Dissimilarity values for species composition: Ochiai index	Dissimilarity values for species composition: Non-nestedness index	Pairwise differentiation index
Solomon Islands					
Russells	47	12	0.72	0.69	0.58
Ugi	49	16	0.63	0.59	0.50
San Cristobal	76	20	0.63	0.49	0.75
Eastern archipelagoes					
New Hebrides	56	22	0.53	0.44	1.27
Santa Cruz	33	20	0.44	0.39	1.50

Values in the right-most three columns are for comparisons of the Rennell avifauna with that of the named archipelago. Data are from table 26.1. Note that Rennell's avifauna is most similar to avifaunas of the eastern archipelagoes in species composition (lowest index values in fourth and fifth columns), but that shared species are most differentiated taxonomically (highest differentiation index value in sixth column) between Rennell and the eastern archipelagoes. See text for discussion of this paradox.

caledonicus, *Aviceda subcristata* (map 2), *Haliastur indus*, *Accipiter albogularis* (map 3), *Haliaeetus sanfordi*, *Megapodius freycinet*, *Gallirallus philippensis*, *Poliolimnas cinereus*, *Amaurornis olivaceus*, *Esacus magnirostris*, *Ptilinopus solomonensis* (map 7), *P. viridis* (map 8), *Ducula rubricera* (map 9), *D. pistrinaria*, *Columba vitiensis* (map 11), *Reinwardtoena crassirostris* (map 13), *Chalcophaps stephani*, *Gallicolumba jobiensis*, *Chalcopsitta cardinalis* (map 15), *Trichoglossus haematodus*, *Charmosyna margarethae*, *Eclectus roratus*, *Cacomantis variolosus*, *Eudynamys scolopacea*, *Ninox jacquinoti* (map 22), *Aerodramus spodiopygius*, *Alcedo atthis*, *Ceyx lepidus*, *Halcyon saurophaga* (map 24), *Eurystomus orientalis*, *Hirundo tahitica*, *Coracina tenuirostris* (map 28), *Rhipidura rufifrons* (map 39), *R. leucophrys*, *Monarcha [melanopsis]* (map 40), *M.* [*manadensis*] (map 41), *Dicaeum* [*erythrothorax*], *Nectarinia jugularis*, *Myzomela lafargei* (map 47), *Aplonis* [*grandis*] (map 51), and *A. metallica*.

Conversely, Rennell supports populations of six species that reached Rennell from the east but that did not go on to reach the San Cristobal group or other Solomon islands. Four of these are Australian species of which at least two or three probably reached Rennell from the New Hebrides Archipelago to the east: *Anas gibberifrons*, *Accipiter fasciatus*, *Chrysococcyx lucidus*, and *Gerygone flavolateralis* (map 34). The other two are Rennell endemic allospecies whose nearest relatives live on Santa Cruz to the east: *Clytorhynchus* [*nigrogularis*] *hamlini* and *Woodfordia* [*superciliosa*] *superciliosa*. However, there are also 11 widespread Northern Melanesian species that reached Rennell from the west but did not spread farther east to Santa Cruz, the New Hebrides, or New Caledonia: *Ardea alba*, *Ixobrychus flavicollis*, *Caloenas nicobarica*, *Lorius* [*lory*], *Micropsitta* [*pusio*], *Geoffroyus heteroclitus*, *Hemiprocne mystacea*, *Coracinea lineata*, *Zosterops* [*griseotinctus*], *Aplonis cantoroides*, and *A.* [*feadensis*].

Similarity to Other Avifaunas

Table 31.6 compares Rennell's avifauna to the avifaunas of three Solomon islands (San Cristobal, Ugi, and the Russells) and of the two nearest archipelagoes to the east (the Santa Cruz Archipelago and the New Hebrides). The absolute number of species shared with Rennell is slightly higher for the eastern archipelagoes (20–22 species) than for Solomon islands (12–20 species), even though the latter are more species rich; hence the percentage of species shared is considerably higher for the two eastern archipelagoes (39% and 61% vs. 26, 26, and 33% for the three Solomon islands). Both of our indices measuring similarity of species composition, the Ochiai index and the non-nestedness index, show Rennell to be more similar to the two eastern archipelagoes (lower index values) than to any of the three Solomon islands compared. In seeming contradiction to this fact, populations of those species actually shared between Rennell and the other islands have much higher differentiation indices between Rennell and either of the eastern archipelagoes (1.27 and 1.50) than between Rennell and any of the three Solomon islands (0.50, 0.58, 0.75).

Since the Solomon Islands are so much closer (168–175 km vs. 574–657 km) to Rennell and are larger than are the eastern archipelagoes, it is not surprising that Rennell's pairwise differentiation index yields lower values for the Solomons than for the eastern archipelagoes. Instead, it is initially perhaps surprising that the eastern archipelagoes make as large a contribution to Rennell's avifauna as they do: between 25% and 37% of the avifauna, in various tabulations (Mayr & Hamlin 1931, Braestrup 1956, Bradley & Wolff 1956, Wolff 1969, 1973, Diamond 1984a).

One factor that may contribute to resolving this paradox is that, since Rennell is the most remote Solomon island except for some very small atolls, Rennell could be reached only by superior overwater colonists and it thus lacks more sedentary Solomon species. Superior colonists—whether from the Solomons, Australia, or Fiji—were also the ones equipped to reach the even more distant eastern archipelagoes, so that the species composition of Rennell is more similar to those of the eastern archipelagoes. However, those archipelagoes are still more than three times farther from Rennell than is any Solomon island. Hence among those species that Rennell shares with other islands, its populations are most similar taxonomically to Solomon populations.

Two other factors that may contribute to resolving the paradox are wind direction and ecology, as noted by Mayr and Hamlin (1931) and Braestrup (1956). The predominant orientation of the trade winds in the neighborhood of Rennell is from the southeast or east (i.e., from the direction of the New Hebrides and Santa Cruz group, rather than from Solomons lying to the north). The fraction of time during which the former (eastern) wind orientation prevails is 4–10 times longer than the fraction of time for prevailing winds from the north. That consideration would preferentially assist colonists from the east. In addition, habitats and ecological conditions are more similar between the Santa Cruz group and Rennell (both being remote and small oceanic islands) than between the larger Solomon islands and Rennell.

We are uncertain whether these three factors—selection for superior colonizing ability, wind direction, and ecology—suffice to resolve the paradox of Rennell's avifauna being disproportionately eastern in species composition, but disproportionately related to the Solomons in taxonomic affinities of shared species. This paradox deserves more attention.

The Smaller Solomon Islands

Among the small Solomon islands that we have not yet considered, Borokua, Bellona, Ramos, Gower, Ulawa, Ontong Java, and Sikaiana have no endemics, while Nissan, Mono, Ugi, Santa Anna/Santa Catalina, Three Sisters, and the Russells have 1–4 endemics each. The resulting low or zero values of the endemism index are generally as expected from these islands' isolations and small areas (eq. 25.3). As expected from their areas, these smaller islands lack most species of the nearest large island. Conversely, the smallest and/or most remote of these islands (Borokua, Bellona, Ramos, Gower, Ontong Java, Sikaiana, and Three Sisters) support one or more populations of the supertramps *Ptilinopus* [*purpuratus*], *Ducula pacifica*, *Monarcha cinerascens*, and *Aplonis* [*feadensis*] lacking on neighboring large islands.

Summary

The four major Solomon Islands groups or islands are the Bukida group, the New Georgia group, Malaita, and San Cristobal. The Bukida group, an island chain united at Pleistocene times of low sea level, includes the largest, highest, and most species-rich Solomon islands. It has high endemism (including four of the Solomon's endemic genera) and lacks no species that are widespread elsewhere in the Solomons.

The New Georgia group's many endemics include two species confined to the mountains of Kulambangra, but only those two species plus one other are absent on the Bukida group. Unlike the larger Bukida group, the New Georgia group lacks some species otherwise widespread in the Solomons, but the smaller Malaita and San Cristobal have even more such absentees. Differentiation of shared species between the Bukida and New Georgia groups is modest because the two groups are aligned parallel to each other for 230 km. Only five species show geographic variation within the New Georgia group, but three of those cases are spectacular: the famous four allospecies of the *Zosterops* [*griseotinctus*] superspecies within a radius of 20 km and the three megasubspecies each of *Monarcha browni* and *Pachycephala pectoralis*.

Malaita, the third largest Solomon island, has its closest affinities with Guadalcanal, with which it shares all but three of its species because the two islands are aligned parallel.

San Cristobal is only the fifth largest Solomon island but has the second most distinctive avifauna after Rennell. It strikingly lacks nine species that are otherwise widespread and common in the Solomons, but it possesses about nine species derived from the east and extending barely or not at all into the Solomons beyond San Cristobal. Nevertheless, those eastern specialties and San Cristobal's endemics do not replace its absentees on a one-for-one basis. San Cristobal's distinctiveness arises from its peripheral location at the end of the Solomon chain, and from the small colonization target presented by its end-on (rather than parallel) orientation to Guadalcanal and Malaita.

Rennell, an isolated, uplifted coral atoll, has the most distinctive and most differentiated avifauna of any Northern Melanesian island. It lacks 42 species otherwise widespread in the Solomons but supports six species derived from the east and absent in the rest of Northern Melanesia. Paradoxically, Rennell is most similar to archipelagos to the east (the New Hebrides and the Santa Cruz group) in its species composition, but is most similar to Solomon islands in the taxonomic affinities of shared species. Factors contributing to the explanation of this paradox may include selection for superior colonizing ability, prevailing wind directions from the east, and ecological similarities.

32

Speciation on Fragmented Pleistocene Islands

Having considered geographic variation across permanent water barriers in Chapters 29–31, we now consider geographic variation across fluctuating water barriers. Six pairs or groups of Northern Melanesian islands are surrounded today by shallow-water shelves that must have been dry land at Pleistocene times of low sea level. Hence these modern island pairs or groups must have been intermittently joined into enlarged Pleistocene islands at such times of low sea level (fig. 1.3). The six are Greater Bukida (now fragmented into the island chain from Buka to Florida and possibly Guadalcanal), Greater Gatumbangra (now fragmented into the chain from Kulambangra to Gatukai), Greater Vellonga (now Vella Lavella, Bagga, and Ganonga), Greater Rendipari (now Rendova and Tetipari), Greater New Ireland (now New Ireland and New Hanover), and Greater New Britain (now New Britain, Umboi, and Duke of York).

At times of high sea level such as the present, these Pleistocene islands broke up again into smaller islands. There are thought to have been dozens of such Pleistocene cycles of sea level rises and falls, resulting in island fissions and fusions. These fission/fusion cycles have had at least five types of consequences for modern distributions. We begin by surveying these five phenomena. We then provide a detailed analysis of subspeciation on fragments of Greater Bukida, the largest Pleistocene island and the one that has undergone fission into the greatest number of modern fragments. We conclude with a brief discussion of incipient speciation on fragments of Northern Melanesia's other five Pleistocene islands. In chapter 33 we discuss differential extinction on fragments of Pleistocene islands.

Five Consequences of Island Fission/Fusion Cycles

The consequences for modern distributions consist of endemics shared among islands, differential extinction, recolonizations from distant sources causing disjunct distributions, differentiation during times of island fission, and erasing of differentiation by island fusion.

Endemics Shared Among Fragments of a Single Pleistocene Island

We have already noted that many Northern Melanesian species have lost the overwater colonizing ability that initially enabled them to reach Northern Melanesia. Many of the thus-derived sedentary populations have differentiated into endemics now confined to (and scattered over) fragments of a single Pleistocene island, fragments that the endemics were able to reach overland when the fragments were joined.

Greater Bukida, the largest of the Pleistocene islands, provides the largest number of examples. Bukida endemics without conspecifics on any other Solomon island include the four monotypic genera and five endemic full species listed on p. 242 of the preceding chapter, plus the endemic allospecies of rail *Nesoclopeus woodfordi*, kingfisher *Halcyon leucopygia* (map 23), pitta *Pitta anerythra*, and fantail flycatcher *Rhipidura drownei* (map 38). In the case of the Bukida endemic weak subspecies of frogmouth *Podargus ocellatus inexpectatus*, the nearest conspecific populations are not in Northern Melanesia but in New Guinea and Australia. Each of these Bukida endemics except *Microgoura meeki* (apparently confined to Choiseul, but probably formerly more widespread on Greater Bukida; p. 32) occurs on 2–10 Bukida islands, but on no other Solomon island.

Similar examples on fragments of other Pleistocene islands include the honey-eater allospecies *Myzomela cineracea* and *Philemon cockerelli* (map 48), confined to the Umboi and New Britain fragments of Pleistocene Greater New Britain; the owl allospecies *Ninox variegata*, confined to the New Hanover and New Ireland fragments of Greater New Ireland (map 22); and three different megasubspecies of the whistler *Pachycephala pectoralis*, each confined to fragments of a different one of the three Pleistocene islands in the New Georgia group (table 31.3).

Differential Extinction

Some populations, formerly widespread over large Pleistocene islands, became stranded on island fragments by rising sea level. Some of those stranded populations then became extinct, especially on the smaller island fragments. In cases of species unable to colonize over water, those gaps persisted until the next cycle of island fusion. In particular, extinctions since the most recent island fissioning at the end of the Pleistocene have produced distributional gaps persisting today. Hence distributions, on modern island fragments, of species unable to disperse across water constitute a study in differential extinction. Examples include the distributional gaps in the ranges of *Nesoclopeus woodfordi*, *Pitta anerythra*, *Zosterops metcalfii*, and *Corvus woodfordi* depicted in figure 32.1, and corresponding gaps in the ranges of *Accipiter imitator*, *Nesasio solomonensis*, and *Halcyon leucopygia* (table 33.1). These gaps will be analyzed in chapter 33.

Recolonizations Resulting in Disjunct Distributions

The preceding paragraph concerned persisting distributional gaps that extinctions on island fragments created in the ranges of species unable to colonize overwater. Of course, extinctions can also create temporary gaps in the ranges of species that do retain the ability to colonize over water, but such gaps are likely to be erased eventually by subsequent overwater recolonization. Those overwater recolonists are most likely to come from the nearest source, usually another fragment of the same Pleistocene island. In a few cases, though, recolonization came from a more distant source, resulting in disjunct distributions of subspecies or even allospecies.

The most striking case is the distribution of thrushes of the *Zoothera* [*dauma*] superspecies on Greater Bukida (map 32). Two endemic Northern Melanesian allospecies, *Z. talaseae* and *Z. margaretae*, live on fragments at opposite ends of Greater Bukida (Bougainville and Guadalcanal, respectively). However, the island of Choiseul in the middle of Greater Bukida is occupied by a weak subspecies of the allospecies *Z. heinei*, which otherwise occurs in New Guinea, eastern Australia, and the St. Matthias group of the Bismarcks. It seems likely that an older colonization wave left founders that evolved into the endemics *Z. talaseae* and *Z. margaretae* on Greater Bukida; that representatives of that wave persisted on Bougainville and Guadalcanal but died out on intervening Bukida fragments; and that Choiseul was subsequently recolonized by thrushes from New Guinea.

The Umboi fragment of Greater New Britain provides six comparable examples: most Umboi populations are identical with New Britain conspecifics, but four instead belong to New Guinea subspecies, while two belong to supertramp subspecies originating from distant small or isolated Bismarck islands (see p. 263 for names). Similarly, the New Hanover fragment of Pleistocene New Ireland shares most of its taxa with modern New Ireland, but two New Hanover populations are instead supertramp subspecies (see p. 263 for names). The anomalous distribution of *Myzomela* honey-

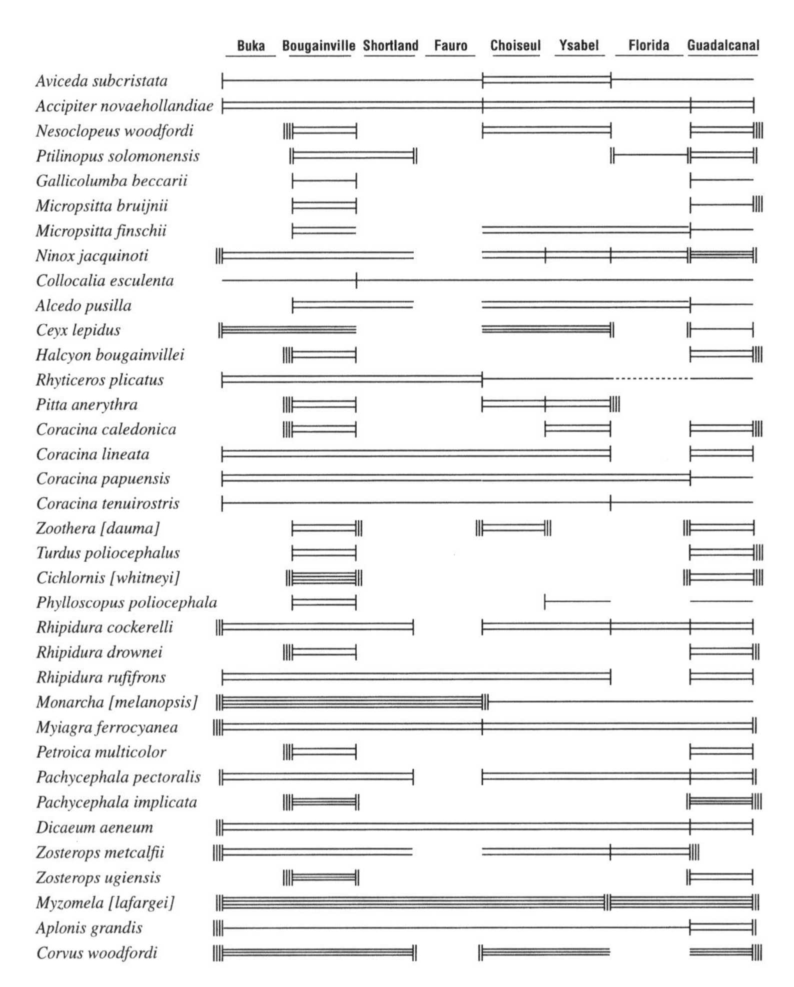

Fig. 32.1. Distributions of all 36 species that vary geographically within the Bukida Group. The group's eight major islands are listed from left to right, in their geographic sequence from northwest to southeast. Residence of a species on an island is indicated by horizontal solid lines; absence, by a gap; and vagrant presence, by a horizontal dashed line. Nonendemic taxa are denoted by a single horizontal line; endemic weak subspecies, megasubspecies, or allospecies, by 2, 3, or 4 horizontal lines, respectively. (Presence on the nearby smaller islands Mono or Savo is ignored in considering a taxon endemic to Bukida islands.) Geographic borders between infraspecific taxa within the Bukida Group are denoted by vertical lines: 1, 2, or 3 vertical lines for the border between weak subspecies, megasubspecies, or allospecies, respectively. Vertical lines to the left of the leftmost or to the right of the rightmost entry denote, respectively, the western or eastern border of a taxon within Northern Melanesia: 1, 2, 3, or 4 vertical lines for a border between weak subspecies, between megasubspecies, or between allospecies, or the range limit of the species itself, respectively. For example, the entry for *Ninox jacquinoti* is to be interpreted as follows: the species occurs on all Bukida islands except Fauro; one endemic weak subspecies is shared by Buka, Bougainville, Shortland, and Choiseul, another is confined to Ysabel, another to Florida, and an endemic megasubspecies is confined to Guadalcanal; a different allospecies occurs to the west of Buka (in the Bismarcks); and a different megasubspecies occurs to the east of Guadalcanal (on Malaita and San Cristobal).

eaters on southeastern fragments of Greater Bukida suggests a recent extinction on Florida, reversed by subsequent recolonization from Guadalcanal (p. 261).

Differentiation Across Intermittent Water Gaps

Almost all intraspecific taxonomic borders in Northern Melanesian birds today coincide with existing sea barriers. This is as expected if differentiation has occurred across sea barriers, and if most earlier barriers were in their present locations. In the cases of endemic subspecies or allospecies replacing each other on modern fragments of the same Pleistocene island, that fact makes it difficult to decide whether the differentiation developed only in the last 10,000 years, after the currently existing sea barriers arose at the end of the Pleistocene, or whether the differentiation instead arose much earlier during the Pleistocene, when sea barriers may have existed in the same places as they do today.

Taxonomic differentiation between conspecific populations on different fragments of the same former Pleistocene island varies from the weak subspecies level to the allospecies level. Within the Bukida Group, there are 36 borders between weak subspecies, 7 borders between megasubspecies, and 5 borders between allospecies (fig. 32.1); corresponding numbers for borders within the same Pleistocene island in the New Georgia Group are 1, 2, and 1, respectively (table 31.3). One might guess that at least some of the weak subspecies arose within the last 10,000 years, while the megasubspecies and especially the allospecies are more likely to have been differentiating since earlier island fissionings during the Pleistocene. When island fragments fused again at Pleistocene times of low sea level, some of the thereby reconnected pairs of differentiated populations already had achieved reproductive isolation. Probable legacies of those earlier divergences are the pairs of allospecies that divide up the modern fragments of Greater Bukida (fig. 32.1; the above-mentioned allospecies pairs of the thrush *Zoothera* [*dauma*], plus allospecies pairs of the thicket-warbler *Cichlornis* [*whitneyi*], the monarch flycatcher *Monarcha* [*melanopsis*], and the honey-eater *Myzomela* [*lafargei*]); and the pair of white-eye allospecies *Zosterops* [*griseotinctus*] *splendidus* and *Zosterops* [*griseotinctus*] *vellalavella* on the modern Ganonga and Vella Lavella/Bagga fragments, respectively, of Greater Vellonga.

Erasing of Differentiation by Island Fusion

The preceding discussion of differentiation between conspecific populations on fragments of the same former Pleistocene island needs to be placed in quantitative perspective. Yes, such differentiation (fig. 26.2) and endemism do exist. But there are much greater differentiation and endemism across permanent water barriers than across the intermittent water barriers separating Pleistocene island fragments. The explanation is obvious: little if any differentiation could occur at times of Pleistocene island fusion, and much of the differentiation that had arisen previously across water barriers at times of island fission must have been erased by gene flow within the fused island. The differentiation least likely to be erased by gene flow at times of island fusion was the above-mentioned differentiation into pairs of allospecies that had already achieved reproductive isolation.

Speciation on Greater Bukida

We now discuss these fission/fusion phenomena in more detail for the largest Pleistocene island, Greater Bukida (see fig. 1.3).

Frequency of Differentiation

In addition to Greater Bukida's numerous smaller modern fragments, there are now eight large fragments ranging in area from 71 to 8,591 km^2: from northwest to southeast, Buka, Bougainville, Shortland, Fauro, Choiseul, Ysabel, Florida, and Guadalcanal. Collectively, 119 species now inhabit those fragments of Pleistocene Greater Bukida. Of those 119 species, 104 occur on more than one Bukida fragment and can therefore be examined for inter-island variation within the Bukida Group. Of those 104 species, 36 are known to vary geographically within Greater Bukida, as depicted in figure 32.1. Of the other 68, 66 do not vary geographically within Greater Bukida, while degree of variation is unknown for two species (*Aerodramus orientalis* and *Erythrura trichroa*) whose Bukida populations have not been analyzed taxonomically.

Most of those 36 species that vary on Greater Bukida also occur on one or more of the other large Solomon islands or island groups: the New Georgia group, Malaita, and the San Cristobal group. Most of those shared species have differentiated not only on the Bukida group but also on the other groups on which they occur: 17 of the 24 species shared with the New Georgia group, 13 of 20 shared with Malaita, and 19 of 21 shared with the San Cristobal group.

Conversely, of the 66 species known not to vary on Greater Bukida, all except two occur elsewhere in Northern Melanesia. Of those 64, 29 do not vary geographically elsewhere in Northern Melanesia, while 35 do. Most of the geographic variation in those 35 species is across Northern Melanesia's major barriers: the barrier between the Bismarcks and Solomons and the barriers surrounding Rennell, San Cristobal, the Northwest Bismarcks, St. Matthias, and the New Georgia group. Only nine of those species have differentiated across minor barriers within Northern Melanesia (e.g., those between New Britain and New Ireland, or those separating New Ireland from its satellites to the east) but failed to differentiate within Bukida.

In short, water gaps between modern Bukida islands rank as minor barriers, for the obvious reason that they were intermittently erased during the Pleistocene. Most species capable of differentiating across those minor intermittent barriers have also been able to differentiate across Northern Melanesia's major permanent barriers. Conversely, many species able to differentiate across the major permanent barriers were unable to differentiate across Bukida's minor barriers. Species able to differentiate across the most significant of the other minor barriers (the New Britain/New Ireland barrier and the barrier between New Ireland and its eastern satellites) have also been able to differentiate within the Bukida Group.

Location of Lowland Barriers

Let us now discuss lowland bird distributions on Bukida islands, which are presumed to have offered an undivided sweep of lowland habitats from Buka to Florida and possibly Guadalcanal in the Pleistocene. Montane bird distributions are considered separately below because Bukida's mountains have always been disjunct from each other.

Today there are 24 lowland bird taxa endemic to a single Bukida island: the endemic pigeon genus *Microgoura* of Choiseul, the owl megasubspecies *Ninox jacquinoti granti* of Guadalcanal, and 22 weak subspecies divided among five islands. *Microgoura*'s restricted modern distribution is probably an artifact of populations on Bukida islands other than Choiseul having been exterminated by cats before ornithological exploration began. The other 23 single-island endemics are distributed unequally over Bukida islands. Buka, Shortland, and Fauro have none; Choiseul has two; Bougainville, Ysabel, and Florida have three each; and Guadalcanal has 11 weak subspecies plus one megasubspecies.

The resulting endemism indices are 0 for Buka, Shortland, and Fauro; 0.03 for Choiseul (omitting *Microgoura*); 0.04 for Bougainville and also for Ysabel; 0.05 for Florida; and 0.16 for Guadalcanal. As expected, all of these values lie below the values predicted for oceanic islands of comparable area and isolation separated by permanent water barriers (eq. 25.3). However, comparison of these numbers among Bukida islands raises the question of why the value for Guadalcanal's lowland avifauna is so much higher than for any other Bukida island. Since Guadalcanal lies at the southeast end of the Bukida chain, these facts suggest that the barrier between Guadalcanal and the nearest Bukida island, Florida, was especially potent (see below).

Further insight can be obtained by calculating pairwise differentiation indices for populations of lowland species shared between adjacent Bukida islands. Figure 32.1 depicts the geographic ranges of differentiated taxa, and table 26.1 calculates the resulting values of pairwise differentiation indices. All index values fall below, or at the lower end of the range of, the values expected for oceanic islands of comparable area and isolations (eq. 26.1). The four weakest barriers (index values 0–0.02) are those separating Bougainville from Buka and Shortland and separating Shortland from Fauro. As for the Buka/Bougainville barrier, even today, when sea level is high and the sea barriers within the Bukida chain are near their maximum width, Buka and Bougainville are very close to each other (1.0 km). As for the other set of weakest barriers, Fauro and Shortland, today 20 km from each other, are both even closer (11.1 and 8.4 km, respectively) to Bougainville, the largest Bukida island. Fauro and Shortland are also the two smallest of the eight main Bukida islands. As a result, Fauro and Shortland share all of their taxa with Bougainville and almost all with each other. Nearly as weak is the barrier between Choiseul and Ysabel (differentiation index only 0.03), dividing only two pairs of weak subspecies despite the large areas of these islands and the considerable distance (65 km) between them.

Conversely, the two strongest barriers within lowland Bukida are those at the northwest ends of Guadalcanal and of Choiseul. Differentiation index values setting off Guadalcanal from Florida (0.17) or from Ysabel (0.30) correspond to the previously mentioned endemism of the Guadalcanal lowland avifauna, greater than for any other Bukida island, though still less than the value for an oceanic island of Guadalcanal's area and modern isolation. The likely explanation is that, as suggested by modern hydrographic charts (fig. 1.3), a water barrier up to 2 km wide separated Guadalcanal from Florida and the rest of the Bukida chain even at Pleistocene times of lowest sea level. Guadalcanal's avifaunal distinctness is testimony to the effectiveness of even such a narrow water gap in inhibiting dispersal of some bird species.

The other strong barrier separates Choiseul from Fauro, Shortland, or Bougainville (differentiation indices 0.12, 0.18, or 0.23, respectively). Hydrographic charts give no hint of any water gap here during the Pleistocene. But they do indicate that the narrowest part of Pleistocene Greater Bukida must have been immediately west of modern Choiseul, where Greater Bukida narrowed to 16 km, then expanded in width to the east to 48 km (in modern Choiseul), and to the west to 76 km (in the vicinity of modern Shortland and Fauro) or 121 km (in modern Bougainville; see fig. 1.3). This former narrow neck evidently restricted gene flow and permitted the avifaunas that it separated to diverge somewhat, as expressed by the allospecies border in the flycatcher *Monarcha* [*melanopsis*] (map 40) and the megasubspecies border in the crow *Corvus woodfordi*.

These two principal barriers in Pleistocene Greater Bukida (the ones at the northwest ends of Guadalcanal and of Choiseul) have left as their modern legacies the accumulations of borders between weak subspecies, and also the few borders between sister megasubspecies or allospecies, in the Bukida chain. Between Guadalcanal and its Bukida neighbors to the northwest lie megasubspecies borders for the owl *Ninox jacquinoti* and the kingfisher *Ceyx lepidus*, while between Choiseul and its neighbors to the northwest lie the allospecies border in the monarch flycatcher superspecies *Monarcha* [*melanopsis*] (map 40) and the megasubspecies border in the crow *Corvus woodfordi*.

Lowland Bukida's sole border above the weak subspecies level that does not coincide with either of these two principal barriers is the allospecies border, lying between Ysabel and Florida, in the honeyeater *Myzomela* [*lafargei*] (map 47). This border is mainly responsible for giving the Ysabel/Florida island pair a higher differentiation index (0.13) than Ysabel/Choiseul (0.03). Hydrographic charts and maps suggest no immediate explanation for this apparent greater potency of the Ysabel/Florida barrier. A plausible alternative explanation is that the *Myzomela* [*lafargei*] population on the small Florida group may have become extinct during the Holocene and then been refounded by colonists from Guadalcanal, which today is closer to Florida than is Ysabel.

In short, geographic variation within the Bukida chain clearly illustrates the influence of Pleistocene land configurations. Legacies of the Pleistocene include the generally low endemism indices and pairwise differentiation indices of Bukida islands, the accumulations of range borders between Choiseul and its northwestern neighbors and between Guadalcanal and its northwestern neighbors, and the great similarity of the Choiseul and Ysabel avifaunas despite the modern distance (65 km) between those two islands. But Bukida also illustrates the influence of modern land configurations, especially in the identity of Buka's populations with those of Bougainville, and in the closer affinity of Florida's populations with those of Guadalcanal than with those of Ysabel.

Bukida's Montane Avifaunas

In contrast to this Pleistocene continuity of Bukida's lowland habitats, Bukida's mountains have always been disjunct. The highest and most extensive mountains are at opposite ends of the Bukida chain: on Bougainville (elevation 2591 km) and Guadalcanal (2448 m), the two highest Solomon islands and also the two with the highest proportion of their total areas lying above 600 m (41% and 33%, respectively). A distance of 600 km separates the high mountains of Bougainville and Guadalcanal. The only other Bukida island exceeding 1,000 m in elevation is Ysabel (1250 m), with only 3% of its area lying above 600 m. As a result, Bougainville and Guadalcanal are the Bukida islands with the best-developed montane avifaunas (18 and 23 species, respectively). Ysabel has only three species confined to its mountains, and no other Bukida island has any montane populations.

Ysabel's three montane populations are not endemic to Ysabel, due undoubtedly to the small area of montane habitat and resulting high extinction rates there. However, endemism in the montane avifaunas of Bougainville and Guadalcanal is very high (endemism indices 1.06 and 0.74, respectively, higher than the values for the whole avifaunas of any Northern Melanesian island except San Cristobal and Rennell). Bougainville's montane endemics consist of the endemic honey-eater genus *Stresemannia* (plate 9), the thicket-warbler allospecies *Cichlornis llaneae*, the whistler megasubspecies *Pachycephala implicata richardsi* (plate 2) and white-eye megasubspecies *Zosterops ugiensis hamlini*, and seven weak subspecies. Guadalcanal has the endemic honey-eater genus *Guadalcanaria* (plate 9), the whistler megasubspecies *Pachycephala implicata implicata* (plate 2), and 10 weak subspecies.

Bougainville and Guadalcanal share 11 montane species that occur on no other Bukida island. (The other montane species of Bougainville and Guadalcanal also occur in the mountains of Ysabel or as lowland populations on other Bukida islands). Of those 11, one (*Cichlornis* [*whitneyi*]) differs between Bougainville and Guadalcanal at the allospecies level; two (*Pachycephala implicata* and *Zosterops ugiensis*), at the megasubspecies level; five differ at the level of weak subspecies (*Micropsitta bruijnii*, *Halcyon bougainvillei*, *Turdus poliocephalus*, *Rhipidura drownei*, and *Petroica multicolor*); two (*Gymnophaps solomonensis* and *Aplonis brunneicapilla*) are subspecifically identical (they also occur on Solomon islands off the Bukida chain without geographic variation); and differentiation in *Erythrura trichroa* cannot be assessed because the Bougainville population has not been identified to subspecies.

Thus, among Bougainville's and Guadalcanal's montane populations, most have differentiated, in marked contrast to the relatively undifferentiated lowland avifaunas of these islands. The montane species of these islands are living on isolated islands of montane habitat, raised into the sky, and separated from other islands of montane habitat by intervening lowlands as well as by sea barriers.

New Georgia Group

Of the New Georgia Group's modern islands, all except Gizo and Simbo were joined in the Pleistocene into three enlarged islands: Greater Vellonga, Greater Gatumbangra, and Greater Rendipari. Within each Pleistocene island, the avifaunas of the modern fragments are highly nested within each other. For instance, within Greater Vellonga, every species of Ganonga and Bagga without exception occurs on the larger Vella Lavella. Within Greater Rendipari, all but one of Tetipari's species are on the larger Rendova. Within the Greater Gatumbangra chain, New Georgia and Wana Wana both share all of their species with Kulambangra; Vangunu and Kohinggo share all except three and one of their species, respectively, with New Georgia; and Gatukai shares all except three of its species with Vangunu.

Regarding geographic variation within modern fragments of these Pleistocene islands, there is no geographic variation whatsoever among the fragments of Greater Gatumbangra. The sole differentiation between Rendova and Tetipari involves the famous megasubspecies of *Zosterops rendovae* (plate 1). Within Greater Vellonga, there is no differentiation between Bagga and Vella Lavella, but Ganonga differs from those two islands in a weak subspecies of *Myzomela eichhorni* and possibly of *Pachycephala pectoralis* (Mayr 1932c), a megasubspecies of *Monarcha browni*, and the famous allospecies of *Zosterops* [*griseotinctus*].

The nesting and nonexistent differentiation among Gatumbangra fragments is expected not only from their Pleistocene connections but also from the proximity of Gatumbangra islands today. No gap between the major islands exceeds 2 km in width except that between Vangunu and Gatukai, which is nearly bridged by a smaller island. The similarity of the Rendova and Tetipari avifaunas is also as expected from their Pleistocene connection and modern distance (3.5 km); the only surprise is that the white-eyes managed to differentiate. As for Greater Vellonga, the perfect nesting of the Bagga avifauna within Vella Lavella's is as expected from their modern distance (3.2 km) and Pleistocene connection.

Ganonga's modern distance from Vella Lavella (9 km) is the greatest separating any two major islands of the New Georgia Group except Gizo, and the Pleistocene land connection between Ganonga and Vella Lavella narrowed to a neck of 5.6 km. Thus, it is not surprising that the greatest differentiation between Pleistocene fragments in the New Georgia Group is that between Ganonga and Vella Lavella. The formation of an endemic allospecies (as well as a megasubspecies) on Ganonga remains striking testimony to the ability of even narrow, intermittent water barriers and narrow, intermittent land bridges to restrict gene flow and thereby permit divergence in *Monarcha browni* and *Zosterops* [*griseotinctus*].

Bismarck Pleistocene Islands

Among Bismarck islands, Pleistocene sea level lowering joined New Hanover to North New Ireland and may also have joined Umboi to West New Britain. (It is uncertain whether the modern 25-km water gap between Umboi and New Britain was completely obliterated during the Pleistocene; hydrographic charts suggest that a very narrow gap may have persisted.) The avifaunas of both New Hanover and Umboi exhibit slight endemism. Their endemism index values of 0.08 and 0.04, respectively, are below expectations for oceanic islands of comparable area and distance (eq. 25.3), as are the New Hanover/New Ireland and Umboi/New Britain pairwise differentiation indices of 0.11 and 0.13, respectively.

Both New Hanover and Umboi are interesting in illustrating recolonizations of a small Pleistocene island fragment from a source other than the nearby larger fragment of the same Pleistocene island. As one might expect, all except three Umboi species (the supertramps *Ptilinopus solomonensis*, *Macropygia mackinlayi*, and *Monarcha cinerascens*) reside on New Britain, and most Umboi populations are subspecifically identical with New Britain populations. Umboi has three endemic weak subspecies (*Tanysiptera sylvia leucura*, *Coracina tenuirostris rooki*, and *Philemon cockerelli umboi*), all of which have their closest relatives on New Britain. However, there are four species whose Umboi populations do not belong to the New Britain race but to the New Guinea race (*Macropygia amboinensis cinereiceps*, *Micropsitta pusio beccarii* (map 19), *Cacomantis variolosus fortior*, and *Hirundo tahitica frontalis*). Umboi populations of two other species belong to a supertramp race of small or remote islands (*Ducula pistrinaria rhodinolaema* and *Halcyon chloris stresemanni* [map 24]), although a different race of the same species lives on New Britain.

Similarly, all except three New Hanover species (*Porphyrio porphyrio*, *Ptilinopus solomonensis*, and *Chalcopsitta cardinalis*) reside on New Ireland, and most New Hanover populations are subspecifically identical with New Ireland populations. New Hanover has six endemic weak subspecies (*Cacomantis variolosus websteri*, *Ninox variegata superior*, *Pitta erythrogaster extima*, *Lalage leucomela albidior*, *Myzomela cruentata lavongai*, and *Lonchura hunsteini nigerrima*), all of which have their closest relatives on New Ireland. However, there are two species whose New Hanover populations belong to a supertramp race (*Ducula pistrinaria rhodinolaema* and *Trichoglossus haematodus flavicans*), not to the New Ireland race.

As discussed previously, a plausible interpretation of the supertramp races on New Hanover and Umboi is that the former New Ireland or New Britain race became extinct on New Hanover or Umboi after post-Pleistocene fragmentation, and that recolonization came from mobile supertramp populations on small islands instead of from the more sedentary New Ireland or New Britain population.

Summary

Six pairs or groups of Northern Melanesian islands repeatedly underwent fission and fusion by cycles of rising and falling sea level during the Pleistocene. These fission/fusion cycles of the past created at least five features of bird distributions today: shared endemic taxa scattered over fragments of a single Pleistocene island, notably Greater Bukida; distributional gaps due to differential extinction, at times of island fission (such as today), of populations of species unable to disperse across water; disjunct infraspecific distributions of populations of some species capable of overwater dispersal, due to recolonization of transient distributional gaps from different directions; infraspecific differentiation across intermittent water gaps; and erasing of differentiation by island fusion.

The two principal taxonomic breaks for lowland birds along the Bukida chain are at the northwest end of Guadalcanal (where Sealark Channel may have persisted as a very narrow water gap even at Pleistocene times of low sea level) and at the northwest end of Choiseul (where Pleistocene Greater Bukida was narrowest). High montane habitats on Greater Bukida were always disjunct, even at Pleistocene times of low sea level, because they are located at opposite ends of the chain, on Bougainville and Guadalcanal. Hence the montane avifaunas of these two islands are sharply differentiated.

Most modern islands of the New Georgia group were fused in the Pleistocene into three enlarged islands. Within each such Pleistocene island, the avifaunas of the modern fragments are almost perfectly nested. Cases of differentiation of shared taxa among fragments are few but spectacular—notably, the white-eye megasubspecies of Rendova and Tetipari and the allospecies of Ganonga and Vella Lavella. Pleistocene fission/fusion cycles in the Bismarcks involved the New Ireland/New Hanover island pair and possibly the New Britain/Umboi island pair.

33

Differential Extinction and Species Occurrences on Fragmented Pleistocene Islands

The preceding chapter considered the development of geographic variation on modern islands derived by fragmentation of a former single Pleistocene island. In this chapter we ask, how uniform are such modern fragments of a single Pleistocene island in their species composition?

For flightless mammals, it is obvious that answers to that question yield insights into differential extinction resulting from habitat fragmentation. For example, the large islands of the Sunda Shelf (Borneo, Sumatra, and Java) support not only volant birds and bats, and rodents easily carried overwater by rafting, but also large, flightless mammals such as rhinoceroses, tigers, orang-utans, and gibbons. Alfred Russel Wallace (1892) recognized that those large mammals had walked out to the Greater Sunda Islands at Pleistocene times of low sea level, when the islands were part of the Asian mainland. Today, those mammals are nonuniformly distributed on the Greater Sunda Islands. For instance, orang-utans occur on Sumatra and Borneo but not on Java, while tigers occur on Java and Sumatra but not on Borneo. However, fossil evidence proves that most of the modern distributional gaps were filled in the Pleistocene (e.g., orang-utans formerly occurred on Java, and tigers on Borneo; Hoojier in Terborgh 1974). That is, the modern patchy distributions result from differential extinction, not from patchy overwater colonization. The primary determinant of long-term survival for Greater Sundan mammals was population size: populations on small islands and populations of species living at low density were especially prone to extinction (Terborgh 1974, Diamond 1984f).

The modern islands in several chains of Northern Melanesian islands were joined to each other during the Pleistocene. As we have seen, numerous Northern Melanesian bird species rarely or never cross water. Their current presence on islands is as unequivocal an indicator of former land connections as are the presences of rhinoceroses and gibbons. Hence the distributions of such bird species on fragmented Pleistocene islands in Northern Melanesia pose problems similar to those recognized by Wallace for distributions of large, flightless mammals on the Greater Sunda islands. Naturally, there are even more Northern Melanesian bird species that do fly across water than species that do not, and that are therefore capable of recolonizing islands and erasing temporary extinctions of their island populations. In this chapter we consider the distributions of non–water-crossing species on modern fragments of Greater Bukida, and we then consider the distributions of all species on modern fragments of Greater New Britain and Greater New Ireland.

Non–water-crossing Species on Fragments of Greater Bukida

The largest Pleistocene island of Northern Melanesia was Greater Bukida, with an area of 46,400 km^2. Four modern fragments of Greater Bukida

Table 33.1. Distribution of relict populations of 13 non–water-crossing bird species on 27 modern islands resulting from post-Pleistocene fragmentation of Greater Bukida.

Island	Area (km^2)	*Halcyon leucopygia*	*Zosterops metcalfii*	*Corvus woodfordi*	*Nesoclopeus woodfordi*	*Pitta anerythra*	*Accipiter imitator*	*Nesasio solomonensis*	*Halcyon bougainvillei*	*Rhipidura drowne*	*Pachycephala iimplicata*	*Stresemannia bougainvillei*	*Guadalcanaria inexpectata*	*Microgoura meeki*
Bougainville	8,591	+	+	+	+	+	+	+	+	+	+	+		
Guadalcanal	5,281	+		+	+				+	+	+		+	
Ysabel	4,095	+	+	+	+	+	+	+						
Choiseul	2,966	+	+	+	+	+	+	+						+
Buka	611	+	+	+										
Florida	368	+	+											
Shortland	232		+	+										
Wagina	91			+										
Fauro	71	+												
Buena Vista	14	+												
Molakobi	7.7		+	+										
Fara	7.3	+												
Arnavon	5.6													
Bates	5.2		+											

The following 13 smaller Bukida islands (area in km^2 in parentheses) support no populations of any of those 13 non–water-crossing bird species: Piru (2.95), Oema (2.85), Nusave (0.53), Bagora (0.33), Nugu (0.15), Samarai (0.0091), New (0.071), Dalakalonga (0.067), Elo (0.055), Kukuvulu (0.040), Tapanu (0.016), Kanasata (0.0091), and Near New (0.0070). Note that number of surviving populations increases with island area; that not even the largest island (Bougainville) retained all 13 species; and that some species (those in the three left-most species columns) were much more successful at persisting on small islands than were other species.

(Bougainville, Guadalcanal, Ysabel, and Choiseul) each still exceed 2900 km^2 in area. The Bukida avifauna includes 13 species that fail to present any type of evidence of overwater colonizing ability (chapter 11). If we assume that these species were formerly distributed uniformly over Greater Bukida, then we can analyze their absences on modern fragments in terms of differential extinction. Of course, the explanation for some of these modern absences may be that the species was already nonuniformly distributed and locally absent in some parts of the expanded Pleistocene island even before it became fragmented by rising sea level. Such postulated local absences could have had either of two causes: lack of suitable habitat (especially in cases of montane species); or the species sampling considerations underlying the so-called continental species/area relation. (That is, if one samples different areas within any continuous land mass, a large area is likely to yield more species than a small area, according to a species/area relation termed the "continental species/area relation"; see p. 269 of this chapter and MacArthur & Wilson 1967, p. 9). We consider both of these alternative explanations below.

In table 33.1 we summarize the distributions of the 13 non–water-crossing species on 27 modern fragments of Greater Bukida, ranging in area from 0.007 km^2 to 8591 km^2, and figure 33.1 plots the numbers of populations of these species against island area. No island smaller than 5 km^2 supports even a single population of these 13 species. Evidently, no bird population can survive in isolation for more than 10,000 years on a central Northern Melanesian island with an area less than 5 km^2. For islands above 5 km^2, the number of populations rises steeply with island area. However, no island supports all 13 species; even the largest fragment, Bougainville, supports only 11 of the 13.

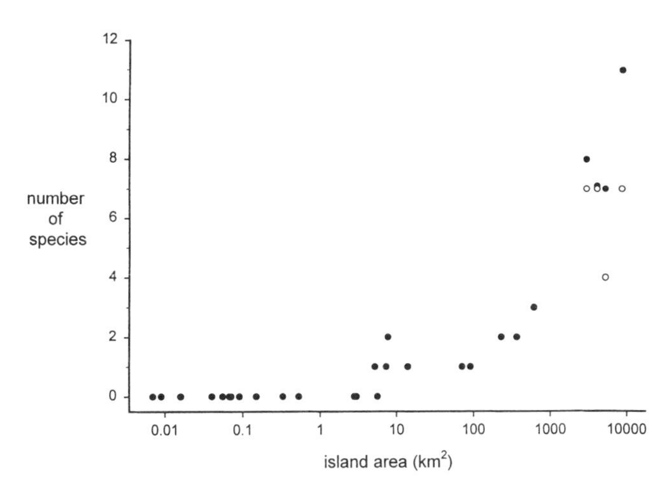

Fig. 33.1. Number of non–water-crossing bird species present as Pleistocene relict populations on each ornithologically explored island that represents a post-Pleistocene fragment of Greater Bukida plotted against island area on a logarithmic scale. The total number of relict species is 13. Filled circles, all species; open circles, the eight lowland species except for *Microgoura meeki*. The second-from-the-right points (seven total species, four lowland species) refer to Guadalcanal. Note that the number of relict populations surviving on an island increases with island area; that no relicts at all have survived on islands smaller than 5 km^2; that even the largest island has retained only 11 of the 13 species; and that Guadalcanal's values are somewhat low for its area (because the possible persistence of a very narrow Sealark Channel may have limited Pleistocene access to Guadalcanal from Greater Bukida).

Some remote Pacific islands of only a few square kilometers, such as Pitcairn and Wake, have endemic bird species that presumably survived there in isolation for much longer than 10,000 years. However, these remote islands have few competing species, and extinction rates for an island of a given area increase with number of competing species and hence decrease with island isolation (Diamond 1980). Thus, our conclusion that no bird population survives more than 10,000 years on a central Northern Melanesian island smaller than 5 km^2 is not incompatible with the existence of endemic bird species on remote Pacific islands of this small area.

Part of the reason that the largest numbers of non–water-crossing species are found on the two largest Bukida islands (Bougainville and Guadalcanal) is their high mountains. The 13 non–water-crossing species include four montane species confined to the high mountains of Bougainville, Guadalcanal, or both (table 33.1). The recent distribution of a fifth species (the lowland pigeon *Microgoura meeki*), apparently confined to Choiseul, may be an artifact of recent extermination by cats (p. 32). If we remove these five species from analysis, the number of populations of the remaining eight non–water-crossing lowland species still rises steeply with island area. The species number for Guadalcanal (four such species) now drops below the general pattern defined by other Bukida fragments (fig. 33.1). The likely explanation is that the modern Sealark Channel between Guadalcanal and the next Bukida fragment (Florida) may not have been entirely obliterated by land during the Pleistocene and may have persisted then as a channel of up to 2 km in width. If so, figure 33.1 suggests that even such a narrow channel sufficed to prevent some species otherwise widely distributed on large Bukida fragments (*Accipiter imitator*, *Nesasio solomonensis*, *Pitta anerythra*, and *Zosterops metcalfii*) from reaching Guadalcanal.

For these eight lowland species, we can reconsider as follows whether our inferences of differential extinction on Bukida fragments are in gross error because of our simplifying assumption that all species initially occurred on all Bukida fragments. The arrangement of Bukida islands is virtually a linear chain running from northwest to southeast. One of the eight species, *Corvus woodfordi*, occurs today on the northwestern-most island (Buka), the southeastern-most island (Guadalcanal), and six intervening islands. Since *Corvus woodfordi* belongs to the group of least vagile species, it must at one time have occurred from end to end of the Bukida chain, and all 19 absences on surveyed intervening Bukida islands must correspond to former extinctions. Another one of the eight species, *Zosterops metcalfii*, occurs on the northwestern-most island (Buka), the third southeastern-most island (Florida), and six intervening islands, but not on the two southeastern-most islands (Nugu and Guadalcanal). We therefore cannot be certain that *Zosterops metcalfii* ever occurred on Nugu and Guadalcanal, which are "terminal" islands beyond the species' modern distribution. However, *Zosterops metcalfii* must have occurred on the 17 surveyed intervening Bukida islands, where its current absences must correspond to former extinctions.

Examining in this way the distributions of all eight species on all 27 surveyed Bukida islands, we find 178 current cases of an island unoccupied by a species. Of those 178 cases, 22 involve terminal Bukida islands beyond the species' current Bukida distribution, while 156 lie within the species' current Bukida distribution. Thus, the great majority (at least 156 of the 178 absences) must correspond to past extinctions. This is a minimum number; it seems likely that many of the 22 absences on terminal islands also reflect former extinctions rather than cases of never having colonized.

Fourteen populations of those eight lowland non–water-crossing species survived on some islands smaller than 700 km^2. All 14 populations belong to just three species: the kingfisher *Halcyon leucopygia*, the white-eye *Zosterops metcalfii*, and the crow *Corvus woodfordi*. All three species are among the more abundant of these non–water-crossing species.

In short, fewer populations survive on small islands than on large islands, and those populations surviving on small islands belong to the most abundant species. That is, risk of extinction decreases steeply with increasing population size, as attested by many other types of evidence (e.g., Diamond 1984f, Pimm et al. 1988).

Greater New Britain and Greater New Ireland

Declines of Total Species Number on Umboi and New Hanover

Having considered Greater Bukida, the largest Pleistocene island of the Solomons, let us now consider the Bismarck islands that became fragmented at the end of the Pleistocene: Greater New Britain and Greater New Ireland. Unlike Greater Bukida, which yielded several large fragments, these two Bismarck Pleistocene islands each yielded only one large fragment (New Britain and New Ireland) and one medium-sized fragment (Umboi and New Hanover, respectively). There are also many small islands between New Ireland and New Hanover derived from the same Pleistocene island, but they have not been surveyed ornithologically.

The two large fragments (New Britain and New Ireland) both have high mountains approximately 2400 m in elevation, but New Britain is much larger than New Ireland (35,742 vs. 7,174 km^2) and has much more extensive nonforest habitats (lakes, rivers, swamps, marshes, and grassland). Of the two medium-sized fragments (Umboi and New Hanover), Umboi is more diverse in habitats than New Hanover, with several small lakes, a marsh, and mountains up to 1655 m. New Hanover is only 1186 m high and is poor in aquatic habitats. New Britain has 127 species, Umboi 84 species, with 81 shared in common; New Ireland has 103 species, New Hanover 75, with 72 shared. New Britain lacks three supertramps of Umboi (*Ptilinopus solomonensis*, *Macropygia mackinlayi*, and *Monarcha cinerascens*). New Ireland lacks two New Hanover supertramps (*Ptilinopus solomonensis* and *Chalcopsitta cardinalis*), and apparently lacks one New Hanover species likely to be present but overlooked on New Ireland (*Porphyrio porphyrio*). Thus, no evidence exists to suggest that either New Britain or New Ireland has lost species as a result of post-Pleistocene shrinkage in area. This is unsurprising, as modern New Britain and New Ireland each account for more than 80% of the area of the Pleistocene island from which each was derived.

Hence let us take the modern avifauna of New Britain or New Ireland as an approximation of the avifauna of Pleistocene Greater New Britain or Greater New Ireland. If one assumes that Umboi or New Hanover started off with most of the species of the Pleistocene island in which they were embedded, then the modern values of their species number mean that both Umboi and New Hanover have lost much of their original avifaunas since the Pleistocene: about 46 (= 127 − 81) of an original 127 species for Umboi, and about 31 (= 103 − 72) of an original 103 species for New Hanover.

However, this estimate of post-Pleistocene species losses can be refined. Umboi, being only a fraction of the area of Pleistocene New Britain, would not have supported the entire Greater New Britain avifauna even at the time when Umboi was still embedded within Greater New Britain. The actual species number of the original Umboi or New Hanover avifaunas can be calculated from the continental species-area relation:

$$S_1/S_2 = (A_1/A_2)^z.$$

This equation gives the relation between species number and census area for areas sampled from the same land mass (p. 266 of this chapter). For sampled areas within New Guinea, the value of the exponent z is 0.055 (Diamond & Gilpin 1983). Using this z value, taking A_2 as the modern area of Umboi or New Hanover, A_1 as the areas of Pleistocene New Britain or New Ireland (about 36,500 and 8,360 km^2 respectively, only slightly greater than the areas of modern New Britain and New Ireland), and approximating S_1 as the modern S value for New Britain or New Ireland, one calculates S_2 (the original S value just before the land bridge was severed) as 103 for Umboi, 93 for New Hanover. Since Umboi now has only 81 of New Britain's 127 species, and since New Hanover has only 72 of New Ireland's 103 species, both islands must have lost about 20 species (103 − 81 = 22 for Umboi, 93 − 72 = 21 for New Hanover) since the end of the Pleistocene. Umboi must already have lacked approximately 127 − 103 = 24 New Britain species, and New Hanover must have lacked approximately 103 − 93 = 10 New Ireland species when still embedded in Greater New Britain or Greater New Ireland, respectively.

Without exhaustive paleontological sampling of Pleistocene sites on Umboi and New Hanover, we cannot know which particular New Britain species now absent on Umboi were already absent at the end of the Pleistocene because of those sampling considerations underlying the continental species/area relation, and which ones instead became extinct since the end of the Pleistocene. The same is true for New Hanover's species deficit compared to New Ireland. Nevertheless, the main conclusion from this analysis is that species totals alone indicate that both Umboi and New Hanover have lost many species as a result of post-Pleistocene fragmentation, even though there is some uncertainty as to which particular species were lost.

Non–water-crossing Species

One can, however, consider those New Britain species considered "non-water crossers" because they exhibit no evidence of overwater colonization (chapter 11). Hence populations of those species on Umboi today are unlikely to have been founded by overwater colonization from New Britain since the Pleistocene land bridge was severed. Instead, they are likely to be relicts of populations distributed over Greater New Britain in the Pleistocene, to have been stranded like flightless mammals on Umboi since the Pleistocene land bridge was severed, and to have survived on Umboi since then. Those four survivors are the goshawk *Accipiter luteoschistaceus*, the pigeon *Henicophaps foersteri*, and the honey-eaters *Myzomela cineracea* and *Philemon cockerelli*. Of New Britain's other nine non–water-crossing species now absent on Umboi, some probably became extinct on Umboi since the Pleistocene, while some others may have already been absent on Umboi before the land bridge was severed. Those nine species are the goshawk *Accipiter princeps*, the hawk *Henicopernis longicauda*, the rail *Gallirallus insignis*, the barn owl *Tyto aurantia*, the kingfisher *Halcyon albonotata*, the thicket-warblers *Ortygocichla rubiginosa* and *Cichlornis grosvenori*, and the honey-eaters *Myzomela erythromelas* and *Melidectes whitemanensis*.

Similarly, New Ireland has six non–water-crossing species, of which only two now occur on New Hanover: the owl *Ninox variegata* and the finch *Lonchura hunsteini*. The other four (the parrot *Lorius albidinucha*, the honey-eaters *Myzomela pulchella* and *Philemon eichhorni*, and the drongo *Dicrurus megarhynchus*) either became extinct on New Hanover since the Pleistocene or else were already absent at the time the land bridge was severed. The latter explanation probably applies to *Lorius albidinucha* and *Myzomela pulchella*, montane species of New Ireland, for which New Hanover's mountains are probably too low and small to offer adequate habitat.

Analysis of All Species

When we restricted comparisons in the preceding section to those New Britain and New Ireland species unable to cross water, we could attribute their modern presences on Umboi or New Hanover straightforwardly to survival since the Pleistocene. Let us now expand our comparison to the entire avifauna of Umboi or New Hanover: species that can disperse across water, as well as species that cannot. Among the water crossers, the modern occurrence of New Britain species on Umboi (or of New Ireland species on New Hanover) could thus reflect either post-Pleistocene overwater colonization or continuous survival since the Pleistocene.

In table 33.2 we summarize which New Britain species are shared with Umboi, and which New Ireland species are shared with New Hanover, as a function of abundance and dispersal ability. For both island pairs, probability of occurrence on the smaller island (Umboi or New Hanover) increases monotonically with dispersal ability and nearly monotonically with abundance, as expected. The former trend arises because dispersal ability measures likelihood of post-Pleistocene overwater colonization. The latter trend arises partly because probability of post-Pleistocene overwater colonization also depends on abundance (more abundant species send out more colonists), and also because the likelihood of extinction of stranded Pleistocene populations decreases with abundance. The sole reversal of these trends is that both Umboi and New Hanover support a lower percentage of species in the highest abundance category (category 5) than in the next highest abundance category (4). Most of the absent species of category 5 are montane species of New Britain or New Ireland, for which Umboi's and especially New Hanover's area of montane habitats may be inadequate.

Instead of this analysis in terms of abundance and dispersal, we can also analyze species by their habitat preference. If we divide the New Britain or New Ireland avifaunas according to preferred habitat, which are the New Britain or New Ireland habitats whose bird species are least well represented on Umboi or New Hanover?

Table 33.2. Comparison of larger and smaller fragments of the same Pleistocene Bismarck island in species composition.

Abundance category	Number of species on larger fragment	Number of those species on smaller fragment	%
Fragments of Greater New Britain: Larger fragment New Britain, smaller fragment Umboi			
1	15	5	33
2	21	10	48
3	53	37	70
4	29	25	86
5	9	4	44
Dispersal category			
1	35	31	89
2	48	33	69
3	30	13	43
4	14	4	29
Fragments of Greater New Ireland: Larger fragment New Ireland, smaller fragment New Hanover			
1	9	3	33
2	12	8	67
3	45	32	71
4	28	23	82
5	9	6	67
Dispersal category			
1	32	30	94
2	39	28	72
3	25	12	48
4	7	2	29

Greater New Britain and Greater New Ireland were Pleistocene islands that became fragmented by rising sea level at the end of the Pleistocene into a larger fragment (New Britain and New Ireland respectively) and a smaller fragment (Umboi and New Hanover respectively). The current species composition of New Britain or New Ireland may be taken to approximate the species composition of the corresponding Pleistocene island, since modern New Britain or New Ireland accounts for more than 80% of the Pleistocene island area. The table lists the number of New Britain or New Ireland species in each category of abundance (1 = rare . . . 5 = abundant; from appendix 5) or dispersal (1 = vagile . . . 4 = sedentary; from column "Bis index" of appendix 6), then lists the number or percentage of those species on Umboi or New Hanover. Note that the most abundant and most vagile species are best represented on the smaller fragments. The sole exception to this trend is the lower percentages for abundance category 5 than 4, because several of the most abundant species (category 5) happen to be sedentary montane species for which the smaller fragments provide only limited suitable habitat.

For both island pairs, the habitat most impoverished in its avifauna on the smaller island is the mountains: only 44% of New Britain's montane species occur on Umboi, and only 20% of New Ireland's montane species occur on New Hanover. This is as expected because New Britain and New Ireland have higher and proportionately more extensive mountains than Umboi and especially than New Hanover. In contrast, habitats whose species were especially successful at establishing or persisting on Umboi or New Hanover include lowland forest (70% of New Britain's species occur on Umboi, 80% of New Ireland's on New Hanover) because it is the predominant habitat on Umboi and New Hanover. Species of the seacoast were also especially successful at establishing or persisting on the smaller island (80% of New Britain's coastal species, and also 80% of New Ireland's, occur on Umboi or New Hanover, respectively) because coastal species are so vagile.

Most absences of New Britain or New Ireland species on Umboi or New Hanover can thus be readily understood in terms of one or several factors: species disadvantaged by limited availability of suitable habitat; species whose distributions are mostly restricted to large islands because of low abundance, low vagility, or both; and species that are apparently absent but are easily overlooked and are likely to be present on Umboi or New Hanover. The most flagrant remaining puzzles are Umboi's lack of New Britain's swiftlet *Collocalia esculenta*, fantail flycatcher *Rhipidura rufiventris*, and flowerpecker *Dicaeum eximium*, all three of which are conspicuous, common (abundance category 4 or 5), and good overwater dispersers.

Summary

At the end of the Pleistocene the large, fused Northern Melanesian islands existing at Pleistocene times of low sea level became fragmented into smaller islands by rising sea level, which isolated populations of bird species that do not fly across water. Study of modern bird distributions on the island fragments constitutes a study of differential extinction because some of those isolated populations have gone extinct since the end of the Pleistocene.

The most extensive data set is for the island fragments of Pleistocene Greater Bukida. Today, the Bukida group supports 13 endemic species unable to fly across water, but no modern Bukida island supports that entire number. Instead, the number of modern populations on each Bukida island increases with island area because smaller fragments lost more species. More abundant species survived on more islands and on smaller islands. No population survived the 10,000 years since the end of the Pleistocene on any Bukida island smaller than 5 km^2.

Post-Pleistocene extinctions also occurred in the Bismarcks, on Umboi and New Hanover (formerly joined to the much larger islands of New Britain and New Ireland, respectively). Each has lost about 20% of its species since it became isolated at the end of the Pleistocene. Among the species remaining, groups of species that are overrepresented include abundant species (because of decreased risk of extinction and increased probability of recolonization), vagile species (because of increased probability of recolonization), species of lowland forest (because that is the islands' predominant habitat), and species of the seacoast (because of high vagility).

PART VII

SYNTHESIS, CONCLUSIONS, AND PROSPECTS

34

Conclusions about Speciation

This last section of the book brings together our conclusions. The present chapter begins by assembling our conclusions about speciation: we reconsider the biological species concept, use the separate snapshots of species distributions to reconstruct the course of speciation with its main trends and variants, and discuss the establishment of isolates and the role of peripheral isolates. Chapter 35 examines species differences as expressed in the evolution of dispersal, the so-called taxon cycle, and the heterogeneity of the montane avifauna. In the final chapter, we suggest some promising directions for future research.

The Biological Species Concept

Consider again the historical background of the biological species concept. When scholars began in the seventeenth and eighteenth centuries to classify, describe, and name natural history specimens, ranging from birds and plants to rocks, they understandably used a phenetic (morphological, typological) species concept for birds and plants just as they did for rocks. That concept could be applied by a museum taxonomist sorting out natural history specimens that someone else had collected and of which the taxonomist had no field experience. For instance, among Northern Melanesian birds, the bright green male and the bright red female of the eclectus parrot, *Eclectus roratus*, were initially described as separate species; the high percentage of missexed specimens in the nineteenth century postponed the day when taxonomists would recognize the true relationship of these two taxa.

But two fatal flaws in this purely phenetic species concept emerged as early as the eighteenth century. One flaw was the one exemplified by this example of the male and female eclectus parrot: individuals belonging to the same population but of different sexes, ages, color morphs, seasonal classes, or parasite life stages often proved to differ from each other in appearance far more than did many generally recognized species. We mentioned previously (chapter 16) how Linnaeus originally gave different names to the very different-looking male and female mallard, and to the adult and immature goshawk, then quietly ignored this morphological species concept and assigned them to the same species when he realized the biological relationship of the male and female or the adult and immature. The other fatal flaw involved cryptic sibling species, which exhibit only very slight phenotypic differences and are difficult to distinguish as dead specimens in museum trays, but which nevertheless exhibit consistent behavioral and ecological differences in the field and do not interbreed. A famous early example involved two European warblers, the willow warbler *Phylloscopus trochilus* and the chiffchaff *P. collybita*, which are extremely similar in plumage and size. By the late 1700s, the British field observer Gilbert White had noticed that the birds that Linnaeus had described from specimens in 1758 as *Motacilla Trochilus* had two different songs, correlated with different preferred habitats and perch heights. That is, these two warblers are otherwise perfectly normal species with normal differences in ecology and vocalizations; they merely do not use plumage differences to distinguish each other. Similar cases have arisen in Northern Melanesian birds among *Myiagra* flycatchers and *Pachycephala* whistlers (see below).

Thus, even museum taxonomists recognized a biological species concept and did not strictly adhere to a phenetic species concept, long before the biological species concept (BSC) was explicitly formulated in the twentieth century. Biologists came increasingly to realize that species taxa are not arbitrary groupings as are those for classifying rocks or art styles, but instead correspond to some reality in nature. The question why this should be the case arose with Darwin: why does the biological world fall into species? Why doesn't nature just present us with an unbroken continuum of individuals, ranging from identical individuals through similar individuals to grossly divergent individuals?

This underlying reality in nature gave rise to the BSC as now explicitly formulated: *a species is a group of interbreeding, natural populations that are reproductively isolated from other such groups.* This definition implies two essential features: species status is a property of populations, not of single individuals; and species are delineated with respect to other species. Because species are delineated by whether they interbreed with each other, the BSC applies to sexually reproducing organisms but not to asexually reproducing organisms, which must be classified on some other basis. Individuals of the latter constitute clones that maintain the genotype from generation to generation without breeding and that therefore have no need of reproductive isolating mechanisms.

The BSC is (at least in principle) straightforwardly applicable to a local biota consisting of sympatric populations that are in reproductive condition and in contact with each other. The taxonomist's task is to classify individuals of the local biota into species, based on purely empirical evidence—not on their degree of phenetic differences, but on whether they belong to interbreeding populations. This principle is simple, although the practical difficulties in assessing reproductive isolation in field situations may be great. Population differences based on the BSC thus correspond to an underlying biological reality that also correlates with other biologically important characteristics, always including separate evolutionary trajectories (because species do not interbreed), and usually including significant ecological, behavioral, physiological, and biochemical differences as well. It is sometimes claimed that the BSC does not apply to plants because of massive hybridization, but a complete analysis of a local flora revealed that the great majority of populations were readily delineated and that only 7% presented taxonomic difficulties (Mayr 1992).

Among the 195 Northern Melanesian bird species discussed in this book, application of the BSC is so straightforward (i.e., delineating species in local avifaunas of individual islands is so unequivocal) that only three uncertainties persisted into the twentieth century. The first uncertainty involved the flycatchers *Myiagra alecto* and *M. hebetior* (plate 5), originally confused by museum taxonomists because adult males are extremely similar in appearance, but later distinguished by museum taxonomists because of the distinct plumage of females (Hartert 1924b). Field biologists then recognized differences in habitat and nesting biology (Meyer 1928). The second uncertainty involved the kingfisher *Halcyon saurophaga*, which exists in two color morphs, green-headed and white-headed, originally described as different species *H. anachoreta* and *H. admiralitatis*, respectively. The two morphs were subsequently found to be otherwise identical in plumage and ecology, to nest as mixed pairs, and to co-occur as an interbreeding population (Mayr 1949b).

The last case involved the golden whistlers (plate 2), formerly all ranked as subspecies but now recognized to belong to two sibling species *P. pectoralis* and *P. melanura*, which are broadly sympatric and which segregate by habitat without any instance of interbreeding from Australia and the Southeastern Papuan Islands throughout the whole of the Bismarcks. The sole cause of ambiguity was one variable hybrid population, *whitneyi*, on a Solomon island at the periphery of *P. melanura's* range (Galbraith 1967, Diamond 1976, Ford 1983). Since Galbraith (1967) recognized the distinctness of *Pachycephala pectoralis* and *P. melanura* in 1967, no further ambiguities in applying the BSC to Northern Melanesian birds have been recognized, and none is now suspected. This happy situation in assessing sympatric populations of Northern Melanesian birds contrasts with our many uncertainties about the species status of allopatric populations (see below).

Thus, the BSC is clear in principle; in an overwhelming percentage of cases involving sympatric populations it is also clear in practice; and the effort to apply it to local biotas can yield significant discoveries (such as the above-mentioned coexistence of two whistlers or of two flycatchers). Nevertheless, controversy persists. The species problem is one of the oldest and most frustrating problems in biology. Even for sexually reproducing species, other species concepts continue to be advocated (e.g., the nominalist, typological, recognition, evolutionary, and phylogenetic species concepts; see Mayr 1988, Mayr & Ashlock 1991 for discussion). Why has such controversy persisted? There are two principal reasons.

First, species differ in many attributes besides those apparent in dead museum specimens. Hence taxonomists who work solely with specimens and who do not do field work lack most of the information relevant to delineating species limits and thereby fail to acquire a sense for the biological reality underlying species. For example, when one of us (J.D.) was first identifying specimens of *Sericornis* warblers from New Guinea, the specimens seemed to form a continuum of individuals varying trivially in body size and color, although museum specimens had somehow been labeled as belonging to either of two species, *S. nouhuysi* and *S. perspicillatus*. It astonished Diamond when New Guinea natives of the Fore tribe, lacking binoculars and glimpsing live warblers in silhouette in dimly lit New Guinea forest at distances of dozens of meters, unhesitatingly assigned them to two categories, termed *mabisena* and *pasagekiyabi* (the Fore names for *S. nouhuysi* and *S. perspicillatus*, respectively; Diamond 1966). Eventually, Diamond learned that New Guineans readily distinguish these species in life by differences not apparent in museum specimens (especially by their different songs, perch heights, perch diameters, foraging modes, foraging substrates, and foraging directions). As another example, New Guinean highlanders sort rat specimens into different folk taxa that museum taxonomists have learned to distinguish as dead specimens only with great difficulty. There remain cases in which New Guineans insist that rat specimens in a museum tray, all appearing to belong to the same taxon, actually belong to two species with different habitat preferences. Based on previous experience, it seems likely that in those cases, too, the New Guineans will turn out to be correct, and that taxonomists will eventually figure out how to distinguish those sibling species from dead study specimens.

The other, more important, reason for controversy is that, while the BSC can in principle be straightforwardly applied to a local biota of sympatric populations, its application to populations isolated in space (e.g., on different islands) or in time (e.g., in different fossil strata at the same site) is not straightforward. Such isolated populations are automatically prevented from interbreeding. To apply the BSC, the taxonomist asks, is it likely that those populations *would* interbreed if they could be brought into contact? Criteria for making that decision include comparing the differences between the two populations with the differences between sympatric species of that same genus, with the differences between intergrading subspecies within geographically widespread species of that genus, and with the differences between occasionally hybridizing populations of species of that genus. But those criteria are not as straightforward in principle as the criteria for delineating sympatric species within a local biota. The greatest practical difficulty is the phenomenon of mosaic evolution: rates of morphological, behavioral, ecological, and molecular divergence may vary independently of rates of acquisition of reproductive isolation. In particular, as already discussed, sibling species may acquire reproductive isolation while maintaining only minimal morphological differences. Conversely, allopatric taxa may diverge greatly in morphology and song without acquiring reproductive isolation, as shown by the stabilized hybrid populations between megasubspecies of *P. pectoralis* that differ greatly in appearance and song (plate 2).

Some taxonomists find these practical difficulties in applying the BSC disconcerting and consider them to weaken the BSC. But, on reflection, the difficulties are entirely to be expected, and the weakness lies with the dissenting taxonomists' unrealistic expectations, not with the BSC. Speciation occurs when a population that initially constituted one species (one interbreeding population) evolves into two non-interbreeding populations. In Northern Melanesian birds and probably in many or most other sexually reproducing organisms, speciation proceeds via the parent population giving rise to two or more geographic isolates (allopatric populations), which may or may not acquire reproductive isolation, and which upon resuming contact may or may not interbreed. The phenomenon of mosaic evolution often makes it difficult to infer, from morphological and other evidence, whether related allopatric taxa would interbreed if they came into contact: one cannot know for sure until they do come into contact.

We encapsulate our own guesses about status along this speciation pathway by assigning the taxonomic ranks of subspecies and megasubspecies (inference of no reproductive isolation) and allospecies (inference of reproductive isolation). We expect that some authors ("lumpers") will instead prefer to reduce some taxa that we rank as allospecies to the rank of megasubspecies, while other authors ("splitters") will prefer to consider many of our megasubspecies as allospecies. Uncertainties about those inferences do affect our quantitative conclusions about the percentage of geographical isolates achieving reproductive isolation, but they do not affect our qualitative conclusions about the pathway in speciation.

We conclude this section by reflecting on how the concepts of subspecies, megasubspecies, and allospecies emerged from the history of efforts to apply the BSC. From Linnaeus's development of classification in the eighteenth century through much of the nineteenth century, each distinguishable local bird population was named binomially (i.e., with a genus name and a species name). By the year 1900, about 30,000 such binomial bird species had been recognized. With the appreciation of the BSC and the development of the new systematics in the twentieth century, it had to be decided on a case-by-case basis which of these binomials should be retained unchanged because they corresponded to monotypic (geographically invariant) species, and which should be reduced to trinomially named subspecies and grouped into polytypic species. In the case of geographically isolated populations, the criterion for the decision was the BSC-based inference of whether they would still interbreed with the main body of the population if they came into contact.

Initially, a consideration in making those decisions was to decrease the large number of binomial species, so the tendency in case of doubt was to reduce allopatric populations to subspecies rank. The result was a reduction from 30,000 to about 8,600 species, with most isolated populations reduced to subspecies rank and with their relationships thereby clearly indicated. The subsequent introduction of the allospecies/superspecies concept permitted one further improvement: isolates so distinct that they probably are reproductively isolated could now be designated as allospecies, thereby conveying both their relationship (as member of a superspecies) and their likely reproductive isolation. Thus, hundreds of Northern Melanesian bird taxa that were described in the nineteenth century as binomial species were reduced in the early twentieth century to the rank of subspecies; but some of these subspecies have subsequently been reconsidered as allospecies. For example, we now rank as allospecies 12 Solomon taxa that Mayr (1945a) ranked as subspecies: *Coracina* [*tenuirostris*] *salamonis*, *Phylloscopus* [*trivirgatus*] *makirensis*, *Monarcha* [*melanopsis*] *erythrostictus*, and nine others. Our designations of megasubspecies will call attention to isolates whose rank merits further consideration. We emphasize again that these decisions about rank are not empty exercises in taxonomy, nor proof of shortcomings in the BSC, but instead are assessments of the most interesting moment in the whole movie of speciation.

The Whole Course of Speciation Assembled from Snapshots

In the preceding section we defined a species, according to the biological species concept, as a group of interbreeding natural populations that are reproductively isolated from other such populations. Speciation is defined as the origin of two or more closely related species from a single ancestral species. When one combines these definitions of a species and of speciation, the problem of speciation becomes, how do the connections among the interbreeding natural populations constituting a single species become broken, so as to yield two or more groups of populations that do not interbreed with each other but that occur together and whose individuals regularly encounter each other while in breeding condition?

The problem of speciation is posed in acute form by so-called species swarms (species-rich sets of closely related, coexisting species), such as Galapagos finches, North American parulid warblers, or Lake Victoria's cichlid fishes. In milder form, the problem is posed by any sympatric pair of closely related species (congeners). More generally, the problem is implicit in any branching evolutionary tree consisting of multiple species descended from a single ancestral species. Thus, the problem of speciation is quite distinct from the problem of evolutionary (phyletic) change with time that receives much more discussion in textbooks of evolution. Evolutionary change with time can proceed without speciation, as exemplified by the development and then the decline of industrial melanism among British moths.

Because completed speciation in birds evidently requires many thousands of years, we have been forced in this book to reconstruct the movie of speciation by considering present-day distributions of different bird species as constituting snapshots or frames of the movie, and by seeking to arrange the snapshots in the correct sequence. The midpoint of the movie is reconstructed from the 19 pairs of closely related species at the crucial stage of reproductive isolation and ecological segregation but only partial sympatry (chapter 22). From that midpoint, we assemble other snapshots backward to the opening scene and forward to finale. Although we describe the movie from beginning to end, we emphasize that its progression is far from inexorable. Most geographic isolates never become full species, for reasons that we discuss (especially extinction, hybridization, and failure to develop ecological segregation). For convenience in describing the continuous action of movie, we divide it into five episodes.

First Episode: Arrival of Colonists in Northern Melanesia

Northern Melanesian islands have never had land connections to New Guinea, Australia, or other large land masses. In that sense, they are truly oceanic islands. Therefore, plant and animal colonists have necessarily arrived over water, presumably already in the distant past in the case of ancestors of endemic genera, and continuing today. For instance, the cormorant *Phalacrocorax carbo* reached Rennell sometime between 1965 and 1976 (Diamond 1984a), the ibis *Threskiornis spinicollis* reached New Ireland in 1963 (Coates 1985), and the ibis *T. moluccus* reached Mono around 1949 (Diamond 2001a).

The ancestors of more than half of the modern resident bird populations of Northern Melanesia arrived from New Guinea, presumably because it is by far the closest large colonization source and has large areas of habitats similar to Northern Melanesian habitats. This preponderance of New Guinean colonists has held for a long time; it emerges from analysis of origins not only of nonendemic populations (presumed recent arrivals) but also of endemic allospecies (presumed early arrivals). Smaller numbers of colonists arrived from the larger but much more distant and ecologically much more distinct Australia, and from the even more distant and much smaller New Hebrides and other eastern archipelagoes. Many New Guinean colonists first occupied the closest (westernmost) Bismarck islands and then spread east through the remaining Bismarcks and Solomons; a few New Guinean colonists and some Australian colonists jumped directly to the Solomons; some New Hebridean colonists first occupied the closest (southeasternmost) Solomon islands of San Cristobal and/or Rennell and then spread west along the Solomon chain; and still other colonists may have had more complex patterns of spread.

For many or most colonizing populations, the movie comes to an abortive end during this first episode, with extinction. For instance, the Australian pelican *Pelecanus conspicillatus* periodically wanders or is blown by cyclones to Northern Melanesia, and occasional individuals breed, but none of these new populations has survived for more than a few years. The pair of the Australian ibis *Threskiornis spinicollis* that reached New Ireland in 1963 never bred; one of the pair soon died, and the other survived alone at least until 1974 (Coates 1985). Similarly, after the ibis *T. moluccus* reached Mono in 1949, only a single individual remained in 1974 (Diamond 2001a).

Second Episode: Geographic Fragmentation of the Gene Pool

On a continent the populations of a single species may become genetically isolated from each other by ecological barriers such as deserts and mountains, or merely by attenuation of gene flow over long distances. That outcome is vanishingly rare in Northern Melanesia, whose largest island is only 470 km long. Instead, geographic fragmentation of Northern Melanesian gene pools occurs across the water gaps between islands. Most fragmentation (called "primary isolation"; see below) develops when colonists from one island cross water to occupy an island previously unoccupied by the species. Less often (secondary isolation; see below), fragmentation develops when a water gap arises (e.g., through eustatic changes of sea level) to divide an island already occupied by a species into two or more separate islands.

Out of Northern Melanesia's 191 analyzed breeding bird species, 26 are confined to a single island, are thus precluded from varying geographically within Northern Melanesia, and cannot be actors in further episodes of the speciation movie unless they should in the future become established on a second island. However, once a species has come to occupy two or more islands, the potential for geographic variation arises, and 100 of the 165 species occupying multiple islands do in fact vary geographically. Their degree of variation in turn varies greatly, ranging from species (such as *Egretta sacra* and *Megapodius freycinet*) in which the populations on all 68 or 69 islands occupied are identical and there is no geographic variation at all, to species in which every island is occupied by a distinct population. Extremes of geographic variation include the whistler *Pachycephala pectoralis* (populations on 34 islands divided among 16 subspecies falling into six megasubspecies groups) and the flycatcher *Monarcha* [*manadensis*] (populations on 32 islands falling into six allospecies, some of which fall, in turn, into 6 megasubspecies and one weak subspecies).

Several factors contribute to these great differences between species in geographic variability. Because genetic differentiation may take some time to develop after populations have become isolated, recently established isolates may not differ from the ancestral stock—such as the populations of the pitta *Pitta sordida* on three islands of the Long Group, colonized from New Guinea after the volcanic defaunation of Long Island a mere three centuries ago. Other species, such as perhaps the cuckoo *Centropus violaceus*, seem to be phenotypically stable and not prone to develop differences among isolates. Highly vagile species are less likely than sedentary species to vary geographically because ongoing dispersal between islands ensures that populations on different islands are not well isolated genetically. Geographic variability increases with abundance because isolated populations of rare species are likely to go extinct before they have existed for enough time to differentiate genetically.

This last point about rare species reemphasizes our comment that the movie of speciation does not proceed inexorably. The fate of most isolates is not to become a distinct species, but to become extinct. Northern Melanesia's great speciators—*Pachycephala pectoralis*, *Monarcha* [*manadensis*], and its other species famous for their geographic variability—tend to be abundant species of moderate vagility. They develop many distinctive isolates because their abundance makes their isolates less likely to go extinct quickly, and because their moderate vagility allows them to colonize many islands but ensures that the new isolate is not swamped by huge numbers of further arriving individuals inhibiting differentiation. The same considerations of abundance and vagility that contribute to explaining differences in geographic variability among species (see part V of this book) also contribute to explaining differences among islands (part VI). The islands with the highest percentages of endemics among their populations are large, remote islands (fig. 25.2), with low rates of population extinctions and low genetic swamping of isolates by further colonists.

Third Episode: Achievement of Reproductive Isolation, the First Moment of Truth

Among the genetic traits that may diverge between isolated populations are traits leading to reproductive isolation (see below). Genetic divergence develops gradually, as illustrated by the many distributional snapshots spanning the whole range from no geographic variation to pronounced geographic variation between island populations. As mentioned above, out of the 165 Northern Melanesian species occupying multiple islands, 65 species (i.e., 39%) nevertheless show no geographic variation. Out of the 385 endemic subspecies that we do recognize, we arbitrarily dichotomize the spectrum of distinctness into 82 more distinctive megasubspecies and 303 less distinctive weak subspecies. Some snapshots from opposite ends of this spectrum include the heron *Egretta sacra* (no geographic variation among 68 islands); the rail *Porphyrio porphyrio* (minor and inconsistent variation in size and hue, insufficient to warrant recognization of more than one subspecies); the fruit dove *Ptilinopus viridis* (the distinctive white-headed San Cristobal race *eugeniae*, which we guess to be not yet reproductively isolated from the green-headed races elsewhere and hence still rank as just a megasubspecies, but which some other authors do guess to be reproductively isolated and rank as an allospecies); and the fruit dove *Ptilinopus solomonensis*, which has two megasubspecies distinctive in ecology (one in the mountains of some islands, the other in the lowlands of other islands) and also distinctive in morphology but bridged by an intermediate subspecies.

There will inevitably be room for disagreement as to whether geographic isolates have achieved reproductive isolation because we cannot know for sure until the former isolates re-expand their ranges to resume geographic contact. At that point comes the first of two discontinuities, or moments of truth, in the otherwise gradually unfolding movie of speciation: do they, or don't they, interbreed? Again, the movie does not proceed inexorably to its finale. The fate of many Northern Melanesian isolates that resume contact is not to demonstrate reproductive isolation and confirm their status as two independent gene pools, but instead to hybridize and to merge again into a single gene pool. Examples of evident merged populations include the swallows *Hirundo tahitica* on New Britain and Long, the cuckoos *Eudynamys scolopacea* of Long, the megapodes *Megapodius freycinet* of Karkar, the cuckoo-shrikes *Coracina tenuirostris nisoria* of Pavuvu (plate 7), and the whistlers *Pachycephala pectoralis christophori* of San Cristobal (plate 2), which appear to be stabilized hybrid populations between colonists arriving from two different source populations belonging to different megasubspecies (chapter 23). But some isolates have diverged sufficiently in their reproductive biology during isolation that they fail to interbreed on resuming geographic contact, or else initially hybridize infrequently until reproductive isolation is perfected or until both stocks find enough potential conspecific mates that they are not forced (by inability to find a conspecific mate) into accepting an allospecific mate. A rare opportunity to observe this moment of truth historically was provided by the two myzomelid honey-eaters of San Cristobal (plate 4): following the recent invasion of San Cristobal by *Myzomela cardinalis*, a few hybrids with the well-established *M. tristrami* were collected in 1908, fewer in 1927, and none in 1953.

The richest material for studying speciation in Northern Melanesia, the golden whistlers *Pachycephala pectoralis* and their relatives (plate 2), illustrates how finely balanced these alternative outcomes of hybridization and perfection of reproduction isolation may be. The dozens of populations of these whistlers attest to several alternative outcomes: one ancient speciation—the two distantly related species *P. pectoralis* and *P. implicata* living sympatrically and segregating by altitude in the mountains of Bougainville and Guadalcanal, with no hybridization at all; one recent speciation—the two closely related species *P. pectoralis* and *P. melanura*, living sympatrically and segregating by island type with no or only very local hybridization throughout most of the Bismarcks, Southeast Papuan Islands, and Northeast Australia, but with the hybrid population *P. whitneyi* in the Northwest Solomons at the edge of *P. melanura*'s range where individuals may have difficulty finding conspecifics as mates; island populations of *P. pectoralis* that are so distinct in plumage that they have often been ranked as allospecies; and stabilized hybrid populations on San Cristobal and Fiji between these very distinctive parental stocks, leading us to rank the stocks as megasubspecies rather than as allospecies.

Allospeciation, the development of reproductive isolation but not yet of sympatry and ecological segregation, is frequent among Northern Melanesian birds, where 35 superspecies are represented by multiple allospecies within Northern Melanesia. The extremes are the superspecies of *Monarcha* [*manadensis*] flycatchers, *Zosterops* [*griseotinctus*] white-eyes, and *Myzomela* [*lafargei*] honey-eaters, each represented by six endemic allospecies within Northern Melenasia. The frequency of allospeciation increases with a species' abundance and peaks at intermediate vagility, as already mentioned for the frequency of subspeciation (of the great speciators) in the previous section and for the same reasons. A whole series of snapshots illustrates the course of allospeciation: sets of allospecies with increasingly wide geographic distributions within Northern Melanesia; increasing numbers of allospecies per superspecies; further differentiation of the allospecies themselves, some of which are represented by multiple subspecies on different islands while other allospecies of the same superspecies have not subspeciated; and distributional gaps suggestive of range retractions and extinctions. Multiple waves of colonization leading to endemic allospecies on some islands and nonendemic allospecies of the same superspecies on other islands are attested by the rail *Gallirallus* [*philippensis*], the pitta *Pitta* [*sordida*], the thrush *Zoothera* [*dauma*] (map 32), and the flycatchers *Rhipidura* [*rufifrons*] and *R.* [*rufidorsa*] (map 39).

In well-studied avifaunas, mate choice within species and reproductive isolation between species are found to depend variously on signals such as vocalizations, plumage signals, and behavioral displays. The Northern Melanesian avifauna offers rich material for studying the evolution of reproductive isolation in allopatry, but such studies have yet to be carried out. Nevertheless, we can mention some starting points for future studies. In regard to vocal divergence in allopatry, Diamond's field observations show that almost every island population has recognizably distinct vocalizations in the kingfisher *Halcyon chloris,* the cuckoo-shrike *Coracina tenuirostris,* the whistler *Pachycephala pectoralis,* and the white-eye superspecies *Zosterops [griseotinctus]* (Diamond 1984a, 1998). Vocalizations differ strikingly between major island populations of the cuckoo *Cacomantis variolosus* and the owl superspecies *Ninox [novaeseelandiae]* (Diamond 1975b) and between Bismarck and New Guinea populations of the starling *Mino dumontii* (Diamond 1975b, Schodde 1977) and the crow *Corvus orru.* In regard to divergence, in allopatry, of plumage signals important in sexual selection, plates 1–9 invite one to speculate whether the swollen red ceres of the pigeon allospecies *Ptilinopus insolitus* (plate 9) and *Ducula rubricera* (plate 8), the crest of the pigeon allospecies *Reinwardtoena crassirostris* (plate 7), and the twisted tail of the drongo allospecies *Dicrurus megarhynchus* (plate 8) function in mate choice.

Fourth Episode: Achievement of Ecological Segregation, the Second Moment of Truth

Even when two geographic isolates have achieved different mate selection criteria and considerable reproductive isolation, that by no means guarantees that they will expand their ranges to coexist sympatrically. Instead, speciation may remain "stuck" for a long time, as exemplified by the 35 Northern Melanesian superspecies whose component allospecies are still completely allopatric. It is initially puzzling to encounter so many cases of different allospecies living to the exclusion of each other on islands separated only by small water gaps, such as the 8-km strait separating the population of the white-eye *Zosterops [griseotinctus] vellalavellae* on Vella Lavella from the population of *Zosterops [griseotinctus] splendidus* on Ganonga. Why does not one allospecies cross over to the other island and establish a breeding population there beside the sister allospecies? Two reasons account for this.

The first reason is that colonists in island archipelagoes often disperse over water as single individuals, rather than spreading as a population over a broad geographic front, as on continents. While a colonist might prefer a mate of its own allospecies over a mate of its sister allospecies if given the choice, the colonist arriving at an island of the sister allospecies may not get the choice: it may encounter no individuals of its own allospecies. If it succeeds in mating with an individual of the resident sister allospecies, its genes will become lost in the huge resident gene pool and may then tend to be eliminated by natural selection.

Should the colonist indeed encounter conspecific mates, either because it disperses in a group or because individuals disperse frequently, the colonists then face a second difficulty: they may be too similar ecologically to the sister allospecies to invade the latter's range. The achievement of reproductive isolation does not automatically imply that sister allospecies have also achieved sufficient ecological differences to permit them to occur sympatrically by filling different ecological niches. Eventually, though, 19 Northern Melanesian superspecies did develop sufficient ecological differences among their component allospecies for the allospecies to achieve partial, though not complete, sympatry (chapter 22). Sympatry in Northern Melanesia is achieved overwhelmingly by means of interarchipelagal invasions: Bismarck, Solomon, New Guinea, and New Hebridean allospecies of the same superspecies invading each other's geographic ranges. In the initial stages of reinvasion, ecological segregation in the zone of sympatry is achieved largely by spatial segregation, with the two taxa occupying different habitats, altitudes, or types of islands. Later, the taxa develop additional ecological differences, such as differences in foraging techniques (especially ones related to differences in body size) or in foraging heights above the ground, that permit the taxa to become syntopic (i.e., to forage in the same space).

Abundant examples illustrate how the niche of a given species can vary between islands as a function of island differences in the competing avifauna and in habitat. Thus, different populations of the same species may with time become adapted to different niches. If those populations also achieve reproductive isolation, they may be preadapted to achieving sympatry when the populations come into contact again. Those observed niche shifts that promote sympatry through spatial segregation are of four types. First, some species that on New Guinea are confined to open, nonforested habitats are able to occupy forest on Northern Melanesian islands, where the forest avifauna is much less species-rich (e.g., the cuckoo *Cacomantis variolosus* and the crow *Corvus orru*; chapter 10). Second, some forest species confined to the canopy on New Guinea are able to forage from the canopy to the ground on Northern Melanesian islands, again because of their lower species diversity (e.g., the cuckoo-shrike *Lalage leucomela* and the honey-eater *Philemon* [*moluccensis*] on New Britain; Diamond 1970). Third, numerous species confined to the mountains on species-rich Northern Melanesian islands descend to sea level on smaller or outlying species-poor islands (e.g., the thrush *Turdus poliocephalus* on Rennell and the warbler *Phylloscopus trivirgatus* on St. Matthias). Finally, some species that occupy large islands in the Solomons are confined to smaller or outlying islands of the Bismarcks (e.g., the pigeons *Ptilinopus solomonensis* and *Macropygia mackinlayi*).

Fifth Episode: Completion of Speciation

Speciation is completed when one allospecies completely overruns the geographic range of its sister allospecies, so that the latter occurs only in sympatry with the former. The result is that one or more taxa can no longer be ranked as component members of the same superspecies but must be considered full species. These are Northern Melanesia's 30 endemic full species and its five endemic monotypic genera (chapter 24). For 8 of the 30 endemic full species we can still identify the most closely related species, occurring in sympatry with the endemic full species. The remaining 22 endemic full species and the 5 endemic monotypic genera are so distinctive that we can no longer definitely identify their closest relative, either because the endemic taxon and its relatives have diverged so far apart with time or because its former close relatives have gone extinct.

Establishment of Primary and Secondary Isolates

As just discussed, the speciation movie begins with birds colonizing Northern Melanesia, followed by the establishment of isolates. Discussions of speciation on continents often invoke the establishment of secondary isolates, when a species' range becomes sundered by the development of a physical or ecological barrier such as a mountain range, low-rainfall area, or water gap. For example, uplift of mountains may split populations of lowland species, while a period of dry climate may fragment forest expanses into forest refugia separated by savannas.

But neither of these two processes appears to have played a role in speciation of Northern Melanesian birds. Among the four highest Northern Melanesian islands, the central mountain chains of Bougainville and Guadalcanal are only 130 km long, and the mountain chains of New Britain and New Ireland are interrupted by lowland gaps, so it is unsurprising that not even subspeciation of birds has been reported from the lowland blocks separated by these mountain chains. The driest areas of Northern Melanesia are the northern watersheds of New Britain and Guadalcanal, but both are sufficiently wet (annual rainfall 1730–2500 mm) that their natural vegetation is rainforest rather than savanna. We do not know whether savannas replaced rainforest in those two rainshadow areas at drier times in the past; modern bird distributions offer no hint of such replacements, unless the few grassland endemics of these two islands (the quail subspecies *Coturnix chinensis lepida,* bustard quail subspecies *Turnix maculosa saturata*, and finch allospecies *Lonchura melaena* of New Britain, and *Turnix maculosa salomonis* of Guadalcanal) constitute such evidence. It is uncertain whether any geographic differentiation has resulted from reduction of gene flow by distance on even the longest Northern Melanesian islands, New Britain (470 km long) and New Ireland (350 km long; chapter 22).

The process of secondary isolation that has been of undoubted significance for the Northern Melanesian avifauna is the repeated fissioning of larger islands into smaller islands by water gaps at Pleistocene and Holocene times of high sea level, including at the present time. In 55 cases those water gaps now separate different subspecies of conspecific populations: 43, six, three, and three cases on islands derived from the fissioning of Pleistocene Greater Bukida, Pleistocene Greater New Ireland, Pleistocene Greater New Britain, and Pleistocene islands in the New Georgia Group, respectively. There are even three pairs of lowland allospecies separated by those water gaps: the white-eyes *Zosterops* [*griseotinctus*] *splendidus* and *Z.* [*g*] *vellalavella* on Ganonga and Vella Lavella, and the honey-eaters *Myzomela* [*lafargei*] *lafargei* and *M.* [*l.*] *melanocephala* and the flycatchers *Monarcha* [*melanopsis*] *erythrostictus* and *M.* [*m.*] *cas-*

taneiventris on the modern islands derived from Greater Bukida. But there have still been much less subspeciation and allospeciation across these intermittent water gaps than across the permanent water gaps of Northern Melanesia (chapter 26). The explanation is that much of the differentiation that developed across the intermittent water gaps at times of high sea level and island fission must have been erased at subsequent times of low sea level and island fusion.

Instead, most geographic differentiation in Northern Melanesian birds has involved primary isolation across permanent water gaps. Such differentiation begins with colonists reaching a previously unoccupied island and establishing a founder population, which may already at its moment of founding, through sampling accidents, differ in gene frequencies from the parent population. Most such populations go extinct soon after founding, while their numbers are still small and before the population has grown to saturate island habitats (e.g., the ephemeral Northern Melanesian breeding populations of the pelican *Pelecanus conspicillatus*). Of those populations that do saturate island habitats, the populations on small islands are still at high risk of extinction, and the populations on islands near source islands are likely to become genetically swamped by the continuing arrival of more colonists. Endemism is highest on remote, large islands because those are the islands on which isolates are least likely to become extinct or to become swamped genetically. These events leading to endemic taxa through primary isolation are illustrated in recent times by the endemic races of the cuckoo *Eudynamys scolopacea* and the thrush *Turdus poliocephalus* that formed on Long and Tolokiwa, respectively, after the defaunation of those islands by a seventeenth-century volcanic explosion (Diamond 1989, 2001a).

Peripheral Isolates

The Existence of Peripheral Isolates

Taxonomists have for a long time realized that, in a group of subspecies or closely related species, the most divergent ones will generally occur at the periphery of the group's distribution. The explanation, of course, is that peripheral populations can diverge because they receive less gene flow from other populations than do centrally located populations. The Northern Melanesian avifauna offers dozens of examples of this phenomenon at several levels. Among the 100 Northern Melanesian species that vary geographically, whenever there is any subspecies or allospecies that is strikingly divergent, it is in every case without exception a geographically peripheral one. Even in those species of which no component taxon is strikingly divergent, the most peripheral taxon is still usually more divergent than, or at least as divergent as, the others.

To begin, we discuss seven wide-ranging superspecies of which the Northern Melanesian allospecies is strikingly divergent. 1. The gray and white sea eagle *Haliaeetus [leucogaster] leucogaster* extends, as a coastal species feeding on fish and carrion, without any geographic variation from India through Australia and New Guinea to the Bismarcks. However, the easternmost (Solomon) allospecies *H. [l.] sanfordi* is uniformally brown, forest-dwelling, and feeds on mammals and large birds. Ornithologists never imagined that the Solomons could harbor an endemic sea eagle, and specimens of *H. sanfordi* were dismissed for 70 years as young individuals of *H. leucogaster* until Mayr (1936) realized that every individual eagle that he observed in the Solomons as a member of the Whitney Expedition was brown. 2. The drongo superspecies *Dicrurus [hottentottus]* extends from India through Australia and New Guinea to Northern Melanesia in 33 named populations. By far the most divergent population is the New Ireland allospecies *D. [h.] megarhynchus*, with its bizarrely twisted tail, large body, and heavy bill (plate 8). 3. The flower-pecker superspecies *Dicaeum [erythrothorax]* extends from Celebes to San Cristobal in nine allospecies and 26 named populations, all but one of them olive or gray to metallic black and with a red-breasted male. The exception is the very different allospecies *Dicaeum [e.] tristrami* of San Cristobal, the southeastern-most Solomon island, which resembles a little brown and white sparrow. 4. The pitta superspecies *Pitta [sordida]* extends from Southeast Asia to the Bismarcks as four allospecies and 18 named populations, of which by far the most divergent is the gigantic *P. [s.] superba* (plate 9) of Manus, the most peripheral Bismarck island. 5. Among the six allospecies and 19 named populations of the thrush superspecies *Zoothera [dauma]*, which ranges from Russia's Ural Mountains to San Cristobal, the most distinctive is the round-winged, long-legged allospecies *Z. [d.] margaretae* of Guadalcanal and San Cristobal. 6.

Among the four allospecies of the pigeon superspecies *Ptilinopus [hyogaster]*, which occurs from the Moluccas to the Bismarcks, the Bismarck allospecies *P. [h.] insolitus* stands out in having its cere inflated into a big, cherrylike, bulbous excrescence (plate 9). 7. Among the three allospecies of the pigeon superspecies *Reinwardtoena [reinwardtii]*, which occurs from the Moluccas to the Solomons, the Solomon allospecies *R. [r.] crassirostris* stands out by its long crest and stout, hooked bill (plate 7).

Several of these peripheral allospecies of Northern Melanesia are so divergent that they were initially diagnosed as monotypic genera (*Dicranostephes, Oedirhinus,* and *Coryphoenas* for *Dicrurus megarhynchus, Ptilinopus insolitus,* and *Reinwardtoena crassirostris*, respectively). Not until later was it realized that these birds are actually the peripheral representatives of wide-ranging superspecies. Similar cases of distinctive peripheral allospecies in very wide-ranging superspecies elsewhere in the Pacific include the flat-billed *Halcyon [chloris]* kingfishers of Polynesia and the long-tailed fruit dove *Ptilinopus [purpuratus] huttoni* of Rapa, originally assigned to the separate genera *Todirhamphus* and *Thyliphaps*, respectively.

The largest Northern Melanesian island, New Britain, has five distinctive peripheral allospecies of superspecies ranging from Celebes, the Moluccas, or New Guinea in the west and reaching their eastern limit on New Britain. They are the hawks *Henicopernis [longicauda] infuscata, Accipiter [poliocephalus] princeps,* and *Accipiter [cirrhocephalus] brachyurus,* the pigeon *Ducula [rufigaster] finschii* (plate 8), and the barn owl *Tyto [novaehollandiae] aurantia.* New Britain also has two distinctive megasubspecies reaching their eastern limit there: the cockatoo *Cacatua galerita ophthalmica,* often ranked as a separate allospecies because of its crest curving backward rather than forward in other races of the same species; and the kingfisher *Tanysiptera sylvia nigriceps.*

All the examples mentioned so far involve a distinctive Northern Melanesian allospecies (or megasubspecies) of a widely ranging superspecies. In other cases the peripheral isolate is only a subspecies, but it is nevertheless noteworthy because the species ranges widely through the Australian region, often as far west as India, with no trace of geographic variation except for the peripheral Northern Melanesian subspecies. These examples include the cormorant *Phalacrocorax melanoleucos brevicauda* of Rennell, the heron *Ixobrychus flavicollis woodfordi* of the Solomons, the ibis *Threskiornis moluccus pygmaeus* of Rennell, the tree-duck *Dendrocygna arcuata pygmaea* of New Britain, and the hawk *Haliastur indus flavirostris* of the Solomons.

We have so far been contrasting peripheral isolates of Northern Melanesia with relatives outside of Northern Melanesia. In addition, even within Northern Melanesia the most peripheral population tends to be the most divergent. Distinctive peripheral allospecies that stand out from more centrally located allospecies elsewhere in Northern Melanesia include the cuckoo-shrike *Coracina [tenuirostris] salomonis* (plate 7), the warbler *Phylloscopus [trivirgatus] makirensis,* the honey-eater *Myzomela [lafargei] tristrami* (plate 4), and the starling *Aplonis [grandis] dichroa,* all of San Cristobal, plus the flycatcher *Rhipidura [rufidorsa] malaitae* of Malata. Sixteen other cases (table 34.1) involve divergent peripheral megasubspecies. Three of these megasubspecies are so distinctive that they are often ranked as allospecies: the white-headed (rather than green-headed) dove *Ptilinopus viridis eugeniae* of San Cristobal, the hen-feathered (rather than sexually dimorphic) whistler *Pachycephala pectoralis feminina* of Rennell (plate 2), and the all-black (rather than rufous and black) flycatcher *Monarcha castaneiventris ugiensis* of Ugi (plate 5).

The island locations of these distinctive peripheral populations within Northern Melanesia are for the most part the most isolated large Northern Melanesian islands, the ones that we have already noted as marked by the highest avifaunal endemism (fig. 25.1). Of the five allospecies mentioned in the preceding paragraph, four are on San Cristobal, one on Malaita. Of the 16 megasubspecies, five each are on San Cristobal and Rennell and one each on Manus, St. Matthias, Ugi, Malaita, the eastern Solomons, and Tabar. Tabar is not peripheral within Northern Melanesia as a whole (see fig. 0.3), but it does constitute the eastern distributional limit of the species *Pitta erythrogaster.*

The Evolutionary Significance of Peripheral Isolates

The question remains as to the evolutionary significance of peripheral isolates. Do they remain mere evolutionary curiosities, confined to the species-poor peripheral locations where they evolved? Or do they instead go on to invade cen-

Table 34.1. Divergent peripheral megasubspecies within Northern Melanesia.

Species	Megasubspecies	Islands
Phalacrocorax melanoleucos	*brevicauda*	Rennell
Ptilinopus viridis	*eugeniae*	San Cristobal Group
Gallicolumba jobiensis	*chalconota*	Eastern Solomons
Geoffroyus heteroclitus	*hyacinthus*	Rennell
Chrysococcyx lucidus	*harterti*	Rennell
Ceyx lepidus	*gentianus*	San Cristobal
Pitta erythrogaster	*splendida*	Tabar
Coracina lineata	*gracilis*	Rennell
Lalage leucomela	*conjuncta*	St. Matthias
Rhipidura cockerelli	*coultasi*	Malaita
Rhipidura rufifrons	*semirubra*	Manus
Monarcha castaneiventris	*ugiensis*	Ugi
Petroica multicolor	*polymorpha*	San Cristobal
Pachycephala pectoralis	*feminina*	Rennell
Myzomela cardinalis	*pulcherrima*	San Cristobal Group
Dicrurus hottentottus	*longirostris*	San Cristobal

The table lists a few examples among the many Northern Melanesian species whose most divergent population is an endemic megasubspecies on a peripheral island or group of islands. See text for discussion.

tral locations and become the important evolutionary novelties of species-rich faunas? For example, among Northern Melanesian birds, have the striking peripheral isolates of San Cristobal, Rennell, and Manus expanded into the rest of Northern Melanesia to become distinctive endemic species widespread through Northern Melanesia? And have Northern Melanesia's distinctive species gone on to invade New Guinea, Australia, Indonesia, and the Indomalayan region?

The answer to these last three questions is clear: very rarely. We can recognize few cases of Northern Melanesian species invading mainland New Guinea, and no cases of San Cristobal, Rennell, or Manus isolates expanding into the rest of Northern Melanesia. As detailed in chapter 14, no more than three species (*Lorius hypoinochrous, Halcyon saurophaga* and *Lonchura spectabilis*) may have spread from Northern Melanesia to mainland New Guinea; the most successful is the kingfisher *Halcyon saurophaga*, which has spread in species-poor beach habitats along the whole North New Guinea coast as far west as the Moluccas. Eight further species (*Megapodius freycinet eremita* and seven others) have spread into the New Guinea region as supertramps to establish themselves on the species-poor island of Karkar off the coast of North New Guinea; the most successful is the monarch flycatcher *Monarcha cineracens*, which has spread on species-poor islets along New Guinea's north coast through most of Wallacea. Also as detailed in chapter 14, about eight Solomon populations (*Accipiter albogularis, Haliastur indus flavirostris,* and six others) have established on the easternmost Bismarck islands nearest the Solomons; three others (*Eclectus roratus solomonensis, Hirundo tahitica subfusca,* and *Coracina tenuirostris remota*) have reached New Ireland; one has reached New Britain (*Halcyon chloris tristrami*); and four (*Ptilinopus solomonensis, Ducula pacifica, Macropygia mackinlayi,* and *Myzomela* [*lafargei*]) have spread as supertramps over small or remote Bismarck islands. Even the largest of the peripheral Northern Melanesian islands, San Cristobal, has contributed negligibly to the avifauna of the rest of Northern Melanesia. A few species (e.g., *Ducula brenchleyi, Coracina caledonica, Petroica multicolor,* and *Rhipidura* [*spilodera*]) have distributions suggesting that they spread from San Cristobal west to Guadalcanal or Bougainville, but all are eastern or Australian species that merely colonized the Solomons through San Cristobal as the southeasternmost Solomon island; they are not distinctive peripheral endemics of San Cristobal.

The explanation for this failure of peripheral isolates to spread lies in the phenomenon of upstream colonization and faunal dominance discussed in chapter 14. The direction of successful colonization is overwhelmingly downstream, from species-rich central locations to species-poor peripheral locations, rather than vice versa. That phenomenon is attributed to the effect of increased interspecific competition in a richer fauna, and increased intraspecific competition within larger populations, on competitive ability. Compared to New Guinea and Australia, Northern Melanesia is far too small and species-poor for its species to be able to compete against New Guinea's and Australia's rich avifauna. By the same token, San Cristobal, Rennell, and Manus are too small for their species to be able to compete in the richer avifaunas of the largest Northern Melanesian islands, New Britain and Greater Bukida. The few successful upstream invaders are mostly confined to species-poor habitats, such as small or remote islands and beach communities.

Summary

We have discussed the biological species concept, which poses no ambiguities in its application to local avifaunas of individual Northern Melanesian islands. Practical difficulties may nevertheless arise in applications to populations isolated in space: it may be difficult to decide whether they could interbreed if brought into contact. We assemble the snapshots, represented by current distributions of Northern Melanesia's 195 species, into the whole movie of speciation—from the arrival of colonists over water, through the geographic fragmentation of their gene pools across water gaps, to their achievement of reproductive isolation, ecological segregation, and sympatry. Speciation in Northern Melanesia has arisen much more often from primary isolation across permanent water gaps than from secondary isolation across the intermittent water gaps produced by Pleistocene island fissioning. Peripheral isolates attract attention because the geographically most peripheral taxon in a species or superspecies is usually the most strikingly divergent one. However, in Northern Melanesia it is rare for peripheral isolates to expand into central locations and become important evolutionary novelties, because of the problems of upstream colonization and faunal dominance discussed in chapter 14.

35

Species Differences, Taxon Cycles, and the Evolution of Dispersal

Throughout this book we have noted many examples of colonizing species changing their habitat preference or dispersal ability in the course of their evolutionary history in Northern Melanesia. For example, many initially vagile species become sedentary (chapter 11), some open-country colonists shift into forest (chapter 10), and some lowland colonists shift into the mountains (chapter 24). Does the course of evolution in Northern Melanesian birds evince any general tendencies? How greatly do colonizing species differ in their evolutionary histories?

We shall begin by summarizing Wilson's (1959, 1961) influential study of these questions for Melanesian ants. Next, we systematically compare, for those bird species that colonized Northern Melanesia from New Guinea, their preferred habitat in New Guinea and Northern Melanesia. We then divide all Northern Melanesian species into four groups distinguished by level of endemism, habitat, and vagility, and we tabulate 12 features of the species in each group. These tabulations permit us to assess evolutionary trajectories and their variation for Northern Melanesian bird species. We specifically examine the origins of one of the most distinctive subsets of the Northern Melanesian avifauna, the montane avifauna. Finally, we reconsider the evolution of dispersal ability.

The Taxon Cycle for Melanesian Ants

Wilson (1959, 1961) surveyed the evolutionary histories of many of the ant species that have colonized Melanesia—not only Northern Melanesia, but also the New Hebrides and Fiji archipelagoes to the southeast and east. As also true of Northern Melanesian birds, Melanesian ants originated from tropical Southeast Asia, New Guinea, and Australia. One large group of ant species was inferred to be in an early stage of colonization (stage I) because they have large, continuous geographic ranges, little geographic variation, and low or no endemism in Melanesia. The New Guinean populations of these species are competitively excluded from the species-rich rainforest interior and are instead concentrated in species-poor marginal habitats such as the coast, savanna, monsoon forest, and open forest. Upon colonizing Melanesian islands, which are species-poor compared to New Guinea, these species undergo ecological release from competition and are able to occupy the rainforest interior. In stage II these colonists evolve into Melanesian endemic species of groups still centered on source regions to the west. They become confined to the interior of rainforest (especially of montane forest) and lose biological attributes adapting them for dispersal. In stage III the colonists evolve further into endemic species belonging to groups endemic to or centered on Melanesia. Some of them undergo contraction of geographic range and become relict species of localized distribution, while others evolve into stage I species again and embark on a new cycle of expansion. Wilson (1961) left it as an open question whether this taxon cycle, as formulated for Melanesian ants, also applies to birds and other taxa.

Habitat Shifts During Colonization by Northern Melanesian Birds

As a first step toward comparing evolutionary histories between Melanesian ants and Northern Melanesian birds, table 35.1 compares habitats occupied within Northern Melanesia and New Guinea for those bird species that colonized Northern Melanesia from New Guinea. For all except one species living in fresh water, swamp, seacoast, mangrove, and aerial habitats, there is not a habitat shift between the New Guinean source population and the Northern Melanesian derived population. (The sole exception is the kingfisher *Halcyon chloris*, confined to mangrove in New Guinea but occurring throughout lowland rainforest on Northern Melanesian islands.) Among the 31 species that colonized Northern Melanesia from New Guinea and exist there in open habitats such as secondary growth, savanna, and grassland, 11 expand into lowland rainforest in Northern Melanesia, as Wilson noted for Melanesian ants, but 20 remain confined to open habitats in Northern Melanesia.

An apparent difference between ants and birds is that 41 bird species of the Northern Melanesian rainforest interior already occupied the rainforest interior in New Guinea. Most of these species also occupy the rainforest edge, rainforest canopy, or secondary growth in New Guinea. However, at least 10 of these 41 species (the megapode *Megapodius freycinet*, rails *Rallina tricolor* and *Gymnocrex plumbeiventris*, pigeons *Ducula* [*rufigaster*] and *Caloenas nicobarica*, kingfishers *Ceyx lepidus* and *Tanysiptera sylvia*, pittas *Pitta* [*sordida*] and *P. erythrogaster*, and flycatcher *Monarcha* [*manadensis*]) are confined in New Guinea to the rainforest ground, lower story, or middle story.

An even more noteworthy difference between ants and birds is the large number of bird colonists of Northern Melanesia that are confined in New Guinea to the mountains: two montane aerial species and 20 montane forest species. Wilson does not mention any New Guinean montane ant species that has colonized Melanesia. Equally striking, while seven Northern Melanesia montane bird species are derived from New Guinean lowland species, as commonly noted for ants, even more Northern Melanesian lowland bird species (10 lowland forest species and two lowland aerial species) are derived from New Guinean montane species—another result apparently unprecedented for Melanesian ants.

Some of these habitat shifts by colonizing New Guinean montane species may be made possible by ecological release from competition due to Northern Melanesia's lack of competing lowland species that excluded these montane species from the New Guinea lowlands. For instance, the pigeon *Gallicolumba beccarii*, parrot *Geoffroyus heteroclitus*, swiftlet *Aerodramus* [*spodiopygius*], honey-eater *Myzomela cruentata*, and wood-swallow *Artamus* [*maximus*] are confined in New Guinea to the mountains and excluded from the New Guinea lowlands by their congeners *G. rufigula*, *G. geof-*

Table 35.1. Do colonizing species change their habitat?

Habitat in New Guinea	Habitat in Northern Melanesia	No. of species
Fresh water	Fresh water	12
Swamp	Swamp	4
Seacoast	Seacoast	5
Mangrove	Mangrove	2
Mangrove	Lowland forest	1
Lowland aerial	Lowland aerial	9
Montane aerial	Lowland aerial	2
Open	Open	20
Open	Lowland forest	11
Lowland forest	Lowland forest	41
Lowland forest	Montane forest	7
Montane forest	Montane forest	10
Montane forest	Lowland forest	10

For those species that colonized Northern Melanesia from New Guinea, the right-most column gives the number of species by habitat in each region. For example, 12 colonists occupy fresh water in New Guinea and also in Northern Melanesia, while 11 colonists occupy open habitats in New Guinea but lowland forest in Northern Melanesia. The analysis is confined to Northern Melanesian species with conspecific or allospecific populations in New Guinea. See text for discussion.

froyi, *A. vanikorensis*, *M. nigrita*, and *A. leucorhynchos*, respectively. These lowland congeners are absent in Northern Melanesia, permitting the New Guinean montane colonists to occupy the Northern Melanesian lowlands. Other colonists, such as the thrushes *Turdus poliocephalus* and *Zoothera heinei*, warbler *Phylloscopus poliocephala*, and white-eye *Zosterops atrifrons*, may expand into the lowlands of at least some Northern Melanesian islands (just outlying species-poor islands in the cases of the first three of these four species) because of relief from diffuse competition (see fig. 42 of Diamond 1975a).

Thus, the evolutionary histories of Melanesian ants and Northern Melanesian birds exhibit both similarities and differences. Similarities include the shifts of some New Guinean open-habitat species into Northern Melanesian lowland rainforest and the shifts of some New Guinean lowland species into Northern Melanesian montane habitats, among both birds and ants. Apparent differences are the direct colonizations of Northern Melanesia by New Guinean lowland forest and montane species among birds but not among ants. We are uncertain whether these differences are apparent or real, because our analysis of birds is based on the entire avifauna, but Wilson's analysis of ants excluded half of the ant fauna that was not well understood taxonomically.

Four Evolutionary Stages among Northern Melanesian Birds

We assigned all analyzed Northern Melanesian bird species to one of four sets. Three of those sets correspond to different stages in evolutionary history: stage 1, consisting of recent colonists that have not yet differentiated into endemic taxa within Northern Melanesia, or have differentiated only to the subspecies level; stage 2, consisting of older colonists that are endemic at the level of the allospecies or higher, but that still retain the overwater dispersal ability that brought them to Northern Melanesia; and stage 3, consisting of older colonists that have now completely lost their overwater colonizing ability. Stage 1 is divided into two subsets by habitat (stages 1a and 1b, nonforest vs. forest species) because some features of biology prove to vary considerably with habitat. We found that 25 superspecies have component allospecies differing in level of endemism or dispersal ability and hence belonging to different stages, so we analyzed such allospecies separately. Table 35.2 tabulates how 12 biological properties vary on the average between stages and also how they vary among individual species of the stage. We call attention to variation among species of the same stage so as to discourage overgeneralizations such as "old endemics are confined to mountain forest of large islands." Such a generalization is true for a disproportionate percentage of old endemics, but they also include supertramps of small islets as well as species of lowland grassland.

Level of endemism necessarily differs between the four stages because of how they were defined (row 2 of table 35.2). However, two other conclusions from the data underlying row 2 of the table are not circular. First, proportionately more forest species than nonforest species (stage 1b vs. 1a) have differentiated to the level of endemic subspecies, as reflected in the higher average level of endemism of the former (1.08 vs. 0.61). Second, all five of the endemic genera are sedentary and have lost their overwater colonizing ability (i.e., all species at endemism level 5 belong to stage 3). But the endemic full species (endemism level 4) include 14 species in the vagile endemic stage 2: three highly vagile species (dispersal class A: the parrot *Chalcopsitta cardinalis*, supertramp honey-eater *Myzomela sclateri*, and supertramp starling *Aplonis feadensis*), as well as 11 moderately vagile species (dispersal class B).

On average, vagility (row 3) does not differ between nonforest nonendemics, forest nonendemics, and vagile endemics (stages 1a, 1b, and 2). But all three of those stages contain both highly vagile and less vagile species (dispersal classes A and B, respectively). Stage 3, by definition, consists only of sedentary species (dispersal class C).

As for habitat (rows 4 and 5), by definition stage 1a contains only nonforest species, and stage 1b contains only forest species. The other differences within rows 4 and 5 are significant and do not result from the definitions of the stages. The vagile endemics include only three nonforest species (the water kingfisher *Alcedo websteri*, the grassland finch *Lonchura melaena*, and the aerial wood-swallow *Artamus insignis*), and the sedentary endemics include only one (the grassland finch allospecies *Lonchura* [*spectabilis*] *hunsteini* and *L.* [*s.*] *forbesi*, confined to modern fragments of Pleistocene Greater New Ireland). The proportion of montane species increases from stage 1 to 2 to 3, from

Table 35.2. Four syndromes/stages among Northern Melanesian bird species.

	Stage 1a: nonendemic, nonforest species	Stage 1b: nonendemic, forest species	Stage 2: vagile endemics	Stage 3: sedentary endemics
1. Number of species or allospecies	54	69	46	49
2. Level of endemism, mean ± SD (range)	0.61 ± 0.64 (0–2)	1.08 ± 0.65 (0–2)	3.3 ± 0.5 (3–4)	3.5 ± 0.7 (3–5)
3. Vagility, mean ± SD	2.4 ± 0.5 (A, B)	2.4 ± 0.5 (A, B)	2.2 ± 0.4 (A, B)	1.0
4. Habitat				
Forest	0	56	32	31
Mountain forest	0	13	11	17
Non-forest	54	0	3	1
5. % in mountains	2%	19%	24%	35%
6. Abundance, mean ± SD (range)	2.4 ± 0.7 (1–4)	3.5 ± 1.0 (1–5)	3.2 ± 1.2 (1–5)	3.3 ± 1.4 (1–5)
7. Smallest island occupied (km^2)	<0.1	<0.1	<0.1	5, 8, 25, 35, 50, 61 . . .
8. % on New Georgia islets	65% (17/26)	50% (16/32)	50% (12/24)	0% (0/4)
9. No. of major islands occupied, mean ± SD (range)	21.4 ± 20.8 (1–67)	18.8 ± 18.7 (1–69)	16.0 ± 14.1 (2–58)	2.3 ± 2.1 (1–9)
10. Geographic variability, mean ± SD (range)	0.17 ± 0.17 (0.01–0.67)	0.29 ± 0.25 (0.01–1.00)	0.29 ± 0.24 (0.03–1.00)	0.62 ± 0.31 (0.20–1.00)
11. Great speciators	3	20	9	13
12. Supertramps	0	7	6	0

All 191 analysed Northern Melanesian bird species are assigned to one of four sets: *Set 1a.* Non-forest species, either not endemic to Northern Melanesia or else endemic only at the subspecies level. *Set 1b.* As set 1a, but forest species. *Set 2.* Endemic at the allospecies, species, or genus level, and capable of overwater dispersal (Chapter 11: dispersal category A or B). *Set 3.* As set 2, but incapable of overwater dispersal (Chapter 11: dispersal category C). For superspecies whose component allospecies differ in dispersal ability or level of endemism (species 29, 37, 54, 63, 67, 78, 83, 96, 97, 120, 127, 132, 141, 144, 148, 152, 162, 167, 175, 177, 184, 187, 188, and 192 of appendix 1), those allospecies are tabulated separately, so that the total number of species in row 1 exceeds 191. Dispersal ability is evaluated at the level of the allospecies, not of the superspecies as in appendix 5.

Row 1: the number of species or allospecies in that set.

Row 2: level of endemism: (0 = nonendemic; 1, 2, 3, 4, 5 = endemic at the level of weak subspecies, strong subspecies, allospecies, species, or genus, respectively).

Row 3: mean value ± SD of vagility, with individual values in parentheses (chapter 11: classes A, B, and C = 3, 2, 1 respectively; A = most vagile, C = sedentary).

Row 4: number of species or allospecies in each habitat category.

Row 5: percentage of species or allospecies in the set with some or all populations confined to the mountains (almost all in montane forest rather than in montane non-forest habitats).

Row 6: abundance (chapter 10: 1 = rarest . . . 5 = most abundant).

Row 7: area (km^2) of the smallest island occupied. Numbers for stage 3 (5, 8, 25, 35, 50, 61 . . .) are the areas of the smallest island occupied for the six species reaching the smallest islands; all other species in the set are confined to islands whose minimum area variously equals 71–35,742 km^2.

Row 8: percentage of species in the stage that occur on any of 83 surveyed islets smaller than 0.39 km^2 in the New Georgia Group of the Solomons (Diamond & Gilpin 1980). This percentage is tabulated only for species that occur at all in the New Georgia group: the two numbers in parentheses are, first, the number of species occurring on islets, followed after a slash by the number of species occurring anywhere in the New Georgia Group.

Row 9: number of major Northern Melanesian islands occupied (the islands listed in appendix 1).

Row 10: the index $(s + A)/i$, a measure of geographic variability (chapter 19: a value of 1.0 means a distinct population on every island occupied).

Row 11: number of species in the stage that are hypervariable "great speciators," as defined in table 19.3 (the total number exceeds the 31 species of table 19.3, because some superspecies of table 19.3 are represented in multiple sets of table 35.2).

Row 12: number of species in the stage that are supertramps (small-island specialists; table 12.2).

See text for discussion.

2–19% to 24–35%. But it is still the case that most species (65%) in the last stage, the sedentary endemics, are not montane species.

Average values of abundance (row 6) vary little among the stages, except that nonforest species (stage 1a) have smaller populations than the other three stages because of the limited area of natural nonforest habitats on Northern Melanesian islands.

Rows 7 and 8 present two alternative measures of size of islands occupied. Row 7 gives the smallest island occupied by any species at that stage (but, of course, there are many other species at that stage that do not occupy that same smallest island). The nonendemics and the vagile endemics (stages 1a, 1b, 2) include species reaching small islets, but not a single one of the sedentary endemics occurs on any island smaller than 5 km^2, and most of the sedentary endemics are confined to much larger islands. Our other measure (row 8) is presence on 83 surveyed small islets in the New Georgia group—a measure that we tabulate only for species present at all in the New Georgia group. In accord with the measure of row 7, half or more of the New Georgia species at stages 1a, 1b, and 2 reach one or more of those small islets, but not a single sedentary endemic does.

The total number of major Northern Melanesian islands occupied (row 9) is related to presence on small islets (row 7 and 8): only species that occupy small as well as large islands can attain a large number of islands occupied. The mean number of islands occupied is similar (21, 19, and 16 islands) for stages 1a, 1b, and 2, but is much lower (only two islands) for the sedentary endemics. Absence from islets and presence on few major islands are in turn related to dispersal ability: populations occasionally go extinct, especially on small islets, so only vagile species capable of reversing those extinctions occur on many islands or on small islets. While confinement to few and large islands characterizes all sedentary species, the reverse is not true for all vagile species, as the ranges of islands occupied in row 9 show. Vagile nonendemics and endemics (stage 1a, 1b, and 2) do, on average, occupy 10 times more islands than sedentary species, but there is great variation within each stage: each of those vagile stages contains species confined to just one or two islands, as well as species on 58–69 islands.

In row 10 we assess geographic variability by the measure introduced in chapter 19, $(s + A)/i$ (the ratio of distinguishable populations to number of islands occupied). On average, this measure is lowest for nonendemic, nonforest species (stage 1a) and highest for sedentary species (stage 3)—again as expected because gene flow from dispersing individuals inhibits differentiation. Again, though, this is only an average trend, and the variation among individual species of each set encompasses almost the whole range of $(s + A)/i$ values. For example, the vagile endemics (stage 2) include two hypervariable species with a distinct population on every island occupied (the swiftlet *Aerodramus orientalis* and white-eye *Zosterops ugiensis*), as well as three species occupying dozens of islands without any geographic variation (the eagle *Haliaeetus sanfordi*, lory *Chalcopsitta cardinalis*, and cockatoo *Cacatua ducorpsi*).

"Great speciators" (row 11) are the 31 hypervariable species defined in table 19.3. They occur at all four stages but are concentrated among the nonendemic forest species and the sedentary endemics (stages 1b and 3).

Finally (row 12), supertramps (highly vagile small-island specialists) consist only of nonendemic forest species (stage 1a) and vagile endemics (most of which are forest species: stage 2). Sedentary endemics, being sedentary and confined to large islands, cannot be supertramps. There are also no nonforest supertramps (stage 1a) because Northern Melanesian small islands lack most nonforest habitats (fresh water, swamp, mangrove, and open habitats), except for seacoast and aerial; the same coastal and aerial species occupy large as well as small islands.

Thus, as Northern Melanesian bird populations advance from early to late stages in their evolutionary history, there is a tendency for the proportion of nonforest populations to decrease, for the proportion of forest and montane populations to increase, for vagility and the number of islands occupied to decrease, for geographic variability to increase, and for populations to become restricted to a few large islands. But these are only weak trends. Most of these characteristics vary greatly among taxa at the same evolutionary stage.

The Montane Avifauna

In many respects the montane avifauna represents one extreme within the Northern Melanesian avifauna: on the average it consists of the most sedentary, oldest, taxonomically most differentiated species. Nevertheless, it is heterogeneous, and quite a few of its component species do not exhibit these extreme characteristics. Hence the montane avifauna is a good subset with which to illustrate a recurrent theme of this book: that our study of 195 species yields not just a mass of details but also generalizations, *and* that there are exceptions to the generalizations, *and* that the exceptions themselves are instructive. We mentioned some of these patterns in previous chapters, and we now synthesize that information.

Montane Populations

We define a montane population as a population of a certain species on a certain island regularly breeding in the mountains of that island but not at sea level. Under this definition we include populations that descend regularly or erratically to the lowlands after breeding, in search of seasonally shifting fruits, flowers, nectar, and seeds (especially populations of the pigeons *Ducula melanochroa* and *Gymnophaps* [*albertisii*], the lorikeets *Charmosyna* [*meeki*] and *C. margarethae*, and the bamboo finch *Erythrura trichroa*). We also include populations of species that are confined to the mountains in some parts of the large islands New Britain and Bougainville but are apparently resident in the lowlands in other parts of the same island (e.g., the whistler *Pachycephala pectoralis*).

By these definitions, we recognize 140 montane populations on Northern Melanesian islands; tables 9.1 and 9.2 give the number of populations on each of the 19 islands high enough to support such populations. The number of populations ranges from just 1 for Vella Lavella to 23 for Guadalcanal, and it increases with island elevation, island area, and area above 600 m (chapter 9; see also Mayr & Diamond 1976). Properties of species with montane populations are tabulated in table 35.3, which adopts a somewhat broader definition of montane species than elsewhere in this book and thereby arrives at 48 species with one or more montane populations.

Dispersal

Among Northern Melanesian habitats, the avifauna of montane habitat has the lowest average dispersal ability (averaged over all its species; table 11.2). The same conclusion applies to the New Guinean source avifauna, as illustrated by the fact that a much lower percentage of New Guinean montane species (12%) than of New Guinean lowland species (44%) have colonized Northern Melanesia (chapter 8). Most of the 48 species with montane populations belong to our dispersal class B or C rather than our most vagile class (column 7 of Table 35.3). This trend of low dispersal is as expected from the fact that montane forest is Northern Melanesia's wettest, most stable, climatically most predictable habitat (chapter 11). Nevertheless, it would be wrong to make a sweeping generalization that "montane species are sedentary." Two fruit pigeon species (*Ptilinopus superbus* and *Ducula melanochroa*) and one flycatcher (*Rhipidura rufifrons*) with montane populations belong to the vagile class A, and a majority of the other montane species (including not just pigeons and parrots but also quite a few territorial forest passerines) belongs to the intermediate class B, not to the most sedentary class C.

Endemism

Among Northern Melanesian habitats, the avifauna of montane habitat also has the highest average level of endemism (chapters 10 and 32): on average, montane species are endemic to Northern Melanesia at the level of an endemic allospecies. Montane species are overrepresented among Northern Melanesia's endemic species and genera (chapter 24); 34 of the 48 montane species are endemic at the level of endemic allospecies, full species, or genus. That still leaves 11 montane species whose Northern Melanesian population belong only to endemic subspecies, and a few are highly vagile (three species in dispersal class A, 11 in class B). Hence even this exception fits a pattern: vagile species tend to have Northern Melanesian populations at lower levels of endemism than sedentary species (chapter 10).

Differentiation of Conspecific Montane Populations Between Islands

As expected from their relatively low vagility, montane populations tend to differentiate taxonomically between Northern Melanesian islands. The upper extreme of differentiation consists of species in which each montane population belongs to a different subspecies, different megasubspecies, or even a different allospecies (column 6 of Table

Table 35.3. Montane bird populations of Northern Melanesia.

Species	Islands	No. of montane populations	No. of nonmontane populations	Endemism	$\frac{(s + A)}{mt.i}$	Overwater dispersal	Abundance	Abutting congener?	Ancestor
Accipiter princeps	NB	1	0	2		C	1		Low (NG)
Pareudiastes sylvestris	SCr	1	0	3		C	1		?
Ptilinopus superbus	Tol	1	36	0		A	3		Low (NG)
Ptilinopus insolitus	Long, Tol, Cr	3	14	2	—	B	4		Low (NG)
Ptilinopus [*rivoli*]	NB, NI, Umb, Boug, Guad, Gan, Kul, NGa, Vang, Gat, Rnd	11	42	2	0.36 s, M, A	B	4		Mt (NG)
Ducula brenchleyi	Guad	1	7	2		B	4		Mt (Heb)
Ducula melanochroa	NB, NI, Umb	3	0	2	—	A	3		Low (NG)
Gymnophaps [*albertisii*]	NB, NI, Boug, Guad, Mal, Kul, Vang, Rnd	8	0	2	0.25 A	B	3		Mt (NG)
Columba vitiensis	NI, Boug, Ys, Guad, Mal, VL, Kul, Vang, Gat	9	2	0	—	B	1		Low (NG)
Columba pallidiceps	Boug, Ch, Guad	3	7	3	—	B	1		Low (NG)
Reinwardtoena [*reinwardtii*]	Boug, Guad, Gan, Kul, Rnd	5	26	2	—	B	1		Low (NG)
Henicophaps foersteri	Umb	1	1	2		C	1		Low (NG)
Lorius [*lory*]	NI	1	19	2		B	4	√	Low (NG)
Charmosyna [*palmarum*]	NB, NI, Boug, Ys, Guad, Mal, Kul, Vang	8	0	2	0.25 A	B	3	√	Mt (NG0
Charmosyna margarethae	Mal, Kul	2	7	3	—	B	3		?
Micropsitta bruijnii	NG, NI, Boug, Guad, Kul	5	0	1	0.6 s	B	3	√	Mt (NG)
Aerodramus orientalis	NI	1	2	3		B	1		?
Ceyx lepidus	Guad	1	24	1		B	4		Low (NG)
Halcyon bougainvillei	Guad	1	1	3		C	1		?
Coarcina caledonica	Boug, Guad	2	4	1	1.0 s	B	3		Low (Heb)
Coracina holopolia	Kul	1	9	3		B	3		?
Zoothera [*dauma*]	NB, Umb, Boug, Ch, Guad, SCr	6	2	2	0.83 s, A	B	3		Mt (NG)
Turdus poliocephalus	NB, NI, Tol, Boug, Guad, Kul	6	2	1	1.0 s	B	5		Mt (NG)
Cichlornis [*whitneyi*]	NB, Boug, Guad	3	0	2	1.0 A	B	3		?
Cettia parens	SCr	1	0	3		C	3		?

1	2	3	4	5	6	7	8	9	10
Phylloscopus [trivirgatus]	NB, NI, Umb, Boug, Kul, Ys, Guad, Mal, SCr	9	1	2	0.67 s, A	B	5		Mt (NG)
Phylloscopus amoenus	Kul	1	0	3		C	3		Mt (NG)
Rhipidura [spilodera]	Boug, Guad	2	2	2	1.0 s	B	5		?
Rhipidura fuliginosa	SCr	1	0	0		B	3		Low (Heb)
Rhipidura rufifrons	Mal	1	26	1		A	5		Low (NG)
Rhipidura [rufidorsa]	NB, NI, Umb, Mal	4	1	2	0.75 s, A	B	4		Low (NG)
Monarcha [manadensis]	Ch	1	31	2		B	4		Low (NG)
Monachella muelleriana	NB	1	0	1		B	2		Mt (NG)
Petroica multicolor	Boug, Guad, Kul, SCr	4	0	1	1.0 s, M	B	4		Low (Heb)
Pachycephala pectoralis	NB, NI, Boug, Gan, Kul, Vang, Gat, Rnd	8	26	1	0.63 s, M	B	5	√	?
Pachycephala implicata	Boug, Guad	2	0	3	1.0 M	C	3	√	?
Zosterops atrifrons	NI, Umb	2	4	1	1.0 s	B	5		Mt (NG)
Zosterops murphyi	Kul	1	0	3		C	5	√	Low (Sol)
Zosterops ugiensis	Boug, Guad	2	1	3	1.0 M	B	5	√	?
Myzomela cruentata	NB	1	4	1		B	5	√	Mt (NG)
Myzomela pulchella	NI	1	0	3		C	5	√	?
Philemon [moluccensis]	NI	1	2	2		B	4		Low (NG)
Melidectes whitemanensis	NB	1	0	3		C	4		Mt (NG)
Stresemannia bougainvillei	Boug	1	0	4		C	3		?
Guadalcanaria inexpectata	Guad	1	0	4		C	3		?
Erythrura trichroa	NB, NI, Long, Umb, Tol, Cr, Guad, Kul	8	4	1	0.25 s	B	4		Mt (NG)
Aplonis brunneicapilla	Guad	1	3	3		B	3	√	Low (NG)
Corvus woodfordi	Guad	1	5	3		C	3		?

This table characterizes the montane bird populations of Northern Melanesia: i.e., island populations confined to the mountains of that island and not normally occurring there at sea level (regardless of whether the species does normally occur at sea level on some other island). For each species or superspecies named in column 1, column 2 names (by the island abbreviations listed in appendix 1) all islands supporting montane populations of that species. Column 3 gives the number of islands with such montane populations; column 4, the number of islands with populations of that species normally occurring at sea level.

Column 5 is that species' level of endemism in Northern Melanesia, from column 2 of appendix 5 (0 = nonendemic, 1 = endemic subspecies, 2 = endemic allospecies, 3 = endemic full species, 4 = endemic genus).

Column 6 is an index of geographic variation among the species' montane populations. The entry is the number of geographically distinct montane populations in Northern Melanesia ([S + A]: distinct subspecies, megasubspecies, or allospecies), divided by the number of islands with montane populations (mt.i: column 3). For example, the fruit dove superspecies *Ptilinopus [rivoli]* has mt. $i = 11$ montane populations, falling into $(s + A)$ = four geographically distinct populations (a Bismarck allospecies with three populations belonging to the same subspecies, and a Solomon allospecies with eight populations belonging to three subspecies), so the index is 4/11 = 0.36. There is no entry for species with only one montane population. A dash indicates species with two or more montane populations that are taxonomically indistinguishable and belong to the same subspecies. The letters s, M, A indicate whether geographic differentiation among the montane populations is at the level of weak subspecies (s), megasubspecies (M), or allospecies (A); some species, including *Ptilinopus [rivoli]*, have distinct montane taxa at more than one level. The geographic variability index $(s + A)$/mt.i in this table refers only to montane populations and is not the same as the index $(s + A)/i$ in the right-most columns of appendix 5, which is calculated for all Northern Melanesian populations of the species, whether montane or not.

Column 7 is the species' overwater dispersal index from column 6 of appendix 5 (A = most vagile . . . C = most sedentary).

Column 8 is the species' abundance from column 7 of appendix 5 (1 = rare . . . 5 = abundant).

Column 9 notes those species whose altitudinal range on at least some islands abuts, and is mutually exclusive of, the altitudinal range of a congener. For example, on three islands (NB, NI, Boug) the lorikeet *Charmosyna [palmarum]* lives at altitudes above its closely related congener *C. placentis*.

Finally, column 10 indicates whether (if known) the ancestral source population from which the Northern Melanesian species originated lives in the lowlands ("low") or mountains ("mt") of the source region (NG = New Guinea, Heb = New Hebrides, Sol = Solomons).

35.3: the cuckoo-shrike *Coracina caledonica*, thrush *Turdus poliocephalus*, thicket warbler *Cichlornis* [*whitneyi*], flycatcher *Rhipidura* [*spilodera*], robin *Petroica multicolor*, whistler *Pachycephala implicata*, and white-eye *Zosterops atrifrons* and *Z. ugiensis*). Hence montane species constitute 35% of Northern Melanesia's great speciators (table 19.3), even though they constitute only 16% of its total avifauna.

The exceptions to this tendency to differentiate again prove to be understandable. The six species that have two or more montane populations but no geographic differentiation at all between these populations (denoted by a dash (—) in column 6 of table 35.3) are, without exception, vagile pigeons or parrots; two (the pigeon *Ducula melanochroa* and lorikeet *Charmosyna margarethae*) are flocking nomads and three of them are rare (column 8: abundance category 1, hence with short-lived local populations likely to go extinct and to be refounded frequently). Conversely, the 13 most differentiated montane species (values of 0.6–1.0 in column 6 of table 35.3) are all abundant or moderately abundant (hence with long-lived local populations); eight of them are among the most abundant Northern Melanesian species (abundance classes 4 and 5: column 8 of table 35.3); and 12 are at least moderately vagile (dispersal class B: column 7 of table 35.3). Thus, both the general tendency of montane species to differentiate, and the exceptions to that tendency, fit our conclusion that proneness to differentiate increases with abundance and decreases with vagility (chapter 19).

Why Are Montane Populations Confined to Mountains?

Nineteen of the species with Northern Melanesian montane populations originated from a lowland species in a source avifauna (last column of Table 35.3; mainly New Guinea, in a few cases the New Hebrides or else the Solomons), as also true for Northern Melanesian montane ants. Fourteen species "jumped" to Northern Melanesia's mountains directly from the mountains of New Guinea (13 species) or the New Hebrides (one species), despite the on-average low vagility of the New Guinea montane avifauna. These findings raise the question of why any population should be confined to mountains and absent at sea level (Mayr & Diamond 1976).

One possible answer is that the species' physiology or diet enables it to survive only among habitats, food sources, temperatures, and humidities in the mountains, and that the species cannot stay alive at sea level. But this interpretation fails to explain why, out of the 37 montane species with populations in the mountains of at least two Northern Melanesian islands (column 3 of table 35.3), 30 occur in the lowlands of at least one Northern Melanesian island—in some cases, of up to 36 Northern Melanesian islands (column 4 of Table 35.3). The other seven species have only montane populations in Northern Melanesia, but even two of those species (*Petroica multicolor* and the superspecies *Ducula* [*pinon*] to which *D. melanochroa* belongs) occur in the lowlands of the New Hebrides and New Guinea, respectively. Evidently, at least 32 of the 37 species can survive in the lowlands, and not more than five species are obligate montane species. Something other than physiology or diet must be barring the 32 species from the lowlands of the islands where their populations are confined to mountains. What, then, accounts for the absences of these 32 species from the lowlands of some islands but not of other islands? Three explanations are apparent.

First, two species are excluded from the lowlands of some islands by a congener but occupy the lowlands of other islands where that congener is absent. Those two species are the white-eye *Zosterops ugiensis*, occurring at sea level on San Cristobal but confined to elevations above 900 m on Bougainville by the presence there of a Bukida endemic lowland congener *Z. metcalfii*; and the honey-eater *Myzomela cruentata*, which occurs in the lowlands of four Bismarck islands but is confined to elevations above 900 m on New Britain by the presence of its New Britain endemic lowland congener *M. erythromelas*. *Myzomela cruentata* is especially interesting because on New Britain it is confined to the mountains at elevations of 900 m up to at least 1720 m and excluded from the lowlands, but on New Ireland it is confined to the lowlands up to 900 m and excluded from the mountains by its New Ireland endemic montane congener *M. pulchella*.

Second, when one compares interisland variation in the altitudinal floor of the montane thrush *Turdus poliocephalus* with interisland variation in bird species diversity as a function of altitude, it becomes apparent that this species descends on each island to an elevation at which it encounters about 30 other species (whatever that elevation happens to be), and that it is excluded from the more species-rich lowlands. For instance, among the islands Bougainville, Guadalcanal, Kulambangra, Tolokiwa, St. Matthias, and Rennell, overall species richness decreases in that sequence, and the altitudinal floor of the thrush decreases to 1200, 1200, 1000, 750, 0, and 0 m, respectively. (That is, the thrush is confined to elevations above 1200 m on Bougainville and Guadalcanal, above 1000 m on Kulambangra, etc., and descends to sea level on St. Matthias and Rennell.) At that altitudinal floor on each island, though, the thrush encounters a community of approximately the same species richness, about 23–36 species, because species diversity on an island decreases with altitude (fig. 42 of Diamond 1975a). Thus, this thrush is excluded from species-rich lowland communities by diffuse competition with many species, rather than by the presence of a single competing congener as in the two cases discussed in the preceding paragraph. A similar explanation may apply to sea-level populations of the otherwise-montane thrush *Zoothera* [*dauma*], the warbler *Phylloscopus* [*trivirgatus*], and the flycatcher *Rhipidura* [*rufidorsa*] on the outlying species-poor Bismarck island St. Matthias.

Third, the two Solomon islands with the highest percentage of their total area lying above 600 m are Bougainville (29%) and Guadalcanal (25%). On one or both of those two islands, numerous species occurring at sea level on smaller and lower islands are virtually confined to elevations above 900 m: *Ptilinopus* [*rivoli*] *solomonensis, Ducula brenchleyi, Columba vitiensis, C. pallidiceps, Reinwardtoena* [*reinwardtii*], *Ceyx lepidus, Coracina caledonica, Rhipidura* [*spilodera*], *Pachycephala pectoralis, Aplonis brunneicapilla, Corvus woodfordi*, and possibly others. Local populations of these species may become adapted to the large area of montane habitat on these islands and may have locally lost their adaptations to lowland habitats.

Finally, some cases remain puzzling and unexplained. As one example, the fruit dove *Ptilinopus insolitus*, derived from *P. iozonus* of the New Guinea lowlands, lives in the lowlands of 14 Bismarck islands but is confined to the mountains of three other Bismarck islands (the now-reforested Long, Tolokiwa, and Crown) that were defaunated in a 17th-century volcanic explosion. Only one, three, and one other species, respectively, are confined to the mountains of these three islands, which otherwise support lowland populations of 31–52 species. We do not know why this fruit dove is virtually the sole lowland species to have shifted into the mountains of these three islands. (The only other example is the fruit dove *Ptilinopus superbus*, confined to the mountains of Tolokiwa but otherwise occurring in the lowlands of dozens of islands of Northern Melanesia and the New Guinea region.) As another puzzling example, the lowland honey-eater *Philemon* [*moluccensis*] *novaeguineae* of New Guinea gave rise to three endemic allospecies in the Bismarcks: *P. eichhorni* on New Ireland, *P. albitorques* on Manus, and *P. cockerelli* on New Britain and Umboi. The latter two species occur in the lowlands of their islands, like their New Guinean relative, and *P. cockerelli* disappears at around 1160 and 760 m on New Britain and Umboi, respectively. But *P. eichhorni* of New Ireland lives up to at least 2200 m and disappears below 900 m, with the result that it remained unknown for a century while collectors were exploring the lowlands of New Ireland and was not discovered until 1923, when Albert Eichhorn at last collected birds in New Ireland's mountains. We cannot explain this upward shift in the altitudinal range of *P. eichhorni* but not of its New Britain and Umboi relative *P. cockerelli*.

Dispersal Ability

Population Differences in Dispersal Ability

Dispersal ability differs enormously among Northern Melanesian bird species. At the one extreme are many species for which there are no records of even a single individual flying across water within the nearly 200-year history of Northern Melanesia's ornithological exploration, and no distributional evidence of crossing water within the last 10,000 years. At the opposite extreme are many species that can be seen flying over water any day in Northern Melanesia. Similar species differences have been noted within other Melanesian vertebrate classes, where the differences in overwater dispersal can often be related to obvious differences in anatomy (e.g., bats have wings but kangaroos do not), in physiology (e.g., the waterproof skins and eggs of geckoes enable them to survive salt spray while rafting overwater), and in reproductive biology (e.g., the direct development of platymantine frogs in terrestrial eggs, and the high percentage of gravid female skinks at any time, predispose both of these groups to colonizing over water; chapter 4).

In a few cases for birds as well, species differences in dispersal ability can be explained by anatomical differences. Out of Northern Melanesia's 195 extant bird species, four are flightless, as were two extinct species known from fossils, and as undoubtedly were other fossil species still to be discovered (chapter 7). Some other extant species are short-winged and probably capable of only short flights. But the overwhelming preponderance of cases of poor overwater dispersal in Northern Melanesian birds has a behavioral explanation: among species that are equally strong fliers over land, some regularly do, and some never do, choose to fly overwater, as measured by our six types of evidence (chapter 11).

Abundant examples attest to the fact that dispersal ability differs not only between full species but also between more closely related taxa (chapter 11). Striking differences between allospecies of the same superspecies include the contrast between the vagile white-eye *Zosterops* [*griseotinctus*] *griseotinctus*, which colonized Long, Tolokiwa, and Crown islands after Long's seventeenth-century volcanic explosion, and the sedentary allospecies *Z.* [*g.*] *vellalavella, Z.* [*g.*] *splendidus, Z.* [*g.*] *luteirostris*, and *Z.* [*g.*] *rennellianus*, never recorded away from their home islands of Vella Lavella and Bagga, Ganonga, Gizo, and Rennell, respectively. Striking differences between subspecies of the same allospecies include the vagile and the sedentary Northern Melanesian races of the pigeon *Ptilinopus solomonensis*, parrot *Trichoglossus haematodus*, swiftlet *Collocalia esculenta*, kingfishers *Halcyon chloris*, and many other species. Striking differences between populations of the same subspecies include the sedentary Bismarck and vagile Solomon populations of the starling *Mino dumontii kreftti.*

Evolutionary Changes in Dispersal Ability with Time

Because Darwin showed that related taxa must have diverged from a common ancestor, these just-mentioned differences of dispersal ability in space between populations of the same group of taxa imply evolutionary changes in dispersal ability, which can either increase or decrease with time. In Europe, where millions of people watch birds, they have noted many examples of range expansions of species associated with a rapid increase in dispersal ability with time, such as the western expansions of the Serin finch *Serinus serinus*, azure tit *Parus cyanus*, Syrian woodpecker *Dendrocopos syriacus*, black-necked grebe *Podiceps nigricollis*, and collared turtle dove *Streptopelia decaocto* beginning around 1790, 1870, 1880, 1900, and 1920, respectively (Mayr 1942, Harrison 1982). Even though there have been far fewer bird observers in Northern Melanesia than in Europe, several range expansions have still been witnessed, notably the twentieth-century invasion of San Cristobal by the honey-eater *Myzomela cardinalis* (chapter 23) and the invasion of the New Georgia group by the hornbill *Rhyticeros plicatus* after 1940 (chapter 13). Many more range expansions in Northern Melanesia's past can be inferred from modern distributions. That inferential evidence, discussed throughout this book, includes the four flightless species and the 100+ behaviorally sedentary populations confined to single, isolated islands, which could only have been reached over water in the past (because Northern Melanesia has never had land connections to other land masses), so that today's sedentary populations must have evolved from vagile ancestors. Sequential waves of colonization in the past can be inferred from the unequal levels of endemism (distinct old endemic populations on some islands, widespread, scarcely differentiated populations on other islands) in the fantail *Rhipidura* [*rufifrons*] (Mayr & Moynihan 1946) and the golden whistlers *Pachycephala pectoralis/P. melanura* (Galbraith 1956).

Thus, the behavioral propensity for overwater dispersal must have a genetic basis and evolve subject to natural selection, just as do anatomical and physiological characters. Our evolutionary cost/benefit analysis of chapter 11 predicted, and the available data confirmed, that dispersal ability is selected by unstable habitats, patchy habitats, in some cases low abundance, and flying in the course of daily foraging. Dispersal ability is penalized by stable habitats, large, coherent habitat blocks, high abundance, and not flying in the course of daily foraging.

Distributional Consequences of Differences in Dispersal Ability

Species differences in dispersal ability, and changes in that ability with evolutionary time, explain a wide range of distributional phenomena in Northern Melanesian birds (see also chapter 36). Those phenomena include the sequence of bird arrivals on volcanic islands after defaunation (chapter 1); differences between the Northern Melanesian representation of New Guinea montane and lowland bird species, and of different groups of New Guinea bird species (chapter 8); differences between small and large islands in how steeply bird species number decreases with isolation (fig. 9.4); now-sedentary, single-island endemics that must have arrived over water (chapter 11); the restriction of sedentary, lowland forest species to a few large islands (chapters 11 and 12), such as the larger fragments of Greater Bukida (table 33.1); the relative sedentariness of higher endemics (tables 11.3 and 24.2); the larger number of islands occupied by vagile species than by sedentary species (fig. 12.2); the decrease in geographic variability with increasing vagility (tables 18.2 and 19.1); the peaks in proneness to speciate (table 19.3) and to form allospecies (chapter 21) at immediate vagility; and convergences in species composition but not in taxonomic affinity between the avifaunas of St. Matthias and Manus, of Biak and the Northwest Bismarcks, of Rennell and Santa Cruz, and among Northern Melanesian islands generally (chapters 30 and 31).

Summary

We assigned Northern Melanesian bird species to four consecutive evolutionary stages beginning with arrival, for comparison with the taxon cycle that E.O. Wilson deduced for Melanesian ants. Bird species at later stages tend to be sedentary, geographically variable endemics of forest and mountains on a few large islands, but there are many exceptions. Analysis of the montane avifauna shows that few of its members are obligate montane species; instead, their confinement to mountains is usually related to competition or to underutilized montane habitats. The behavioral basis for overwater dispersal ability varies greatly among Northern Melanesian bird populations (even among populations of the same subspecies), evolves with time under natural selection, and has many distributional consequences.

36

Promising Directions for Future Research

In this final chapter we call attention to five promising sets of directions for future research. We begin with directions that depend on gathering new data: molecular evidence of relationships, bird fossils, archival searches for early European observations, refined measurements of vagility and abundance, searches for undiscovered bird populations and species, and surveys within the two longest Northern Melanesian islands. The second set involves improved analytical techniques for quantifying effects of island area, distance, and elevation. The third set consists of approaches for estimating the absolute time scale of speciation. Fourth is the effort to recognize which, among our attempts to explain bird distributions parsimoniously, are parsimonious but wrong, and to replace them with more complicated but correct explanations. Finally, generalizing about speciation will require comparing our study of Northern Melanesian birds with studies of other biotas, such as biotas of remote archipelagoes, continental biotas, temperate-zone biotas, and taxa other than birds.

Priorities for Gathering New Data

Molecular Evidence

Our wish list for new data begins with molecular evidence of bird relationships. Such evidence is now becoming available for many avifaunas, constitutes a growing focus of ornithology, and often yields dramatic new insights. At the time of this writing (December 2000), no molecular studies have been reported for Northern Melanesian birds. Now that so much nonmolecular evidence is available, there is a plethora of problems ripe for study at the molecular level.

At taxonomic levels above the species level, molecular evidence could clarify the relationships of any of Northen Melanesia's five endemic monotypic genera and 30 endemic full species (chapter 24). What are the origins of Northern Melanesia's most distinctive bird, the Choiseul crested pigeon *Microgoura meeki* (#74, plate 8)? or of the big owl *Nesasio solomonensis* (#98, plate 8)? (Numbers here and in the following discussion refer to the number sequence of species in appendix 1.) Where, within the large genera *Halcyon* and *Centropus*, lie the affinities of the Bukida mountain kingfisher *Halcyon bougainvillei* (#115, plate 8) and the big Bismarck coucal *Centropus violaceus* (#94)? Is the Kulambangra mountain white-eye *Zosterops murphyi* (#168, plate 1) really derived from the *Z.* [*griseotinctus*] superspecies (#167, plate 1), implying a double invasion of Kulambangra? Is the parrot *Chalcopsitta cardinalis* (#76, plate 8) really related to New Guinea *Chalcopsitta* species? Is the New Britain mountain honey-eater *Melidectes whitemanensis* (#179, plate 9) really related to the many New Guinean species of *Melidectes*? And is the San Cristobal mountain warbler *Cettia parens*

(#137) really related to Asian species of *Cettia*? Did the monarch flycatchers *Myiagra alecto* (#154, plate 5) and *M. hebetior* (#155, plate 5) really originate as the New Guinea and Bismarck representatives of the *M.* [*rubecula*] superspecies (#156, plate 5)?

Other questions arise at lower taxonomic levels. Will molecular evidence help sort out the multiple colonization waves that we suspect gave rise to the diverse Northern Melanesian populations of the rail *Gallirallus* [*philippensis*] (#37), kingfisher *Halcyon chloris* (#113, plate 6), cuckoo-shrike *Coracina* [*tenuirostris*] (#127, plate 7), fantail-flycatcher *Rhipidura rufifrons* (#147), monarch flycatcher *Monarcha manadensis* (#152, plate 3), whistler *Pachycephala pectoralis* (#159, plate 2), and honeyeater *Philemon* [*moluccensis*] (#178, plate 7)? Which of Manus's endemic populations are derived independently from New Guinea rather than from the Bismarcks? The hawk *Aviceda subcristata coultasi* (#18Aa) and the pigeon *Macropygia amboinensis admiralitatis* (#66Ac) are candidates (see chapter 30). Where among the populations of the fantail flycatcher *Rhipidura rufrifrons* did the distinctive Manus race *R. rufrifrons semirubra* (#147Aa) come from?

Of course, calculations of molecular clocks may help date the age of Northern Melanesian endemic taxa. For instance, how long ago did the endemics of Rennell Island, that remarkable uplifted coral atoll thought to be a few million years old, begin to diverge?

Fossil Evidence

Humans arrived in at least parts of Northern Melanesia by 33,000 years ago. On other Pacific islands, arrival of humans was associated with mass extinctions of birds and other species. If this were also true in Northern Melanesia, its modern avifauna, or even the avifauna encountered by the first European collectors in the nineteenth century, might have been quite different from the original avifauna. As discussed in chapter 7, the sole collections of Northern Melanesian bird fossils at early archeological sites reported to date are from two large Bismarck islands, New Ireland and St. Matthias. Those collections suggest that, while those islands indeed suffered bird extinctions, about 75% of the original avifauna survived: a far higher percentage than on remote Pacific islands like Hawaii. We attributed the apparent relative robustness of the Northern Melanesian avifauna to its having co-evolved with native mammalian and reptilian predators, but this conclusion is preliminary.

Hence another high priority will be to seek bird fossils on many more islands. We hypothesize that other major central islands that supported native mammals and reptiles—such as New Britain, Manus, the Bukida chain, the New Georgia Group, Malaita, and San Cristobal—suffered only modest extinctions. We also hypothesize that outliers without native mammals or reptiles may have suffered heavy extinctions for the same reason as Hawaii: Rennell is a prime candidate. These predictions await testing.

Archival Evidence

Fossils can identify Northern Melanesian bird populations that became extinct prehistorically due to the arrival of the first human colonists thousands of years ago, long before Europeans arrived. In addition, Hawaii and other Pacific islands provide examples of bird species that survived those first humans, only to succumb soon after the arrival of Europeans with their associated vermin and weapons. In chapter 7 we summarized the first six sets of bird observations in Northern Melanesia by Europeans who landed between 1568 and 1838. Those observations failed to provide any hint of bird species extant then but extinct today.

Archives may hold other accounts of bird observations by Europeans who landed before ornithological exploration began seriously. We mention three promising leads. First, as we noted in chapter 7, the French navigator Louis-Antoine de Bougainville landed for two weeks on New Ireland in 1768. Bougainville's (1771) journals mention only one identifiable bird species. However, Bougainville's accompanying naturalist, Philibert Commerson, also landed with him on New Ireland, made collections which unfortunately became lost or dispersed, and wrote manuscripts that remain unpublished (Dunmore 1965). The manuscripts are still in Paris, in the Muséum National d'Histoire Naturelle there (Laissus 1978). Laissus remarks that the manuscripts that he lists as numbers 42 and 73 describe and illustrate birds; we have not examined them. Second, Marist missionaries lived on Umboi (then known as Rooke) in 1848–1849 and again in 1852–1856 (volume 2A pp. 68–69 and 80–81 of Wichmann (1902–1912), pp. 478–487 and 574–574 of Wiltgen (1979)). Many of those missionaries' unpublished letters and journals are preserved in Rome (chapter 31 of Wiltgen 1979). These might provide valuable information not only about birds, but also about posteruption conditions on nearby Long Island after its volcanic defaunation (believed to have occurred in the seventeenth century). Finally, whaling museums and libraries in the United States, Britain, Australia, and New Zealand contain diaries of whalers, many of whom visited Northern Melanesia between 1799 and 1887, and some of whom may have commented on birds (see pp. 24–33 and Appendix 1 of Bennett (1987)). Will any of these three sets of accounts provide evidence for bird populations that subsequently became extinct?

Improved Measures of Abundance and Vagility

Our estimates of abundance and vagility for each species turned out to account for many distributional and evolutionary characteristics that vary among species. For example, the number of islands occupied (chapter 12), geographic variability (table 19.1), proneness to become a "great speciator" (table 19.4), allospeciation (chapter 21), and survival of isolated populations of non–water-crossing species on land-bridge islands (chapter 33, including table 33.2) all increase with abundance, while vagility among forest passerines decreases with abundance (chapter 11). Number of islands occupied (fig. 12.2) increases with vagility; geographic variability (chapter 18, table 19.1) and allospeciation (chapter 21) decrease with vagility; vagility varies among habitats (chapter 11); sedentary groups of New Guinea species are underrepresented in Northern Melanesia (chapter 8); sedentary species are overrepresented on large islands (chapters 12 and 33) and among higher endemics (chapters 11 and 24); vagile species are overrepresented on remote islands, whose species compositions therefore converge (chapter 30); and species of intermediate vagility are overrepresented among great speciators (table 19.4).

It is pleasantly surprising that we found abundance and vagility to have such high predictive value even with our relatively crude estimates of them. We merely categorized species abundances into five classes, and we synthesized six types of evidence to categorize species vagility into three classes. Hence improved estimates may yield even more and stronger predictions. Abundance should be quantified from transect counts of species population densities, multiplied by areas of different habitats on a Northern Melanesian island. More data to estimate species vagilities could be gathered for the Bismarcks by surveying birds of very small islands, and by sitting in a canoe between islands and watching birds fly over water (chapter 12).

One major type of species difference that deserves study is that some species will probably prove to disperse infrequently but in large pulses as adults (wind-driven large water birds? flocking nomadic frugivores and nectarivores?), while other species will disperse regularly but singly as juveniles (species maintaining pair territories?). A second major type of species difference in dispersal involves dispersal distance: some species have longer mean dispersal distances than other species (Diamond et al. 1976). That is, dispersal rates could be represented by an equation $I = I_0 \ exp \ (-d/d_0)$, where d_0 is a species-specific mean dispersal distance and I_0 is a product of population size times the per capita rate of production of dispersing colonists (eq. 7 of Diamond et al. 1976). Values of d_0 could be estimated by using the species/island lists of our appendix 1 to calculate species incidences (chapter 12) as a function of distance (i.e., to group islands of a given area class according to their isolation, and to calculate what fraction of the islands in a given isolation class are occupied by a given species). Diamond et al. (1976, p. 2162) give an example of such data for the *Monarcha* [*manadensis*] superspecies in the Solomons.

Further Ornithological Exploration

There are still interesting things to be learned from making carefully targeted collections of bird specimens, as opposed to the general collecting prevalent in the early days of ornithological exploration. At the end of chapter 6 we listed some undescribed taxa that have been reported or observed but not collected, and we also listed some habitats that may yield novelties. In addition, collections, or study of specimens, at opposite ends of Northern Melanesia's two longest islands (New Ireland and New Britain) could tell us what distances are required for the operation of continental speciation (i.e., completed speciation within a single land mass) in birds (chapter 22). Continental speciation has proceeded to completion among birds of New Guinea (2400 km long), Madagascar (1600 km long), and perhaps New Zealand (1500 km long), but not on New Caledonia (400 km long; Diamond 1977). Are

New Britain (470 km long) and New Ireland (340 km long) long enough for geographic differentiation to have at least developed between opposite ends of the island? For New Britain, all that is required is analysis of already collected bird study skins in the Department of Ornithology at the American Museum of Natural History (New York), which houses one large collection from the west end (Mt. Talawe, Cape Gloucester, and Malisonga) and many collections from the east end. For New Ireland, the kingfisher *Halcyon chloris* and the grass-finch *Lonchura* [*spectabilis*] are already known to be differentiated between North and Central New Ireland (p. 168 of Chapter 22), but very few specimens have been collected at the south end.

Improved Analyses of Area and Distance Effects

Our analyses of island area and distance turned out to account for many features of avifaunas that vary among islands. Increasing island area is associated with increasing values of island species number (figs. 9.1–9.3), of the island endemism index (fig. 25.2), of the pairwise differentiation index (fig. 26.2), and (island mean area) of the Ochiai dissimilarity index (chapter 27). Each species is restricted to a certain species-specific range of island areas (chapter 12, figs. 12.5 and 12.6). For example, higher endemics (chapter 24), sedentary species (chapter 33), and isolated populations of non–water-crossing species on land-bridge islands (chapter 33) are concentrated on the largest islands. Archipelagal area affects interarchipelagal faunal dominance (chapter 14).

The number of montane species increases with the proportion of an island's area above 600 m (chapter 9). As for distance, increasing island distances are associated with decreasing values of island species number (figs. 9.1 and 9.4) and with increasing values of island endemism index (fig. 25.2), pairwise differentiation index (fig. 26.2), Ochiai dissimilarity index (chapter 27), and nonnestedness index (chapter 27). Increasing archipelagal isolation is associated with increasing values of archipelagal endemism (chapter 10).

We suggest at least four ways to extend these analyses. First, the relation between montane area and number of montane species has been analyzed for the Solomon Islands but not for the Bismarck Islands (Mayr & Diamond 1976), but six Bismarck islands harbor montane populations (table 9.1): does montane area also predict their number? Second, our distance measure is the usual one in island biogeography: the straight-line distance between the two nearest points of two islands. This is an oversimplified description of distance-dependent dispersal between islands. A more realistic model might consider all locations on two islands and the distances between them, integrated over island areas. Such a model seems likely to yield improved predictions of distance effects, because we already noted (chapter 31) that the distance effect on the pairwise differentiation index is greater for a pair of two long islands pointed end to end at each other (San Cristobal vs. Malaita) than for long island pairs aligned parallel to each other (Malaita vs. Guadalcanal, and the Bukida group vs. the New Georgia group).

Third, we suggested earlier in this chapter how to estimate the effect of distance on the incidences of individual species; incidence of some species is likely to decrease much more steeply with distance than for other species. Finally, species also differ greatly in the effect of area on incidence: some species require much larger islands than do other species (figs. 12.5 and 12.6, or else Figs. 12.3 and 12.4 with the abscissa of species number converted to an area abscissa). Species-specific curves of incidence versus area could be fitted to equations derived from simple models in order to extract the area dependence of extinction and immigration for each species (Diamond & Mayr 1977b, Gilpin & Diamond 1981, Diamond 1984e).

The Time Scale of Speciation

How rapidly do geographic differentiation and speciation proceed in Northern Melanesian birds? To mention some completely arbitrary and purely hypothetical numbers, is the mean duration of isolation required for geographic differentiation to produce an endemic taxon 5,000 years for a subspecies, 50,000 years for a megasubspecies, 500,000 years for an allospecies, and 2,000,000 years for a full species?

At least five approaches are possible to address this time question. First, four or five subspecifically distinct populations have developed on Long and its neighbor Tolokiwa since the volcanic defaunation of Long Island in the seventeenth century (p. 141 of chapter 19, p. 200 of chapter 25). This suggests that at least some endemic subspecies develop very quickly through the founder effect, character displacement, and formation of stabilized hybrid populations.

Second, most populations of non–water-crossing species on Holocene island fragments of Pleistocene expanded islands are not subspecifically distinct. For instance, the seven non–water-crossing species occupying the lowlands of two or more Bukida islands (table 33.1) have 30 island populations that have been analyzed taxonomically. If there had been no subspeciation since Greater Bukida became fragmented by rising sea level around 10,000 years ago, these seven species would constitute only seven taxa; if there had been subspeciation of every one of those species on every analyzed Bukida fragment, there would be 30 taxa. There are actually only 13 taxa, meaning that no more than six distinct subspecies have formed in 10,000 years (it could have been fewer, because some of those subspecies may pre-date the last Ice Age) and that at least three-quarters of the isolates failed to differentiate.

Third, many Northern Melanesian islands are Late Cenozoic volcanoes (e.g., Crown, Feni, Hermit, Kulambangra, Lihir, Long, Rambutyo, Sakar, Savo, Simbo, Tabar, Tanga, Tolokiwa, Umboi, Witu, and Vuatom). Others are uplifted coral (e.g., Bellona, Emirau, Gower, Nissan, Rennell, and Three Sisters). Of these 22 islands, two (Kulambangra and Rennell) have endemic full species, and 10 others have endemic subspecies. Dating the origins of these islands and their termination of volcanism would help fix the ages of their endemic taxa. For islands of a given age and isolation, what is the relation between endemism and island age? Some relation will probably emerge; for instance, we already noted (chapter 30) that the New Ireland satellite of Dyaul, whose volcanism ended in the Miocene, has more than double the endemism of the other four New Ireland satellites, where volcanism continued until more recently.

Fourth and fifth, fossils and molecular clocks offer two more approaches to studying the time scale of speciation.

Distributions with Complex Histories

Throughout this book, we have presented parsimonious interpretations of the available data. For instance, the distributions of 10 bird species summarized in figure 13.1 are most simply interpreted as snapshots of 10 successive stages in the linear spread of New Guinean colonists through the Bismarcks and Solomons. Nevertheless, in any field of science, an interpretation that parsimoniously fits the available data at one time may prove wrong when more data become discovered and require more complex interpretation.

An example from the field of island biogeography is that newer taxonomic studies using molecular markers revealed expansion histories of West Indian yellow warblers and bananaquits more complex than could have been realized before those markers were studied (Klein & Brown 1994, Seutin et al. 1994). We assume that molecular evidence, fossils, and other discoveries will similarly demonstrate that some of our interpretations of Northern Melanesian bird histories are too simple. At least four types of evidence already exist to contradict hypotheses of neat linear spreads of colonists through Northern Melanesia.

First, even if colonization begins with a smooth wave of spread, the resulting distribution may become disrupted by local extinctions on certain islands. The most flagrant such distributional gap in Northern Melanesia is the absence of the monarch flycatcher *Monarcha chrysomela* on the large Bismarck island of New Britain near New Guinea, despite the flycatcher's presence throughout the New Guinea region and on five Bismarck islands beyond New Britain (p. 229 of chapter 30). This suggests the extinction of a former New Britain population: if such a gap were to be filled by a recolonization from New Guinea, one would be deceived into assuming a single wave of expansion when there had actually been two waves.

Second, subspecific relationships show us that such distributional gaps may be recolonized not from an adjacent island but from a more distant island, creating a mosaic distribution that we can sometimes recognize even without molecular markers. Examples include six recognizable colonizations of Umboi and two recognizable colonizations of New Hanover from more distant islands rather than from the nearby islands of New Britain and New Ireland, to which these two islands were respectively joined during the Pleistocene (p. 257 of chapter 32).

Third, we have noted examples in which one superspecies is represented within Northern Melanesia by several populations at different levels of endemism and with different capacities for overwater dispersals, strongly suggesting multiple waves of colonization (chapter 21).

Finally, the affinities of the avifaunas of Manus, St. Matthias, and Biak (chapter 33) warn us that islands similar in species composition may have become so as the result not of one wave of colonization, but of community convergence caused by repeated independent colonizations by the same vagile members of the colonist source pool.

Comparisons and Generalization

One of the main themes of this book is that, even among the 195 bird species of Northern Melanesia, speciation has followed quite different courses for different species. Hence we expect, even more so, that speciation in other taxa and in other parts of the world may follow different courses. Nevertheless, our book's theme is not the nihilistic one that every situation is different and that there can be no generalizations. Instead, we have seen that much variation in speciation can be understood in terms of the different biology of species and the differing geographic configurations of their habitats. We conclude by mentioning some differences that are likely to emerge when our findings are compared with speciation studies on other biotas.

Remote versus Close Archipelagoes

As measured against the dispersal ability of birds, Northern Melanesia is sufficiently close to its major colonization sources (New Guinea, Australia, and eastern archipelagoes) that arrivals of colonists from outside Northern Melanesia are frequent. Hence most cases of completed speciation (18 out of 19 recognized cases) have involved double invasions of Northern Melanesia across water gaps by colonists from outside. In contrast, in remote archipelagoes where colonists rarely arrive from outside, most speciation across water gaps is likely to involve multiple invasions between islands of the remote archipelago, leading to sympatric swarms of closely related species—a phenomenon conspicuously lacking among Northern Melanesian birds (fig. 22.1). The most famous avian examples are the intra-archipelagal invasions that produced the much-studied sympatric swarms of Darwin's finches (Geospizinae) in the Galapagos (Lack 1947, Grant & Grant 1996, 1997, Grant 1999) and of honey-creepers (Drepanididae) in the Hawaiian archipelago (Amadon 1950, Sibley & Ahlquist 1982, James & Olson 1999).

It is no accident that these two famous species swarms of birds occupy the world's two most remote, island-rich tropical archipelagoes. All other tropical archipelagoes of many islands have been too accessible to birds for such radiations to have occurred; available niches have instead been preempted by bird colonists arriving from the outside. In fact, few other archipelagal avifaunas provide any examples of intra-archipelagal speciation in birds; there is a modest number of cases each from Fiji, the Marquesas, the Societies, possibly New Zealand, the Philippines, and the West Indies (see Diamond 1977 for details). But we emphasize that isolation must be measured by the dispersal abilities of the taxa involved. Species less vagile than birds, such as some plants, reptiles, amphibia, and invertebrates, can radiate by intra-archipelagal speciation in archipelagoes much less remote than the Galapagos and Hawaii, because those other less vagile taxa arrive so infrequently. Examples include species swarms of plants on the Canary and Juan Fernandez archipelagoes (Carlquist 1974), and possibly of skinks and platymantine frogs in Northern Melanesia (chapter 4).

Continents versus Islands

When isolates expand and reencounter each other, three outcomes are possible: they may hybridize, they may remain allopatric and not hybridize; or they may become sympatric and not hybridize. The relative likelihoods of these outcomes, and the form of hybridization if it occurs, differ between continents and islands.

What, really, is the difference between a continent and an island? From a biogeographer's perspective, a continent could be defined as a land mass large enough that speciation can proceed to completion within the land mass, because distances within the land mass are large enough to isolate populations through reduction of gene flow by distance alone or by ecological barriers (fig. 22.1). The formation of western and eastern allospecies and subspecies of birds across the Great Plains of North America is a typical example. Speciation in such cases is termed "continental speciation" (Diamond 1977). In contrast, an island is too small for isolates to develop within the island; isolation must instead be across water barriers between islands. By this definition, continental speciation is the rule for birds within the conventionally recognized continents and also within the world's largest "islands" (cf. the radiation of species swarms of birds of paradise, Vangidae, and moas in New Guinea, Madagascar, and New Zealand, respectively).

There is no such example of continental speciation of birds in Northern Melanesia, simply because Northern Melanesian islands are too small as measured against the dispersal abilities of birds. In fact, only on the second longest Northern Melanesian island (New Ireland) is there any documented example of geographic variation of a bird species within the island (chapter 22). Instead, all bird speciation in Northern Melanesia has been across water gaps.

This difference between continental speciation and insular speciation (regardless of whether that insular speciation is intra-archipelagal, as in the Galapagos, or interarchipelagal, as in Northern Melanesia) has at least two important consequences. First, expanding continental isolates are likely to meet in large numbers over a long and broad front (e.g., over much of the Great Plains in North America). If the isolates have not become reproductively isolated, hybridization will occur over that front, creating an entire belt populated by hybrids, with the two parental populations occupying either side of the hybrid belt. One of the best-studied hybrid belts in birds is that between two megasubspecies of crow, the carrion crow *Corvus corone corone* and the hooded crow *C. c. cornix*, across Europe (Meise 1928). Hybrid belts are relatively common in continental birds; for instance, they have been noted in about 6% of North American species (Mayr & Short 1970). They have been described for birds of all the conventionally recognized continents: North America (Mayr & Short 1970), South America (Haffer 1974), the Palaearctic (Vaurie 1959, 1965), the western Palaearctic (Harrison 1982), Africa (Hall & Moreau 1970), and Australia (Keast 1961).

Not a single bird hybrid belt has been discovered on any Northern Melanesian island. This is because interisland water barriers prevent birds from meeting in large numbers over a long, broad front. Instead, when individual colonists from one island arrive on another island with an established population of a sister neo-species, the colonists are greatly outnumbered by the established population, which acts as a genetic sink. The colonists are likely either to be absorbed by hybridization, or to die without leaving offspring due to the absence of appropriate mates, or to succumb to competition from the established population. In chapter 23 we noted hybrid populations, some of them now stabilized and uniform, on some Northern Melanesian islands. Those insular hybrid populations differ from the hybrid belts of continents in two ways: they are not geographically directly adjacent to parental populations but are separated by water barriers from both parents; and the potential exists on islands, but much less so in a continental hybrid belt, for the hybrid population to stabilize into a uniform and disjunct population that happens to have hybrid origins but is otherwise no more variable than any other disjunct island population (e.g., the San Cristobal whistler population *Pachycephala pectoralis christophori* [plate 2], which is of hybrid origins but is not especially variable).

Thus, one consequence of the difference between continental and insular speciation is that hybrid belts are frequent in the former case, nonexistent in the latter case. A second difference between continental and insular speciation is that, when continental isolates that have developed reproductive isolation and ecological differences meet in large numbers on a broad front, sympatry can develop easily. On islands the smaller number of colonists and their dilution by an established host population make the likelihood of developing sympatry much smaller: each isolate is unlikely to be able to invade the other island and is likely to remain confined to its own island. Consider the following tabulations for superspecies, which as usual we take to be sets of closely related taxa with either no or only modest geographic overlap. Northern Melanesian birds include representatives of 57 superspecies having either two or more allospecies within Northern Melanesia (33 cases) or having one allospecies within Northern Melanesia and a different allospecies in 1 of the source areas adjacent to Northern Melanesia (35 cases: New Guinea in 27 cases, Australia or eastern archipelagoes but not New Guinea in eight cases). (Eleven of these 57 superspecies have multiple allospecies within Northern Melanesia and also have a different allospecies in an adjacent source area.) Of these 57 superspecies, only three (the fruit-doves *Ptilinopus* [*rivoli*] *rivoli* and *P.* [*r.*] *solomonensis*, cuckoo-doves *Macropygia* [*ruficeps*] *nigrirostris* and *M.* [*r.*] *mackinlayi*, and parrots *Lorius* [*lory*] *hypoinochrous* and *L.* [*l.*] *albidinucha*) exhibit any sympatry between the allospecies. We can recognize only 16 other cases of sympatric taxa that probably originated as members of a superspecies but that are now too distinct for us to group them in the same superspecies (e.g., the pigeons *Columba vitiensis* and *C. pallidiceps*, the parrots *Charmosyna* [*palmarum*] and *C. placentis*, and the 14 other cases discussed in chapter 21). We do not have comparable figures for the continents, but we suspect that the frequency of completion of speciation there will prove to be higher. This conjecture could be tested by analyzing the above-cited reference works for the birds of North America, South America, the Palaearctic, Africa, and Australia.

In other words, in island environments more often than in continental environments, isolates may remain "stuck" at the stage of sister allospecies having exclusive ranges, without completion of speciation. This may be the outcome even more often in Northern Melanesia than in other insular environments. Because Northern Melanesian islands exhibit such modest interisland variation in rainfall and habitat, isolates of different islands are likely to be ecologically similar and hence to have difficulty in evolving ecological differences sufficient to permit co-existence. Northern Melanesia lacks the xeric habitats encountered along with wet habitats in the West Indies, Hawaii, and the Galapagos.

Temperate Zones versus the Tropics

Tropical bird species are, on average, much more sedentary and much less prone to disperse over water than are bird species of the temperate zones. For instance, 50% of analyzable New Guinean lowland bird species (115 out of 228 analyzable species) have never been recorded dispersing over water. They have never been observed as residents or even as individual vagrants on any satellite island of New Guinea (even on islands only a few kilometers from New Guinea) lacking recent land bridges to New Guinea and accessible only by overwater colonization. The corresponding figure for southern California birds is only 17% (16 out of 93 species; Diamond 1971b). Associated with their sedentariness, tropical birds may have on average smaller geographical ranges than temperate birds.

For this reason we hypothesize that, all other things being equal, speciation can develop over smaller distances in tropical birds than in temperate birds. This is exemplified by the dozens of examples of well-differentiated bird populations on islands separated by only narrow water gaps in Northern Melanesia, such as the allospecies and megasubspecies of white-eyes separated by only 9 km and 3.5 km, respectively, in the New Georgia group (plate 1, map 45). Such profusion of distinctive geographic isolates within such small distances is unprecedented among temperate birds.

Birds versus Other Taxa

Chapters 10–24 were devoted to the differences among Northern Melanesian bird species in aspects of speciation, such as their propensity to geographic differentiation and to forming higher endemics. We were able to interpret many of these species differences in terms of differences in dispersal and abundance. Differentiation and endemism decrease with increasing vagility (tables 11.3 and 19.1, chapters 11, 19, 21), for the reason that dispersing individuals inhibit differentiation of target populations by injecting their own genes into the target gene pool. Differentiation and endemism increase with abundance because the risk of extinction decreases with population size (hence with abundance), and extinction would terminate the process of differentiation.

While bird species vary in abundance and dispersal, other taxa have abundances and dispersal abilities extending the range of values for birds both upward and downward. Among other vertebrates, arthropods, and plants, one can find taxa living at much higher population densities than birds (e.g., nematode worms) as well as at much lower densities (e.g., whales); and one can find taxa much better adapted to overwater dispersal (e.g., ferns) as well as much more poorly adapted to overwater dispersal (e.g., land snails). Hence one might expect taxon differences in speciation for these two reasons of differences in dispersal and abundance, as well as for other reasons.

Let us start with taxon differences within Northern Melanesia. In chapter 4, which compared different vertebrate classes, we noted big differences in endemism and also in radiations of sympatric endemic species. Relative numbers of endemic species in Northern Melanesia decrease in the sequence: frogs > rodents > birds > bats (table 4.5). (Data for lizards and snakes are too incomplete to arrange them in this sequence.) Ability to disperse over water varies in the reverse sequence (table 4.5): it is highest for birds and bats, intermediate for rodents, and lowest for frogs. Hence dispersal differences are in the correct direction to explain differences in differentiation among vertebrate classes. Abundance differences may also contribute; frogs, on average, live at much higher population densities than do rodents, birds, or bats.

These differences in dispersal and abundance among vertebrate classes are also in the correct direction to explain class differences in radiations of endemic sympatric species within Northern Melanesia. There are few such examples in birds: just 19 species pairs in a fauna of 195 species (table 22.1 and chapter 22), with one or two of those examples actually constituting species triplets (the whistlers *Pachycephala pectoralis, P. melanura*, and *P. implicata* on Bougainville and its offshore islets, and possibly the hawks *Accipiter novaehollandiae, A. albogularis*, and *A. imitator* on Bukida). Bats provide even fewer examples of radiations, perhaps only the species pair *Pteralopex atrata* and *P. pulchra* replacing each other altitudinally on Guadalcanal (Flannery 1995). But two radiations account for more than one-third of Northern Melanesia's native rodent species: the three sympatric species of *Uromys* on Guadalcanal and the five species of the endemic genus *Solomys* on Bukida. Radiations of sympatric species reach their peak in frogs, the class that has the highest abundance and lowest overwater dispersal ability, with three radiations of endemic genera on Bukida: four species of *Discodeles*, eight species of *Batrachylodes*, and nine species of *Platymantis*. Favoring radiations of frogs are their high abundance and low dispersal even over land, and their very low dispersal over water, ensuring that few outside invaders establish successfully.

Such differences among vertebrate classes and among phyla are familiar outside Northern Melanesia. New Caledonia, which is even smaller than Pleistocene Bukida, has nevertheless supported radiations of gymnosperms, angiosperms, lizards, and carabid beetles, but not of bats or birds (Carlquist 1974). To the former taxa, New Caledonia is effectively a continent, not an island. The Canary, Madeira, and Juan Fernandez archipelagoes have had radiations of angiosperms but not of birds (Carlquist 1974). Opposite extremes in Pacific biotas are land snails, which have radiated and speciated profusely on many Pacific islands (e.g., 40 species of the genus *Achatinella* on Oahu alone; Carlquist 1970), contrasting with ferns, which have undergone intra-archipelagal speciation only on Hawaii, the most remote Pacific archipelago (Tryon 1971). Just as within Northern Melanesia, these taxon differences fit considerations based on dispersal and abundance: high endemism and radiations characterize abundant, sedentary groups of taxa. One hectare supports far more individual daisies, land snails, and beetles than birds. In addition, land snails are highly sedentary (individuals may remain on the same bush for years), while ferns with their windborne spores are highly adapted to overwater dispersal. To land snails, Oahu (area 1574 km^2) is a continent, but to ferns the entire Pacific is a single archipelago.

Thus, taxon differences in dispersal and abundance not only among birds, but also among groups of plants and animals, contribute to taxon differences in geographic differentiation, endemism, and the relative frequencies of interarchipelagal, intra-archipelagal, and continental speciation.

While dispersal and abundance may be the most important variables affecting species differences in speciation among birds, other biological attributes that do not vary among bird species may also contribute to differences in speciation among other taxa. Genetic and behavioral differences, including the potential for disruptive speciation or for instant speciation, could make sympatric speciation possible in other taxa (Bush 1975, Coyne 1992). For instance, among Mediterranean annual plants, Zohary (1999) notes that syntopic local swarms of closely related species suggestive of sympatric speciation are common in groups where self-pollination is the rule, but not in predominantly cross-pollinated groups, as one might expect from the potential of predominantly self-pollinated plants for sympatric speciation. As another example, there has been much discussion about whether the monophyletic species swarms of cichlid fishes in African lakes arose sympatrically. For Lake Victoria and other large lakes, geographic isolation of populations within the same lake seems readily possible. But Schliewen et al. (1994) studied the cichlids of two small volcanic crater lakes in Cameroon. One lake of only 4.2 km^2 has a monophyletic swarm of 11 endemic cichlid species; an even smaller lake of only 0.6 km^2 has a swarm of nine species. These lakes are too small, and too deficient in geographic barriers, to make geographic isolation possible. Evidently, these species flocks are the result of sympatric speciation. This is apparently made possible by females having a simultaneous preference for mating in a specific niche and for the special characteristics of certain males that have the same niche preference. Such sympatric speciation is unknown in birds.

Thus, as the most important promising direction for future research, we encourage studies of geographic variation and of superspecies, similar to those that we have reported for Northern Melanesian birds, but for frogs, lizards, arthropods, plants, and other taxa. We hope that the Northern Melanesian avifauna will serve as a fruitful example and comparison.

Summary

We conclude our book by calling attention to five promising sets of directions for future research. First are directions that involve obtaining new data; second, improved analytical procedures to measure effects of island area, distance, and elevation on bird distributions and differentiation. The third set comprises at least five methods to estimate the absolute time scale of speciation. Fourth are several warning signs of bird distributions that may actually have arisen from more complex histories (involving local extinctions, mosaic colonization, multiple waves of colonization, and community convergence) than the most parsimonious histories inferred from currently available data. Finally, comparisons of our study of Northern Melanesian birds with studies of other biotas are needed to formulate generalizations about speciation. Available evidence suggests differences in speciation between remote and close archipelagoes, between continents and islands, between the temperate zones and the tropics, and between birds, other vertebrate classes, and various groups of invertebrates and plants.

MAPS

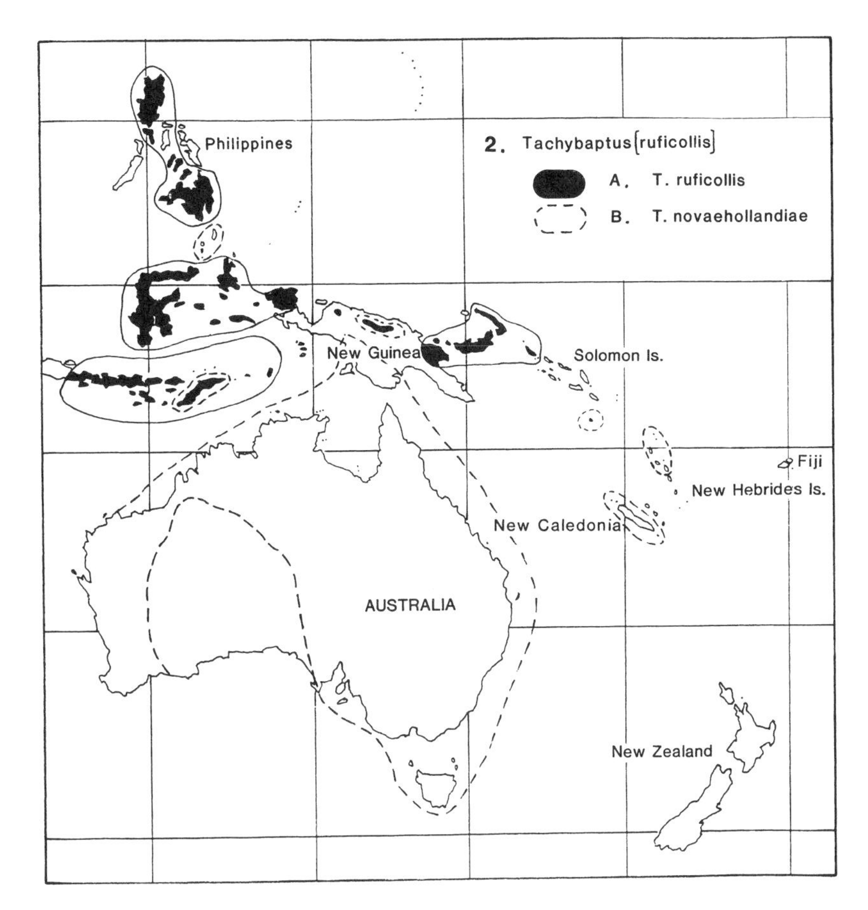

Map 1. 2 = grebe superspecies *Tachybaptus* [*ruficollis*]. A = *T. ruficollis*; B = *T. novaehollandiae*. *T. ruficollis*, the Eurasian allospecies, invaded Northern Melanesia from New Guinea to occupy the Bismarcks and Northwest Solomons. *T. novaehollandiae*, the Australasian allospecies, invaded from the southeast to occupy only Rennell, the southeasternmost Northern Melanesian island. The two allospecies co-occur only in North New Guinea and perhaps Java and Timor.

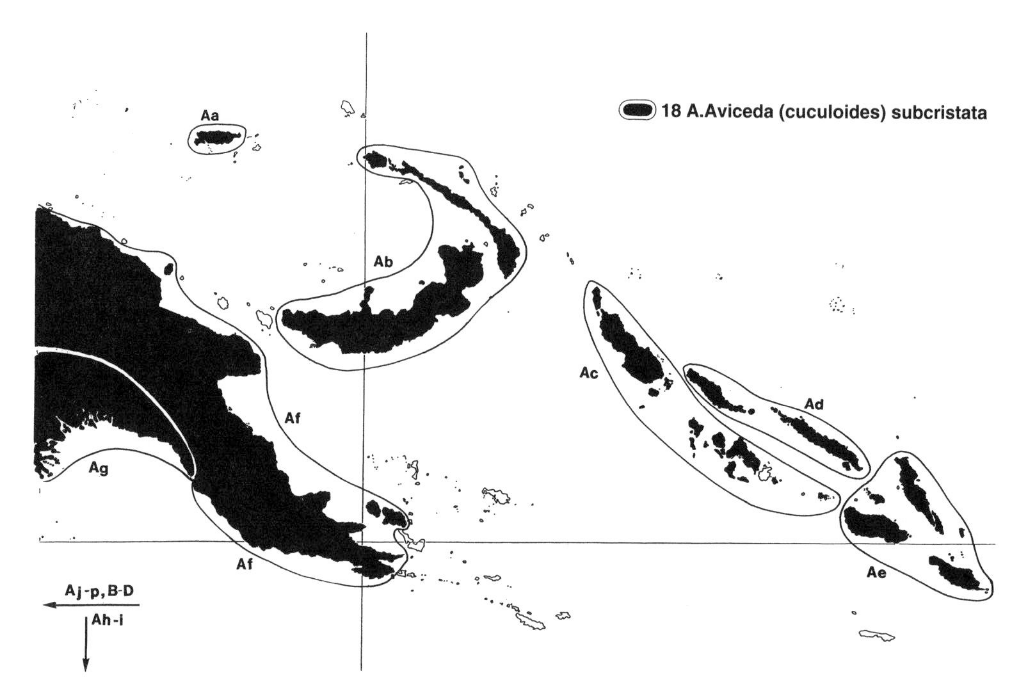

Map 2. 18 = hawk superspecies *Aviceda* [*cuculoides*]. A = *A. subcristata* (Aa = *A. s. coultasi*, Ab = *A. s. bismarckii*, Ac = *A. s. proxima*, Ad = *A. s. robusta*, Ae = *A. s. gurneyi*, Af = *A. s. megala*, Ag = *A. s. stenozona*, Ah–Ap = other subspecies); B = *A. jerdoni*; C = *A. madagascariensis*; D = *A. cuculoides*. The Australian allospecies *A. subcristata* (18A) invaded Northern Melanesia from New Guinea, occupied all of the larger islands except the outliers St. Matthias and Rennell, and differentiated into five weak subspecies. Other subspecies live in Australia and Wallacea. The three other allospecies (18B, C, D) live in Asia, Africa, and Madagascar.

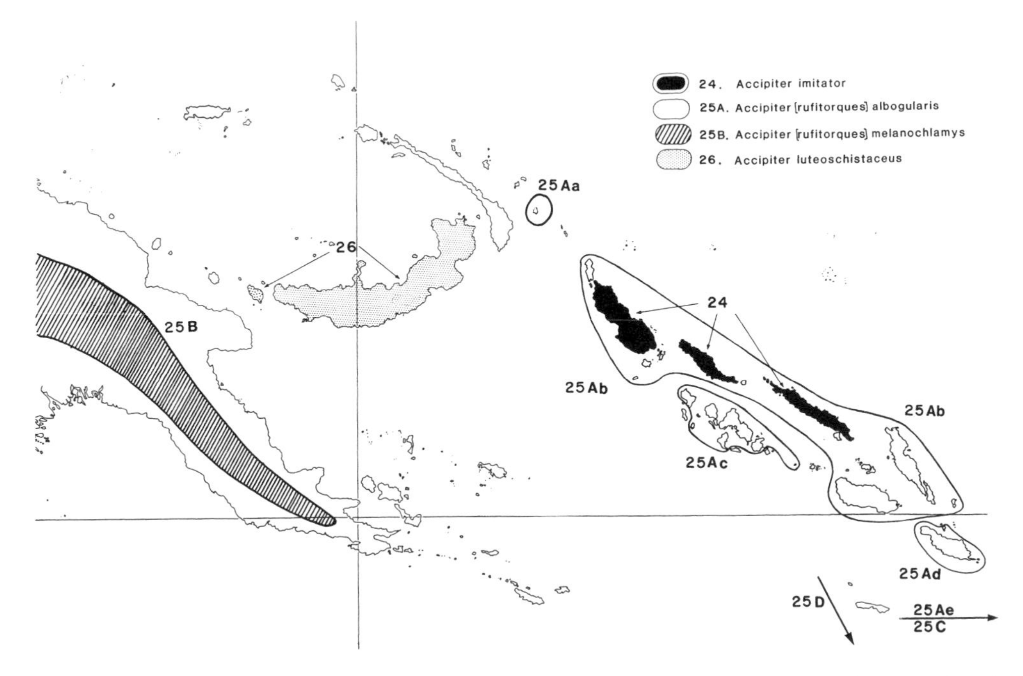

Map 3. 24–26 = three goshawks, the superspecies *Accipiter* [*rufitorques*] and two relatives. 24 = *A. imitator*. 25A = *A.* [*rufitorques*] *albogularis* (Aa = *A. albogularis eichhorni*, Ab = *A. a. woodfordi*, Ac = *A. a. gilvus*, Ad = *A. a. albogularis*, Ae = *A. a. sharpei*); 25B = *A. melanochlamys*; 25C = *A. rufitorques*; 25D = *A. haplochrous*. 26 = *A. luteoschistaceus*. The *A.* [*rufitorques*] superspecies (species 25) ranges as four allospecies from New Guinea (25B) to New Caledonia (25D) and Fiji (25C). Its Solomon allospecies, *A. albogularis* (25A), varies geographically in the Solomons in a pattern shared with many other species: one race (25Ad) in the San Cristobal group, another race (25Ac) in the New Georgia group, and a third race (25Ab) in the Bukida group and adjacent islands. *A. albogularis* also extends to Feni (25Aa), the Bismarck island nearest the Solomons. Possible relatives are two sedentary and monotypic goshawk species: *A. imitator* (24) on the three largest post-Pleistocene fragments of Pleistocene Greater Bukida, and *A. luteoschistaceus* (26) on the two largest fragments of Pleistocene Greater New Britain. See p. 174 of chapter 22 for discussion.

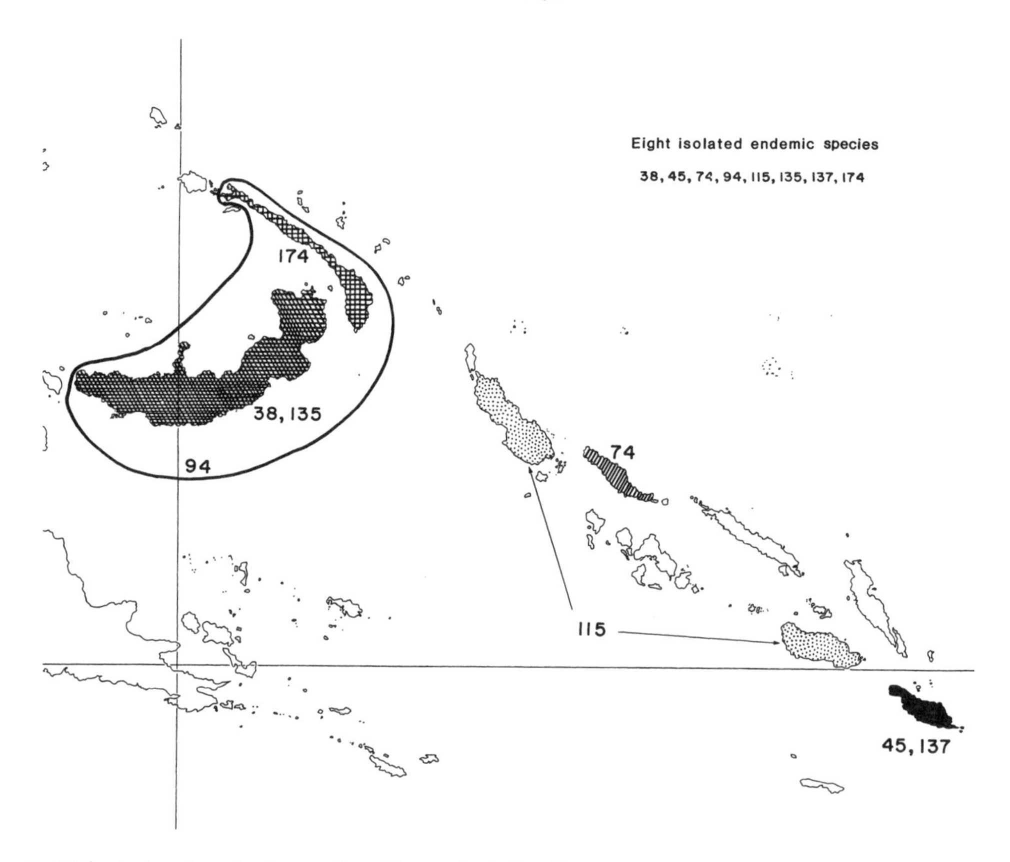

Map 4. Eight isolated endemic species. 38 = rail *Gallirallus insignis* (plate 8). 45 = gallinule *Pareudiastes sylvestris* (plate 8). 74 = ground pigeon *Microgoura meeki* (plate 8). 94 = cuckoo *Centropus violaceus* (plate 8). 115 = kingfisher *Halcyon bougainvillei* (plate 9). 135 = thicket warbler *Ortygocichla rubiginosa*. 137 = warbler *Cettia parens*. 174 =honey-eater *Myzomela pulchella* (plate 4). These sedentary, old endemics, for none of which except *Centropus violaceus* is there any evidence of overwater dispersal, are confined to the largest Northern Melanesian islands.

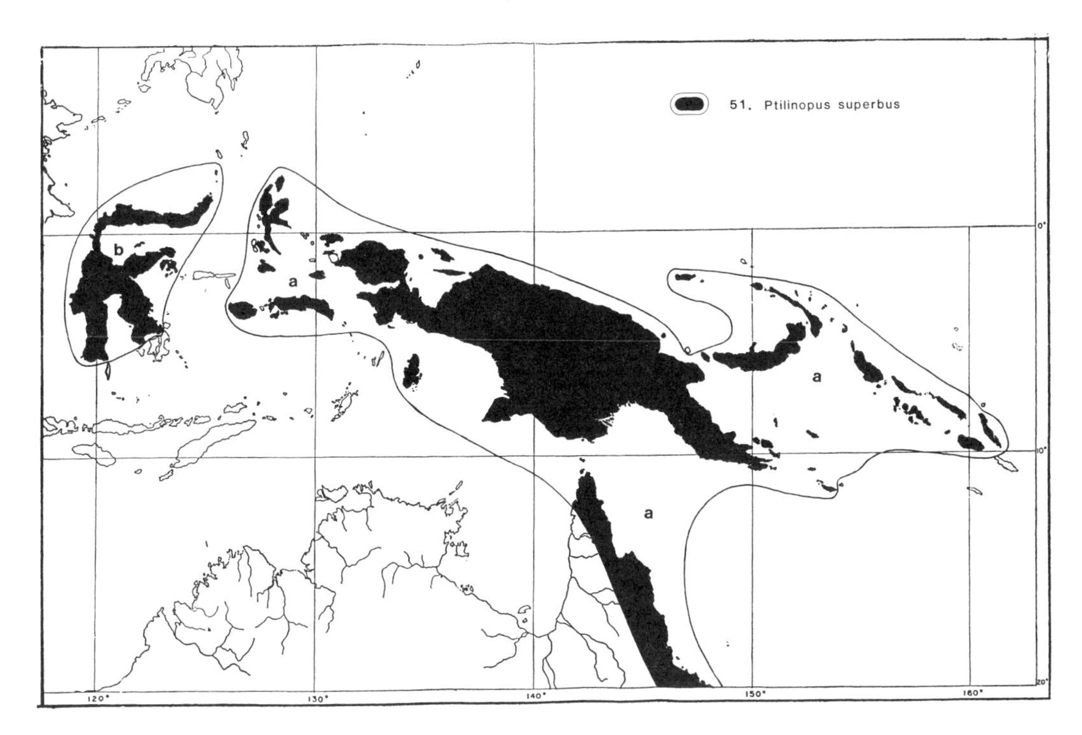

Map 5. 51 = fruit-dove *Ptilinopus superbus* (a = *P. s. superbus*, b = *P. s. temminckii*). This vagile pigeon occupies virtually all of Northern Melanesia (except for the peripheral San Cristobal group and St. Matthias group and other outliers) and also occupies all of the New Guinea region and Moluccas and eastern Australia, without geographic variation. Only the peripheral Celebes population (51b) has differentiated subspecifically.

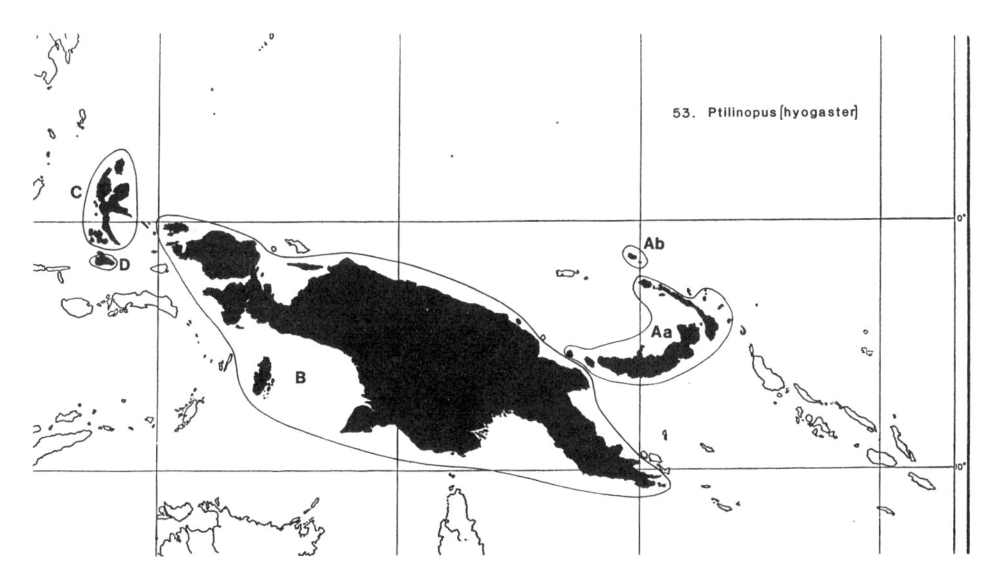

Map 6. 53 = fruit-dove superspecies *Ptilinopus* [*hyogaster*]. A = P. *insolitus* (Aa = *P. i. insolitus*, Ab = *P. i. inferior*); B = P. *iozonus*; C = *P. hyogaster*; D = *P. granulifrons*. The superspecies consists of four allospecies, of which the distinctive Bismarck allospecies (see plate 9) has differentiated subspecifically only on the peripheral St. Matthias group (53Ab) and has not reached the Northwest Bismarcks.

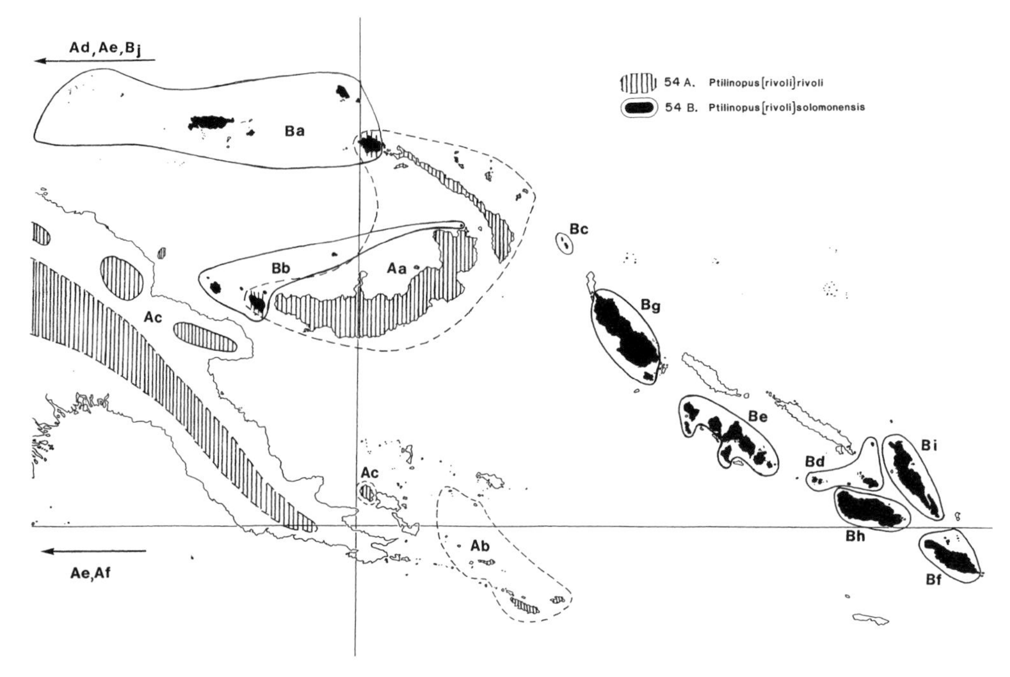

Map 7. 54 = fruit-dove superspecies *Ptilinopus* [*rivoli*] (see plate 6). A = *P.* [*rivoli*] *rivoli* (Aa = *P. r. rivoli*, Ab = *P. r. strophium*, Ac = *P. r. bellus*, Ad = *P. r. miquelii*, Ae = *P. r. prasinorrhous*, Af = *P. r. buruanus*); B = *P.* [*rivoli*] *solomonensis* (Ba = *P. s. johannis*, Bb = *P. s. meyeri*, Bc = *P. s. neumanni*, Bd = *P. s.* subsp.?, Be = *P. s. vulcanorum*, Bf = *P. s. solomonensis*, Bg = *P. s. bistictus*, Bh = *P. s. ocularis*, Bi = *P. s. ambiguus*, Bj = *P. s. speciosus*). This superspecies is among Northern Melanesia's hypervariable "great speciators" (see chapter 19). The ranges of the two allospecies *P. rivoli* (54A) and *P. solomonensis* (54B) interdigitate complexly in the Bismarcks, where the former (54Aa) mainly occupies the larger central islands and the latter (54Bb, 54Bc) occupies smaller or remote or volcanically disturbed islands. Resident populations of both species share only one or two islands, but vagrants of *P. solomonensis* may reach islands occupied by *P. rivoli*. Probably *P. solomonensis* differentiated as the Solomon allospecies of the superspecies, then spread westward as a supertramp into the Bismarck (Aa) and Geelvink Bay (Ad, Ae, Bj) island range of *P. rivoli* but remains excluded from most islands occupied by *P. rivoli*. See p. 175 of chapter 22 for discussion of this case of speciation. *P. solomonensis* has montane subspecies (Bg and Bh), constituting a distinct megasubspecies (see plate 6), on the two largest and highest Solomon islands (p. 148 of chapter 20).

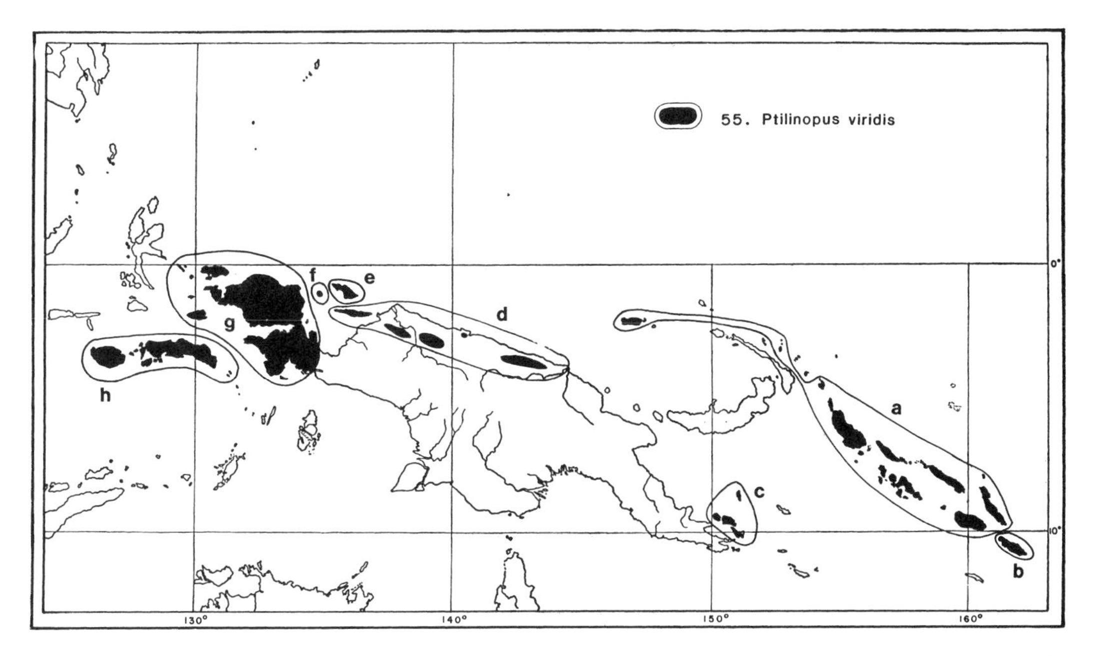

Map 8. 55 = fruit-dove *Ptilinopus viridis* (a = *P. v. lewisi*, b = *P. v. eugeniae*, c = *P. v. vicinus*, d = *P. v. salvadorii*, e = *P. v. geelvinkianus*, f = *P. v. pseudogeelvinkianus*, g = *P. v. pectoralis*, h = *P. v. viridis*). The most distinctive race of this species is the white-headed, peripheral isolate *P.v. eugeniae* (55b) of the San Cristobal group, sometimes considered a separate allospecies. All other Northern Melanesian populations belong to a single subspecies (55a). The species may have colonized the Solomons directly from New Guinea. It has spread west from the Solomons only to the nearest Bismarck islands east of New Ireland, plus (within the last few decades) Manus (which it may alternatively have colonized directly from New Guinea). See p. 96 of chapter 13 and p. 147 of chapter 20 for discussion.

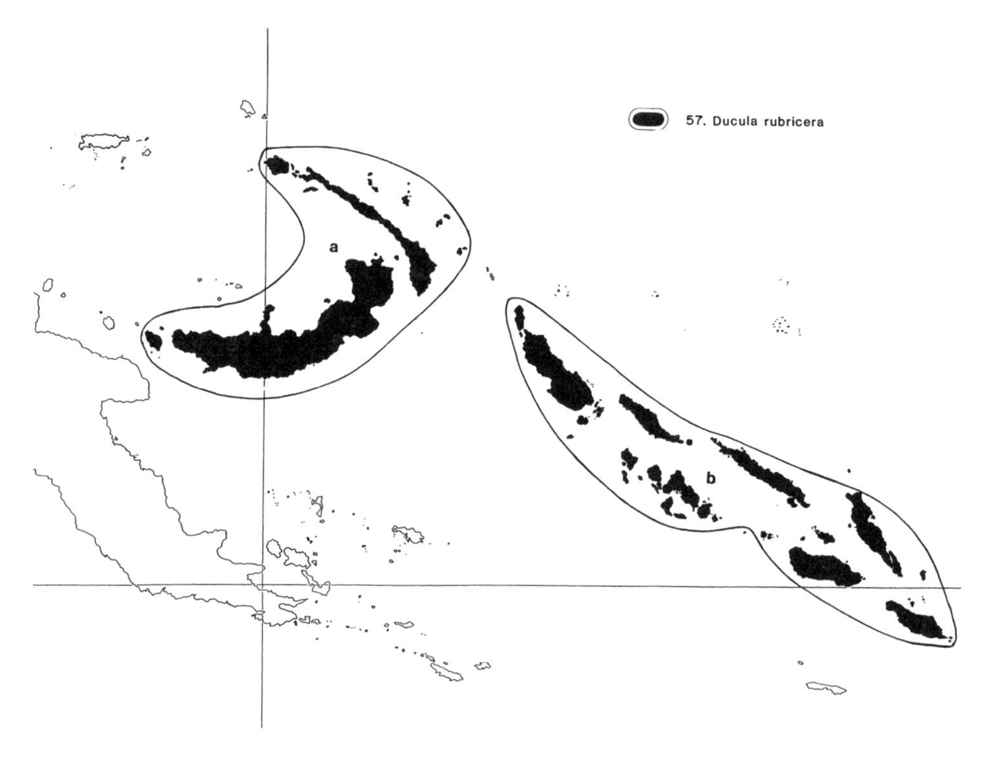

Map 9. 57 = fruit-pigeon *Ducula rubricera* (a = *D. r. rubricera*, b = *D. r. rufigula*) (see plate 8). The Bismarcks and Solomons are occupied by distinct megasubspecies, but there is no further subspeciation of this vagile pigeon within either archipelago. It is absent from all outlying islands, such as Rennell, St. Matthias, and the Northwest Bismarcks.

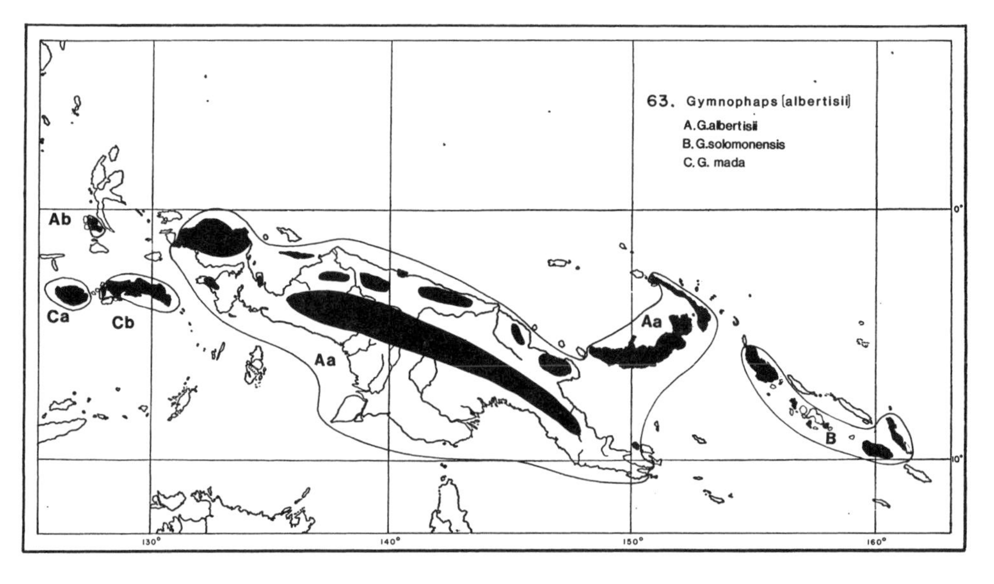

Map 10. 63 = pigeon superspecies *Gymnophaps* [*albertisii*]. A = *G. albertisii* (Aa = G. *a. albertisii*, Ab = *G. a. exsul*); B = *G. solomonensis*; C = *G. mada* (Ca = *G. m. mada*, Cb = *G. m. stalkeri*). This vagile montane species occupies many high mountain ranges of New Guinea and many islands, with very little subspecific variation. One subspecies occupies New Guinea and the two highest Bismarck islands (63Aa), while a separate monotypic allospecies occupies six of the seven highest Solomon islands (63B).

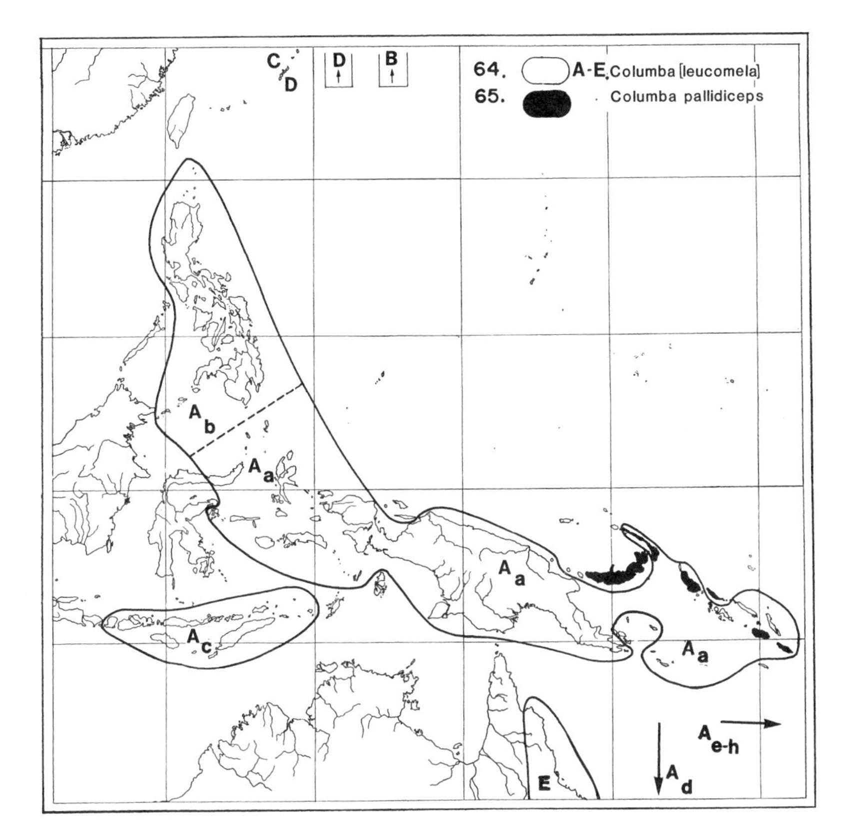

Map 11. 64 = pigeon superspecies *Columba* [*leucomela*]. A = *C. vitiensis* (Aa = *C. v. halmaheira*, Ab = *C. v. griseogularis*, Ac = *C. v. metallica*, Ad = *C. v. godmanae*, Ae = *C. v. leopoldi*, Af = *C. v. hypoenochroa*, Ag = *C. v. vitiensis* , Ah = *C. v. castaneiceps*); B = *C. versicolor*; C = *C. jouyi*; D = *C. janthina*; E = *C. leucomela*. 65 = *Columba pallidiceps* (see plate 8). The superspecies *C.* [*leucomela*] has a wide range from Japan to Samoa, with most of this range occupied by the arboreal allospecies *C. vitiensis* (64A). Its terrestrial relative *C. pallidiceps* (65) is endemic to Northern Melanesia and may have originated as the Northern Melanesian allospecies, whose range was subsequently overrun by *C. vitiensis* expanding from the west. See p. 175 of chapter 22.

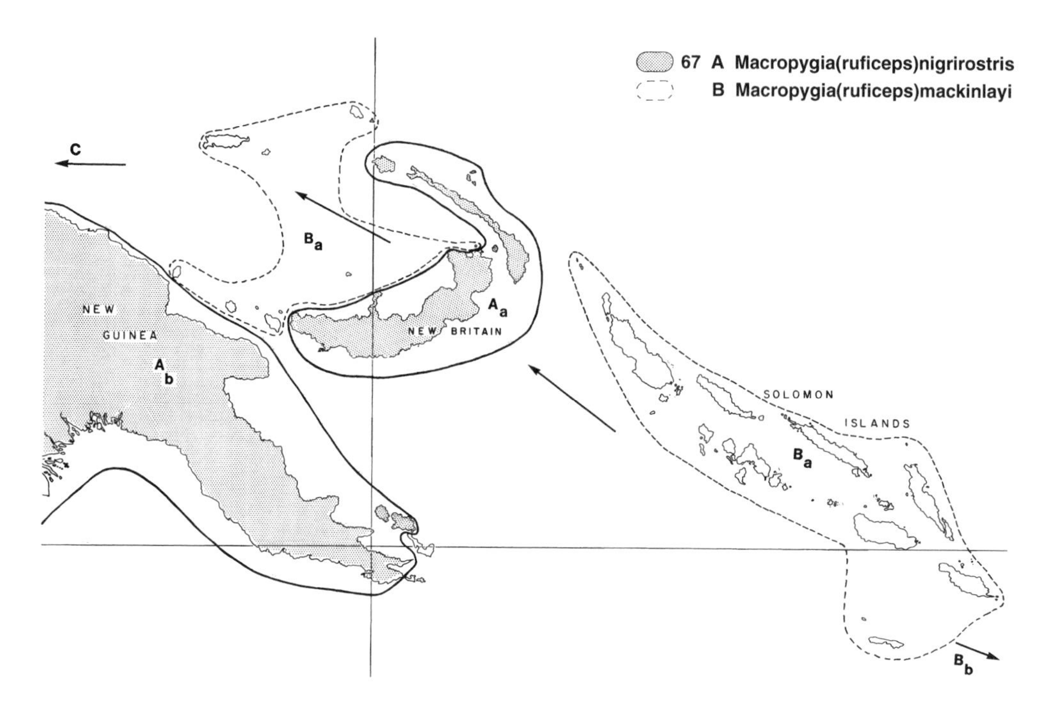

Map 12. 67 = cuckoo-dove superspecies *Macropygia* [*ruficeps*]. A = *M.* [*ruficeps*] *nigrirostris* (Aa = *M. n. major*, Ab = *M. n. nigrirostris*); B = *M.* [*ruficeps*] *mackinlayi* (Ba = *M. m. arossi*, Bb = *M. m. mackinlayi*); C = *M.* [*ruficeps*] *ruficeps*. The two allospecies *M. nigrirostris* (67A) and *M. mackinlayi* (67B) interdigitate in the Bismarcks, where the former mainly occupies large, high, central islands while the latter occupies small or remote or volcanically disturbed islands. On four islands one of the species is resident while the other has been recorded as a vagrant. In the Solomons, where *M. mackinlayi* occurs alone without *M. nigrirostris*, it occupies large as well as small islands. Probably *M. mackinlayi* arose as the superspecies' Solomon allospecies and expanded westward into the Bismarcks and to Karkar Island off New Guinea's north coast as a supertramp, excluded from islands occupied by *M. nigrirostris*. See p. 163 of chapter 22 for discussion.

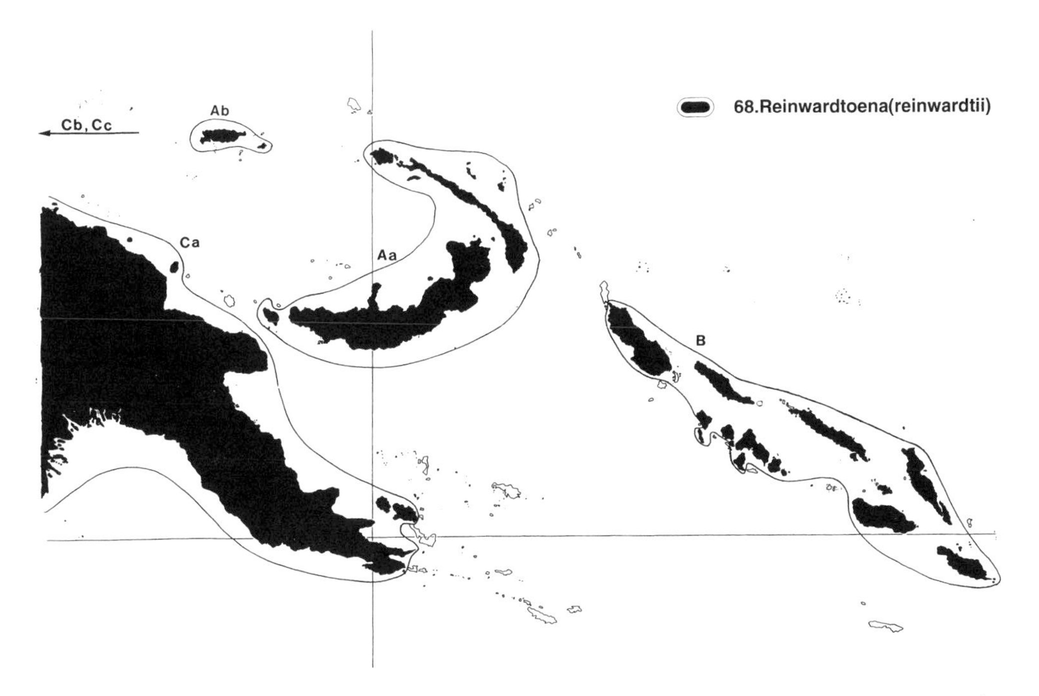

Map 13. 68 = giant cuckoo-dove superspecies *Reinwardtoena* [*reinwardtii*] (plate 7). A = *R. browni* (Aa = *R. b. browni*, Ab = *R. b. solitaria*); B = *R. crassirostris*; C = *R. reinwardtii* (Ca = *R. r. griseotincta*, Cb = *R. r. brevis*, Cc = *R. r. reinwardtii*). The Bismarcks, Solomons, and New Guinea each support a different allospecies (68A, B, and C, respectively). Within the superspecies, the most distinctive allospecies is the peripheral isolate of the Solomons, *R. crassirostris* (68B; see plate 7). Within the Bismarck allospecies, the only distinctive population is the peripheral isolate of the Admiralties (68Ab). The superspecies is absent from the outliers Rennell and St. Matthias.

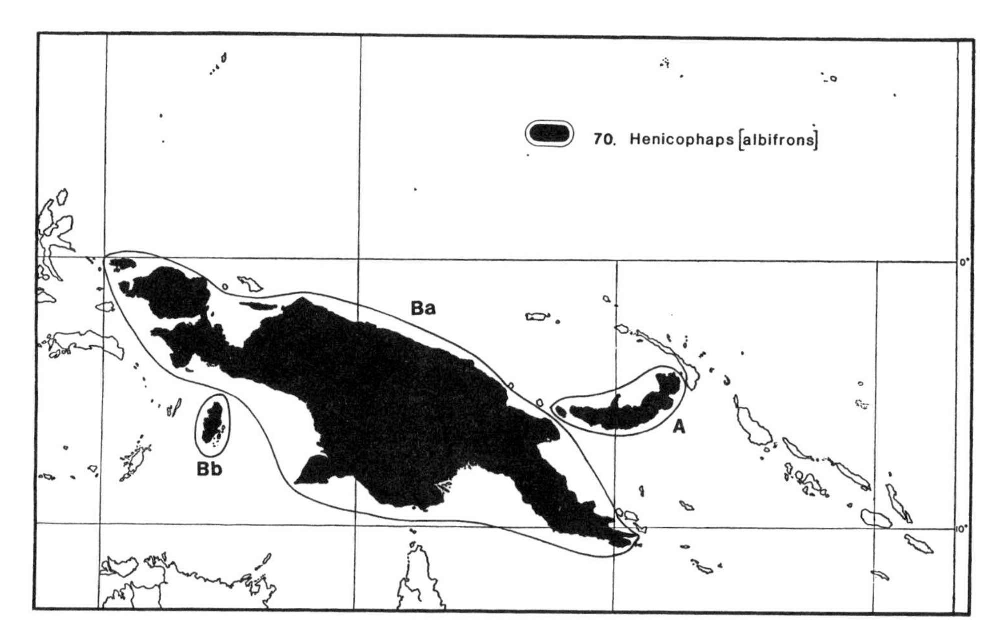

Map 14. 70 = ground-pigeon superspecies *Henicophaps* [*albifrons*]. A = *H. foersteri*; B = *H. albifrons* (Ba = *H. a. albifrons*, Bb = *H. a. schlegeli*). The sedentary Bismarck allospecies *H. foersteri* (70A) is confined to post-Pleistocene island fragments of Pleistocene Greater New Britain, while the sedentary New Guinea allospecies *H. albifrons* (70Ba, 70Bb) is similarly confined to post-Pleistocene island fragments of Pleistocene New Guinea.

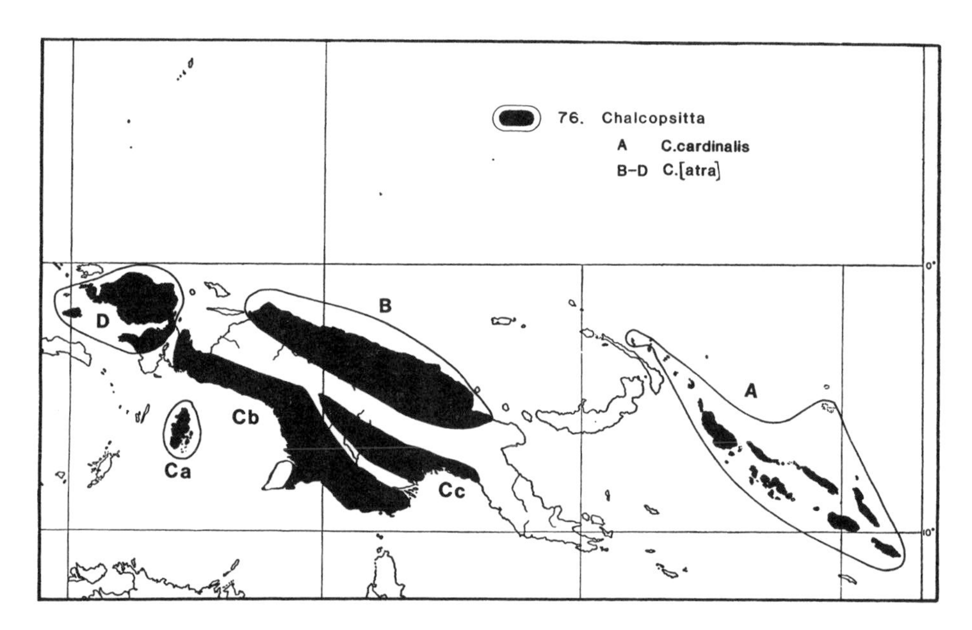

Map 15. 76 = lories of genus *Chalcopsitta*. A = *C. cardinalis* (see plate 8); B = *C.* [*atra*] *duivenbodei*; C = *C.* [*atra*] *scintillata* (Ca = *C. s. rubrifrons*, Cb = *C. s. scintillata*, Cc = *C. s. chloroptera*); D = *C.* [*atra*] *atra*. *C. cardinalis* (76A), one of the most vagile Northern Melanesian endemics, occupies virtually all central Solomon islands plus adjacent outlying Bismarck islands of the New Ireland group. It can be seen flying over water any day in the Solomons. In contrast, the related *C.* [*atra*] superspecies of New Guinea (76B–D) never flies overwater and is confined to land-bridge islands that had Pleistocene connections to New Guinea.

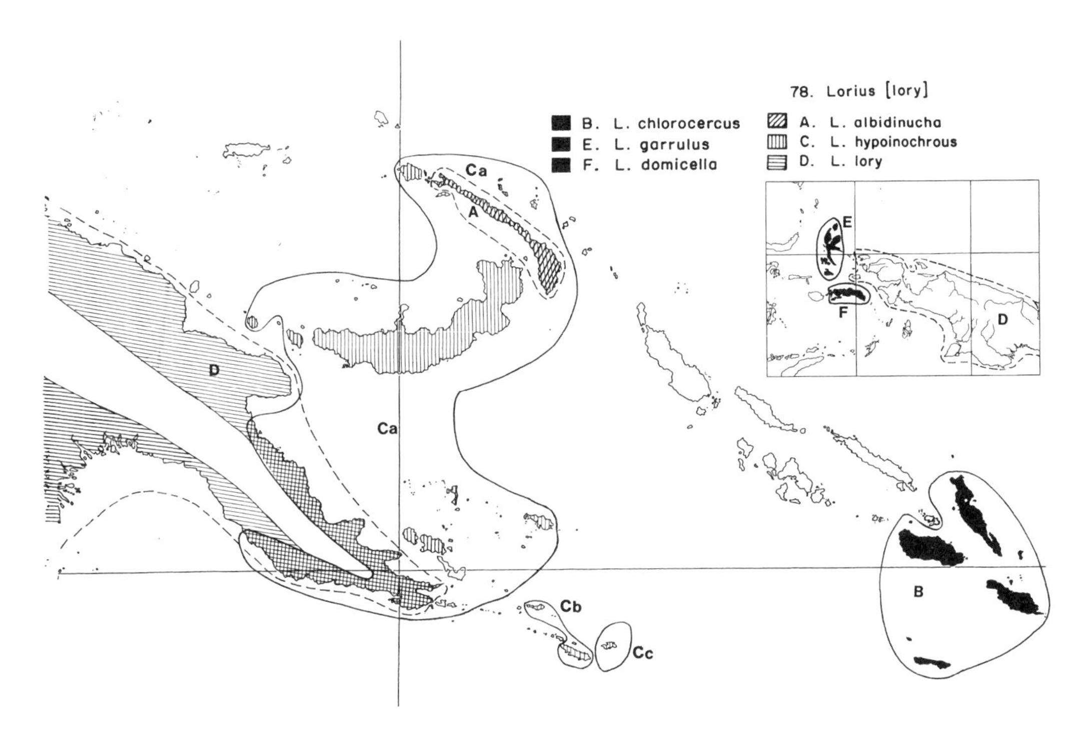

Map 16. 78 = *Lorius* [*lory*] lory superspecies (see plate 6). A = *L.* [*lory*] *albidinucha*; B = *L.* [*lory*] *chlorocercus*; C = *L.* [*lory*] *hypoinochrous* (Ca = *L. h. devittatus*, Cb = *L. h. hypoinochrous*, Cc = *L. h. rosselianus*); D = *L.* [*lory*] *lory*; E = *L.* [*lory*] *garrulus*; F = *L.* [*lory*] *domicella*. The six allospecies are allopatric, except that *L. hypoinochrous* (78Ca) shares its range in Southeast New Guinea with *L. lory* (78D) and that *L. albidinucha* (78A), endemic to New Ireland, shares that island with *L. hypoinochrous* (78Ca), where the two taxa segregate by altitude, body size, and bill size. Thus, this superspecies illustrates stages in completion of speciation. See p. 159 of chapter 21 and p. 165 of chapter 22 for discussion. The allospecies 78E and 78F occur in the Moluccas, to the west of the islands depicted on this map.

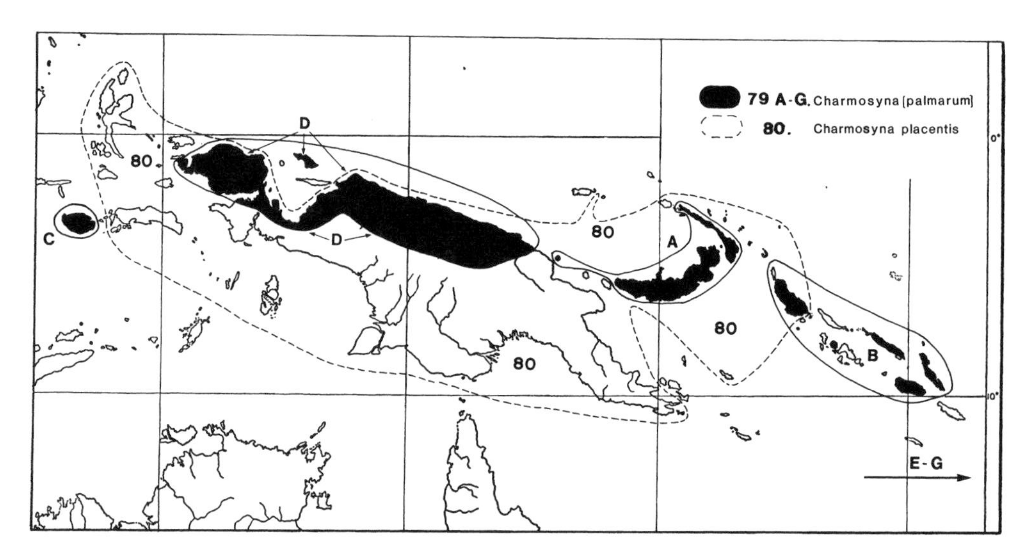

Map 17. 79 = lorikeet superspecies *Charmosyna* [*palmarum*]. 79A = *C. rubrigularis*; 79B = *C. meeki*; 79C = *C. toxopei*; 79D = *C. rubronotata*; 79E = *C. palmarum*; 79F = *C. diadema*; 79G = *C. amabilis*. 80 = *C. placentis*. The *C.* [*palmarum*] superspecies consists of seven allospecies (79A–G) that are commonest in montane forest and distributed from Buru in the west (79C) to Fiji in the east (79G). One of the Northern Melanesian allospecies (79A) is confined to the two highest Bismarck islands, the other (79B) to the six highest Solomon islands. *C. placentis* (80) is related to this superspecies, especially closely to the allospecies *C. rubronotata* (79D), but lives in the lowlands and occurs sympatrically with three or four allospecies (79A, B, D, and possibly C) of the *C.* [*palmarum*] superspecies. *C. placentis* may have originated as the South New Guinea representative of the superspecies, in which case the relations of species 79 and 80 represent an intermediate stage of speciation. *C. placentis* is also at an intermediate stage of expansion from New Guinea over Northern Melanesia, having occupied most of the Bismarcks but only the northwestmost Solomon island (Bougainville). See p. 97 of chapter 13, figure 13.1, and p. 164 of chapter 22 for discussion.

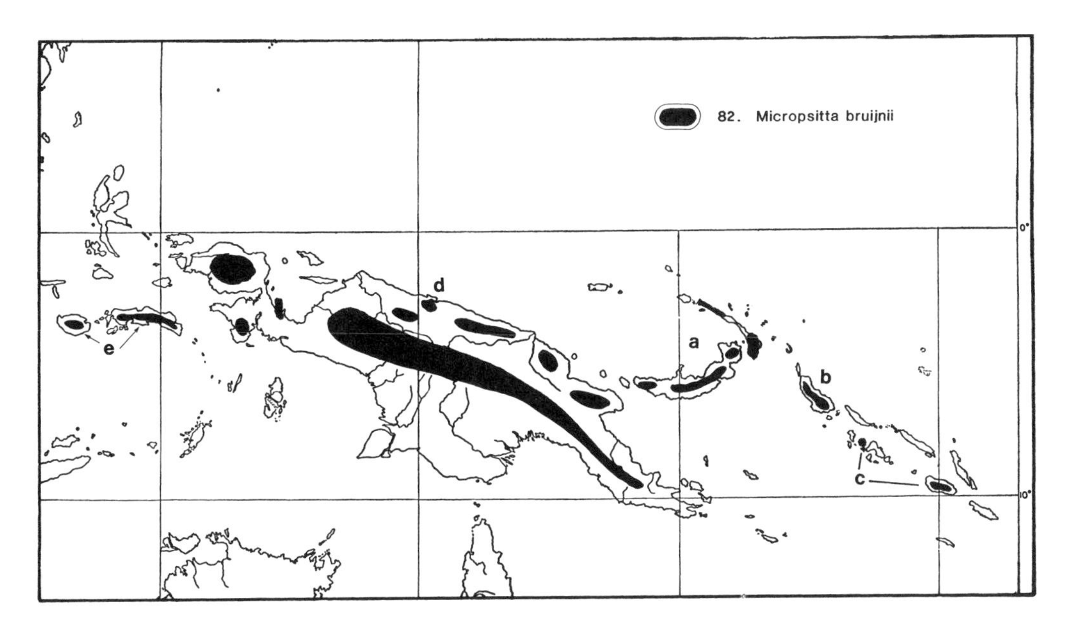

Map 18. 82 = pygmy parrot *Micropsitta bruijnii* (a = *M. b. necopinata*, b = *M. b. brevis*, c = *M. b. rosea*, d = *M. b. bruijnii*, e = *M. b. pileata*). This parrot of montane forest is patchily distributed on high, isolated mountain ranges from the Moluccas in the west (82e), through New Guinea (82d), to Northern Melanesia (82a, b, c). In Northern Melanesia it occupies all five islands exceeding 1700 m in elevation, but no other island. Its five subspecies, of which three are endemic to Northern Melanesia, are only weakly distinct, perhaps because this is one of the more vagile montane species. *M. bruijnii* shares almost its whole geographical range with the *M.* [*pusio*] superspecies (species 83, compare Map 19), from which it segregates altitudinally: *M.* [*pusio*] in the lowlands, *M. bruijnii* in the mountains. Thus, these two species represent a late stage in completed speciation.

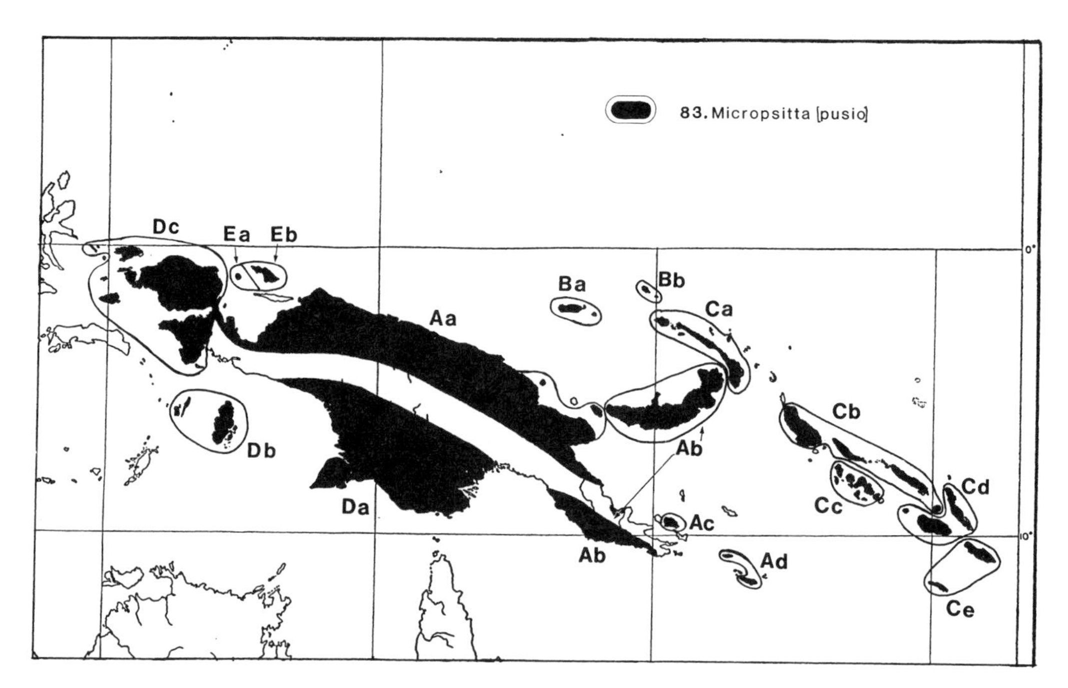

Map 19. 83 = pygmy parrot superspecies *Micropsitta* [*pusio*]. A = *M.* [*pusio*] *pusio* (Aa = *M. p. beccarii*, Ab = *M. p. pusio*, Ac = *M. p. harterti*, Ad = *M. p. stresemanni*); B = *M.* [*pusio*] *meeki* (Ba = *M. m. meeki*, Bb = *M. m. proxima*); C = *M.* [*pusio*] *finschii* (Ca = *M. f. viridifrons*, Cb = *M. f. nanina*, Cc = *M. f. tristrami*, Cd = *M. f. aolae*, Ce = *M. f. finschii*); D = *M.* [*pusio*] *keiensis* (Da = *M. k. viridipectus*, Db = *M. k. keiensis*, Dc = *M. k. chloroxantha*); E = *M.* [*pusio*] *geelvinkiana* (Ea = *M. g. geelvinkiana*, Eb = *M. g. misoriensis*). This superspecies is among Northern Melanesia's "great speciators" (chapter 19), occupying almost all major islands but with especially marked geographic variation. Some features of its variation are shared with many other species. For example, populations of the New Ireland group (83Ca) are distinct from the New Britain Group (83Ab); subspecifically distinct populations on the Bismarck outliers Manus (83Ba) and St. Matthias (83Bb) are more closely related to each other than either is to the New Ireland population, which belongs to a different allospecies (83Ca); and separate subspecies exist in the Solomons on the Bukida group (83Cb), New Georgia group (83Cc), San Cristobal group (83Ce), and Guadalcanal and Malaita (83Cd). Unusual features are that, despite all this differentiation, the New Britain population is not even subspecifically distinct from that of Southeast New Guinea (83Ab); the Rennell population is not subspecifically distinct from that of San Cristobal (83Ce); and the New Ireland population (83Ca) belongs to the same allospecies as the Solomon populations (83Cb–Ce), rather than to the same allospecies as the New Britain population (83Ab). See legend to map 18 for relations with *M. bruijnii*.

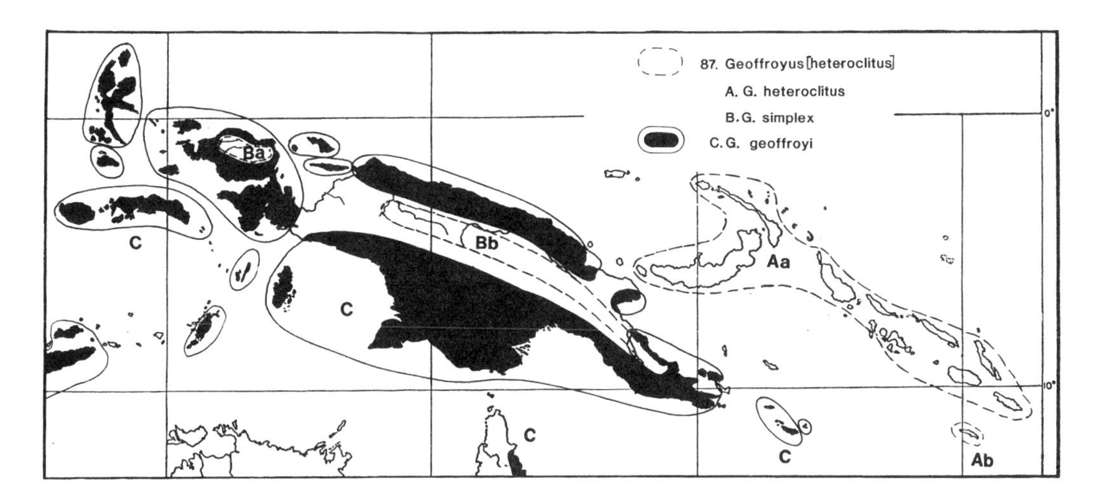

Map 20. 87 = parrot superspecies *Geoffroyus* [*heteroclitus*] (plate 9). A = *G.* [*heteroclitus*] *heteroclitus* (Aa = *G. h. heteroclitus*, Ab = *G. h. hyacinthus*); B = *G.* [*heteroclitus*] *simplex* (Ba = *G. s. simplex*, Bb = *G. s. buergersi*). C = *G. geoffroyi*. The sole local race of *G. heteroclitus* is the distinctive peripheral isolate *G. h. hyacinthus* (87Ab) on Rennell. *G. heteroclitus* appears to be derived from *G. simplex* (87B) of the New Guinea mountains rather than from *G. geoffroyus* (C) of the New Guinea lowlands.

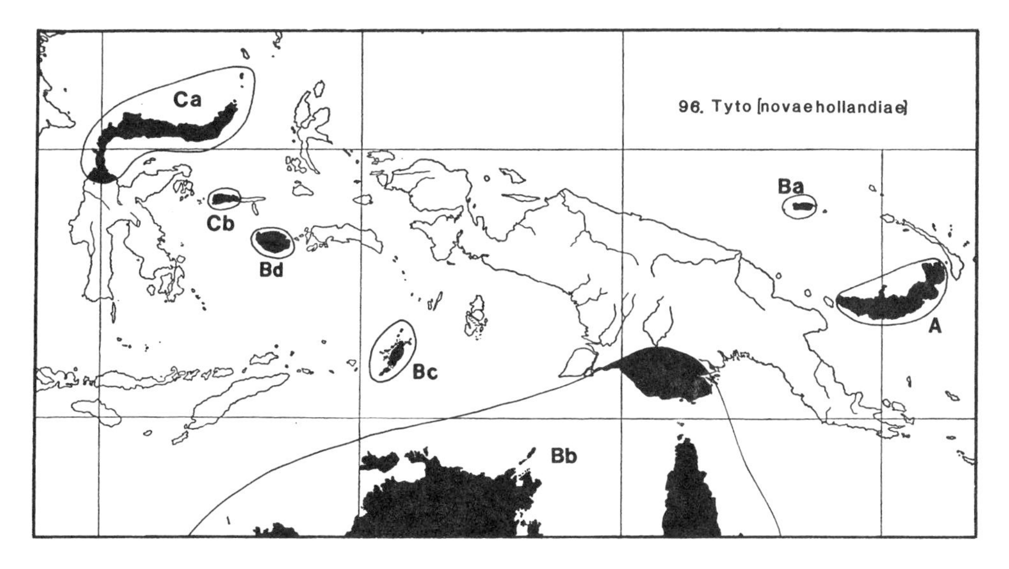

Map 21. 96 = barn owl superspecies *Tyto* [*novaehollandiae*]. A = *T. aurantia*; B = *T. novaehollandiae* (Ba = *T. n. manusi*, Bb = *T. n. kimberli*, Bc = *T. n. sorocula*, Bd = *T. n. cayelii*); C = *T. inexspectata* (Ca = *T. i. inexspectata*, Cb = *T. i. nigrobrunnea*). The New Britain allospecies *T. aurantia* (96A) is very distinct.

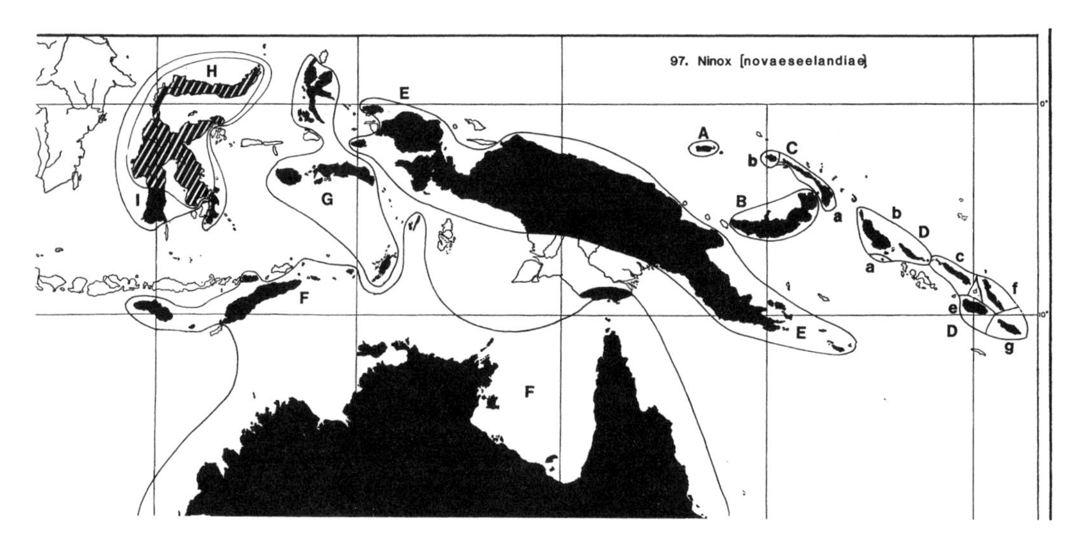

Map 22. 97 = owl superspecies *Ninox* [*novaeseelandiae*] (plate 7). A = *N. meeki*; B = *N. odiosa*; C = *N. variegata* (Ca = *N. v. variegata*, Cb = *N. v. superior*); D = *N. jacquinoti* (Da = *N. j. mono*, Db = *N. j. eichhorni*, Dc = *N. j. jacquinoti*, Dd = *N. j. floridae*, De = *N. j. granti*, Df = *N. j. malaitae*, Dg = *N. j. roseoaxillaris*); E = *N. theomacha*; F = *N. novaeseelandiae*; G = *N. squamipila*; H = *N. ochracea*; I = *N. punctulata*. This owl is another of Northern Melanesia's "great speciators" (chapter 19) with marked geographic variation. Its pattern of variation among large islands is shared with many other species: endemic allospecies on New Britain (97B), New Ireland (97C), Manus (97A), and the Solomons (97D); and, within the Solomons, different megasubspecies on Greater Bukida (97 Da–d), Guadalcanal (97De), and Malaita and San Cristobal (97Df–g). It is among the otherwise widespread Solomon species strikingly absent from the New Georgia Group (see p. 243 of chapter 31); it is also absent from the outliers St. Matthias and Rennell and from almost all small islands. 97H and 97I occur sympatrically as a doublet over most of Celebes (barred area of map), but only 97I occurs in South Celebes (solid black area of map).

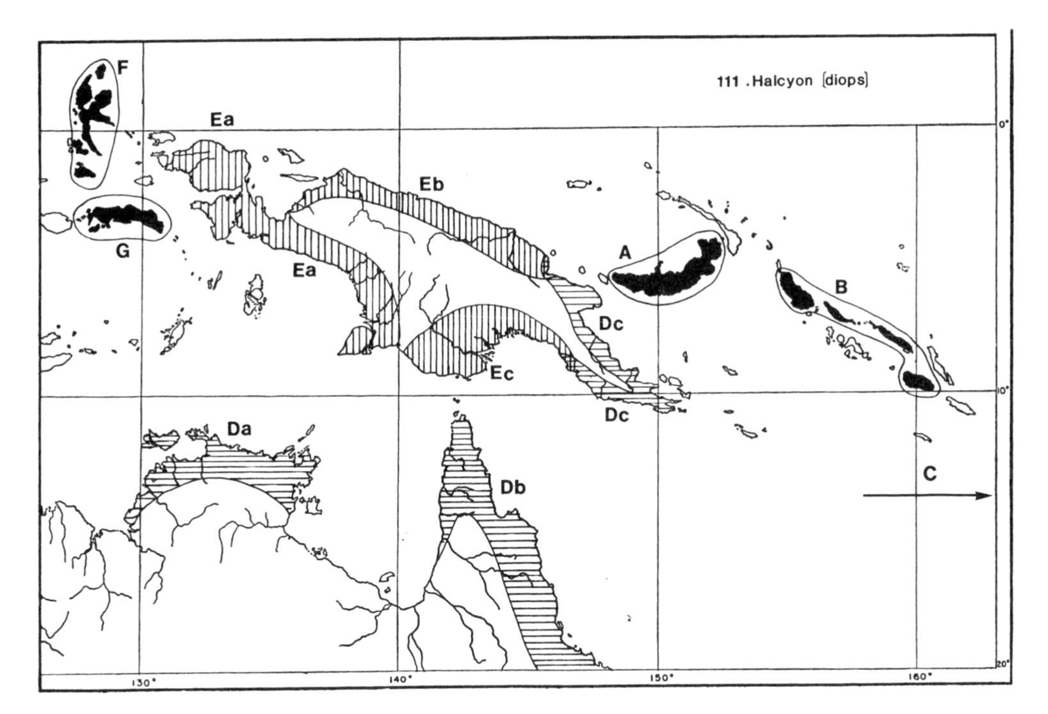

Map 23. 111 = kingfisher superspecies *Halcyon* [*diops*]. A = *H. albonotata*; B = *H. leucopygia*; C. = *H. farquhari*; D = *H. macleayii* (Da = *H. m. macleayii*, Db = *H. m. incincta*, Dc = *H. m. elisabeth*); E = *H. nigrocyanea* (Ea = *H. n. nigrocyanea*, Eb = *H. n. quadricolor*, Ec = *H. n. stictolaema*); F = *H. diops*; G = *H. lazuli*. Despite the wide distribution of the superspecies from the Moluccas to the New Hebrides, both Northern Melanesian populations are sedentary endemic allospecies with restricted distributions: a Bismarck allospecies *H. albonotata* (111A) confined to New Britain and a Solomon allospecies *H. leucopygia* (111B) confined to the Bukida group and Guadalcanal.

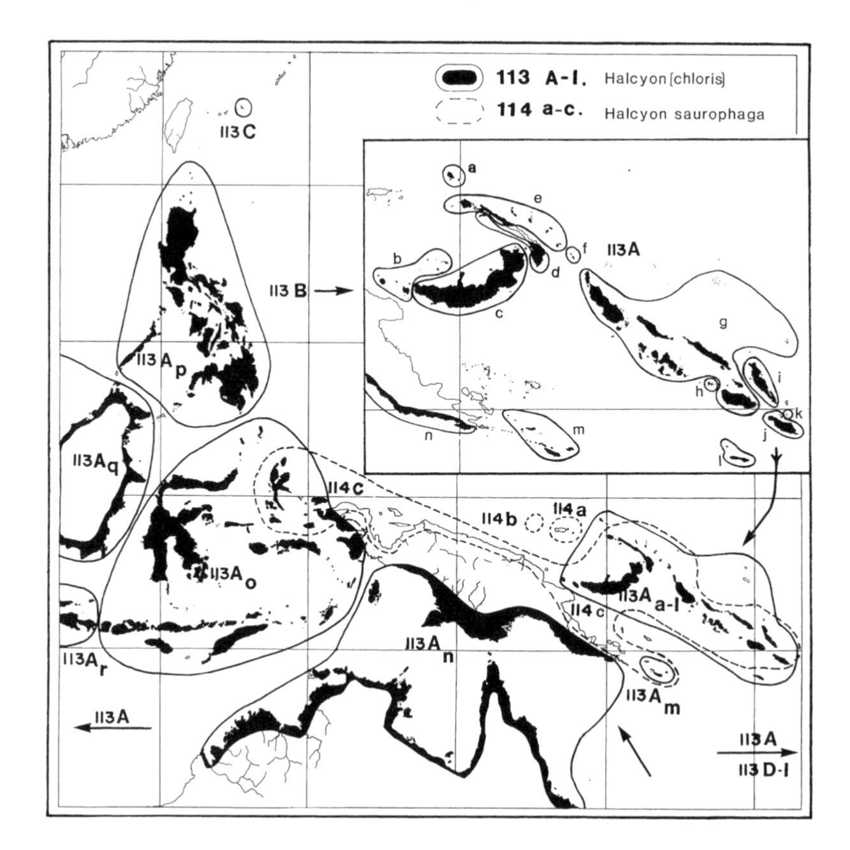

Map 24. 113 = kingfisher superspecies *Halcyon* [*chloris*] (plate 6). 113A = *H. chloris* (Aa = *H. c. matthiae*, Ab = *H. c. stresemanni*, Ac = *H. c. tristrami*, Ad = *H. c. novaehiberniae*, Ae = *H. c. nusae*, Af = *H. c. bennetti*, Ag = *H. c. alberti*, Ah = *H. c. pavuvu*, Ai = *H. c. mala*, Aj = *H. c. solomonis*, Ak = *H. c. sororum*, Al = *H. c. amoena*, Am = *H. c. colona*, An = H. *c. sordida*, Ao = *H. c. chloris*, Ap = *H. c. collaris*, Aq = *H. c. laubmanniana*, Ar = *H. c. palmeri*); 113B = *H. cinnamomina*; 113C = *H. miyakoensis*; 113D = *H. ruficollaris*; 113E = *H. recurvirostris*; 113F = *H. venerata*; 113G = *H. tuta*; 113H = *H. gambieri*; 113I = *H. godeffroyi*. 114 = *Halcyon sauropphaga* (a = *H. s. admiralitatis*, b = *H. s. anachoretae*, c = *H. s. saurophaga*). Over its wide range from the Red Sea to Samoa, the *H.* [*chloris*] superspecies, another one of Northern Melanesia's "great speciators," falls into 9 allospecies and 47 subspecies. Within Northern Melanesia alone, it falls into 3 megasubspecies and 12 subspecies. Of those three megasubspecies, one occupies the easternmost Solomon islands plus the New Hebrides to the southeast, and includes subspecies 113 Aj, k, l. The second comprises the remaining Solomon races (113Ag, h, i), plus (surprisingly) the New Britain race (113Ac). All the other Bismarck races (113Aa, b, d, e, f) comprise the remaining megasubspecies, within which the peripheral isolate of St. Matthias (113Aa) is especially distinct in its white head.

H. chloris is absent from both North New Guinea and the Northwest Bismarcks, suggesting that the related *H. saurophaga* (114a–c, plate 6) may have originated as the *H. chloris* subspecies of that region before acquiring species rank, spreading westward to the Northern Moluccas and eastward into the Bismarck Archipelago and the Solomon Islands, and coming to overlap widely with *H. chloris*. See page 175 of chapter 22 for discussion.

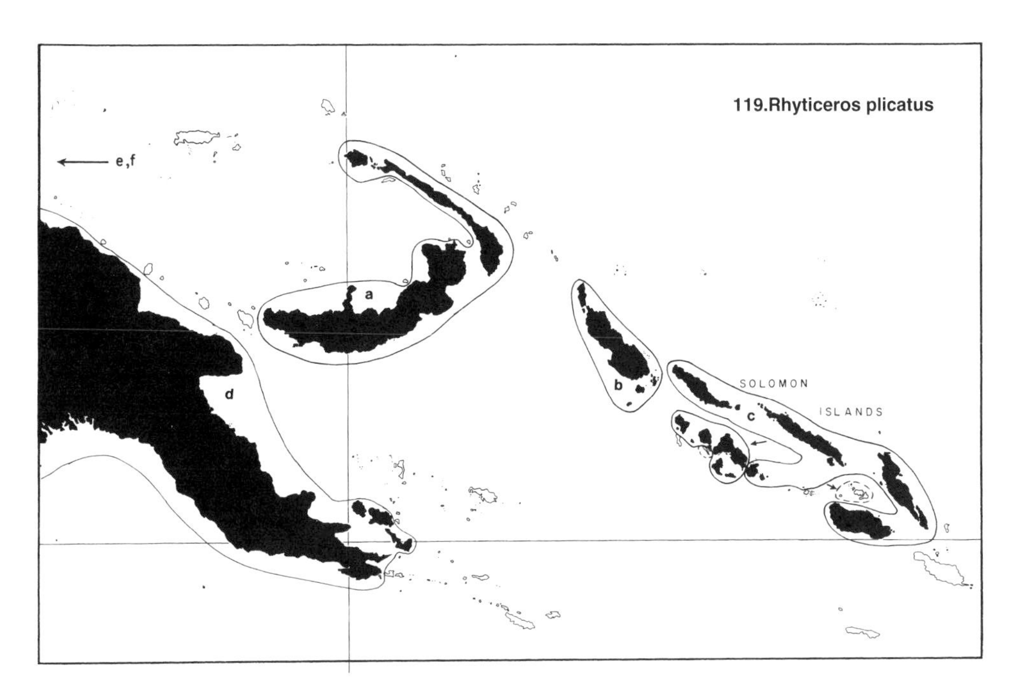

Map 25. 119 = hornbill *Rhyticeros plicatus* (a = *R. p. dampieri*, b = *R. p. harterti*, c = *R. p. mendanae*, d = *R. p. jungei*, e = *R. p. ruficollis*, f = *R. p. plicatus*). *R. p. mendanae* colonized most islands of the New Georgia Group only in recent decades (arrow) and still occurs only as a vagrant on Wana Wana, Savo, and the Florida Group (dashed lines, arrow). Because this is a vagile species, the three Northern Melanesian subspecies (119a, b, c) are only weakly distinct from each other and from the ancestral New Guinea population (119d). Note the absences from outlying islands, not only from the smaller islands east of New Ireland but also from Northern Melanesia's four largest outliers, San Cristobal and Rennell in the eastern Solomons and St. Matthias and Manus in the Bismarcks — perhaps because this species' seasonal movements in search of fruiting trees restrict it to the largest islands or else to medium-sized islands near other islands. See p. 147 of chapter 20 for discussion.

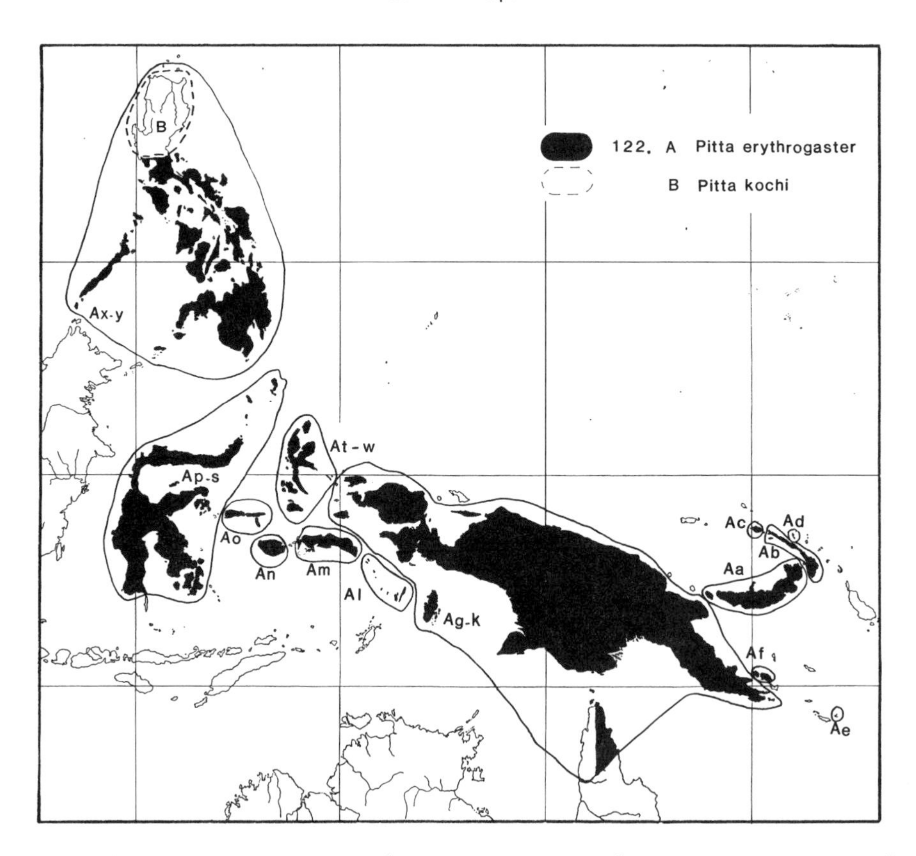

Map 26. 122 = pitta superspecies *Pitta* [*erythrogaster*]. A = *P. erythrogaster* (Aa = *P. e. gazellae*, Ab = P. *e. novaehibernicae*, Ac = *P. e. extima*, Ad = *P. e. splendida*, Ae = *P. e. meeki*, Af = *P. e. finschii*, Ag = *P. e. macklotii*, Ah = *P. e. loriae*, Ai = *P. e. oblita*, Aj = *P. e. habenichti*, Ak = *P. e. aruensis*, Al = *P. e. kuehni*, Am = *P. e. piroensis*, An = *P. e. rubrinucha*, Ao = *P. e. dohertyi*, Ap = *P. e. celebensis*, Aq = *P. e. palliceps*, Ar = *P. e. caeruleitorques*, As = *P. e. inspeculata*, At = *P. e. bernsteini*, Au = *P. e. cyanonota*, Av = *P. e. obiensis*, Aw = *P. e. rufiventris*, Ax = *P. e. propinqua*, Ay = *P. e. erythrogaster*); B = *Pitta kochi*. This pitta is confined in Northern Melanesia to the Bismarcks. Within the Bismarcks it is absent from the outlying Admiralty group, St. Matthias group, and from all the islands east of New Ireland except for Tabar, where it has developed a megasubspecies (122 Ad), one of the two most distinctive endemics of the islands east of New Ireland.

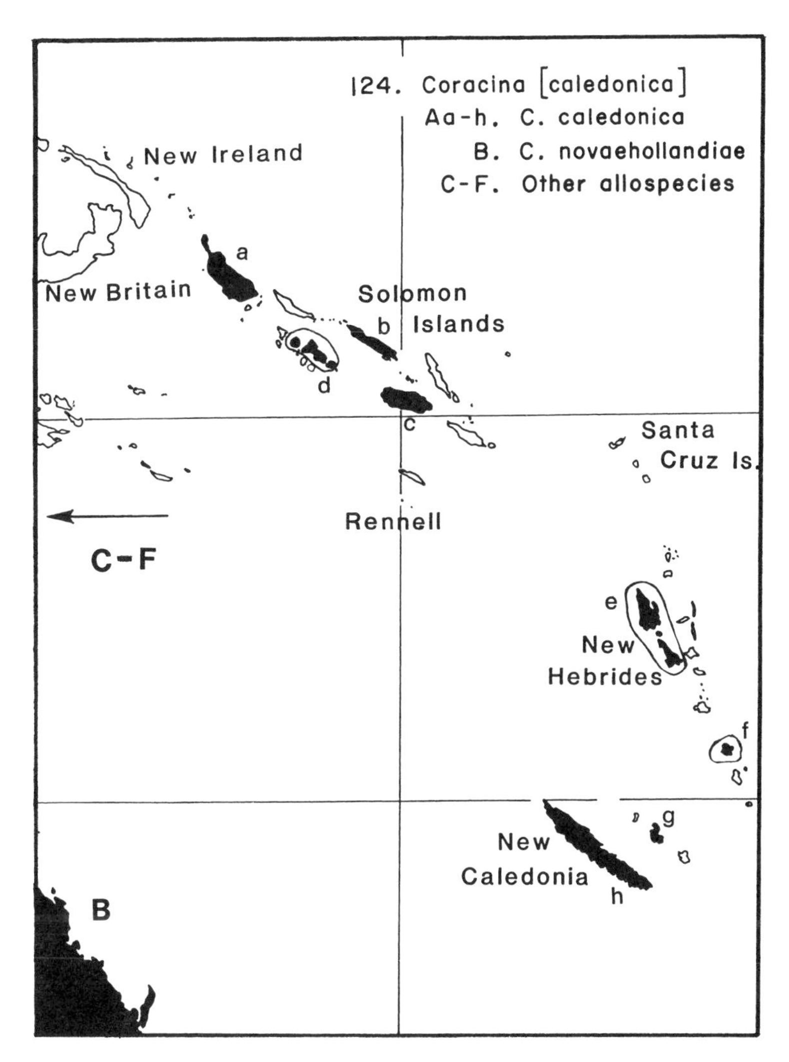

Map 27. 124 = cuckoo-shrike superspecies *Coracina* [*caledonica*]. A = *C. caledonica* (Aa = *C. c. bougainvillei*, Ab = *C. c. welchmani*, Ac = *C. c. amadonis*, Ad = *C. c. kulambangrae*, Ae = *C. c. thilenii*, Af = *C. c. seiuncta*, Ag = *C. c. lifuensis*, Ah = *C. c. caledonica*); B = *C. novaehollandiae;* C = *C. fortis;* D = *C. atriceps;* E = *C. pollens;* F = *C. schistacea*. This superspecies may have reached the Solomons (124 Aa–d) from populations of the same allospecies *C. caledonica* in the New Hebrides and New Caledonia (124 Ae–h), derived in turn from the Australian allospecies *C. novaehollandiae* (124B). It has not spread farther west from the Solomons to the Bismarcks. The Solomon populations are confined to mountains above 1000 m on the highest islands (Bougainville (124Aa) and Guadalcanal (124Ac)), descend to lower elevations or sea level on smaller islands, and fall into a different subspecies on each island or island group occupied (124Aa, Ab, Ac, Ad).

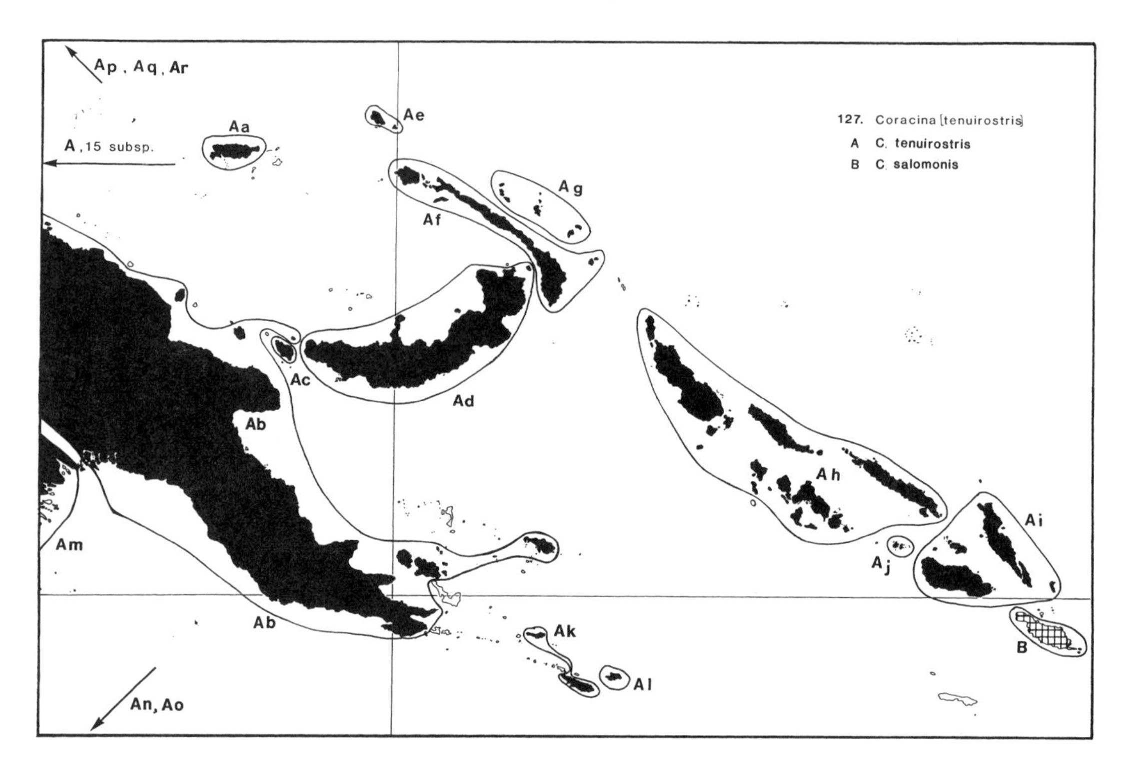

Map 28. 127 = cuckoo-shrike superspecies *Coracina* [*tenuirostris*] (plate 7). A = *C. tenuirostris* (Aa = *C. t. admiralitatis*, Ab = *C. t. muellerii*, Ac = *C. t. rooki*, Ad = *C. t. heinrothi*, Ae = *C. t. matthiae*, Af = *C. t. remota*, Ag = *C. t. ultima*, Ah = *C. t. saturatior*, Ai = *C. t. erythropygia*, Aj = *C. t. nisoria*, Ak = *C. t. tagulana*, Al = *C. t. rostrata*, Am = *C. t. aruensis*, An = *C. t. tenuirostris*, Ao = *C. t. melvillensis*, Ap = *C. t. monacha*, Aq = *C. t. nesiotis*, Ar = *C. t. insperata*); B = *C. salomonis*. This superspecies is highly prone to geographic variation, with two allospecies, three megasubspecies, and nine weak subspecies in Northern Melanesia alone. The most distinctive population of the superspecies is the allospecies *C. salamonis* (127B, plate 7) of San Cristobal, a sedentary peripheral isolate derived from a vagile ancestor. Long Island, the southwesternmost large Bismarck island, was recolonized from New Guinea (127Aa) after the seventeenth-century volcanic defaunation of Long (see p. 180 of chapter 23 for discussion).

Some other patterns of geographic variation are unusual: the populations of the New Ireland group (127Af,g) are related not to the other Bismarck populations (122 Aa, b, c, d, e) but to the widespread Solomon populations (127 Ah, i), with which they constitute one megasubspecies; the subspecies of the Bismarck outliers Manus (127 Aa) and St. Matthias (127 Ae) are not especially distinctive, as true for many other species, but are closely related to the subspecies of New Britain (127Ad) and its neighbor Umboi (127Ac); and the subspecies of the Russell group in the Solomons (127Aj) is unexpectedly distinctive, perhaps because of hybrid origins between colonists from the population 127Ab and 127Ah–i.

Note that this vagile superspecies has colonized almost every Melanesian island, with the exception of Rennell in the southeastern Solomons plus a few small islands.

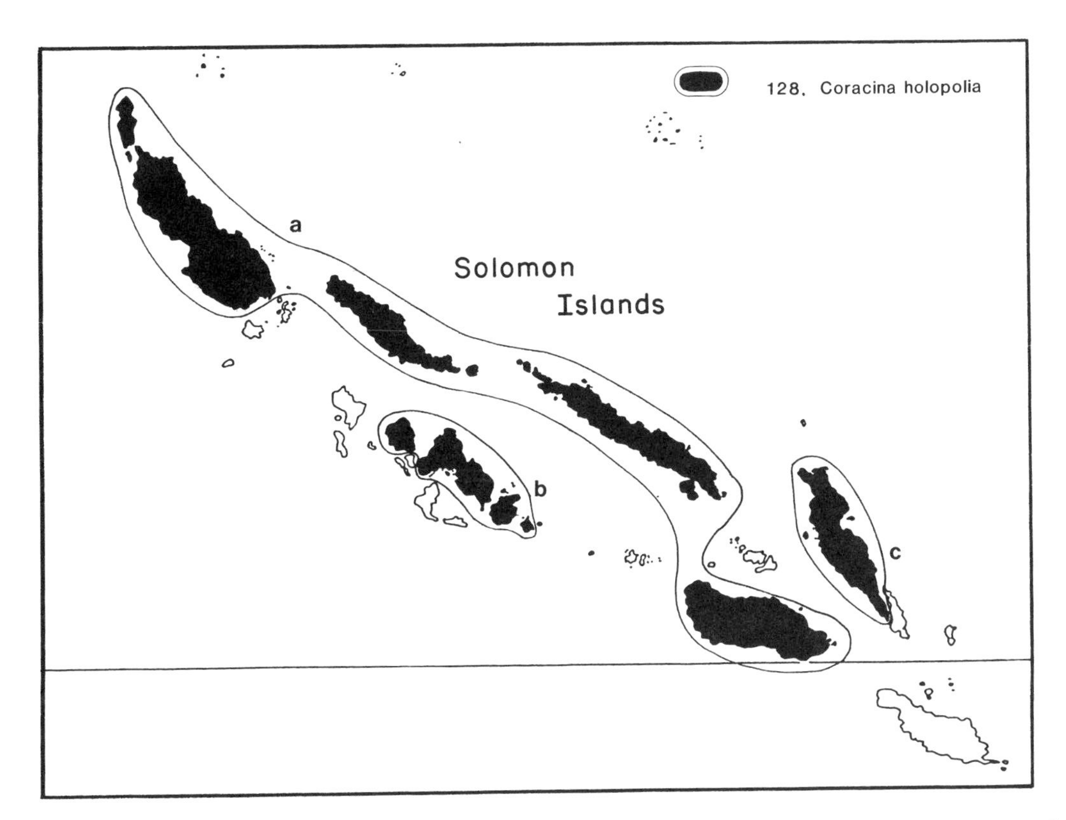

Map 29. 128 = cuckoo-shrike *Coracina holopolia* (a = *C. h. holopolia*, b = *C. h. pygmaea*, C = *C. h. tricolor*). This sedentary, isolated endemic species without close relatives is confined to the Solomons, where it has notable absences on San Cristobal, Florida group, Russells, Rendipari group, Vellonga group, and Shortland group. The three subspecies are very distinct (megasubspecies) and have a familiar distributional pattern: one race (128a) on the Bukida group and Guadalcanal, another race (128b) on the New Georgia group (where it is confined to modern island fragments of Pleistocene Greater Gatumbangra), and the third race (128c) on Malaita.

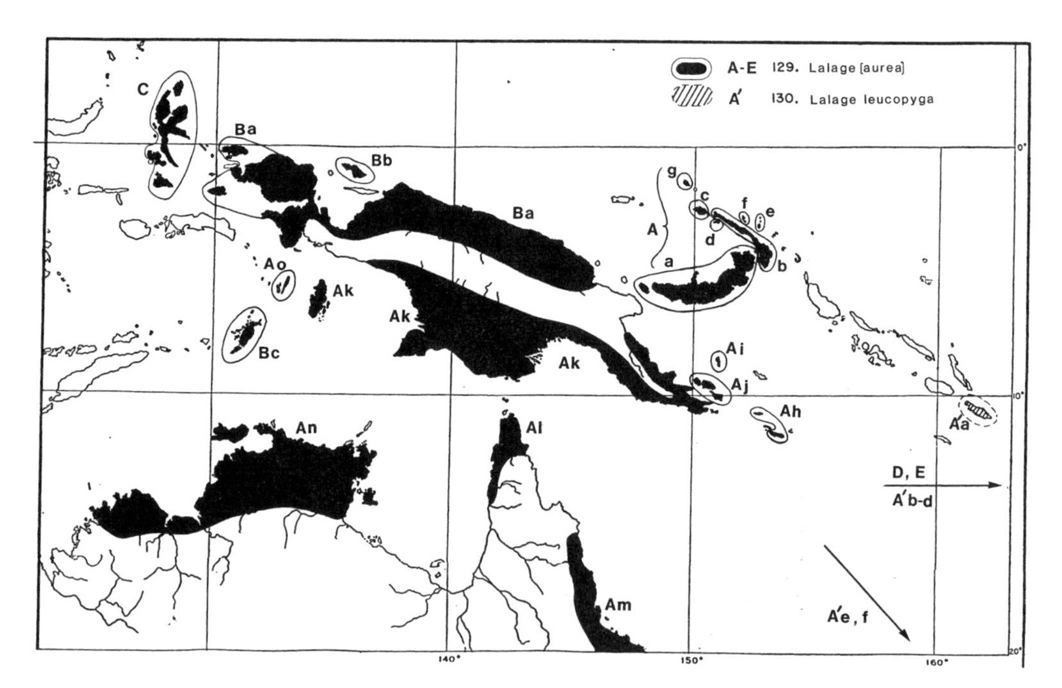

Map 30. 129 = cuckoo-shrike superspecies *Lalage* [*aurea*]. A = *L.* [*aurea*] *leucomela* (Aa = *L. l. falsa*, Ab = *L. l. karu*, Ac = *L. l. albidior*, Ad = *L. l. sumunae*, Ae = *L. l. ottomeyeri*, Af = *L. l. tabarensis*, Ag = *L. l. conjuncta*, Ah = *L. l. pallescens*, Ai = *L. l. trobriandi*, Aj = *L. l. obscurior*, Ak = *L. l. polygrammica*, Al = *L. l. yorki*, Am = *L. l. leucomela*, An = *L. l. rufiventer*, Ao = *L. l. keyensis*); B = *L.* [*aurea*] *atrovirens* (Ba = *L. a. atrovirens*, Bb = *L. a. leucoptera*, Bc = *L. a. moesta*); C = *L.* [*aurea*] *aurea*; D = *L.* [*aurea*] *maculosa*; E = *L.* [*aurea*] *sharpei*. 130A′ = *L. leucopyga* (A′a = *L. l. affinis*, A′b–f = other subspecies). Within the genus *Lalage*, *L. leucopyga* (130A′) is an isolated species, distant taxonomically from the *L. leucomela* superspecies (129A–E) and sympatric with it (130A′b–d, 129D) in the New Hebrides. *L. leucopyga* is confined in the Solomons to the San Cristobal group (130A′a) and may have arrived from the New Hebrides.

L. leucomela (129A) reached the Bismarcks (129Aa–g) from Southeast New Guinea (129Ah–k) but failed to reach the isolated Northwest Bismarcks or to spread east to the Solomons. Its Bismarck populations are geographically variable and fall into three megasubspecies groups: the peripheral isolate of St. Matthias (129Ag); the two races on Lihir and Tabar islands (129Ae and Af) east of New Ireland, which together with the Tabar pitta *Pitta erythrogaster splendida* (population 122Ad on map 26) are the most distinctive endemics of the islands east of New Ireland; and the populations of the rest of the New Ireland group (129Ab–d) and of the New Britain group (129Aa), which are closely related to the populations of the New Guinea region (129Ah–k). Note that the Bismarcks were colonized by the South New Guinea allospecies *L. leucomela* (129A), not by the North New Guinea allospecies *L. atrovirens* (129B).

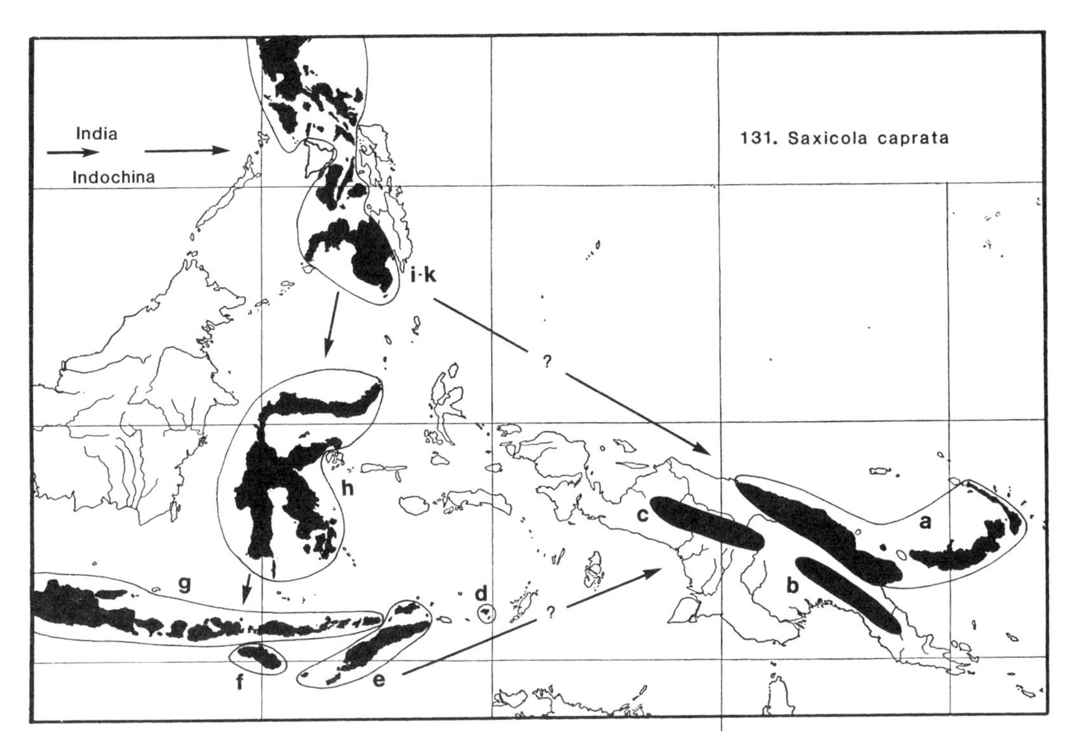

Map 31. 131 = chat *Saxicola caprata* (a = *S. c. aethiops*, b = *S. c. wahgiensis*, c = *S. c. belensis*, d = *S. c. cognata*, e = *S. c. pyrrhonota*, f = *S. c. francki*, g = *S. c. fruticola*, h = *S. c. albonotata*, i = *S. c. anderseni*, j = *S. c. randi*, k = *S. c. caprata*). From North New Guinea this open-country species of Asian origins has reached only the Bismarck islands of New Britain and New Ireland and the nearby recently defaunated volcanic islands of Long and Vuatom. Like most other open-country colonists of Northern Melanesia, it has scarcely differentiated: the Bismarck populations still belong to the same subspecies as the New Guinea source population (131a).

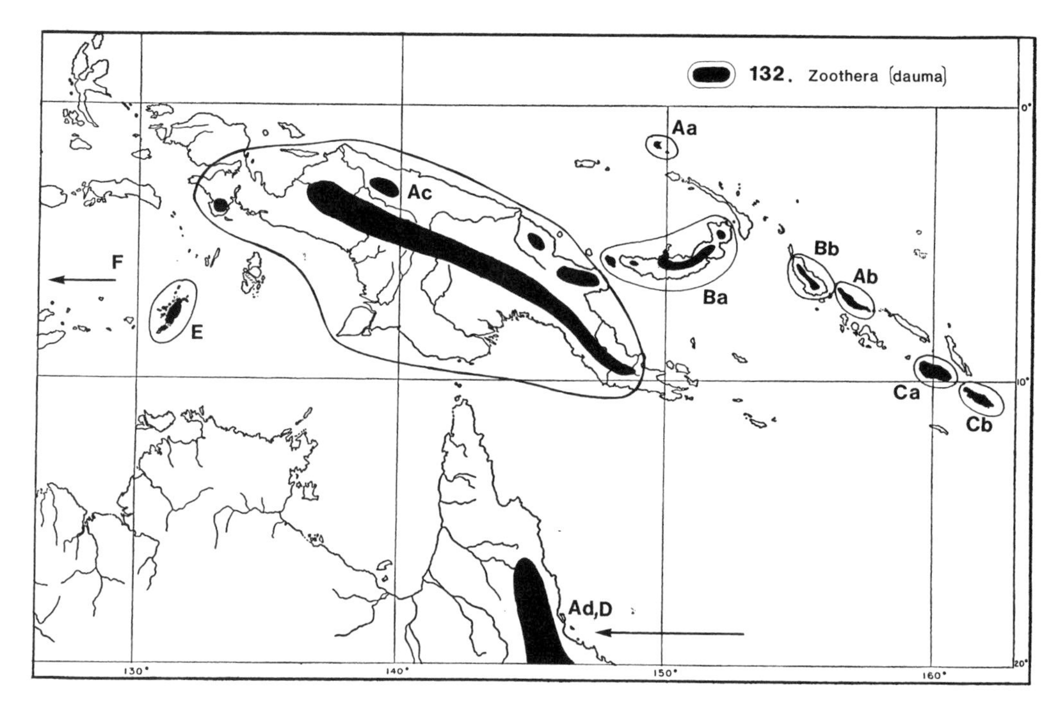

Map 32. 132 = thrush superspecies *Zoothera* [*dauma*]. A = Z. [*dauma*] *heinei* (Aa = *Z. h. eichhorni*, Ab = *Z. h. choiseuli*, Ac = *Z. h. papuensis*, Ad = *Z. h. heinei*); B = Z. [*dauma*] *talaseae* (Ba = *Z. t. talaseae*, Bb = *Z. t. atrigena*); C = Z. [*dauma*] *margaretae* (Ca = *Z. m. turipavae*, Cb = *Z. m. margaretae*); D = Z. [*dauma*] *lunulata*; E = Z. [*dauma*] *machiki*; F = Z. [*dauma*] *dauma*. This superspecies has a complex distribution within Northern Melanesia on seven large islands and one small one, indicating multiple waves of colonization (see p. 160 of chapter 21 for discussion). The populations of New Guinea (132Ac) and of high islands (132 Ba, Bb, Ca, Cb) are confined to mountains, whereas those of low islands occur at sea level (132 Aa, E, and probably Ab). The superspecies originated in Eurasia, where congeners and the allospecies *Z. dauma* (132F) occur. Two other allospecies represent old invaders of Northern Melanesia: *Z. margaretae* (139 Ca, Cb) on two high islands of the eastern Solomons, and the melanistic *Z. talaseae* on two high islands of the Bismarcks (139Ba) plus the western Solomon high island of Bougainville (139Bb). In addition, the allospecies *Z. heinei* of New Guinea (139Ac) and Australia (139Ad) extends to three low islands, St. Matthias and Emirau in the Bismarcks (139Aa) and Choiseul in the Solomons (139Ab), as an evidently more recent invader. *Z heinei* has speciated in Australia to produce a very similar sympatric sibling species, *Z. lunulata* (139D).

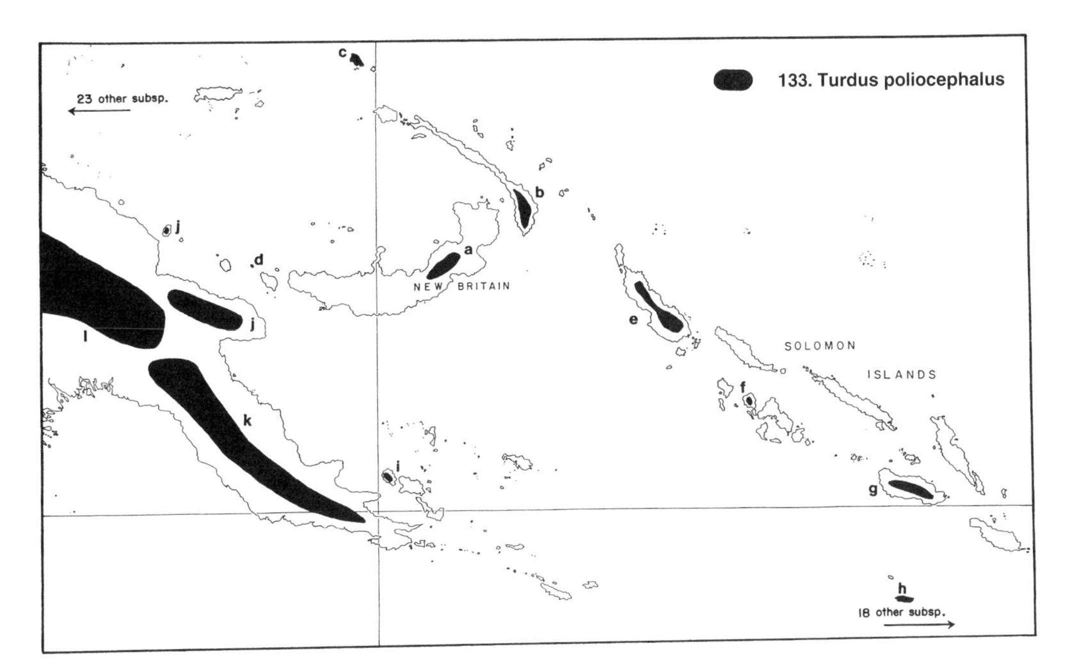

Map 33. 133 = thrush *Turdus poliocephalus* (a = *T. p.* subsp.?, b = *T. p. beehleri*, c = *T. p. heinrothi*, d = *T. p. tolokiwae*, e = *T. p. bougainvillei*, f = *T. p. kulambangrae*, g = *T. p. sladeni*, h = *T. p. rennellianus*, i = *T. p. canescens*, j = *T. p. keysseri*, k = *T. p. papuensis*, l = *T. p. carbonarius*). One of Northern Melanesia's "great speciators" (chapter 19), this thrush breaks up into 53 subspecies over its far-flung island range from Sumatra east to Samoa, including a different subspecies on each of its eight Northern Melanesian islands of occurrence. Six of those populations (133a, b, d, e, f, g) are confined to high elevations of high islands, while the other two populations (133c, h) live at sea level on low islands.

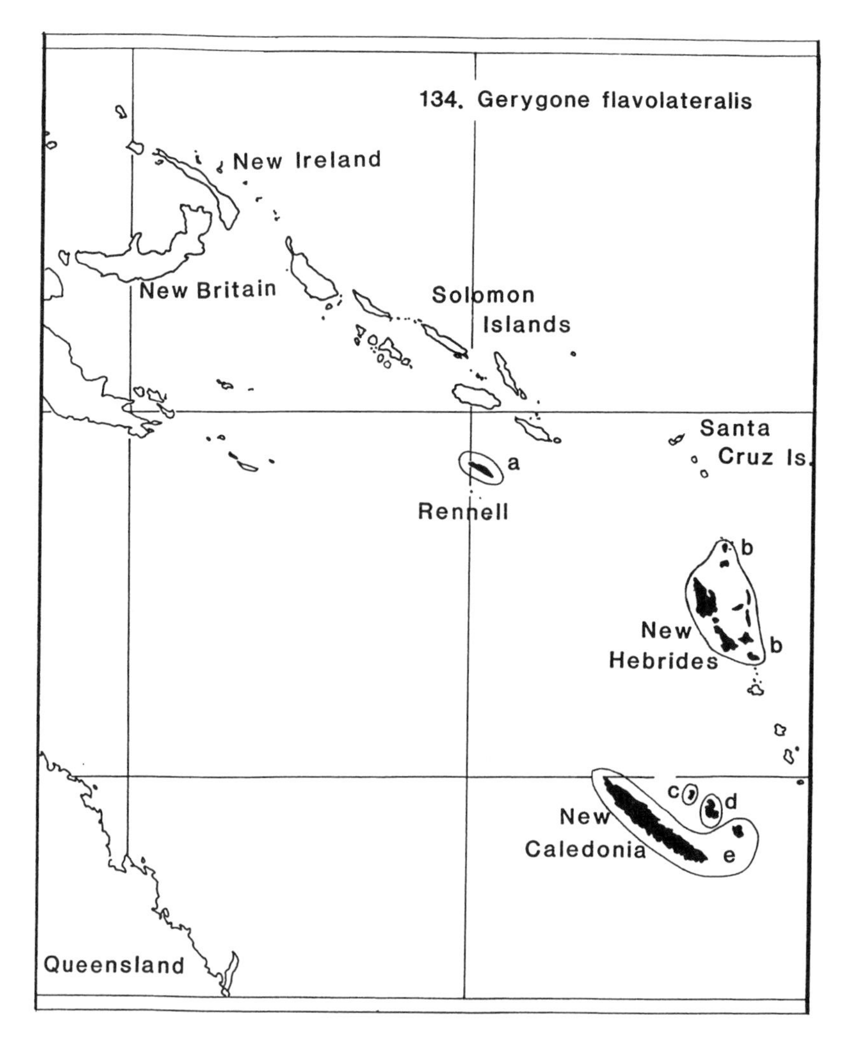

Map 34. 134 = warbler *Gerygone flavolateralis* (a = *G. f. citrina*, b = *G. f. correiae*, c = *G. f. rouxi*, d = *G. f. lifuensis*, e = *G. f. flavolateralis*). This warbler is the sole Northern Melanesian representative of the Australasian warbler family Acanthizidae, which has 17 representatives in New Guinea and 38 in Australia. *G. flavolateralis* reached Northern Melanesia from archipelagoes to the southeast (134b–e) and occupied only the southeasternmost Solomon island, Rennell, where it evolved the most distinctive race of the species (134a).

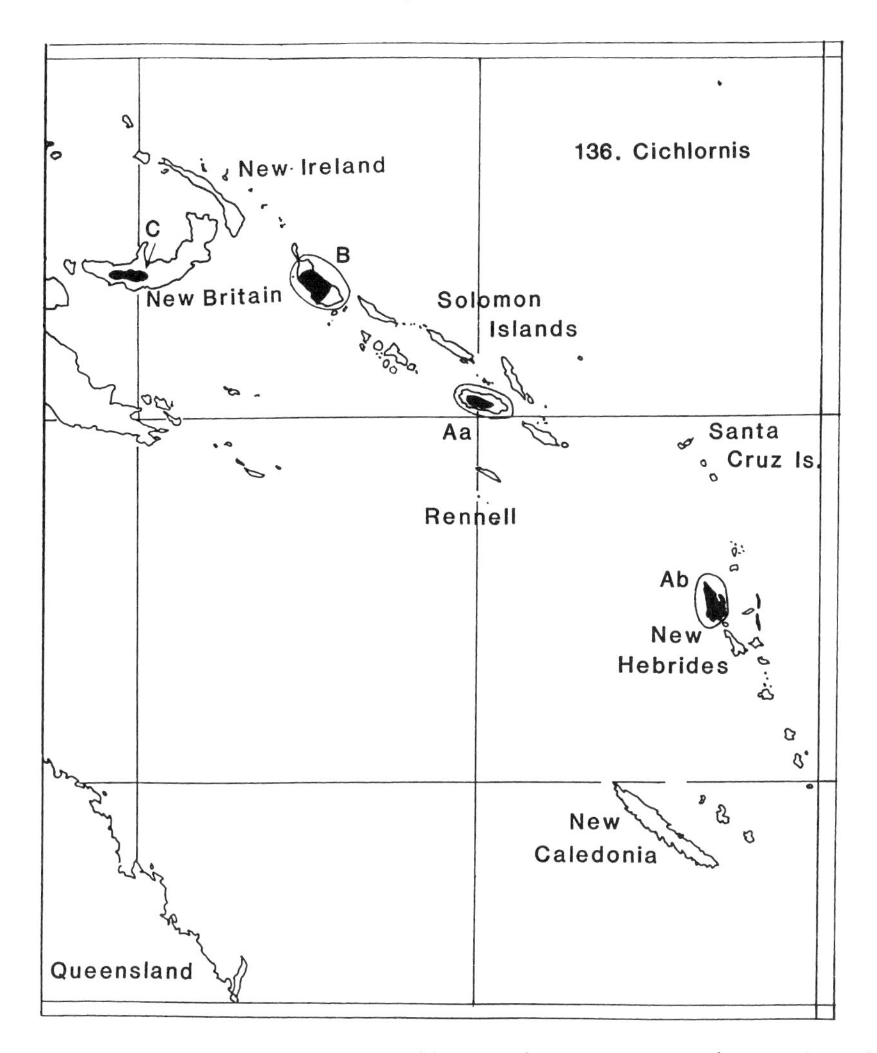

Map 35. 136 = thicket-warbler superspecies *Cichlornis* [*whitneyi*]. A = *C. whitneyi* (Aa = *C. w. turipavae*, Ab = *C. w. whitneyi*); B = *C. llaneae*; C = *C. grosvenori* (plate 9). Each of the four allospecies or subspecies of these secretive, sedentary warblers is confined to the mountains of one large high island.

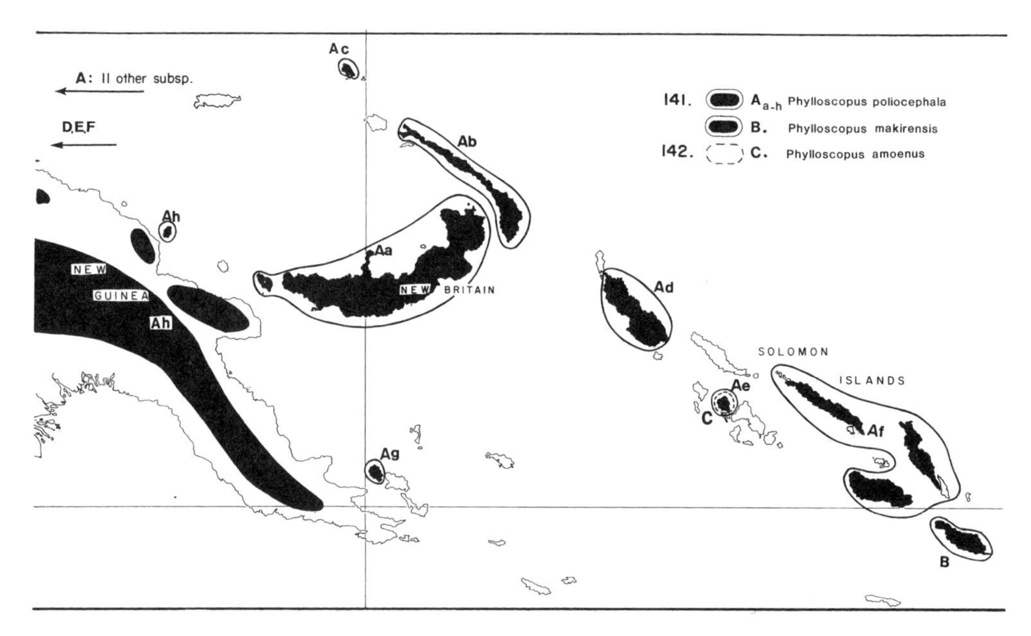

Map 36. 141 = warbler superspecies *Phylloscopus* [*trivirgatus*] and the related warbler species 142 = *P. amoenus*. A = *Phylloscopus* [*trivirgatus*] *poliocephala* (Aa = *P. p. moorhousei*, Ab = *P. p. leletensis*, Ac = *P. p. matthiae*, Ad = *P. p. bougainvillei*, Ae = *P. p. pallescens*, Af = *P. p. becki*, Ag = *P. p. hamlini*, Ah = *P. p. giulianettii*); B = *P.* [*trivirgatus*] *makirensis*; C = *Phylloscopus amoenus*; D = *Phylloscopus* [*trivirgatus*] *presbytes*; E = *P.* [*trivirgatus*] *sarasinorum*; F = *P.* [*trivirgatus*] *trivirgatus*. Of the five allospecies in the superspecies *P.* [*trivirgatus*], three live in Wallacea and west of Wallace's line (141D, E, F). The fourth, *P. poliocephala* (141A), occurs from the Moluccas in the west to the Solomons in the east and falls into 19 subspecies, of which six (141Aa–f) occupy nine Northern Melanesian islands, but none of those subspecies is very distinct. Instead, the most distinctive population of Northern Melanesia is a peripheral isolate, the allospecies *P. makirensis* (141B) of San Cristobal. Most populations of the superspecies, and eight of the nine Northern Melanesian populations, are confined to mountains of high islands, but the race *P.p. matthiae* (141Ac) lives in the lowlands of a low island, St. Matthias. *P. amoenus* (species 142), a related warbler species endemic to the top of the highest peak of Kulambangra, may represent the product of a double invasion of that island by *P. poliocephala*, whose race *P.p. pallescens* (141Ae) now occurs sympatrically on Kulambangra with *P. amoenus* (see p. 176 of chapter 22 for discussion).

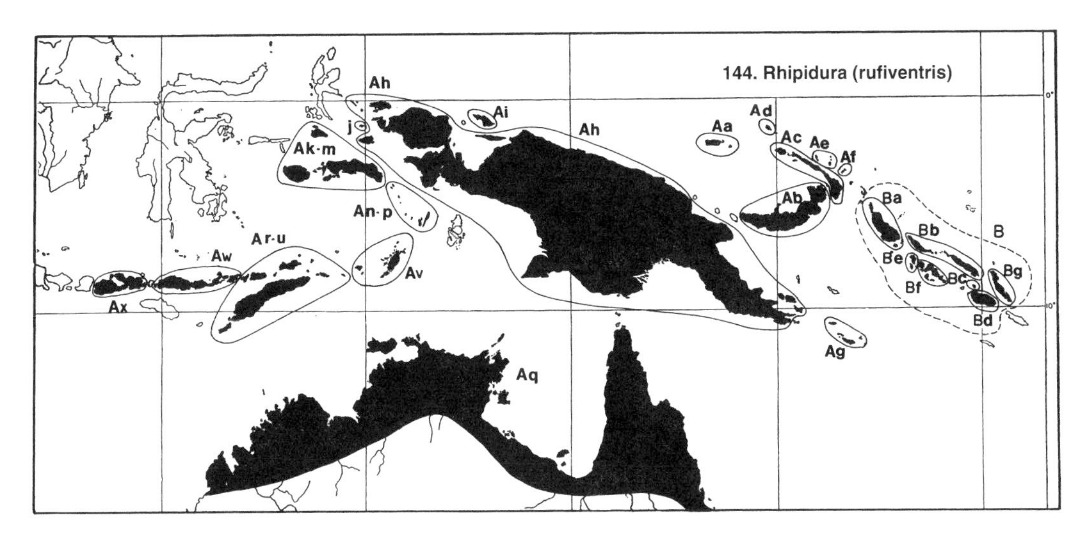

Map 37. 144 = fantail-flycatcher superspecies *Rhipidura* [*rufiventris*]. A = *Rhipidura* [*rufiventris*] *rufiventris* (Aa = *R. r. niveiventris*, Ab = *R. r. finschi*, Ac = *R. r. setosa*, Ad = *R. r. mussai*, Ae = *R. r. gigantea*, Af = *R. r. tangensis*, Ag = *R. r. nigromentalis*, Ah = *R. r. gularis*, Ai = *R. r. kordensis*, Aj = *R. r. vidua*, Ak = *R. r. obiensis*, Al = *R. r. bouruensis*, Am = *R. r. cinerea*, An = *R. r. perneglecta*, Ao = *R. r. finitima*, Ap = *R. r. assimilis*, Aq = *R. r. isura*, Ar = *R. r. buettikoferi*, As = *R. r. pallidiceps*, At = *R. r. tenkatei*, Au = *R. r. rufiventris*, Av = *R. r. fuscorufa*, Aw = *R. r. diluta*, Ax = *R. r. sumbawensis*); B = *R.* [*rufiventris*] *cockerelli* (Ba = *R. c. septentrionalis*, Bb = *R. c. interposita*, Bc = *R. c. floridana*, Bd = *R. c. cockerelli*, Be = *R. c. lavellae*, Bf = *R. c. albina*, Bg = *R. c. coultasi*). Within the superspecies *R.* [*rufiventris*], the allospecies *R. rufiventris* (144A) falls into 24 subspecies, of which 6 (144Aa–f) are endemic to the Bismarcks. The peripheral Solomon populations at the eastern edge of the superspecies' range are more distinct and constitute a separate allospecies, *R. cockerelli* (144B), of which the most distinct race is the easternmost on Malaita (144Bg). Note the absence of this otherwise widespread superspecies from San Cristobal, the large island at the southeastern end of the Solomon chain.

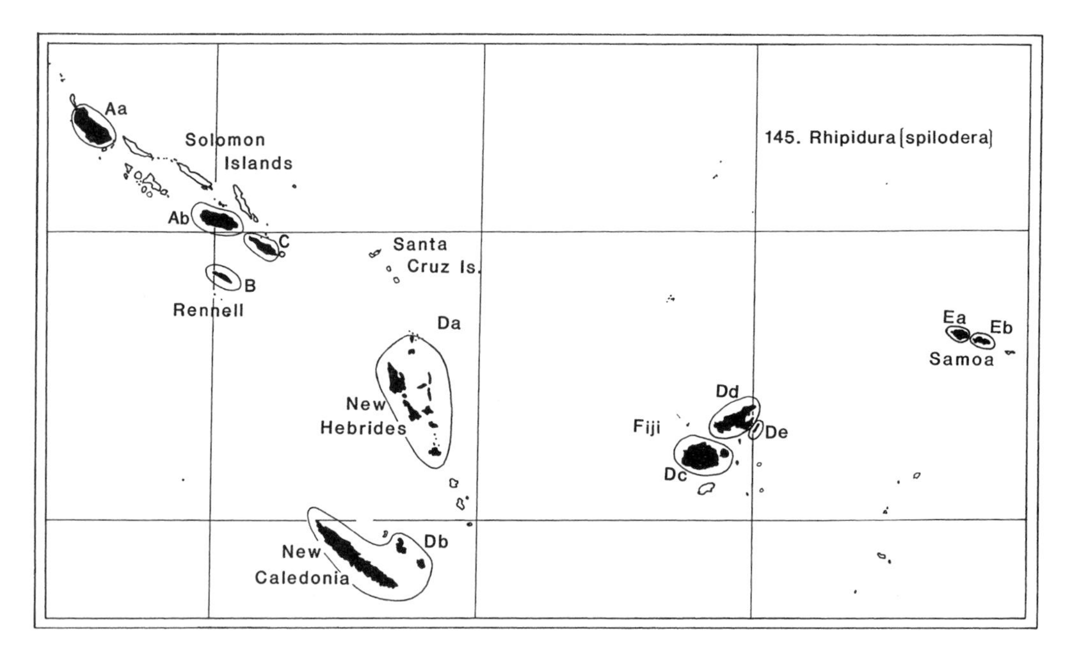

Map 38. 145 = fantail flycatcher superspecies *Rhipidura* [*spilodera*]. A = *R. drownei* (Aa = *R. d. drownei*, Ab = *R. d. ocularis*); B = *R. rennelliana*; C = *R. tenebrosa*; D = *R. spilodera* (Da = *R. s. spilodera*, Db = *R. s. verreauxi*, Dc = *R. s. layardi*, Dd = *R. s. erythronota*, De = *R. s. rufilateralis*); E = *R. nebulosa* (Ea = *R. n. altera*, Eb = *R. n. nebulosa*). The superspecies occupies a wide range from Samoa west to the Solomons, where each of the four islands occupied harbors a distinct population (three allospecies [145 A, B, C] and two subspecies [145Aa, b]). Three of the four Solomon populations (145Aa, 145Ab, 145C) are confined to the mountains of high islands, while the fourth population (145B) occurs at sea level on a low island, Rennell. Although ancestral taxa of the superspecies must have been sufficiently vagile to spread 3500 km over water between the Solomons and Samoa, all of the Solomons and Samoan populations are now confined to single islands and must have undergone evolutionary complete loss of overwater colonizing ability (see pp. 70–72 and 298 for discussion). *R. tenebrosa* (145C) occurs sympatrically with the related *R. fuliginosa* (species 146) on San Cristobal: see p. 165 of chapter 22.

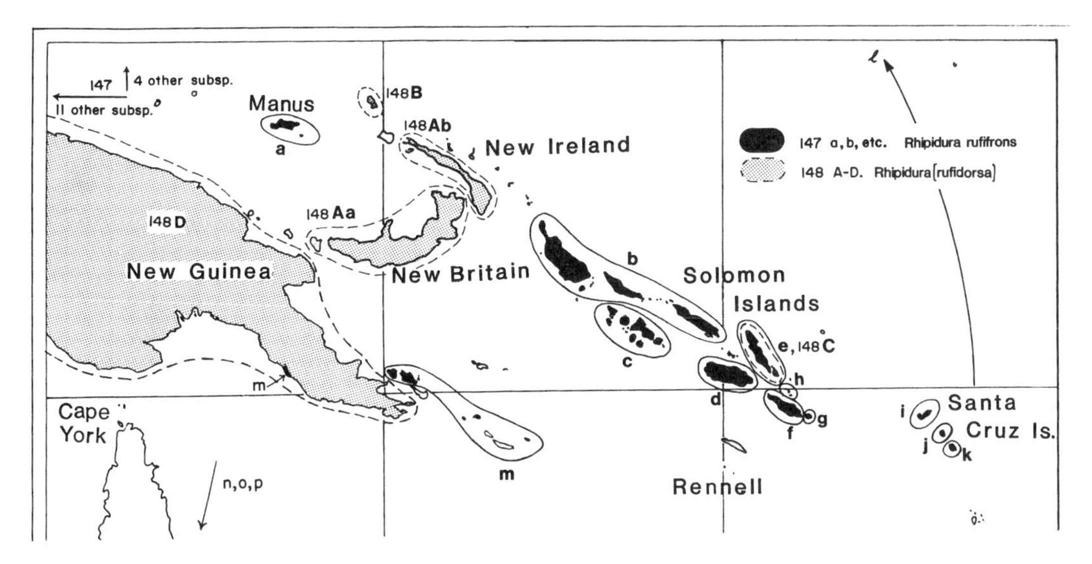

Map 39. Two related fantail-flycatcher superspecies, 147 = *Rhipidura* [*rufifrons*] and 148 = *R.* [*rufidorsa*]. a–p = *R. rufifrons* (a = *R. r. semirubra*, b = R. *r. commoda*, c = *R. r. granti*, d = *R. r. rufofronta*, e = *R. r. brunnea*, f = *R. f. russata*, g = *R. r. kuperi*, h = *R. r. ugiensis*, i = *R. r. agilis*, j = *R. r. utupuae*, Ak = *R. r. melanolaema*, l = *R. r. kubaryi*, m = *R. r. louisiadensis*, n = *R. r. dryas*, o = *R. r. intermedia*, p = *R. r. rufifrons*). 148A = *R.* [*rufidorsa*] *dahli* (148Aa = *R. d. dahli*, 148Ab = *R. d. antonii*); 148B = *R.* [*rufidorsa*] *matthiae*; 148C = *R.* [*rufidorsa*] *malaitae;* 148D = [*rufidorsa*] *rufidorsa.*

The *Rhipidura* [*rufifrons*] superspecies is another "great speciator." Its Northern Melanesian allospecies *R. rufifrons* has eight subspecies within Northern Melanesia and 23 elsewhere, while about six other allospecies or derivatives live in Wallacea, Fiji, and Palau. Although *A. rufifrons* occupies virtually every central Solomon island, the species is strangely absent from the Bismarcks except for the very distinctive Manus race (147a). All three of the distinctive megasubspecies of Northern Melanesia are peripheral isolates: that Manus race, the Ugi race (147h), and the two other races of the San Cristobal group (147f, g). The remaining races of the Solomons (147b, c, d, e) are related to (and probably derived from) the race of the Southeast Papuan Islands (147m).

As for the *R.* [*rufidorsa*] superspecies, consisting of four allospecies, it is a *R. rufrifrons* offshoot that now lives sympatrically with *R. rufifrons* on Malaita (148C and 147e; see p. 176 of chapter 22 for discussion) and in parts of New Guinea and its western islands (148D and 147). Thus, the *R.* [*rufifrons*]–*R.* [*rufidorsa*] complex is at an intermediate stage of speciation, with achievement of reproductive isolation but only limited sympatry. The lowland New Guinea allospecies *R. rufidorsa* (148D) was probably the source of the four Northern Melanesian population of the *R.* [*rufidorsa*] superspecies, of which three populations (148Aa, Ab, C) now live in the mountains of high islands, while the fourth lives at sea level on the low island St. Matthias (148B).

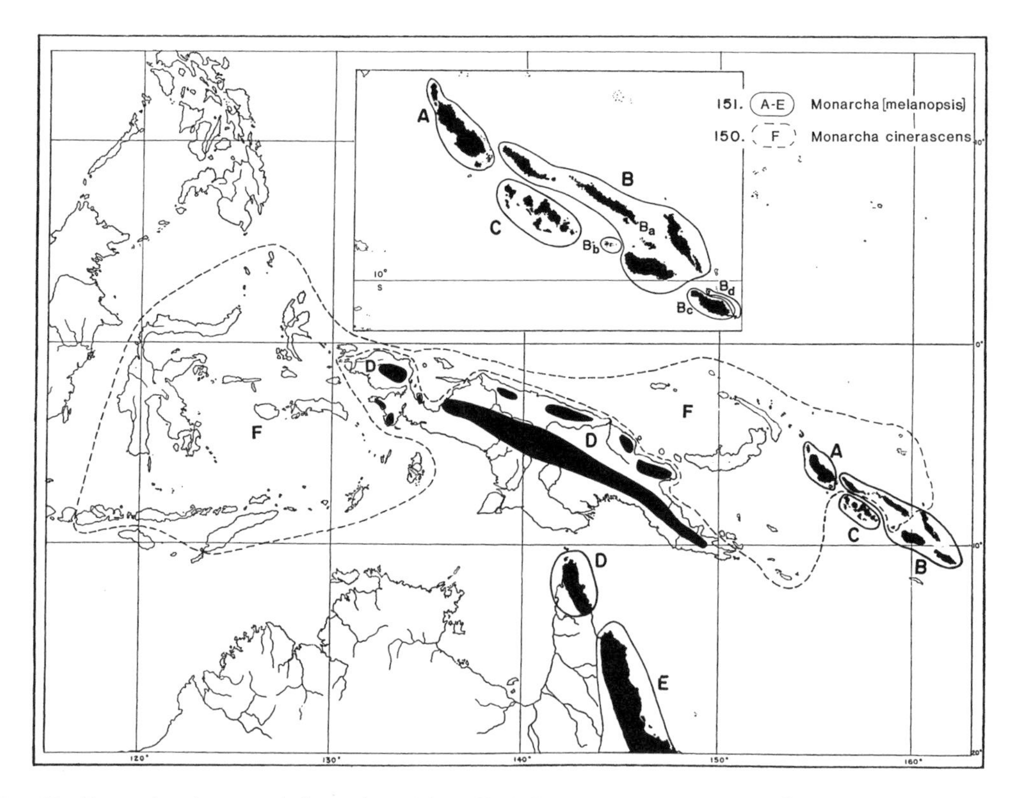

Map 40. Two related monarch flycatchers (plate 5), 150 = superspecies *Monarcha cinerascens* and 151 = *Monarcha* [*melanopsis*]. A = *M.* [*melanopsis*] *erythrostictus*; B = *M.* [*melanopsis*] *castaneiventris* (Ba = *M. c. castaneiventris*, Bb = *M. c. obscurior*, Bc = *M. c. megarhynchus*, Bd = *M. c. ugiensis*); C = *M.* [*melanopsis*] *richardsii*; D = *M.* [*melanopsis*] *frater*; E = *M.* [*melanopsis*] *melanopsis*; F = *M. cinerascens*. The superspecies *M.* [*melanopsis*] consists of two allospecies (151D, E) in New Guinea and Australia, plus three allospecies (151A, B, C) on Solomon islands within short distances of each other. One of those three Solomon allospecies (151B) breaks up into two megasubspecies (151Ba–c and Bd), one of which in turn consists of three subspecies (151Ba, b, c).

The locations of the boundaries between those Solomon taxa are interesting. As usual, the New Georgia group has a separate allospecies (151C). Somewhat unusually, there is another allospecies border between islands that were formerly joined in Pleistocene Greater Bukida, at the modern strait between Choiseul and the Shortland group separating the allospecies 151A and 151B. This border may be a legacy of the restrictions to gene flow even when the strait was dry land during the Pleistocene, because the modern location of the strait was even then the narrowest neck within the Pleistocene island of Greater Bukida (p. 261 of chapter 32). Allospecies 151B also has weak subspecies on the Russell group (151Bb) and San Cristobal (151Bc). The remaining Solomon taxon is the melanistic subspecies 151 Bd (see plate 5) on the islands Ugi, Santa Anna, and Santa Catalina very near San Cristobal — one of the three most distinctive endemics of those islands (p. 247 of chapter 31). The superspecies is absent from the large southeastern outlier Rennell.

The *M.* [*melanopsis*] superspecies is absent from the Bismarcks, which are occupied instead by the closely related supertramp *Monarcha cinerascens* (species 150), living on small, remote, or volcanically disturbed islands. *M. cinerascens* may have originated as the Bismarck allospecies of the *M.* [*melanopsis*] superspecies. Its range now extends far west of the Bismarcks into Wallacea, and east of the Bismarcks into the Solomons, where it lives on islets within at least a few kilometers of larger islands occupied by *M.* [*melanopsis*] and shares three islets with *M.* [*melanopsis*]. Thus, *M.* [*melanopsis*] and *M. cinerascens* are at an intermediate stage of speciation, with extensive geographic interdigitation and slight local sympatry. See p. 176 of chapter 22 for discussion.

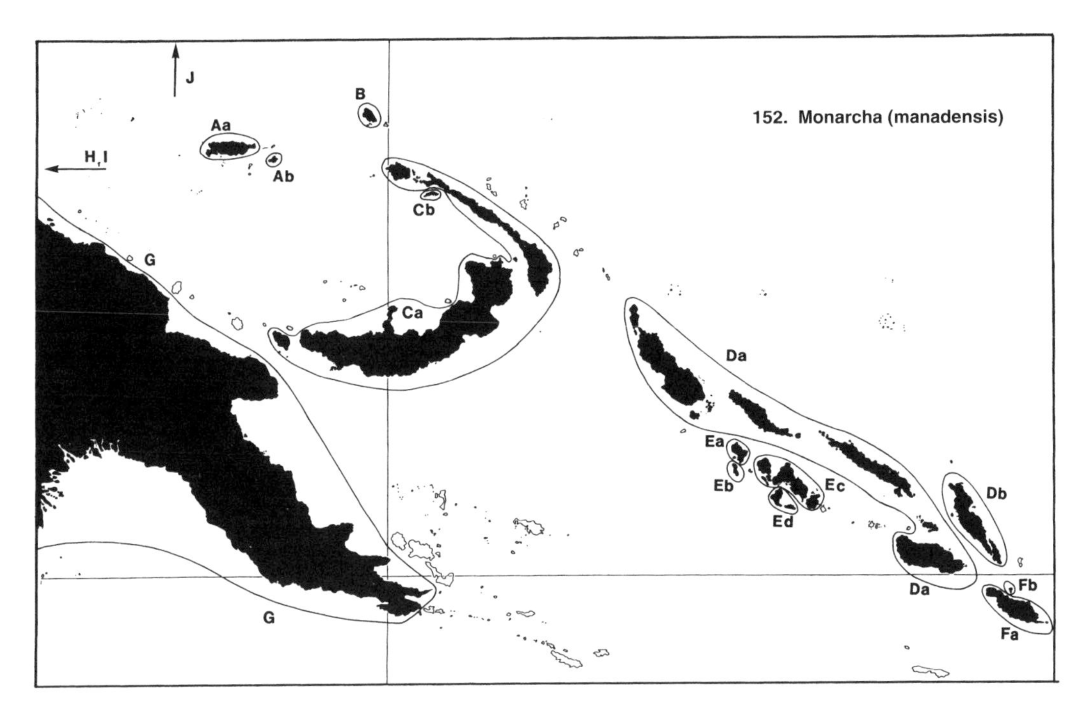

Map 41. 152 = monarch flycatcher superspecies *Monarcha* [*manadensis*] (see plate 3). A = *M. infelix* (Aa = *M. i. infelix*, Ab = *M. i. coultasi*); B = *M. menckei*; C = *M. verticalis* (Ca = *M. v. verticalis*, Cb = *M. v. ateralbus*); D = *M. barbatus* (Da = *M. b. barbatus*, Db = *M. b. malaitae*); E = *M. browni* (Ea = *M. b. nigrotectus*, Eb = *M. b. ganongae*, Ec = *M. b. browni*, Ed = *M. b. meeki*); F = *M. viduus* (Fa = *M. v. viduus*, Fb = *M. v. squamulatus*); G = *M. manadensis*; H = *M. brehmii*; I = *M. leucurus*; J = *M. godeffroyi*. This superspecies is one of the extremes among "great speciators," with six allospecies, 11 megasubspecies, and two weak subspecies in Northern Melanesia, plus four allospecies (152G, H, I, J) outside Northern Melanesia. The superspecies thus comprises forms of a wide range of distinctness, from weak subspecies to very distinct allospecies, with many borderline forms about which it is difficult to decide whether to classify them as allospecies or subspecies. All Northern Melanesian taxa are allopatric, but speciation has produced sympatric related species on New Guinea and several other Papuan and Lesser Sundan islands.

In Northern Melanesia the superspecies occupies all large islands except Rennell, but only one small island (Tong in the Northwest Bismarcks). Almost all taxon borders in Northern Melanesia lie at locations shared by many other species—for example, different allospecies in the Bismarcks (152A–C) and Solomons (152D–F); within the Bismarcks, distinct allospecies as peripheral isolates on St. Matthias (152B) and in the Admiralities (152A), plus a megasubspecies on Dyaul (152Cb); within the Solomons, distinct allospecies on the Bukida group (152D), New Georgia group (152E), and San Cristobal group (152F); within the New Georgia group, three distinct megasubspecies on Vella Lavella and Bagga (152Ea), Ganonga (152Eb), and the Gatumbangra and Rendipari groups (152Ec, d), plus two separate weak subspecies on the latter two groups; and endemic megasubspecies on Malaita (152Db) and Ugi (152Fb). The only unusual taxon border is the separate megasubspecies on Rambutyo (152Ab), the most distinctive endemic taxon of that island. See p. 150 of chapter 20 for discussion.

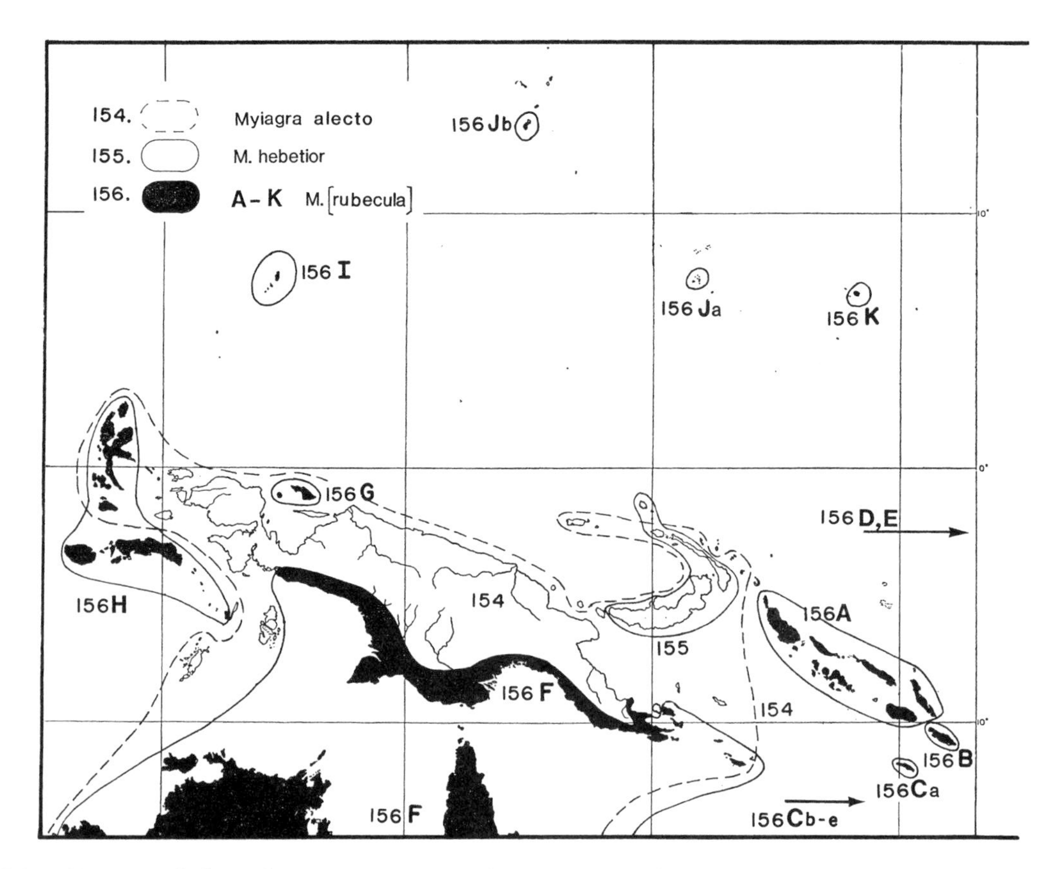

Map 42. monarch flycatchers 154 = *Myiagra alecto*, 155 = *M. hebetior*, and 156 = superspecies *Myiagra* [*rubecula*] (plate 5). 156A = *M. ferrocyanea*; 156B = *M. cervinicauda*; 156C = *M. caledonica* (Ca = *M. c. occidentalis*, Cb = *M. c. caledonica*, Cc = *M. c. viridinitens*, Cd = *M. c. melanura*, Ce = *M. c. marinae*); 156D = *M. vanikorensis*; 156E = *M. albiventris*; 156F = *M. rubecula*; 156G = *M. atra*; 156H = *M. galeata*; 156I = *M. erythrops*; 156J = *M. oceanica* (Ja = *M. o. oceanica*, Jb = *M. o. freycineti*); 156K = *M. pluto*.

The *M.* [*rubecula*] superspecies consists of 11 allospecies (156A–K), among which 156 D and 156F occur sympatrically with closely related species in Fiji (156D) and New Guinea and Australia (156F), products of completed speciations. The superspecies splits up within the Solomons at the usual places: a distinct allospecies in the San Cristobal group (156B); the Rennell subspecies (156Ca) derived from the New Hebrides and New Caledonia to the southeast (156Cb-e) rather than from the Central Solomons; and separate megasubspecies (not distinguished on this map) on the Bukida group (156Aa, b), Malaita (156Ac), and the New Georgia group (156Ad).

Despite its wide geographic range from the Moluccas (156H) east to Samoa (156E) and from South Australia (156F) north to Guam (156Jb), the superspecies is absent from the Bismarcks, occupied instead by its distinct congeners *M. alecto* (154) and *M. hebetior* (155), which may have arisen as Bismarck representatives of *M.* [*rubecula*]. *M. alecto* and *M. hebetior* are sibling species (see plate 5) at a late stage in speciation: all *M. hebetior* populations except the one on St. Matthias occur sympatrically with *M. alecto*, but *M. alecto* also occurs allopatrically in a wide range encompassing New Guinea, Australia, and the Moluccas. See p. 148 of chapter 20 and p. 176 of chapter 22 for discussion.

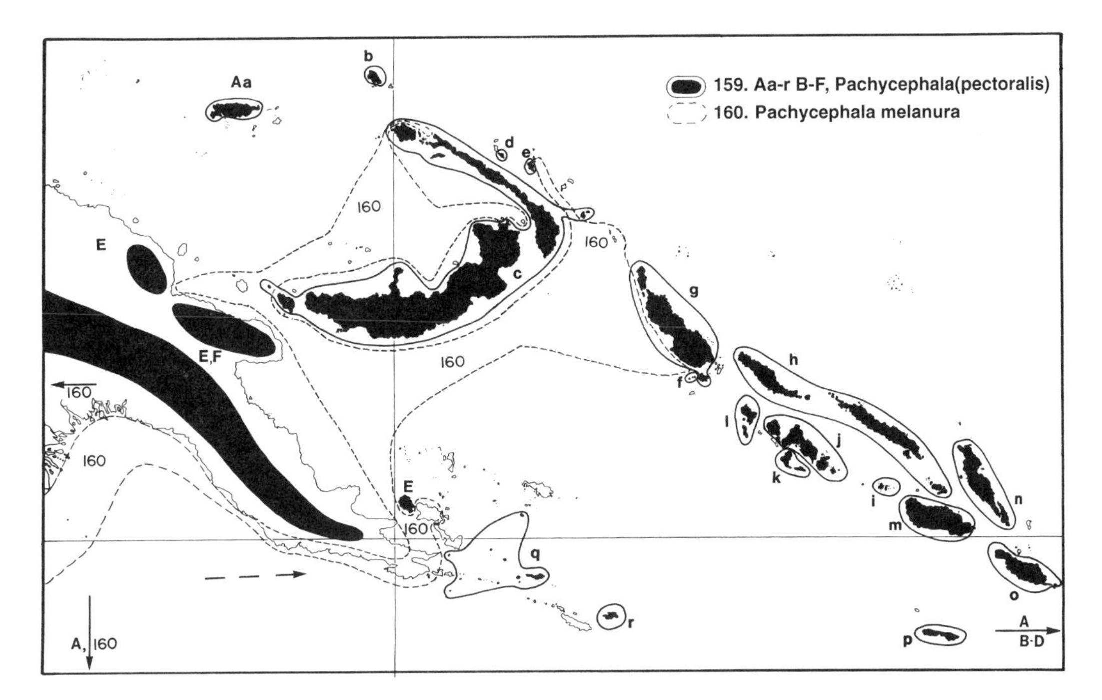

Map 43. 159 = whistler superspecies *Pachycephala* [*pectoralis*] (Plate 2). 159A = *P. pectoralis* (Aa = *P. p. goodsoni*, Ab = *P. p. sexuvaria*, Ac = *P. p. citreogaster*, Ad = *P. p. tabarensis*, Ae = *P. p. ottomeyeri*, Af = *P. p. whitneyi*, Ag = *P. p. bougainvillei*, Ah = *P. p. orioloides*, Ai = *P. p. pavuvu*, Aj = *P. p. centralis*, Ak = *P. p. melanoptera*, Al = *P. p. melanonota*, Am = *P. p. cinnamomea*, An = *P. p. sanfordi*, Ao = *P. p. christophori*, Ap = *P. p. feminina*, Aq = *P. p. collaris*, Ar = *P. p. rosseliana*); 159B = *P. caledonica*; 159C = *P. melanops*; 159D = *P. flavifrons*; 159E = *P. soror*; 159F = *P. schlegelii*. 160 = *P. melanura*.

These golden whistlers provide the most complicated examples of speciation within Northern Melanesia, as well as the richest examples of geographic variation of any bird species in the world. Within Northern Melanesia, speciation has proceeded to completion to produce two pairs of sympatric species. First, *P. implicata* (species 161, range not depicted) is endemic to the mountains of the two highest Solomon islands, Bougainville and Guadalcanal, where it coexists by altitudinal segregation with the races 159 Ag and Am, respectively, of the widespread *P. pectoralis*. Second, the small-island specialist *P. melanura* (species 160) extends from Australia to the Northwest Solomons and lives in close proximity to *P. pectoralis* large-island races 159 Ac, d, e, and g. Speciation has also proceeded to completion to produce pairs of sympatric species on Australia and four Lesser Sunda islands and to produce about six sympatric species (including 159E, 159F, and 160) on New Guinea.

The *P.* [*pectoralis*] superspecies falls into six allospecies (159A–F) over its range from Java to Samoa. Its Northern Melanesian allospecies *P. pectoralis* (159A) occupies most of this range (Java to Fiji) and falls into about 66 subspecies, of which 16, constituting six very different megasubspecies (see plate 2), live within Northern Melanesia and occupy most major islands. Each of the Solomon races 159Ak, Al, An, and Ap, living on the Rendipari group, the Vellonga group, Malaita, and Rennell, respectively, constitutes a separate megasubspecies. Most remaining Solomon races (159 Ag, h, i, j, m) constitute a fifth megasubspecies, which has hybridized with the New Hebrides megasubspecies from the southeast to produce a stabilized hybrid race (159Ao) on San Cristobal, and which has hybridized with *P. melanura* (species 160) at the eastern periphery of the latter's range to produce a variable hybrid race (159 Af) on islets of the Shortland group. All Bismarck races (159 Aa–e) belong to a widespread sixth megasubspecies distributed from the Lesser Sundas to Fiji. See pp. 149, 177, and 181 of chapters 20, 22, and 23, respectively, for discussion.

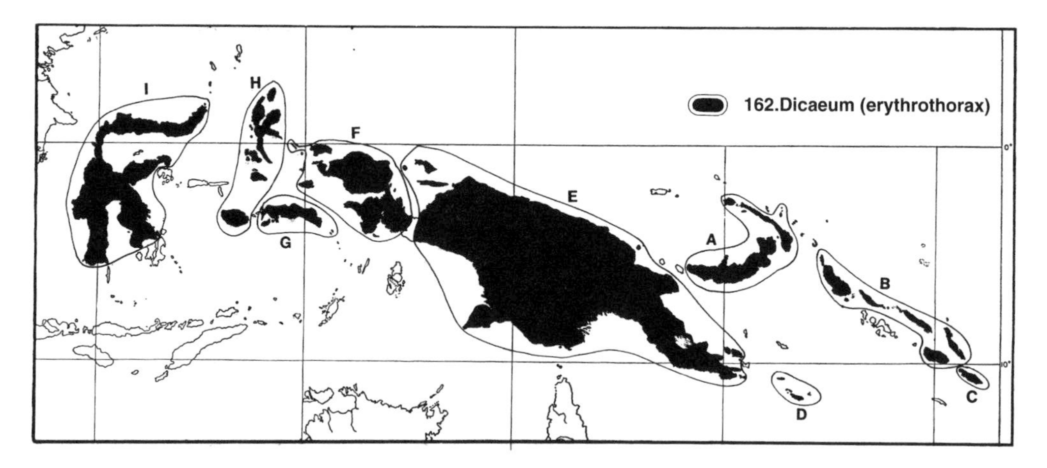

Map 44. 162 = flower-pecker superspecies *Dicaeum* [*erythrothorax*]. A = *D. eximium*; B = *D. aeneum*; C = *D. tristrami*; D = *D. nitidum*; E = *D. geelvinkianum*; F = *D. pectorale*; G = *D. vulneratum;* H. = *D. erythrothorax*; I = *D. nehrkorni*. The superspecies falls into nine allospecies, three of which (162 A, B, C) are endemic to Northern Melanesia. The Bismarck allospecies 162A and the widespread Solomon allospecies 162B, in turn, each fall into three subspecies, but those subspecies are weakly distinct compared to the large differences between the allospecies. The most peripheral allospecies, *D. tristrami* of San Cristobal (162C), resembles a plump, brown sparrow and is so bizarre that its placement in this superspecies is debated. Note the striking absence from the New Georgia group, as well as from the outlying Northwest Bismarcks, St. Matthias, and Ugi.

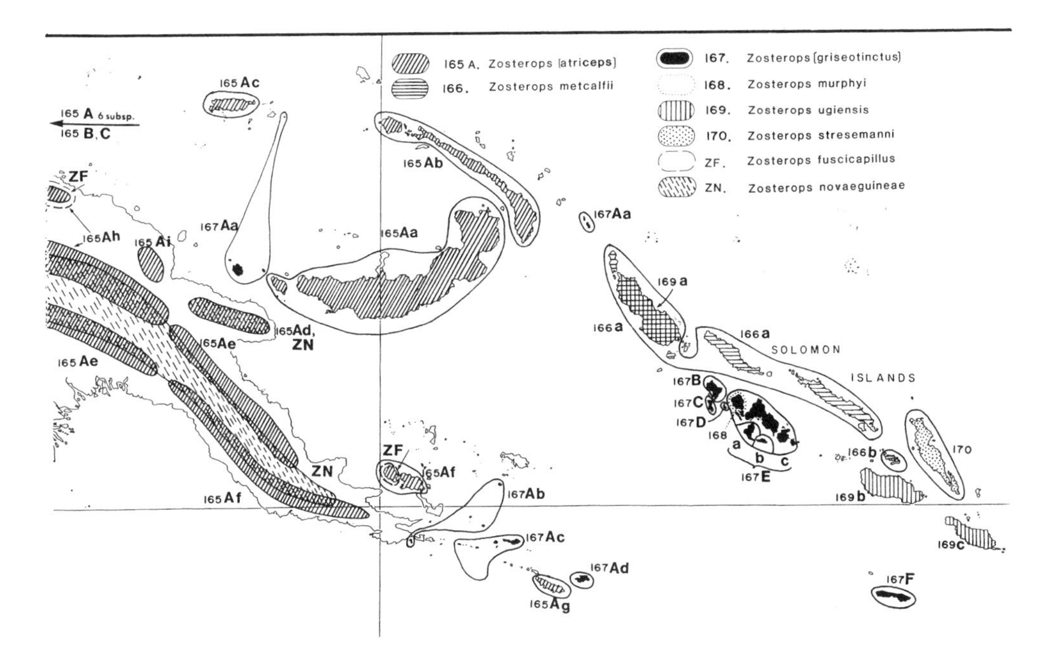

Map 45. 165–170 = Species of *Zosterops* white-eyes (plate 1). 165 = Z. [*atriceps*]. 165A = Z. [*atriceps*] *atrifrons* (Aa = *Z. a. hypoxanthus*, Ab = *Z. a. ultimus*, Ac = *Z. a. admiralitatis*, Ad = *Z. a. gregaria*, Ae = *Z. a. chrysolaema*, Af = *Z. a. delicatula*, Ag = *Z. a. meeki*, Ah = *Z. a. minor*, Ai = *Z. a.* subsp); 165B = Z. [*atriceps*] *atriceps*; 165C = Z. [*atriceps*] *mysorensis*. 166 = *Z. metcalfii* (a = *Z. m. metcalfii*, b = *Z. m. floridanus*). 167 = Z. [*griseotinctus*]. 167A = Z. [*griseotinctus*] *griseotinctus* (Aa = *Z.g. eichhorni*, Ab = *Z. g. longirostris*, Ac = *Z. g. griseotinctus*, Ad = *Z. g. pallidipes*); 167B = Z. [*griseotinctus*] *vellalavella*; 167C = Z. [*griseotinctus*] *spendidus*; 167D = Z. [*griseotinctus*] *luteirostris*; 167E = Z. [*griseotinctus*] *rendovae* (Ea = *Z. r. rendovae*, Eb = *Z. r. tetiparius*, Ec = *Z. r. kulambangrae*); 167F = Z. [*griseotinctus*] *rennellianus*. 168 = *Z. murphyi*. 169 = *Z. ugiensis* (a = *Z. u. hamlini*, b = *Z. u. oblitus*, c = *Z. u. ugiensis*). 170 = *Z. stresemanni*. ZF = *Z. fuscicapillus*. ZN = *Z. novaeguineae*.

Depicted are the ranges of all six *Zosterops* species of Northern Melanesia (165–170) and the two additional species of eastern New Guinea (ZF, ZN). Pairs of sympatric species coexist by altitudinal segregation at five sites: in the mountains of Bougainville (169a and 166a), Kulambangra (168 and 167 Ec), Goodenough (ZF and 165Af), New Guinea's North Coastal Range (ZF and 165Ah) and other mountains of eastern New Guinea (ZN and 165Ad, e, f, h).

Within Northern Melanesia, *Z. stresemanni* (170) and *Z. murphyi* (168) are each endemic to a single island, Malaita and Kulambangra, respectively. Six large Bismarck islands are occupied by an endemic megasubspecies (divided further into the weak subspecies 165Aa–c) of the allospecies *Z. atrifrons*, which extends west to New Guinea (165Ad–i) and Wallacea, with two related allospecies on Biak (165C) and the Moluccas (165B). *Z. metcalfii*, divided into two weak subspecies (166a, b), is endemic to islands formed by post-Pleistocene rising sea levels from Pleistocene Greater Bukida. *Z. ugiensis* is endemic to the mountains of three high Solomon islands, Bougainville (169a) and Guadalcanal (169b) and San Cristobal (169c).

The most complicated distribution is of the superspecies Z. [*griseotinctus*] (167). Its western allospecies *Z. griseotinctus* is a supertramp, with four subspecies (167Aa–d) on small or volcanically disturbed islands. Its eastern allospecies (167F) is endemic to the Solomon outlier Rennell. The other four allospecies are endemic to islands within short distances of each other in the New Georgia group (167B, C, D, E). One of those allospecies, *Z. rendovae* (167E), is further divided into three megasubspecies (167 Ea, b, c) separated by water gaps as narrow as 3.5 km. The Kulumbangra endemic *Z. murphyi* (168) may belong to the *Z. griseotinctus* superspecies, in which case the superspecies would be represented by a doublet on Kulambangra (168 *Z. murphyi* and 167 *Z. rendovae kulambangrae*). See pp. 158 and 177 of chapters 21 and 22 for discussion.

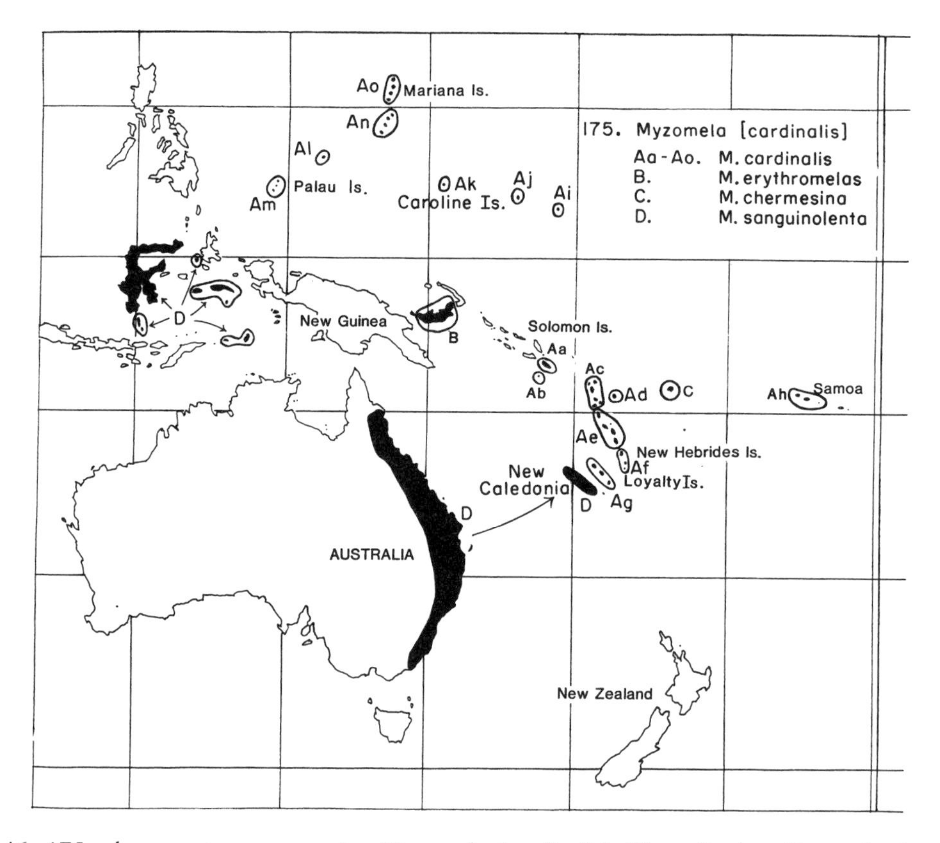

Map 46. 175 = honey-eater superspecies *Myzomela* [*cardinalis*] (Plate 4). A = *M. cardinalis* (Aa = *M. c. pulcherrima*, Ab = *M. c. sanfordi*, Ac = *M. c. sanctaecrucis*, Ad = *M. c. tucopiae*, Ae = *M. c. tenuis*, Af = *M. c. cardinalis*, Ag = *M. c. lifuensis*, Ah = *M. c. nigriventris*, Ai = *M. c. rubratra*, Aj = *M. c. dichromata*, Ak = *M. c. major*, Al = *M. c. kurodai*, Am = *M. c. kobayashii*, An = *M. c. saffordi*, Ao = *M. c. asuncionis*); B = *M. erythromelas*; C. = *M. chermesina*; D = *M. sanguinolenta*. This wide-ranging superspecies encompasses four allospecies: two localized endemics (175B on New Britain, 175C on Rotuma), *M. sanguinolenta* (175D) in the west and south, and *M. cardinalis* (175A) in the east and north. Three other closely related species live in New Guinea, North Australia, and the Lesser Sundas. The eastern allospecies *M. cardinalis* invaded the eastern Solomons, where it differentiated into one megasubspecies on Rennell (175Ab) and another on the San Cristobal Group (175Aa). *M. cardinalis* may have colonized San Cristobal from smaller islands of the San Cristobal group only within the past century, and it hybridized there with its congener *M. tristrami* (species 177F, map 47) in the first decades of the invasion (p. 180 of Chapter 23).

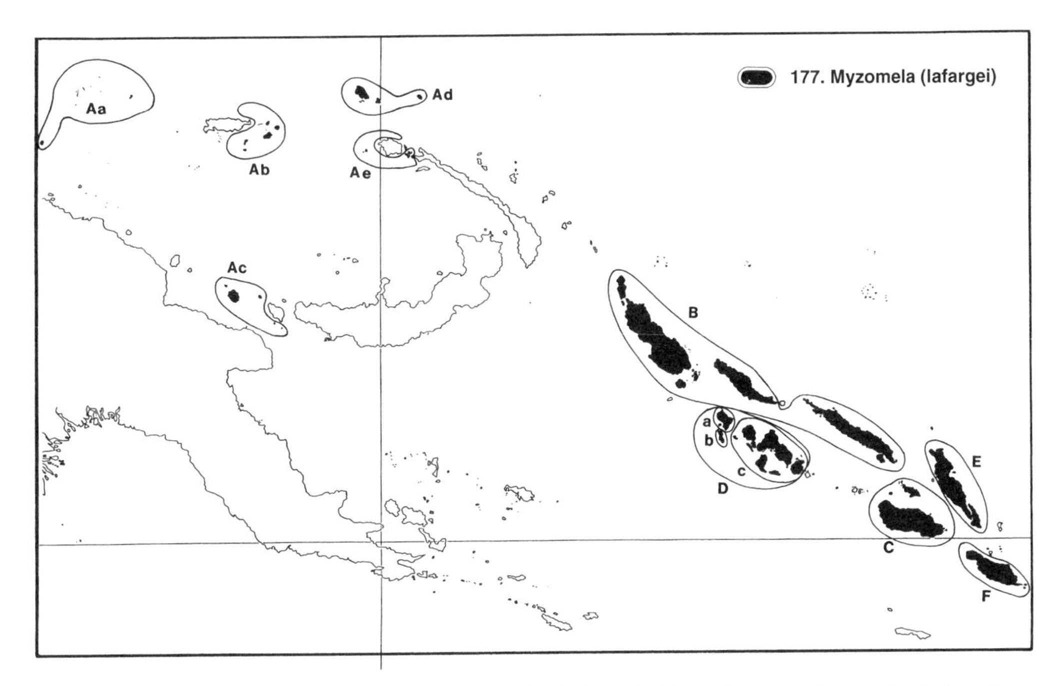

Map 47. 177 = honey-eater superspecies *Myzomela [lafargei]* (Plate 4). A = *Myzomela [lafargei] pammelaena* (Aa = *M. p. ernstmayri*, Ab = *M. p. pammelaena*, Ac = *M. p. nigerrima*, Ad = *M. p. hades*, Ae = *M. p. ramsayi*; B = *M. [lafargei] lafargei;* C = *M. [lafargei] melanocephala*; D = *M. [lafargei] eichhorni* (Da = *M. e. atrata*, Db = *M. e. ganongae*, Dc = *M. e. eichhorni*); E = *M. [lafargei] malaitae;* F = *M. [lafargei] tristrami*. This superspecies, endemic to Northern Melanesia, consists of five allospecies in the Solomons (177B–F), which evidently invaded the Bismarcks to give rise to a sixth there (*M. pammelaena*, 177A). The latter is among the Solomon colonists that became supertramps on invading the richer Bismarck avifauna, and it is now confined to small (races Aa, Ab, Ae), outlying (Ad), or volcanically disturbed (Ac) islands (see p. 85 of chapter 12). Race Ac arose recently by recolonization of Long and its neighbors from populations Aa or Ab, following the seventeenth-century volcanic defaunation of the Long Group. Within this short time, Ac diverged from Aa and Ab in the direction of larger size through character displacement with its smaller congener *M. sclateri* (species 176), with which *M. pammelaena* occurs sympatrically in the Long Group but nowhere else.

Geographic variation in the Solomons mostly follows the usual pattern: different allospecies on the Bukida group (177B), Guadalcanal (177C), New Georgia group (177D), Malaita (177E), and San Cristobal group (177F); absent from Rennell. The only exceptional feature is that the easternmost island of the Bukida group, Florida, is occupied by the Guadalcanal allospecies rather than the widespread Bukida allospecies (see p. 261 of chapter 32 for discussion). *M. tristrami*, the San Cristobal allospecies (177F), has hybridized there with its congener *M. cardinalis* (species 175Aa, map 46), apparently within the first decades of the latter's invasion of San Cristobal. See p. 180 of chapter 23.

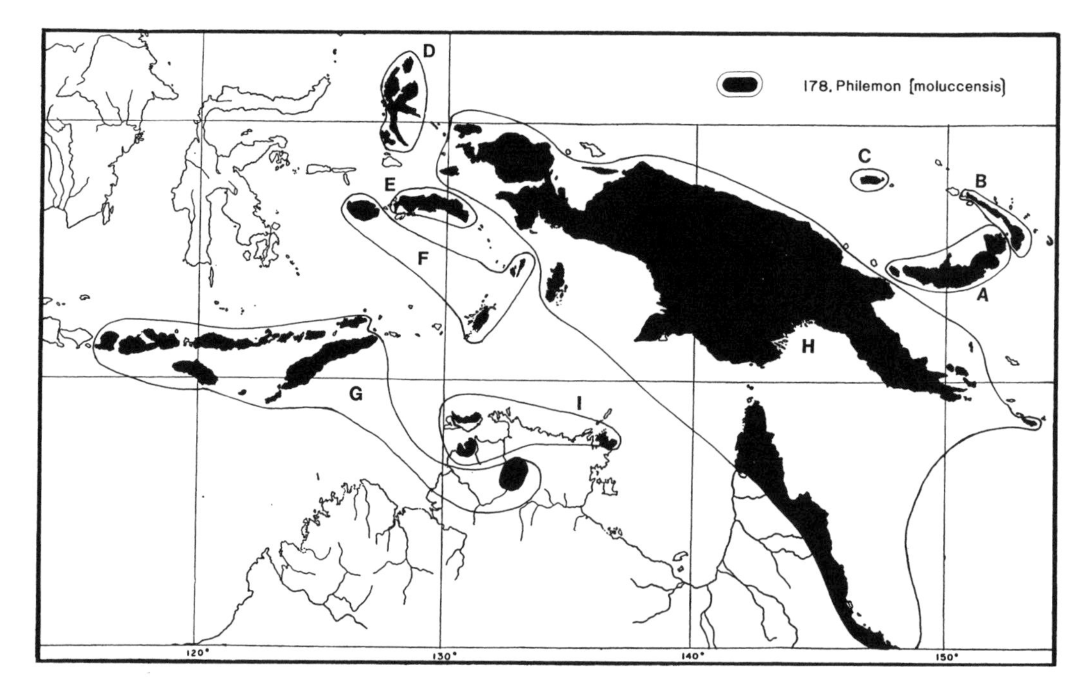

Map 48. 178 = honey-eater superspecies *Philemon* [*moluccensis*] (plate 7). A = *P.* [*moluccensis*] *cockerelli*; B = *P.* [*moluccensis*] *eichhorni*; C = *P.* [*moluccensis*] *albitorques*; D = *P.* [*moluccensis*] *fuscicapillus*; E = *P.* [*moluccensis*] *subcorniculatus*; F = *P.* [*moluccensis*] *moluccensis*; G = *P.* [*moluccensis*] *buceroides*; H = *P.* [*moluccensis*] *novaeguineae*; I = *P.* [*moluccensis*] *gordoni*. The superspecies consists of six allospecies outside Northern Melanesia (178D–I, of which D–H serve as models for the famous case of visual mimicry by orioles discovered by Alfred Russel Wallace; see Diamond 1982b), plus three allospecies endemic to three Bismarck islands or island groups (178A–C). The latter may have arisen by multiple waves of invasion from the ancestral allospecies 178H of lowland New Guinea, because the New Britain allospecies 178A is still fairly similar to the New Guinea source population, while the New Ireland allospecies 178B is distinct in plumage (see plate 7) and confined to mountains over 1000 m. Although all three Bismarck allospecies must have been founded by overwater colonists, all three have now completely lost that original overwater colonizing ability: none has been recorded as crossing water, and each is now confined either to a single island (178B) or else to a pair of islands that were joined at Pleistocene times of low sea level (178A, also 178C).

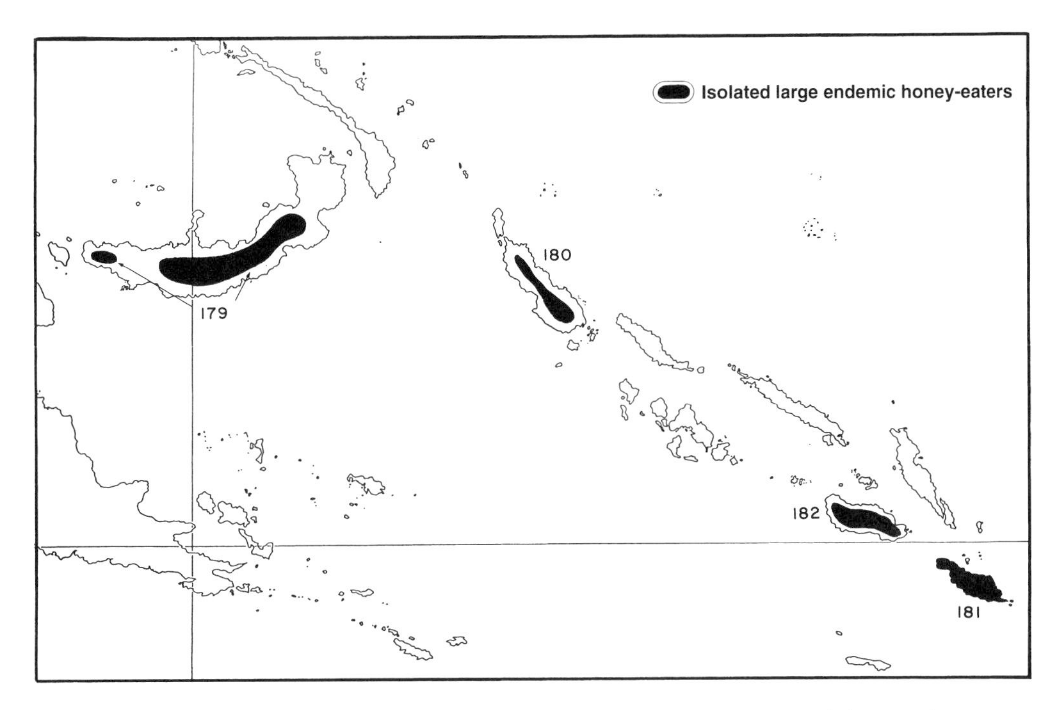

Map 49. 179–182 = Isolated, large endemic honey-eaters (plate 9). 179 = *Melidectes whitemanensis.* 180 = "*Stresemannia*" *bougainvillei.* 181 = *Meliarchus sclateri.* 182 = "*Guadalcanaria*" *inexpectata.* Each of these four honey-eaters is endemic to one large Northern Melanesian island, has never been recorded dispersing overwater, and must have lost the overwater colonizing ability that originally brought it to Northern Melanesia. Species 181 occurs in the lowlands of San Cristobal, while the other three species (179, 180, 182) are confined to the mountains of Northern Melanesia's three highest islands. The New Britain honey-eater (179) belongs to genus *Melidectes*, whose other nine species are confined to the mountains of New Guinea. The other three species (180–182) are assigned to monotypic genera out of ignorance of their relationships, not because they are distinctive (they are not).

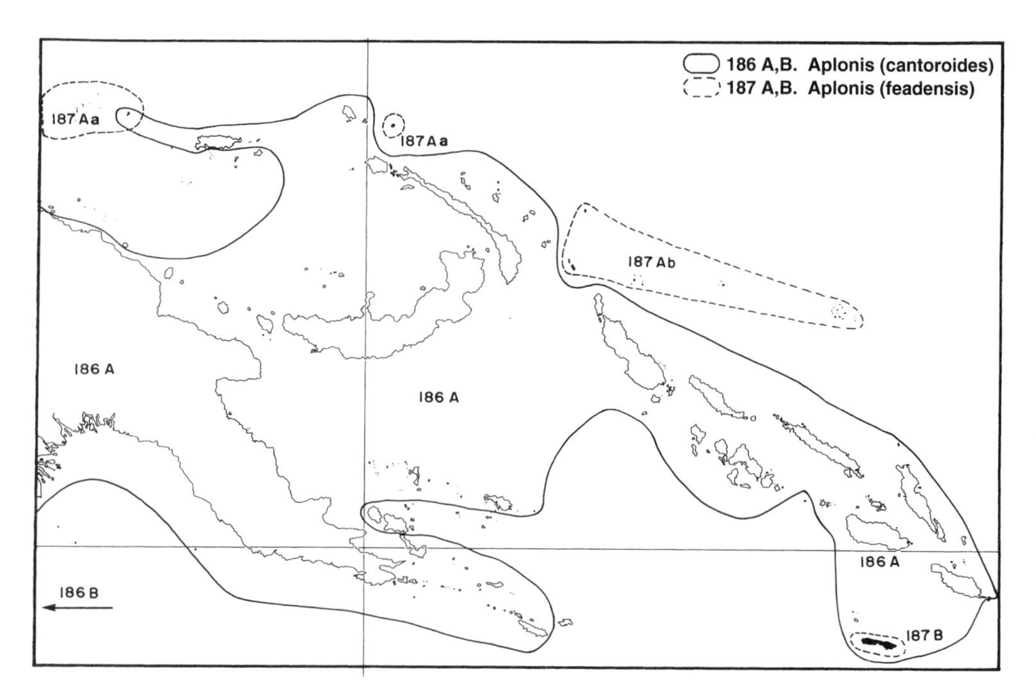

Map 50. 186 = starling superspecies *Aplonis* [*cantoroides*]. 186A = *A. cantoroides*; 186B = *A. crassa*. 187 = *Aplonis* [*feadensis*]. 187A = *A. feadensis* (Aa = *A. f. heureka*, Ab = *A. f. feadensis*); 187B = *A. insularis*. The vagile allospecies *A. cantoroides* (186A) ranges throughout almost the whole of Northern Melanesia and the New Guinea region without any geographic variation, but there is a sedentary endemic allospecies (186B) on Tenimbar in eastern Wallacea. Derived from *A.* [*cantoroides*] is the endemic Northern Melanesian supertramp superspecies *A.* [*feadensis*], with one allospecies endemic to remote islets of the Bismarcks (187Aa) and Northern Solomons (187Ab), another allospecies endemic to Rennell (187B), and sympatric with *A. cantoroides* on Rennell and Hermit and possibly Wuvulu. See p. 178 of chapter 22.

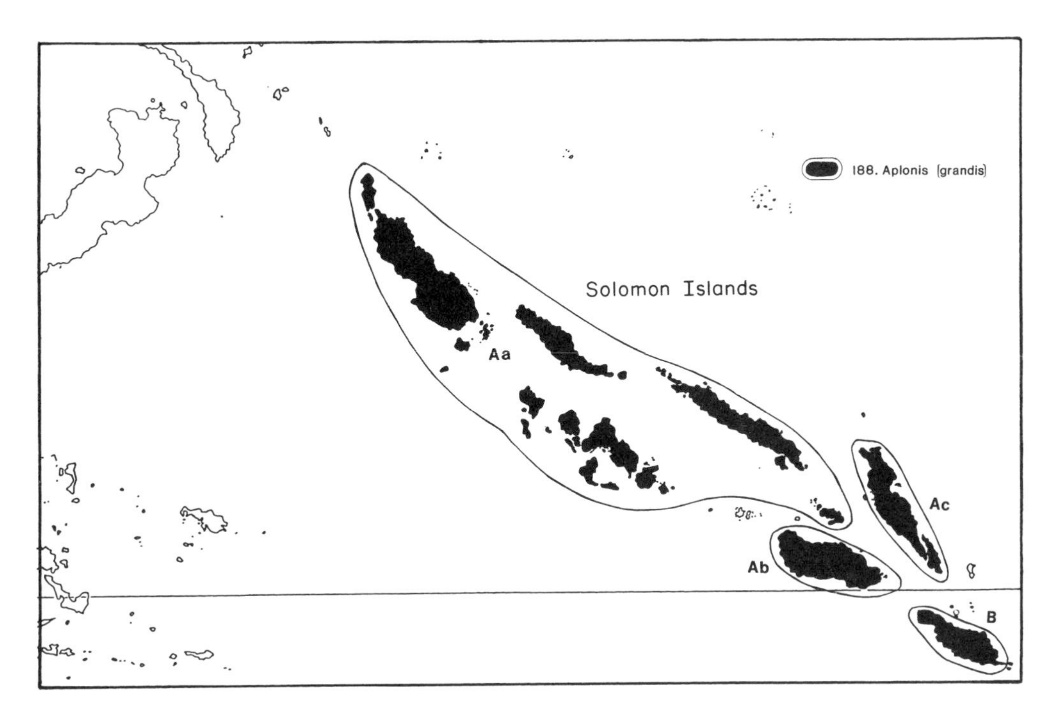

Map 51. 188 = starling superspecies *Aplonis* [*grandis*]. A = *A. grandis* (Aa = *A. g. grandis,* Ab = *A. g. macrura*, Ac = *A. g. malaitae*); B = *A. dichroa.* The superspecies is endemic to the Solomons and has no close relatives. Geographic variation conforms to a common pattern: a distinctive endemic allospecies on San Cristobal (188B), a distinctive megasubspecies on Malaita (189Ac), and the remaining megasubspecies (188Aa, b) widely distributed over the Bukida group, New Georgia group, and Guadalcanal, with only a weakly differentiated subspecies (188 Ab) on Guadalcanal; absent from Rennell. The widespread race 188Aa is a good overwater colonist that maintains itself even on small islands of the New Georgia Group, but the San Cristobal allospecies 188B is sedentary and has never been recorded off the large island of San Cristobal. See p. 149 of chapter 20 for discussion.

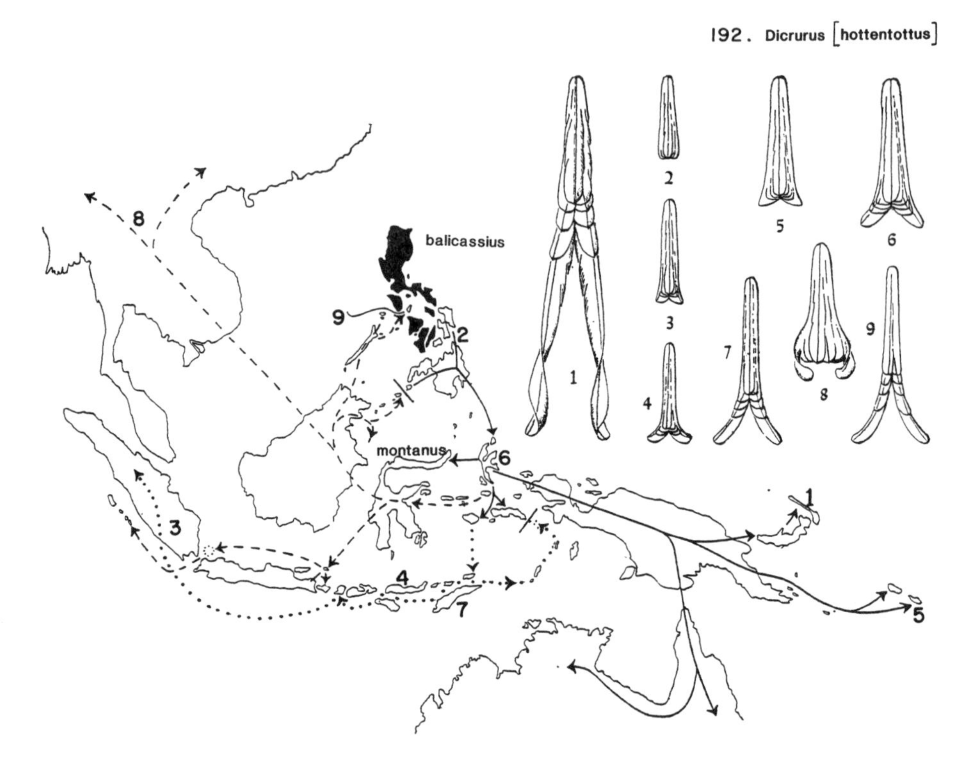

Map 52. 192 = drongo superspecies *Dicrurus* [*hottentottus*]. Tail forms and inferred dispersal routes of three branches of this highly variable superspecies are shown. Solid line: the earliest branch to disperse; dotted line: the next branch; dashed line: the most recently dispersing branch. *D. balicassius*, a closely related species, occupies a range in the northern Philippines allopatric to *D. hottentottus* in the southern Philippines. The closely related *D. montanus*, the product of a double invasion of Celebes by *D. hottentottus,* now lives in the mountains of Celebes above *D. hottentottus* in the lowlands. The nine numbers on the map indicate the geographic ranges of nine taxa whose tails are depicted at upper right. 1 = the allospecies *D.* [*hottentottus*] *megarhynchus* (New Ireland). 2–9 are subspecies of the allospecies *D.* [*hottentottus*] *hottentottus:* 2 = *samarensis* (central Philippines), 3 = *sumatranus* (Sumatra), 4 = *renschi* (Sumbawa), 5 = *longirostris* (San Cristobal), 6 = *atrocaeruleus* (Halmahera), 7 = *densus* (Timor), 8 = *hottentottus* (Southeast Asia), 9 = *menagei* (Tablas). The tails of races 4 and 6, which are near the ancestral geographic location of their respective branches, are typical for *D.* [*hottentottus*] and for drongos in general: forked, and without modifications of the outermost tail feathers. The tails of the other depicted races, which occupy geographically more peripheral locations within the distributions of their respective branches, are more aberrant and specialized in various directions. These trends reach their peaks in the geographically most peripheral populations: *sumatranus* (3) and *longirostis* (5), with their flat, almost square, nearly unforked tails; conversely, *menagei* (9), with its very deeply forked tail; *hottentottus* (8), with the tips of its outer tail feathers curled and twisted completely around; and *D. megarhynchus* (1), with its enormously prolonged and spirally twisted outer tail feathers (see plate 8). Based on Vaurie (1949) and Mayr and Vaurie (1948).

APPENDICES

Appendix 1

Systematic List

Breeding Land and Fresh-water Native Birds of Northern Melanesia

This appendix provides taxonomic information and island distributions for all 195 recent land and freshwater native bird species that have been recorded from any Northern Melanesian island and that are believed to breed or to have bred recently. We have reappraised the taxonomic status and island distribution of each species.

Our principal data sources have been as follows: (1) The bird collections of the American Museum of Natural History, which are by far the largest collections of Northern Melanesian birds. (2) Publications from the eighteenth century to the present. (3) The unpublished diary accounts, preserved as manuscripts in the American Museum of Natural History, that the collectors of the Whitney South Sea Expedition wrote about each island visited and its birds. These diaries provide detailed documentation for island occurrences of many bird species of which specimens were not collected on that island. (4) Unpublished observations by J. D. on most ornithologically significant Solomon islands, and on Bismarck islands from New Britain west to Crown, in the years 1969–1976. (5) Unpublished reports and personal communications of field records by other observers since 1942. (6) Study of Northern Melanesian specimens in the bird collections of the Zoological Museum of the Copenhagen Museum, and of the Bernice Bishop Museum (Honolulu).

Most of the 195 species tabulated are believed to be resident and regular breeders within Northern Melanesia. Borderline cases are three species (the pelican *Pelecanus conspicillatus*, the cuckoo *Scythrops novaehollandiae*, and the kingfisher *Halcyon sancta*) of which most Northern Melanesian records represent winter visitors or vagrants, and for which there are few breeding records; and the herons *Ardea intermedia* and *A. alba* and the spoonbill *Platalea regia*, for which the relative contributions of visitors and of resident breeders to records in Northern Melanesia are unknown. Winter visitors and vagrants for which there is no evidence of breeding are listed separately in appendix 2. The five non-native introduced species are listed separately in appendix 3, except that appendix 1 includes one species (the cassowary *Casuarius bennetti*) whose sole Northern Melanesian population (on New Britain) may result from a prehistoric introduction. We do not list marine bird species. We also do not list the few Northern Melanesian species or populations known only from fossils, which are discussed instead in chapter 7.

In the course of our reappraisal we rejected records that we did not find fully convincing, for either of two reasons: uncertainty about locality or about species identity. First, for some specimens collected in the nineteenth century, it is uncertain whether the locality listed on the specimen label or as stated by the describer is correct. This uncertainty especially affects specimens collected by nineteenth-century exploring expeditions (e.g., see tables 7.2 and 7.3) before the significance of accurate locality data became appreciated; specimens reported in the 1870s and 1880s by E. P. Ramsay, who was especially careless about specimen localities; and specimens reported in papers by Sclater and Ramsay and obtained for the Reverend George Brown by native hunters who collected on New Britain and New Ireland and brought back their specimens to Brown's home on Duke of York Island, with the result that many of those records attributed to Duke of York now appear in light of subsequent collections to be erroneous. Second, most records before 1940 were based on specimens, but increasing numbers since 1940 have been based instead on field identification without collection of specimens. Some of these field records (especially, but not only, some records by observers without lengthy experience in Northern Melanesia) are evident misidentifications. We have neither rejected nor accepted field records en masse but have evaluated them individually, based on the supporting details.

A practical problem that arose while finalizing the data we present is that we used island distributions for many laborious quantitative analyses and for constructing maps. Some new distributional records continued to accumulate during finalization of the manuscript and during galley proof stage. It was easy to incorporate such records into this appendix, but it would have been prohibitively difficult and expensive to redo the analyses and maps. Rather than deprive readers of information, we have incorporated the records, but only in this appendix. As a result, there are minor numerical discrepancies between distributions as given in this appendix, and as analyzed in the text and as depicted in the maps. None of the discrepancies affect our main conclusions; they instead involve numbers such as additions of one or two species to a few of the *S* values of tables 9.1 and 9.2, and additions of one or two islands to a few of the *i* values of appendix 5. The most interesting additions are the recent arrival of the cormorant *Phalacrocorax carbo* as a breeder at Lake Tegano on Rennell, and of the moorhen *Gallinula tenebrosa* at two lakes on New Britain; and the addition to the avifauna of Dyaul of eight otherwise widespread species not previously recorded for that island.

Distributional ecologists and biogeographers who want to use the data of this appendix for quantitative analyses should bear in mind the following caveats. In analyses of resident populations, island records followed in this appendix by a question mark should be omitted because it is uncertain whether that species really does occur on that island; island records followed by "(vagrant)" should be omitted because that species is evidently a vagrant rather than a resident on that island; but island records followed by "(subsp?)" should be included because that species does occur on that island, and only its subspecific affiliation is uncertain. Some islands have been surveyed less completely than other islands; we expect that further studies could reveal a couple of additional resident populations on Buka, Dyaul, Santa Anna, Santa Catalina, Tabar, Tanga, and Wuvulu, more additional resident populations on Lolobau and Unea, and many more on Bagga. This appendix does not list the species of some islets whose species totals are reported in table 9: islets 9–16 of table 9.1 and islets 10–21, 28, and 29 of table 9.2.

Format of This Appendix

Each of the 195 zoogeographic species (isolated species, or else superspecies) of Northern Melanesian birds is assigned a number. Superspecies names are indicated by square brackets: for instance, species 1, *Casuarius bennetti*, does not belong to a superspecies, but species 2, *Tachybaptus [ruficollis]*, does consist of a superspecies. As discussed in the text, a superspecies is a monophyletic group of closely related and largely or entirely allopatric species (i.e., a group of taxa that have achieved reproductive isolation but little or no geographic overlap). A practical question then arises: how much geographic overlap among component taxa do we accept before dismembering one superspecies into two or more species or superspecies? In practice, within our superspecies numbers 54, 67, and 78 we accept some limited sympatry among the component taxa. We recommend that community ecologists using our data analyze taxa 54A and 54B as separate units, taxa 67A and 67B as separate units, and taxa 78A and 78B–C as separate units.

Within a superspecies, allospecies are assigned uppercase letters (e.g., the allospecies 2B *T. novaehollandiae* under the superspecies 2. *T. [ruficollis]*). Within each allospecies, or within each species not constituting a superspecies, subspecies are assigned lowercase letters: cf. the subspecies 2Ba *T. novaehollandiae rennellianus* under the allospecies *T. novaehollandiae*. For each subspecies, or for each allospecies not divided into subspecies, or for each monotypic species not divided into allospecies or subspecies, we list Northern Melanesian islands of occurrence by abbreviations (explained below). If the allospecies or species also includes some extralimital subspecies (i.e., species confined to outside Northern Melanesia), we then summarize briefly their ranges without naming those extralimital subspecies (e.g., the entry "other subspecies: Afr, Eurasia to NG" under 2A. *T. ruficollis* following its sole Northern Melanesian subspecies, 2Aa. *T. ruficollis collaris*).

If a species or superspecies is depicted in one of our nine color plates, or if its range is depicted in one of our 52 distributional maps, the plate or map number is cited after the taxon name. For each taxon whose type locality lies within Northern Melanesia (but not for taxa whose type locality lies outside Northern Melanesia), the taxon name is immediately followed in parentheses by the author and year of the description. For instance, the entry A1. *Casuarius bennetti* (Gould 1857)" means that the name and description of that cassowary were published by Gould in 1857 and that the type locality lies within Northern Melanesia. The full references for almost all of these descriptions will be found in the 16-volume reference, *Check-list of Birds of the World* (1931–1987) by J. L. Peters and his successors. For dozens of such references that we cite only as author/year descriptions, we save space by omitting the full citation in our list of references because the full citation can instead be looked up in Peters.

Following those paragraphs with island distributions, each species' or superspecies' entry continues with a paragraph specifying which, if any, of the subspecies mentioned are ranked as megasubspecies (i.e., as exceptionally distinctive subspecies, which some readers may prefer to rank as allospecies). An example is the paragraph about the distinctive megasubspecies *brevicauda* of species 4. *Phalocrocorax melanoleucos*. This paragraph will make it easier for taxonomists inclined to splitting to modify our taxonomy according to their preferences.

The next paragraph ("Other allospecies") lists extralimital allospecies, if there are any, of the same superspecies. For instance, the superspecies 2. *Tachybaptus* [*ruficollis*] includes two allospecies that occur in Northern Melanesia, plus a third allospecies (*T. rufolavatus*) confined to Madagascar.

The next paragraph ("Comments") provides more details, references, and taxonomic opinions about the distributional and taxonomic data summarized in the preceding paragraphs.

Island Names and Abbreviations

Island distributions (using abbreviations given below) for each subspecies, monotypic allospecies, or monotypic species are listed beginning with Bismarck islands of occurrence, followed after a semicolon by Solomon islands of occurrence, followed in parentheses by the extralimital range, if any. For instance, the entry for species 1. *Casuarius bennetti* reads "NB (and NG)," meaning that this species occurs on the Bismarck island of New Britain, on no Solomon island, and outside Northern Melanesia only on New Guinea. The entry for subspecies 2Aa. *Tachybaptus ruficollis collaris* reads "NB, NI, Long, Umb, Lol, Witu; Boug (and Huon Pensinsula of NG)," meaning that this subspecies occurs on those six named Bismarck islands, on the Solomon island of Bougainville, and on New Guinea's Huon Peninsula.

We now summarize abbreviations and alternative names for Bismarck islands, then for Solomon islands, then for extralimital islands. The island names that we use are those employed in most of the taxonomic literature on Northern Melanesian birds, which peaked in the years 1875–1957. Some alternative island names were used in the older literature (especially the older geographic literature), while some different names are now current in the independent nations of Papua New Guinea and Solomon Islands that constitute Northern Melanesia. Our choice of names is not intended to disacknowledge usage current today, but merely to facilitate use of most of the relevant taxonomic literature. Each abbreviation is followed by the full name used in the taxonomic literature (if the full name differs from the abbreviation), followed in parentheses by alternative names, with the current name italicized if it differs from our choice of name. For instance, the entry "3. NH = New Hanover (*Lavongai*)" means that the island cited in the 1875–1957 literature as New Hanover is abbreviated as NH, and is termed Lavongai in modern Papua New Guinea.

Bismarck Islands

We provide complete distributional records for the following 31 Bismarck islands. An entry that reads

"All Bismarcks" or "All islands of Northern Melanesia" or "Most Bismarck islands except" or "Most Bismarck islands (no records for)" means that that species has been recorded from all 31 of these Bismarck islands, with the exceptions noted (e.g., see species 6, 9, 20, 29, and 30 below). The term "Northwest Bismarcks" (abbreviated NW Bismarcks) refers to the islands numbered 5–12; "Admiralties" refers to the islands numbered 5–8.

1. NB = New Britain (Neu Pommern)
2. NI = New Ireland (Neu Mecklenburg)
3. NH = New Hanover (*Lavongai*)
4. SMt = St. Matthias (*Mussau*)
5. Man = Manus
6. Ramb = Rambutyo
7. Nau = Nauna (La Vandola)
8. SMig = San Miguel
9. Anch = Anchorites (Kanit, Kaniet)
10. Her = Hermits
11. Nin = Ninigos (Schachbrett, Echiquier)
12. Wuv = Wuvulu (Matty, Mary)
13. Long = Long (Ahrup)
14. Umb = Umboi (Rooke)
15. Vu = Vuatom
16. DY = Duke of York (Neu Lauenburg)
17. Cred = Credner (Palakuuro, Pigeon), an islet near Duke of York
18. Em = Emirau (Squally, Storm, Kerue)
19. Dyl = Dyaul (Djaul, Sandwich)
20. Ting = Tingwon (Portland)
21. Lih = Lihir (Gardenijs); a group consisting of Lihir, Mali, Masahet, Sinabiet, and Mahur islands
22. Feni = Feni (Ambitle, Anir, St. Jan)
23. Tab = Tabar; a group consisting of Simberi (Fischer), Tabar (Gardner), and Tatau islands
24. Tang = Tanga (Anthony, Caens, Kaan)
25. Lol = Lolobau
26. Tol = Tolokiwa (Lottin)
27. Sak = Sakar (Tupinier)
28. Cr = Crown
29. Witu = Witu (Garowe)
30. Unea = Unea (Mérite)
31. Tench = Tench

Solomon Islands

We provide complete distributional records for the following 45 Solomon islands. An entry that reads "All Solomons" or "All islands of Northern Melanesia" or "Most Solomon islands except" or "Most Solomon islands (no records for)" means that the species has been recorded from all of the islands numbered 1–15 and 17–45; these expressions ignore Bagga (island 16) because it was incompletely surveyed. The term "All Central Solomons" refers to the islands numbered 1–15 and 17–35, omitting Borokua (island 36) even though its geographical location is central within the Solomons, but it has few species so we would frequently have to say "All Central Solomon islands except Borokua" if "All Central Solomon islands" were defined as including Borokua. The term "New Georgia group" refers to the islands numbered 15–27. The term "San Cristobal Group" refers to the islands numbered 28–32. The term "outliers" refers to the islands numbered 37–45.

1. Buka = Buka
2. Boug = Bougainville
3. Shtl = Shortland (Alu)
4. Faur = Fauro
5. Ch = Choiseul
6. Ys = Ysabel (Bugotu)
7. Fla = Florida (*Nggela*)
8. BV = Buena Vista (Vatilau)
9. Savo = Savo
10. Guad = Guadalcanal
11. Pav = Pavuvu (in Russell group)
12. Ban = Banika (in Russell group)
13. Mal = Malaita
14. Ul = Ulawa
15. VL = Vella Lavella
16. Bag = Bagga
17. Gan = Ganonga (*Ranongga*, Ronongo)
18. Simb = Simbo (Narovo, Eddystone)
19. Gizo = Gizo (*Ghizo*)
20. Kul = Kulambangra (*Kolombangara*)
21. Wana = Wana Wana
22. Koh = Kohinggo (Arundel)
23. NGa = New Georgia
24. Vang = Vangunu
25. Gat = Gatukai (*Nggatokae*)
26. Rnd = Rendova (Hammond)
27. Tet = Tetipari (*Tetepare*, Montgomery)
28. SCr = San Cristobal (*Makira*, Bauro)
29. Ugi = Ugi (Golfe)
30. 3S = Three Sisters
31. SAn = Santa Anna (Awa Raha)
32. SCat = Santa Catalina (Awa Kaba)
33. Mono = Mono (Treasury, Stirling)
34. Rmos = Ramos
35. Gow = Gower (*Ndai*, Inattendue)
36. Bor = Borokua (Murray)
37. Ren = Rennell (Mu Nggava)
38. Bel = Bellona

39. Nis = Nissan (Green, Sir Charles Hardy)
40. Fead = Fead (Nuguria, Abgarris)
41. OJ = Ontong Java (Lord Howe, Liuaniua, Luanguia)
42. Sik = Sikaiana (Stewart)
43. Kili = Kilinailau (Carteret)
44. Tauu = Tauu (Taku, Mortlock, Marcken, Marqueen)
45. Nuku = Nukumanu (Tasman)

Extralimital Islands or Continents

We use the following abbreviations:

Afr = Africa
Au = Australia
CAm = Central America
Cel = Celebes (*Sulawesi*)
Eu = Europe
GS = Greater Sundas
LS = Lesser Sundas
Mad = Madagascar
Marq = Marquesas
Mol = Moluccas
NAm = North America
NCal = New Caledonia
NG = New Guinea
NHeb = New Hebrides
NZ = New Zealand
Phil = Philippines
SAm = South America

Compass directions are abbreviated N, S, E, W, SE, etc. For instance, SE Asia = Southeast Asia, N NG = North New Guinea, E Au = East Australia, SE Papuan Islands = Southeast Papuan Islands, and W Papuan Islands = West Papuan Islands.

Species List

CASUARIIDAE (cassowaries)

1. *Casuarius bennetti* (Gould 1857). NB (and NG). COMMENTS: The cassowary was probably introduced to NB by humans (White 1976). In the absence of adequate series from any locality, we take this species as monotypic (Mayr 1940a).

PODICIPEDIDAE (grebes)

2. *Tachybaptus* [*ruficollis*] (map 1)
   A. *T. ruficollis.* (a) *collaris* (Mayr 1945b). NB, NI, Long, Umb, Lol, Witu; Boug (and Huon Peninsula of NG). Other subspecies: Afr, Eurasia to NG.
   B. *T. novaehollandiae.* (a) *rennellianus* (Mayr 1931b). Ren. Guad: a sight record of a pair of undetermined subspecies and breeding status (Lees et al. 1991). Other subspecies: Java, Timor (?), Au, NG, NCal, NHeb

   MEGASUBSPECIES: none.
   OTHER ALLOSPECIES: *T. rufolavatus* (Mad).
   COMMENTS: *T. ruficollis* and *T. novaehollandiae* coexist on NG, and doubtfully on Java and Timor (Mayr 1943, White & Bruce 1986). See Storer (1963) for generic classification and Mayr (1945b) for geographic variation in *T. ruficollis.*

PELECANIDAE (pelicans)

3. *Pelecanus conspicillatus.* NB, Feni; Buka, Boug, Guad, Pav, Mal, VL, Gan, Gizo, Kul, Wana, NGa, Gat, Rnd, SCr, SAn, Mono, Ren, Nis (and Au, and vagrants to NG, Indonesia, NHeb, NZ). COMMENTS: The Northern Melanesian records represent wind-blown vagrants, some of which bred, after storms in 1952 and subsequent years (Bradley & Wolff 1956, Cain & Galbraith 1956, Galbraith & Galbraith 1962, Filewood 1969, 1972). The closest relatives are *P. onocrotalus* (Eurasia, Afr), *P. rufescens* (Afr), and *P. philippensis* (Eurasia).

PHALACROCORACIDAE (cormorants)[a]

4. *Phalacrocorax melanoleucos.* (a) *melanoleucos.* NB; Boug, Ch, Guad, Gan, Gizo, Kul, Wana, NGa, Gat, SCr, 3S, SAn (and Indonesia, Au, NG, Palau, Santa Cruz group, NCal). (b) *brevicauda* (Mayr 1931b). Ren. Other subspecies: NZ
   MEGASUBSPECIES: *brevicauda,* distinctive in its very small size (no size overlap with nominate *melanoleucos*).

[a]*Phalacrocorax carbo* has been observed on Ren in 1976 (Diamond 1984a), 1994 (Gibbs 1996), 1995 (M. LeCroy & P. Webb unpublished observations), and 1997 (Filardi et al. 1999). In 1976 Rennellese informants stated that the species did not breed and merely visited occasionally; in 1994 and 1995 they stated that it had begun breeding since further arrivals with Cyclone Nina in 1993. We have not included this species in our analyses of Northern Melanesian breeding species.

COMMENTS: Most records from the NGa group represent wind-blown vagrants after the cyclone of 1952. The closest relatives are *P. niger* (SE Asia and GS), *P. pygmaeus* (Eu), and *P. africanus* (Afr). See Amadon (1942a) for geographic variation.

ARDEIDAE (herons)

See Bock (1956), Payne & Risley (1976), Hancock & Elliott (1978), Payne in Peters (1979, vol. 1), and Sheldon (1987) for discussion of classification.

5. *Butorides striatus.* (a) *solomonensis* (Mayr 1940a). NB, NI; almost all central Solomon islands (no records for BV, Ul, SCat, and Rmos). Other subspecies: New World, Afr, Asia, east to Au, NG, NCal, NHeb, Fiji, Tahiti
MEGASUBSPECIES: none.
COMMENTS: See Mayr (1940a), Salomonsen (1966b), and Schodde et al. (1980) for geographic variation. Contrary to Payne (in Peters 1979, vol. 1), we separate the population of Santa Cruz, NHeb, and Fiji as *diminutus* (Mayr 1940a). We consider *virescens* and *sundevalli* as conspecific.

6. *Egretta sacra.* (a) *sacra.* Probably all islands of Northern Melanesia (no records for Ting, Fead, Kili, Tauu, Nuku) (and SE Asia, east to NZ and Tahiti). Other subspecies: NCal, Loyalty Is. COMMENTS: See Mayr & Amadon (1941) for geographic variation and polymorphism.

7. *Ardea intermedia.* (a) *intermedia.* NB, Man, Umb; Buka, Boug, SCr (and India to NG, Au). Other subspecies: Afr, Asia through Indonesia. COMMENTS: Breeding status in Northern Melanesia not established. We follow Mees (1975) and Payne (in Peters 1979, vol. 1) in not recognizing *plumifera.* We follow Sheldon (1987) in placing this species in *Ardea,* not (as traditionally) in *Egretta.*

8. *Ardea alba.* (a) *modesta.* NB, Man (vagr), Unea; Buka, Boug, Ys (vagr), Ren (vagr), Bel (vagr); also reported as vagrant on many islands of the New Georgia group (and India to Au, NG, NZ). Other subspecies: cosmopolitan. COMMENTS: Breeding status in Northern Melanesia not established. We follow Payne (in Peters 1979, vol. 1) and Sheldon (1987) in placing this species in *Ardea,* not in *Egretta.*

9. *Nycticorax* [*nycticorax*]
A. *N. caledonicus.* (a) *hilli.* NW Bismarcks (not recorded from Nau, Her) and Long (and NZ, Au, NG, Mol, LS). (b) *mandibularis* (Ogilvie-Grant 1888). Most Bismarck islands except for Long and the NW Bismarcks (no records for SMt, Cred, Em, Ting, Witu, Unea, Tench); most central Solomon islands (no records for Rmos) plus Bor, Nis, Fead. It is not known whether the populations of Umb, Tol, Sak, and Cr belong to 9Aa or 9Ab. Other subspecies: Bonin Islands, Borneo, Java, Cel, Phil, Micronesia, NCal
OTHER ALLOSPECIES: *N. nycticorax* (NAm, SAm, Afr, Eurasia to GS and Cel).
MEGASUBSPECIES: none.
COMMENTS: See Amadon (1942a) for geographic variation, and White (1973) and White and Bruce (1986) for the relations of *N. caledonicus* and *N. nycticorax* in Wallacea. Hybridization between those two allospecies has been reported for Java and possibly for Cel and Phil.

10. *Ixobrychus* [*minutus*]
A. *I. sinensis.* NB, Nin (vagr), Vu (vagr); Boug (and Asia to NG and Micronesia)
OTHER ALLOSPECIES: *I. exilis* (New World), *I. minutus* (Eurasia, Afr, Au).
COMMENTS: See Hadden (1981) for proof of breeding on Boug.

11. *Ixobrychus flavicollis.* (a) *australis.* NB, NI, NH, SMt, Man, Her, Nin, Umb, Vu, DY, Em, Dyl, Lih, Feni, Tab, Witu (and Au, NG, Mol, Timor). (b) *woodfordi* (Ogilvie-Grant 1888). Boug, Shtl (?), Ch, Ys, Guad, Ul, VL, Kul, Wana, Koh, NGa, Vang, Tet, Ren. Other subspecies: SE Asia through Phil and GS to Cel
MEGASUBSPECIES: *woodfordi,* distinct in plumage of the female and in the relatively long bill and legs.
COMMENTS: We do not recognize the races *gouldi* nor *pallidior* (Mayr 1931b), nor the genus *Dupetor.* See Mayr (1945b) and Bradley (1962).

THRESKIORNITHIDAE (ibises and spoonbills)

12. *Threskiornis* [*aethiopicus*]
    A. *T. moluccus*. (a) *pygmaeus* (Mayr 1931b). Ren, Bel. Vagrants unidentified to subspecies have reached Mono. Other subspecies: Au, NG, Mol
    MEGASUBSPECIES: *pygmaeus*, distinct in its very small size and short bill.
    OTHER ALLOSPECIES: *T. aethiopicus* (Afr, Asia).
    COMMENTS: *T. moluccus* and *T. aethiopicus* may be conspecific (e.g., Payne in Peters 1979, vol. 1). The Mad (*bernieri*) and SE Asian (*melanocephalus*) populations are variously considered as allospecies (e.g., White & Bruce 1986) or as subspecies of *T. aethiopicus* (e.g., Payne in Peters 1979, vol. 1). See Mayr (1931b) and Lowe & Richards (1991) for geographic variation in *T. moluccus* and in the superspecies, respectively.

13. *Platalea* [*leucorodia*]
    A. *P. regia*. 3S (vagr), Ren, Bel (vagr) (and Au, NZ, Java)
    OTHER ALLOSPECIES: *P. leucorodia* (Eurasia), sometimes considered conspecific.
    COMMENTS: Breeding status on Ren not established. *P. minor* (Korea and China) and *P. alba* (Afr) are related.

ANATIDAE (ducks, geese, and swans)

See Johnsgard (in Peters 1979, vol. 1).

14. *Dendrocygna* [*bicolor*]
    A. *D. arcuata*. (a) *pygmaea* (Mayr 1945b). NB, Umb (and formerly Fiji). Other subspecies: Phil, Indonesia, Au, NG, NCal
    MEGASUBSPECIES: none.
    OTHER ALLOSPECIES: *D. bicolor* (NAm, SAm, Afr, Asia).
    COMMENTS: See Mayr (1945b) for geographic variation. While White & Bruce (1986) suggest that this species may be monotypic, we recognize *pygmaea* on the basis of its small size. *D. javanica* is related.

15. *Dendrocygna guttata*. NB, Umb (and Phil and Cel to NG).

16. *Anas* [*platyrhynchos*]
    A. *A. superciliosa*. (a) *pelewensis*. NB, NI, NH, Man, Ramb, Long, Umb, Lih, Feni, Tab; most central Solomon islands (no records for Shtl, Savo, Ul, SCat, Rmos, Gow), plus Ren (and N NG, Micronesia, to Polynesia). Other subspecies: Indonesia, S NG, Au, NZ
    OTHER ALLOSPECIES: *A. poecilorhyncha* (SE Asia), *A. luzonica* (Phil), *A. melleri* (Mad), *A. undulata* (Afr), *A. platyrhynchos* (NAm, Eurasia), *A. rubripes* (NAm).
    COMMENTS: See Amadon (1943a) for geographic variation in *A. superciliosa* and Livezey (1991) for relationships. *A. poecilorhyncha* and *A. superciliosa* may be conspecific. *A. oustaleti* (Marianas) probably represents hybrid *A. superciliosa* x *A. platyrhynchos* (Johnsgard in Peters 1979, vol. 1). *A. laysanensis* (Laysan), *A. wylvilliana* (Hawaii), *A. fulvigula* (NAm), and *A. diazi* (Mexico) are sometimes considered separate allospecies, rather than as subspecies of *A. platyrhynchos*.

17. *Anas* [*gibberifrons*]
    A. *A. gibberifrons*. (a) *remissa* (Ripley 1942a). Ren (extinct: Diamond 1984a). Other subspecies: Andaman, Java, LS, Cel, Au, NG, NZ, NHeb, NCal
    MEGASUBSPECIES: none.
    OTHER ALLOSPECIES: *A. bernieri* (Mad).
    COMMENTS: *A. castanea* (Au) is closely related; *A. chlorotis* (NZ), *A. aucklandica* (Auckland Island), and *A. nesiotis* (Campbell Island) are less close. See Ripley (1942a) for geographic variation and Livezey (1991) and Kennedy and Spencer (2000) for relationships.

ACCIPITRIDAE (hawks)[b]

See Brown and Amadon (1968) Stresemann and Amadon (in Peters 1979, vol. 1), and Amadon and Bull (1988) for the whole family and Mayr (1957) and Wattel (1973) for *Accipiter*.

[b] *Milvus migrans* has never been reported for NB since the passing mention by Ramsay (1877) in his report on a collection secured for the Australian Museum. That museum has no record of the specimen, and we doubt Ramsay's claim because of his notorious unreliability with localities.

18. *Aviceda* [*cuculoides*]. (map 2)
A. *A. subcristata*. (a) *coultasi* (Mayr 1945b). Man, SMig. (b) *bismarckii* (Sharpe 1888b). NB, NI, NH, Dyl, Tab, Lol. (c) *proxima* (Mayr 1945b). Buka, Boug, Shtl, Fauro, Pav, New Georgia Group (except no records for Vang), Mono. (d) *robusta* (Mayr 1945b). Ch, Ys. (e) *gurneyi* (Ramsay 1882f). Fla (subsp?), Guad, Mal, Ul, SCr, Ugi, SAn, SCat. Other subspecies: LS, Mol, NG, Au
MEGASUBSPECIES: none.
OTHER ALLOSPECIES: *A. jerdoni* (SE Asia to GS, Phil, Cel), *A. cuculoides* (Afr), *A. madagascarensis* (Mad).
COMMENTS: See Mayr (1945b) for subspecies and Schodde (1977) for a different view.

19. *Henicopernis* [*longicauda*]
A. *H. infuscata* (Gurney 1882). NB, Lol
OTHER ALLOSPECIES: *H. longicauda* (NG).

20. *Haliastur indus*. (a) *girrenera*. Virtually all Bismarck islands except Feni (no records for Anch, Her, Cred, Witu, Unea, Tench) (and Mol, NG, Au). (b) *flavirostris* (Condon & Amadon 1954). Feni; all central Solomon islands plus Bor, Nis. Other subspecies: SE Asia through Indonesia
MEGASUBSPECIES: none.
COMMENTS: See Condon & Amadon (1954) for geographic variation. We doubt the provenance of the specimen of *flavirostris* reported for DY by Galbraith and Galbraith (1962).

21. *Accipiter* [*gentilis*]
A. *A. meyerianus*. NB, Long (?), Umb, Vu (vagr); Buka (?), Guad, VL, Kul (and Mol, NG)
OTHER ALLOSPECIES: *A. gentilis* (NAm and Eurasia), and perhaps *A. melanoleucus* (Afr) and *A. henstii* (Mad). *A. buergersi* of NG may be related (Wattel 1973). See Mayr (1934a) for discussion of plumages.

22. *Accipiter novaehollandiae*. (a) *manusi* (Mayr 1945b). Man, Ramb, Nau, SMig. (b) *matthiae* (Mayr 1945b). SMt. (c) *lihirensis* (Stresemann 1933). Lih, Tang. (d) *lavongai* (Mayr 1945b). NI, NH, Dyl, Tab (subsp?). (e) *dampieri* (Gurney 1882). NB, Umb, Vu (vagr), DY, Lol. (f) *bougainvillei* (Rothschild & Hartert 1905). Buka, Boug, Shtl, Faur, Mono. (g) *rufoschistaceus* (Rothschild & Hartert 1902a). Ch, Ys, Fla. (h) *rubianae* (Rothschild & Hartert 1905). VL, Bag, Simb, Gizo, Kul, Wana, Koh, NGa, Vang, Rnd, Tet. (i) *pulchellus* (Ramsay 1882f). Guad. (j) *malaitae* (Mayr 1931e). Mal. Other subspecies: LS, Mol, NG, Au
MEGASUBSPECIES: none.
COMMENTS: We do not follow Schodde (1977), who separates the Papuan forms as an allospecies (*A. hiogaster*) from Australian *A. novaehollandiae*. For geographic variation in Northern Melanesia, see Mayr (1945b); for outside Northern Melanesia, see Condon and Amadon (1954), Wattel (1973), and White and Bruce (1986). The nearest relatives are *A. fasciatus* and *A.* [*rufitorques*], and possibly (more distantly) *A. imitator* and *A. luteoschistaceus* (Wattel 1973).

23. *Accipiter fasciatus*. (a) *fasciatus*. Ren, Bel (and S Au). Other subspecies: LS, Buru, NG, N Au, NCal, NHeb. COMMENTS: See Condon and Amadon (1954) and White and Bruce (1986) regarding subspecies. See under species 22, *A. novaehollandiae*, for relationships.

24. *Accipiter imitator* (Hartert 1926a). (map 3). Boug, Ch, Ys. COMMENTS: An isolated species (Mayr 1957). See Wattel (1973) and Schodde (1977) for possible relationships, explained under species 22, *A. novaehollandiae*.

25. *Accipiter* [*rufitorques*]. (map 3)
A. *A. albogularis*. (a) *eichhorni* (Hartert 1926a). Feni. (b) *woodfordi* (Sharpe 1888a). Buka, Boug, Shtl, Faur, Ch, Ys, Fla, Guad, Mal, Ul, Mono. (c) *gilvus* (Mayr 1945b). VL, Simbo, Gizo, Kul, Wana, Koh, NGa, Vang, Rnd, Tet, Bor (subsp?). (d) *albogularis* (Gray 1870). SCr, Ugi, SAn, SCat. Other subspecies: Santa Cruz Islands
MEGASUBSPECIES: none.
OTHER ALLOSPECIES: *A. melanochlamys* (NG), *A. rufitorques* (Fiji), *A. haplochrous* (NCal).
COMMENTS: See Mayr (1957) and under species 22, *A. novaehollandiae*, for relationships, and Mayr (1945b) for geographic variation.

26. *Accipiter luteoschistaceus* (Rothschild & Hartert 1926b). (Map 3). NB, Umb. COMMENTS: An isolated species with a distinctive, kestrel-like juvenal plumage (Mayr 1957). It may be the ecological equivalent of species 21 (*A.* [*rufitorques*]) in the Bismarcks (Diamond 1976). See Wattel (1973) for possible relationships as explained under species 25 (*A. novaehollandiae*), and see Diamond (1976) for discussion. There is also a claimed sighting for NI (Finch & McKean 1987), which we consider insufficiently substantiated because of the subsequent collection of the very similar *A. brachyurus* on NI.

27. *Accipiter* [*poliocephalus*].
    A. *A. princeps* (Mayr 1934a). NB
    OTHER ALLOSPECIES: *A. poliocephalus* (NG).
    COMMENTS: See Mayr (1934a).

28. *Accipiter* [*cirrhocephalus*].
    A. *A. brachyurus* (Ramsay 1879d). NB, NI.
    OTHER ALLOSPECIES: *A. cirrhocephalus* (NG, Au), *A. erythrauchen* (Mol), and possibly *A. rhodogaster* (Cel)
    COMMENTS: *A. nisus* and other sparrowhawks may be related (Wattel 1973).

29. *Haliaeetus* [*leucogaster*]
    A. *H. leucogaster*. Most Bismarck islands (no records for Ramb, Anch, Her, Nin, Cred, Dyl, Ting, Sak, Unea, Tench); Nis (and India to NG, Au).
    B. *H. sanfordi* (Mayr 1935). All central Solomon islands except Rmos, Gow; Nis (?)
    COMMENTS: See Mayr (1936) for discussion of *H. sanfordi*.

30. *Pandion haliaetus*. (a) *melvillensis*. Most Bismarck islands (no records for SMig, Anch, Her, Cred, Ting, Lol, Sak, Unea); all central Solomon islands plus Bor, Ren, Bel, Nis (and Phil, Indonesia, NG, tropical Au, Palau, NCal). Other subspecies: *cristatus* (nontropical Au) and three other subspecies; cosmopolitan except most of non-Mediterranean Afr. COMMENTS: Contrary to the statement by White (1975), *melvillensis* is significantly smaller than *cristatus* and distinct from it (see Brown & Amadon 1968, Stresemann & Amadon in Peters 1979, vol. 1, White & Bruce 1986).

FALCONIDAE (falcons)

See Brown and Amadon (1968), Stresemann and Amadon (in Peters 1979, vol. 1), Cade and Digby (1982), Amadon and Bull (1988), and Olsen et al. (1989).

31. *Falco* [*peregrinus*]
    A. *F. peregrinus*. (a) *ernesti*. NB, Long, Lih, Feni, Witu; Guad, 3S (vagr), Anusagaru (islet off Boug) (and Phil, GS, NG). Other subspecies: cosmopolitan
    OTHER ALLOSPECIES: *F. deiroleucus* (CAm, SAm).
    COMMENTS: See Cade and Digby (1982) for status of *pelegrinoides* and *babylonicus* (N Afr and central Asia), *F. fasciinucha* (Afr), and *F.* "*kreyenborgi*" (S SAm: Ellis *et al.* 1981).

32. *Falco* [*subbuteo*]
    A. *F. severus*. (a) *papuanus*. NB, Vu (vagr), DY, Tang; Buka, Boug, Ch, Ys, Guad, Gizo, Kul, SCr (and NG, Mol, Cel). Other subspecies: SE Asia, GS, Phil
    OTHER ALLOSPECIES: *F. subbuteo* (Eurasia), *F. cuvieri* (Afr), *F. longipennis* (Au, LS), and possibly *F. rufigularis* (Neotropics).
    COMMENTS: See Mayr (1945b), Brown and Amadon (1968), and White and Bruce (1986) for subspecies. *F. longipennis* and *F. severus* may be convergent on the other hobbies, rather than closely related to them (Olsen et al. 1989).

33. *Falco berigora*. (a) *novaeguineae*. Long (and NG). Other subspecies: Au. COMMENTS: An isolated species, possibly related to *F. novaeseelandiae* (Olsen et al. 1989). See Diamond (1976) for occurrence on Long.

MEGAPODIIDAE (megapodes)

34. *Megapodius* [*freycinet*].
    A. *M. freycinet*. (a) *eremita* (Hartlaub 1867). Probably all islands of Northern Melanesia except Ren and Bel (no records for Unea, Kili, Tauu). Other subspecies: Nicobar, Phil, Indonesia, Au, NG, NHeb
    MEGASUBSPECIES: *eremita*, hybridizing with the N NG race *affinis* on Karkar (Mayr 1938a).
    OTHER ALLOSPECIES: *M. laperouse* (Micronesia), *M. pritchardii* (Niuafou).

COMMENTS: We consider *eremita* and *freycinet* conspecific for the reasons discussed by Mayr (1938a), contrary to Schodde (1977), White and Bruce (1986), and earlier authors, who break up *M. freycinet* into six or seven allospecies.

PHASIANIDAE (quail and pheasants)

35. *Coturnix* [*chinensis*]
    A. *C. chinensis*. (a) *lepida* (Hartlaub 1879). NB, NI, NH, Vu, DY, Lih, Tab, Tang. Other subspecies: India to Au, NG
    MEGASUBSPECIES: none.
    OTHER ALLOSPECIES: *C. adansonii* (Afr) (unless conspecific: Snow 1978).
    COMMENTS: For subspecies see Mayr (1949a). We do not recognize the genus *Excalfactoria* for this superspecies or species.

TURNICIDAE (bustard quail)

36. *Turnix* [*sylvatica*]
    A. *T. maculosa*. (a) *saturata* (Forbes 1882). NB, DY. (b) *salomonis* (Mayr 1938b). Guad. Other subspecies: Phil, LS, Au, NG
    MEGASUBSPECIES: none.
    OTHER ALLOSPECIES: *T. sylvatica* (Afr, Asia, to Phil, Java).
    COMMENTS: *T. sylvatica* and *T. maculosa* may be sympatric in Phil (Sutter 1955; but see White & Bruce 1986). See also Rensch (1931) and Mayr (1938b, 1949a) for the relations of these two species.

RALLIDAE (rails)[c]

We follow the generic revision by Olson (1973). See also Ripley (1977).

37. *Gallirallus* [*philippensis*]
    A. *G. philippensis*. (a) *anachoretae* (Mayr 1949a). Anch, Her, Nin, Wuv (subsp?). (b) *praedo* (Mayr 1949a). Skoki (Admiralty Group). (c) *admiralitatis* (Stresemann 1929). Papenbush, Pityili (Admiralty Group). (d) *lesouefi* (Mathews 1911). NI, NH, Ting, Tab, Tang. (e) *meyeri* (Hartert 1930). NB, Vu, DY, Witu. (f) *reductus* (Mayr 1938b). Long (and northeast NG). (g) *christophori* (Mayr 1938b). Fla(?), Guad, Ban, Mal, SCr, Ugi, 3S, SAn, SCat. Other subspecies: Phil to Au, NG, NZ, NCal, NHeb, Fiji, Samoa.
    B. *G. rovianae* (Diamond 1991). Kul, Wana, Koh, NGa, Rnd
    MEGASUBSPECIES: some subspecies or subspecies groups of *G. philippensis* may constitute megasubspecies.
    ALLOSPECIES: *G. dieffenbachi* (Chatham, possibly just a subspecies)
    COMMENTS: *G. wakensis* (Wake) and *G. owstoni* (Guam) are derivatives (Mayr 1949a), as are *G. modestus* (Chatham; Olson 1973) and possibly *G. pacificus* (Tahiti) and numerous unnamed extinct taxa on other Pacific islands. *G. australis* (NZ) and *G. sylvestris* (Lord Howe) may be more remote derivatives. Reasons for including these species in *Gallirallus* have been stated by Olson (1973). For geographic variation, see Mayr (1938b, 1949a) and Schodde and de Naurois (1982).

38. *Gallirallus insignis* (Sclater 1880a). (plate 8, map 4). NB. COMMENTS: A derivative of *G. torquatus* (Olson 1973), as is *G. okinawae* (described as *Rallus okinawae*; Yamashina & Mano 1981). Vuilleumier et al. (1992) place these three taxa in the same superspecies. We do not recognize the monotypic genus *Habropteryx* for *insignis*. See Mayr (1949a).

39. *Rallina* [*eurizonoides*]
    A. *R. tricolor*. (a) *convicta* (Stresemann 1925). NI, NH. (b) *laeta* (Mayr 1949a). SMt. Other subspecies: Damar, Tenimbar, NG, Waigeu, Cape York
    MEGASUBSPECIES: none.
    OTHER ALLOSPECIES: *R. eurizonoides* (SE Asia to Phil, Cel), *R. canningi* (Andaman).
    COMMENTS: *Rallicula* (NG) may be a derivative. See Mayr (1949a) for subspecies. Ripley (1977) placed the SMt population in the Tenimbar race *victa*, which is of doubtful validity (White & Bruce 1986).

40. *Nesoclopeus* [*poecilopterus*]
    A. *N. woodfordi*. (a) *woodfordi* (Ogilvie-Grant 1889). Guad. (b) *immaculatus*

[c]In 1997 *Gallinula tenebrosa* was found breeding at two lakes on NB (Dutson 2001). We have not included this species in our analyses of Northern Melanesian breeding species.

(Mayr 1949a). Ch (?), Ys. (c) *tertius* (Mayr 1949a). Boug
MEGASUBSPECIES: none.
OTHER ALLOSPECIES: *N. poecilopterus* (Fiji), considered conspecific by Ripley (1977).
COMMENTS: We retain the genus *Nesoclopeus*, contrary to Greenway (1967) and Ripley (1977), who sink it in *Rallina* and *Rallus*, respectively. For subspecies, see Mayr (1949a).

41. *Gymnocrex* [*plumbeiventris*]
A. *G. plumbeiventris*. NI(?) (and Mol, NG)
OTHER ALLOSPECIES: *G. rosenbergii* (Cel).
COMMENTS: *Aramides* may be related; Ripley (1977) sinks *Gymnocrex* in *Aramides*. The sole putative Northern Melanesian record of this species is the type of *Rallus intactus* (Sclater 1869), from a collection supposedly made in the Solomon Islands. However, most identifiable species in this collection are actually from New Ireland (Mayr 1933d). The type differs in some plumage characteristics from otherwise monotypic *G. plumbeiventris* and may be subspecifically distinct (M. Walters, personal communication). Thus, *G. plumbeiventris* may occur on New Ireland, but not with certainty.

42. *Porzana* [*tabuensis*]
A. *P. tabuensis*. (a) *tabuensis*. Wuv, Vu; Guad, SCr, Ren (and Polynesia to NG, Phil). Other subspecies: NZ, Au, mountains of NG
OTHER ALLOSPECIES: *P. atra* (Henderson), *P. monasa* (Kusaie).
COMMENTS: See Baker (1951), Olson (1973), and Ripley (1977) for relationships, and Amadon (1942a), Mayr (1949a), and White and Bruce (1986) for geographic variation.

43. *Poliolimnas cinereus*. (a) *leucophrys*. NB, NI, NH, Umb, Lih; Buka, Boug, Guad, SCr, 3S (and Mol, NG, Au). (b) *meeki* (Hartert 1924b). SMt, Em. Other subspecies: Malay Peninsula through LS, Micronesia, NCal, NHeb, Fiji, Samoa
MEGASUBSPECIES: none.
COMMENTS: This genus may be near *Porzana* and is considered congeneric by Ripley (1977a), Mees (1982), and White and Bruce (1986). See Olson (1970, 1973), Ripley (1977), and Mees (1982) for relationships and Mayr (1949a) and Mees (1982) for subspecies.

44. *Amaurornis* [*akool*]
A. *A. olivaceus*. (a) *nigrifrons* (Hartert 1926c). NB, NI, NH, Umb, Vu, DY, Lih, Tab, Tang, Tol, Sak, Witu; almost all Central Solomon islands (subsp?) except range of *ultimus* (no records for Savo, Ul, Tet, 3S, Rmos). (b) *moluccanus*. Long (and Mol, NG) (subsp?). (c) *ultimus* (Mayr 1949a). SCr, Ugi, SAn, SCat, Gow. Other subspecies: Phil, Au
MEGASUBSPECIES: none.
OTHER ALLOSPECIES: *A. akool* (SE Asia), *A. isabellina* (Cel).
COMMENTS: See Stresemann (1939) for relationships and Mayr (1949a) for subspecies. Revision of subspecies based on more material is needed.

45. *Pareudiastes sylvestris* (Mayr 1933a). (plate 8, map 4). SCr. COMMENTS: This species is distantly related to *Pareudiastes pacificus* (Samoa), and both may be related to *Gallinula*, in which Ripley (1977) sinks *Pareudiastes*. We do not recognize the monotypic genus *Edithornis* (Mayr 1933a, 1949a, Olson 1973, 1975).

46. *Porphyrio porphyrio*. (a) *samoensis*. NB, NH, Man, Ramb, Umb, Vu, Tab; virtually all Solomon islands (no records for Ul, 3S, Rmos, Bor, Nis, Fead, Sik, Kili, Tauu, Nuku) (and NHeb, NCal, Fiji, Samoa). Other subspecies: Eu, Afr, Asia, to NG, Au, NZ. COMMENTS: *Porphyrio mantelli* (NZ: "*Notornis*") is a derivative, as is *P. albus* (Lord Howe) if considered distinct from *P. porphyrio* (Ripley 1977). For geographic variation, see Mayr (1949a).

JACANIDAE (jacanas)

47. *Irediparra gallinacea*. (a) subsp? NB. Other populations: Borneo, Phil, Cel, Mol, LS, NG, Au. COMMENTS: Recorded on NB in 1978 and 1997, including chicks (*Papua New Guinea Bird Soc. Newsletter* no. 145: 21 (1978) and no. 291:3 (1997)). The three named races of this species are of doubtful validity (Mees 1982, White & Bruce 1986).

CHARADRIIDAE (plovers)

48. *Charadrius dubius*. (a) *dubius*. NB, NI (and Phil, NG). Other subspecies: Eurasia, N Afr. COMMENTS: *C.* [*hiaticula*] is closely related. For subspecies, see Mayr (1949a) and Mees (1982).

RECURVIROSTRIDAE (avocets and stilts)

49. *Himantopus* [*himantopus*]
    A. *H. leucocephalus*. NB, Long, Umb (and Phil, Indonesia, to NG, Au, NZ)
    OTHER ALLOSPECIES: *H. himantopus* (Eurasia, Afr), *H. melanurus* (south SAm), *H. mexicanus* (NAm, north SAm), *H. knudseni* (Hawaii), *H. novaezelandiae* (NZ). We follow Mayr and Short (1970) in considering these populations as allospecies, but some of them may be conspecific with others. See Diamond (1976) and Bishop (1983) for Bismarck records.

BURHINIDAE (stone-curlews and thick-knees)

50. *Esacus* [*magnirostris*]
    A. *E. magnirostris*. NB, NI, NH, SMt, Man, Long, Vu (vagr), DY, Cred, Em, Lih, Tol, Sak, Cr; almost all central Solomon islands and Bor (no records for Buka, Savo, Ul, SCr, Gow) (and coastal SE Asia to NG, Au, NCal)
    OTHER ALLOSPECIES: *E. recurvirostris* (India). COMMENTS: If *Esacus* were lumped with *Burhinus*, the species name would become *giganteus* (White & Bruce 1986). See Mayr (1949a) for distribution and measurements.

COLUMBIDAE (pigeons)

Species relationships within the whole family are discussed by Goodwin (1983), within *Ptilinopus* by Cain (1954), and within *Ducula* by Goodwin (1960).

51. *Ptilinopus superbus*. (map 5). (a) *superbus*. NB, NI, NH, Man, Ramb, Umb, Vu (vagr), DY (?), Lih, Feni, Tab, Tang, Lol, Tol; Nis (?), and almost all central Solomon islands except SCr group (no records for BV, Savo, Ul, Rmos, Gow) (and NG, Mol, E Au). Other subspecies: Cel. COMMENTS: Cain (1954) places this species in a superspecies with *P. perousii*, and Goodwin (1983) considers it related to *P. wallacii*, but we consider it an isolated species on the basis of its distinctive vocalizations and other characters.

52. *Ptilinopus* [*purpuratus*]
    A. *P. greyii*. Gow (and Santa Cruz, NHeb, NCal).
    B. *P. richardsii*. (a) *richardsii* (Ramsay 1882g). Ugi, 3S, SAn, SCat. (b) *cyanopterus* (Mayr 1931b). Ren, Bel
    MEGASUBSPECIES: none.
    OTHER ALLOSPECIES: *P. regina* (LS, Au), *P. monacha* (Mol), *P. pelewensis* (Palau), *P. roseicapilla* (Marianas), *P. purpuratus* (Societies and Tuamotus), *P. huttoni* (Rapa), *P. rarotongensis* (Rarotonga), *P. insularis* (Henderson), plus the doublets *P. pulchellus* and *P. coronulatus* (NG), *P. mercieri* and *P. dupetithouarsi* (Marquesas), and *P. perousii* and *P. porphyraceus* (Fiji, Samoa; *P. porphyraceus* also in Micronesia). See Ripley and Birckhead (1942) and Goodwin (1983).

53. *Ptilinopus* [*hyogaster*]. (plate 9, map 6).
    A. *P. insolitus*. (a) *insolitus* (Schlegel 1863). NB, NI, NH, Long, Umb, Vu, DY, Dyl, Lih, Feni, Tab, Lol, Tol, Sak, Cr. (b) *inferior* (Hartert 1924b). SMt, Em
    MEGASUBSPECIES: none.
    OTHER ALLOSPECIES: *P. iozonus* (NG), *P. hyogaster* (N Mol), *P. granulifrons* (Obi). COMMENTS: *P. melanospila* (Phil, Cel, Java, Bali, LS) and possibly *P. nanus* (NG) have been suggested as relatives (Cain 1954, Goodwin 1983).

54. *Ptilinopus* [*rivoli*]. (plate 6, map 7)
    A. *P. rivoli*. (a) *rivoli* (Prevost 1843). NB, NI, NH, Umb, DY (?), Dyl, Lih, Tab, Tang, Lol. Other subspecies: Mol, NG.
    B. *P. solomonensis*. (a) *johannis* (Sclater 1877b). NH, SMt, Man, Ramb, Nau, SMig, Anch, Her, Nin, Wuv (subsp?), Em, Ting, Tench. (b) *meyeri* (Hartert 1926c). NB (?), NI (?), Long, Umb, Vu, Tol, Sak, Cr, Witu, Ritter (a small island east of Umboi). (c) *neumanni* (Hartert 1926a). Nis. (d) subsp? Fla, BV, Savo, Pav, Rmos. (e) *vulcanorum* (Mayr 1931e). VL, Gan,

Kul, Koh, NGa, Vang, Gat, Rnd. (f) *solomonensis* (Gray 1870). SCr, Ugi, 3S. (g) *bistictus* (Mayr 1931e). Boug, Shtl. (h) *ocularis* (Mayr 1931e). Guad. (i) *ambiguus* (Mayr 1931e). Mal. Other subspecies: Numfor, Biak, and smaller islands of Geelvink Bay

MEGASUBSPECIES: Plumage characters separate *P. solomonensis* populations into three distinct megasubspecies, one consisting of the NMel montane races *bistictus* and *ocularis*; a second, of all the other NMel races, which occur in the lowlands (*ambiguus* is intermediate between these first two megasubspecies); and a third, of the race *speciosus* on the islands of Geelvink Bay.

COMMENTS: *P. rivoli* and *P. solomonensis* are largely allopatric but coexist on Numfor and Traitor's islands in Geelvink Bay (Mayr 1941c), on Umboi, and marginally on NH and perhaps NI and NB. *P. solomonensis* occurs on small islets near NH and NI (Salomonsen 1976); it rarely occurs on NH itself, and it is uncertain whether it similarly occurs rarely on NI and NB (no definite records). See Mayr (1931e) and Galbraith and Galbraith (1962) for geographic variation in *P. solomonensis*.

55. *Ptilinopus viridis*. (map 8). (a) *lewisi* (Ramsay 1882g). Man, Lih, Tang; all Central Solomons except SCr group and Rmos; Nis. (b) *eugeniae* (Gould 1856). SCr, Ugi, 3S, SAn, SCat. Other subspecies: S Mol, N NG, islands of Geelvink Bay, SE Papuan Islands

MEGASUBSPECIES: *eugeniae*.

COMMENTS: The relationships of *P. viridis* are uncertain. *P. v. eugeniae*, with its white head, is sometimes considered a separate allospecies (Goodwin 1983).

56. *Ducula* [*pacifica*]
    A. *D. pacifica*. (a) *pacifica*. Fla, BV, Savo, Mal (vagr), 3S, Rmos, Gow, Ren, Bel, OJ, Sik, Nuku (and Louisiades, NHeb, NCal, Fiji, Samoa, Cooks). (b) *sejuncta* (Amadon 1943a). Anch, Her, Nin, Wuv, Tench (subsp?) (and Tarawai and Seleo off N NG)

MEGASUBSPECIES: none.

OTHER ALLOSPECIES OR RELATED SPECIES: *D. oceanica* (Micronesia), *D. aurorae* (Societies), *D. galeata* (Marq). These species are allopatric today, but subfossil remains indicate past sympatry. Subfossil bones indistinguishable from bones of *D. pacifica* or *D. aurorae* are known from Henderson, while bones indistinguishable from those of *D. galeata* are known from Henderson and from Mangaia in the Cooks (Steadman 1985, Steadman & Olson 1985). The related extinct *D. david* is known as a subfossil from Uvea (Wallis Group), where *D. pacifica* also occurs (Balouet & Olson 1987).

COMMENTS: See Mayr (1931b) and Amadon (1943a) for subspecies and relationships. *D.* [*myristicivora*] (species 57) might be related.

57. *Ducula* [*myristicivora*]. (plate 8, map 9)
    A. *D. rubricera*. (a) *rubricera* (Bonaparte 1854). NB, NI, NH, Umb, Vu, DY, Cred, Dyl, Lih, Feni, Tab, Tang, Lol, Sak. (b) *rufigula* (Salvadori 1878). Virtually all central Solomons (no records for Rmos, Gow)

MEGASUBSPECIES: the two races are distinct in plumage and constitute megasubspecies.

OTHER ALLOSPECIES: *D. myristicivora* (W Papuan islands, islands of Geelvink Bay).

COMMENTS: See under superspecies 56, *D.* [*pacifica*].

58. *Ducula* [*rufigaster*]. (plate 8)
    A. *D. finschii* (Ramsay 1882f). NB, NI, Umb, Vu (vagr), Dyl

OTHER ALLOSPECIES: *D. basilica* (Mol), the doublet *D. rufigaster* and *D. chalconota* (NG).

59. *Ducula* [*rosacea*]. (plate 8)
    A. *D. pistrinaria*. (a) *rhodinolaema* (Sclater 1877b). NH, SMt, Man, Ramb, Nau, SMig, Her, Wuv, Long, Umb, Em, Tol, Cr (and Manam, Karkar). (b) *vanwyckii* (Cassin 1862). NB, NI, Vu, DY, Cred, Dyl, Lih, Lol, Witu, Unea. (c) *pistrinaria* (Bonaparte 1855). Feni, Tab, Tang; virtually all central Solomons (no records for Gow) plus Bor, Nis. Other subspecies: SE Papuan Islands

MEGASUBSPECIES: none.

OTHER ALLOSPECIES: *D. rosacea* (LS and islets of Java Sea), *D. whartoni* (Christmas Island in Indian Ocean), *D. pickeringii* (islets near Borneo and Phil).

COMMENTS: See Ripley (1947) for subspecies.

60. *Ducula* [*latrans*]. (plate 8)
 A. *D. brenchleyi* (Gray 1870). Guad, Mal, Ul, SCr, Ugi, 3S, SAn, SCat
 OTHER ALLOSPECIES: *D. latrans* (Fiji), *D. bakeri* (NHeb), *D. goliath* (NCal).
 COMMENTS: See Amadon (1943a) and Goodwin (1960) for relationships.

61. *Ducula* [*pinon*]. (plate 8)
 A. *D. melanochroa* (Sclater 1878b). NB, NI, Umb, Vu (vagr), DY (vagr)
 OTHER ALLOSPECIES: the doublet *D. pinon* and *D. muellerii* (NG).
 COMMENTS: This superspecies may be related to superspecies 60, *D.* [*latrans*] (Goodwin 1960).

62. *Ducula* [*bicolor*]
 A. *D. spilorrhoa*. (a) *subflavescens* (Finsch 1886). NB, NI, NH, Man, Nau, Long, Umb, Vu (vagr), DY (?), Lol, Tol, Sak, Cr. Other subspecies: NG, Au
 MEGASUBSPECIES: none.
 OTHER ALLOSPECIES: *D. bicolor* (islets from India to NG), *D. luctuosa* (Cel), *D. melanura* (Mol).
 COMMENTS: Johnstone (1981) treats all allospecies as subspecies of *D. bicolor*, but see Siebers (1930), Goodwin (1960), and White and Bruce (1986). Diamond and LeCroy (1979) give the characters of the Bismarck race.

63. *Gymnophaps* [*albertisii*]. (map 10)
 A. *G. albertisii*. (a) *albertisii*. NB, NI (and NG). Other subspecies: Batjan.
 B. *G. solomonensis* (Mayr 1931e). Boug, Guad, Mal, Kul, Vang, Rnd (?)
 OTHER ALLOSPECIES: *G. mada* (Buru, Ceram).
 COMMENTS: See Goodwin (1963) regarding the affinities of this superspecies, and Mayr (1931e) for variation in *G. solomonensis*.

64. *Columba* [*leucomela*]. (map 11)
 A. *C. vitiensis*. (a) *halmaheira*. NB (vagr), NI; Boug, Ys, Fla, Guad, Mal, VL, Kul, Vang, Gat, SCr (and Mol, NG). Other subspecies: Phil, LS, NHeb, NCal, Fiji, Samoa
 OTHER ALLOSPECIES: *C. janthina* (Japan), *C. versicolor* (Bonin, extinct), *C. jouyi* (Riu Kiu), *C. leucomela* (Au), some of them extremely different.
 COMMENTS: See Amadon (1943a) for geographic variation. For use of the name *leucomela* instead of *norfolciensis*, see Hindwood (1965) and Goodwin (1983).

65. *Columba pallidiceps* (Ramsay 1878b). (plate 8, map 11). NB, NI, DY (?); Boug, Ch, Fla, Guad, VL, SCr, Rmos, Gow. COMMENTS: A derivative of *C. vitiensis*, sympatric with it on Boug, Fla, Guad, VL, SCr, and marginally on NB. See Mayr (1934a) for variation.

66. *Macropygia* [*amboinensis*]
 A. *M. amboinensis*. (a) *cinereiceps*. Long, Umb, Tol, Sak (and SE NG). (b) *carteretia* (Bonaparte 1850). NB, NI, NH, Vu, DY, Dyl, Lih, Feni, Tab, Tang, Lol. (c) *admiralitatis* (Mayr 1937b). Man. Other subspecies: Cel, Mol, NG
 MEGASUBSPECIES: *admiralitatis* is distinct in plumage from the other two NMel races and most similar to the heavily barred *amboinensis* and *doreya* of west New Guinea and the southern Moluccas. However, further study of this geographically highly variable species would be required to divide it into megasubspecies.
 OTHER ALLOSPECIES: *M. tenuirostris* (Phil), *M. emiliana* (GS, Lombok, Sumbawa, Flores), *M. magna* (Timor, Southwest Islands, Tenimbar), *M. phasianella* (Au), *M. rufipennis* (Andaman, Nicobar).
 COMMENTS: We do not recognize the subspecies *hueskeri* for NH birds (Hartert 1925, Mayr 1937b). See Mayr (1944a) and Goodwin (1983) for relationships and Rothschild and Hartert (1901b), Mayr (1937b), and Goodwin (1983) for geographic variation. *M. unchall* (SE Asia to Sumatra, Java, Lombok) is closely related but occurs sympatrically with *M. emiliana* on Sumatra, Java, and Lombok. Mayr (1944a) considered the members of the superspecies more closely related to each other than to *M. unchall* and excluded *M. unchall* from the superspecies, which would then consist of largely or strictly allopatric forms. Alternatively, since *M. amboinensis* is intermediate between *M. unchall* and *M. emiliana*, this group could be regarded as a superspecies (including *M. unchall*) with overlap or a doublet on Sumatra, Java, and Lombok. It is uncertain

whether *M. magna* and *M. amboinensis* occur sympatrically on the south tip of Celebes (Mees 1972, White & Bruce 1986). Other arrangements are treating *emiliana* and *tenuirostris* as subspecies of *M. phasianella* (Frith 1982, Goodwin 1983) and treating *phasianella*, *amboinensis*, and *magna* as conspecific (Frith 1982), or treating all the forms as conspecific (Condon 1975).

67. *Macropygia* [*ruficeps*]. (map 12)
    A. *M. nigrirostris*. (a) *major* (van Oort 1908). NB, NI, NH, Vu (vagr), DY, Tab. Other subspecies: NG. COMMENTS: For geographic variation, see Gilliard and LeCroy (1967).
    B. *M. mackinlayi*. (a) *arossi* (Tristram 1879). NB (vagr), NI (vagr), SMt, Man, Ramb, Nau, SMig, Long, Umb, Vu, Em, Lih, Tol, Sak, Cr, Witu, Tench; almost all Central Solomons (no records for Ul, Tet, Rmos) plus Ren, Bel, Nis (and Karkar). Other subspecies: NHeb

    MEGASUBSPECIES: none.

    OTHER ALLOSPECIES: *M. ruficeps* (Burma to GS, LS).

    COMMENTS: *M. nigrirostris* and *M. mackinlayi* both occur on Karkar, Vu, NB, and NI, but in each case one of the two species is very uncommon and probably a vagrant (*M. nigrirostris* on Karkar and Vu, *M. mackinlayi* on NB and NI) (Diamond & LeCroy 1979). See Gilliard and LeCroy (1967) for geographic variation in *M. nigrirostris*. We do not consider the Bismarck ("*goodsoni*") or Karkar ("*krakari*") populations of *M. mackinlayi* separable from *M. m. arossi* (Diamond & LeCroy 1979).

68. *Reinwardtoena* [*reinwardtii*]. (plate 7, map 13)
    A. *R. browni*. (a) *browni* (Sclater 1877a). NB, NI, NH, Umb (subsp?), Vu (vagr), DY, Dyl, Lih, Tab, Lol. (b) *solitaria* (Salomonsen 1972). Man, Ramb, Nau.
    B. *R. crassirostris* (Gould 1856). Boug, Ch, Ys, Fla, Guad, Mal, VL, Gan, Kul, Koh, NGa, Vang, Gat, Rnd, SCr, Ugi, 3S, SAn, SCat

    MEGASUBSPECIES: *solitaria* is distinct from *browni* in its small size and dark color; the Umboi population may be intermediate.

    OTHER ALLOSPECIES: *R. reinwardtii* (NG, Mol).

    COMMENTS: See Salomonsen (1972) for subspecies of *R. browni*.

69. *Chalcophaps stephani*. (a) *stephani*. Almost all Bismarcks (no records for SMig, Anch, Her, Nin, Wuv, Cred, Ting, Tench) (and NG). (b) *mortoni* (Ramsay 1882g). All central Solomons plus Nis. Other subspecies: Cel

    MEGASUBSPECIES: none (*mortoni* differs from nominate *stephani* only in its slightly larger size; see Schodde 1977, p. 48).

    COMMENTS: *C. indica* (India to Au, NG, NHeb, NCal) is related.

70. *Henicophaps* [*albifrons*]. (map 14)
    A. *H. foersteri* (Rothschild & Hartert 1906). NB, Umb, Lol

    OTHER ALLOSPECIES: *H. albifrons* (NG).

71. *Gallicolumba* [*canifrons*].
    A. *G. beccarii*. (a) *johannae* (Sclater 1877a). NB, NI, NH, Long, Umb, DY, Lih (subsp?), Feni (subsp?), Tab (?), Tang, Lol, Tol, Sak, Cr, Witu (and Karkar). (b) *eichhorni* (Hartert 1924b). SMt, Em. (c) *admiralitatis* (Rothschild & Hartert 1914b). Man. (d) *masculina* (Salomonsen 1972). Nis. (e) *intermedia* (Rothschild & Hartert 1905). Boug, Gizo, Kul, NGa. (f) *solomonensis* (Ogilvie-Grant 1888). Guad, SCr, 3S, SAn, Gow, Ren, Bel. Other subspecies: NG

    MEGASUBSPECIES: *masculina*, distinct in the cock-feathered female. The remaining NMel races are fairly similar to each other and constitute a further megasubspecies somewhat distinct from nominate *beccarii* of NG.

    OTHER ALLOSPECIES: *G. canifrons* (Palau).

    COMMENTS: Among *G. hoedtii* (Wetar), *G. salamonis* (Solomons: species 72), *G. rubescens* (Marq), and *G. sanctaecrucis* (Santa Cruz) and *G. stairii* (Fiji, Samoa) (closely related to each other), it is uncertain which is closer to this superspecies, which to the *G. erythroptera* superspecies (species 73). Extinct taxa from Norfolk (originally described as *Columba norfolciensis*; see Hindwood 1965, Goodwin 1983, Schodde et al. 1983), the Cooks (Steadman 1985), and

Marq (Steadman 1992) may also have been related. See Mayr (1936), Amadon (1943a), Baker (1951), and Goodwin (1983) for relationships and Salomonsen (1972) for subspecies of *G. beccarii*.

72. *Gallicolumba salamonis* (Ramsay 1882e). SCr, Rmos. COMMENTS: See under species 71, *G.* [*canifrons*]. Known only from two specimens. Possibly extinct (Parker 1968, Diamond 1987).

73. *Gallicolumba* [*erythroptera*]
   A. *G. jobiensis*. (a) *jobiensis*. NB, NI, Her, Umb, Vu, DY (?), Lih, Tab, Lol, Sak, Tong (an islet in the Admiralty group) (and NG). (b) *chalconota* (Mayr 1935). Guad, VL, SCr

   MEGASUBSPECIES: possibly nominate *jobiensis* and *chalconota* (see comments below).
   OTHER ALLOSPECIES: *G. erythroptera* (Societies and Tuamotus), *G. xanthonura* (Marianas), *G. kubaryi* (Carolines, possibly conspecific with *G. xanthonura*).
   COMMENTS: See under species 71, *G.* [*canifrons*], for relationships, and see Holyoak (1978) for discussion of the Solomon race *chalconota* and of its possibly distinctive female plumage, known only from a single specimen.

74. *Microgoura meeki* (Rothschild 1904a). (plate 8, map 4). Ch. COMMENTS: An isolated species, apparently extinct (Parker 1967a, 1972, Diamond 1987).

75. *Caloenas nicobarica*. (a) *nicobarica*. Almost all Bismarcks (no records for SMig, Anch, Nin, Ting, Witu) and Solomons (no records for Buka, SCr, Kili, Tauu) (and islets from Nicobars through Indonesia to NG). Other subspecies: Palau. COMMENTS: An isolated species.

PSITTACIDAE (parrots)

See Forshaw & Cooper (1973). Smith (1975) discusses relationships between genera.

76. *Chalcopsitta cardinalis* (Gray 1859). (plate 8, map 15). NH (vagr?), Lih (only on Mahir and Masahet), Feni, Tab, Tang; virtually all Central Solomons (no records for SAn, Rmos, Gow), Bor, Nis, OJ. COMMENTS: We follow Peters (1937), Auber (1938), and Forshaw & Cooper (1973) in placing this species in *Chalcopsitta*, if that genus is kept separate from *Eos*. The closest relative is the *Chalcopsitta* [*atra*] superspecies of NG, not the red *Eos* species of Indonesia and W NG islands.

77. *Trichoglossus* [*ornatus*].
   A. *T. haematodus*. (a) *nesophilus* (Neumann 1929). Nin. (b) *flavicans* (Cabanis & Reichenow 1876). NH, SMt, Man, Ramb, Nau, SMig, Her, Em, Ting (no records for Anch, Wuv, Tench). (c) *massena*. All Bismarcks except ranges of Aa and Ab (no records for Cred); virtually all central Solomons (no records for 3S, SAn, SCat, Rmos), Nis, Fead (and Karkar, NHeb). Other subspecies: Bali, LS, Mol, NG, Au, NCal

   MEGASUBSPECIES: following Cain (1955), we consider all populations of NMel, NHeb, NCal, NG, and Mol to belong to the same megasubspecies, within which *flavicans* and *nesophilus* form a separate subgroup.
   OTHER ALLOSPECIES: *T. ornatus* (Cel), *T. rubiginosus* (Ponape).
   COMMENTS: See Cain (1955) for subspecies. *T. chlorolepidotus* (E Au), *T. flavoviridis* (Cel), *T. johnstoniae* (Mindanao), and *T. euteles* (LS) are related.

78. *Lorius* [*lory*]. (plate 6, map 16)
   A. *L. albidinucha* (Rothschild & Hartert 1924b). NI.
   B. *L. chlorocercus* (Gould 1856). Savo, Guad, Mal, Ul, SCr, Ugi, SCat, Ren, Bel (vagr).
   C. *L. hypoinochrous*. (a) *devittatus*. NB, NI, NH, Long, Umb, Vu (vagr), Dyl, Lih, Tab, Lol, Sak, Witu (and SE NG, D'Entrecasteaux group, Woodlark). Other subspecies: Louisiades

   OTHER ALLOSPECIES: *L. lory* (NG), *L. domicella* (Ceram and Amboina), *L. garrula* (NMol), *L. tibialis* (range unknown, possibly just an aberrant specimen of *L. domicella*; Forshaw 1973).
   COMMENTS: *L. hypoinochrous* overlaps *L. albidinucha* on NI, *L. lory* in SE NG. *L. amabilis* (Stresemann 1931) is evidently a plumage variant of 78 Ca (Mayr 1941c, Forshaw 1971). See Hartert (1925) for discussion of relationships, and Mayr (1969) for a map of ranges.

79. *Charmosyna* [*palmarum*]. (map 17)
   A. *C. rubrigularis* (Sclater 1881). NB, NI (and Karkar).
   B. *C. meeki* (Rothschild & Hartert 1901c). Boug, Ys, Guad, Mal, Kul, Vang (?)
   OTHER ALLOSPECIES (see Amadon 1942B): *C. toxopei* (Buru) and *C. palmarum* (NHeb); possibly also *C. rubronotata* (N NG, closely related to species 80, *C. placentis*), *C. diadema* (NCal), *C. amabilis* (Fiji; see Mayr 1945c for priority of this name over *C. aureicinctus*).
   COMMENTS: Amadon (1942b) considered *Charmosyna* as a subgenus of *Vini*, but Forshaw & Cooper (1973), Schodde (1977), and most other authors recognize *Charmosyna*.

80. *Charmosyna placentis*. (map 17) (a) *pallidior* (Rothschild & Hartert 1905). NB, NI, NH, Long, Umb, Vu, DY, Cred, Dyl, Lih, Feni, Tab, Tang, Lol, Tol, Sak, Cr, Witu, Lou and Pak (Admiralty group); Buka, Boug, Nis, Fead (and Woodlark). Other subspecies: NG, Mol
   MEGASUBSPECIES: none.
   COMMENTS: *C. placentis* originated as the NG-Mol representative of the *C. palmarum* superspecies and is now sympatric with the superspecies on Boug (*C. meeki*), on NB and NI (*C. rubrigularis*), on N NG and Salawati (*C. rubronotata*), and possibly on Buru (*C. toxopei*; see Siebers 1930, p. 255; Mayr 1940b; Jepson 1993). *C. rubronotata* is the closest relative of *C. placentis*. See Mees (1965a) for geographic variation.

81. *Charmosyna margarethae* (Tristram 1879). (plate 9). Boug, Ys, Guad, Mal, Gizo, Kul, Gat, SCr, SAn. COMMENTS: The nearest relatives are *C. papou*, *C. josefinae*, and *C. pulchella* of New Guinea. The superficial resemblance in color pattern to the larger lory, *Lorius chlorocercus* (78B), is striking, but it is uncertain whether this resemblance is coincidental or biologically significant (Moynihan 1968, Diamond 1982b).

82. *Micropsitta bruijnii*. (map 18). (a) *necopinata* (Hartert 1925). NB, NI. (b) *brevis* (Mayr 1940b). Boug. (c) *rosea* (Mayr 1940b). Guad, Kul. Other subspecies: NG, Buru, Ceram
   MEGASUBSPECIES: none.
   COMMENTS: For geographic variation, see Mayr (1940b) and Gilliard and LeCroy (1967). Forshaw and Cooper (1973) and Schodde (1977) sink *brevis* in *rosea*.

83. *Micropsitta* [*pusio*]. (map 19)
   A. *M. pusio*. (a) *beccarii*. Umb, Tol, Sak (and N NG). (b) *pusio* (Sclater 1866). NB, Vu, DY, Lol, Witu, Unea (and SE NG). Other subspecies: Fergusson, Misima, Tagula.
   B. *M. meeki*. (a) *meeki* (Rothschild & Hartert 1914c). Man, Ramb. (b) *proxima* (Rothschild & Hartert 1924a). SMt, Em.
   C. *M. finschii*. (a) *viridifrons* (Rothschild & Hartert 1899). NI, NH, Dyl, Lih, Tab. (b) *nanina* (Tristram 1891). Buka, Boug, Ch, Ys, Fla. (c) *tristrami* (Rothschild & Hartert 1902a). VL, Gan, Simb, Gizo, Kul, Wana, Koh, NGa, Vang, Rnd, Tet. (d) *aolae* (Ogilvie-Grant 1888). Guad, Pav, Ban, Mal. (e) *finschii* (Ramsay 1881). SCr, Ugi, SAn, Ren
   MEGASUBSPECIES: none.
   OTHER ALLOSPECIES: *M. keiensis* (SW NG, Kei), *M. geelvinkiana* (Numfor, Biak).
   COMMENTS: See Hartert (1924a) for geographic variation in *M. finschii*.

84. *Cacatua* [*alba*]
   A. *C. galerita*. (a) *ophthalmica* (Sclater 1864). NB. Other subspecies: NG, Au
   MEGASUBSPECIES: *ophthalmica*, distinct in its backward-curving rather than forward-curving crest.
   OTHER ALLOSPECIES: *C. sulphurea* (Cel and LS), *C. alba* (NMol), *C. moluccensis* (SMol). Some recent authors (e.g., Forshaw & Cooper 1973, Coates 1985, White & Bruce 1986, Brown & Toft 1999) recognize *C. ophthalmica* as a distinct allospecies.

85. *Cacatua* [*tenuirostris*]. (plate 8)
   A. *C. ducorpsi* (Bonaparte 1850). All Central Solomons except Ul, Gan, Simb, Rmos, SCr group
   OTHER ALLOSPECIES: *C. sanguinea* (Au and S NG), *C. pastinator* (SW Au), *C. goffini* (Tenimbar), *C. tenuirostris* (SE Au).
   COMMENTS: *C. haematuropygia* is sister taxon to this superspecies (Brown & Toft 1999). See Schodde et al. (1979b) and Ford (1985) for revision of the Au populations. *C. sanguinea* occurs sympatrically with *C. pastinator* and with *C. tenuirostris*.

86. *Eclectus roratus.* (a) *goodsoni* (Hartert 1924d). NB, Man, Ramb, Umb, Vu, DY, Lol, Witu, Unea. (b) *solomonensis* (Rothschild & Hartert 1901a). NI, NH, Dyl, Lih, Feni, Tab, Tang; all Central Solomons except 3S, SAn, SCat, Rmos, Gow; Bor. Other subspecies: Sumba, Tenimbar, Mol, NG, Cape York
MEGASUBSPECIES: none.
COMMENTS: This genus is closest to *Geoffroyus.* Unlike Forshaw & Cooper (1973), we consider *goodsoni* separable from *solomonensis.*

87. *Geoffroyus* [*heteroclitus*]. (plate 9, map 20)
A. *G. heteroclitus.* (a) *heteroclitus* (Hombron & Jacquinot 1841). NB, NI, NH, Umb, DY, Dyl, Lih, Tab, Lol, Witu; all Central Solomons except Gan, Simb, Ugi, 3S, SAn, SCat, Mono, Rmos, Gow. (b) *hyacinthinus* (Mayr 1931b). Ren
MEGASUBSPECIES: *heteroclitus, hyacinthinus.*
OTHER ALLOSPECIES: *G. simplex* (NG).
COMMENTS: In voice and plumage *G. heteroclitus* is more similar to *G. simplex* (NG) than to *G. geoffroyi* (LS, Mol, NG, Cape York). Racial characters suggest that *geoffroyi* is a relatively recent colonizer of NG (from LS and Mol), making a *simplex–heteroclitus* relationship more likely than a *geoffroyi–heteroclitus* relationship.

88. *Loriculus* [*aurantiifrons*].
A. *L. tener* (Sclater 1877a). NB, NI, NH, DY
OTHER ALLOSPECIES: *L. aurantiifrons* (NG), and possibly *L. flosculus* (Flores) and *L. exilis* (Cel).
COMMENTS: This would be an interesting genus for study of character variation. The other species are *L. vernalis* (India and SE Asia), *L. beryllinus* (Ceylon), *L. pusillus* (Java, Bali), *L. philippensis* (Phil), *L. galgulus* (Malay Peninsula, Sumatra, Borneo), *L. stigmatus* (Cel), and *L. amabilis* (Mol and Sula). All forms are allopatric except on Cel, which is shared by *L. exilis* and *L. stigmatus. L. stigmatus* is close to *L. amabilis* (Stresemann 1940–41, White & Bruce 1986), while *L. vernalis* is close to *L. beryllinus.* The *L. aurantiifrons* group is connected to *L. stigmatus* through *L. amabilis amabilis*; White and Bruce (1986) group these taxa as a superspecies. The *L. aurantiifrons* group may also be connected to *L. vernalis* and *L. beryllinus* through *L. flosculus* and *L. exilis* (Rensch 1931). Peters (1931–87/1937) and Forshaw & Cooper (1973) consider *tener* to be a subspecies of *L. aurantiifrons.*

CUCULIDAE (cuckoos)[d]

89. *Cacomantis variolosus.* (a) *fortior.* Long, Umb (and Manam, Karkar, Fergusson, Goodenough). (b) *macrocercus* (Stresemann 1921). NB, NI, Vu, DY, Dyl, Lih, Tang, Lol, Unea. (c) *websteri* (Hartert 1898). NH. (d) *tabarensis* (Amadon 1942b). Tab. (e) *blandus* (Rothschild & Hartert 1914b). Man, Ramb, Nin. (f) *addendus* (Rothschild & Hartert 1901c). Boug, Ys, Guad, Mal, Ul, Kul, Wana, Koh, NGa, SCr, Ugi. Other subspecies: Malay Peninsula, Phil, Indonesian Archipelago, NG, Au
MEGASUBSPECIES: uncertain, pending reevaluation of subspecies.
COMMENTS: The subspecific affinities of some of the Bismarck populations (Long, Umb, Lih, Tab, Tang, Unea, Nin) are uncertain (Amadon 1942b). Further uncertainties include whether the Malay, Philippine, and Indonesian populations included in *C. variolosus* (e.g., *sepulcralis* and *virescens*) are actually conspecific; which of the other species of *Cacomantis* is mostly closely related to *C. variolosus*; and whether *Cacomantis* should be merged with *Cuculus* (see White & Bruce 1986 for discussion of these questions and for references).

90. *Chrysococcyx lucidus.* (a) *harterti* (Mayr 1932b). Ren, Bel. Other subspecies: Au, NZ, NCal, NHeb
MEGASUBSPECIES: *harterti*, characterized by much more marked sexual dimorphism than in extralimital subspecies.
COMMENTS: *C. basalis* (Au) and the *C. "malayanus"* group (Malay Peninsula, Indonesian Archipelago, NG, Au) may be related (Friedmann 1968, Marchant 1972,

[d] *Cacomantis pyrrhophanus meeki* (Rothschild and Hartert 1902a), recorded for Ys, Guad, and Bel, seems to consist of wintering individuals of nominate *pyrrhophanus*, which breeds in NCal (Amadon 1942b, Galbraith & Galbraith 1962).

Harrison 1973, Parker 1981). The breeding races *C. l. lucidus* of NZ and *C. l. plagosus* of Au are winter visitors to the Solomons and Bismarcks (Mayr 1932b; Galbraith & Galbraith 1962).

91. *Eudynamys scolopacea*. (a) Subsp? Long, Tol, Cr. (b) *salvadorii* (Hartert 1900b). NB, NI, Umb, Vu, DY, Lol, Sak. (c) *alberti* (Rothschild & Hartert 1907). Almost all Central Solomons (no records for Ul, Ugi, Rmos, Gow). Other subspecies: India to NG, Au
MEGASUBSPECIES: *salvadorii* and *alberti*, differing from each other and from *rufiventer* of NG in size (large, small, and medium respectively) and in female ventral coloration (whitish, ochraceous, and pale ochraceous respectively).
COMMENTS: The population of Long, Cr, and Tol appears to be a stable hybrid population between *salvadorii* and *rufiventer*, formed in the three centuries since the volcanic defaunation of those islands (Diamond 2001a). See Mayr (1944b) for discussion of the genus. It is uncertain whether all populations from India to the Solomons, grouped under *E. scolopacea*, are conspecific or constitute a superspecies (Siebers 1930, Rand 1941a, White & Bruce 1986).

92. *Scythrops novaehollandiae*. NB, NI, NH (?), SMt, Man, SMig, Long, Umb, Vu, DY, Dyl, Lih, Feni, Tab, Tang, Lol, Sak; Savo, Simb, Ren (and Cel, LS, Mol, NG, Au)
COMMENTS: The great majority of records from Northern Melanesia, and from outside Au generally, are of wintering Australian birds. There are three inadequately documented breeding records for NB, based on one *Scythrops* chick reported third-hand to have been taken from a nest of the crow *Corvus orru* (Meyer 1927b), plus two finds of eggs identified as *Scythrops* eggs and collected in *Corvus orru* nests (Meyer 1933, Schönwetter 1935; see also Hartert 1926b, Mason & Forrester 1996). In part because of this uncertainty as to breeding status, we do not recognize the Bismarck race *S.n, schoddei*, described on the basis of few specimens and slight distinctions (Mason & Forrester 1996), but further study is required.

93. *Centropus* [*ateralbus*]
A. *C. ateralbus* (Lesson 1826). NB, NI, Umb, Dyl, Lol.
B. *C. milo*. (a) *milo* (Gould 1856). Fla, BV, Guad. (b) *albidiventris* (Rothschild 1904). VL, Bag, Gan, Simb, Gizo, Kul, Koh, NGa, Vang, Gat, Rnd, Tet
MEGASUBSPECIES: the two races are distinct in ventral coloration and constitute megasubspecies.
OTHER ALLOSPECIES: *C. goliath* (Mol).
COMMENTS: Mason et al. (1984) instead argue that *C. milo*, *C. goliath*, and *C. violaceus* (species 94) form a species group together with *C. menbeki* (NG) and *C. chalybeus* (Biak), while *C. ateralbus* is related to *C. phasianinus* (Au, NG), *C. spilopterus* (Kei), and possibly *C. bernsteini* (NG).

94. *Centropus violaceus* (Quoy & Gaimard 1830). (map 4). NB, NI. COMMENTS: *C. chalybeus* (Biak) may be related. The affinities of this species are otherwise unclear. See discussion of relationships under species 93.

TYTONIDAE (barn owls)

See Stresemann (1940–41), Burton (1973), Eck and Busse (1973), Schodde and Mason (1980), and Amadon and Bull (1988).

95. *Tyto* [*alba*]
A. *T. alba*. (a) *crassirostris* (Mayr 1935). Long (subsp?), Tang; Buka, Boug, Ys, Guad, Mal, VL, NGa and possibly other islands of the NGa group, SCr, Ugi, 3S, SAn, SCat, Ren, Bel, Nis. Other subspecies: worldwide
MEGASUBSPECIES: none.
OTHER ALLOSPECIES: *T. rosenbergii* (Cel).
COMMENTS: The race *bellonae* (Bradley 1962) is not recognized (Galbraith & Galbraith 1962). See Mayr (1936) and Galbraith & Galbraith (1962) for geographic variation. The NMel race *crassirostris* is much more similar to *delicatula* of Au than to *meeki* of NG. Unlike Stresemann (1940–41) or Schodde and Mason (1980), Eck and Busse (1973) and White and Bruce (1986) consider *T. rosenbergii* closer to species 96 (*T.* [*novaehollandiae*]) than to *T. alba*.

96. *Tyto* [*novaehollandiae*]. (map 21)
   A. *T. aurantia* (Salvadori 1881). NB.
   B. *T. novaehollandiae*. (a) *manusi* (Rothschild & Hartert 1914b). Man. Other subspecies: Tenimbar, Buru, Au, S NG
   MEGASUBSPECIES: none.
   OTHER ALLOSPECIES: *T. inexspectata* (Cel), *T. nigrobrunnea* (Taliabu; may be conspecific with *T. inexspectata*).
   COMMENTS: White & Bruce (1986) treat the Tenimbar and Buru populations as an allospecies (*T. sorocula*). Eck & Busse (1973) and White & Bruce (1986) consider *T. rosenbergii* (Cel) related to this superspecies rather than to *T.* [*alba*].

STRIGIDAE (owls)

See Burton (1973), Schodde and Mason (1980), and Amadon and Bull (1988).

97. *Ninox* [*novaeseelandiae*]. (plate 7, map 22)
   A. *N. meeki* (Rothschild & Hartert 1914c). Man, Los Negros.
   B. *N. odiosa* (Sclater 1877a). NB, Vu.
   C. *N. variegata*. (a) *variegata* (Quoy & Gaimard 1830). NI. (b) *superior* (Hartert 1925). NH.
   D. *N. jacquinoti*. (a) *mono* (Mayr 1935). Mono. (b) *eichhorni* (Hartert 1929). Buka, Boug, Shtl, Ch. (c) *jacquinoti* (Bonaparte 1850). Ys. (d) *floridae* (Mayr 1935). Fla. (e) *granti* (Sharpe 1888a). Guad. (f) *malaitae* (Mayr 1931e). Mal. (g) *roseoaxillaris* (Hartert 1929). SCr, Ugi, SCat
   MEGASUBSPECIES: populations of *N. jacquinoti* fall into three groups — *roseoaxillaris*/*malaitae*, *granti*, and *mono*/*eichhorni*/*jacquinoti*/*floridae* — differing markedly in plumage.
   OTHER ALLOSPECIES: *N. theomacha* (NG), *N. novaeseelandiae* (Sumba to Au, NZ, S NG), *N. squamipila* (Mol, Tenimbar, Christmas Island in Indian Ocean), *N. ochracea* and *N. punctulata* (doublet on Cel).
   COMMENTS: *N. philippensis* (Phil), *N. scutulata* (SE Asia to GS), *N. affinis* (Nicobar and Andaman, doublet with *N. scutulata obscura*), and even *N. superciliaris* (Mad) are related. The status of some populations as subspecies or allospecies is controversial (see Eck & Busse 1973, Schodde & Mason 1980, White & Bruce 1986). The nineteenth-century record of *N. variegata* and the resulting supposed sympatry of *N. variegata* and *N. odiosa* on NB, repeated in some modern literature (e.g., Peters 1931–87/1940, Amadon & Bull 1988), are erroneous. See Hartert (1929) and Mayr (1931e, 1936) for geographic variation in *N. jacquinoti*.

98. *Nesasio solomonensis* (Hartert 1901). (plate 8). Boug, Ch, Ys. COMMENTS: The relationships of this monotypic genus are unknown.

PODARGIDAE (frogmouths)

See Schodde and Mason (1980).

99. *Podargus ocellatus*. (a) *inexpectatus* (Hartert 1901). Boug, Ch, Ys. Other subspecies: NG, E Au
   MEGASUBSPECIES: none.
   COMMENTS: The extralimital race most similar to the Solomon race is *intermedius* of the D'Entrecasteaux Archipelago.

CAPRIMULGIDAE (nightjars)

See Schodde and Mason (1980).

100. *Eurostopodus mystacalis*. (a) *nigripennis* (Ramsay 1882c). Boug, Faur, Ys, VL, Gizo, Kul, Wana, NGa, Vang, Gat, Rnd, Tet. Other subspecies: Au, NCal
   MEGASUBSPECIES: possibly *nigripennis*.
   COMMENTS: *E. argus* (Au) is related, possibly allospecific. See Schodde & Mason (1980), who also discuss reasons for the name *argus* rather than *guttatus*. The inclusion of NI in the range of *E. argus*, cited by authors such as Peters (1931–87/1940), is based solely on a pullus unidentifiable to species (Sclater 1879).

101. *Caprimulgus macrurus*. (a) *yorki*. NB, NI, NH, Long, Umb, Vu, Lih, Tab, Lol, Tol (and NG, N Au, Mol, LS). Other subspecies: India to GS, Cel; Tagula
   COMMENTS: See Mees (1977), Greenway (1978), Schodde and Mason (1980), and White and Bruce (1986) for range and nomenclature. We follow Greenway's (1978) rejection of the racial name *schlegelii* in favor of *yorki*.

APODIDAE (swifts)

See Chantler and Driessens (1995). For *Aerodramus* and *Collocalia*, see Mayr (1937a), Peters (1931–87/1940), Medway (1966), Brooke (1970, 1972), Medway and Pye (1977), and Salomonsen (1983). Cytochrome b mitochondrial DNA sequences (Lee et al. 1996) support Brooke's (1972) earlier suggestion that the large, dull-plumaged, echo-locating swiftlet species now assigned to genus *Aerodramus* and formerly lumped under *Collocalia* (including species 102–104 below) are not closely related to the small, glossy-plumaged, non–echo-locating swiftlet species retained in genus *Collocalia* (including species 105 below). Our treatment mostly follows that of Salomonsen (1983).

102. *Aerodramus orientalis*. (a) *orientalis* (Mayr 1935). Guad. (b) *leletensis* (Salomonsen 1962). NI. (c) Subsp? Boug
MEGASUBSPECIES: none.
COMMENTS: See Salomonsen (1962, 1983), Somadikarta (1967), and Ripley (1983). The taxa *papuensis* (NG) and *nuditarsus* (NG) may be conspecific or allospecific, and all these taxa have variously been treated as subspecies or allospecies of *whiteheadi* (Phil), *maxima* (= *lowi*, GS), or *brevirostris* (SE Asia), some of which may in turn be conspecific with each other.

103. *Aerodramus* [*vanikorensis*]
A.*A. vanikorensis*. (a) *coultasi* (Mayr 1937a). SMt, Man, Ramb, Em, Los Negros (in Admiralty group). (b) *pallens* (Salomonsen 1983). NB, NI, NH, Long, Umb, Vu, DY, Dyl, Ting, Tol, Cr, Witu, Unea. (c) *lihirensis* (Mayr 1937a). Lih, Feni, Tab, Tang; Fead. (d) *lugubris* (Salomonsen 1983). Most Central Solomon islands (no records for Faur, BV, Ul, Gan, Simb, Tet, Rmos, Gow), Ren, Bel, Nis. Other subspecies: Cel, Mol, NG, Cape York (vagr), NHeb, NCal (?)
MEGASUBSPECIES: none.
OTHER ALLOSPECIES: *A. inquietus* (Carolines).
COMMENTS: See Mayr (1937a), Baker (1951), Medway (1966, 1975), Holyoak and Thibault (1978), and Salomonsen (1983). *A. bartschi* (Marianas, Palau), *A. salanganus* (GS), *A. mearnsi* (Phil), and *A. leucophaeus* (Societies, Marquesas, Cooks) have variously been treated as subspecies, allospecies, or relatives.

104. *Aerodramus* [*spodiopygius*].
A. *A. spodiopygius*. (a) *delichon* (Salomonsen 1983). Man, Los Negros (in Admiralty Group). (b) *eichhorni* (Hartert 1924b). SMt, Em. (c) *noonaedanae* (Salomonsen 1983). NB, NI, Vu, Lih, Tab. (d) *reichenowi* (Stresemann 1912). Buka, Boug, Wagina (near Ch), Ch, Ys, Guad, Mal, Kul, Wana, SCr, Ugi. Other subspecies: Cel, Mol, Cape York, NCal, NHeb, Fiji, Samoa
MEGASUBSPECIES: the NMel races may constitute a group distinct from the populations of Cape York (*terrae-reginae*) and Mol and Cel, and possibly also distinct from populations to the east.
OTHER ALLOSPECIES: *A. hirundinaceus* (NG), treated as conspecific by Salomonsen (1983).
COMMENTS: See Mayr (1937a), Medway (1966), Salomonsen (1983), and White and Bruce (1986) for discussion of this species and possibly related populations west of Wallace's line.

105. *Collocalia esculenta*. (a) *stresemanni* (Rothschild & Hartert 1914b). Man, Ramb, Nau, Tench, Los Negros (in Admiralty group), Nusa (islet off N NI). (b) *tametamele* (Stresemann 1921). NB, Long, Vu (vagr), Tol, Cr, Witu; Buka, Boug. (c) *kalili* (Salomonsen 1983). NI, NH, Dyl. (d) *spilogaster* (Salomonsen 1983). Lih, Feni, Tab, Tang; Nis. (e) *becki* (Mayr 1931b). Shtl, Faur, Ch, Wagina (near Ch), Ys, Fla, Guad, Pav, Ban, Mal, VL, Gan, Gizo, Kul, NGa, Vang, Gat, Rnd, Tet, Gow. (f) *makirensis* (Mayr 1931b). SCr, Ugi, 3S. (g) *desiderata* (Mayr 1931b). Ren, Bel. Other subspecies: Malay Peninsula and Phil to NG, Cape York, NHeb, NCal
MEGASUBSPECIES: none. Northern Melanesian populations with white rumps (e.g. *tametamele*, *desiderata*, *stresemanni*) may have evolved that trait independently (Salomonsen 1983).
COMMENTS: See Mayr (1931b), Medway (1966), Salomonsen (1983), White and Bruce (1986), and Somadikarta (1986). Among populations west of Wallace's line included in *C. esculenta*, the *linchi* group and others may comprise one or more

sibling species. The subspecific allocation of the Long, Tol, Cr, Feni, Tang, and Buka populations of *C. esculenta* are uncertain. We do not separate the Nis population ("*hypogrammica*": Salomonsen 1983) from *spilogaster*.

HEMIPROCNIDAE (tree swifts)

See Chantler and Driessens (1995).

106. *Hemiprocne mystacea*. (a) *macrura* (Salomonsen 1983). Man, Ramb, Los Negros. (b) *aeroplanes* (Stresemann 1921). NB, NI, NH, SMt, Long, Umb, Vu, DY, Em, Dyl, Lih, Tab, Tang, Lol. (c) *woodfordiana* (Hartert 1896). Feni; Ren plus all Central Solomons except Fla, BV, Savo, Ul, Mono, Rmos, Gow, and SCr group. (d) *carbonaria* (Salomonsen 1983). SCr, SAn, SCat (no record for Ugi or 3S). Other subspecies: Mol, NG
MEGASUBSPECIES: none.
COMMENTS: We follow Salomonsen (1983) in considering *aeroplanes* and *macrura* separable from *mystacea*, contrary to Mees (1964, 1965a) and Gilliard and LeCroy (1967).

ALCEDINIDAE (kingfishers)

See Fry (1980a, b), Fry et al. (1992), and Forshaw and Cooper (1983–94).

107. *Alcedo* [*atthis*]
A. *A. atthis*. (a) *hispidoides*. NB, NI, NH, SMt, Man, Long, Umb, Vu, DY, Em, Dyl, Lih, Feni, Tab, Tang, Tol, Sak (and Cel, Mol, NG). (b) *salomonensis* (Rothschild & Hartert 1905). Buka, Boug, Ch, Ys, Guad, Mal, VL, Bag, Gan, Gizo, Kul, Wana, Koh, NGa, Vang, Gat, Rnd, Tet, SCr, Mono, Nis. Other subspecies: Eurasia to LS
MEGASUBSPECIES: none.
OTHER ALLOSPECIES: *A. semitorquata* (Afr), sometimes considered conspecific with *A. atthis*.
COMMENTS: See Rothschild & Hartert (1905), Schodde (1977), and Forshaw and Cooper (1983–94) for geographic variation. Fry (1980a) considers this superspecies, the *A.* [*azurea*] superspecies (species 108, including *A. meninting* and *A. quadribrachys*), and *A. hercules* (SE Asia) to be related.

108. *Alcedo* [*azurea*]
A. *A. websteri* (Hartert 1898). NB, NI, NH, Umb, Vu (vagr), Lih
OTHER ALLOSPECIES: *A. azurea* (NG, Mol, Tenimbar, Au), and possibly (Fry 1980a) *A. meninting* (SE Asia and GS) and *A. quadribrachys* (Afr).
COMMENTS: This species was formerly placed in *Alcyone* or *Ceyx*. It is uncertain whether to recognize these genera as distinct from *Alcedo*, and, if so, which species to place in them. Schodde and Mason (1976) and Fry (1980a) place *azurea* in *Alcedo*, while Forshaw and Cooper (1983) retain it in *Alcyone*.

109. *Alcedo pusilla*. (a) *masauji* (Matthews 1927). NB, NI, NH, Umb, Dyl, Tab. (b) *bougainvillei* (Ogilvie-Grant 1914). Buka, Boug, Shtl, Ch, Ys, Fla. (c) *aolae* (Ogilvie-Grant 1914). Guad, Pav (subsp?), Mal (subsp?). (d) *richardsi* (Tristram 1882). VL, Bag, Gan, Gizo, Kul, Wana, Koh, NGa, Vang, Rnd, Tet. Other subspecies: Mol, NG, Au
MEGASUBSPECIES: none.
COMMENTS: Fry (1980a) considers this species and *A. coerulescens* (Sumatra to Sumbawa) to form a superspecies. This species was formerly placed in *Alcyone* or *Ceyx* (see under species 108 *A.* [*azurea*]); Forshaw and Cooper (1983–94) retain it in *Alcyone*.

110. *Ceyx lepidus*. (a) *dispar* (Rothschild & Hartert 1914c). Man. (b) *sacerdotis* (Ramsay 1882f). NB, Umb, Vu, Lol. (c) *mulcatus* (Rothschild & Hartert 1914d). NI, NH, Dyl, Lih, Tab. (d) *meeki* (Rothschild 1901). Buka, Boug, Ch, Ys. (e) *collectoris* (Rothschild & Hartert 1901c). VL, Gan, Kul, NGa, Vang, Gat, Rnd, Tet. (f) *nigromaxilla* (Rothschild & Hartert 1905). Guad. (g) *malaitae* (Mayr 1935). Mal. (h) *gentianus* (Tristram 1879). SCr. Other subspecies: Phil, Mol, NG.
MEGASUBSPECIES: *dispar*, *meeki*, and *gentianus* are each very distinct, while the other NMel races belong in a group with *solitarius* of NG (Forshaw & Cooper 1983–94).

COMMENTS: Fry (1980a) considers *C. argentatus* (Phil) the closest relative. Stresemann (1940–41) believed instead that *C. fallax* (Cel) connects *C. lepidus* to the superspecies *C. erithacus* (including *rufidorsum*)–*C. melanurus* of SE Asia to Phil. See Mayr (1936), Schodde (1977), Fry (1980a), and Forshaw and Cooper (1983–94) for geographic variation in plumage and in bill color and shape. The Lih and Tab populations may belong to undescribed subspecies.

111. *Halcyon* [*diops*]. (map 23)
A. *H. albonotata* (Ramsay 1885). NB.
B. *H. leucopygia* (Verreaux 1858). Buka, Boug, Faur, Ch, Ys, Fla, BV, Guad, Fara (near Ys), Fatura (near Ys)
OTHER ALLOSPECIES: *H. farquhari* (NHeb), *H. macleayii* (E NG, Au), *H. nigrocyanea* (NG), *H. diops* (NMol), *H. lazuli* (S Mol, sometimes treated as a subspecies of *H. diops*).
COMMENTS: *H. winchelli* (Phil) is related. The Au breeding race *H. macleayii incincta* occurs as a winter visitor on NB.

112. *Halcyon* [*australasia*]. (plate 6)
A. *H. sancta*. (a) *sancta*. Breeds in Au. Widespread throughout NMel as a winter visitor; some individuals remain through the year. Recorded from all Bismarck islands except SMig, Anch, Nin, Dyl, Ting, Tang, and Lol, and from all Solomon islands except Ul, Gan, Simb, SAn, SCat, Bor, OJ, Sik, and Nuku. Breeding records for 3S (French 1957) and possibly Guad (Cain & Galbraith 1956) and other Solomon islands (Diamond 2001a). Other subspecies: NZ, NCal, Loyalty Islands
OTHER ALLOSPECIES: *H. australasia* (LS, Southwest Islands, Tenimbar). Fry (1980a) considers the Polynesian *H. tuta*, *H. gambieri*, *H. godeffroyi*, and *H. venerata* to be further allospecies, and *recurvirostris* (Samoa) to be a race of *H. sancta*; Pratt et al. (1987) consider *vitiensis* (Fiji), *eximia* (Fiji), and *regina* (Futuna) also to be races of *H. sancta*; we consider these forms closer to species 113 (*H.* [*chloris*]). The relations of these forms to *H. sancta* and *H. chloris* require further study, especially as *H. sancta* and *H. chloris* are undoubtedly closely related and seem each to have spread in several waves. *H. saurophaga* (species 114) and *H. pyrrhopygia* (Au) are also related. Clarity can be achieved only through a thorough revision of the entire group.

113. *Halcyon* [*chloris*]. (plate 6, map 24)
A. *H. chloris*. (a) *matthiae* (Heinroth 1902). SMt, Em. (b) *stresemanni* (Laubmann 1923). Long, Umb, Tol, Cr, Witu, Unea. (c) *tristrami* (Layard 1880). NB, Vu, Lol. (d) *novaehiberniae* (Hartert 1925). S NI, Dyl. (e) *nusae* (Heinroth 1902). N NI, NH, Ting, Lih, Feni, Tab, Tang. (f) *bennetti* (Ripley 1947). Nis. (g) *alberti* (Rothschild & Hartert 1905). Buka, Boug, Shtl, Faur, Ch, Ys, Fla, BV, Savo, Guad, VL, Bag, Gan, Simb, Gizo, Kul, Wana, Koh, NGa, Vang, Gat, Rnd, Tet, Gow (subsp?), OJ (subsp?), Sik (subsp?). (h) *pavuvu* (Mayr 1935). Pav, Ban. (i) *mala* (Mayr 1935). Mal. (j) *solomonis* (Ramsay 1882b). SCr, Ugi, SAn, SCat. (k) *sororum* (Galbraith & Galbraith 1962). 3S. (l) *amoena* (Mayr 1931b). Ren, Bel. The sole Northern Melanesian islands on which this species is absent are the NW Bismarcks (except possibly present on Wuv), Sak, DY, Cred, Tench, Ul, Mono, Rmos, Bor, Fead, Kili, Tauu, and Nuku. Other subspecies: Red Sea, south coast of Asia, Indonesian Archipelago, NG, Au, Micronesia, Santa Cruz, NHeb, Fiji, Tonga, Samoa
MEGASUBSPECIES: the NMel races may be grouped into at least three megasubspecies. First, the races *solomonis*, *sororum*, and *amoena* of the eastern Solomons are closest to the NHeb races; *amoena* is distinct within this group. Second, the remaining Sol races (*alberti*, *mala*, and *pavuvu*) are very close to each other and (surprisingly) to *tristrami* of NB. The third group consists of the remaining Bis races, among which *novaehiberniae*, *nusae*, and *bennetti* are close to each other, *stresemanni* approaches *sordida* of NG, and *matthiae* is distinct.
OTHER ALLOSPECIES: *H. cinnamomina* (Micronesia), *H. miyakoensis* (Miyako, Riu Kiu Islands, possibly a race of *H. cinnamomina*, Greenway 1967), *H. ruficollaris* (Cooks, possibly a race of *H.*

tuta), *H. recurvirostris* (Samoa), *H. venerata* (Societies), *H. tuta* (Societies), *H. gambieri* (Tuamotus), *H. godeffroyi* (Marquesas, possibly a race of *H. tuta*). The last five allospecies form the *Todirhamphus* group, which Fry (1980a) considers instead to be allospecies of species 112 (*H.* [*australasia*]).
COMMENTS: The Talaud population *enigma* may be a race of *H. chloris* (Eck 1978) or a distinct species (White & Bruce 1986). *H. cinnamomina* is connected to *H. chloris* via *H. chloris matthiae* and is sympatric with *H. chloris* only on Palau. Among sympatric forms *H. saurophaga* (species 114 below), *H.* [*australasia*] (species 112 above), and *H. funebris* (Mol) are fairly close, *H. pyrrhopygia* (Au) less close. *H. chloris* is curiously absent from the NW Bismarcks and N NG, and some uncertainty remains whether races 114a and 114b below belong to *H. chloris* or to *H. saurophaga*, though the latter seems more likely. See Mayr (1931a, 1941b, 1949b), Baker (1951), Holyoak (1974), Diamond (1976), and Fry (1980a) for discussion of species lines, and see under species 112 (*H.* [*australasia*]) for further comments. For geographic variation within NMel, see Hartert (1926b), Mayr (1936), Galbraith and Galbraith (1962), and Forshaw and Cooper (1983–94/1985).

114. *Halcyon saurophaga*. (plate 6, map 24). (a) *admiralitatis* (Sharpe 1892). Man, Ramb, Nau, SMig. (b) *anachoretae* (Reichenow 1898). Anch, Her, Nin, Wuv (subsp?). (c) *saurophaga*. Virtually all Bismarcks except NW Bismarcks (ranges of 114a and 114b) (no records for Sak, Unea); virtually all Central Solomons plus Nis, Fead (no records for BV, Savo, Rmos) (and Mol, N NG, SE Papuan Islands)
MEGASUBSPECIES: the NW Bis races *admiralitatis* and *anachoretae* constitute one megasubspecies, while nominate *saurophaga* (occupying all the rest of the species' range) is the other megasubspecies.
COMMENTS: See under species 113 above.

115. *Halcyon bougainvillei*. (plate 8, map 4). (a) *bougainvillei* (Rothschild 1904b). Boug. (b) *excelsa* (Mayr 1941b). Guad
MEGASUBSPECIES: none.
COMMENTS: Fry (1980a) transferred this species from *Halcyon* to *Actenoides*, along with *H. monacha* (Cel), *H. princeps* (Cel), and the *H. lindsayi* (Phil)–*H. concreta* (Malay Peninsula, Sumatra, Borneo)–*H. hombroni* (Phil) superspecies. Forshaw and Cooper (1983–94) and White and Bruce (1986) concurred, but we prefer to retain this species in *Halcyon* until relationships within *Halcyon* become better understood. The New Guinea species *Melidora* (Lesson 1830) *macrorhina* also seems to belong to this group, and this generic name is older than *Actenoides* (Bonaparte 1850). See Mayr (1941b), Galbraith and Galbraith (1962), and duPont and Niles (1980) for plumage and races.

116. *Tanysiptera sylvia*. (a) *leucura* (Neumann 1915). Umb. (b) *nigriceps* (Sclater 1877a). NB, Vu (vagr), DY, Lol. Other subpsecies: Au, SE NG
MEGASUBSPECIES: the two Bis races constitute one megasubspecies, while the Au race *sylvia* and NG race *salvadoriana* constitute the other (see Hartert 1926b for the plumage differences).
COMMENTS: The Bismarck populations are sometimes separated as an allospecies *T. nigriceps*. The *T. galatea* (Mol, NG)–*T. ellioti* (Kofiau)–*T. riedelii* (Biak)–*T. carolinae* (Numfor)–*T. hydrocharis* (Aru and S NG, sympatric with *T. galatea*) superspecies and the *T. nympha* (N NG)–*T. danae* (SE NG) superspecies are closely related. *Tanysiptera* is close to *Halcyon* (Sibley & Ahlquist 1983).

MEROPIDAE (bee-eaters)

See Fry (1969, 1984), Fry et al. (1992), Forshaw and Cooper (1983-1994).

117. *Merops* [*superciliosus*]
A. *M. philippinus*. NB, Long, Umb, Sak (and India to SE Asia, Cel, NG)
OTHER ALLOSPECIES: *M. superciliosus* (SW Asia, Afr).
COMMENTS: We agree with Fry (1984) and Forshaw and Cooper (1983–94/1987) that the race *salvadorii* (Meyer 1891) does not deserve recognition. Sympatry between the taxa *persicus* and *philippinus* in India has

variously been interpreted to mean that *M. superciliosus* (including *M. s. persicus*) and *M. philippinus* are a superspecies with slight overlap (Fry 1969) or that *M. superciliosus* (excluding *persicus*) and *M. philippinus* are conspecific while *M. persicus* is a separate species (Fry 1984), or that *M. superciliosus*, *M. philippinus*, and *M. persicus* are each separate species (Forshaw & Cooper 1983–94/1987). See also Mees (1982) and White and Bruce (1986). *M. ornatus* (Au) is related.

CORACIIDAE (rollers)

See Forshaw and Cooper (1983–94) and Fry et al. (1992).

118. *Eurystomus orientalis*. (a) *crassirostris* (Sclater 1869). NB, NI, NH, SMt, Umb, Vu (vagr), DY, Lih, Tab, Lol, Tol, Sak, Witu, Unea. (b) *solomonensis* (Sharpe 1890). Feni; virtually all Central Solomons (no records for Rmos). Other subspecies: E Asia to NG, Au
MEGASUBSPECIES: none.
COMMENTS: See Mayr (1934a) and Ripley (1942b) for subspecies. The two NMel races resemble *waigiouensis* of NG in their brightly colored, dark plumage and differ from the dull pale *pacificus* of Au. Mees (1965a) and White and Bruce (1986) treat the N Mol population, *azureus*, as an allospecies. *E. o. pacificus*, which breeds in Au, occurs as a winter visitor in the Bismarcks.

BUCEROTIDAE (hornbills)

See Sanft (1960), Forshaw and Cooper (1983–94), and Kemp (1995).

119. *Rhyticeros plicatus*. (map 25). (a) *dampieri* (Mayr 1934a). NB, NI, NH. (b) *harterti* (Mayr 1934a). Buka, Boug, Shtl, Faur, Mono. (c) *mendanae* (Hartert 1924e). Ch, Wagina (near Ch), Ys, Fla (vagr), BV (vagr), Savo (vagr), Guad, Mal, VL, Gan, Gizo, Kul, Wana (vagr), NGa, Koh, Vang, Gat, Rnd, Tet, Leru of the Russell group, Bor. Other subspecies: SE Asia, Sumatra, Mol, NG
MEGASUBSPECIES: none.
COMMENTS: The population of SE Asia and Sumatra (*subruficollis*) has alternatively been treated as an allospecies or separate species (White & Bruce 1986, Kemp 1995). *R. undulatus* (SE Asia, GS), *R. everetti* (Sumba), and *R. narcondami* (Narcondam in the Bay of Bengal) are related, and some of these may be allospecies. Peters (1931–87/1945) sinks *Rhyticeros* in *Aceros*, but most recent authors retain *Rhyticeros*. Most populations on islands of the NGa group were founded by an invasion after 1940 (Diamond 2001a). For geographic variation, see Mayr (1934a).

PASSERINES

Passerine family-level and subfamily-level taxonomy is in a state of flux, due especially to recent results of DNA/DNA hybridization studies (Sibley & Ahlquist 1983, 1985, 1990). Of the volumes of Peters's *Check-list of Birds of the World*, all except volume 11 were produced prior to Sibley and Ahlquist's reclassification of Australo-Papuan passerines. We do not attempt to provide a revised or new family- and subfamily-level classification of the passerines, as do Sibley and Monroe (1990). Instead, our arrangement is closer to the traditional Peters arrangement (merely so that readers can more easily find taxa by looking in the traditional place) but incorporates some features of Sibley and Ahlquist's classification. See Sibley and Ahlquist (1983, 1985, 1990) and Mayr and Bock (1994) for further discussion.

PITTIDAE (pittas)

See Mayr (in Peters 1979) and Lambert and Woodcock (1996).

120. *Pitta* [*sordida*]
A. *P. superba* (Rothschild & Hartert 1914c). (plate 9). Man.
B. *P. sordida*. (a) *novaeguineae*. Long, Tol, Cr (and NG). Other subspecies: SE Asia, Phil, Borneo, Sumatra, Cel, islands off NG
OTHER ALLOSPECIES: *P. maxima* (N Mol) and probably *P. steerei* (Phil).
COMMENTS: *P.* [*brachyura*] (species 121 below) is related. See Mayr (1955).

121. *Pitta* [*brachyura*]. (plate 9)

A. *P. anerythra.* (a) *pallida* (Rothschild 1904b). Boug. (b) *nigrifrons* (Mayr 1935). Ch. (c) *anerythra* (Rothschild 1901). Ys
MEGASUBSPECIES: none.
OTHER ALLOSPECIES: *P. versicolor* (Au, LS, Mol), *P. moluccensis* (SE Asia), *P. brachyura* (India), *P. nympha* (Japan, Korea, China), *P. angolensis* (including *reichenowi* (Afr). *P. elegans* (LS, Mol) and *P. iris* (N Au) are sometimes separated from *P. versicolor* (E Au). *P. moluccensis megarhyncha* is sometimes treated as a separate species (Medway & Wells 1976). See Mayr (1936, 1955) for geographic variation in *P. anerythra.*

122. *Pitta* [*erythrogaster*]. (map 26)
A. *P. erythrogaster.* (a) *gazellae* (Neumann 1908). NB, Umb, Vu, DY (?), Lol, Tol. (b) *novaehibernicae* (Ramsay 1878c). NI, Dyl (subsp?). (c) *extima* (Mayr 1955). NH. (d) *splendida* (Mayr 1955). Tab. Other subspecies: Phil, Cel, Mol, NG, Cape York
MEGASUBSPECIES: *splendida* constitutes one megasubspecies, while the other three NMel races together constitute another.
OTHER ALLOSPECIES: *P. kochi* (Luzon), a doublet.
COMMENTS: *P. arcuata* (Borneo), *P. granatina* (Malay Peninsula, Sumatra, Borneo), and (if it is a separate species from *P. granatina*) *P. venusta* (Borneo, Sumatra) are related. See Mayr (1955) for geographic variation.

HIRUNDINIDAE (swallows)

123. *Hirundo* [*rustica*]
A. *H. tahitica.* (a) *frontalis.* Man (subsp?), Long, Umb, Tol, Sak, Cr (and Cel, LS, Mol, NG). (b) *ambiens* (Mayr 1934a). NB, Vu, DY. (c) *subfusca.* NI, NH, Lih, Feni, Tab, Tang; virtually all Central Solomons (no records for Savo, Rmos, Gow) plus Nis (and NHeb, NCal, Fiji). Other subspecies: India to GS, Phil, Au, Societies
MEGASUBSPECIES: *subfusca.*
OTHER ALLOSPECIES (Hall & Moreau 1970): *H. rustica* (Eurasia, N Afr, NAm), *H. lucida* (W Afr), *H. angolensis* (Central Afr), *H. aethiopica* (central Afr), *H. albigularis* (S Afr). *H. neoxena* (Au) is sometimes separated from *H. tahitica.*
COMMENTS: See Mayr (1934a, 1955) for geographic variation. The race *ambiens* is intermediate between *frontalis* and *subfusca* and may have arisen as a hybrid between them (Mayr 1934a), while the population of Long and its neighbors arose from *frontalis* and *ambiens*, mainly *frontalis*, after the volcanic explosion of Long (Mayr 1955).

CAMPEPHAGIDAE (cuckoo-shrikes)

See Ripley (1941) and Voous & van Marle (1949) for *Coracina*; see Mayr & Ripley (1941) for *Lalage.*

124. *Coracina* [*caledonica*]. (map 27)
A. *C. caledonica.* (a) *bougainvillei* (Mathews 1928). Boug. (b) *welchmani* (Tristram 1892). Ys, Fara (islet near Ys). (c) *amadonis* (Cain & Galbraith 1955). Guad. (d) *kulambangrae* (Rothschild & Hartert 1916a). Kul, NGa, Vang. Other subspecies: NHeb, NCal
MEGASUBSPECIES: the four Solomon races differ from the NHeb and NCal races in possessing marked sexual dimorphism and form a megasubspecies.
OTHER ALLOSPECIES: *C. novaehollandiae* (India to Java, Bali, LS, Au), *C. fortis* (Buru), *C. atriceps* (Mol), *C. pollens* (Tenimbar and Kei), *C. schistacea* (Sula).
COMMENTS: See Ripley (1941), Voous & van Marle (1949), Peters (1931–87/1960), and White & Bruce (1986) for species limits in the superspecies and Mayr (1955) for geographic variation in *C. caledonica. C. macei* (SE Asia), *C. javensis* (Java, Bali), and *C. personata* (LS, including *C. pollens*) are sometimes separated from *C. novaehollandiae.*

125. *Coracina lineata.* (a) *sublineata* (Sclater 1879). NB, NI, Lol. (b) *nigrifrons* (Tristram 1892). Buka, Boug, Shtl, Faur, Ch, Ys, Wagina. (c) *ombriosa* (Rothschild & Hartert 1905). VL, Gizo, Kul, Wana, Koh, NGa, Vang, Gat, Rnd, Tet. (d) *pusilla* (Ramsay 1879b). Guad. (e) *malaitae* (Mayr

1931e). Mal. (f) *makirae* (Mayr 1935). SCr. (g) *gracilis* (Mayr 1931b). Ren, Bel. Other subspecies: NG, Au
MEGASUBSPECIES: *gracilis*, which lacks sexual dimorphism and resembles *lineata* of Au. The race *makirae* is somewhat intermediate between *gracilis* and the other NMel races, which in turn resemble *axillaris* of NG.
COMMENTS: See Mayr (1931b, 1936, 1955), Galbraith and Galbraith (1962), and Schodde (1977) for geographic variation. The relationships of this species within *Coracina* are debated (Ripley 1941, Voous & van Marle 1949).

126. *Coracina* [*papuensis*]
A. *C. papuensis*. (a) *ingens* (Rothschild & Hartert 1914c). Manus. (b) *sclateri* (Salvadori 1878). NB, NI, NH, Umb, Vu, Lol. (c) *perpallida* (Rothschild & Hartert 1916a). Buka, Boug, Shtl, Faur, Ch, Ys, Wagina, Fla, BV, Mono (?). (d) *elegans* (Ramsay 1881). Guad, Pav, and all islands of NGa group except Simb. (e) *eyerdami* (Mayr 1931e). Mal, Ul. Other subspecies: Mol, NG, Au
MEGASUBSPECIES: none.
OTHER ALLOSPECIES: *C. leucopygia* (Cel).
COMMENTS: The Australian *robusta* appears to be a race of *C. papuensis* (Galbraith 1969). See Rothschild and Hartert (1916a) and Mayr (1955) for subspecies.

127. *Coracina* [*tenuirostris*]. (plate 7, map 28)
A. *C. tenuirostris*. (a) *admiralitatis* (Rothschild & Hartert 1914c). Man. (b) *muellerii*. Long, Sak (and NG). (c) *rooki* (Rothschild & Hartert 1914c). Umb. (d) *heinrothi* (Stresemann 1922). NB, DY (?), Lol. (e) *matthiae* (Sibley 1946). SMt, Em. (f) *remota* (Sharpe 1878). NI, NH, Dyl, Feni. (g) *ultima* (Mayr 1955). Lih, Tab, Tang. (h) *saturatior* (Rothschild & Hartert 1902a). Buka, Boug, Shtl, Faur, Ch, Ys, Wagina (near Ch), Mono, and all islands of NGa group except Simb. (i) *erythropygia* (Sharpe 1888a). Fla, BV, Savo, Guad, Mal, Ul (subsp?). (j) *nisoria* (Mayr 1950). Pav, Ban. Other subspecies: Timor, Cel, Mol, NG, Au, Micronesia.
B. *C. salomonis* (Tristram 1879). SCr
MEGASUBSPECIES: *nisoria* is one megasubspecies; the group with ventrally unbarred females, consisting of *remota*, *ultima*, *saturatior*, and *erythropygia*, is another. The remaining five races of the western Bismarcks and NG (*admiralitatis*, *muellerii*, *rooki*, *heinrothi*, *matthiae*) have ventrally barred females and belong to a third megasubspecies.
COMMENTS: *C. morio* (Phil, Cel, Mol, NG) and *C. dohertyi* (Sumba, Flores) are the nearest relatives. We follow Galbraith and Galbraith (1962) in treating the distinctive population *salomonis* as an allospecies. Species limits within this group are uncertain, and van Bemmel (1948), Mees (1982), and White and Bruce (1986) give different arrangements. See Mayr (1955) for geographic variation.

128. *Coracina holopolia*. (map 29) (a) *holopolia* (Sharpe 1888a). Buka, Boug, Ch, Ys, Guad. (b) *pygmaea* (Mayr 1931e). Kul, NGa, Vang, Gat. (c) *tricolor* (Mayr 1931e). Mal
MEGASUBSPECIES: the races are very distinct and constitute three megasubspecies.
COMMENTS: It is uncertain whether the superficially similar *C. montana* (NG) is closely related. See Mayr (1931e) for geographic variation.

129. *Lalage* [*aurea*]. (map 30)
A. *L. leucomela*. (a) *falsa* (Hartert 1925). NB, Umb, Vu, DY, Lol, Sak. (b) *karu* (Lesson & Garnot 1827). NI. (c) *albidior* (Hartert 1924a). NH. (d) *sumunae* (Salomonsen 1964). Dyl. (e) *ottomeyeri* (Stresemann 1933). Lih. (f) *tabarensis* (Mayr 1935). Tab. (g) *conjuncta* (Rothschild & Hartert 1924a). SMt. Other subspecies: NG, Au, Kei
MEGASUBSPECIES: *conjuncta* is one megasubspecies; the group consisting of *ottomeyeri* and *tabarensis* is another. The remaining Bismarck races belong to the same megasubspecies as the NG races.
OTHER ALLOSPECIES: *L. aurea* (Mol), *L. atrovirens* (Tenimbar, N NG), *L. maculosa* (NHeb, Fiji, Samoa), *L. sharpei* (Samoa).
COMMENTS: See Mayr & Ripley (1941) for relationships and Mayr (1955) and Salomonsen (1964) for subspecies.

130. *Lalage leucopyga*. (map 30). (a) *affinis* (Tristram 1879). SCr, Ugi. Other subspecies: NHeb, NCal
MEGASUBSPECIES: none.
COMMENTS: See Mayr & Ripley (1941) for geographic variation.

TURDIDAE (thrushes)

131. *Saxicola caprata*. (map 31). (a) *aethiops* (Sclater 1880a). NB, NI, Long, Vu (and N NG). Other subspecies: Asia to Phil, Cel, Java, LS, NG.
COMMENTS: See Mayr (1944a) for distribution and Mayr (1955) for geographic variation.

132. *Zoothera* [*dauma*]. (Map 32).
A. *Z. heinei*. (a) *eichhorni* (Rothschild & Hartert 1924a). SMt, Em. (b) *choiseuli* (Hartert 1924b). Ch. Other subspecies: NG (*papuensis*), E Au (*heinei*).
B. *Z. talaseae*. (a) *talaseae* (Rothschild & Hartert 1926b). NB, Umb. (b) *atrigena* (Ripley & Hadden 1982). Boug.
C. *Z. margaretae*. (a) *turipavae* (Cain & Galbraith 1955). Guad. (b) *margaretae* (Mayr 1935). SCr
MEGASUBSPECIES: none.
OTHER ALLOSPECIES: *Z. lunulata* (E Au, including *Z. l. cuneata*), *Z. machiki* (Tenimbar), *Z. dauma* (Eurasia to Java, Sumatra, Lombok).
COMMENTS: See Mayr (1936, 1955), Ripley and Hadden (1982), Ford (1983), and White and Bruce (1986) for discussion. Ford (1983) showed that *lunulata* and *heinei*, formerly considered races of *Z. dauma*, occur sympatrically in E Au and must be distinct species. *Z. lunulata* and *Z. heinei* may be more similar to each other than either is to Asian *Z. dauma* (R. Schodde, personal communication). We list them as separate allospecies and consider the superspecies to be represented by a doublet in Au. *Z. machiki* is closest to *Z. heinei*, but we follow White and Bruce (1986) in considering it as a separate allospecies. Three of the four Solomon races are known only from unique types, making assessment of allospecies lines uncertain. We do not follow Ripley and Hadden (1982), who rank *margaretae* and *turipavae* as subspecies of *talaseae*.

133. *Turdus poliocephalus*. (map 33). (a) subsp? NB. (b) *beehleri* (Ripley 1977). NI. (c) *heinrothi* (Rothschild & Hartert 1924a). SMt. (d) *tolokiwae* (Diamond 1989). Tol. (e) *bougainvillei* (Mayr 1941b). Boug. (f) *kulambangrae* (Mayr 1941b). Kul. (g) *sladeni* (Cain & Galbraith 1955). Guad. (h) *rennellianus* (Mayr 1931b). Ren. Other subspecies: Phil, Indonesia, NG, NHeb, NCal, Fiji, Samoa
MEGASUBSPECIES: none.
COMMENTS: *T. merula* (Eurasia) and many other Asian, African, and New World species appear related. See Mayr (1941b, 1955) for geographic variation.

ACANTHIZIDAE (Australasian warblers)

134. *Gerygone flavolateralis*. (map 34). (a) *citrina* (Mayr 1931b). Ren. Other subspecies: NCal, NHeb
MEGASUBSPECIES: *citrina*, the most distinct race of this species (Ford 1986).
COMMENTS: Ford's (1986) reanalysis indicates that this species is closest to *G. insularis* (Lord Howe), *G. fusca* (Au), and *G. laevigaster* (N and E Au, S NG) rather than to *G. olivacea* (Diamond & Marshall 1976) or *G. chrysogaster* (Ford 1981a). See Meise (1931) and Mayr (1931b) for geographic variation in *G. flavolateralis*.

SYLVIIDAE (Old World warblers)

135. *Ortygocichla* [*rubiginosa*]. (plate 9, map 4)
A. *O. rubiginosa* (Sclater 1881). NB
OTHER ALLOSPECIES: questionably, *O. rufa* (Fiji).
COMMENTS: The interrelations and family affiliations of the warblers *O. rubiginosa* and *O. rufa*, *Cichlornis* [*whitneyi*] (species 136 below), *Megalurulus mariei* (NCal) and *M.* (*Buettikoferella*) *bivittata* (Timor), and possibly *Eremiornis carteri* (Au) are uncertain. The latter three species resemble each other, while *Cichlornis* resembles *Ortygocichla* but differs by an attenuated tail and a much heavier bill. See Mayr (1933a, 1944a, 1955), Cain and Galbraith

(1955), Gilliard (1960b), Galbraith and Galbraith (1962), Hadden (1983), and Ripley (1985).

136. *Cichlornis* [*whitneyi*]. (map 35)
   A. *C. whitneyi*. (a) *turipavae* (Cain & Galbraith 1955). Guad. Other subspecies: Santo.
   B. *C. llaneae* (Hadden 1983). Boug.
   C. *C. grosvenori* (Gilliard 1960b). (Plate 9). NB
   MEGASUBSPECIES: none (*turipavae* is very similar to nominate *whitneyi* of Santo).
   COMMENTS: See under species 135 for relationships of the genus, and Mayr (1933a), Cain & Galbraith (1955), Gilliard (1960b), Hadden (1983), and Ripley (1985) for discussion of the species. We disagree with Ripley's (1985) conclusion that the four populations of *Cichlornis* should be considered conspecific.

137. *Cettia parens* (Mayr 1935). (map 4). SCr. COMMENTS: See Mayr (1936) for possible relationship to *C. ruficapilla* (Fiji). We agree with Orenstein and Pratt (1983) in placing both species in *Cettia* (rather than segregating them in *Vitia*) and considering them related to *C. diphone* (Manchuria, Korea, Japan, China, Luzon), *C. annae* (Palau), and *C. carolinae* (Tenimbar; Rozendaal 1987).

138. *Acrocephalus* [*arundinaceus*]
   A. *A. stentoreus*. (a) *sumbae*. NB, NI, Long, Umb; Buka, Boug, Ys, Guad, Ban, Gijunabena (and Sumba, Timor, Mol, Cape York, NG). Other subspecies: Egypt and Asia to Java, Phil, Cel, LS, Au
   OTHER ALLOSPECIES: *A. arundinaceus* (Eurasia) (marginal overlap with *A. stentoreus*), *A. orientalis* (E Asia), *A. luscinia* (Micronesia), *A. aequinoctialis* (Line Islands), *A. familiaris* (Hawaii), *A. caffer* (Marq, Societies), *A. atyphus* (Tuamotus), *A. vaughani* (Pitcairn, Henderson, Rimitara, Cooks).
   COMMENTS: For distinction between *A. arundinaceus* and *A. stentoreus*, see Stresemann and Arnold (1949). See Mayr (1948) and White and Bruce (1986) for geographic variation.

139. *Cisticola exilis*. (a) *polionota* (Mayr 1934a). NB, NI, NH, Long, Umb, Vu, DY, Lih, Tab, Sak. Other subspecies: India to Phil, Indonesia, NG, Au
   MEGASUBSPECIES: *polionota*, differing from *diminuta* of NG in the lack of an eclipse plumage.
   COMMENTS: See Mayr (1934a, 1955) for geographic variation.

140. *Megalurus timoriensis*. (a) *interscapularis* (Sclater 1880a). NB, NI, NH, Vu, Tol. Other subspecies: Phil, Cel, Timor, Sumba, Ambon, NG, Au
   MEGASUBSPECIES: none (*interscapularis* is very similar to *macrurus* of NG).
   COMMENTS: See Hartert (1925, 1930) for geographic variation.

141. *Phylloscopus* [*trivirgatus*]. (map 36)
   A. *P. poliocephala*. (a) *moorhousei* (Gilliard & LeCroy 1967). NB, Umb (subsp?). (b) *leletensis* (Salomonsen 1965). NI. (c) *matthiae* (Rothschild & Hartert 1924a). SMt. (d) *bougainvillei* (Mayr 1935). Boug. (e) *pallescens* (Mayr 1935). Kul. (f) *becki* (Hartert 1929). Ys, Guad, Mal. Other subspecies: Mol, Kei, NG.
   B. *P. makirensis* (Mayr 1935). SCr
   MEGASUBSPECIES: none.
   OTHER ALLOSPECIES: *P. trivirgatus* (Malaya, GS, Phil, Lombok, Sumbawa), *P. sarasinorum* (Cel), *P. presbytes* (Flores, Timor).
   COMMENTS: *P. olivaceus* (Phil) and *P. cebuensis* (Phil) are closely related but are sympatric with each other and with *P. trivirgatus* on Negros. The populations that we consider to be allospecies have also been considered subspecies of *P. trivirgatus* but may represent several independent invasion waves. See Mayr (1944a, b), Parkes (1971), and White and Bruce (1986) for discussion. *P. amoenus* (species 142) is a derivative (Mayr 1944b, 1955). For geographic variation, see Mayr (1936), Salomonsen (1965), and Gilliard and LeCroy (1967).

142. *Phylloscopus amoenus* (Hartert 1929). (map 36). Kul. COMMENTS: See under species 141 for relationships, and Mayr (1944b) for reasons for rejecting the monotypic genus *Mochthopoeus*.

RHIPIDURIDAE (fantails)

See Mayr (1931d) for Solomon forms.

143. *Rhipidura leucophrys*. (a) *melaleuca*. Most Bismarcks except NW Bismarcks (no records for Feni, Tang, Cr, Tench); almost all Central Solomons (no records for Savo, Pav, Rmos, Gow) (and Mol, NG). Other subspecies: Au. COMMENTS: See Mayr (1931d) for geographic variation.

144. *Rhipidura* [*rufiventris*]. (map 37)
   A. *R. rufiventris*. (a) *niveiventris* (Rothschild & Hartert 1914c). Man, Ramb. (b) *finschii* (Salvadori 1882). NB, Vu, DY, Lol. (c) *setosa* (Quoy & Gaimard 1830). NI, NH, Dyl (subsp?). (d) *mussai* (Rothschild & Hartert 1924a). SMt. (e) *gigantea* (Stresemann 1933). Lih, Tab. (f) *tangensis* (Mayr 1955). Tang. Other subspecies: Sumbawa, Flores, Timor, Southwest Islands, Tenimbar, Mol, NG, Au.
   B. *R. cockerelli*. (a) *septentrionalis* (Rothschild & Hartert 1916b). Buka, Boug, Shtl. (b) *interposita* (Rothschild & Hartert 1916b). Ch, Ys. (c) *floridana* (Mayr 1931d). Fla. (d) *cockerelli* (Ramsay 1879e). Guad. (e) *lavellae* (Rothschild & Hartert 1916b). VL, Gan. (f) *albina* (Rothschild & Hartert 1901c). Kul, Koh, NGa, Vang, Rnd, Tet. (g) *coultasi* (Mayr 1931d). Mal
   MEGASUBSPECIES: *coultasi* (see Mayr 1931d, Galbraith & Galbraith 1962).
   COMMENTS: The populations of Tenimbar and of Flores and Sumbawa have also been considered allospecies distinct from *R. rufiventris* (*R. fuscorufa* and *R. diluta*, respectively; Rensch 1931, White & Bruce 1986). *R. perlata* (Malay Peninsula, GS), *R. javanica* (SE Asia, Phil, GS), and *R. euryura* (GS) may be related. See Mayr (1931d, 1955) for geographic variation, and Rensch (1931) and Galbraith and Galbraith (1962) for relationships.

145. *Rhipidura* [*spilodera*]. (map 38)
   A. *R. drownei*. (a) *drownei* (Mayr 1931d). Boug. (b) *ocularis* (Mayr 1931d). Guad.
   B. *R. rennelliana* (Mayr 1931b). Ren.
   C. *R. tenebrosa* (Ramsay 1882b). SCr
   MEGASUBSPECIES: none.
   OTHER ALLOSPECIES: *R. spilodera* (NHeb, NCal, Fiji), *R. nebulosa* (Samoa).
   COMMENTS: *R. fuliginosa* (species 146 below) is closely related but shares SCr with *R. tenebrosa*, NHeb and NCal with *R. spilodera*. *R. tenebrosa* may not be so closely related to this group. See Mayr (1931d) for geographic variation and relationships.

146. *Rhipidura fuliginosa*. (a) *brenchleyi* (subsp?). SCr (and NHeb). Other subspecies: Au, S NG, NZ, NCal
   COMMENTS: Relatives are *R.* [*spilodera*] (species 145 above, q.v.) and possibly *R. albolimbata* (NG) and *R. hyperythra* (NG) (Schodde 1977, Ford 1981b). The mangrove population of N Au and S NG may be a distinct species, *R. phasiana* (Ford 1981b). See Mayr (1931d) for geographic variation.

147. *Rhipidura* [*rufifrons*]. (map 39)
   A. *R. rufifrons*. (a) *semirubra* (Sclater 1877b). Man, SMig. Tong. (b) *commoda* (Hartert 1918). Buka, Boug, Shtl, Faur, Ch, Ys. (c) *granti* (Hartert 1918). VL, Bag, Gan, Simb, Gizo, Kul, Wana, Koh, NGa, Vang, Gat, Rnd, Tet. (d) *rufofronta* (Ramsay 1879e). Guad. (e) *brunnea* (Mayr 1931d). Mal. (f) *russata* (Tristram 1879). SCr. (g) *kuperi* (Mayr 1931d). SAn, SCat (subsp?). (h) *ugiensis* (Mayr 1931d). Ugi. Other subspecies: LS from Flores eastwards, Mol, W and SE Papuan Islands, coastal NG, Au, Micronesia, Santa Cruz Islands
   MEGASUBSPECIES: *semirubra* is one, *ugiensis* another, *russata* plus *kuperi* a third. All other NMel races belong to the same megasubspecies as *louisiadensis* (SE Papuan Islands).
   OTHER ALLOSPECIES OR MEMBERS OF THE SPECIES GROUP: *R. superflua* (Buru), *R. teysmanni* (Cel), *R. lepida* (Palau), *R. dedemi* (Ceram), *R. opistherythra* (Tenimbar), *R. personata* (Kandavu), and the *R.* [*rufidorsa*] superspecies (species 148 below).
   COMMENTS: *R. rufifrons* overlaps *R. opistherythra* on Tenimbar and overlaps *R. rufidorsa* marginally on Misol and S NG. *R. nigrocinnamomea* (Mindanao) and *R. brachyrhyncha* (NG) are older derivatives. See Mayr (1931d) and Mayr & Moynihan (1946).

148. *Rhipidura* [*rufidorsa*]. (map 39)
A. *R. dahli*. (a) *dahli* (Reichenow 1897). NB, Umb (subsp?). (b) *antonii* (Hartert 1926b). NI.
B. *R. matthiae* (Heinroth 1902). SMt.
C. *R. malaitae* (Mayr 1931d). Mal
MEGASUBSPECIES: none.
OTHER ALLOSPECIES: *R. rufidorsa* (NG).
COMMENTS: See under species 147 above.

MONARCHIDAE (monarch flycatchers)

149. *Clytorhynchus* [*nigrogularis*]. (plate 9)
A. *C. hamlini* (Mayr 1931b). Ren
OTHER ALLOSPECIES: *C. nigrogularis* (Fiji, Santa Cruz). See Mayr (1933b).

150. *Monarcha cinerascens*. (plate 5, map 40). (a) *fulviventris* (Hartlaub 1868). Ramb, Nau, SMig, Anch, Her, Nin, Wuv (subsp?). (b) *perpallidus* (Neumann 1924). SMt, Em, Dyl (vagr?), Ting, Lih, Tab, Tench, islets off NH and NI. (c) *impediens* (Hartert 1926a). Malie and Sinabiet (Lihir group), Feni, Tang, Witu, Unea; Rmos, Gow, Bor, Nis, Fead, OJ, Sik, Kili, Tauu, Nuku; islets near Buka, Boug, Shtl, Ch, and Ys but not these main islands. (d) subsp? Long, Umb (vagr?), Vu, Cred, Tol, Sak, Cr. Other subspecies: islets around the Banda Sea, and islets west, north, and east of NG
MEGASUBSPECIES: none.
COMMENTS: See Meise (1929c), Mayr (1955), and Mees (1965a) for subspecies. Species 151 below (*M.* [*melanopsis*]) is closely related. The genus name, *Monarcha*, is masculine.

151. *Monarcha* [*melanopsis*]. (plate 5, map 40)
A. *M. erythrostictus* (Sharpe 1888a). Buka, Boug, Shtl, Faur.
B. *M. castaneiventris*. (a) *castaneiventris* (Verreaux 1858). Ch, Ys, Fla, BV, Savo, Guad, Mal. (b) *obscurior* (Mayr 1935). Pav and other islands of Russell group (Ban, Moie, Kiome). (c) *megarhynchus* (Rothschild & Hartert 1908b). SCr. (d) *ugiensis* (Ramsay 1882f). Ugi, 3S, SAn, SCat.
C. *M. richardsii* (Ramsay 1881). All islands of NGa group
MEGASUBSPECIES: *ugiensis*, unique in its uniformly black plumage.
OTHER ALLOSPECIES: *M. melanopsis* (Au), *M. frater* (NG, Cape York).
COMMENTS: *M. cinerascens* is closely related. The form *ugiensis* might be considered an allospecies distinct from *M. castaneiventris*.

152. *Monarcha* [*manadensis*]. (plate 3, map 41)
A. *M. infelix*. (a) *infelix* (Sclater 1877b). Man. (b) *coultasi* (Mayr 1955). Ramb, Tong (islet in Admiralty Group).
B. *M. menckei* (Heinroth 1902). SMt.
C. *M. verticalis*. (a) *verticalis* (Sclater 1877a). NB, NI, NH, Umb, DY. (b) *ateralbus* (Salomonsen 1964). Dyl.
D. *M. barbatus*. (a) *barbatus* (Ramsay 1879e). Buka, Boug, Shtl, Faur, Ch, Ys, Fla, BV, Guad. (b) *malaitae* (Mayr 1931e). Mal.
E. *M. browni*. (a) *nigrotectus* (Hartert 1908). VL, Bag. (b) *ganongae* (Mayr 1935). Gan. (c) *browni* (Ramsay 1883a). Kul, Wana, Koh, NGa, Vang. (d) *meeki* (Rothschild & Hartert 1905). Rnd, Tet.
F. *M. viduus*. (a) *viduus* (Tristram 1879). SCr, SAn. (b) *squamulatus* (Tristram 1882). Ugi
MEGASUBSPECIES: *infelix*, *coultasi*, *verticalis*, *ateralbus*, *barbatus*, *malaitae*, *nigrotectus*, *ganongae*, *viduus*, *squamulatus*, and the group consisting of *browni* plus *meeki*.
OTHER ALLOSPECIES: *M. manadensis* (NG), *M. brehmii* (Biak), *M. leucurus* (Kei, Buru, Tanahdjampea), *M. godeffroyi* (Yap), possibly *M. takatsukasae* (Tinian).
COMMENTS: Many of the Northern Melanesian forms are on the borderline between subspecies and allospecies. The relationships of the Northern Melanesian forms to each other, to the extralimital taxa that we place in the same superspecies, and to the other pied monarchs (the so-called *trivirgatus* group: *M. guttula* [NG], *M. julianae* [Kofiau], *M. mundus* [Tenimbar, Babar, Damar], *M. sacerdotum* [Flores], *M. boanensis* [Boano], and *M. trivirgatus* [Mol, Flores, Timor, Damar, E Au, S NG, SE Papuan Islands]) are controversial and require careful study. There is no sympatry among the forms that we group in a superspecies. Within the *trivirgatus* group, *M. trivirgatus* shares Timor with *M. sacerdotum*, Damar with *M. mundus*, and S NG and the SE Papuan Islands with *M. guttula*. The *trivirgatus* group is sympatric with the *M. manadensis* superspecies only

on NG (*M. manadensis* and *M. guttula*). Some authors variously consider the Northern Melanesian forms and *M. leucurus* related to the *trivirgatus* group, not to *M. manadensis*. See Meise (1929c), Mayr (1931e, 1944a, 1955, 1971), van Bemmel (1948), Ripley (1959), Galbraith and Galbraith (1962), Salomonsen (1964), Schodde (1977), White and Bruce (1986), and Moeliker and Heij (1995).

153. *Monarcha chrysomela*. (a) *chrysomela* (Garnot 1827). NI, NH. (b) *whitneyorum* (Mayr 1955). Lih. (c) *tabarensis* (Mayr 1955). Tab. (d) *pulcherrimus* (Salomonsen 1964). Dyl. Other subspecies: NG
MEGASUBSPECIES: *pulcherrimus*, the most distinct race of the species.
COMMENTS: See Mayr (1955) and Salomonsen (1964) for subspecies.

154. *Myiagra alecto*. (plate 5, map 42). (a) *chalybeocephala* (Garnot 1828). NB, NI, NH, Man, Ramb, Umb, Vu, DY, Cred, Dyl, Feni, Tab, Tang, Lol, Sak, Unea (and NG). Other subspecies: Tenimbar, Mol, Au
COMMENTS: *M. hebetior* is closely related. Whether these two species, formerly placed in *Monarcha*, belong in *Myiagra* or in a separate genus *Piezorhynchus* has been debated (Keast 1958, Schodde & Hitchcock 1968, Mees 1982); we consider them greatly modified derivatives of the *Myiagra* [*rubecula*] superspecies (species 156 below, q.v. for discussion). See Mayr (1941a, 1955), Salomonsen (1964), and Ford (1983) for geographic variation.

155. *Myiagra hebetior*. (plate 5, map 42). (a) *hebetior* (Hartert 1924b). SMt. (b) *eichhorni* (Hartert 1924b). NB, NI, NH, Vu. (c) *cervinicolor* (Salomonsen 1964). Dyl
MEGASUBSPECIES: the three races are very distinct in female plumage, and each constitutes a separate megasubspecies.
COMMENTS: See under species 154 for generic position; Mayr (1955) and Salomonsen (1964) for geographic variation and distinction from *M. alecto*; and Mayr (1969) for distributional map.

156. *Myiagra* [*rubecula*]. (plate 5, map 42)
A. *M. ferrocyanea*. (a) *cinerea* (Mathews 1928). Buka, Boug, Shtl, Faur, Mono. (b) *ferrocyanea* (Ramsay 1879e). Ch, Wagina, Ys, Fla, BV, Guad. (c) *malaitae* (Mayr 1931e). Mal. (d) *feminina* (Rothschild & Hartert 1901c). All islands of NGa group except Simbo.
B. *M. cervinicauda* (Tristram 1879). SCr, Ugi, SAn.
C. *M. caledonica*. (a) *occidentalis* (Mayr 1931b). Ren. Other subspecies: NCal, NHeb
MEGASUBSPECIES: *malaitae*, *feminina*, and the group consisting of *cinerea* plus *ferrocyanea*.
OTHER ALLOSPECIES: *M. galeata* (Mol), *M. vanikorensis* (Vanikoro, Fiji), *M. erythrops* (Palau), *M. oceanica* (Guam, Truk), *M. pluto* (Ponape), *M. rubecula* (N and E Au, S NG), *M. atra* (Numfor, Biak), *M. albiventris* (Samoa).
COMMENTS: Several of these allospecies could be combined. Indeed, if one wanted to be extreme, one could treat all of them as subspecies of *M. rubecula*. *M. ruficollis* and *M. cyanoleuca* differ more from *M. rubecula* than do the allospecies of *rubecula*, but are nevertheless closely related. *M. ruficollis*, *M. cyanoleuca*, and *M.* [*rubecula*] are all absent from the Bismarck Archipelago and are also absent from NG except for a population of *M. ruficollis* in the mangroves of S NG and a population of *M. rubecula* in the savanna of SE and S NG. These species are apparently replaced in most of the Bismarck Archipelago and NG by 154. *M. alecto* and 155. *M. hebetior*. This pattern of distribution suggests that *M. alecto* and *M. hebetior* are ancient and greatly modified allospecies of *M.* [*rubecula*]. See Mayr (1933c) and Galbraith and Galbraith (1962) for discussion of the superspecies and Mayr (1945a) for geographic variation in Northern Melanesia.

EOPSALTRIIDAE (Australasian robins)

157. *Monachella muelleriana*. (a) *coultasi* (Mayr 1934a). NB. Other subspecies: NG
MEGASUBSPECIES: *coultasi* COMMENTS: This monotypic genus is most closely related to *Microeca* (Mayr 1941a). For a different

view see Orenstein (1975). The distinctive NB race is known only from the type series (Mayr 1934a).

158. *Petroica multicolor*. (a) *septentrionalis* (Mayr 1934b). Boug. (b) *kulambangrae* (Mayr 1934b). Kul. (c) *dennisi* (Cain & Galbraith 1955). Guad. (d) *polymorpha* (Mayr 1934b). SCr. Other subspecies: Au, NHeb, Fiji, Samoa
MEGASUBSPECIES: *polymorpha*.
COMMENTS: This species is nearest *P. phoenicea* (Au) and *P. goodenovii* (Au). See Mayr (1934b) for geographic variation.

PACHYCEPHALIDAE (whistlers)

159. *Pachycephala* [*pectoralis*]. (plate 2, map 43)
A. *P. pectoralis*. (a) *goodsoni* (Rothschild & Hartert 1914b). Man. (b) *sexuvaria* (Rothschild & Hartert 1924a). SMt. (c) *citreogaster* (Ramsay 1876). NB, NI, NH, Umb, Dyl (subsp?), Feni, Tol. (d) *tabarensis* (Mayr 1955). Tab. (e) *ottomeyeri* (Stresemann 1933). Lih. (f) *whitneyi* (Hartert 1929). Small islets of Shortland group, and islets near Boug. (g) *bougainvillei* (Mayr 1932c). Buka, Boug, Shtl. (h) *orioloides* (Pucheran 1853). Ch, Ys, Fla. (i) *pavuvu* (Mayr 1932c). Pav and other islands of Russell group (Ban, Moie, Kiome). (j) *centralis* (Mayr 1932c). Kul, Koh, NGa, Vang, Gat. (k) *melanoptera* (Mayr 1932c). Rnd, Tet. (l) *melanonota* (Hartert 1908). VL, Bag, Gan. (m) *cinnamomea* (Ramsay 1879e). Guad. (n) *sanfordi* (Mayr 1931e). Mal. (o) *christophori* (Tristram 1879). SCr, SAn. (p) *feminina* (Mayr 1931b). Ren. Other subspecies: Au, LS, Mol, SE Papuan Islands, NHeb, Fiji
MEGASUBSPECIES: *melanoptera*, *melanonota*, *sanfordi*, and *feminina*. The Solomon races *bougainvillei*, *orioloides*, *pavuvu*, *centralis*, and *cinnamomea* collectively form a fifth megasubspecies, to which the former four megasubspecies are related ("group C" of Galbraith [1956]). The race *christophori* may have arisen as a hybrid between group C and the megasubspecies of NHeb and NCal ("group G" of Galbraith [1956]). The five Bismarck races (*goodsoni* through *ottomeyeri*) belong to a widespread megasubspecies also represented in LS, Mol, SE Papuan Islands, and Fiji ("group H" of Galbraith [1956]).
OTHER ALLOSPECIES: *P. caledonica* (NCal), *P. melanops* (Tonga, possibly only a subspecies of *P. pectoralis*), *P. flavifrons* (Samoa), *P. soror* and *P. schlegelii* (NG, doublet).
COMMENTS: *P. melanura* (species 160) is a closely related derivative; 159 Af, *P. pectoralis whitneyi*, is a hybrid population between 159 Ag and 160 a. Other relatives of this group may include *P. meyeri* (NG), *P. orpheus* (Wetar, Timor), *P. implicata* (species 161), *P. lorentzi* (NG), *P. aurea* (NG), and *P. nudigula* (Sumbawa, Flores). See Galbraith (1956) for monographic treatment and also Mayr (1932c, 1955).

160. *Pachycephala melanura*. (plate 2, map 43). (a) *dahli* (Reichenow 1897). NB (vagr), Long, Umb (vagr), Vu, Cred, Ting, Malie and Sinabiet (Lihir group), Tang (vagr), Tol (vagr), Cr, Witu, islets near NB and NI and NH and Umb; Nis, and islets near Buka and Boug and Shtl (and SE NG, Fergusson). Other subspecies: coasts of Au and S NG, Balim Valley of NG. COMMENTS: See under species 159. *P. melanura* and *P. pectoralis* are largely allopatric but occasionally meet on the coasts of Queensland and when vagrants of *P. melanura* reach Tol, NB, Umb, and probably many other islands on which *P. pectoralis* is resident (see comments under species 159 about the hybrid population *P. p. whitneyi*). See Galbraith (1967), Diamond (1976), and Ford (1983).

161. *Pachycephala implicata*. (plate 2). (a) *implicata* (Hartert 1929). Guad. (b) *richardsi* (Mayr 1932d). Boug
MEGASUBSPECIES: the two races are very distinct and constitute megasubspecies.
COMMENTS: This species is a derivative of the group that includes *P. pectoralis* (see comments under species 159), though not necessarily of *P. pectoralis* itself. See Mayr (1932d).

DICAEIDAE (flower-peckers)

See Mayr and Amadon (1947), Salomonsen (1960a, b).

162. *Dicaeum* [*erythrothorax*]. (map 44)
   A. *D. eximium*. (a) *layardorum* (Salvadori 1880). NB, Vu, Lol. (b) *eximium* (Sclater 1877a). NI, NH, Lih. (c) *phaeopygium* (Salomonsen 1964). Dyl.
   B. *D. aeneum*. (a) *aeneum* (Pucheran 1853). Buka, Boug, Shtl, Faur, Ch, Ys, Fla, BV. (b) *becki* (Hartert 1929). Guad. (c) *malaitae* (Salomonsen 1960b). Mal.
   C. *D. tristrami* (Sharpe 1884). SCr
   MEGASUBSPECIES: none.
   OTHER ALLOSPECIES: *D. erythrothorax* (Mol), *D. nehrkorni* (Cel), *D. vulneratum* (Ceram, Ambon), *D. pectorale* (W Papuan Islands and W NG), and *D. geelvinkianum* (NG except west) and *D. nitidum* (Louisiades) (sometimes lumped with *D. pectorale*).
   COMMENTS: Although *D. tristrami* is highly aberrant, its distribution suggests that it is an outpost of *D.* [*erythrothorax*] (Mayr & Amadon 1947, Mayr 1955). However, Salomonsen (1960b) and Galbraith and Galbraith (1962) consider *tristrami* a relict species of unknown affinities. Relatives of *D.* [*erythrothorax*] include *D. igniferum* and *D. maugei* (LS), the *D. hirundinaceum* superspecies (India to Phil, GS, LS, Cel, Au), *D. cruentatum* (India to Sumatra, Borneo), and *D. trochileum* (Java, Borneo, Lombok). See Mayr (1955) and Salomonsen (1964) for geographic variation in Northern Melanesia.

NECTARINIIDAE (sunbirds)

See Delacour (1944).

163. *Nectarinia* [*sperata*]
   A. *N. sericea*. (a) *corinna* (Salvadori 1878). NB, NI, NH, Umb, Vu, DY, Cred, Dyl, Lih, Tab, Lol, Sak, Unea. (b) *eichhorni* (Rothschild & Hartert 1926a). Feni. Other subspecies: Cel, Mol, NG
   MEGASUBSPECIES: none.
   OTHER ALLOSPECIES: *N. sperata* (SE Asia, GS, Phil).
   COMMENTS: *N. zeylonica* (India), *N. minima* (India), and *N. calcostetha* (SE Asia, GS) may be related. See Delacour (1944), Mayr (1955), and Mees (1965b). We do not recognize *caeruleogula* (Mees 1965b) for the NB and Umb population.

164. *Nectarinia* [*jugularis*].
   A. *N. jugularis*. (a) *flavigaster* (Gould 1843). All Bismarcks except Nau, SMig, Anch, Wuv, Cred, Ting, Tol, Witu; Bor, Nis, and all Central Solomons except SCr. Other subspecies: SE Asia through Phil and Indonesia to NG
   MEGASUBSPECIES: none.
   OTHER ALLOSPECIES: *N. buettikoferi* (Sumba), *N. solaris* (LS).
   COMMENTS: *N. jugularis* and *N. solaris* overlap on Sumbawa, Flores, and Lomblen. *N. asiatica* (Afghanistan to SE Asia) is related. See Rensch (1931), Mayr (1944a), and Delacour (1944). The Northern Melanesian race *flavigaster* differs weakly from the race *frenata* (NG, N Mol, Cape York).

ZOSTEROPIDAE (white-eyes)

See Mees (1957, 1961, 1969).

165. *Zosterops* [*atriceps*]. (plate 1, map 45)
   A. *Z. atrifrons*. (a) *hypoxanthus* (Salvadori 1881). NB, Umb, Vu. (b) *ultimus* (Mayr 1955). NI, NH. (c) *admiralitatis* (Rothschild & Hartert 1914c). Man. Other subspecies: Cel, Ceram, NG
   MEGASUBSPECIES: the three NMel races collectively constitute a megasubspecies.
   OTHER ALLOSPECIES: *Z. atriceps* (N Mol), *Z. mysorensis* (Biak).
   COMMENTS: Species 165 Aa, b, and c are sometimes separated as *Z. hypoxanthus*, and the Tagula race as *Z. meeki*. See Mayr (1955, 1965b) and Mees (1961).

166. *Zosterops metcalfii*. (plate 1, map 45). (a) *metcalfii* (Tristram 1894). Buka, Boug, Shtl, Ch, Ys, Molakobi, Bates. (b) *floridanus* (Rothschild & Hartert 1901c). Fla, Tulagi.
   MEGASUBSPECIES: none.
   COMMENTS: We do not recognize *exiguus* for the Buka, Boug, Shtl, and Ch population. The affinities of this species are unclear (Mees 1961).

167. *Zosterops* [*griseotinctus*]. (plate 1, map 45)
   A. *Z. griseotinctus*. (a) *eichhorni* (Hartert 1926a). Nau, Long, Tol, Cr; Nis. Other subspecies: SE Papuan islands.
   B. *Z. vellalavella* (Hartert 1908). VL, Bag.

C. *Z. splendidus* (Hartert 1929). Gan.
D. *Z. luteirostris* (Hartert 1904). Gizo.
E. *Z. rendovae*. (a) *rendovae* (Tristram 1882). Rnd. (b) *tetiparius* (Murphy 1929). Tet. (c) *kulambangrae* (Rothschild & Hartert 1901c). Kul, Wana, Koh, NGa, Vang, Gat.
F. *Z. rennellianus* (Murphy 1929). Ren
MEGASUBSPECIES: *rendovae*, *tetiparius*, and *kulambangrae*.
COMMENTS: We agree with Galbraith (1957), Galbraith and Galbraith (1962), and Mayr (1965b) in rejecting Mees's (1955, 1961) arguments for transferring the name *rendovae* from 167 Ea to 169 c and thereby renaming species 167 Ea and 169. See Murphy (1929) and Mees (1961) for geographic variation and relationships.

168. *Zosterops murphyi* (Hartert 1929). (plate 1, map 45). Kul. COMMENTS: This species may belong to the *Z.* [*griseotinctus*] superspecies, in which case the superspecies would be represented by a doublet on Kul (Mees 1961).

169. *Zosterops ugiensis*. (plate 1, map 45). (a) *hamlini* (Murphy 1929). Boug. (b) *oblitus* (Hartert 1929). Guad. (c) *ugiensis* (Ramsay 1881). SCr
MEGASUBSPECIES: *hamlini*, and the group consisting of *oblitus* plus *ugiensis*.
COMMENTS: The affinities of this species are unclear (see Mees 1961, who uses the name *Z. rendovae* for this species, as discussed under species 167).

170. *Zosterops stresemanni* (Mayr 1931e). (plate 1, map 45). Mal. COMMENTS: The affinities of this species are unclear (Mayr 1931e).

171. *Woodfordia* [*superciliosa*]. (plate 1)
A. *W. superciliosa* (North 1906). Ren.
OTHER ALLOSPECIES: *W. lacertosa* (Santa Cruz). See Mees (1969).

MELIPHAGIDAE (honey-eaters)

See Koopman (1957) for *Myzomela* and Mayr (1932a) for Solomon taxa.

172. *Myzomela* [*eques*]. (plate 4)
A. *M. cineracea* (Sclater 1879). NB, Umb
OTHER ALLOSPECIES: *M. blasii* (Ceram, Ambon), *M. albigula* (Louisiades), *M. eques* (NG), *M. obscura* (Au, N Mol, Biak, overlapping *M. eques* in S NG).
COMMENTS: See Mayr (1955) for geographic variation and Koopman (1957) for relationships. We do not recognize the race *rooki* (Hartert 1926b) from Umboi.

173. *Myzomela cruentata*. (plate 4). (a) *coccinea* (Ramsay 1878a). NB, DY (?). (b) *erythrina* (Ramsay 1878a). NI. (c) *lavongai* (Salomonsen 1966a). NH. (d) *cantans* (Mayr 1955). Tab. (e) *vinacea* (Salomonsen 1966a). Dyl. Other subspecies: NG
MEGASUBSPECIES: the four E Bis races (*erythrina*, *lavongai*, *cantans*, *vinacea*) collectively form one megasubspecies; the other consists of *coccinea* plus the NG race *cruentata*.
COMMENTS: For subspecies, see Mayr (1955), Salomonsen (1966a), and Gilliard and LeCroy (1967).

174. *Myzomela pulchella* (Salvadori 1891). (plate 4, map 4). NI. COMMENTS: The relationships of this species are uncertain (Koopman 1957).

175. *Myzomela* [*cardinalis*]. (plate 4, map 46)
A. *M. cardinalis*. (a) *pulcherrima* (Ramsay 1881). SCr, Ugi, 3S, SAn (vagr). (b) *sanfordi* (Mayr 1931b). Ren. Other subspecies: Santa Cruz, NHeb, Loyalties, Samoa, Micronesia.
B. *M. erythromelas* (Salvadori 1881). NB
MEGASUBSPECIES: *pulcherrima*, *sanfordi*.
OTHER ALLOSPECIES: *M. chermesina* (Rotuma), *M. sanguinolenta* (Cel, Mol, Tenimbar, E Au, NCal). *M. erythrocephala* (Sumba, N Au, S NG), *M. kuehni* (Wetar), and *M. adolphinae* (NG) are closely related, *M. vulnerata* (Timor) possibly related. The limits of this group, and the question as to whether it includes *M. erythromelas*, *M. rosenbegii* of New Guinea, and the *M.* [*lafargei*] superspecies (species 177), need further study (Koopman 1957, LeCroy & Peckover 1999). *M. c. pulcherrima* has hybridized with *M. tristrami* (Mayr 1932a). See Mayr (1931b, 1932a) for geographic variation in *M. cardinalis*.

176. *Myzomela sclateri* (Forbes 1879). (plate 4). Long, Vu, Cred, Tol, Cr, Witu, Unea, and islets off the north coast of NB (and Karkar)
COMMENTS: The relationships of this species are uncertain (Koopman 1957). The song resembles that of *M. adolphinae* (Diamond 1972b).

177. *Myzomela* [*lafargei*]. (plate 4, map 47)
A. *M. pammelaena*. (a) *ernstmayri* (Meise 1929a). Anch, Her, Nin, Manu, Wuvulu (subsp?). (b) *pammelaena* (Sclater 1877b). Ramb, Nau, SMig, other small islands of Admiralty Group, occasionally on coast of Manus. (c) *nigerrima* (Salomonsen 1966c). Long, Tol, Cr, islets near Umb. (d) *hades* (Meise 1929a). SMt, Em, Tench (subsp?). (e) *ramsayi* (Finsch, in Finsch & Meyer 1886). Ting, and islets off NH and off N NI.
B. *M. lafargei* (Pucheran 1853). Buka, Boug, Shtl, Faur, Ch, Ys.
C. *M. melanocephala* (Ramsay 1879e). Fla, Savo, Guad.
D. *M. eichhorni*. (a) *atrata* (Hartert 1908). VL, Bag. (b) *ganongae* (Mayr 1932a). Gan. (c) *eichhorni* (Rothschild & Hartert 1901c). Gizo, Kul, Wana, Koh, NGa, Vang, Rnd, Tet.
E. *M. malaitae* (Mayr 1931e). Mal.
F. *M. tristrami* (Ramsay 1881). SCr, Ugi (vagr), SAn, SCat
MEGASUBSPECIES: *M. pammelaena* falls into two megasubspecies, one consisting of the small races *hades* and *ramsayi* with whitish underwings, the other of the large races *ernstmayri*, *pammelaena*, and *nigerrima* with dark underwings. The three races of *M. eichhorni* are quite similar to each other and do not constitute megasubspecies.
COMMENTS: *M. tristrami* (Galbraith & Galbraith 1962) and *M. pammelaena* (Diamond 1976) belong to this superspecies, not to *M. nigrita*. *M. malaitae* may instead belong to the *M.* [*cardinalis*] superspecies, species 175 (Mayr 1931e, 1932a). Further study is needed to evaluate whether the *M.* [*cardinalis*] superspecies and the *M.* [*lafargei*] superspecies are closely related to each other; they are allopatric except on SCr, where *M. cardinalis* may be a recent invader that has hybridized with *M. tristrami*. *M. jugularis* (Fiji) is related (Koopman 1957), and perhaps also *M. chermesina* (Rotuma) (Galbraith & Galbraith 1962). See Mayr (1932a) and Diamond (1976) for geographic variation and relationships. The race *M. pammelaena nigerrima* of Long and its neighbors, differing from *ernstmayri* and *pammelaena* by larger size, evidently arose since the seventeenth century volcanic defaunation of Long as a result of competition and character displacement with the smaller *M. sclateri* (Diamond et al. 1989).

178. *Philemon* [*moluccensis*]. (plate 7, map 48)
A. *P. cockerelli*. (a) *umboi* (Hartert 1926b). Umb. (b) *cockerelli* (Sclater 1877a). NB.
B. *P. eichhorni* (Rothschild & Hartert 1924b). NI.
C. *P. albitorques* (Sclater 1877b). Man, Los Negros
MEGASUBSPECIES: none (*umboi* differs from *cockerelli* only by slightly larger size and heavier bill).
OTHER ALLOSPECIES: *P. fuscicapillus* (N Mol), *P. subcorniculatus* (Ceram), *P. moluccensis* (Buru, Tenimbar, Kei), *P. buceroides* (LS and N Au), *P. novaeguineae* (NG and Cape York), *P. gordoni* (N Au, possibly conspecific with *P. buceroides*; see Parker 1971, Schodde et al. 1979a). For discussion of the superspecies, see Mayr (1944a) and Diamond (1982b).

179. *Melidectes whitemanensis* (Gilliard 1960a). (plate 9, map 49). NB. COMMENTS: The closest relative is probably *M. fuscus* of NG (Diamond 1971). We do not recognize the monotypic genus *Vosea* (Gilliard 1960a).

180. "*Stresemannia*" *bougainvillei* (Mayr 1932a). (plate 9, map 49). Boug. COMMENTS: We reluctantly place this undistinctive species in the monotypic genus *Stresemannia* (Meise 1950) until its relationships are known. It has been variously placed in *Lichmera*, *Melilestes*, *Meliphaga*, and *Stresemannia* (Mayr 1932a, 1955, Meise 1950, Salomonsen in Peters 1967, Schodde 1977).

181. *Meliarchus sclateri* (Gray 1870). (plate 9, map 49). SCr. COMMENTS: Relationships

of this monotypic genus are unclear. Two hypotheses are that it is near *Acanthagenys rufogularis* (Parkes 1980) or that it belongs in *Melidectes* (Sibley & Monroe 1990).

182. "*Guadalcanaria*" *inexpectata* (Hartert 1929). (plate 9, map 49).
Guad. COMMENTS: The relationships of this species, often separated as *Guadalcanaria*, are unclear (Hartert 1929, Mayr 1932a). It may belong to *Meliphaga* (Salomonsen in Peters 1967).

ESTRILDIDAE (waxbills, mannikins, and grassfinches)

See Delacour (1943) and Goodwin (1982) for the family and Restall (1996) for *Lonchura*.

183. *Erythrura* [*trichroa*]
A. *E. trichroa*. (a) *sigillifera*. NB, NI, Long, Umb, Feni, Tol, Cr (and NG, Cape York). (b) *eichhorni* (Hartert 1924b). SMt, Em. (c) *woodfordi* (Hartert 1900a). Boug (subsp?), Guad, Kul (subsp?). Other subspecies: Cel, Mol, NHeb, Micronesia
MEGASUBSPECIES: none.
OTHER ALLOSPECIES: *E. coloria* (Phil); possibly *E. tricolor* (Timor, Tenimbar, Southwest Islands).
COMMENTS: *E. papuana* (NG) is closely related. See Hartert (1900a), Mayr (1931c), Ziswiler *et al.* (1972), and Goodwin (1982) for geographic variation and relationships.

184. *Lonchura* [*spectabilis*]
A. *L. spectabilis*. (a) *spectabilis* (Sclater 1879). NB, Long, Umb, Vu, Tol. Other subspecies: NE NG.
B. *L. forbesi* (Sclater 1879). NI (except far north?).
C. *L. hunsteini*. (a) *hunsteini* (Finsch 1886). NI (Kavieng district in far north). (b) *nigerrima* (Rothschild & Hartert 1899). NH. Other subspecies: Ponape (possibly introduced rather than native; Pratt et al. 1987)
MEGASUBSPECIES: none.
OTHER ALLOSPECIES: possibly *L. nevermanni* (S NG) (Delacour 1943, Mayr 1955).
COMMENTS: Whether *L. forbesi* and *L. h. hunsteini* actually are sympatric in N NI is uncertain. *L. nevermanni* of S NG may (Delacour 1943, Mayr 1955) or may not (Goodwin 1982, Mees 1982) be another allospecies. This superspecies belongs to the *castaneothorax* species group (see under species 185; Goodwin 1982).

185. *Lonchura* [*castaneothorax*]
A. *L. melaena* (Sclater 1880a). NB, NI (subsp?); Buka (subsp?)
OTHER ALLOSPECIES: *L. flaviprymna* (NW Au), *L. castaneothorax* (parts of Au and NG).
COMMENTS: For relations between *L. flaviprymna* and *L. castaneothorax* in areas of sympatry, see Immelmann (1962). Other relatives include *L. teerinki* (NG), the *L. montana–L. monticola* superspecies (NG), *L. stygia* (S NG), *L. caniceps* (NG), *L. vana* (NG), *L. nevermanni* (NG), *L. grandis* (NG), the *L. spectabilis* superspecies (species 184), and possibly *L. quinticolor* (LS) (Delacour 1943, Mayr 1968, Mees 1982, Goodwin 1982).

STURNIDAE (starlings)

See Amadon (1943b, 1956) and Feare and Craig (1999). Finch (1986) discusses Solomon populations of *Aplonis*.

186. *Aplonis* [*cantoroides*]. (map 50)
A. *A. cantoroides*. Most Bismarck islands (no records for Ramb, Anch, Nin, Cred, and Tench); most Central Solomon islands plus Ren, Bel (no records for Faur and SCat) (and NG). Probably actually absent on some of the islands for which there are no records
OTHER ALLOSPECIES: *A. crassa* (Tenimbar).
COMMENTS: The *A. feadensis* superspecies is related but is sympatric with *A. cantoroides* on Her and Ren and possibly Wuv. Other possible relatives include the superspecies consisting of *A. cinerascens* (Rarotonga), *A. tabuensis* (Santa Cruz, Fiji, Samoa), *A. opaca* (Micronesia), and *A. fusca* (Norfolk, Lord Howe) (Amadon 1943b, Baker 1951).

187. *Aplonis* [*feadensis*]. (map 50)
A. *A. feadensis*. (a) *heureka* (Meise 1929b). Her, Nin, Wuv, Tench (subsp?). (b) *feadensis* (Ramsay 1882f). Nis, Fead, OJ, Kili, Tauu.

B. *A. insularis* (Mayr 1931b). Ren
MEGASUBSPECIES: none.
COMMENTS: See under species 186 for relationships; see Finch (1986) for habits.

188. *Aplonis* [*grandis*]. (map 51)
A. *A. grandis*. (a) *grandis* (Salvadori 1881). Buka, Boug, Shtl, Faur, Ch, Wagina, Ys, Fla, Mono, and all islands of NGa group. (b) *macrura* (Mayr 1931e). Guad. (c) *malaitae* (Mayr 1931e). Mal.
B. *A. dichroa* (Tristram 1895). SCr
MEGASUBSPECIES: *malaitae*, distinct in its small size, green breast, and white iris.
COMMENTS: See Mayr (1931e) for geographic variation, and Finch (1986) for habits. The affinities of this superspecies are unclear.

189. *Aplonis metallica*. (a) *purpureiceps* (Salvadori 1878). Man, Los Negros (Admiralty Group). (b) *nitida* (Gray 1858). Most Bismarck islands except Man, Ramb, and Los Negros (no records for Nau, SMig, Anch, Her, Nin, Wuv, and Tench); most Central Solomon islands plus Nis, Kili (no records for Gow). Probably absent on most of the islands for which there are no records. (c) *metallica*. Ramb (and Karkar, NG, Cape York, Mol). Other subspecies: Biak, Numfor, Tenimbar, Damar
MEGASUBSPECIES: none.
COMMENTS: *A. mystacea* (NG) and *A. brunneicapilla* (species 190) represent independent derivatives (Galbraith & Galbraith 1962). See Rothschild and Hartert (1914b) and Amadon (1956) for geographic variation.

190. *Aplonis brunneicapilla* (Danis 1938). Boug, Ch, Guad, Rnd. COMMENTS: See under species 189 for relationships, and see Cain & Galbraith (1956) and Finch (1986) for habits.

191. *Mino dumontii*. (a) *kreffti* (Sclater 1869). NB, NI, NH, Umb, Tang, Lol; all Central Solomon islands except SCr group, Rmos, and Gow. Other subspecies: NG
MEGASUBSPECIES: *kreffti*, very distinct in behavior and voice.
COMMENTS: We do not recognize the races *sanfordi* and *giliau*. The races *kreffti* and *dumontii* differ in morphology (Amadon 1956) and in behavior and voice (Schodde 1977), but we do not follow Schodde in considering them allospecies. See Amadon (1956) for geographic variation.

DICRURIDAE (drongos)

See Vaurie (1949).

192. *Dicrurus* [*hottentottus*]. (map 52)
A. *D. hottentottus*. (a) *laemostictus* (Sclater 1877a). NB, Umb, Lol. (b) *meeki* (Rothschild & Hartert 1903). Guad. (c) *longirostris* (Ramsay 1882e). SCr. Other subspecies: India to NG, Au.
B. *D. megarhynchus* (Quoy & Gaimard 1830). (plate 8). NI
MEGASUBSPECIES: *longirostris*, distinct in its very long bill and its flat, nearly square tail. (See map 52 for illustration.)
OTHER ALLOSPECIES: *D. montanus* (Cel, doublet with *D. hottentottus*).
COMMENTS: *D. balicassius* (Phil) is closely related. *D. hottentottus* itself may also be divided into allospecies, in which case 192 Aa, b, and c would belong to *D. bracteatus* (other subspecies in Phil, Cel, Mol, NG, Au). See Mayr and Vaurie (1948), Vaurie (1949), Mayr (1955), and White & Bruce (1986).

ARTAMIDAE (wood-swallows)

See Etchécopar & Hüe (1977).

193. *Artamus* [*maximus*]. (plate 9)
A. *A. insignis* (Sclater 1877a). NB, NI, Vu (vagr)
OTHER ALLOSPECIES: *A. maximus* (NG).
COMMENTS: See Mayr (1955). *A. monachus* (Cel) may be related (Etchécopar & Hüe 1977).

CORVIDAE (crows)

See Goodwin (1976) and Jollie (1978).

194. *Corvus orru*. (a) *insularis* (Heinroth 1903). NB, NI, NH, Umb, Vu, Dyl, Lol, Sak, Witu. Other subspecies: Mol, NG, Au, Tenimbar, Babar
MEGASUBSPECIES: none.
COMMENTS: The *C. coronoides* group (Au) and *C. macrorhynchos* (E Asia, GS, LS,

Phil) seem to be related (Stresemann 1943, Rowley 1970, Goodwin 1976, Jollie 1978).

195. *Corvus woodfordi.* (a) *meeki* (Rothschild 1904). Buka, Boug, Shtl. (b) *woodfordi* (Ogilvie-Grant 1887). Ch, Ys, Guad, Molakobi, Wagina
MEGASUBSPECIES: *meeki*, *woodfordi.*
COMMENTS: See Mayr (1955) and Vaurie (1958) for geographic variation. *C. meeki* has also been treated as a distinct allospecies (see Goodwin 1976, Jollie 1978). *C. fuscicapillus* (W Papuan islands, Aru, NG), *C. tristis* (NG), and *C. moneduloides* (NCal) have been suggested as relatives (Goodwin 1976, Jollie 1978).

Appendix 2

Nonbreeding Visitors to Northern Melanesia

This appendix lists all land and freshwater bird taxa that breed outside Northern Melanesia and have been recorded as visiting Northern Melanesia but are known or believed not to breed there. Some of these species arrive regularly in considerable numbers each year as winter visitors from either Australia (e.g., the cuckoo *Chrysococcyx lucidus plagosus*, bee-eater *Merops ornatus*), New Zealand (e.g., *Chrysococcyx lucidus lucidus*), or the Palaearctic (e.g., numerous species of waders). Others arrive infrequently as vagrants from Australia, New Guinea, or Indonesia, and their months of occurrence have not been systematically studied (e.g., several species of cormorants, egrets, and ibises).

The format of the list is as follows. Each entry consists of the Latin name, followed by the generally accepted vernacular name (generally from Beehler et al. 1986). The list then notes separately whether that species has been recorded from the Bismarcks ("Bis") and from the Solomons ("Sol"). If there are records from several islands within the Bismarcks or Solomons, the islands are not noted individually, nor are literature references cited. References to Northern Melanesian occurrences of these frequently recorded species can instead be found in the following works: Mayr (1945a) and Galbraith and Galbraith (1962) for the Solomons, Coates (1985, 1990) for the Bismarcks and Bougainville, Hadden (1981) for Bougainville, Diamond (1984a) for Rennell and Bellona, Mayr (1949a) for waders in the Bismarcks and Solomons, Bull (1948) for waders in the Solomons, and Stickney (1943) for six species of waders occurring in the Bismarcks and Solomons. If the species has been recorded on only a few occasions or from only a few islands, the islands are named and literature references are provided. The island abbreviations are the same as those used in appendix 1. Finally, the breeding range of the species is given. The 13 most frequently recorded visitors to Northern Melanesia are marked by asterisks.

In addition to the taxa that visit Northern Melanesia without breeding there, there are also some taxa which we included in appendix 1 that lists taxa known or believed to breed in Northrn Melanesia, but whose breeding numbers are supplemented or even greatly outnumbered by nonbreeding visitors. These taxa are the pelican *Pelecanus conspicillatus*, cormorant *Phalacrocorax melanoleucos*, spoonbill *Platalea regia*, cuckoo *Scythrops novaehollandiae*, and kingfisher *Halcyon sancta*; and possibly the herons *Ardea intermedia intermedia*, *Ardea alba modesta*, *Ixobrychus sinensis*, and the stilt *Himantopus leucocephalus*. Of these, the commonest is *Halcyon sancta*, which is an abundant winter visitor from Australia to both the Bismarcks and Solomons and is not known to breed in the Bismarcks and for which there is only limited evidence of breeding in the Solomons.

Tachybaptus novaehollandiae novaehollandiae. Australasian Grebe. Bis (one record: Man (Heinroth 1903)). Breeds in Au, NG.

Phalacrocorax sulcirostris. Little black cormorant. Bis (two records: NB, Man (Dutson 2001)). Sol (one record: Boug (Hadden 1981)). Breeds from GS to NG, Au.

Phalacrocorax carbo novaehollandiae. Great cormorant. Sol (Ren, possibly breeding there; (Diamond 1984a, Filardi et. al 1999). Breeds in Au.

Bubulcus ibis subsp. Cattle egret. Bis (three records: NB, NI, Man (Finch 1985, B. Coates, unpublished observation)). Breeding cosmopolitan, including Indonesia, Au, NG.

Egretta novaehollandiae. White-faced heron. Sol (two records: Boug, Guad). Breeds in LS, Au, NG, NZ, NCal.

Egretta garzetta nigripes. Little egret. Bis (one record: NB (Coates 1985)). Sol (sole records are from Buka, Boug, Wana (Hadden 1981, Blaber 1990)). Breeds in Indonesia, Au, NZ, NCal.

Threskiornis spinicollis. Straw-necked ibis. Bis (one record: NI (Coates 1973a, 1985)). Breeds in Au.

Plegadis falcinellus. Glossy ibis. Sol (sole records are from Boug, Kul (Hadden 1981, Blaber 1990)). Breeds from Eurasia, Afr, NAm to Au.

Anas gibberifrons gracilis. Gray teal. Bis (one record: NB (Bishop 1983)). Breeds in Au, NG.

Anas acuta acuta. Northern pintail. Bis (one record: Umb (Diamond 1976)). Sol (one record: Boug (Skyrme 1981)). Breeds in Palaearctic, NAm.

Anas querquedula. Garganey. Bis (two records: NB, NI (Lindgren 1978, Bishop 1983)). Breeds in Palaearctic.

Anas clypeata. Northern shoveler. Bis (one record: NB (Bishop 1983)). Breeds in Palaearctic.

Aythya fuligula. Tufted duck. Bis (one record: NB (Dutson 2001)). Breeds in Palaearctic.

Dendrocygna eytoni. Plumed whistling duck. Sol (two records: SCr, Ren (Buckingham et al. 1996, Dutson 2001)).

Falco longipennis longipennis. Australian hobby. Bis (two records: NB, Vu; Stresemann 1934, Meyer 1937). Breeds in Au.

Circus approximans gouldi. Swamp harrier. Sol (three records: Guad, 3S (Beecher 1945, French 1957, Buckingham et al. 1996)). Breeds in Au, NZ, S NG.

Stiltia isabella. Australian pratincole. Bis (one record: NB (Coates 1985)). Breeds in Au.

Glareola maldivarum. Oriental pratincole. Sol (one record: Boug (Hadden 1981)). Breeds in SE Asia.

Vanellus miles novaehollandiae. Masked lapwing. Sol (one record: NGa (Blaber 1990)). Breeds in Au.

Pluvialis squatarola. Gray plover. Bis, Sol. Breeds in Holarctic.

**Pluvialis fulva.* Pacific golden plover. Bis, Sol (Stickney 1943, Bull 1948). Breeds in N Siberia, W Alaska.

Charadrius mongolus mongolus. Mongolian plover. Bis, Sol. Breeds in NE Asia.

Charadrius leschenaultii leschenaultii. Large sand-plover. Bis, Sol. Breeds in W China, Mongolia, and adjacent Russia.

Charadrius veredus. Oriental plover. Sol (two records: Boug, Ren (Hadden 1981, Diamond 1984a)). Breeds in Mongolia and N China.

**Arenaria interpres interpres.* Ruddy turnstone. Bis, Sol (Stickney 1943, Bull 1948). Breeds in N Eurasia, Alaska.

Numenius madagascariensis. Eastern curlew. Bis, Sol. Breeds in E Siberia.

**Numenius phaeopus variegatus.* Whimbrel. Bis, Sol (Bull 1948, Mayr 1949a). Breeds in E Siberia.

Numenius minutus. Little curlew. Bis. Sol (recorded only from Boug (Hadden 1981)). Breeds in E Siberia.

**Tringa brevipes.* Gray-tailed tattler. Bis, Sol (Stickney 1943, Bull 1948). Breeds in E Siberia.

Tringa incana. Wandering tattler. Bis, Sol (Stickney 1943). Breeds in NW NAm.

**Tringa hypoleucos.* Common sandpiper. Bis, Sol (Bull 1948, Mayr 1949a). Breeds in Palaearctic.

Tringa nebularia. Common greenshank. Bis, Sol. Breeds in Palaearctic.

Tringa stagnatilis. Marsh sandpiper. Bis (NB). Sol (recorded only from Boug (Hadden 1981, Bluff & Skyrme 1984)). Breeds in Palaearctic.

Tringa glareola. Wood sandpiper. Bis (one record: NB (Dutson 2001)). Sol: (one record: Boug (Skyrme 1984)). Breeds in Palaearctic.

Tringa terek. Terek sandpiper. Bis, Sol. Breeds in Palaearctic.

Gallinago megala. Swinhoe's snipe. Bis, Sol. Breeds in NE Asia.

?*Gallinago stenura.* Pin-tailed snipe. Bis (two unconfirmed reports: NB (Dutson 2001), NI (Finch 1985)). Breeds in NE Asia.

Limosa limosa melanuroides. Black-tailed godwit. Bis, Sol. Breeds in NE Asia.

**Limosa lapponica baueri.* Bar-tailed godwit. Bis, Sol (Stickney 1943, Bull 1948). Breeds in NE Asia, NW NAm.

**Calidris acuminata.* Sharp-tailed sandpiper. Bis, Sol (Bull 1948, Mayr 1949a). Breeds in E Siberia.

Calidris melanotos. Pectoral sandpiper. Sol (two records: Boug, Ren (Hadden 1981, Diamond 1984a)). Breeds in E Siberia.

Calidris ruficollis. Red-necked stint. Bis, Sol. Breeds in NE Siberia, Alaska.

Calidris subminuta. Long-toed stint. Bis (two records: NB (Mayr 1949a; K.D. Bishop, unpublished observation)). Breeds in E Siberia.

Calidris ferruginea. Curlew sandpiper. Bis, Sol. Breeds in N Siberia.

Calidris alba. Sanderling. Bis (one record: NB (M. Lecroy & W. Peckover, unpublished observation)). Sol (recorded only from Boug, 3S (French 1957, Hadden 1981)). Breeds in Holarctic.

Limicola falcinellus sibirica. Broad-billed sandpiper. Sol (one record: 3S (French 1957)). Breeds in N Siberia.

Philomachus pugnax. Ruff. Bis (two records: NB, Manus (Finch 1985; K.D. Bishop, unpublished observation)). Sol (two records: Boug (Hadden 1981)). Breeds in Palaearctic.

Phalaropus lobatus. Red-necked phalarope. Bis (Mayr 1949a, Bishop 1983, Coates 1985). Breeds in Holarctic.

Cuculus saturatus subsp. Oriental cuckoo. Bis, Sol. Breeds in Asia.

Cacomantis pyrrhophanus pyrrhophanus. Fan-tailed cuckoo. Sol (recorded from Ys, Guad, Bel (Amadon 1942b, Galbraith & Galbraith 1962)). Breeds in NCal.

**Chrysococcyx lucidus lucidus.* Shining bronze-cuckoo. Bis, Sol (Mayr 1932c). Breeds in NZ.

**Chrysococcyx lucidus plagosus.* Shining bronze-cuckoo. Bis, Sol (Mayr 1932c, Galbraith & Galbraith 1962). Breeds in Au.

Eudynamys taitensis. Long-tailed koel. Bis, Sol (Bogert 1937). Breeds in NZ.

Hirundapus caudacutus caudacutus. White-throated needletail. Bis (one record: Umb (Dutson 2001)). Breeds in E Siberia.

Halcyon macleayii incincta. Forest kingfisher. Bis (recorded only from NB, NI). Sol (Kul?; Buckingham et al. 1996). Breeds in Au.

**Merops ornatus*. Rainbow bee-eater. Bis. Sol (two records: Nis (Cooke 1971, and specimens collected by Hamlin in 1929 and examined in American Museum of Natural History; apparently the sole other records for Sol are the questionable ones of Ramsay 1882d)). Breeds in Au.

**Eurystomus orientalis pacificus*. Dollarbird. Bis. Breeds in Au.

Hirundo rustica gutturalis. Barn swallow. Bis (one record: Hornos in the Admiralties (Mackay 1977)). Breeds in NE Asia.

Hirundo daurica subsp. Red-rumped swallow. Bis (NB, NI (Finch 1985, Coates 1990)). Sol (one record: Boug (Hicks 1990)). Breeds in Eurasia, Afr.

**Hirundo nigricans nigricans*. Tree martin. Bis, Sol. Breeds in Au.

Motacilla flava simillima. Yellow wagtail. Bis (two records: NB (Diamond 1971), NI (B. Coates, unpublished observation)). Sol (one record: Boug (Skyrme 1984)). Breeds in NE Siberia.

Motacilla cinerea caspica. Gray wagtail. Bis (one record: Wuv (Coates & Swainson 1978)). Breeds in Asia.

**Coracina novaehollandiae melanops*. Black-faced cuckoo-shrike. Bis. Sol (recorded only from Nis, Boug, Ren). Breeds in Au.

Phylloscopus borealis kennicotti or *P. b. borealis*. Arctic warbler. Bis (one record: Anch (Mayr 1955)). Breeds in Alaska or NE Siberia.

Locustella fasciolata. Gray's grasshopper warbler. Bis (one record: NI (Diamond 1986b)). Breeds in NE Asia.

Myiagra rubecula subsp. Leaden flycatcher. Bis? (two records: NB (specimen from Liverpool Museum examined; collected by G.E. Richards in 1879), NI (questionable locality: Gray 1859, Mayr 1955)). Breeds in Au, SE NG.

Myiagra cyanoleuca. Satin flycatcher. Bis. Breeds in Au.

Lonchura tristissima subsp. Streak-headed mannikin. Bis (one record: Umb (Dutson 2001)). Breeds in NG.

Oriolus sagittatus subsp. Olive-backed oriole. Bis (one record: NB (Dutson 2001)). Breeds in Au.

Appendix 3

Introduced Bird Species in Northern Melanesia

There have been only five documented introductions of exotic bird species in Northern Melanesia. Of these five, only one (*Acridotheres tristis*) has become firmly established and is spreading. Three (*Streptopelia chinensis* on New Britain, *Gymnorhina tibicen* on Guadalcanal, and *Myzantha melanocephala* on Three Sisters) have certainly gone extinct. The fifth, *Columba livia* in the Bismarcks, is apparently present in just one town.

The few successful establishments of introduced species in Northern Melanesia contrast with the many successes in Hawaii, Tahiti, and many other oceanic islands. On such islands many or most of the bird individuals that one sees, especially in disturbed habitats, are exotic species. In contrast, in Northern Melanesia all the bird individuals that one sees, even in disturbed habitats, are native species, except that *Acridotheres tristis* is abundant in disturbed habitats of about six Solomon islands. The low success in Northern Melanesia compared with remote oceanic islands fits a general pattern—namely, that species introductions have the greatest probability of success in habitats with few native species (Elton 1958, Case 1996). The Bismarcks and Solomons are much richer in native bird species and have lost far fewer of their original native bird species than more remote oceanic islands such as Hawaii. Details of the five exotic species are given below.

Streptopelia chinensis tigrina. Spotted dove. Native of Southeast Asia to the Greater Sundas, and introduced to many other areas of Indonesia and to other parts of the world. It was evidently introduced to New Britain, where Coultas collected three specimens at Rabaul in 1933 (Mayr 1934a). There have been no reports of it since then, and the numerous bird watchers in Rabaul in recent decades have never seen it there.

Columba livia. Rock dove. Probably originally from the Palaearctic. Some pigeon fanciers in New Guinea keep rock doves, and feral populations exist in a few New Guinea cities (Port Moresby, Lae, Madang, Goroka, and Popondetta). Rock doves have been reported in the town of Kimbe on New Britain, associated with buildings. In the Admiralty Islands in 1977 three individuals were seen at Lorengau on Manus, four on Lou Island, and a flock of six was seen on Pak Island (Bell 1986). Those at Lou were brought from a town on New Guinea. None of these pigeons in the Admiralties seemed feral, and their present status is unknown (Mackay 1977).

Manorina melanocephala. Noisy miner. From Australia. French (1957) recorded this species as breeding on one of the Three Sisters Islands in the Solomons during his stay there, probably in the 1930s. Details of its introduction are unknown. Buckingham et al. (1996) did not observe this conspicuous species on the Three Sisters in 1990, so it may be presumed to have died out.

Gymnorhina tibicen. Black-backed magpie. From Australia. Earlier observers on Guadalcanal did not observe this species. Three specimens were collected out of one group of birds observed in a coconut grove on Guadalcanal in 1944 (Beecher 1945, Baker 1948), but at that time the species was not observed elsewhere on Guadalcanal, and signs of a disease often found in caged birds in one of the specimens suggested that it may have been an imported, captive bird. In 1953 Cain and Galbraith (1956) observed only two individuals. The species had disappeared from Guadalcanal by the late 1960s.

Acridotheres tristis. Indian myna. Originally from Southeast Asia and introduced to Australia, New Zealand, many Pacific islands, and other areas. The first record in the Solomons was a specimen collected on Guadalcanal in 1920 (Davidson 1929). When the Whitney Expedition collected in the Solomons in the late 1920s, they found the myna only in the Russell group (Pavuvu and Banika) (Mayr 1955). It may have been introduced to the Russells by Lever Plantations in the belief that it would control insects in their coconut plantations. French (1957) reported it as breeding on Three Sisters, probably in the 1930s, and it was still there in 1990 (Buckingham et al. 1996). When Cain and Galbraith (1956) were in the Solomons in 1953, they found the myna on Guadalcanal but not on the other islands that they visited (San Cristobal, Ugi, and Ulawa). When Diamond was in the Solomons (1969, 1972, 1974, 1976), it was common on Guadalcanal and in the Russells, and reported present in the town of Auki on Malaita, but absent on Bougainville and all other islands visited (he did not visit Three Sisters). It was first observed in 1976 on Bougainville in Toniva and the town of Arawa (Hadden 1981), and a few years later it was reported to be only at Arawa and not at the towns of Kieta or Panguna (Bell 1986). Residents of New Georgia told Diamond in 1976 that an attempt had been made to introduce the myna there from the Russells, but the attempt had failed. On Guadalcanal and the Russells the species is now well established in disturbed habitats (towns, roadside, coconut plantations).

Appendix 4

Chronologies of Ornithological Exploration

Each entry in the list below consists of the following: the year of the field work; the observer or collector; the islands visited (using the abbreviations explained in appendix 1); and the source for the resulting distributional records (a publication, report, unpublished account, or specimens in museums). We give first the chronology for the Solomons, then for the Bismarcks. For the chronology of ornithological description (as opposed to exploration), see appendix 1, which gives the author and year of description of every bird taxon for which the type locality falls within Northern Melanesia.

Solomon Islands

1838. Pôle Sud Expedition of Dumont d'Urville. San Jorge (near Ys). Jacquinot and Pucheran (1853).

ca. 1850. J. MacGillivray (voyage of the H. M. S. Rattlesnake). Guad, SCr. Gould (1856), Gray (1859), Sclater (1869).

1858. Novara Expedition. Sik. von Pelzeln (1865).

? Collector?. Guad? Verreaux (1858).

Summary of Exploration to Date: Gray (1859)

1860's. J. Buttray (yacht "Chance"). SCr. Sclater (1869), Mayr (1933d).

1865. J. Brenchley (H. M. S. Curacoa). Ys, Guad, SCr, Ugi. Gray (1870), Brenchley (1870).

Summary of Exploration to Date: Sclater (1869)

1878. E. Layard and G. Richards. Guad, SCr. Tristram (1879), Layard (1880), Salvadori (1880).

ca. 1879. J. Cockerell. Savo, Guad. Ramsay (1879b,c, 1882d), Salvadori (1880).

1880. G. Richards. Guad, Russell, Rnd, SCr, Ugi. Ramsay (1881, 1882d, 1883a), Tristram (1882).

ca. 1881. A. Morton. Ys, Fla, Mal, SCr, Ugi. Ramsay (1882a).

ca. 1881. J. Farrie. Fla, NGa. Ramsay (1882c,d).

1882. J. Stephens. SCr, Ugi. Ramsay (1882b, d,e, 1883b), Parker (1968).

Summary of Exploration to Date: Salvadori (1880–82), Ramsay (1882d), Tristram (1882)

1882. T. Heming. SAn. Ramsay (1883b).

ca. 1882. G. Brown. Fead. Ramsay (1883a), Sclater (1883).

1886–87. C. Woodford. Shtl, Faur, Guad, Mal. Ogilvie-Grant (1887), Woodford (1890).

1886–87. C. Woodford. Guad, NGa. Ogilvie-Grant (1888), Sharpe (1888a), Woodford (1890).

early 1890s. H. Welchman. Ys. Tristram (1892, 1894, 1895).

ca. 1891. J. Plant. Fla. Tristram (1892).

1893–95. C. Wahnes and C. Ribbe. Faur, Munia. Specimens examined in American Museum of Natural History, and scattered mentions by Rothschild and Hartert (e.g., 1901a).

1894. H. Cayley-Webster and Cotton. NGa. Scattered mentions of specimens by Rothschild and Hartert (e.g., 1901a, 1905).

1900–01. A. Meek. Ys, Fla, Guad, Kul, Mono. Rothschild and Hartert (1901b,c,d, 1902a,b). See Meek (1913) and Parker (1967b) for the itineraries of this and the following two expeditions of Meek to the Solomons.

1903–04. A. Meek. Boug, Ch, Gizo, NGa, Rnd. Rothschild and Hartert (1905).

1906. C. Woodford. Ren. North (1906), Woodford (1916), Kinghorn (1937).

1908. A. Meek. Boug, VL, SCr. Rothschild and Hartert (1908a,b).

1910. E. Sarfert. OJ, Nuku. Sarfert and Damm (1929).

1920–21. J. Kusche. Guad. Davidson (1929).

1924. A. Eichhorn. Nis. Hartert (1926a).

1927. G. Stanley and J. Hogbin. Ren. Kinghorn (1937).

1927–30. Whitney South Sea Expedition. All Central Solomons except Wana, Koh, and 3S; plus Bag, Bor, Ren, Bel, Nis, and OJ. Mayr (1931b,e, 1945a), plus many papers by Mayr and others on particular taxonomic groups in *American Museum Novitates*; specimens, specimen registers, and unpublished journals examined in the American Museum of Natural History

1929. Crane Pacific Expedition. Ys, Mal, Kul, Ugi. Mayr and Camras (1938).

1933. Templeton Crocker Expedition. Guad, Mal, SCr, Ugi, SAn, SCat, Ren, Bel. Davidson (1934).

ca. 1934–39. W. French. 3S. French (1957)

1936–37. J. Poncelet. Boug. Danis (1937a,b, 1938), White (1937)

1942–43. W. Donaggho. Fla, Guad. Donaggho (1950)

1943. F.E. Ludwig and K.W. Prescott. Boug, Gla, Guad, Pav, Banika. Specimens in the University of Michigan Museum of Zoology

1943–45. W. Beecher. Boug, Guad. Beecher (1945)

1944. L. Bennett. Nis. Ripley (1947)

1944–45. P. Bull. Boug, Fla, Guad, Pav, NGa. Bull (1948)

1944–45. U.S. Naval Medical Research Unit no. 2. Boug, Guad. Baker (1948)

1944–45. R. Virtue. Boug. Virtue (1947)

1944–45. C. Sibley. Simb, NGa. Sibley (1951)

Summary of Exploration to Date: Mayr (1945a)

1951. T. Wolff. Ren. Bradley and Wolff (1956)

1953. D. Bradley. Ren, Bel, OJ. Bradley and Wolff (1956), Bradley (1957)

1953. N. Laird and E. Laird. Ren, Bel. Laird (1954), Laird and Laird (1956), Bradley and Wolff (1956)

1953–54. Oxford University Expedition (A. Cain and I. Galbraith). Guad, Ul, SCr, Ugi. Cain and Galbraith (1955, 1956, 1957), Galbraith and Galbraith (1962)

1958–59. T. Monberg and S. Willer-Andersen. Ren, Bel. Bradley (1962)

1960s, 1970s. G. Stevens and J. Tedder. Guad. Stevens and Tedder (1973)

1962. T. Monberg. Bel. Wolff (1973)

1962. Noona Dan Expedition (zoologist, F. Salomonsen). Fead, Kili, Tauu, Nuku. Wolff (1966); specimens examined in the University of Copenhagen Zoological Museum

1962, 1965. T. Wolff. Ren. Wolff (1973)

1963–64. P. Temple and Shanahan. Shtl, Faur, Ch, Ys, Fla, Guad, Mal, VL, Kul. Specimens examined in the Bishop Museum

1964. R. Schodde. Boug. Schodde (1977)

1965–67. A. Lang. Buka, Nis. A. Lang & S. Parker (unpublished observations)

1966. H. Officer. Gizo. Officer (unpublished observations)

1968. A. Mirza. Boug. Specimens examined in the Bishop Museum

1968. S. Parker. Buka, Ch, Wagina, Guad, Mal, VL, Gizo. Parker (1972, unpublished observations)

1969–70. L. Filewood. Boug. Filewood (1969, 1972)

1970–71. T. Bayliss-Smith. OJ, Sik. Bayliss-Smith (1972, 1973)

1971. R. Cooke. Nis. Cooke (1971)

1972. J. Diamond. Boug. Diamond (1975b)

1974, 1976. J. Diamond. All Central Solomons except Buka, Mal, Ul, 3S, Rmos, Gow; Bor, Ren, Bel. Diamond (1984a, 2001a, unpublished observations)

1976–80. D. Hadden. Buka, Boug. Hadden (1981)

1981. D. Roper. Savo. Roper (1983)

1981. L. Howell. Nis. Howell (1981)

1982. E. Harding. Nis. Harding (1982)

1984. G. Bluff and T. Skyrme. Boug. Bluff and Skyrme (1984)

1985. B. Finch. Boug, Guad, Mal, VL, Gizo, Kul, NGa, Rnd, SCr, Ren, Nis. Finch (1985)

1985–88. S. Blaber. NGa, Wana, Koh. Blaber (1990)

1986–88. H.P. Webb. Ys. Webb (1992)

1989. R. Loyn, D. Tully. Guad, Mal. Unpublished.

1990. A. Lees, M. Garnett, S. Wright. Ch, Guad, Mal, NGa, Gat, SCr, Ren. Lees et al. (1991)

1990. Scofield, P. Guad, Mal, Gizo, Kul, NGa, Rnd, Ren, Bel. Unpublished

1990. D. Buckingham, G. Dutson, J. Newman. Guad, Rus, Gizo, Kul, SCr, 3S, Ren, Bel. Buckingham et al. (1996)

1991. J. Engbring. Pav, Ban. Unpublished

1992. H. P. Webb. Ch, Ys, Guad, VL. Unpublished

1993. D. Gibbs. Ch, Ys, Guad, Mal, VL, Gan, Gizo, Kul, SCr, Ren. Gibbs (1996)

1994. M. LeCroy, H. P. Webb. Ys, Guad, Ren. Unpublished

1997. C. Filardi, A. Kratter, C. Smith, D. Steadman, H. P. Webb. Ys, Guad, Mal, Ren. Filardi et al. (1999), Kratter et al. (2001)

1997–98. G. Dutson. Buka, Ch, Ys, Guad, Mal, VL, Gan, Gizo, Kul, NGa, Vang, Rnd, Tet, Ulawa, SCr, Ugi, SAn, Ren, Nis. Dutson (2001)

Bismarck Islands

1823. Coquille Expedition (leader, L. Duperrey; zoologists, R. Lesson and P. Garnot). NI. Lesson and Garnot (1826–28)

1827. Astrolabe Expedition (leader, J. Dumont d'Urville; zoologists, J. Quoy and J. Gaimard). NI. Quoy and Gaimard (1830).

1840. Sulphur Expedition (leader, E. Belcher; zoologist, R. Hinds). NI. Gould (1843)

Summary of Exploration to Date: Gray (1859)

1860s. Collectors for Haus Godeffroy. Nin. Hartlaub (1867)

1860s. J. Buttray (yacht "Chance"). NI. Sclater (1869), Mayr (1933d)

1872. Captain Fergusson. NB, NI. Sclater (1873), Ramsay (1876)

1875. Challenger Expedition. Man. Sclater (1877b, 1880b)

1875. Gazelle Expedition (Hüsker and T. Studer, collectors). NB, NI, NH, Anch. Cabanis and Reichenow (1876), Studer (1889)

1875–ca.1883. G. Brown (partly with J. Cockerell). NB, NI, DY, Cred. Sclater (1877a, 1878a,b, 1879, 1880a, 1883), Ramsay (1877, 1878a,b,c, 1879a, 1883a)

1877–78. F. Hübner. NB, DY. Finsch (1879)

1879. E. Layard and G. Richards. NB, DY, Cred. Layard (1880), Gurney (1882)

1880–81. O. Finsch. NB, NI. Finsch (1881, 1884, 1886)

1881. T. Kleinschmidt. NB, DY, Cred. Salvadori (1881), Sclater (1881), Forbes (1882), Reichenow (1899), Rothschild and Hartert (1901c)

1880s. R. Parkinson. NB, DY. North (1888)

1886. J. Kubary. NB. A. Meyer (1890, 1891)

1891. H. and B. Geisler. NB. A. Meyer (1891), Reichenow (1899)

1896–97. F. Dahl. NB, Vu, DY, Cred. Dahl (1899), Reichenow (1899)

1897. H. Cayley-Webster. NH. Hartert (1898, 1899), Rothschild and Hartert (1899)

Summary of Exploration to Date: Reichenow (1899)

1900–01. I. Deutsche Südsee Expedition von Br. Mencke (zoologist, O. Heinroth). NB, NI, SMt, DY, Cred. Heinroth (1902, 1903)

1902–1937. O. Meyer. NB, NI, NH, Vu, Lih. O. Meyer (1906, 1909, 1927a,b, 1928, 1929a,b, 1930, 1933, 1934a,b, 1936, 1937), W. Meyer (1909), Stresemann (1929, 1933, 1938), Schönwetter (1935)

ca. 1905. C. Wahnes. NB. Hartert (1911)

1908–09. Südsee-Expedition der Hamburger Wissenschaftlichen Stiftung. NB, Em, Tench. Martens (1922)

1913. A. Meek. Man, Umb. Rothschild and Hartert (1914a,b)

1923–25. A. Eichhorn. NB, NI, NH, SMt, Feni, Witu, Unea. Hartert (1924a,b,c, 1925, 1926a,b,c).

1928. E. Mayr. Nin, Wuv, Manu. Meise (1929a,b).

1928. R. Beck. NB, NI. Specimens examined in the American Museum of Natural History

1932–35. W. Coultas (Whitney South Sea Expedition). NB, Man, Ramb, Nau, SMig, Anch, Her, Nin, Long, Umb, Lih, Feni, Tab, Tang, Lol. Mayr (1934a), LeCroy and Peckover (1983); specimens and unpublished diaries examined in the American Museum of Natural History

1943. K.W. Prescott. Em. Specimens in the University of Michigan Museum of Zoology.

1944–45. U. S. Naval Medical Research Unit no. 2. SMt, Man. Baker (1948).

1944. C. Sibley. SMt, Em, Tench. Unpublished.

1945. L. Bennett. Man, Los Negros. Ripley (1947).

1958–59. E. Gilliard. NB. Gilliard and LeCroy (1967).

1962. H. Clissold. NB. Thompson (1964); specimens examined in the Bishop Museum.

1962. Noona Dan Expedition (zoologists, F. Salomonsen and T. Wolff). NB, NI, NH, SMt, Man, Her, DY, Cred, Dyl, Ting, Feni. Salomonsen (1962, 1964, 1965, 1966a,b, 1976, 1983), Wolff (1966); specimens examined in the University of Copenhagen Zoological Museum.

1966–67. R. Donaghey. NB. Unpublished.

1969. J. Diamond. NB. Diamond (1971a, 1975a).

1970. H. Bell. Her, Nin. Bell (1970, 1975).

1972. J. Diamond. Long, Umb, Tol, Sak, Cr. Diamond (1974, 1975a, 1976).

1972. E. Lindgren and K. Kisokau. NB, Cred. Unpublished.

1972–76. B. Coates. Em, Feni, Witu, Unea, Tench. Coates (1972, 1973b, 1974, 1977, 1985, 1990, unpublished observations).

1973. K. Silva. Feni, Tench. Silva (1973a,b).

1973. R. Orenstein. NB. Orenstein (1976).

1975. K. Silva. Man, SMt, Em. Silva (1975).

1975. B. Coates and G. Swainson. Wuv. Coates and Swainson (1978, unpublished observations).

1976. B. Beehler. NI. Beehler (1978).

1976. H. Childs. Wuv. Unpublished.

1976. W. Peckover. Em. Unpublished.

1977. R. Mackay. NI, Man, Ramb. Mackay (1977).

1978–80. K. Bishop. NB. Bishop (1983; Bishop & Jones 2001).

1979. M. LeCroy and W. Peckover. NB. LeCroy and Peckover (1983).

1979, 1981. J. Smith. NB, NI, SMt, Man, Los Negros, DY. Specimens examined in the Los Angeles County Museum of Natural History.

1984. B. Finch and J. McKean. NB, NI, Finch (1984), Finch and McKean (1987).

1985. B. Finch. NB, NI, SMt, Man. Finch (1985).

1986. D. Jones and P. Lambley. NI. Jones and Lambley (1987).

1987. I. Burrows. Lih. Burrows (1987).

1990. D. Buckingham, G. Dutson, and J. Newman. Man. Dutson and Newman (1991).

1993. L.. Tolhurst. Man, SMig. Tolhurst (1993).

1994. C. Eastwood. Man. Eastwood (1995a).

1994. B. Beehler. NI. Unpublished.

1996. F. Crome. Tab (Simberi Island). Unpublished.

1996. C. Eastwood. NI, SMt, Dyl. Eastwood (1996).

1997–98. G. Dutson. NB, NI, NH, SMt, Man, Umb, Vu, Dyl, Lih, Tab, Tang, Tench. Dutson (1997, 2001).

1999. P. Gregory. NI, NH, SMt, Dyl, Tench.

1999. C. Schipper, M. Shanahan, S. Cook, I. Thornton. Long. Unpublished.

Appendix 5

Attributes of Each Bird Species

This appendix lists some evolutionary, ecological, and distributional attributes for each of the 195 breeding land and fresh-water bird species and superspecies of Northern Melanesia. In 12 superspecies we separately list, for individual allospecies, several attributes that vary among allospecies. For four species (# 1, 3, 92, 112) not analyzed further because they breed only occasionally or may have been introduced prehistorically, we tabulate only certain attributes. Attributes tabulated in the colums are as follows.

Level of endemism identifies for each species whether its Northern Melanesian populations are not endemic to Northern Melanesia at all (because they belong to the same subspecies as an extralimital population: denoted by 0), or else are endemic at the subspecies (1), allospecies (2), full species (3), or genus (4) level. If a Northern Melanesian population extends to just a few nearby extralimital islands (e.g., Karkar, Louisiades, Santa Cruz), adding just a small fraction to its Northern Melanesian range, that extension is ignored in determining level of endemism. Two numbers indicate species whose populations are variously at two different levels of endemism; for example, "0,1" means that some populations belong to an endemic subspecies while others belong to the same subspecies as an extralimital population. In analyses in the text we allocate such species to the higher of the two levels of endemism. If, however, most Northern Melanesian populations of a species are subspecifically distinct and the populations of just a few islands near New Guinea (e.g., Long) are conspecific with the New Guinea population, the latter few populations are ignored.

Habitat denotes habitat preference by the following codes. f = lowland forest (regardless of whether the species also occurs in montane forest or in lowland nonforest habitats). All other codes refer to species generally absent from lowland forest: c = seacoast; a = lowland aerial; w = fresh water; man = mangrove; o = open country, gardens, secondary growth, and forest edge; s = swamps and marshes; mt = mountains (montane forest, except for one montane aerial and one montane riverine species).

Incidence category describes a species' distribution over small and large islands, as defined in table 12.3: HS = high-*S* species; ST = supertramp; A, B, C, or D = A tramp, B tramp, C tramp, or D tramp, respectively; — = absent from that archipelago; n = confined to one or a few geographically peripheral islands in that archipelago, and hence not classified into an incidence category.

Dispersal is graded from A to C, with A being the 56 most vagile species and C being the 28 most sedentary species. That Northern Melanesian dispersal index is based on the separate archipelago dispersal indices for Bismarck and Solomon populations in appendix 6, as follows. A Northern Melanesian dispersal index of C applies to a species or superspecies confined to a single island or to modern fragments of a single Pleistocene island, with no records of overwater dispersal. An index of A applies to a species or superspecies confined to either the Bismarcks or Solomons and with an archipelago index of 1 (appendix 6) in that archipelago; or present in both archipelagoes and with an archipelago index of 1 in both, or of 1 in one archipelago and 2 in the other, or of 1 in the Bismarcks and 3* in the Solomons. In the case of a superspecies represented by multiple allospecies within the Bismarcks or Solomons, the archipelago index is taken as 1 (or 2) only if all allospecies have an index of 1 (or 2). All species not qualifying for a Northern Melanesian dispersal index of A or C are assigned an index of B.

Abundance is graded from 1 to 5, with 1 being the rarest species (average population density <1 pair/10 km^2) and 5 being the most abundant species (average population density ≥100 pairs/km^2).

The variable i is the number of Northern Melanesian islands occupied by the species, out of 74 islands analysed.

Allospecies is the number of distinct allospecies in Northern Melanesia if greater than 1, followed by the number of endemic allospecies in parentheses.

The sum $(s + A)$ is the number of geographically distinct populations in Northern Melanesia (number of subspecies plus monotypic allospecies).

The ratio $(s + A)/i$ is not calculated for species confined to a single island ($i = 1$).

	Species	Level of endemism	Habitat	Incidence category		Dispersal	Abundance	*i*	Allospecies	(*s* + *A*)	(*s* + *A*)/*i*
				Bis	Sol						
1	*Casuarius bennetti*	0	f	HS	—	—	—				
2	*Tachybaptus [ruficollis]*	1				B	2	8	2(0)	2	0.25
	T. ruficollis	0	w	A	HS						
	T. novaehollandiae	1	w	—	n						
3	*Pelecanus conspicillatus*	0	c	n	n	—	—				
4	*Phalacrocorax melanoleucos*	0,1	w	HS	A	A	2	14		2	0.14
5	*Butorides striatus*	1	man	HS	D	B	3	31		1	0.03
6	*Egretta sacra*	0	c	D	D	A	2	67		1	0.01
7	*Ardea intermedia*	0	o	HS	HS	B	2	6		1	0.17
8	*A. alba*	0	o	HS	HS	A	2	4		1	0.25
9	*Nycticorax caledonicus*	1	f	C	D	A	3	55		2	0.04
10	*Ixobrychus sinensis*	0	s	HS	HS	B	2	2		1	0.50
11	*I. flavicollis*	0,1	f	C	A	A	3	30		2	0.07
12	*Threskiornis moluccus*	1	o	—	n	A	3	2		1	0.50
13	*Platalea regia*	0	w	—	n	A	2	1		1	—
14	*Dendrocygna arcuata*	1	w	HS	—	B	2	2		1	0.50
15	*D. guttata*	0	w	HS	—	B	2	2		1	0.50
16	*Anas superciliosa*	0	w	B	D	A	3	38		1	0.03
17	*A. gibberifrons*	1	w	—	n	B	2	1		1	—
18	*Aviceda subcristata*	1	f	A	D	B	3	34		5	0.15
19	*Henicopernis infuscata*	2	f	HS	—	C	1	2		1	0.50
20	*Haliastur indus*	0,1	o	D	D	A	3	58		2	0.03
21	*Accipiter meyerianus*	0	f	HS	HS	B	1	5		1	0.20
22	*A. novaehollandiae*	1	f	C	C	A	3	35		10	0.29
23	*A. fasciatus*	0	o	—	n	B	3	2		1	0.50
24	*A. imitator*	3	f	—	HS	C	1	3		1	0.33
25	*A. albogularis*	2	f	n	D	A	3	26		4	0.15
26	*A. luteoschistaceus*	3	f	HS	—	C	1	2		1	0.50
27	*A. princeps*	2	mt	HS	—	C	1	1		1	—
28	*A. brachyurus*	2	f	HS	—	B	1	2		1	0.50
29	*Haliaeetus [leucogaster]*	2				A	1	53	2(1)	2	0.04
	H. leucogaster	0	c	D	—						
	H. sanfordi	2	f,c	—	D						
30	*Pandion haliaetus*	0	c	D	D	A	2	60		1	0.02
31	*Falco peregrinus*	0	a	A	HS	B	1	6		1	0.17
32	*F. severus*	0	a	A	A	B	1	9		1	0.11

33	*F. berigora*	0	o	n	—	B	1	1		1	—
34	*Megapodius freycinet*	1	f	D	D	A	3	69		1	0.01
35	*Coturnix chinensis*	1	o	A	—	B	2	8		1	0.13
36	*Turnix maculosa*	1	o	HS	HS	B	2	3		2	0.67
37	*Gallirallus [philippensis]*	2				B	3	21	2(1)	8	0.38
	G. philippensis	1	o	B	B						
	G. rovianae	2	f	—	B						
38	*G. insignis*	3	f	HS	—	C	3	1		1	—
39	*Rallina tricolor*	1	f	HS	—	B	2	3		2	0.67
40	*Nesoclopeus woodfordi*	2	f	—	HS	C	3	4		3	0.75
41	*Gymnocrex plumbeiventris*	0	f	HS	—	B	2	1		1	—
42	*Porzana tabuensis*	0	s	n	n	B	2	4		1	0.25
43	*Poliolimnas cinereus*	0,1	s	A	A	B	2	12		2	0.17
44	*Amaurornis olivaceus*	1	o	B	D	B	3	40		3	0.08
45	*Pareudiastes sylvestris*	3	mt	—	n	C	1	1		1	—
46	*Porphyrio porphyrio*	0	o	A	D	A	3	40		1	0.03
47	*Irediparra gallinacea*	0	w	HS	—	B	2	1		1	—
48	*Charadrius dubius*	0	w	HS	—	B	2	2		1	0.50
49	*Himantopus leucocephalus*	0	w	HS	—	B	2	3		1	0.33
50	*Esacus magnirostris*	0	c	C	D	A	2	42		1	0.02
51	*Ptilinopus superbus*	0	f	C	C	A	3	37		1	0.03
52	*P. [purpuratus]*	2				B	4	7	2(1)	3	0.43
	P. greyii	0	f	—	ST						
	P. richardsi	2	f	—	ST						
53	*P. insolitus*	2	f	C	—	B	4	17		2	0.12
54	*P. [rivoli]*	2				B	4	52	2(1)	10	0.19
	P. rivoli	1	mt	A	—						
	P. solomonensis	2	f, mt	ST	C						
55	*P. viridis*	1	f	n	D	A	4	35		2	0.06
56	*Ducula pacifica*	0,1	f	ST	ST	A	4	16		2	0.13
57	*D. rubricera*	2	f	B	D	A	4	45		2	0.04
58	*D. finschii*	2	f	HS	—	B	3	4		1	0.25
59	*D. pistrinaria*	2	f	ST	D	A	4	58		3	0.05
60	*D. brenchleyi*	2	f	—	A	B	4	8		1	0.13
61	*D. melanochroa*	2	mt	HS	—	A	3	3		1	0.33
62	*D. spilorrhoa*	1	f	ST	—	A	3	12		1	0.08
63	*Gymnophaps [albertisii]*	2				B	3	7	2(1)	2	0.29
	G. albertisii	0	mt	HS	—						
	G. solomonensis	2	mt	—	A						
64	*Columba vitiensis*	0	f, mt	HS	B	B	1	11		1	0.09
65	*C. pallidiceps*	3	f, mt	HS	A	B	1	10		1	0.10

continued

Species		Level of endemism	Habitat	Incidence category		Dispersal	Abundance	i	Allospecies	$(s + A)$	$(s + A)/i$
				Bis	Sol						
66	*Macropygia amboinensis*	1	f	C	—	B	4	16		3	0.19
67	*M. [ruficeps]*	1				A	4	52	2(0)	2	0.04
	M. nigrirostris	1	f	A	—						
	M. mackinlayi	1	f	ST	D						
68	*Reinwardtoena [reinwardtii]*	2	f, mt	B	C	B	1	31	2(2)	3	0.10
69	*Chalcophaps stephani*	0,1	f	D	D	A	4	56		2	0.04
70	*Henicophaps foersteri*	2	f	HS	—	C	1	3		1	0.33
71	*Gallicolumba beccarii*	1	f	C	B	B	4	31		6	0.19
72	*G. salamonis*	3	f	—	n	B	3	2		1	0.50
73	*G. jobiensis*	0,1	f	B	HS	B	3	13		2	0.15
74	*Microgoura meeki*	4	f	—	HS	C	1	1		1	—
75	*Caloenas nicobarica*	0	f	ST	ST	A	3	64		1	0.02
76	*Chalcopsitta cardinalis*	3	f	n	D	A	4	36		1	0.03
77	*Trichoglossus haematodus*	1	f	D	D	A	4	58		3	0.05
78	*Lorius [lory]*	2				B	4	20	3(2)	3	0.15
	L. albidinucha	2	mt	HS	—						
	L. chlorocercus	2	f	—	A						
	L. hypoinochrous	1	f	B	—						
79	*Charmosyna [palmarum]*	2				B	3	7	2(2)	2	0.29
	C. rubrigularis	2	mt	HS	—						
	C. meeki	2	mt	—	HS						
80	*C. placentis*	1	f	C	HS	A	3	23		1	0.04
81	*C. margarethae*	3	f	—	B	B	3	9		1	0.11
82	*Micropsitta bruijnii*	1	mt	HS	HS	B	3	5		3	0.60
83	*Micropsitta [pusio]*	2				B	4	41	3(2)	9	0.22
	M. pusio	0	f	C	C						
	M. meeki, M. finschii	2	f	C	C						
84	*Cacatua galerita*	1	f	HS	—	B	3	1		1	—
85	*C. ducorpsi*	2	f	—	C	A	3	24		1	0.04
86	*Eclectus roratus*	1	f	C	D	A	3	45		2	0.04
87	*Geoffroyus heteroclitus*	2	f	A	C	A	3	35		2	0.06
88	*Loriculus tener*	2	f	HS	—	B	1	4		1	0.25
89	*Cacomantis variolosus*	1	f	B	B	B	3	27		5	0.19
90	*Chrysococcyx lucidus*	1	f	—	n	B	4	2		1	0.50
91	*Eudynamys scolopacea*	1	f	B	D	A	3	38		3	0.08
92	*Scythrops novaehollandiae*	0	f	B	—	—	—				

93	*Centropus [ateralbus]*	2	f	HS	B	B	3	20	2(2)	3	0.15
94	*C. violaceus*	3	f	HS	—	B	1	2		1	0.50
95	*Tyto alba*	1	o	n	C	B	1	16		1	0.06
96	*T. [novaehollandiae]*	2				B	1	2	2(1)	2	1.00
	T. aurantia	2	f	HS	—						
	T. novaehollandiae	1	f	n	—						
97	*Ninox [novaeseelandiae]*	2	f	HS	B	B	3	18	4(4)	11	0.61
98	*Nesasio solomonensis*	4	f	—	HS	C	1	3		1	0.33
99	*Podargus ocellatus*	1	f	—	HS	B	1	3		1	0.33
100	*Eurostopodus mystacalis*	1	a	—	B	A	2	12		1	0.08
101	*Caprimulgus macrurus*	0	a	B	—	B	3	10		1	0.10
102	*Aerodramus orientalis*	3	mt	HS	HS	B	1	3		3	1.00
103	*A. vanikorensis*	1	a	D	D	A	4	49		4	0.08
104	*A. spodiopygius*	1	a	A	A	B	3	19		4	0.21
105	*Collocalia esculenta*	1	a	C	C	A	4	44		7	0.16
106	*Hemiprocne mystacea*	1	a	C	C	A	3	42		4	0.10
107	*Alcedo atthis*	0,1	w	C	C	A	2	38		2	0.05
108	*A. websteri*	2	w	HS	—	B	3	5		1	0.20
109	*A. pusilla*	1	man	HS	C	B	3	24		4	0.17
110	*Ceyx lepidus*	1	f	A	C	B	4	25		8	0.32
111	*Halcyon [diops]*	2	f	HS	A	B	3	8	2(2)	2	0.25
112	*H. sancta*	0	o	—	D	—	—				
113	*H. chloris*	1	f	D	D	A	4	56		12	0.21
114	*H. saurophaga*	0,1	c	D	D	A	2	61		3	0.05
115	*H. bougainvillei*	3	f	—	HS	C	1	2		2	1.00
116	*Tanysiptera sylvia*	1	f	A	—	A	4	4		2	0.50
117	*Merops philippinus*	0	a	A	—	B	3	4		1	0.25
118	*Eurystomus orientalis*	1	a	C	D	A	3	46		2	0.04
119	*Rhyticeros plicatus*	1	f	HS	C	B	1	24		3	0.13
120	*Pitta [sordida]*	2				B	4	4	2(1)	2	0.50
	P. superba	2	f	n	—						
	P. sordida	0	f	n	—						
121	*P. anerythra*	2	f	—	HS	C	4	3		3	1.00
122	*P. erythrogaster*	1	f	B	—	B	4	10		4	0.40
123	*Hirundo tahitica*	1	a	C	D	A	3	46		3	0.07
124	*Coracina caledonica*	1	mt	—	A	B	3	6		4	0.67
125	*C. lineata*	1	f	HS	C	B	3	25		7	0.28
126	*C. papuensis*	1	o	A	C	A	3	32		5	0.16
127	*C. [tenuirostris]*	1,2	f	C	D	B	3	44	2(1)	11	0.25
128	*C. holopolia*	3	f	—	B	B	3	10		3	0.30
129	*Lalage leucomela*	1	f	B	—	B	5	12		7	0.58

continued

Species	Level of endemism	Habitat	Incidence category		Dispersal	Abundance	i	Allospecies	$(s + A)$	$(s + A)/i$
			Bis	Sol						
130 *L. leucopyga*	1	f	—	n	B	5	2		1	0.50
131 *Saxicola caprata*	0	o	A	—	B	2	4		1	0.25
132 *Zoothera [dauma]*	2				B	3	8	3(2)	6	0.75
Z. heinei	1	f	A	HS						
Z. talaseae, Z. margaretae	2	mt	A	HS						
133 *Turdus poliocephalus*	1	mt	A	HS	B	5	8		7	0.88
134 *Gerygone flavolateralis*	1	f	—	n	B	5	1		1	—
135 *Ortygocichla rubiginosa*	3	f	HS	—	C	3	1		1	—
136 *Cichlornis [whitneyi]*	1,2	mt	HS	HS	B	3	3	3(2)	3	1.00
137 *Cettia parens*	3	mt	—	n	C	3	1		1	—
138 *Acrocephalus stentoreus*	0	s	HS	A	B	3	8		1	0.13
139 *Cisticola exilis*	1	o	B	—	B	3	10		1	0.10
140 *Megalurus timoriensis*	1	o	A	—	B	3	5		1	0.20
141 *Phylloscopus [trivirgatus]*	1,2	mt	HS	HS	B	5	10	2(1)	7	0.70
142 *P. amoenus*	3	mt	—	HS	C	3	1		1	—
143 *Rhipidura leucophrys*	0	o	D	D	A	4	48		1	0.02
144 *R. [rufiventris]*	2				B	5	29	2(1)	13	0.45
R. rufiventris	1	f	B	—						
R. cockerelli	2	f	—	B						
145 *R. [spilodera]*	2	mt	—	HS	B	5	4	3(3)	4	1.00
146 *R. fuliginosa*	0	mt	—	n	B	3	1		1	—
147 *R. rufifrons*	1	f	n	D	A	5	27		8	0.26
148 *R. [rufidorsa]*	2	mt, f	HS	HS	B	4	5	3(3)	4	0.80
149 *Clytorhynchus hamlini*	2	f	—	n	C	4	1		1	—
150 *Monarcha cinerascens*	1	f	ST	ST	A	3	42		4	0.10
151 *M. [melanopsis]*	2	f	—	D	A	5	30	3(3)	6	0.20
152 *M. [manadensis]*	2	f	A	C	B	4	32	6(6)	13	0.38
153 *M. chrysomela*	1	f	A	—	B	4	5		4	0.80
154 *Myiagra alecto*	0	f	C	—	B	4	16		1	0.06
155 *M. hebetior*	3	f	A	—	B	4	6		3	0.50
156 *M. [rubecula]*	1,2	f	—	D	B	4	27	3(2)	6	0.22
157 *Monachella muelleriana*	1	mt	HS	—	B	2	1		1	—
158 *Petroica multicolor*	1	mt	—	HS	B	4	4		4	1.00
159 *Pachycephala pectoralis*	1	f, mt	B	C	B	5	34		16	0.47
160 *P. melanura*	0	f	ST	ST	A	3	12		1	0.08
161 *P. implicata*	3	mt	—	HS	C	3	2		2	1.00

162	*Dicaeum [erythrothorax]*	2	f	A	B	B	5	18	3(3)	7	0.39
163	*Nectarinia sericea*	1	f	C	—	A	3	14		2	0.15
164	*N. jugularis*	1	o	D	D	A	3	57		1	0.02
165	*Zosterops atrifrons*	1	f	A	—	B	5	6		3	0.50
166	*Z. metcalfii*	3	f	—	A	C	5	6		2	0.33
167	*Z. [griseotinctus]*	3	f	ST	B	B	5	18	6(5)	8	0.44
168	*Z. murphyi*	3	mt	—	HS	C	5	1		1	—
169	*Z. ugiensis*	3	mt	—	HS	B	5	3		3	1.00
170	*Z. stresemanni*	3	f	—	HS	C	5	1		1	—
171	*Woodfordia superciliosa*	2	f	—	n	C	5	1		1	—
172	*Myzomela cineracea*	2	f	HS	—	C	4	2		1	0.50
173	*M. cruentata*	1	f, mt	A	—	B	5	7		5	0.71
174	*M. pulchella*	3	mt	HS	—	C	5	1		1	—
175	*M. [cardinalis]*	1,2	f	HS	n	B	5	5	2(1)	3	0.60
176	*M. sclateri*	3	f	ST	—	A	3	8		1	0.13
177	*M. [lafargei]*	3	f	ST	C	B	5	42	6(6)	12	0.29
178	*Philemon [moluccensis]*	2	f	HS	—	B	4	5	3(3)	4	0.80
179	*Melidectes whitemanensis*	3	mt	HS	—	C	4	1		1	—
180	*Stresemannia bougainvillei*	4	mt	—	HS	C	3	1		1	—
181	*Meliarchus sclateri*	4	f	—	n	C	4	1		1	—
182	*Guadalcanaria inexpectata*	4	mt	—	HS	C	3	1		1	—
183	*Erythrura trichroa*	0,1	mt	A	HS	B	4	12		3	0.25
184	*Lonchura [spectabilis]*	1,2	o	A	—	B	3	7	3(2)	4	0.57
185	*L. melaena*	2	o	HS	n	B	2	2		1	0.50
186	*Aplonis cantoroides*	0	o	D	D	A	3	58		1	0.02
187	*A. [feadensis]*	3	f	ST	ST	A	3	10	2(2)	3	0.30
188	*A. [grandis]*	3	f	—	C	B	4	24	2(2)	4	0.17
189	*A. metallica*	1	f	D	D	A	4	59		3	0.05
190	*A. brunneicapilla*	3	mt	—	HS	B	3	4		1	0.25
191	*Mino dumontii*	1	f	A	D	B	4	36		1	0.03
192	*Dicrurus [bottentottus]*	1,2	f	HS	HS	B	4	6	2(1)	4	0.67
193	*Artamus insignis*	2	a	HS	—	B	3	2		1	0.50
194	*Corvus orru*	1	f	HS	—	A	3	9		1	0.11
195	*C. woodfordi*	3	f, mt	—	A	C	3	6		2	0.33

Appendix 6

Evidence of Overwater Dispersal Ability of Each Species and Allospecies in the Bismarcks and Solomons

This appendix summarizes all available types of evidence of overwater dispersal, for each Northern Melanesian allospecies or species, for the Bismarcks and Solomons separately. Consecutive entries under the same species number are different allospecies of the same superspecies. The right-most two columns distill this evidence into dispersal indices for each taxon in each archipelago.

The available evidence for the Bismarcks is as follows. *Seen* is the number of instances in which Diamond observed the species flying over water in the Bismarcks, or in which similar records are contained in publications, reports, or diaries of other observers (H.L. Bell, W.F. Coultas, F. Dahl, W. King, E. Lindgren, and O. Meyer). *Islets* = number of Bismarck islets of area ≤25 ha on which the species has been recorded, from surveys of eight islets near Umboi or Long (J. Diamond), seven near Northeast New Britain (E. Lindgren), one near New Hanover (Noona Dan Expedition), one in the Ninigo group (H. L. Bell), Tench (B. Coates), one in the Admiralities (Whitney Expedition), and Credner (several observers). *Vagrant*: r = repeatedly observed as a stray on Vuatom (O. Meyer); "1" or "2" = 1 or 2 records of straying to Vuatom or other Bismarck islands (W. Coultas, O. Meyer). *Colonized* means the species colonized Vuatom as a breeding resident between 1910 and 1936 (O. Meyer), or Manus after 1960 (*Ptilinopus viridis*: several observers). *CLRT* means now resident on Crown, Long, Ritter, and/or Tolokiwa, which were defaunated by seventeenth- or nineteenth-century volcanic explosions. *Volcanoes* is the number of other Bismarck Holocene volcanoes or volcanically defaunated islands on which the species is now resident (Duke of York, Garowe, Sakar, Unea, Vuatom; post-1910 colonizations of Vuatom are instead counted under the column "colonized").

The available evidence for the Solomons is similar, except as follows. *Seen* is observations by J. Diamond and T. Zinghite. *Islets* refers to 81 islets surveyed by Diamond, of which 71 were in the New Georgia group, 7 in the Shortland group, and 3 near Florida or Guadalcanal. *Vagrant* indicates records of vagrants or new colonists (by J. Diamond and other observers). *Volcanoes* is the number of Solomon Holocene volcanoes on which the species is now resident (Borokua, Savo, Simbo).

Ext indicates whether the species is resident on islands separated by permanent water gaps, either within Northern Melanesia or else between Northern Melanesia and its colonization sources (mainly New Guinea).

Taxa are separately assigned a dispersal index in each of the two archipelagoes (right-most two columns), as follows. Index 4 = confined and endemic to a single Northern Melanesian island, or else to several islands derived from the same larger Pleistocene island; hence there is no evidence of overwater dispersal during the species' evolutionary lifetime. Index 3 = resident on islands separated by permanent water gaps; hence the species is evidently capable of crossing water within the species' evolutionary lifetime, but there is no evidence specifically of Holocene or twentieth-century overwater dispersal. Index 2 = single records of overwater dispersal during the twentieth century (columns "seen," "islets," "vagrant," or "colonized"), or else resident on Holocene volcanoes or else on volcanoes that erupted in the seventeenth or nineteenth centuries (columns "CLRT" or "volcanoes"). Index 1 = two or more records of overwater dispersal during the twentieth century. Index 3* for the Solomons = four species widespread in the Bismarcks but barely extending to the Solomons; hence little or no evidence of overwater dispersal could be obtained for the Solomons.

		Bismarcks						Solomons				Index		
		Seen	Islets	Vagrant	Colonized	CLRT	Volcanoes	Seen	Islets	Vagrant	Volcanoes	Ext	Bismarcks	Solomons
(1	*Casuarius bennetti)*							—	—	—	—		—	—
2	*Tachybaptus ruficollis*					1						√	2	3
	T. novaehollandiae	—	—	—	—	—	—					√	—	3
(3	*Pelecanus conspicillatus)*	1										√	—	—
4	*Phalacrocorax melanoleucos*									2		√	2	1
5	*Butorides striatus*							5	40	3	2	√	3	1
6	*Egretta sacra*	1	10	r		3	4	7	39	1	3	√	1	1
7	*Ardea intermedia*											√	3	3
8	*A. alba*			1						2		√	2	1
9	*Nycticorax caledonicus*		1	r,1		3	2		5		3	√	1	1
10	*Ixobrychus sinensis*			1								√	2	3
11	*I. flavicollis*		1				3	1	4			√	2	1
12	*Threskiornis moluccus*	—	—	—	—	—	—			2		√	—	1
13	*Platalea regia*	—	—	—	—	—	—			2		√	—	1
14	*Dendrocygna arcuata*							—	—	—	—	√	3	—
15	*D. guttata*							—	—	—	—	√	3	—
16	*Anas superciliosa*					1		3			1	√	2	1
17	*A. gibberifrons*	—	—	—	—	—	—					√	—	3
18	*Aviceda subcristata*	—	—	—	—	—	—	4			1	√	3	1
19	*Henicopernis infuscata*							—	—	—	—		4	—
20	*Haliastur indus*					3	3	8	25	1	3	√	2	1
21	*Accipiter meyerianus*			r								√	1	3
22	*A. novaehollandiae*		1	r				2	4		1	√	1	1
23	*A. fasciatus*	—	—	—	—	—	—					√	—	3
24	*A. imitator*	—	—	—	—	—	—						—	4
25	*A. albogularis*							2			2	√	3	1
26	*A. luteoschistaceus*							—	—	—	—		4	—
27	*A. princeps*							—	—	—	—		4	—
28	*A. brachyurus*							—	—	—	—		3	—
29	*Haliaeetus leucogaster*	1	2	r		3	2					√	1	3*
	H. sanfordi	—	—	—	—	—	—	6	12	3	2	√	—	1
30	*Pandion haliaetus*		2			3	3	6	15		3	√	1	1
31	*Falco peregrinus*			1		1	1					√	2	3
32	*F. severus*			r								√	1	3
33	*F. berigora*					1		—	—	—	—	√	2	—
34	*Megapodius freycinet*		9			3	3	2	53	1	3	√	1	1
35	*Coturnix chinensis*						2	—	—	—	—	√	2	—

36	*Turnix maculosa*						1					√	2	3
37	*Gallirallus philippensis*					1	2			2		√	2	1
	G. rovianae	—	—	—	—	—	—						—	4
38	*G. insignis*							—	—	—	—		4	—
39	*Rallina tricolor*							—	—	—	—	√	3	—
40	*Nesoclopeus woodfordi*	—	—	—	—	—	—						—	4
41	*Gymnocrex plumbeiventris*							—	—	—	—	√	3	—
42	*Porzana tabuensis*				1							√	2	3
43	*Poliolimnas cinereus*											√	3	3
44	*Amaurornis olivaceus*					2	4				1	√	2	2
45	*Pareudiastes sylvestris*	—	—	—	—	—	—						—	4
46	*Porphyrio porphyrio*						1	2	23		2	√	2	1
47	*Irediparra gallinacea*							—	—	—	—	√	3	—
48	*Charadrius dubius*							—	—	—	—	√	3	—
49	*Himantopus leucocephalus*							—	—	—	—	√	3	—
50	*Esacus magnirostris*		3	r		3	2	4	14	2	2	√	1	1
51	*Ptilinopus superbus*			r		1			6		1	√	1	1
52	*P. greyii*	—	—	—	—	—	—					√	—	3
	P. richardsiii	—	—	—	—	—	—					√	—	3
53	*P. insolitus*					3	3	—	—	—	—	√	2	—
54	*P. rivoli*							—	—	—	—	√	3	—
	P. solomonensis		3			4	3				1	√	1	2
55	*P. viridis*				1				24	1	2	√	2	1
56	*Ducula pacifica*		2								1	√	1	2
57	*D. rubricera*	1	1				3	2	13		2	√	1	1
58	*D. finschii*			1				—	—	—	—	√	2	—
59	*D. pistrinaria*	4	8	1		4	4	8	61	2	3	√	1	1
60	*D. brenchleyi*	—	—	—	—	—	—					√	—	3
61	*D. melanochroa*			r			1	—	—	—	—	√	1	—
62	*D. spilorrhoa*		3	r, 1		3	1	—	—	—	—	√	1	—
63	*Gymnophaps albertisii*							—	—	—	—	√	3	—
	G. solomonensis	—	—	—	—	—	—					√	—	3
64	*Columba vitiensis*											√	3	3
65	*C. pallidiceps*											√	3	3
66	*Macropygia amboinensis*					2	3	—	—	—	—	√	2	—
67	*M. nigrirostris*			r			1	—	—	—	—	√	1	—
	M. mackinlayi		4		1	3	2	1			2	√	1	2
68	*Reinwardtoena browni*			r			1	—	—	—	—	√	1	—
	R. crassirostris	—	—	—	—	—	—					√		
69	*Chalcophaps stephani*	3				3	5		15		2	√	1	1
70	*Henicophaps foersteri*							—	—	—	—		4	—

continued

		Bismarcks						Solomons				Index		
		Seen	Islets	Vagrant	Colonized	CLRT	Volcanoes	Seen	Islets	Vagrant	Volcanoes	Ext	Bismarcks	Solomons
71	*Gallicolumba beccarii*					3	3					√	2	3
72	*G. salamonis*	—	—	—	—	—	—					√	—	3
73	*G. jobiensis*				1		1					√	2	3
74	*Microgoura meeki*	—	—	—	—	—	—						—	4
75	*Caloenas nicobarica*		5	r,2		3	3	2	11	1	3	√	1	1
76	*Chalcopsitta cardinalis*				1			10	56	4	3	√	2	1
77	*Trichoglossus haematodus*	3	2	r		3	3	7	37	2	2	√	1	1
78	*Lorius albidinucha*							—	—	—	—		3	—
	L. chlorocercus	—	—	—	—	—	—			2	1	√	—	1
	L. hypoinochrous			r,2		1	2	—	—	—	—	√	1	—
79	*Charmosyna rubrigularis*							—	—	—	—	√	3	—
	C. meeki	—	—	—	—	—	—					√	—	3
80	*C. placentis*		2			3	4					√	1	3*
81	*C. margarethae*	—	—	—	—	—	—	1				√	—	2
82	*Micropsitta bruijnii*											√	3	3
83	*M. pusio*					1	5	—	—	—	—	√	2	—
	M. meeki							—	—	—	—	√	3	—
	M. finschii										1	√	3	2
84	*Cacatua galerita*							—	—	—	—	√	3	—
85	*C. ducorpsi*	—	—	—	—	—	—	10	37	3	1	√	—	1
86	*Eclectus roratus*	1		r			3	8	11	2	3	√	1	1
87	*Geoffroyus heteroclitus*						2	3	5	1	1	√	2	1
88	*Loriculus tener*						1	—	—	—	—	√	2	1
89	*Cacomantis variolosus*					1	3					√	2	3
90	*Chrysococcyx lucidus*	—	—	—	—	—	—					√	—	3
91	*Eudynamys scolopacea*		1			3	3	2	4		2	√	2	1
(92	*Scythrops novaehollandiae*)												—	—
93	*Centropus ateralbus*							—	—	—	—	√	3	—
	C. milo	—	—	—	—	—	—				1	√	—	2
94	*C. violaceus*							—	—	—	—	√	3	—
95	*Tyto alba*					1			1			√	2	2
96	*T. aurantia*							—	—	—	—		4	—
	T. novaehollandiae							—	—	—	—	√	3	—
97	*Ninox meeki*							—	—	—	—		4	—
	N. odiosa						1	—	—	—	—	√	2	—
	N. variegata							—	—	—	—		4	—
	N. jacquinoti	—	—	—	—	—	—					√	—	3

98	*Nesasio solomonensis*	—	—	—	—	—	—						—	4
99	*Podargus ocellatus*	—	—	—	—	—	—					√	—	3
100	*Eurostopodus mystacalis*	—	—	—	—	—	—		2			√	—	1
101	*Caprimulgus macrurus*					2	1	—	—	—	—	√	2	—
102	*Aerodramus orientalis*											√	3	3
103	*A. vanikorensis*	2	1			3	4	7	5		2	√	1	1
104	*A. spodiopygius*	1					1					√	2	3
105	*Collocalia esculenta*	1		r		3	1	1				√	1	2
106	*Hemiprocne mystacea*					1	2	1	7		1	√	2	1
107	*Alcedo atthis*		1			2	3	3	5			√	2	1
108	*A. websteri*			1				—	—	—	—	√	2	—
109	*A. pusilla*								4			√	3	1
110	*Ceyx lepidus*						1					√	2	3
111	*Halcyon albonotata*							—	—	—	—		4	—
	H. leucopygia	—	—	—	—	—	—						—	4
(112	*H. sancta)*												—	—
113	*H. chloris*		3			3	3	1	35	1	2	√	1	1
114	*H. saurophaga*		14	r		3	2	9	35	1	1	√	1	1
115	*H. bougainvillei*	—	—	—	—	—	—						—	4
116	*Tanysipteria sylvia*			r			1	—	—	—	—	√	1	—
117	*Merops philippinus*					1	1	—	—	—	—	√	2	—
118	*Eurystomus orientalis*			r		2	4	1	3	1	2	√	1	1
119	*Rhyticeros plicatus*							5	3	7	2	√	3	1
120	*Pitta sordida*					3		—	—	—	—	√	2	—
	P. superba							—	—	—	—		4	—
121	*P. anerythra*	—	—	—	—	—	—						—	4
122	*P. erythrogaster*					1	1	—	—	—	—	√	2	—
123	*Hirundo tahitica*	1		1		4	3	5	5		1	√	1	1
124	*Coracina caledonica*	—	—	—	—	—	—					√	—	3
125	*C. lineata*							1	17			√	3	1
126	*C. papuensis*						1	2	51	1		√	2	1
127	*C. tenuirostris*					1	1	1	7		1	√	2	1
	C. salamonis	—	—	—	—	—	—						—	4
128	*C. holopolia*	—	—	—	—	—	—					√	—	3
129	*Lalage leucomela*						3	—	—	—	—	√	2	—
130	*L. leucopyga*	—	—	—	—	—	—					√	—	3
131	*Saxicola caprata*					1	1	—	—	—	—	√	2	—
132	*Zoothera heinei*											√	3	3
	Z. talaseae											√	3	3
	Z. margaretae	—	—	—	—	—	—					√	—	3
133	*Turdus poliocephalus*					1						√	2	3

continued

		Bismarcks						Solomons				Index		
		Seen	Islets	Vagrant	Colonized	CLRT	Volcanoes	Seen	Islets	Vagrant	Volcanoes	Ext	Bismarcks	Solomons
134	*Gerygone flavolateralis*	—	—	—	—	—	—					√	—	3
135	*Ortygocichla rubiginosa*							—	—	—	—		4	—
136	*Cichlornis whitneyi*	—	—	—	—	—	—					√	—	3
	C. llaneae	—	—	—	—	—	—						—	4
	C. grosvenori							—	—	—	—		4	—
137	*Cettia parens*	—	—	—	—	—	—						—	4
138	*Acrocephalus stentoreus*					1						√	2	3
139	*Cisticola exilis*					1	3	—	—	—	—	√	2	—
140	*Megalurus timoriensis*					1	1	—	—	—	—	√	2	—
141	*Phylloscopus poliocephala*											√	3	3
	P. makirensis	—	—	—	—	—	—						—	4
142	*P. amoenus*	—	—	—	—	—	—						—	4
143	*Rhipidura leucophrys*		14			2	5	8	64	1	1	√	1	1
144	*R. rufiventris*		1				2	—	—	—	—	√	2	—
	R. cockerelli	—	—	—	—	—	—					√	—	3
145	*R. drownei*	—	—	—	—	—	—						—	4
	R. rennelliana	—	—	—	—	—	—						—	4
	R. tenebrosa	—	—	—	—	—	—						—	4
146	*R. fuliginosa*	—	—	—	—	—	—					√	—	3
147	*R. rufifrons*		1						6		1	√	2	1
148	*R. dahli*							—	—	—	—	√	3	—
	R. matthiae						·	—	—	—	—	√	4	—
	R. malaitae	—	—	—	—	—	—						—	4
149	*Clytorhynchus hamlini*	—	—	—	—	—	—						—	4
150	*Monarcha cinerascens*		10			3	4				1	√	1	2
151	*M. erythrostictus*	—	—	—	—	—	—		5			√	—	1
	M. castaneiventris	—	—	—	—	—	—		3		1	√	—	1
	M. richardsii	—	—	—	—	—	—		40	1	1	√	—	1
152	*M. infelix*							—	—	—	—		3	—
	M. menckei							—	—	—	—	√	4	—
	M. verticalis						1	—	—	—	—	√	2	—
	M. barbatus	—	—	—	—	—	—		1			√	—	2
	M. browni	—	—	—	—	—	—					√	—	3
	M. viduus	—	—	—	—	—	—					√	—	3
153	*M. chrysomela*							—	—	—	—	√	3	—
154	*Myiagra alecto*		1				4	—	—	—	—	√	2	—
155	*M. hebetior*						1	—	—	—	—	√	2	—

156	*M. ferrocyanea*	—	—	—	—	—	—	1	15			√	—	1
	M. cervinicauda	—	—	—	—	—	—					√	—	3
	M. caledonica	—	—	—	—	—	—					√	—	3
157	*Monachella muelleriana*							—	—	—	—	√	3	—
158	*Petroica multicolor*	—	—	—	—	—	—					√	—	3
159	*Pachycephala pectoralis*					1						√	2	3
160	*P. melanura*		8			4	2		3			√	1	1
161	*P. implicata*	—	—	—	—	—	—						—	4
162	*Dicaeum eximium*						1	—	—	—	—	√	2	—
	D. aeneum	—	—	—	—	—	—		1			√	—	2
	D. tristrami	—	—	—	—	—	—						—	4
163	*Nectarinia sericea*		5				4	—	—	—	—	√	1	—
164	*N. jugularis*		1	r	1	2	3	4	78		3	√	1	1
165	*Zosterops atrifrons*						1	—	—	—	—	√	2	—
166	*Z. metcalfii*	—	—	—	—	—	—						—	4
167	*Z. griseotinctus*					3						√	2	3*
	Z. vellalavella	—	—	—	—	—	—						—	4
	Z. splendidus	—	—	—	—	—	—						—	4
	Z. luteirostris	—	—	—	—	—	—						—	4
	Z. rendovae	—	—	—	—	—	—	1	1	2		√	—	1
	Z. rennellianus	—	—	—	—	—	—						—	4
168	*Z. murphyi*	—	—	—	—	—	—						—	4
169	*Z. ugiensis*	—	—	—	—	—	—					√	—	3
170	*Z. stresemanni*	—	—	—	—	—	—						—	4
171	*Woodfordia superciliosa*	—	—	—	—	—	—						—	4
172	*Myzomela cineracea*							—	—	—	—		4	—
173	*M. cruentata*							—	—	—	—	√	3	—
174	*M. pulchella*							—	—	—	—		4	—
175	*M. cardinalis*	—	—	—	—	—	—					√	—	3
	M. erythromelas							—	—	—	—		4	—
176	*M. sclateri*		4			3	3	—	—	—	—	√	1	—
177	*M. pammelaena*		12			3		—	—	—	—	√	1	—
	M. lafargei	—	—	—	—	—	—						—	4
	M. melanocephala	—	—	—	—	—	—				1		—	2
	M. eichhorni	—	—	—	—	—	—	1	11	1		√	—	1
	M. malaitae	—	—	—	—	—	—						—	4
	M. tristrami	—	—	—	—	—	—						—	4
178	*Philemon cockerelli*							—	—	—	—		4	—
	P. eichhorni							—	—	—	—		4	—
	P. albitorques							—	—	—	—		4	—
179	*Melidectes whitemanensis*							—	—	—	—		4	—

continued

		Bismarcks						Solomons				Index		
		Seen	Islets	Vagrant	Colonized	CLRT	Volcanoes	Seen	Islets	Vagrant	Volcanoes	Ext	Bismarcks	Solomons
180	*Stresemannia bougainvillei*	—	—	—	—	—	—						—	4
181	*Meliarchus sclateri*	—	—	—	—	—	—						—	4
182	*Guadalcanaria inexpectata*	—	—	—	—	—	—						—	4
183	*Erythrura trichroa*					3						√	2	3
184	*Lonchura spectabilis*	1				2	1	—	—	—	—	√	2	—
	L. forbesi							—	—	—	—		4	—
	L. hunsteini							—	—	—	—		4	—
185	*Lonchura melaena*											√	3	3
186	*Aplonis cantoroides*		2		1	3	4	10	46	3	2	√	1	1
187	*A. feadensis*	1	2									√	1	3*
	A. insularis	—	—	—	—	—	—						—	4
188	*A. grandis*	—	—	—	—	—	—	1	10	1	1	√	—	1
	A. dichroa	—	—	—	—	—	—					√	—	4
189	*A. metallica*	1	2			3	5	3	15	1	2	√	1	1
190	*A. brunneicapilla*	—	—	—	—	—	—					√	—	3
191	*Mino dumontii*								14	2	2	√	3	1
192	*Dicrurus hottentottus*		1									√		
	D. megarhynchus							—	—	—	—		4	—
193	*Artamus insignis*			1				—	—	—	—	√	2	—
194	*Corvus orru*		3				3	—	—	—	—	√	1	—
195	*C. woodfordi*	—	—	—	—	—	—						—	4

Appendix 7

Distributions and Origins of Northern Melanesian Bird Populations

7.1. Distributions and Origins of Endemic Subspecies and Nonendemic Populations

This appendix lists each species with Northern Melanesian populations belonging to an endemic subspecies confined to Northern Melanesia (abbreviated s, column 5 *Endemism in NMel*), or else not endemic (belonging to the same subspecies as an extralimital population: 0, column 5). Numbers before each species' name are that species' number in appendix 1. In this appendix table and also in the following appendices 7.2 and 7.3, taxa confined to Northern Melanesia except for occurring on just one or a few neighboring islands (e.g., *Accipiter albogularis* in appendix 7.2h, extending to the Santa Cruz group) are still considered endemics.

Species are grouped in appendices 7.1a through 7.1i according to their inferred route of proximate colonization of Northern Melanesia (from Australia to any part of Northern Melanesia, from New Guinea to the Solomons, etc.). For the species of Appendices 7.1f and 7.1g, it is not possible to decide whether they arrived from New Guinea, from Australia, or from both on a broad front. Appendix 7.1h ("Eastern origins") consists of species inferred to have arrived from the east, generally from the New Hebrides, or at least to be most closely related to New Hebridean taxa.

Columns 3 and 4 (*Resident in*) indicate whether the species occurs in the Bismarcks or Solomons, respectively. If the species occurs within an archipelago only on one island, that island is listed by the abbreviation used in appendix 1. + = on more than one island of an archipelago; ( ) = only marginally extending into that archipelago from the other archipelago; v = vagrant; m = migrant winter visitor from Australia; – = not present.

Columns 6–8 (*Species or allospecies in*) denote whether that species (if it does not break up into allospecies) or that allospecies occurs in New Guinea (NG), Australia (Au), or southwest Pacific archipelagoes east of Northern Melanesia (Z = Santa Cruz group, H = New Hebrides, C = New Caledonia and Loyalty Islands, F = Fiji).

Column 9 (*Proximate*) denotes the proximate source of colonists of Northern Melanesia, insofar as it can be inferred by the criteria discussed in the text (see chapter 13 for discussion). Superscript s = inference based on subspecific characters. Superscript r = inference based on distribution being permanent and widespread in one potential source region but local or transient in another potential source region (e.g., species widespread in New Guinea but confined to tropical Australia). Superscript g = species present both in New Guinea and Australia, but confined in Northern Melanesia to a Bismarck island lying north of New Guinea and shielded from Australian colonists by New Guinea.

Column 10, denoting the ultimate source of colonists of Northern Melanesia, indicates whether the ancestry of the Northern Melanesian population can be traced back ultimately to Asia (at the level of the subspecies itself = 0, allospecies = 1, superspecies or full species = 2, group of species = 3, genus = 4, group of genera = 5, tribe = 6, subfamily = 7, family = 8, parvorder = 9); whether, instead, the ancestry can be traced back within the Australian region up to a certain taxonomic level (same codes 2 . . . 9) and cannot be traced back farther outside the Australian region; or whether ultimate origins cannot be classified at all (no entry; see chapter 15 for discussion).

Taxon		Resident in		Endemism in NMel	Species or allospecies in			Origins	
		Bismarcks	Solomons		NG	Au	East	Proximate	Ultimate
Appendix 7.1a. Australia → Northern Melanesia									
2B	*Tachybaptus novaehollandiae*	–	Ren	s	+	+	H,C	Au^{r} or H	Asia(2)
3	*Pelecanus conspicillatus*	v	+,v	0	m	+	–	Au	Asia(4)
12	*Threskiornis moluccus*	–	Ren	s	+	+	–	Au^{r}	
13	*Plalatea regia*	–	Ren	0	m	+	–	Au	
17	*Anas gibberifrons*	–	Ren	s	+	+	H,C	Au^{r}	
23	*Accipiter fasciatus*	–	Ren	0	+	+	H,C	Au^{s}	Au(3)
92	*Scythrops novaehollandiae*	+,v	v	0	m	+	–	Au	Au(4)
95	*Tyto alba*	( )	+	s	+	+	Z,H,C,F	Au^{s}	
100	*Eurostopodus mystacalis*	–	+	s	–	+	C	Au^{r}	Au(3)
112	*Halcyon sancta*	m	m,+	0	m	+	C	Au	Au(3)
Appendix 7.1b. New Guinea → Solomons									
55	*Ptilinopus viridis*	(Lih)	+	s	+	–	–	NG	Au(5)
64	*Columba vitiensis*	(NI)	+	0	+	–	H,C,F	NG	Asia(4)
99	*Podargus ocellatus*	–	+	s	+	+	–	NG	Au(4)
Appendix 7.1c. New Guinea → Bismarcks									
1	*Casuarius bennetti*	NB	–	0	+	–	–	NG	Au(8)
15	*Dendrocygna guttata*	+	–	0	+	–	–	NG	
33	*Falco berigora*	Long	–	0	+	+	–	NG^{g}	Au(2)
35	*Coturnix chinensis*	+	–	s	+	+	–	NG^{s}	Asia(2)
39	*Rallina tricolor*	+	–	s	+	+	–	NG^{r}	Au(2)
41	*Gymnocrex plumbeiventris*	NI?	–	0	+	–	–	NG	Au(4)
48	*Charadrius dubius*	+	–	0	+	–	–	NG	Asia(1)
54A	*Ptilinopus rivoli*	+	–	s	+	–	–	NG	Au(5)
62	*Ducula spilorrhoa*	+	–	s	+	+	–	NG^{r}	Au(5)
63A	*Gymnophaps albertisii*	+	–	0	+	–	–	NG	Au(4)
66	*Macropygia amboinensis*	+	–	s	+	–	–	NG	A u(5)
67A	*Macropygia nigrirostris*	+	–	s	+	–	–	NG	Au(5)
83A	*Micropsitta pusio*	+	–	0	+	–	–	NG	Au(7)
96B	*Tyto novaehollandiae*	Man	–	s	+	+	–	NG^{g}	Au(2)
101	*Caprimulgus macrurus*	+	–	s	+	+	–	NG^{r}	Asia(1)

continued

Taxon		Resident in: Bismarcks	Resident in: Solomons	Endemism in NMel	Species or allospecies in: NG	Species or allospecies in: Au	Species or allospecies in: East	Origins: Proximate	Origins: Ultimate
Appendix 7.1c. New Guinea → Bismarcks (*continued*)									
117	*Merops philippinus*	+	–	0	+	–	–	NG	Asia(1)
120B	*Pitta sordida*	+	–	0	+	–	–	NG	
122	*Pitta erythrogaster*	+	–	s	+	+	–	NG^{s,r}	
129	*Lalage leucomela*		–	s	+	+	–	NG^{s}	Au(5)
131	*Saxicola caprata*	+	–	0	+	–	–	NG	Asia(1)
139	*Cisticola exilis*	+	–	s	+	+	–	NG^{s}	Asia(1)
140	*Megalurus timoriensis*	+	–	s	+	+	–	NG^{s}	Asia(8)
144A	*Rhipidura rufiventris*	+	–	s	+	+	–	NG^{s}	Au(6)
153	*Monarcha chrysomela*	+	–	s	+	–	–	NG	Au(5)
154	*Myiagra alecto*	+	–	0	+	+	–	NG^{s}	Au(5)
157	*Monachella muelleriana*	NB	–	s	+	–	–	NG	Au(9)
163	*Nectarinia sericea*	+	–	s	+	–	–	NG	Asia(3)
165	*Zosterops atrifrons*	+	–	s	+	–	–	NG	Au(2)
173	*Myzomela cruentata*	+	–	s	+	–	–	NG	Au(9)
194	*Corvus orru*	+	–	s	+	+	–	NG^{s}	Asia(4)
Appendix 7.1d. New Guinea → Bismarcks, western Solomons									
2A	*Tachybaptus ruficollis*	+	Boug	0	+	–	–	NG	Asia(1)
10	*Ixobrychus sinensis*	NB	Boug	0	+	–	–	NG	Asia(1)
29A	*Haliaeetus leucogaster*	+	Nis	0	+	+	–	NG or Au	Asia(4)
80	*Charmosyna placentis*	+	+	0	+	–	–	NG	Au(7)
160	*Pachycephala melanura*	+	+	0	+	+	–	NG^{s}	Au(9)
Appendix 7.1e. New Guinea → Bismarcks, Solomons									
5	*Butorides striatus*	( )	+	s	+	+	Z,H,C,F	NG^{s}	Asia(1)
16	*Anas superciliosa*	+	+	0	+	+	Z,H,C,F	NG^{s}	Asia(2)
21	*Accipiter meyerianus*	+	+	0	+	–	–	NG	Asia(2)
22	*Accipiter novaehollandiae*	+	+	s	+	+	–	NG^{s}	Au(3)
30	*Pandion haliaetus*	+	+	0	+	+	C	NG^{s}	
31	*Falco peregrinus*	+	+	0	+	+	H,C,F	NG^{s}	Asia(0)
32	*Falco severus*	+	+	0	+	–	–	NG	Asia(1)
34	*Megapodius freycinet*	+	+	s	+	+	H,F	NG^{r}	Au(8)

36	*Turnix maculosa*	+	Guad	s	+	+	–	NG^{s}	Au(1)
37Aa–f	*Gallirallus philippensis*	+	+	0,s	+	+	H,C,F	NG^{s}	Au(3)
42	*Porzana tabuensis*	+	+	0	+	+	Z,H,C,F	NG^{s}	Asia(5)
43	*Poliolimnas cinereus*	+	+	0,s	+	+	H,C,F	NG^{r}	Asia(5)
44	*Amaurornis olivaceus*	+	+	s	+	+	–	NG^{s}	Au(1)
69	*Chalcophaps stephani*	+	+	0,s	+	–	–	NG	Au(5)
71	*Gallicolumba beccarii*	+	+	s	+	–	–	NG	Au(4)
73	*Gallicolumba jobiensis*	+	+	0,s	+	–	–	NG	Au(4)
75	*Caloenas nicobarica*	+	+	0	+	–	–	NG	
77	*Trichoglossus haematodus*	+	+	s	+	+	Z,H,C	NG^{s}	Au(7)
82	*Micropsitta bruijnii*	+	+	s	+	–	–	NG	Au(7)
86	*Eclectus roratus*	+	+	s	+	+	–	NG^{r}	Au(5)
89	*Cacomantis variolosus*	+	+	s	+	+	–	NG^{s}	
91	*Eudynamys scolopacea*	+	+	s	+	+	–	NG^{s}	Au(4)
103	*Aerodramus vanikorensis*	+	+	s	+	v	Z,H	NG	Au(2)
105	*Collocalia esculenta*	+	+	s	+	+	Z,H,C	NG^{r}	
106	*Hemiprocne mystacea*	+	+	s	+	–	–	NG	Au(2)
107	*Alcedo atthis*	+	+	0,s	+	+	–	NG^{s}	Asia(1)
109	*Alcedo pusilla*	+	+	s	+	+	–	NG^{r}	Au(2)
110	*Ceyx lepidus*	+	+	s	+	–	–	NG	Au(2)
118	*Eurystomus orientalis*	+	+	s	+	+	–	NG^{s}	Asia(1)
119	*Rhyticeros plicatus*	+	+	s	+	–	–	NG	Asia(4)
125	*Coracina lineata*	+	+	s	+	+	–	NG^{s}	Au(5)
126	*Coracina papuensis*	+	+	s	+	+	–	NG^{s}	Au(5)
127A	*Coracina tenuirostris*	+	+	0,s	+	+	–	NG^{s}	Au(5)
133	*Turdus poliocephalus*	+	+	s	+	–	Z,H,C,F	NG	Asia(3)
138	*Acrocephalus stentoreus*	+	+	0	+	–	–	NG^{r}	Asia(1)
141A	*Phylloscopus poliocephalus*	+	+	s	+	–	–	NG	Asia(3)
143	*Rhipidura leucophrys*	+	+	0	+	+	–	NG^{s}	Au(6)
164	*Nectarinea jugularis*	+	+	s	+	+	–	NG^{r}	Asia(3)
183	*Erythrura trichroa*	+	+	0,s	+	+	H,C	NG^{r}	
186	*Aplonis cantoroides*	+	+	0	+	–	–	NG	Asia(6)
189	*Aplonis metallica*	+	+	0,s	+	+	–	NG^{r}	Asia(6)
191	*Mino dumontii*	+	+	s	+	–	–	NG	Asia(6)
192A	*Dicrurus hottentottus*	+	+	s	+	+	–	NG^{s}	Asia(4)

Appendix 7.1f. New Guinea or Australia → Bismarcks

14	*Dendrocygna arcuata*	+	–	s	+	+	C,F	NG or Au	
47	*Irediparra gallinacea*	NB	–	0	+	+	–	NG or Au	Au(4)

continued

Taxon		Resident in		Endemism in NMel	Species or allospecies in			Origins	
		Bismarcks	Solomons		NG	Au	East	Proximate	Ultimate
Appendix 7.1f. New Guinea or Australia → Bismarcks (*continued*)									
49	*Himantopus leucocephalus*	+	–	0	+	+	–	NG or Au	Asia(2)
84	*Cacatua galerita*	NB	–	s	+	+	–	NG or Au	Au(7)
116	*Tanysiptera sylvia*	+	–	s	+	+	–	NG or Au	Au(4)
Appendix 7.1g. New Guinea or Australi → Bismarcks, Solomons									
4	*Phalacrocorax melanoleucos*	NB	+	0,s	+	+	Z,C		Asia(3)
6	*Egretta sacra*	+	+	0	+	+	Z,H,C,F	Au^r, NG^r, or H	Asia(1)
7	*Ardea intermedia*	+	+	0	+	+	–		
8	*Ardea alba*	+	+	0	+	+	–		
9	*Nycticorax caledonicus*	+	+	0,s	+	+	C		
11	*Ixobrychus flavicollis*	+	+	0,s	+	+	–		Asia(1)
18	*Aviceda subcristata*	+	+	s	+	+	–		Asia(2)
20	*Haliastur indus*	+	+	0,s	+	+	–		
46	*Porphyrio porphyrio*	+	+	0	+	+	Z,H,C,F		Asia(1)
50	*Esacus magnirostris*	+	+	0	+	+	C		Asia(1)
51	*Ptilinopus superbus*	+	+	0	+	+	–		Au(5)
113Aa–i	*Halcyon chloris*	+	+	s	+	+	Z,H,F		Au(3)
132A	*Zoothera heinei*	+	Ch	s	+	+	–		Asia(2)
Appendix 7.1h. Eastern origins									
37Ag	*Gallirallus philippensis*	–	+	s	+	+	H,C,F	H^s	Au(3)
52A	*Ptilinopus greyi*	–	Gow	0	–	–	Z,H.C	H	Au(5)
56	*Ducula pacifica*	+	+	0,s	–	–	Z,H,C,F	H	Au(5)
90	*Chrysococcyx lucidus*	–	Ren	s	–	+	Z,H,C	H^s	Au(3)
113Aj–l	*Halcyon chloris*	–	+	s	+	+	Z,H,F	H^s	Au(3)
124	*Coracina caledonica*	–	+	s	–	–	H,C	H or Au	Au(5)
130	*Lalage leucopyga*	–	SCr	s	–	–	H,C	H	Au(5)
134	*Gerygone flavolateralis*	–	Ren	s	–	–	H,C	H	Au(9)
146	*Rhipidura fuliginosa*	–	SCr	0?	–	+	H,C	H^s	Au(6)
156C	*Myiagra caledonica*	–	Ren	s	–	–	H,C	H	Au(5)
158	*Petroica multicolor*	–	+	s	–	+	H,F	H^s	Au(9)
159Ao	*Pachycephala pectoralis*	–	+	s	(+)	+	Z,H,F	H^s	Au(9)
175A	*Myzomela cardinalis*	–	SCr,Ren	s	–	–	Z,H,C	H	Au(9)

Appendix 7.1i. Interpretation complex or difficult

78C	*Lorius hypoinochrous*	+	–	0	+	–	–	NG?	Au(7)
104	*Aerodramus spodiopygius*	+	+	s	–	+	Z,H,C,F		Au(2)
114	*Halcyon saurophaga*	+	+	0,s	+	–	–	NG?	Au(3)
123	*Hirundo tahitica*	+	+	0,s	+	+	Z,H,C,F	NG(+H?)	Asia(1)
147	*Rhipidura rufifrons*	Man	+	s	+	+	Z		Au(6)
150	*Monarcha cinerascens*	+	+	s	–	–	–		Au(5)
159Aa–n,p	*Pachycephala pectoralis*	+	+	s	–	+	Z,H,F	?,+H	Au(9)
167A	*Zosterops griseotinctus*	+	–	s	(+)	–	–		Au(2)
184A	*Lonchura spectabilis*	+	–	s	+	–	–	NG?	Au(3)

7.2. Distributions and Origins of Endemic Allospecies

Entries are as for appendix 7.1, but for species with Northern Melanesian populations belonging to an endemic allospecies confined to Northern Melanesia (abbreviated a, column 5, *Endemism*). [0] in columns 3 or 4 (*Resident in*) notes the occurrence of a nonendemic allospecies. For instance, in Appendix 7.2b the endemic allospecies *Dicrurus megarhynchus* is confined to New Ireland, but the same superspecies is also represented both in the Bismarcks and Solomons by an allospecies extending outside Northern Melanesia. Columns 6–8 (*Superspecies*) denote occurrences of the superspecies (regardless of whether the same allospecies) in New Guinea, Australia, or southwest Pacific archipelagoes east of Northern Melanesia.

Taxon		Resident in: Bismarcks	Resident in: Solomons	Endemism in NMel	Superspecies: NG	Superspecies: Au	Superspecies: East	Origins: Proximate	Origins: Ultimate
Appendix 7.2a. Australia → Northern Melanesia									
85	*Cacatua ducorpsi*	–	+	a	–	+	–	Au	Au(7)
121	*Pitta anerythra*	–	+	a	–	+	–	Au	
Appendix 7.2b. New Guinea → Bismarcks									
19	*Henicopernis infuscata*	+	–	a	+	–	–	NG	Au(4)
27	*Accipiter princeps*	NB	–	a	+	–	–	NG	Au(2)
53	*Ptilinopus insolitus*	+	–	a	+	–	–	NG	Au(5)
58	*Ducula finschii*	+	–	a	+	–	–	NG	Au(5)
61	*Ducula melanochroa*	+	–	a	+	–	–	NG	Au(5)
70	*Henicophaps foersteri*	+	–	a	+	–	–	NG	Au(5)
88	*Loriculus tener*	+	–	a	+	–	–	NG	Au(2)
108	*Alcedo websteri*	+	–	a	+	+	–	NG[s]	Au(1)
120A	*Pitta superba*	Man	–	a	+	–	–	NG	
172	*Myzomela cineracea*	+	–	a	+	+	–	NG[s]	Au(9)
178	*Philemon [moluccensis]*	+	–	a	+	+	–	NG[r]	Au(9)
184B,C	*Lonchura [spectabilis]*	+,[0]	–	a	+	–	–	NG	Au(3)
192B	*Dicrurus megarhynchus*	NI,[0]	[0]	a	+	+	–	NG[s]	Asia(4)
193	*Artamus insignis*	+	–	a	+	–	–	NG	Au(9)
Appendix 7.2c. New Guinea → Bismarcks, Solomons									
54B	*Ptilinopus solomonensis*	+,[0]	+	a	+	–	–	NG	Au(5)
63B	*Gymnophaps solomonensis*	[0]	+	a	+	–	–	NG	A u(4)
67B	*Macropygia mackinlayi*	+,[0]	+	a	+	–	Z,H	NG	Au(5)
68	*Reinwardtoena [reinwardtii]*	+	+	a	+	–	–	NG	Au(5)
78A,B	*Lorius [lory]*	NI,[0]	+	a	+	–	–	NG	Au(7)
83B,C	*Micropsitta [pusio]*	+,[0]	+	a	+	–	–	NG	Au(7)
87	*Geoffroyus heteroclitus*	+	+	a	+	(+)	–	NG[r]	Au(5)
97	*Ninox [novaeseelandiae]*	+	+	a	+	+	–	NG[s]	Au(4)
141B	*Phylloscopus makirensis*	[0]	SCr	a	+	–	–	NG	Asia(3)
148	*Rhipidura [rufidorsa]*	+	+	a	+	–	–	NG	Au(6)
152	*Monarcha [manadensis]*	+	+	a	+	–	–	NG	Au(5)
162	*Dicaeum [erythrothorax]*	+	+	a	+	–	–	NG	Asia(4)

continued

Taxon		Resident in		Endemism in NMel	Superspecies			Origins	
		Bismarcks	Solomons		NG	Au	East	Proximate	Ultimate
Appendix 7.2d. New Guinea or Australia → Bismarcks									
28	*Accipiter brachyurus*	NB	–	a	+	+	–	NG or Au	Au(2)
96A	*Tyto aurantia*	NB,[0]	–	a	+	+	–	NG or Au	Au(2)
Appendix 7.2e. New Guinea or Australia → Bismarcks, Solomons									
29B	*Haliaeetus sanfordi*	[0]	+	a	+	+	–	NG or Au	Asia(4)
111	*Halcyon [diops]*	NB	+	a	+	+	H	NG or Au	Au(2)
127B	*Coracina salamonis*	[0]	SCr,[0]	a	+	+	–	NG or Au	Au(5)
132B,C	*Zoothera [dauma]*	+,[0]	+,[0]	a	+	+	–	NG or Au	Asia(2)
144B	*Rhipidura cockerelli*	[0]	+	a	+	+	–	NG or Au	Au(6)
185	*Lonchura melaena*	NB	Buka	a	+	+	–	NG or Au	Au(3)
Appendix 7.2f. New Guinea or Australia → Solomons									
151	*Monarcha [melanopsis]*	–	+	–	+	+	–	NG or Au	Au(5)
Appendix 7.2g. Eastern origins or affinities									
40	*Nesoclopeus woodfordi*	–	+	a	–	–	F	East?	Au(4)
52B	*Ptilinopus richardsi*	–	+	a	+	+	Z,H,C,F	East?	Au(5)
60	*Ducula brenchleyi*	–	+	a	–	–	H,C,F	East	Au(5)
145	*Rhipidura [spilodera]*	–	+	a	–	–	H,C,F	East?	Au(6)
149	*Clytorhynchus hamlini*	–	Ren	a	–	–	Z,F	East	Au(5)
171	*Woodfordia superciliosa*	–	Ren	a	–	–	Z	East?	Au(4)
Appendix 7.2h. Interpretation complex or difficult									
25	*Accipiter albogularis*	(Feni)	+	a	+	–	Z,C,F	NG or E	Au(3)
37B	*Gallirallus rovianae*	[0]	+,[0]	a	+	+	H,C,F	Ng, Au, or E	Au(3)
57	*Ducula rubricera*	+	+	a	(+)	–	–		Au(5)
59	*Ducula pistrinaria*	+	+	a	(+)	–	–		Au(5)
79	*Charmosyna [palmarum]*	+	+	a	+	–	Z,H,C,F	NG or E	Au(7)
93	*Centropus [ateralbus]*	+	+	a	–	–	–		Au(2)
156A,B	*Myiagra [rubecula]*	–	+	a	+	+	Z,H,C,F	Ng, Au, or E	Au(5)
167B–F	*Zosterops [griseotinctus]*	[0]	+	a	(+)	–	–		Au(2)
175B	*Myzomela erythromelas*	NB	–	a	+	+	Z,H,C		Au(9)

7.3. Distributions and Origins of Endemic Species and Genera

Entries are as for appendix 7.1, but for species endemic to Northern Melanesia at the level of full species or superspecies (S, column 5, Endemism) or genus (G). Column 6 (*Nearest relatives*) gives the geographic range of the most closely related species (if known), according to the abbreviations used in appendix 1. Proximate origins, if they can be inferred, are classified as either New Guinea (NG), New Guinea or other islands to the west ("west"), or islands to the east ("east").

Taxon		Resident in		Endemism in NMel	Nearest relatives	Origins	
		Bismarcks	Solomons			Proximate	Ultimate
24	*Accipiter imitator*	–	+	S	LS, NG, Au to NHeb, NCal, Fiji	?	Au(3)
26	*Accipiter luteoschistaceus*	+	–	S	LS, NG, Au to NHeb, NCal, Fiji	?	Au(3)
38	*Gallirallus insignis*	NB	–	S	Wallacea, Okinawa	West	Au(3)
45	*Pareudiastes sylvestris*	–	SCr	S	Samoa	?	
65	*Columba pallidiceps*	+	+	S	Japan, Phil to Samoa	?	Asia(4)
72	*Gallicolumba salamonis*	–	+	S	LS to Marq, Micronesia	?	Au(4)
74	*Microgoura meeki*	–	Ch	G	?	?	
76	*Chalcopsitta cardinalis*	(+)	+	S	NG	NG	Au(7)
81	*Charmosyna margaretae*	–	+	S	NG	NG	Au(7)
94	*Centropus violaceus*	+	–	S	?	?	
98	*Nesasio solomonensis*	–	+	G	?	?	
102	*Aerodramus orientalis*	NI	+	S	NG?	NG?	
115	*Halcyon bougainvillei*	–	+	S	GS, Phil, Cel, NG	West	
128	*Coracina holopolia*	–	+	S	?	?	Au(5)
135	*Ortygocichla rubiginosa*	NB	–	S	Fiji?	East?	Au(5)
136	*Cichlornis [whitneyi]*	NB	+	G	LS to Fiji	?	Au(5)
137	*Cettia parens*	–	SCr	S	Fiji?	East?	Asia(4)
142	*Phylloscopus amoenus*	–	Kul	S	GS to Sol	West	Asia(3)
155	*Myiagra hebetior*	+	–	S	Mol, NG, Au, Bis	NG?	Au(5)
161	*Pachycephala implicata*	–	+	S	LS to Samoa	?	Au(9)
166	*Zosterops metcalfii*	–	+	S	?	?	
168	*Zosterops murphyi*	–	Kul	S	Sol?	?	
169	*Zosterops ugiensis*	–	+	S	?	?	
170	*Zosterops stresemanni*	+	Mal	S	?	?	
174	*Myzomela pulchella*	NI	–	S	?	?	Au(9)
176	*Myzomela sclateri*	+	–	S	?	?	Au(9)
177	*Myzomela [lafargei]*	+	+	S	?	?	Au(9)
179	*Melidectes whitemanensis*	NB	–	S	NG	NG	Au(9)
180	*Stresemannia bougainvillei*	–	Boug	G	?	?	Au(9)
181	*Meliarchus sclateri*	–	SCr	G	?	?	Au(9)
182	*Guadalcanaria inexpectata*	–	Guad	G	?	?	Au(9)
187	*Aplonis feadensis*	+	+	S	NG to Sol	NG	Asia(6)
188	*Aplonis grandis*	–	+	S	?	?	Asia(6)
190	*Aplonis brunneicapilla*	–	+	S	Mol to Sol	NG	Asia(6)
195	*Corvus woodfordi*	–	+	S	?	?	Asia(4)

References

Abbott, L.D., E.I. Silver, P.R. Thompson, M.V. Filewicz, C. Schneider & Abdoerrias. 1994. Stratigraphic constraints on the development and timing of arc-continent collision in northern Papua New Guinea. *J. Sedimentary Research* B64: 169–183.

Allen, J. & R. C. Green. 1972. Mendana 1595 and the fate of the lost "Almiranta": an archaeological investigation. *J. Pac. Hist.* 7: 73–91.

Allison, A. 1982. Distribution and ecology of New Guinea lizards. In: J.L. Gressitt (ed.), *Biogeography and Ecology of New Guinea*, pp. 803–814. Junk, The Hague.

Allison, A. 1996. Zoogeography of amphibians and reptiles of New Guinea and the Southwestern Pacific. In: A. Keast & S.E. Miller (eds.), *The Origin and Evolution of Pacific Island Biotas, New Guinea to Eastern Polynesia: Patterns and Processes*, pp. 407–436. SPB Academic Publishing, Amsterdam.

Amadon, D. 1942a. Birds collected during the Whitney South Sea Expedition. 49. Notes on some non-passerine genera, 1. *Amer. Mus. Novit.* 1175.

Amadon, D. 1942b. Birds collected during the Whitney South Sea Expedition. 50. Notes on some non-passerine genera, 2. *Amer. Mus. Novit.* 1176.

Amadon, D. 1943a. Birds collected during the Whitney South Sea Expedition. 52. Notes on some non-passerine genera, 3. *Amer. Mus. Novit.* 1237.

Amadon, D. 1943b. The genera of starlings and their relationships. *Amer. Mus. Novit.* 1247.

Amadon, D. 1950. The Hawaiian Honeycreepers (Drepaniidae). *Bull. Amer. Mus. Nat. Hist.* 95: 157–257.

Amadon, D. 1956. Remarks on the starlings, family Sturnidae. *Amer. Mus. Novit.* 1803.

Amadon, D. & J. Bull. 1988. Hawks and owls of the world: an annotated list of species. *Proc. Western Foundation Vert. Zool.* 3: 295–357.

Amadon, D. & L.L., Short. 1992. Taxonomy of lower categories: suggested guidelines. *Bull. Brit. Orn. Club*, Suppl. 112A: 11–38.

Amherst, Lord & B. Thomson. 1901. *The Discovery of the Solomon Islands by Alvaro de Mendaña in 1568*. Hakluyt Society, London.

Atkinson, I.A.E. 1985. The spread of commensal species of *Rattus* to oceanic islands and their effects on island avifaunas. In: P.J. Moors (ed.), *Conservation of Island Birds*, pp. 35–81. International Council for Bird Preservation, Cambridge.

Auber, L. 1938. Die Rassen- und Artenkreise des Genus *Eos* Wagler (Aves). *Festschr. Embrik Strand* 4: 673–782.

Bain, J.H.C., H.L. Davies, P.D. Hohnen, R.J. Ryburn, I.E. Smith, R. Grainger, R.J. Tingey & M.R. Moffat. 1972. *Geology of Papua New Guinea*. 1:1,000,000 map. Bureau of Mineral Resources, Canberra.

Baker, R. H. 1948. Report on collections of birds made by United States Naval Medical Research Unit No. 2 in the Pacific war area. *Smithson. Misc. Coll.* 107: 1–74.

Baker, R.H. 1951. The avifauna of Micronesia, its origin, evolution, and distribution. *Univ. Kansas Publ. Mus. Nat. Hist.* 3: 1–359.

Balgooy, M.M.J. van. 1960. Preliminary plant-geographical analysis of the Pacific. *Blumea 10*: 385–430.

Balgooy, M.M.J. van. 1971. Plant-geography of the Pacific. *Blumea* 6 (suppl.): 1–222.

Balgooy, M.M.J. van., P.H. Hovenkamp & P.C. van Welzen. 1996. Phytogeography of the Pacific: floristic and historical distribution patterns in plants. In: A. Keast & S.E. Miller (eds.), *The Origin and Evolution of Pacific Island Biotas, New Guinea to Eastern Polynesia: Patterns and Processes*, pp. 191–214. SPB Academic Publishing, Amsterdam.

Balouet, J.C. & S.L. Olson. 1987. A new extinct species of giant pigeon (Columbidae: *Ducula*) from archaeological deposits on Wallis (Uvea) Island, South Pacific. *Proc. Biol. Soc. Wash.* 100: 769–775.

Bayliss-Smith, T. P. 1972. The birds of Ontong Java and Sikaiana, Solomon Islands. *Bull. Brit. Orn. Club* 92: 1–10.

Bayliss-Smith, T. P. 1973. A recent immigrant to Ontong Java atoll, Solomon Islands. *Bull. Brit. Orn. Club* 93: 52–53.

Beaglehole, J. C. 1966. *The Exploration of the Pacific*, 3rd ed. Stanford University Press, Stanford, CA.

Beecher, W. J. 1945. A bird collection from the Solomon Islands. *Fieldiana* (*Zool.*) 31: 31–37.

Beehler, B. 1978. Notes on the mountain birds of New Ireland. *Emu* 78: 65–70.

Beehler, B., T. Pratt & D. Zimmerman. 1986. *Birds of New Guinea*. Princeton University Press, Princeton, NJ.

Bell, H. L. 1970. Ninigo and Hermit Islands. *New Guinea Bird Soc. Newsletter* no. 57: 2–3.

Bell, H. L. 1975. Avifauna of the Ninigo and Hermit Islands, New Guinea. *Emu* 75: 77–84.

Bell, H.L. 1986. Occupation of urban habitats by birds in New Guinea. *Proc. West. Found. Vert. Zool.* 3: 1–48.

Bellwood, P. 1985. *Prehistory of the Indo-Malaysian Archipelago*. Academic Press, Sydney.

Bellwood, P. 1987. *The Polynesians*, revised ed. Thames and Hudson, London.

Bellwood, P., J. J. Fox & D. Tryon, eds. 1995. *The Austronesians*. Australian National University, Canberra.

Bennett, J. A. 1987. *Wealth of the Solomons*. University of Hawaii Press, Honolulu.

Bishop, K. D. 1983. Some notes on non-passerine birds of west New Britain. *Emu* 83: 235–241.

Bishop, K.D. & J.D. Jones. 2001. The montane avifauna of the Nakanai Mountains, West New Britain (PNG). *Emu*, in press.

Blaber, S.J.M. 1990. A checklist and notes on the current status of the birds of New Georgia, Western Province, Solomon Islands. *Emu* 90: 205–214.

Blake, D.H. & Y. Miezitis. 1967. Geology of Bougainville and Buka Islands, New Guinea. *Bureau of Mineral Resources, Geology and Geophysics Bulletin*, no. 93.

Bluff, G. & T. Skyrme. 1984. Birds of the Jaba delta. *Papua New Guinea Bird Soc. Newsletter* no. 211: 3–6.

Bock, W. J. 1956. A generic review of the family Ardeidae (Aves). *Amer. Mus. Novit.* 1779.

Bock, W. J. 1986. Species concepts, speciation, and macroevolution. In: K. Iwatsuki, P. H. Raven & W. J. Bock, *Modern Aspects of Species*, pp. 31–57. University of Tokyo Press, Tokyo.

Bock, W. J. 1995. The species concept versus the species taxon: their roles in biodiversity analyses and conservation. In: R. Arai, M. Kato & Y. Doi (eds.), *Biodiversity and Evolution*, pp. 47–72. National Science Museum Foundation, Tokyo.

Bogert, C. 1937. Birds collected during the Whitney South Sea Expedition. 34. The distribution and the migration of the Long-tailed Cuckoo (*Urodynamis taitensis* Sparrman). *Amer. Mus. Novit.* no. 933.

Bonaparte, C.L.J.L. 1850. *Conspectus Generum Avium*, vol. 1. Brill, Lugduni Batavorum.

Bonaccorso, F. J. 1998. *Bats of Papua New Guinea*. Conservation International, Washington, DC.

Bougainville, L.A. de. 1771. *Voyage autour du monde par la fregate du Roi La Boudeuse et la flute L'Étoile en 1766–69*. Le Breton, Paris.

Bradley, D. 1957. Birds of the Solomon Islands. *Ibis* 99: 352–353.

Bradley, D. 1962. Additional records of birds from Rennell and Bellona Islands. *Nat. Hist. Ren. Isl., Br. Solomon Isl.* 4: 11–12.

Bradley, D. & T. Wolff. 1956. The birds of Rennell Island. *Nat. Hist. Ren. Isl., Br. Solomon Isl.* 1: 85–120.

Braestrup, F.W. 1956. The significance of the strong "oceanic" affinities of the vertebrate fauna on Rennell Island. *Nat. Hist. Ren. Isl., Br. Solomon Isl.* 1: 135–148.

Brenchley, J. L. 1873. *Jottings During the Cruise of H.M.S. Curaçoa Among the South Sea Islands in 1865*. Longmans, Green, and Co., London.

Briffa, K.R., P. D. Jones, F.H. Schweingruber, and T.J. Osborn (1998). Influence of volcanic eruptions on Northern Hemisphere summer temperature over the past 600 years. *Nature* 393: 450–455.

Brooke, R.K. 1970. Taxonomic and evolutionary notes on the subfamilies, tribes, genera, and subgenera of the swifts (Aves: Apodidae). *Durban Mus. Novit.* 9: 13–24.

Brooke, R.K. 1972. Generic limits in Old World Apodidae and Hirundinidae. *Bull. Brit. Orn. Club* 92: 53–57.

Brookfield, H.C. & D. Hart. 1966. *Rainfall in the Tropical Southwest Pacific*. Australian National University, Canberra.

Brown, D.M. & C.A. Toft, 1999. Molecular systematics and biogeography of cockatoos (Psittaciformes: Cacatuidae). *Auk* 116: 141–157.

Brown, L. & D. Amadon. 1968. *Eagles, Hawks, and Falcons of the World*. Country Life Books, Middlesex, UK.

Brown, W.C. 1952. The amphibians of the Solomon Islands. *Bull. Mus. Comp. Zool.* 107: 1–64.

Brown, W.C. 1991. Lizards of the genus *Emoia* (Scincidae) with observations on their evolution and biogeography. *Memoirs Cal. Acad. Sci.* no. 15.

Brown, W.C. 1997. Biogeography of amphibians in the islands of the Southwest Pacific. *Proc. Cal. Acad. Sci.* 50: 21–38.

Brown, W.C. & J.I. Menzies. 1978. A new *Platymantis* (Amphibia: Ranidae) from New Ireland, with notes on the amphibians of the Bismarck Archipelago. *Proc. Biol. Soc. Wash.* 91: 965–971.

Brown, W.C. & G.S. Myers. 1949. A new frog of the genus *Cornufer* from the Solomon Islands, with notes on the endemic nature of the Fijian frog fauna. *Amer. Mus. Novitates* no. 1418.

Brown, W.C. & F. Parker. 1970. New frogs of the genus *Batrachylodes* (Ranidae) from the Solomon Islands. *Mus. Comp. Zool. Breviora* no. 346.

Brown, W.C. & F. Parker. 1977. Lizards of the genus *Lepidodactylus* (Gekkonidae) from the Indo-Australian Archipelago and the islands of the Pacific, with descriptions of new species. *Proc. Cal. Acad. Sci.*, ser. 4, 41: 253–265.

Brown, W.C. & M.J. Tyler. 1968. Frogs of the genus *Platymantis* (Ranidae) from New Britain with descriptions of new species. *Proc. Biol. Soc. Washington* 81: 69–86.

Buckingham, D.L., G.C.L. Dutson, and J.L. Newman. 1996. Birds of Manus, Kolombangara and Makira (San Cristobal) with notes on mammals and records from other Solomon Islands (self-published report).

Bull, P. C. 1948. Field notes on waders in the south-west Pacific with special reference to the Russell Islands. *Emu* 47: 165–176.

Burrows, I. 1987. Some notes on the birds of Lihir. *Muruk* 2: 40–42.

Burton, J.A. 1973. *Owls of the World: Their Evolution, Structure and Ecology*. Peter Lowe, London.

Bush, G.L. 1975. Modes of animal speciation. *Ann. Rev. Ecol. Syst.* 6: 339–364.

Cabanis, J. & A. Reichenow. 1876. Uebersicht der auf der Expedition Sr. Maj. Schiff "Gazelle" gesammelten Vögel. *J. f. Orn.* 24: 319–330.

Cade, T.J. & R.D. Digby. 1982. *The Falcons of the World*. Collins, London.

Cain, A.J. 1954. Subdivisions of the genus *Ptilinopus*. *Bull. Brit. Mus. (Nat. Hist.) Zool.* 2: 267–284.

Cain, A.J. 1955. A revision of *Trichoglossus haematodus* and of the Australian Platycercine parrots. *Ibis.* 97: 432–479

Cain, A. J. & I. C. J. Galbraith. 1955. Five new subspecies from the mountains of Guadalcanal (British Solomon Islands). *Bull. Brit. Orn. Club* 75: 90–93.

Cain, A. J. & I. C. J. Galbraith. 1956. Field notes on birds of eastern Solomon Islands. *Ibis* 98: 100–134, 262–295.

Cain, A. J. & I. C. J. Galbraith. 1957. [Birds of the Solomon Islands]. *Ibis* 99: 128–130.

Carlquist, S. 1974. *Island Biology*. Columbia University Press, New York.

Case, T.J. 1996. Global patterns in the establishment and distribution of exotic birds. *Biol. Conservation* 78: 69–96.

Cavalli-Sforza, L. L., P. Menozzi, & A. Piazza. 1994. *The History and Geography of Human Genes*. Princeton University Press, Princeton, NJ.

Chantler, P. & G. Driessens (1995). *Swifts: A Guide to Swifts and Treeswifts of the World*. Pica, Sussex, U.K.

Chappell, J. & H. Polach. 1991. Post-glacial sea-level rise from a coral record at Huon Peninsula, Papua New Guinea. *Nature* 349: 147–149.

Chappell, J. & M.J. Shackleton. 1986. Oxygen isotopes and sea level. *Nature* 324: 137–140.

Chowning, A. 1982. Physical anthropology, linguistics, and ethnology. In: J. L. Gressitt (ed.), *Biogeography and Ecology of New Guinea*, pp. 131–168. Junk, The Hague.

Coates, A. 1970. *Western Pacific Islands*. Her Majesty's Stationery Office, London.

Coates, B. J. 1972. List of birds recorded on Witu and Bali Islands (Bismarck Sea). *New Guinea Bird Soc. Newsletter* no. 82: 3–4.

Coates, B.J. 1973a. [Straw-necked ibis *Threskiornis spinicollis*]. *New Guinea Bird Soc. Newsletter* no. 84: 2.

Coates, B. J. 1973b. Additions to the list of Anir (Feni Island, New Ireland District). *New Guinea Bird Soc. Newsletter* no. 87: 2–5.

Coates, B. J. 1974. Anir Island, New Ireland District. *New Guinea Bird Soc. Newsletter* no. 101: 2.

Coates, B. J. 1977. Observations on the birds of Tench Island, New Ireland Province. *New Guinea Bird Soc. Newsletter* no. 136: 8–10.

Coates, B. J. 1985. *The Birds of Papua New Guinea, vol. 1, Non-passerines*. Dove Publications, Alderley, Australia.

Coates, B.J. 1990. *The Birds of Papua New Guinea. Vol. 2. Passerines*. Dove Publications, Alderley, Australia.

Coates, B. J. & G. W. Swainson. 1978. Notes on the birds of Wuvulu Island. *New Guinea Bird Soc. Newsletter* no. 145: 8–10.

Cogger, H.G. 1972. A new scincid lizard of the genus *Tribolonotus* from Manus Island, New Guinea. *Zool. Meded.* 47: 202–210.

Coleman, P.J. 1966. The Solomon Islands as an island arc. *Nature* 211: 1249–1251.

Coleman, P.J. 1970. Geology of the Solomon and New Hebrides Islands, as part of the Melanesian re-entrant, Southwest Pacific. *Pacific Science* 24: 289–314.

Collar, N.J. & P. Andrew. 1988. *Birds to Watch*. International Council for Bird Preservation, Cambridge.

Collar, N.J., M.L. Crosby & A.J. Stattersfield. 1991. *Birds to Watch 2: The World List of Threatened Birds*. BirdLife International, Cambridge.

Comrie, B., S. Matthews & M. Polinsky. 1996. *The Atlas of Languages. Facts on File*, New York.

Condon, H.T. 1975. *Checklist of the Birds of Australia*. Part I. Non-passerines. Royal Australasian Ornithologists Union, Melbourne.

Condon, H. T. & D. Amadon. 1954. Taxonomic notes on Australian hawks. *Rec. S. R. Aust. Mus.* 11: 189–246.

Cooke, R. 1971. Notes on species observed at Nissan Island. *New Guinea Bird Soc. Newsletter* no. 69: 2.

Cooper, P. & B. Taylor. 1987. Seismotectonics of New Guinea: a model for arc reversal following arc-continent collision. *Tectonics* 6: 53–67.

Coyne, J. A. 1992. Genetics and speciation. *Nature* 355: 511–515.

Cracraft, J. 1984. The terminology of allopatric speciation. *Syst. Zool.* 33:115–116.

Crook, K.A.W. & L. Belbin. 1978. The Southwest Pacific area during the last 90 million years. *J. Geol. Soc. Australia* 25: 23–40.

Dahl, F. 1899. Das Leben der Vögel auf den Bismarckinseln. *Mitteil. Zool. Mus. Berlin* 1, no. 3: 108–222.

Dampier, W. 1939. *A Voyage to New Holland*. Argonaut Press, London.

Danis, V. 1937a. Étude d'une collection d'oiseaux de l'Ile Bougainville. *Bull. Mus. Hist. Nat., Paris* (2) 9: 119–123.

Danis, V. 1937b. Étude d'une nouvelle collection d'oiseaux de l'Ile Bougainville. *Bull. Mus. Hist. Nat, Paris* (2) 9: 362–365.

Danis, V. 1938. Étude d'une nouvelle collection d'oiseaux de l'Ile Bougainville. *Bull. Mus. Hist. Nat, Paris* (2) 10: 43–47.

Darlington, P.J., Jr. 1957. *Zoogeography*. Wiley, New York.

Darwin, C. 1859. *On the Origin of Species by Means of Natural Selection*. Murray, London.

Davidson, M. E. M. 1929. On a small collection of birds from Torres Strait islands, and from Guadalcanal Island, Solomon group. *Proc. Calif. Acad. Sci.* (4) 18: 245–260.

Davidson, M. E. M. 1934. The Templeton Crocker Expedition to western Polynesian and Melanesian islands, 1933. No. 16. Notes on the birds. *Proc. Calif. Acad. Sci.* (4) 21: 189–198.

Davies, H.L., E. Honza, D.L. Tiffin, J. Lock, Y. Okuda, J.B. Keene, F. Murakami, & K. Kisimoto. 1987. Regional setting and structure of the Western Solomon Sea. *Geo-Marine Letters* 7: 153–160.

Delacour, J. 1943. A revision of the subfamily Estrildinae of the family Ploceidae. *Zoologica* 28: 69–86.

Delacour, J. 1944. A revision of the family Nectariniidae (sunbirds). *Zoologica* 29: 17–38.

Diamond, J.M. 1966. Zoological classification system of a primitive people. *Science* 151: 1102–1104.

Diamond, J.M. 1970. Ecological consequences of island colonization by Southwest Pacific birds. I. Types of niche shifts. *Proc. Natl. Acad. Sci. USA* 67: 529–536.

Diamond, J. M. 1971a. Bird records for western New Britain. *Condor* 73: 481–483.

Diamond, J.M. 1971b. Comparison of faunal equilibrium turnover rates on a tropical island and a temperate island. *Proc. Natl. Acad. Sci.* USA 68: 2742–2745

Diamond, J. M. 1972a. Biogeographic kinetics: estimation of relaxation times for avifaunas of southwest Pacific islands. *Proc. Natl. Acad. Sci. USA* 69: 3199–3203.

Diamond, J.M. 1972b. *Avifauna of the Eastern Highlands of New Guinea.* Nuttall Ornithological Club, Cambridge, MA.

Diamond, J.M. 1974. Colonization of exploded volcanic islands by birds: the supertramp strategy. *Science* 183: 803–806.

Diamond, J. M. 1975a. Assembly of species communities. In: M. L. Cody and J. M. Diamond (eds.), *Ecology and Evolution of Species Communities*, pp. 342–444. Harvard University Press, Cambridge, MA.

Diamond, J. M. 1975b. Distributional ecology and habits of some Bougainville birds (Solomon Islands). *Condor* 77: 14–23.

Diamond, J. M. 1976. Preliminary results of an ornithological exploration of the islands of Vitiaz and Dampier Straits, Papua New Guinea. *Emu* 76: 1–7.

Diamond, J.M. 1977. Continental and insular speciation in Pacific land birds. *Syst. Zool.* 26: 263–268.

Diamond, J. M. 1980. Species turnover in island bird communities. *Proc. Int. Orn. Congr.* 17: 777–782.

Diamond, J. M. 1982a. Effect of species pool size on species occurrence frequencies: musical chairs on islands. *Proc. Natl. Acad. Sci. USA* 79: 2420–2424.

Diamond, J.M. 1982b. Mimicry of friarbirds by orioles. *Auk* 99: 187–196.

Diamond, J. M. 1983. Survival of bird populations stranded on land-bridge islands. *Natl. Geog. Soc. Res. Reports* 15: 127–141.

Diamond, J. M. 1984a. The avifaunas of Rennell and Bellona Islands. *Nat. Hist. Ren. Isl., Br. Solomon Isls.* 8: 127–168.

Diamond J. M. 1984b. Historic extinctions: a Rosetta Stone for understanding prehistoric extinctions. In: P. Martin & R. Klein (eds.), *Quaternary Extinctions*, pp. 824–862. University of Arizona Press, Tucson.

Diamond, J. M. 1984c. Biogeographic mosaics in the Pacific. In: F.J. Radovsky, P.H. Raven & S.H. Sohner (eds.), *Biogeography of the Tropical Pacific*, pp. 1–14. Association of Systematics Collections, Lawrence, Kansas.

Diamond, J. M. 1984d. Distributional patchiness in birds of tropical Pacific islands. *Natl. Geog. Soc. Res. Reports* 16: 319–327.

Diamond, J. M. 1984e. Distribution of New Zealand birds on real and virtual islands. *New Zealand J. Ecol.* 7: 37–55.

Diamond, J.M. 1984f. "Normal" extinctions of isolated populations. In: M.H. Nitecki (ed.), *Extinctions*, pp. 191–246. University of Chicago Press, Chicago.

Diamond, J. 1986a. Evolution of ecological segregation in the New Guinea montane avifauna. In: J. Diamond & T.J. Case (eds.), *Community Ecology*, pp. 98–125. Harper and Row, New York.

Diamond, J.M. 1986b. First record of the Large Grasshopper Warbler *Locustella fasciolata* from islands east of New Guinea. *Emu* 86: 249.

Diamond, J.M. 1987. Extant unless proven extinct? Or, extinct unless proven extant? *Conservation Biol.* 1: 77–79.

Diamond, J. M. 1989. A new subspecies of the Island Thrush *Turdus poliocephalus* from Tolokiwa Island in the Bismarck Archipelago. *Emu* 89: 58–60.

Diamond, J.M. 1991. A new species of rail from the Solomon Islands and convergent evolution of insular flightlessness. *Auk* 108: 461–470.

Diamond, J. 1998. Geographic variation in vocalizations of the white-eye superspecies *Zosterops* [*griseotinctus*] in the New Georgia group. *Emu* 98: 70–74.

Diamond, J.M. 2001a. New records and observations of birds from Northern Melanesia. *Pacific Science*, in press.

Diamond, J. M. 2001b. Contributions of convergent community assembly to homogeneity of biogeographic regions. *Proc. Natl. Acad. Sci. USA,* in press.

Diamond, J. M. & M. E. Gilpin. 1980. Contribution of turnover noise to variance in species number. *Am. Nat.* 115: 884–889.

Diamond, J. M. & M. E. Gilpin. 1981. Examination of the 'null' model of Connor and Simberloff for species co-occurrences on islands. *Oecologia* 52: 64–74.

Diamond, J.M. & M.E. Gilpin. 1983. Biogeographic umbilici and the evolution of the Philippine avifauna. *Oikos* 41: 307–321.

Diamond, J. M., M. E. Gilpin & E. Mayr. 1976. The species-distance relation for birds of the Solomon Archipelago, and the paradox of the great speciators. *Proc. Natl. Acad. Sci. USA* 73: 2160–2164.

Diamond, J.M. & M. LeCroy. 1979. Birds of Karkar and Bagabag Islands, New Guinea. *Bull. Amer. Mus. Nat. Hist.* 164: 467–531.

Diamond, J.M. & A.G. Marshall. 1976. Origin of the New Hebridean avifauna. *Emu* 76: 187–200.

Diamond, J. M. & A. G. Marshall. 1977a. Niche shifts in New Hebridean birds. *Emu* 77: 61–62.

Diamond, J. M. & A. G. Marshall. 1977b. Distributional ecology of New Hebridean birds: a species kaleidoscope. *J. Anim. Ecol.* 46: 703–727.

Diamond, J. M. & E. Mayr. 1976. Species-area relation for birds of the Solomon Archipelago. *Proc. Natl. Acad. Sci. USA* 73: 262–266.

Diamond, J.M., S.L. Pimm, M.E. Gilpin & M. LeCroy. 1989. Rapid evolution of character displacement in myzomelid honeyeaters. *Amer. Natur.* 134: 675–708.

Diamond, J.M. & C.R. Veitch. 1981. Extinctions and introductions in the New Zealand avifauna: cause and effect? *Science* 211: 499–501.

Donaghho, W. R. 1950. Observations on some birds of Guadalcanal and Tulagi. *Condor* 52: 127–132.

Dow, D.B. 1977. A geological synthesis of Papua New Guinea. *Bureau of Mineral Resources, Geology and Geophysics Bulletin* 201.

Dumont d'Urville, M. J. 1841–46. *Voyage au Pôle Sud et dans l'Océanie sur les Corvettes l'Astrolabe et la Zelée.* Gide, Paris.

Dunmore, J. 1965. *French Explorers in the Pacific.* Oxford University Press, Oxford.

duPont, J.E. & D.M. Niles. 1980. Redescription of *Halcyon bougainvillei excelsa* Mayr, 1941. *Bull. Br. Orn. Club* 100: 232–233.

Dutson, G. 1997. Interesting sightings from the Bismarcks. *Papua New Guinea Bird Soc. Newsletter* no. 291: 2–3.

Dutson, G. 2001. New distributional ranges for Melanesian birds. *Emu*, in press.

Dutson, G. and J. Newman. 1991. Observations on the Superb Pitta *Pitta superba* and other Manus endemics. *Bird Conservation Int.* 1: 215–222.

Eastwood, C. 1995a. Manus: a trip report. *Muruk* 7: 53–55.

Eastwood, C. 1995b. Interesting sightings during 1993 & 1994. *Muruk* 7: 128.

Eastwood, C. 1996. Kavieng, Djaul Island and Mussau Island, New Ireland: a trip report. *Muruk* 8: 28–32.

Eck, S. 1978. Die blaugrünen *Halcyon-Formen* der Talaut-Inseln (Aves, Coraciiformes, Alcedinidae). *Zool. Abh. Staatl. Mus. Tierk. Dresden* 35: 275–283.

Eck, S. & H. Busse. 1973. Eulen. Die rezenten und fossilen Formen. Aves, Strigidae. *Die Neue Brehm Büch.* 469: 1–196.

Ellis, D.H., C.M. Anderson & T. B. Roundy. 1981. *Falco kreyenborgi*: more pieces for the puzzle. *Raptor Research* 15 (2): 42–45.

Elton, C.S. 1958. *The Ecology of Invasion by Animals and Plants.* Chapman and Hall, London.

Etchécopar, R.-D. & F. Hüe. 1977. Les Artamidés. *L'Oiseau et Rev. fr. Orn.* 47: 381–410.

Faaborg, J. 1977. Metabolic rates, resources, and the occurrence of non-passerines in terrestrial avian communities. *Amer. Natur.* 111: 903–916.

Feare, C. & A. Craig. 1999. *Starlings and Mynas.* Princeton University Press, Princeton, NJ.

Filardi, C.E., C.E. Smith, A.W. Kratter, D.W. Steadman, & H.P. Webb. 1999. New behavioral, ecological, and biogeographic data on the avifauna of Rennell, Solomon Islands. *Pacific Science* 53: 319–340.

Filewood, L. W. C. 1969. New avifaunal sight-recordings for Bougainville. *Proc. Papua New Guinea Sci. Soc.* 21: 20–22.

Filewood, L. W. C. 1972. Notes on the birds of Bougainville Island. *Emu* 72: 32.

Finch, B. W. 1984. South-east New Ireland. *Papua New Guinea Bird Soc. Newsletter* no. 209: 9.

Finch, B.W. 1986. The *Aplonis* starlings of the Solomon Islands. *Muruk* 1: 3–16.

Finch, B. W. 1985. Noteworthy observations in Papua New Guinea and Solomons. *Papua New Guinea Bird Soc. Newsletter* no. 215: 6–12.

Finch, B.W. & J.L. McKean. 1987. Some notes on the birds of the Bismarcks. *Muruk* 2: 3–28.

Finsch, O. 1879. On a collection of birds made by Mr. Hübner on Duke-of-York Island and New Britain. *Proc. Zool. Soc. London* 9–17.

Finsch, O. 1881. Ornithological letters from the Pacific. VIII. New Britain. *Ibis* (4) 5: 532–540.

Finsch, O. 1884. Ueber Vögel der Südsee. I. Neu Britannien. *Mitt. Orn. Ver. Wien* 54–94.

Finsch, O. 1886. On two new species of birds from New Ireland. *Ibis* 5 (4): 1–2.

Finsch, O., and A. B. Meyer. 1886. Vögel von Neu Guinea. *Zeitschr. ges. Orn.* 3: 1–29.

Fisher, N.H. 1957. *Catalog of the Active Volcanoes of the World.* Part V. Melanesia. International Volcanological Association, Naples.

Fisher, N.H. & L.C. Noakes. 1942. Geological reports on New Britain. *Territory of New Guinea Geological Bulletin* no. 3.

Flannery, T.F. 1990. *Mammals of New Guinea.* Robert Brown, Carina, Queensland, Australia.

Flannery, T.F. 1992. New Pleistocene marsupials (Macropodidae, Diprotodontidae) from subalpine habitats in Irian Jaya, Indonesia. *Alcheringa* 16: 323–331.

Flannery, T.F. 1994. The fossil land mammal record of New Guinea: a review. *Science in New Guinea* 20: 39–48.

Flannery, T.F. 1995. *Mammals of the Southwest Pacific and Moluccan Islands.* Reed Books, Chatswood.

Flannery, T.F., E. Hoch & K. Aplin. 1989. Macropodines from the Pliocene Otibanda Formation, Papua New Guinea. *Alcheringa* 13: 145–152.

Flannery, T.F., P. Kirch, J. Specht & M. Spriggs. 1988. Holocene mammal faunas from archaeological sites in Island Melanesia. *Archaeol. Oceania* 23: 89–94.

Flannery, T.F., R. Martin, & A. Szalay. 1996. *Tree Kangaroos.* Reed Books, Port Melbourne.

Flannery, T.F., M.-J. Mountain & K. Aplin. 1983. Quaternary kangaroos (Macropodidae: Marsupialia) from Nombe Rock Shelter, Papua New Guinea, with comments on the nature of megafaunal extinction in the New Guinea highlands. *Proc. Linn. Soc. NSW* 107: 77–99.

Flannery, T.F. & M. Plane. 1986. A new late Pleistocene diprotodontid (Marsupialia) from Pureni, Southern Highlands Province, Papua New Guinea. *BMR J Austr. Geol. Geophysics* 10: 65–76.

Flannery, T.F. & J.P. White. 1991. Animal translocation. *Nat. Geogr. Res. Exploration* 7: 96–113.

Flannery, T.F. & S. Wickler. 1990. Quaternary murids (Rodentia: Muridae) from Buka Island, Papua New Guinea, with descriptions of two new species. *Aust. Mammal.* 13: 127–139.

Foley, W.A. 1986. *The Papuan Languages of New Guinea.* Cambridge Universlty Press, Cambridge.

Forbes, W. A. 1882. On a new species of hemipode from New Britain. *Ibis* 4 (6): 428–431.

Ford, J. 1981a. Morphological and behavioural evolution in populations of the *Gerygone fusca* complex. *Emu* 81: 57–81.

Ford, J. 1981b. Evolution, distribution and stage of speciation in the *Rhipidura fuliginosa* complex in Australia. *Emu* 81: 128–144.

Ford, J. 1983. Taxonomic notes on some mangrove-inhabiting birds in Australia. *Rec. West. Aust. Mus.* 10: 381–415.

Ford, J. 1985. Species limits and phylogenetic relationships in corellas of the *Cacatua pastinator* complex. *Emu* 85: 163–180.

Ford, J. 1986. Phylogeny of the acanthizid warbler genus *Gerygone* based on numerical analyses of morphological characters. *Emu* 86: 12–22.

Forshaw, J.M. 1971. Status of *Lorius amabilis* Stresemann. *Bull. Brit. Orn. Club* 91: 64–65.

Forshaw, J.M. & W.T. Cooper. 1973. *Parrots of the World.* Lansdowne, Melbourne.

Forshaw, J.M. & W.T. Cooper. 1983–94. *Kingfishers and Related Birds.* Lansdowne, Melbourne.

French, W. 1957. Birds of the Solomon Islands. *Ibis* 99: 126–127.

Friedman, H. 1968. The evolutionary history of the avian genus *Chrysococcyx. Bull. U. S. Nat. Mus.* 265: 1–137.

Frith, H.J. 1982. *Pigeons and Doves of Australia.* Rigby, Adelaide, Australia.

Fry, C.H. 1969. The evolution and systematics of bee-eaters (Meropidae). *Ibis* 111: 557–592.

Fry, C.H. 1980a. The evolutionary biology of kingfishers (Alcedinidae). *Living Bird* 18: 113–160.

Fry, C.H. 1980b. The origin of Afrotropical kingfishers. *Ibis.* 122: 57–74.

Fry, C.H. 1984. *The Bee-eaters.* Poyser, Calton.

Fry, C.H., K. Fry & A. Harris. 1992. *Kingfishers, Bee-eaters and Rollers.* Christopher Helm, London.

Galbraith, I.C.J. 1956. Variation, relationships and evolution in the *Pachycephala pectoralis* superspecies (Aves, Muscicapidae). *Bull. Brit. Mus. (Nat. Hist.) Zool.* 4: 133–222.

Galbraith, I.C.J. 1957. On the application of the name *Zosterops rendovae* Tristram, 1882. *Bull. Brit. Orn. Club* 77: 10–16.

Galbraith, I.C.J. 1967. The Black-tailed and robust whistlers *Pachycephala melanura* as a species distinct from the golden whistler *P. pectoralis. Emu* 66: 289–294.

Galbraith, I.C.J. 1969. The Papuan and little Cuckoo-shrikes, *Coracina papuensis* and *robusta* as races of a single species. *Emu* 69: 9–29.

Galbraith, I. C. J. & E. H. Galbraith. 1962. Land birds of Guadalcanal and the San Cristoval group, eastern Solomon Islands. *Bull. Brit. Mus. (Nat. Hist.) Zool.* 9: 1–86.

Gallup, C.D., R.L. Edwards & R.G. Johnson. 1994. The timing of high sea levels over the past 200,000 years. *Science* 263: 796–800.

Gash, N. & J. Whittaker. 1975. *A Pictorial History of New Guinea.* Brown, Brisbane, Australia.

Gibbons, J.R.H. 1985. The biogeography and evolution of Pacific island reptiles and amphibians. In: G. Grigg, R. Shine & H. Ehmann (eds.), *Biology of Australasian Frogs and Reptiles*, pp. 125–142. Surrey Beatty, Sydney.

Gibbs, D. 1996. Notes on Solomon Island birds. *Bull. Brit. Orn. Club* 116: 18–25.

Gill, F.B. 1998. Hybridization in birds. *Auk* 115: 281–283.

Gilliard, E.T. 1960a. Results of the 1958–1959 Gilliard New Britain Expedition. 1. A new genus of honeyeater (Aves). *Amer. Mus. Novit.* 2001.

Gilliard, E.T. 1960b. Results of the 1958–1959 Gilliard New Britain expedition. 2. A new thicket warbler (Aves, *Cichlornis*) from New Britain. *Amer. Mus. Novit.* 2008.

Gilliard, E. T. & M. LeCroy. 1967. Results of the 1958–59 Gilliard New Britain expedition. 4. Annotated list of birds of the Whiteman Mountains, New Britain. *Bull. Am. Mus. Nat. Hist.* 135: 173–216.

Gilpin, M. E. & J. M. Diamond. 1976. Calculation of immigration and extinction curves from the species-distance relation. *Proc. Natl. Acad. Sci. USA* 73: 4130–4134.

Gilpin, M. E. & J. M. Diamond. 1981. Immigration and extinction probabilities for individual species: relation to incidence functions and species colonization curves. *Proc. Natl. Acad. Sci. USA* 78: 392–396.

Goodenough, W. H., ed. 1996. Prehistoric settlement of the Pacific. *American Philosophical Society Transactions*, vol. 86, pt. 5. Philadelphia.

Goodwin, D. 1960. Taxonomy of the genus *Ducula*. *Ibis* 102: 526–535.

Goodwin, D. 1963. On the affinities of *Gymnophaps*. *Ibis* 105: 116–118.

Goodwin, D. 1976. *Crows of the World*. Brit. Mus. (Nat. Hist.), London.

Goodwin, D. 1982. *Estrildid Finches of the World*. Brit. Mus. (Nat. Hist.), London.

Goodwin, D. 1983. *Pigeons and Doves of the World*, 3rd ed. Cornell University Press, Ithaca, NY.

Gould,J. 1843. [Birds of the H.M.S. Sulphur Expedition]. *Proc. Zool. Soc. London* 103–104.

Gould, J. 1856. [Birds collected by Mr. John MacGillivray . . .]. *Proc. Zool. Soc. London* 135–138.

Gould, S.J. & N. Eldridge 1977. Punctuated equilibria: the tempo and mode of evolution reconsidered. *Paleobiology* 3: 115–151.

Grant, B.R. & P.R. Grant. 1989. *Evolutionary Dynamics of a Natural Population*. University of Chicago Press, Chicago.

Grant, P. R. 1981. Speciation and the adaptive radiation of Darwin's finches. *Amer. Sci.* 69: 653–663.

Grant, P.R. 1984. Recent research on the evolution of land birds on the Galapagos. *Biol. J. Linn. Soc.* 21: 113–136.

Grant, P.R. 1999. *Ecology and Evolution of Darwin's Finches*. 2nd ed. Princeton University Press, Princeton, NJ.

Grant, P.R. & B.R. Grant. 1996. Speciation and hybridization in island birds. *Phil. Trans. R. Soc. Lond. B* 351: 765–772.

Grant, P.R. & B.R. Grant. 1997. Genetics and the origin of bird species. *Proc. Natl. Acad. Sci. USA* 94: 7768–7775.

Gray, G. R. 1859. *Catalogue of the Birds of the Tropical Islands of the Pacific Ocean, in the Collection of the British Museum*. British Museum, London.

Gray, G. R. 1870. Descriptions of new species of birds from the Solomon and Banks's groups of islands. *Ann. Mag. Nat. Hist.* (4) 5: 327–331.

Greenberg, J. H. 1971. The Indo-Pacific hypothesis. *Current Trends in Linguistics*, vol. 8, pp. 807–871.

Greenway, J. C. 1967. *Extinct and Vanishing Birds of the World*. Dover, New York.

Greenway, J.C. 1978. Type specimens of birds in the American of Natural History. Part II. *Bull. Amer. Mus. Nat. Hist.* 161: 1–306.

Greer, A.E. 1982. A new species of *Geomyersia* (Scincidae) from the Admiralty Islands, with a summary of the genus. *J. Herpetol.* 16: 61–66.

Gressitt, J.L. 1982. *Biogeography and Ecology of New Guinea*. Junk, The Hague.

Grimes, B.F., ed. 1988. *Ethnologue: Languages of the World*, 11th ed. Summer Institute of Linguistics, Dallas, TX.

Groube, L., J. Chappell, J. Muke & D. Price. 1986. A 40,000 year-old human occupation site at Huon Peninsula, Papua New Guinea. *Nature* 324: 453–455.

Grover, J.C. 1955. Geology, mineral deposits and prospects of mining development in the British Solomon Islands Protectorate. *Interim Geological Survey of the British Solomon Islands Memoir* no. 1. Dept. of Geological Surveys, Honiara, British Solomon Islands.

Grover, J.C., R.B. Thompson, P.J. Coleman, R.L. Stanton & J.D. Bell. 1959–62. *The British Solomon Islands Geological Record*, vol. 2.

Guppy, H.B. 1887a. *The Solomon Islands and Their Natives*. Swan Sonnenschein, Lowrey, London.

Guppy, H.B. 1887b. *The Solomon Islands: Their Geology, General Features, and Suitability for Colonization*. Swan Sonnenschein, Lowrey, London.

Gurney, J. H. 1882. Notes on the raptorial birds collected in New Britain by Lieut. G. E. Richards, R. N. *Ibis* 6: 126–133.

Hadden, D. 1981. *Birds of the North Solomons*. Wau Ecology Institute, Wau.

Hadden, D. 1983. A new species of Thicket Warbler *Cichlornis* (Sylviinae) from Bougainville Island, North Solomons

Province, Papua New Guinea. *Bull. Brit. Orn. Club* 103: 22–25.

Haffer, J. 1974. *Avian Speciation in Tropical South America*. Nuttall Ornithological Club, Cambridge, MA.

Hall, B.P. & R.E. Moreau. 1970. *An Atlas of Speciation in African Passerine Birds*. British Museum of Natural History, London.

Hammer, K.L. 1907. *Die geographische Verbreitung der vulkanischen Gebilde und Erscheinungen im Bismarckarchipel und auf den Salomonen*. Universitäts-Druckerei, Giessen.

Hancock, J. & H. Elliott. 1978. *The Herons of the World*. London Editions, London.

Harding, E. 1982. Birds of Nissan and Pinipel Island, North Solomons Province. *Papua New Guinea Bird Soc. Newsletter* no. 195–196: 4–12.

Harrison, C.J.O. 1973. The zoogeographical dispersal of the genus *Chrysococcyx*. *Emu*. 73: 129–133.

Harrison, C. 1982. *An Atlas of the Birds of the Western Palaearctic*. Princeton University Press, Princeton, NJ.

Hartert, E. 1896. On the forms of *Macropteryx mystacea*. *Novit. Zool.* 3: 19–20.

Hartert, E. 1898. List of the birds collected on New Hanover. In: H. Cayley-Webster, *Through New Guinea and the Cannibal Countries*, pp. 369–375. T. Fisher, London.

Hartert, E. 1899. On the birds of New Hanover. *Ibis* (5): 277–281.

Hartert, E. 1900a. The birds of Ruk in the Central Carolines. *Novit. Zool.* 7: 1–7.

Hartert, E. 1900b. The birds of Buru, being a list of collections made on that island by Messrs. William Doherty and Dumas. *Novit. Zool.* 7: 226–242.

Hartert, E. 1901. [Descriptions of new species of birds]. *Bull. Brit. Orn. Club* 12: 24–25.

Hartert, E. 1904. [Description of a new *Zosterops*]. *Bull. Brit. Orn. Club* 14: 61.

Hartert, E. 1908. [Descriptions of new forms from the Solomon Islands]. *Bull. Brit. Orn. Club* 21: 105–107.

Hartert, E. 1924a. The birds of New Hanover. *Novit. Zool.* 31: 194–213.

Hartert, E. 1924b. The birds of St. Matthias Island. *Novit. Zool.* 31: 261–275.

Hartert, E. 1924c. The birds of Squally or Storm Island. *Novit. Zool.* 31: 276–278.

Hartert, E. 1924d. Types of birds in the Tring Museum. *Novit. Zool.* 31: 112–134.

Hartert, E. 1924e. [Description of a new Hornbill]. *Bull. Brit. Orn. Club* 45: 46.

Hartert, E. 1925. A collection of birds from New Ireland (Neu Mecklenburg). *Novit. Zool.* 32: 115–136.

Hartert, E. 1926a. On the birds of Feni and Nissan Islands, east of South New Ireland. *Novit. Zool.* 33: 33–48.

Hartert, E. 1926b. On the birds of the district of Talasea in New Britain. *Novit. Zool.* 33: 122–145.

Hartert, E. 1926c. On the birds of the French Islands, north of New Britain. *Novit. Zool.* 33: 171–178.

Hartert, E. 1929. Birds collected during the Whitney South Sea Expedition. 8. Notes on birds from the Solomon Islands. *Amer. Mus. Novit.* 364.

Hartert. E. 1930. List of the birds collected by Ernst Mayr. *Novit. Zool.* 36: 27–128.

Hartlaub, G. 1867. On a collection of birds from some less-known localities in the western Pacific. *Proc. Zool. Soc. London* 828–832.

Heinroth, O. 1902. Ornithologische Ergebnisse der "I. Deutschen Südsee Expedition von Br. Mencke". *J. f. Orn.* 50: 390–457.

Heinroth, O. 1903. Ornithologische Ergebnisse der "I. Deutschen Südsee Expedition von Br. Mencke". *J. f. Orn.* 51:65–126.

Hempenstall, P. J. 1978. *Pacific Islanders under German Rule*. Australian National University Press, Canberra.

Hicks, R. 1990. Recent observations January–March 1989. *Muruk* 4: 76–84.

Hindwood, K.A. 1965. John Hunter: a naturalist and artist of the First Fleet. *Emu* 65: 83–95.

Hohnen, P.D. 1978. Geology of New Ireland, Papua New Guinea. *Bureau of Mineral Resources, Geology and Geophysics Bulletin* 194.

Holyoak, D.T. 1974. Les oiseaux des Iles de Société. *L'Oiseau et Rev. fr. Orn.* 44: 153–184.

Holyoak, D.T. 1978. A female specimen of *Gallicolumba jobiensis* from San Cristoval, Solomon Islands. *Bull. Brit. Orn. Club* 98: 98–99.

Holyoak, D.T. & J.T. Thibault. 1978. Notes on the biology and systematics of Polynesian swiftlets, *Aerodramus*. *Bull. Brit. Orn. Club* 98: 59–65.

Hombron, J. B. & C.H. Jacquinot. 1841. Description de plusieurs oiseaux nouveaux ou peu connus, provenant de l'expedition autour du monde faite sur les corvettes l'Astrolabe et la Zélée. *Ann Sci. Nat. Zool.* (2) 16: 312–320.

Honza, E., H.L. Davies, J.B. Keene, & D.L. Tiffin. 1987. Plate boundaries and evolution of the Solomon Sea region. *Geo-Marine Letters* 7: 161–168.

Howard, D.J. & S.H. Berlocher. 1998. *Endless Forms: Species and Speciation*. Oxford University Press, New York.

Howell, L. 1981. Birds of Nissan Island, North Solomons Province. *Papua New Guinea Bird Soc. Newsletter* no. 185–186: 28–36.

Immelmann, K. 1962. Besiedlungsgeschichite und Bastardierung von *Lonchura castaneothorax* und *Lonchura flaviprymna* in Nordaustralien. *J. f. Orn. 103:* 344–357.

Jacquinot, C.H. & J. Pucheran. 1853. *Voyage au Pole Sud et dans l'Oceanie sur les corvettes L'Astrolabe et La Zélée exécuté par ordre du Roi pendant les années 1837–1840. Zoologie,* vol. 3, Paris.

James, H. F. & S. L. Olson. 1991. *Descriptions of 32 new species of birds from the Hawaiian Islands*: Part II Passeriformes. *Ornithological Monographs* no. 46. American Ornithologists Union, Washington, D.C.

Jaques, A.L. & G.P. Robinson. 1977. Continent/island-arc collision in northern Papua New Guinea. *BMR J. Australian Geol. Geophysics* 2: 289–303.

Jepson, P. 1993. Recent ornithological observations from Buru. *Kukila* 6: 85–109.

Johnson, R.W. 1976. *Volcanism in Australasia.* Elsevier, New York.

Johnson, R.W. 1979. Geotectonics and volcanism in Papua New Guinea: a review of the late Cenozoic. *BMR J. Australian Geol. Geophysics* 4: 181–207.

Johnson, R.W. 1981. *Cooke-Ravian Volume of Volcanological Papers.* Geological Survey of Papua New Guinea, Port Moresby, Papua New Guinea.

Johnson, R.W. & A.L. Jaques. 1980. Continent-arc collision and reversal of arc polarity: new interpretations from a critical area. *Tectonophysics* 63: 111–124.

Johnson, T. & P. Molnar. 1972. Focal mechanisms and plate tectonics of the Southwest Pacific. *J. Geophysical Research* 77: 5000–5032.

Johnson, R.W., G. A. M. Taylor & R. A. Davies. 1972. Geology and petrology of Quaternary volcanic islands off the north coast of New Guinea. *Bureau of Mineral Resources, Geology and Geophysics Record* 1972/21.

Johnstone, R. E. 1981. Notes on the distribution, ecology and taxonomy of the Red-crowned Pigeon (*Ptilinopus regina*) and Torres Strait Pigeon *Ducula bicolor* in Western Australia. *Rec. W. Aust. Mus.* 9: 7–22.

Jollie, M. 1978. Phylogeny of the species of *Corvus. The Biologist* 60: 73–108.

Jones, D. & P. Lambley. 1987. Notes on the birds of New Ireland. *Muruk* 2: 29–33.

Keast, J.A. 1958. Variation and speciation in the Australian flycatchers. *Rec. Austr. Mus.* 24: 73–108.

Keast, A. 1961. Bird speciation on the Australian continent. *Bull. Mus. Comp. Zool.* 123: 305–495.

Kemp, A. 1995. *The Hornbills.* Oxford University Press, Oxford.

Kennedy, M. & H.G. Spencer. 2000. Phylogeny, biogeography, and taxonomy of Australian teals. *Auk* 117: 154–163.

King, W.B. 1981. *Endangered Birds of the World.* Smithsonian Institution Press, Washington, D.C.

Kinghorn, J. R. 1937. Notes on some Pacific Island birds. *Proc. Zool. Soc. Lond. B* 107: 177–184.

Kirch, P. V. 1988. The Talepakemali Lapita site and oceanic prehistory. *Nat. Geog. Res.* 4: 328–342.

Kirch, P. V. 1997. *The Lapita Peoples.* Blackwell, Oxford.

Kirch, P. V. 2000. *On the Road of the Winds: an Archaeological History of the Pacific Islands before European Contact.* University of California Press, Berkeley.

Kirch, P. V. & T. L. Hunt. 1988. *Archaeology of the Lapita Cultural Complex: A Critical Review.* Burke Museum, Seattle, NA.

Klein, N.K. & W.M. Brown 1994. Intraspecific molecular phylogeny in the yellow warbler (*Dendroica petechia*)and implications for avian biogeography in the West Indies. *Evolution* 48: 1914–1932.

Koopman, K.F. 1957. Evolution in the genus *Myzomela* (Aves: Meliphagidae). *Auk* 74: 49–72.

Kratter, A.W., D.W., Steadman, C.E. Smith, C.E. Filardi & H.P. Webb. 2001. The avifauna of a lowland forest site on Isabel, Solomon Islands. *Pacific Science,* in press.

Kroenke, L.W. 1996. Plate tectonic development of the western and southwestern Pacific: mesozoic to the present. In: A. Keast & S.E. Miller (eds.), *The Origin and Evolution of Pacific Island Biotas, New Guinea to Eastern Polynesia: Patterns and Processes,* pp. 19–34. SPB Academic Publishing, Amsterdam.

Lack, D. 1940. Evolution of the Galapagos finches. *Nature* 146: 324–327.

Lack, D. 1944. Ecological aspects of species-formation in passerine birds. *Ibis.* 1944: 260–286.

Lack, D. 1945. The Galapagos finches (*Geospizinae)*: a study in variation. *Occ. Papers Calif. Acad. Sci.* 21: 1–159.

Lack, D. 1947. *Darwin's Finches.* Cambridge University Press, Cambridge.

Lack, D. 1949. The significance of ecological isolation. In: G. L. Jepsen, E. Mayr & G. G. Simpson (eds.), *Genetics, Paleontology, and*

Evolution, pp. 299–308. Princeton University Press, Princeton, NJ.

Lack, D. 1971. *Ecological Isolation in Birds.* Harvard University Press, Cambridge, MA.

Laird, M. 1954. Australian Pelicans in the Solomon Islands and New Hebrides. *Notornis* 6: 11–13.

Laird, M. & E. Laird. 1956. Account of a visit to Bellona and Rennell in August 1953. *Nat. Hist. Ren. Isl., Br. Solomon Isl.* 1: 65–71.

Laissus, Y. 1978. Catalogue des manuscrits de Philibert Commerson (1727–1773) conservés à la Bibliothèque centrale du Muséum national d'Histoire naturelle (Paris). *Rev. Hist. Sci.* 31: 131–162.

Lambert, F. & M. Woodcock. 1996. *Pittas, Broadbills and Asities.* Pica, Sussex, UK.

Laracy, H. 1976. *Marists and Melanesians.* University Press of Hawaii, Honolulu.

Laurance, W.F., J. Garesche & C.W. Payne. 1993. Avian nest predation in modified and natural habitats in tropical Queensland: an experimental study. *Wildlife Research* 20: 711–723.

Laurance, W.F. & J.D. Grant. 1994. Photographic identification of ground-nest predators in Australian tropical rainforest. *Tropical Research* 21: 241–248.

Laurie, E.M.O. & J.E. Hill. 1954. *List of Land Mammals of New Guinea, Celebes, and Adjacent Islands 1758–1952.* British Museum of Natural History, London.

Layard, E. L. C. 1880. Notes of a collecting-trip in the New Hebrides, the Solomon Islands, New Britain, and the Duke-of-York Islands. *Ibis* 4: 290–309.

Layard, E. L. & E. L. C. Layard. 1880. Note on *Pachycephala assimilis* of J. Verreaux and O. Des Murs. *Ibis* 4: 460–461.

LeCroy, M. & W. S. Peckover. 1983. Birds of the Kimbe Bay area, west New Britain, Papua New Guinea. *Condor* 85: 297–304.

LeCroy, M. & W.S. Peckover. 1999. Plumages of the Red-collared Honeyeater *Myzomela rosenbergii longirostris* from Goodenough Island, D'Entrecasteaux Islands, Papua New Guinea. *Bull. Brit. Orn. Club* 119: 62–65.

Lee, P.L.M., D.H. Clayton, R. Griffiths & R.D.M. Page. 1996. Does behavior reflect phylogeny in swiftlets (Aves: Apodidae)? A test using cytochrome b mitochondrial DNA sequences. *Proc. Nat. Acad. Sci. USA* 93: 7091–7096.

Lees, A., M. Garnett & S. Wright. 1991. *A Representative Protected Forests System for the Solomon Islands.* Maruia Society, Nelson, New Zealand.

Lesson, R.P. & P. Garnot. 1826–28. *Voyage autour du Monde, Exécuté par Ordre du Roi, sur la Corvette de Sa Majesté La Coquille, pendant les Années 1822, 1823, 1824, 1825. Zoologie*, vol. 1. Bertrand, Paris.

Lever, R.J.A.W. 1937. The geology of the British Solomon Islands Protectorate. *Geological Magazine* 74: 271–277.

Lindgren, E. 1978. [Observations]. *Papua New Guinea Bird Soc. Newsletter* no. 146: 16.

Livezey, B.C. 1991. A phylogenetic analysis and classification of recent dabbling ducks (tribe Anatini) based on comparative morphology. *Auk* 108: 471–507.

Löffler, E. 1977. *Geomorphology of Papua New Guinea.* Commonwealth Scientific and Industrial Research Organization, Canberra.

Löffler, E. 1982. Land forms and land form development. In: J.L. Gressitt (ed.), *Biogeography and Ecology of New Guinea*, vol. 1, pp. 57–72. Junk, The Hague.

Loveridge, A. 1948. New Guinean reptiles and amphibians in the Museum of Comparative Zoology and United States National Museum. *Bull. Mus. Comp. Zool.* 101: 305–430.

Lowe, K.W. & G.C. Richards. 1991. Morphological variation in the Sacred Ibis *Threskiornis aethiopicus* superspecies complex. *Emu* 91: 41–45.

Ludwig, J.A. & J.F. Reynolds. 1988. *Statistical Ecology.* Wiley, New York.

MacArthur, R.H., J.M. Diamond & J. Karr. 1972. Density compensation in island faunas. *Ecology* 53: 330–342.

MacArthur, R. H. & E. O. Wilson. 1967. *The Theory of Island Biogeography.* Princeton University Press, Princeton, NJ.

Mackay, R. D. 1977. Birds recorded in Manus and New Ireland Provinces between 21/2/77 and 11/3/77. *New Guinea Bird Soc. Newsletter* no. 137: 4–6.

Malinote, E.V. & G. Underwood. 1988. Australasian natricine snakes of the genus *Tropidonophis. Proc. Acad. Nat. Sci. Phil.* 140: 59–201.

Marchant, S. 1972. Evolution of the genus *Chrysococcyx. Ibis.* 114: 219–233.

Marlow, M.S., S.V. Dadisman, and N.F. Exon. 1988. *Geology and Offshore Resources of Pacific Island Arcs: New Ireland and Manus Region, Papua New Guinea.* Circum-Pacific Council for Energy and Mineral Resources, Houston, TX.

Martens, G. H. 1922. Die Vögel der Südsee-Expedition der Hamburger Wissenschaftlichen Stiftung 1908–1909. *Archiv f. Naturgeschichte* 88A, no. 7: 44–54.

Mason, I.J. & R.I. Forrester. 1996. Geographical differentiation in the Channel-billed cuckoo

Scythrops novaehollandiae Latham, with description of two new subspecies from Sulawesi and the Bismarck Archipelago. *Emu* 96: 217–233.
Mason, I.J., J.L. McKean & M.L. Dudzinski. 1984. Geographical variation in the pheasant coucal *Centropus phasianinus* (Latham) and a description of a new subspecies from Timor. *Emu.* 84: 1–15.
Mayr, E. 1931a. Birds collected during the Whitney South Sea Expedition. 12. Notes on *Halcyon chloris* and some of its subspecies. *Amer. Mus. Novit.* 469.
Mayr, E. 1931b. Birds collected during the Whitney South Sea Expedition. 13. A systematic list of the birds of Rennell Island with descriptions of new species and subspecies. *Amer. Mus. Novit.* 486.
Mayr, E. 1931c. The parrot finches (genus *Erythrura*). *Amer. Mus. Novit.* 489.
Mayr, E. 1931d. Birds collected during the Whitney South Sea Expedition. 16. Notes on fantails of the genus *Rhipidura. Amer. Mus. Novit.* 502.
Mayr, E. 1931e. Birds collected during the Whitney South Sea Expedition. 17. The birds of Malaita Island (British Solomon Islands). *Amer. Mus. Novit.* 504.
Mayr, E. 1932a. Birds collected during the Whitney South Sea Expedition. 18. Notes on Meliphagidae from Polynesia and the Solomon Islands. *Amer. Mus. Novit.* 516.
Mayr, E. 1932b. Birds collected during the Whitney South Sea Expedition. 19. Notes on the Bronze Cuckoo *Chalcites lucidus* and its subspecies. *Amer. Mus. Novit.* 520.
Mayr, E. 1932c. Birds collected during the Whitney South Sea Expedition. 20. Notes on thickheads (*Pachycephala*) from the Solomon Islands. *Amer. Mus. Novit.* 522.
Mayr, E. 1932d. Birds collected during the Whitney South Sea Expedition. 21. Notes on thickheads (*Pachycephala*) from Polynesia. *Amer. Mus. Novit.* 531.
Mayr, E. 1933a. Birds collected during the Whitney South Sea Expedition. 22. Three new genera from Polynesia and Melanesia. *Amer. Mus. Novit.* 590.
Mayr, E. 1933b. Birds collected during the Whitney South Sea Expedition. 24. Notes on Polynesian flycatchers and a revision of the genus *Clytorhynchus* (Elliot). *Amer. Mus. Novit.* 628.
Mayr, E. 1933c. Birds collected during the Whitney South Sea Expedition. 25. Notes on the genera *Myiagra* and *Mayrornis*. *Amer. Mus. Novit.* 651.
Mayr, E. 1933d. On a collection of birds, supposedly from the Solomon Islands. *Ibis* 13 (3): 549–552.
Mayr, E. 1934a. Birds collected during the Whitney South Sea Expedition. 28. Notes on some birds from New Britain, Bismarck Archipelago. *Amer. Mus. Novit.* 709.
Mayr, E. 1934b. Birds collected during the Whitney South Sea Expedition. 29. Notes on the genus *Petroica. Amer. Mus. Novit.* 714.
Mayr, E. 1935. Birds collected during the Whitney South Sea Expedition. 30. Descriptions of 25 new species and subspecies. *Amer. Mus. Novit.* 820.
Mayr, E. 1936. Birds collected during the Whitney South Sea Expedition. 31. Descriptions of 25 species and subspecies. *Amer. Mus. Novit.* 828.
Mayr, E. 1937a. Birds collected during the Whitney South Sea Expedition. 33. Notes on New Guinea birds. I. *Amer. Mus. Novit.* 915.
Mayr, E. 1937b. Birds collected during the Whitney South Sea Expedition. 36. Notes on New Guinea birds. III. *Amer. Mus. Novit.* 947.
Mayr, E. 1938a. Birds collected during the Whitney South Sea Expedition. 39. Notes on New Guinea birds. IV. *Amer. Mus. Novit.* 1006.
Mayr, E. 1938b. Birds collected during the Whitney South Sea Expedition. 40. Notes on New Guinea birds. V. *Amer. Mus. Novit.* 1007.
Mayr, E. 1940a. Birds collected during the Whitney South Sea Expedition. 41. Notes on New Guinea birds. VI. *Amer. Mus. Novit.* 1056.
Mayr, E. 1940b. Birds collected during the Whitney South Sea Expedition. 43. Notes on New Guinea birds. VII. *Amer. Mus. Novit.* 1091.
Mayr, E. 1941a. Birds collected during the Whitney South Sea Expedition. 45. Notes on New Guinea birds. VIII. *Amer. Mus. Novit.* 1133.
Mayr, E. 1941b. Birds collected during the Whitney South Sea Expedition. 47. Notes on the genera *Halcyon*, *Turdus*, and *Eurostopodus*. *Amer. Mus. Novit.* 1152.
Mayr, E. 1941c. *List of New Guinea Birds.* American Museum of Natural History, New York.
Mayr, E. 1941d. The origin and history of the bird fauna of Polynesia. *Proc. 6th Pac. Sci. Congr.* 4: 197–216.
Mayr, E. 1942. *Systematics and the Origin of Species.* Columbia University Press, New York.
Mayr, E. 1943. Notes on Australian birds (II). *Emu* 43: 3–17.
Mayr, E. 1944a. The birds of Timor and Sumba. *Bull. Amer. Mus. Nat. Hist.* 83: 123–194.

Mayr, E. 1944b. Birds collected during the Whitney South Sea Expedition. 54. Notes on some genera from the southwest Pacific. *Amer. Mus. Novit.* 1269.

Mayr, E. 1944c. Wallace's line in the light of recent zoogeographic studies. *Quart. Rev. Biol* 29: 1–14.

Mayr, E. 1945a. *Birds of the Southwest Pacific.* Macmillan, New York.

Mayr, E. 1945b. Birds collected during the Whitney South Sea Expedition. 55. Notes on the birds of Northern Melanesia. 1. *Amer. Mus. Novit.* 1294.

Mayr, E. 1945c. The correct name of the Fijian Mountain Lorikeet. *Auk* 62: 139.

Mayr, E. 1948. Geographic variation in the Reed-Warbler. *Emu* 47: 205–210.

Mayr, E. 1949a. Notes on the birds of northern Melanesia. 2. *Amer. Mus. Novit.* 1417.

Mayr, E. 1949b. Artbildung und Variation in der *Halcyon-chloris*-Gruppe. In: E. Mayr & E. Schüz (eds.), *Ornithologie als Biologische Wissenschaft*, pp. 55–60. Carl Winter, Heidelberg.

Mayr, E. 1950. A new cuckoo-shrike from the Solomon Islands. *Auk* 67: 104.

Mayr, E. 1953. Fragments of a Papuan ornithogeography. *Proc. 1949 Pac. Sci. Congr.* 4: 11–19.

Mayr, E. 1955. Notes on the birds of northern Melanesia. 3. Passeres. *Amer. Mus. Novit.* 1707.

Mayr, E. 1954. Change of genetic environment and evolution. In: J. Huxley, A. C. Hardy & E. B. Ford (eds.), *Evolution as a Process*, pp. 157–180. Allen & Unwin, London.

Mayr, E. 1957. Notes on the birds of northern Melanesia. 4. The genus *Accipiter*. *Amer. Mus. Novit.* 1823.

Mayr, E. 1963. *Animal Species and Evolution.* Harvard University Press, Cambridge, MA.

Mayr, E. 1965a. Avifauna: turnover on islands. *Science* 150: 1587–1588.

Mayr, E. 1965b. Relationships among Indo-Australian Zosteropidae (Aves). *Breviora* 228.

Mayr, E. 1965c. The nature of colonizations in birds. In: H.G. Baker & G.L. Stebbins (eds.), *The Genetics of Colonizing Species*, pp. 29–47. Academic Press, New York.

Mayr, E. 1968. The sequence of genera in the Estrildidae (Aves). *Breviora* 287.

Mayr, E. 1969. Bird speciation in the tropics. *Biol. J. Linn. Soc.* 1: 1–17.

Mayr, E. 1971. New species of birds described from 1956 to 1965. *J. f. Ornithologie* 112: 302–316.

Mayr, E. 1982a. Speciation and macroevolution. *Evolution* 36: 1119–1132.

Mayr, E. 1982b. Processes of speciation in animals. In: C. Barigozzi (ed.), *Mechanisms of Speciation*, pp. 1–19. New York, Liss

Mayr, E. 1988. *Toward a New Philosophy of Biology*. Harvard University Press, Cambridge, MA.

Mayr, E. 1991. *One Long Argument: Charles Darwin and the Genesis of Modern Evolutionary Thought.* Harvard University Press, Cambridge, MA.

Mayr, E. 1992. A local flora and the biological species concept. *Am. J. Bot.* 79: 222–238.

Mayr, E. 1997. *This Is Biology*. Harvard University Press, Cambridge, MA.

Mayr, E. & D. Amadon. 1941. Birds collected during the Whitney South Sea Expedition. 46. Geographical variation in *Demigretta sacra* (Gmelin). *Amer. Mus. Novit.* 1144.

Mayr, E. & D. Amadon. 1947. A review of the Dicaeidae. *Amer. Mus. Novit.* 1360.

Mayr, E. & P.D. Ashlock, 1991. *Principles of Systematic Zoology*, 2nd ed., McGraw Hill, New York.

Mayr, E. & W. Bock. 1994. Provisional classifications *v* standard avian sequences: heuristics and communication in ornithology. *Ibis* 136: 12–18.

Mayr, E. & S. Camras. 1938. Birds of the Crane Pacific Expedition. *Publ. Field Mus.* (*Zool*). 20: 453–473.

Mayr, E. & J. M. Diamond. 1976. Birds on islands in the sky: origin of the montane avifauna of Northern Melanesia. *Proc. Natl. Acad. Sci. USA* 73: 1765–1769.

Mayr, E. & H. Hamlin. 1931. Birds collected during the Whitney South Sea Expedition. 14. With notes on the geography of Rennell Island and the ecology of its bird life. *Amer. Mus. Novit.* 488.

Mayr, E. & M. Moynihan. 1946. Evolution in the *Rhipidura rufifrons* group. *Amer. Mus. Novit.* 1321.

Mayr, E. & W. B. Provine. 1980. *The Evolutionary Synthesis.* Harvard University Press, Cambridge, MA.

Mayr, E. & S. D. Ripley. 1941. Birds collected during the Whitney South Sea Expedition. 44. Notes on the genus *Lalage* (Boie). *Amer. Mus. Novit.* 1116.

Mayr, E. & L.L. Short. 1970. *Species Taxa of North American Birds.* Nuttall Ornithological Club, Cambridge, MA.

Mayr, E. & C. Vaurie. 1948. Evolution in the family Dicruridae (birds). *Evolution* 2: 238–265.

McAlpine, J.R., G. Keig & K. Short. 1975. Climatic tables for Papua New Guinea. *CSIRO Div. Land Use Res. Tech. Paper* no. 37.

McCoy, M. 1980. *Reptiles of the Solomon Islands*. Wau Ecology Institute, Wau.

McDowell, S. B. 1970. On the status and relationships of the Solomon Island elapid snakes. *J. Zool. Lond.* 161: 145–190.

McDowell, S.B. 1974. A catalogue of the snakes of New Guinea and the Solomons, with special reference to those in the Bernice P. Bishop Museum, Part I. Scolecophidia. *J. Herpetol.* 8: 1–57.

McDowell, S.B. 1975. A catalogue of the snakes of New Guinea and the Solomons, with special reference to those in the Bernice P. Bishop Museum. Part II. Anilioidea and Pythoninae. *J. Herpetol.* 9:1–79.

McDowell, S.B. 1979. A catalogue of the snakes of New Guinea and the Solomons, with special reference to those in the Bernice P. Bishop Museum. Part III. Boinae and Acrochordoidea (Reptilia, Serpentes). *J. Herpetol.* 13: 1–92.

McDowell, S.B. 1984. Results of the Archbold Expeditions. No. 112. The snakes of the Huon Peninsula, Papua New Guinea. *Amer. Mus. Novitates* no. 2775.

Medway, L. 1966. Field characters as a guide to the specific relations of swiftlets. *Proc. Linn. Soc. Lond.* 177: 151–172.

Medway, L. 1975. The nest of *Collocalia v. vanikorensis,* and taxonomic implications. *Emu* 75: 154–155.

Medway, L. & J.D. Pye. 1977. Echolocation and the systematics of swiftlets. In: B. Stonehouse & C. M. Perrins (eds.), *Evolutionary Ecology*. pp. 225–238. Macmillan, London.

Medway, L. & D. R. Wells. 1976. *The Birds of the Malay Peninsula,* vol. 5. Witherby, London.

Meek, A. S. 1913. *A Naturalist in Cannibal Land*. Unwin, London.

Mees, G.F. 1955. The name of the white-eye from Rendova Island (Solomon Islands). *Zool. Meded.* 33: 299–300.

Mees, G.F. 1957. A systematic review of the Indo-Australian Zosteropidae (Part I). *Zool. Verh.* 35: 1–204.

Mees, G.F. 1961. A systematic review of the Indo-Austalian Zosteropidae (Part II). *Zool. Verh.* 50: 1–168.

Mees, G.F. 1964. Notes on two small collections of birds from New Guinea. *Zool. Verhandel.* no. 65.

Mees, G.F. 1965a. The avifauna of Misool. *Nova Guinea, Zool.* 31: 139–203.

Mees, G.F. 1965b. Revision of *Nectarinia sericea* (Lesson). *Ardea* 53: 38–56.

Mees, G.F. 1969. A systematic review of the Indo-Australian Zosteropidae (Part III). *Zool. Verh.* 102: 1–390.

Mees, G.F. 1975. A list of the birds known from Roti and adjacent islets (Lesser Sunda Islands). *Zool. Meded.* 49: 115–140.

Mees, G.F. 1977. Geographical variation of *Caprimulgus macrurus* Horsfield (Aves, Caprimulgidae). *Zool. Verh.* 155: 1–47.

Mees, G.F. 1982. Birds from the lowlands of southern New Guinea (Merauke and Koembe). *Zool. Verhandel.* No. 191.

Meise, W. 1928. Die Verbreitung der Aaskrähe (Formenkreis *Corvus corone* L.). *J. f. Orn.* 76: 1–203.

Meise, W. 1929a. Zwei neue Rassen von *Myzomela nigrita*. *Orn. Monatsber.* 37: 84–85.

Meise, W. 1929b. Ueber den Formenkreis *Aplonis cantoroides*. *Orn. Monatsber.* 37: 111–113.

Meise, W. 1929c. Die Vögel von Djampea und benachbarten Inseln nach einer Sammlumg Baron Plessens. *J. f. Orn.* 77: 431–479.

Meise, W. 1931. Zur Systematik der Gattung *Gerygone*. *Novit. Zool.* 36: 317–379.

Meise, W. 1950. *Stresemannia*, eine neue Meliphagidengattung von den Salomon-Inseln.. *Orn. Berichte* 2: 118.

Menzies, J.I. 1976. *Handbook of Common New Guinea Frogs*. Wau Ecology Institute, Wau, Papua New Guinea.

Menzies, J.I. 1982. Systematics of *Platymantis papuensis* (Amphibia: Ranidae) and related species of the New Guinea region. *Brit. J. Herp.* 6: 236–240.

Menzies, J. 1991. *A Handbook of New Guinea Marsupials and Monotremes*. Kristen Press, Madang.

Menzies, J.I. & E. Dennis. 1979. *Handbook of New Guinea Rodents*. Wau Ecology Institute, Wau, Papua New Guinea.

Meyer, A. B. 1890. Notes on birds from the Papuan region, with descriptions of some new species. *Ibis* 6 (2): 412–424.

Meyer, A. B. 1891. Ueber Vögel von Neu Guinea und Neu Britannien. *Abh. Ber. Königl. Zool. Mus. Dresden* 1890/91, no. 4: 1–17.

Meyer, O. 1906. Die Vögel der Insel Vuatom. *Natur. u. Offenbarung*. 52: 513–617.

Meyer, O. 1909. Das Leben der Vögel im Urwald Neu-Pommerns. *Natur. u. Offenbarung* 55: 464–475.

Meyer, O. 1927a. Zur Ornithologie des Bismarckarchipels. *Orn. Monatsber.* 35: 112–113.

Meyer, O. 1927b. Zur Lebensweise zweier Vogelarten des Bismarck-Archipels. *Orn. Monatsber.* 35: 139–140.

Meyer, O. 1928. Meine Beobachtungen an *Monarcha hebetior eichhorni* Hartert. *J. f. Orn.* 76: 654–660.

Meyer, O. 1929a. Beiträge zur Biologie der Vögel von Vuatom (Bismarck-Archipel). *Orn. Monatsber.* 37: 106–108.

Meyer, O. 1929b. Zur Brutbiologie einiger Vögel des Bismarckarchipels. *J. f. Orn.* 77: 21–35.

Meyer, O. 1930. Uebersicht ueber die Brutzeiten der Vögel auf der Insel Vuatom (New Britain). *J. f. Orn.* 78: 19–38.

Meyer, O. 1933. Vogeleier und Nester aus Neubritannien, Südsee. *Beiträge z. Fortpflanzungsbiol. der Vögel* 9: 122–136, 182–185.

Meyer, O. 1934a. Die Vogelwelt auf der Inselgruppe Lihir. *J. f. Orn.* 82: 294–308.

Meyer, O. 1934b. Seltene Vögel auf Neubritannien. *J. f. Orn.* 82: 568–578.

Meyer, O. 1936. *Die Vögel des Bismarckarchipel.* Katholische Mission, Vunapope.

Meyer, O. 1937. Australische Zugvögel im Bismarckarchipel. *Orn. Monatsber.* 45: 48–51.

Meyer, W. 1909. Zur Vogel-Fauna des Bismarck-Archipels. *Orn. Monatsber.* 17: 33–38.

Moeliker, C. W. & C. J. Heij. 1995. The rediscovery of *Monarcha boanensis* (Aves: Monarchidae) from Boano Island, Indonesia. *Deinsia* 2: 123–143.

Moynihan, M. 1968. Social mimicry: character convergence versus character displacement. *Evolution* 22: 315–331.

Murphy, R. C. 1929. Birds collected during the Whitney South Sea Expedition. 9. Zosteropidae from the Solomon Islands. *Amer. Mus. Novit.* 365.

North, A. J. 1888. Notes on the nests and eggs of certain Australian birds. *Proc. Linn. Soc. N.S.W.* 2 (2): 405–411.

North, A. J. 1906. [Description of *Woodfordia superciliosa*]. *Victoria Nat.* 23: 104.

Ogilvie-Grant, W. R. 1887. A list of the birds collected by Mr. Charles Morris Woodford in the Solomon Archipelago. *Proc. Zool. Soc. London* 328–333.

Ogilvie-Grant, W. R. 1888. Second list of the birds collected by Mr. C. M. Woodford in the Solomon Archipelago. *Proc. Zool. Soc. London* 185–204.

Oliver, D. 1973. *Bougainville.* University Press of Hawaii, Honolulu.

Olsen, P.D., R.C. Marshall, and A. Gaal. 1989. Relationships within the genus *Falco*: a comparison of the electrophoretic patterns of feather proteins. *Emu* 89: 193–203.

Olson, S.L. 1970. The relationships of *Porzana flaviventer. Auk* 87: 805–808.

Olson, S.L. 1973. A classification of the Rallidae. *Wilson Bull.* 85: 381–416.

Olson, S.L. 1975. Paleornithology of St. Helena Island, South Atlantic Ocean. *Smithsonian Contributions to Paleobiology* no. 23.

Olson, S.L. & H.F. James. 1982. Fossil birds from the Hawaiian Islands: evidence for wholesale extinction by man before Western contact. *Science* 217: 633–635.

Orenstein, R.I. 1975. Observations and comments on two stream-adapted birds of Papua New Guinea. *Bull. Brit. Orn. Club* 95: 161–165.

Orenstein, R. I. 1976. Birds of the Plesyumi area, central New Britain. *Condor* 78: 370–374.

Orenstein, R.I. & H.D. Pratt. 1983. The relationships and evolution of the Southwest Pacific warbler genera *Vitia* and *Psamathia* (Sylviinae). *Wilson Bull.* 95: 184–198.

O'Shea, M. 1996. *A Guide to the Snakes of Papua New Guinea.* Independent Publishing, Port Moresby.

Palfreyman, W.D., R.W. Johnson, R.J.S. Cooke, & R.J. Bultitude. 1986. *Volcanic activity in Papua New Guinea before 1944: an annotated bibilography of reported observations.* Bureau of Mineral Resources, Geology and Geophysics Report no. 254.

Papua New Guinea Bird Society Newsletter. 1978. No. 145: 21.

Papua New Guinea Bird Society Newsletter. 1997. No. 291: 3.

Parker, S. 1967a. New information on the Solomon Islands Crown Pigeon, *Microgoura meeki* (Rothschild). *Bull. Brit. Orn. Club* 87: 86–89.

Parker, S. 1967b. A. S. Meek's three expeditions to the Solomon Islands. *Bull. Brit. Orn. Club* 87: 129–135.

Parker, S. 1968. On the thick-billed ground dove *Gallicolumba salamonis* (Ramsay). *Bull. Brit. Orn. Club.* 88: 58–59.

Parker, S.A. 1971. Taxonomy of the Northern Territory friarbirds known as *Philemon buceroides gordoni. Emu* 71: 54–56.

Parker, S. 1972. An unsuccessful search for the Solomon Islands crowned pigeon. *Emu* 72: 24–26.

Parker, S.A. 1981. Prolegomenon to further studies in the *Chrysococcyx "malayanus"* group (Aves, Cuculidae). *Zool. Verh.* 187: 1–56.

Parkes, K.C. 1971. Taxonomic and distributional notes on Philippine birds. *Nemouria* 4: 1–67.

Parkes, K.C. 1980. A new subspecies of the Spiny-cheeked Honey-eater *Acanthagenys rufogularis*, with notes on generic relationships. *Bull. Brit. Orn. Club* 100: 143–147.

Parkinson, R. 1926. *Dreissig Jahre in der Südsee.* Strecker und Schröder, Stuttgart.

Patterson, B.D. & W. Atmar. 1986. Nested subsets and the structure of insular mammalian faunas and archipelagos. *Biol. J. Linn. Soc.* 28: 65–82.

Payne, R.B. & C.J. Risley. 1976. Systematics and evolutionary relationships among the herons (Ardeidae). *Misc. Publ. Mus. Zool. Univ. Mich.* 150: 1–115.

Pelzeln, A. von. 1865. *Reise der Österreichischen Fregatte "Novara" um die Erde in den Jahren 1857, 1858, 1859. Zoologisches Theil. Vol. 1: Vögel.* Geroldssohn, Vienna.

Percival, M. & J.S. Womersley. 1975. *Floristics and Ecology of the Mangrove Vegetation of Papua New Guinea.* Botany Bulletin no. 8, Division of Botany, Dept. of Forests, Lae, Papua New Guinea.

Peters, J.L. 1931–87. *Check-list of Birds of the World*, 16 vols. Museum of Comparative Zoology, Cambridge, MA.

Peterson, A.T. 1998. New species and new species limits in birds. *Auk* 115: 555–558.

Pieters, P.E. 1982. Geology of New Guinea. In: J.L. Gressitt (ed.), *Biogeography and Ecology of New Guinea*, vol. 1, pp. 15–38. Junk, The Hague.

Pigram, C.J. & H.L. Davies. 1987. Terranes and the accretion history of the New Guinea orogen. Bureau of Mineral Resources J. *Australian Geology & Geophysics* 10: 193–211.

Pimm, S.L., H.L. Jones & J. Diamond. 1988. On the risk of extinction. *Amer. Natur.* 132: 757–785.

Pippet, J.R. 1975. The marine toad, *Bufo marinus*, in Papua New Guinea. *Papua New Guinea Agricultural J.* 26: 23–30.

Polhemus, J.T. & D.A. Polhemus. 1993. The Trepobatinae (Heteroptera: Gerridae) of New Guinea and surrounding regions, with a review of the world fauna. Part 1. Tribe Metrobatini. *Entomologica Scandinavica* 24: 241–284.

Pratt, H.D., P.L. Bruner & D.G. Berrett. 1987. *A Field Guide to the Birds of Hawaii and the Tropical Pacific.* Princeton University Press, Princeton, NJ.

Pregill, G.K., D.W. Steadman, S.L. Olson, & F.V. Grady. 1988. Late Holocene fossil vertebrates from Burma Quarry, Antigua, Lesser Antilles. *Smithsonian Contributions to Zoology* no. 463.

Quoy, J.R.C. & J.P. Gaimard. 1830. *Voyage de Decouvertes de L'Astrolabe, Exécuté par Ordre du Roi pendant les Années 1826-1827-1828-1829, sous le Commandemant de M. J. Dumont d'Urville, Zoologie, vol. 1.* Tastu, Paris.

Ramsay, E. P. 1876. Description of a supposed new species of *Pachycephala*, from New Britain, proposed to be called *Pachycephala citreogaster. Proc. Linn. Soc. N.S.W.* (1) 1: 66–67.

Ramsay, E. P. 1877. Notes of a collection of birds from New Britain, New Ireland, and the Duke of York Islands, with some remarks on the zoology of the group. *Proc. Linn. Soc. N.S.W.* (1) 1: 369–378.

Ramsay, E. P. 1878a. Description of some new species of birds from New Britain, New Ireland, Duke of York Island, and the south-east coast of New Guinea. *Proc. Linn. Soc. N.S.W.* (1) 2: 104–107.

Ramsay, E. P. 1878b. Description of a new species of *Ianthoenas*, from the Duke of York Islands. *Proc. Linn. Soc. NSW* (1) 2: 248–249.

Ramsay, E. P. 1878c. Descriptions of five species of new birds, from Torres Straits and New Guinea, etc. *Proc. Linn. Soc. NSW* (1) 3: 72–75.

Ramsay, E. P. 1879a. Contributions to the zoology of New Guinea. Parts I and II. *Proc. Linn. Soc. N.S.W.* (1) 3: 241–305.

Ramsay, E. P. 1879b. Notes on the zoology of the Solomon Islands. I. Aves. *Proc. Linn. Soc. N.S.W.* (1) 4: 65–84.

Ramsay, E. P. 1879c. Notes on some recently described birds from the Solomon Islands. *Proc. Linn. Soc. N.S.W.* (1) 4: 313–315.

Ramsay, E. P. 1879d. On some new and rare birds, from South East Coast of New Guinea, etc. *Proc. Linn. Soc. NSW* (1) 4: 464–470.

Ramsay, E. P. 1879e. Notes on the fauna of the Solomon Islands. *Nature* 20: 125–126.

Ramsay, E. P. 1881. Notes on the zoology of the Solomon Islands, with descriptions of some new birds. *Proc. Linn. Soc. N.S.W.* (1) 6: 176–181.

Ramsay, E. P. 1882a. Notes on the zoology of the Solomon Islands, with descriptions of some new birds. *Proc. Linn. Soc. N.S.W.* (1) 6: 718–727.

Ramsay, E. P. 1882b. Descriptions of two new birds from the Solomon Islands. *Proc. Linn. Soc. N.S.W.* (1) 6: 833–835.

Ramsay, E. P. 1882c. On a new species of *Eurystopodus. Proc. Linn. Soc. N.S.W.* (1) 6: 843–845.

Ramsay, E. P. 1882d. Notes on the zoology of the Solomon Islands. *Proc. Linn. Soc. N.S.W.* (1) 7: 16–43.

Ramsay, E. P. 1882e. Descriptions of two new birds from the Solomon Islands. *Proc. Linn. Soc. N.S.W.* (1) 7: 299–301.

Ramsay, E. P. 1882f. Descriptions of some new birds from the Solomon Islands and New Britain. *J. Linn. Soc. (London) Zool.* 16: 128–131.

Ramsay, E. P. 1882g. New birds from the Solomon Islands. *Nature* 25: 282.

Ramsay, E. P. 1883a. *Proc. Zool. Soc. London,* p. 711.

Ramsay, E. P. 1883b. Notes on birds from the Solomon Islands. *Proc. Linn. Soc. N.S.W.* (1) 7: 665–673.

Ramsay, E. P. 1885. Descriptions of two new species of birds from the Austro-Malayan region. *Proc. Linn. Soc. NSW* (1) 9: 863–864.

Rand, A.L. 1941a. Results of the Archbold Expeditions. No. 32. New and interesting birds from New Guinea. *Amer. Mus. Novitates* no. 1102.

Rand, A.L. 1941b. Results of the Archbold Expeditions. No. 33. A new race of quail from New Guinea, with notes on the origin of the grassland avifauna. *Amer. Mus. Novitates* no. 1122.

Reichenow, A. 1899. Die Vögel der Bismarckinseln. *Mitteil. Zool. Mus. Berlin* 1, no. 3: 1–106.

Rensch, B. 1931. Die Vogelwelt von Lombok, Sumbawa und Flores. *Mitt. Zool. Mus. Berlin* 17: 451–637.

Restall, R. 1996. *Munias and Mannikins.* Pica, Sussex, UK.

Ricklefs, R.E. & E. Bermingham. 1997. Molecular phylogenetics and conservation of Caribbean birds. *El Pitirre* 10: 85–92.

Ricklefs, R.E. & E. Bermingham. 1999. Taxon cycles in the Lesser Antillean avifauna. *Ostrich* 70: 49–59.

Ricklefs, R.E. & G.C. Cox. 1972. Taxon cycles in the West Indian avifauna. *Amer. Natur.* 106: 195–219.

Ricklefs, R.E. & G.C. Cox. 1978. Stage of taxon cycle, habitat distribution, and population density in the avifauna of the West Indies. *Amer. Natur.* 112: 875–895.

Ripley, S.D. 1941. Notes on the genus *Coracina. Auk* 58: 381–395.

Ripley, S.D. 1942a. A review of the species *Anas castanea. Auk* 59: 90–99.

Ripley, S.D. 1942b. The species *Eurystomus orientalis. Proc. Biol. Soc. Wash.* 55: 169–176.

Ripley, S. D. 1947. A report on the birds collected by Logan J. Bennett on Nissan Island and the Admiralty Islands. *J. Washington Acad. Sci.* 37: 95–102.

Ripley, S.D. 1959. Comments on birds from the Western Papuan Islands. *Postilla* 38: 1–17.

Ripley, S.D. 1977. *A Monograph of the Family Rallidae.* Godine, Boston.

Ripley, S.D. 1983. A record of Whitehead's swiftlet *Collocalia whiteheadi* from Bougainville Island. *Bull. Brit. Orn. Club.* 103: 82–84.

Ripley, S.D. 1985. Relationships of the Pacific warbler *Cichlornis* and its allies. *Bull. Brit. Orn. Club* 105: 109–112.

Ripley, S. D. & H. Birckhead. 1942. Birds collected during the Whitney South Sea Expedition. 51. On the fruit pigeons of the *Ptilinopus purpuratus* group. *Amer. Mus. Novit.* 1192.

Ripley, S.D. & D. Hadden. 1982. A new subspecies of *Zoothera* (Aves: Muscicapidae: Turdinae) from the Northern Solomon Islands. *J. Yamashina Inst. Ornith.* 14: 103–107.

Roberts, R.G., R. Jones & M. A. Smith. 1994. Beyond the radiocarbon barrier in Australian prehistory. *Antiquity* 68: 611–616.

Rohwer, S. & C. Wood. 1998. Three hybrid zones between Hermit and Townsend's warblers in Washington and Oregon. *Auk* 115: 284–310.

Roper, D. S. 1983. Egg incubation and laying behaviour of the Incubator Bird *Megapodius freycinet* on Savo. *Ibis* 125: 384–389.

Rothschild, W. 1901. [Descriptions of new species from Isabel, Solomon Islands]. *Bull. Brit. Orn. Club* 12: 22–23.

Rothschild, W. 1904a. [Description of a new pigeon]. *Bull. Brit. Orn. Club* 14: 77–78.

Rothschild, W. 1904b. [Descriptions of two new kingfishers]. *Bull. Brit. Orn. Club* 15: 5–7.

Rothschild, W. & E. Hartert. 1899. Ein kleiner Beitrag zur ferneren Kenntnis der Ornis von Neu-Hannover. *Ornith. Monatsber.* 7: 138–139.

Rothschild, W. & E. Hartert. 1901a. Notes on Papuan birds. *Novit. Zool.* 8: 55–88.

Rothschild, W. & E. Hartert. 1901b. Notes on Papuan birds. *Novit. Zool.* 8: 102–162.

Rothschild, W. & E. Hartert. 1901c. List of a collection of birds from Kulambangra and Florida Islands, in the Solomons group. *Novit. Zool.* 8: 179–189.

Rothschild, W. & E. Hartert. 1901d. List of a collection of birds from Guadalcanar Island in the Solomon group. *Novit. Zool.* 8: 373–382.

Rothschild, W. & E. Hartert. 1902a. List of a collection of birds made on Ysabel Island in the Solomon group by Mr. A. S. Meek. *Novit. Zool.* 9: 581–594.

Rothschild, W. & E. Hartert. 1902b. List of a small collection of birds made by Mr. A. S. Meek on Treasury Island, Solomon Islands. *Novit. Zool.* 9: 594.

Rothschild, W. & E. Hartert. 1903. Notes on Papuan birds. *Novit. Zool.* 10: 65–116.

Rothschild, W. & E. Hartert. 1905. Further contributions to our knowledge of the ornis of the Solomon Islands. *Novit. Zool.* 12: 243–268.

Rothschild, W. & E. Hartert. 1906. [Description of a new pigeon]. *Bull. Brit. Orn. Club* 19: 28.
Rothschild, W. & E. Hartert. 1907. Notes on Papuan birds. *Novit. Zool.* 14: 433–446.
Rothschild, W. & E. Hartert. 1908a. The birds of Vella Lavella, Solomon Islands. *Novit. Zool.* 15: 351–358.
Rothschild, W. & E. Hartert. 1908b. On a collection of birds from San Cristoval, Solomon Islands. *Novit. Zool.* 15: 359–365.
Rothschild, W. & E. Hartert. 1914a. On the birds of Rook Island, in the Bismarck Archipelago. *Novit. Zool.* 21: 207–218.
Rothschild, W. & E. Hartert. 1914b. The birds of the Admiralty Islands, north of German New Guinea. *Novit. Zool.* 21: 281–298.
Rothschild, W. & E. Hartert. 1914c. [Descriptions of new birds from the Admiralty Islands]. *Bull. Brit. Orn. Club* 33: 104–109.
Rothschild, W. & E. Hartert. 1916a. On some forms of *Coracina* (*Graucalus* Auct.) from the Solomon Islands. *Novit. Zool.* 23: 289–291.
Rothschild, W. & E. Hartert. 1916b. [Description of new forms of *Rhipidura cockerelli*]. *Bull. Brit. Orn. Club* 36: 73–74.
Rothschild, W. & E. Hartert. 1924a. [Descriptions of new forms from St. Matthias Island]. *Bull. Brit. Orn. Club* 44: 50–53.
Rothschild, W. & E. Hartert. 1924b. [Descriptions of a new species of lory and honey-eater from New Ireland]. *Bull. Brit. Orn. Club* 45: 7–8.
Rothschild, W. & E. Hartert. 1926a. In Hartert (1926a), p. 41.
Rowley, I. 1970. The genus *Corvus* (Aves: Corvidae) in Australia. *CSIRO Wildlife Res.* 15: 27–71.
Rozendaal, F.G. 1987. Description of a new species of bush warbler of the genus *Cettia* Bonaparte, 1834 (Aves: Sylviidae) from Yamdena, Tanimbar Islands, Indonesia. *Zool. Meded.* 61: 177–202.
Ruhlen, M. 1987. *A Guide to the World's Languages*, vol. I. Stanford University Press, Stanford, CA.
Salomonsen, F. 1960a. Notes on flowerpeckers (Aves, Dicaeidae). 2. The primitive species of the genus *Dicaeum*. *Amer. Mus. Novit.* 1991.
Salomonsen, F. 1960b. Notes on flowerpeckers (Aves, Dicaeidae). 3. The species group *Dicaeum concolor* and the superspecies *Dicaeum erythrothorax*. *Amer. Mus. Novit.* 2016.
Salomonsen, F. 1962. Whitehead's swiftlet (*Collocalia whiteheadi* Ogilvie-Grant) in New Guinea and Melanesia. *Vidensk. Medd. Dansk naturh. Foren.* 125: 509–512.
Salomonsen, F. 1964. Some remarkable new birds from Dyaul Island, Bismarck Archipelago, with zoogeographical notes. *Biol. Skr. Dan. Vid. Selsk.* 14: 1–37.
Salomonsen, F. 1965. Notes on the Mountain Leaf-Warbler (*Phylloscopus trivirgatus* Strickland) in the Bismarck Archiplago. *Vidensk. Medd. Dansk naturhist. Foren.* 128: 77–83.
Salomonsen, F. 1966a. *Myzomela cruentata* Meyer (Aves, Meliphagidae) in the Bismarck Archipelago. *Dansk. Orn. Foren. Tidsskr.* 60: 118–122.
Salomonsen, F. 1966b. Notes on the Green Heron (*Butorides striatus* (Linnaeus)) in Melanesia and Papua. *Vidensk. Medd. fra Dansk Naturhist. Foren.* 129: 279–283.
Salomonsen, F. 1966c. Preliminary descriptions of new honeyeaters (Aves, Meliphagidae). *Breviora* 254.
Salomonsen, F. 1972. New pigeons from the Bismarck Archipelago (Aves, Columbidae). *Steenstrupia* 2: 183–189.
Salomonsen, F. 1976. The main problems concerning avian evolution on islands. *Proc. 16th Intern. Orn. Congr.* 585–602.
Salomonsen, F. 1983. Revision of the Melanesian swiftlets (Apodes, Aves) and their conspecific forms in the Indo-Australian and Polynesian Region. *Biol. Skr. Dan. Vid. Selsk.* 23: 1–112.
Salvadori, 1880–1882. Ornitologia della Papuasia e delle Molocche. Vols. 1–3. Stamperia Reale, Torino.
Salvadori, T. 1880. Remarks on two recently published papers on the ornithology of the Solomon Islands. *Ibis* 4: 126–131.
Salvadori, T. 1881. Descrizione di alcune specie nuove o poco conosciute di uccelli della Nuova Britannia, della Nuova Guinea e delle Isole del Duca di York. *Atti R. Accad. Sci. Torino* 16: 619–625.
Sanft, K. 1960. Aves/Upupae: Bucerotidae. *Das Tierreich* 76: 1–174.
Sarfert, E. & H. Damm. 1929. Luangiua and Nukumanu. In: G. Thilenius (ed.), *Ergebnisse der Südsee-Expedition 1908-10, 2. Ethn., B. Micronesien*, vol. 12. Friederichsen, Hamburg.
Savidge, J. 1987. Extinction of an island forest avifauna by an introduced snake. *Ecology* 68: 660–668.
Schliewen, U.K., D. Tautz & S. Pääbo. 1994. Sympatric speciation suggested by monophyly of crater lake cichlids. *Nature* 368: 629–632.
Schodde, R. 1977. Contributions to Papuasian ornithology. VI. Survey of the birds of

southern Bougainville Island, Papua New Guinea. *CSIRO Div. Wildlf. Res. Tech. Paper* no. 34.

Schodde, R. & R. De Naurois. 1982. Patterns of variation and dispersal in the Buff-banded Rail (*Gallirallus philippensis)* in the south-west Pacific with description of new subspecies. *Notornis* 29: 131–142.

Schodde, R., P. Fullager & N. Hermes. 1983. *A Review of Norfolk Island Birds: Past and Present.* Australian National Parks and Wildlife Service, Canberra.

Schodde, R. & W.B. Hitchcock. 1968. Contributions to Papuasian ornithology. I. Report on the birds of the Lake Kutubu area, Territory of Papua and New Guinea. *CSIRO Div. Wildlf. Res. Tech. Paper* no. 13.

Schodde, R. & I.J. Mason. 1976. Infra-specific variation in *Alcedo azurea* Latham (Alcedinidae). *Emu* 76: 161–166.

Schodde, R. & I.J. Mason. 1980. *Nocturnal Birds of Australia.* Lansdowne, Melbourne.

Schodde, R., I.J. Mason, M.L. Dudzinski & J.L. McKean. 1980. Variation in the Striated Heron *Butorides striatus* in Australasia. *Emu* 80: 203–212.

Schodde, R., I.J. Mason & J.L. McKean. 1979a. A new subspecies of *Philemon buceroides* from Arnhem Land. *Emu* 79: 24–30.

Schodde, R., G.T. Smith, I.J. Weatherly & R.G. Weatherly. 1979b. Relationships and speciation in the Australian corellas (Psittacidae). *Bull. Brit. Orn. Club* 99: 128–137.

Schönwetter, N. 1935. Vogeleier aus Neubritannien. *Beiträge z. Fortpflanzungsbiologie der Vögel* 11: 129–136.

Sclater, P. L. 1869. On a collection of birds from the Solomon Islands. *Proc. Zool. Soc. London* 118–126.

Sclater, P. L. 1873. [Exhibit of skins collected by Capt. Fergusson]. *Proc. Zool. Soc. London* 3.

Sclater, P. L. 1877a. On the birds collected by Mr. George Brown, C.M.Z.S., on Duke-of-York Island, and on the adjoining parts of New Ireland and New Britain. *Proc. Zool. Soc. London* 96–114.

Sclater, P. L. 1877b. Report on the collection of birds made during the voyage of H. M. S. "Challenger". III. On the birds of the Admiralty Islands. *Proc. Zool. Soc. London* 551–557.

Sclater, P. L. 1878a. [A second collection of birds from the Rev. G. Brown]. *Proc. Zool. Soc. London* 289–290.

Sclater, P. L. 1878b. On a third collection of birds made by the Rev. G. Brown, C.M.Z.S., in the Duke-of-York Group of Islands and its vicinity. *Proc. Zool. Soc. London* 670–673.

Sclater, P. L. 1879. On a fourth collection of birds made by the Rev. G. Brown, C.M.Z.S., on Duke-of-York Island and in its vicinity. *Proc. Zool. Soc. London* 446–451.

Sclater, P. L. 1880a. On a fifth collection of birds made by the Rev. G. Brown, C.M.Z.S., on Duke-of-York Island and in its vicinity. *Proc. Zool. Soc. London* 65–67.

Sclater, P. L. 1880b. On the birds collected in the Admiralty Islands. In: *Report on the Scientific Results of H.M.S. Challenger During the Years 1873-76, Zoology* 2: 25–34.

Sclater, P. L. 1881. [Birds obtained in New Britain by Mr. Kleinschmidt]. *Proc. Zool. Soc. London* 451–453.

Sclater, P. L. 1883. [A collection of birds by the Rev. George Brown]. *Proc. Zool. Soc. London* 347–348.

Scott, R. M., P. B. Heyligers, J. R. McAlpine, J. C. Saunders & J. G. Speight. 1967. Lands of Bougainville and Buka Islands, Territory of Papua and New Guinea. CSIRO Land Research series no. 20. Commonwealth Scientific and Industrial Research Organization, Melbourne, Australia.

Seutin, G., N.K. Klein, R.E. Ricklefs & E. Bermingham. 1994. Historical biogeography of the bananaquit (*Coereba flaveola*) in the Caribbean region: a mitochondrial DNA assessment. *Evolution* 48: 1041–1061.

Sharpe, R. B. 1884. [title]. *Proc. Zool. Soc. London*, p. 579.

Sharpe, R. B. 1888a. Descriptions of some new species of birds from the island of Guadalcanar in the Solomon Archipelago, discovered by Mr. C. M. Woodford. *Proc. Zool. Soc. Lond.* 182–185.

Sheldon, F.H. 1987. Phylogeny of herons estimated from DNA-DNA hybridization data. *Auk* 104: 97–108.

Sibley, C. G. 1951. Notes on the birds of New Georgia, central Solomon Islands. *Condor* 53: 81–92.

Sibley, C. G. & J. E. Ahlquist. 1982. The relationships of the Hawaiian honeycreepers (Drepaninini) as indicated by DNA hybridization. *Auk* 99: 130–140.

Sibley, C.G. & J.E. Ahlquist. 1983. The phylogeny and classification of birds based on the data of DNA-DNA hybridization. *Current Orn.* 1: 245–292.

Sibley, C.G. & J.E. Ahlquist. 1985. The phylogeny and classification of the passerine birds, based on comparisons of the genetic material, DNA. *Proc. Int. Orn. Congr.* 18: 83–121.

Sibley, C.G. & J.E. Ahlquist. 1990. *Phylogeny and Classification of Birds: a Study in Molecular Evolution.* Yale University Press, New Haven, CT.

Sibley, C.G. & B.L. Monroe, Jr. 1990. *Distribution and Taxonomy of Birds of the World.* Yale University Press, New Haven, CT.

Siebers, H.C. 1930. Fauna Buruana. Aves. *Treubia 7 (suppl.)*: 165–303.

Silva, K. 1973a. Observations. *New Guinea Bird Soc. Newsletter* no. 89: 1–2.

Silva, K. 1973b. Observations. *New Guinea Bird Soc. Newsletter* no. 90: 2–3.

Silva, K. 1975. Observations. *New Guinea Bird Soc. Newsletter* no. 112: 4–6.

Silver, E.A., L.D. Abbott, K.S. Kirchoff-Stein, D.L. Reed, and B. Bernstein-Taylor. 1991. Collision propagation in Papua New Guinea and the Solomon Sea. *Tectonics* 10: 863–874.

Skyrme, A.C. 1984. Further observations on Bougainville. *Papua New Guinea Bird Soc. Newsletter* no. 207: 18–19.

Smith, G.A. 1975. Systematics of parrots. *Ibis* 117: 18–68.

Snow, D. W. 1978. *An Atlas of Speciation in African Non-passerine Birds.* British Museum of Natural History, London.

Somadikarta, S. 1967. A recharacterization of *Collocalia papuensis Rand*, the three-toed swiftlet. *Proc. U. S. Nat. Mus.* 124: 1–8.

Somadikarta, S. 1986. *Collocalia linchi* Horsfield & Moore C—a revision. *Bull. Brit. Orn. Club* 106: 32–40.

Spriggs, M. 1997. *The Island Melanesians.* Blackwell, Oxford.

Steadman, D.W. 1985. Fossil birds from Mangaia, southern Cook Islands. *Bull. Brit. Orn. Club* 105: 58–66.

Steadman, D.W. 1989. Extinction of birds in Eastern Polynesia: a review of the record, and comparisons with other Pacific island groups. *J. Archaeol. Sci.* 16: 177–205.

Steadman, D.W. 1992. New species of *Gallicolumba* and *Macropygia* (Aves: Columbidae) from archeological sites in Polynesia. *Los Angeles County Mus. Nat. Hist. Sci. Ser.* 36: 329–348.

Steadman, D.W. 1993. Biogeography of Tongan birds before and after human impact. *Proc. Natl. Acad. Sci. USA* 90: 818–822.

Steadman, D.W. 1995. Prehistoric extinctions of Pacific Island birds: biodiversity meets zooarchaeology. *Science* 267: 1123–1131.

Steadman, D.W. & P.V. Kirch. 1998. Biogeography and prehistoric exploitation of birds in the Mussau Islands, Bismarck Archipelago, Papua New Guinea. *Emu* 98: 13–22.

Steadman, D.W. & S.L. Olson. 1985. Bird remains from an archaeological site on Henderson Island, South Pacific: man-caused extinctions on an "uninhabited" island. *Proc. Natl. Acad. Sci. USA* 82: 6191–6195.

Steadman, D. W., J. P. White & J. Allen. 1999. Prehistoric birds from New Ireland, Papua New Guinea: extinctions on a large Melanesian island. *Proc. Natl. Acad. Sci. USA* 96: 2563–2568.

Stevens, G. W. & J. L. O. Tedder. 1973. *A Honiara Bird Guide.* British Solomon Islands Scout Association, Honiara.

Stickney, E.H. 1943. Birds collected during the Whitney South Sea Expedition. 53. Northern shore birds in the Pacific. *Amer. Mus. Novit.* no. 1248.

Storer, R.W. 1963. Courtship and mating behavior and the phylogeny of grebes. *Proc. Int. Orn. Congr.* 13: 562–569.

Stresemann, E. 1929. *Himantopus himantopus leucocephalus* Gould im Bismarck-Archipel. *Orn. Monatsber.* 37: 47.

Stresemann, E. 1931. *Lorius amabilis* species nova. *Orn. Monatsb.* 39: 182–183.

Stresemann, E. 1933. Neue Vogelrassen von Lihir (Bismarck-Archipel). *Orn. Monatsber.* 41: 114–116.

Stresemann, E. 1934. *Falco longipennis longipennis* Swainson im Bismarckarchipel. *Orn. Monatsber.* 42: 157.

Stresemann, E. 1938. P. Otto Meyer. *J. f. Orn.* 86: 166–169.

Stresemann, E. 1939. Die Vögel von Celebes. Teil I und II. *J. f. Orn.* 87: 299–425.

Stresemann, E. 1940–41. Die Vögel von Celebes. Teil III. *J. f. Orn.* 88: 1–135, 389–487; 89: 1–102.

Stresemann, E. 1943. Die Gattung *Corvus* in Australien und Neuguinea. *J. f. Orn.* 91: 121–135.

Stresemann, E. 1950. Birds collected during Capt. James Cook's last expedition (1776–1780). *Auk* 67: 66–88.

Stresemann, E. & J. Arnold. 1949. Speciation in the group of Great Reed Warblers. *J. Bombay Nat. Hist. Soc.* 48: 428–443.

Studer, T. 1889. *Die Forschungsreise S.M.S. "Gazelle" in den Jahren 1874 bis 1876 unter Kommando des Kapitan zur See Freiherrn von Schleinitz. Teil 3. Zoologie und Geologie.* Mittler, Berlin.

Sutter, E. 1955. Über die Mauser einiger Laufhuhnchen und die Rassen von *Turnix maculosa* und *sylvatica* im indo-australischen Gebiet. *Verh. Naturf. Ges. Basel* 66: 85–139.

Terborgh, J. W. 1974. Preservation of natural diversity: the problem of extinction prone species. *BioScience* 24: 715–722.

Terrell, J. 1986. *Prehistory in the Pacific Islands.* Cambridge University Press, Cambridge.

Thompson, M. C. 1964. Two new distributional records of birds from the Southwest Pacific. *Ardea* 52: 121.

Thorne, R.F. 1969. Floristic relationships between New Caledonia and the Solomon Islands. *Phil. Trans. Roy. Soc.* B 255: 595–602.

Thornton, I. 1996. *Krakatau.* Harvard University Press, Cambridge, MA.

Tolhurst, L. 1993. Observations from Manus Province. *Muruk* 6 (1): 15–18.

Tregenza, T. & R.K. Butlin. 1999. Speciation without isolation. *Nature* 400: 311–312.

Tristram, H. B. 1879. On a collection of birds from the Solomon Islands and New Hebrides. *Ibis* 4 (3): 437–444.

Tristram, H. B. 1882. Notes on a collection of birds from the Solomon Islands, with descriptions of new species. *Ibis* (4) 6: 133–146.

Tristram, H. B. 1891. On an apparently new species of pigmy parrot of the genus *Nasiterna. Ibis* (6) 3: 608.

Tristram, H. B. 1892. On two small collections of birds from Bugotu and Florida, two of the smallest Solomon Islands. *Ibis* (6) 4: 293–297.

Tristram, H. B. 1894. On some birds from Bugotu, Solomon Islands, and Santa Cruz. *Ibis* (6) 6: 28–31.

Tristram, H. B. 1895. Further notes on birds from Bugotu, Solomon Islands, with description of a new species. *Ibis* (7) 1: 373–376.

Tryon, R. 1971. Development and evolution of fern floras of oceanic islands. In: W. L. Stern (ed.), *Adaptive Aspects of Insular Evolution*, pp. 54–62. Washington State University Press, Pullman.

van Bemmel, A.C.V. 1948. A faunal list of the birds of the Moluccan Islands. *Treubia* 19: 323–402.

Vaurie, C. 1949. A revision of the bird family Dicruridae. *Bull. Amer. Mus. Nat. Hist.* 93: 203–342.

Vaurie, C. 1958. Remarks on some Corvidae of Indo-Malaya and the Australian Region. *Amer. Mus. Novit.* 1915.

Vaurie, C. 1959. *The Birds of the Palearctic Fauna: Passeriformes.* Witherby, London.

Vaurie, C. 1965. *The Birds of the Palearctic Fauna: Non-passeriformes.* Witherby, London.

Verreaux, J. 1858. Description d'oiseaux nouveaux. *Rev. Mag. Zool., Paris* 10: 304–306.

Virtue, R. M. 1947. Birds observed at Torokina, Bougainville Island. *Emu* 46: 324–331.

Voous, K.H. & J.G. van Marle. 1949. The distributional history of *Coracina* in the Indo-Australian Archipelago. *Bijgr. Dierk.* 28: 513–529.

Vuilleumier, F., M. LeCroy & E. Mayr. 1992. New species of birds described from 1981 to 1990. *Bull. Brit. Orn. Club* (Suppl.) 112A: 267–309.

Wallace, A. R. 1892. *Island Life*, 2nd ed. Macmillan, London.

Wattel, J. 1973. Geographical differentiation in the genus *Accipiter.* Nuttall Ornithological Club, Cambridge, MA.

Webb, H.P. 1992. Field observations of the birds of Santa Isabel, Solomon Islands. *Emu* 92: 52–56.

White, C. M. N. 1937. [Notes on some Solomon Island birds.] *Bull. Brit. Orn. Club* 58: 46–48.

White, C.M.N. 1973. Night herons in Wallacea. *Bull. Brit. Orn. Club* 93: 175–176.

White, C.M.N. 1975. Further notes on birds of Wallacea. *Bull. Brit. Orn. Club* 95: 106–109.

White, C.M.N. 1976. The problem of the cassowary in New Britain. *Bull. Brit. Orn. Club* 96: 66–68.

White, C.M.N. & M.D. Bruce. 1986. *The Birds of Wallacea.* British Ornithologists' Union, London.

White, J. P. & J. F. O'Connell. 1982. *A Prehistory of Australia, New Guinea and Sahul.* Academic Press, Sydney.

Whitmore, T.C. 1966. *Guide to the Forests of the British Solomon Islands.* Oxford University Press, London.

Whitmore, T.C. 1969a. The vegetation of the Solomon Islands. *Phil. Trans. Roy. Soc.* B 255: 259–270.

Whitmore, T.C. 1969b. Geography of the flowering plants. *Phil. Trans. Roy. Soc. B* 255: 549–566.

Whitmore, T.C. 1974. Change with time and the role of cyclones in tropical rain forest on Kolombangra, Solomon Islands. Institute Paper no. 46. Commonwealth Forestry Institute, Oxford.

Whitmore, T.C. 1981. Wallace's line and some other plants. In: T.C. Whitmore (ed.), *Wallace's Line and Plate Tectonics*, pp. 70–80. Oxford University Press, Oxford.

Whitmore, T.C. 1984. *Tropical Rain Forests of the Far East*, 2nd ed. Oxford University Press, Oxford.

Whitmore, T.C. 1990. *An Introduction to Tropical Rain Forests.* Oxford University Press, Oxford.

Wichmann, A. 1909–12. Entdeckungsgeschichte von Neu-Guinea. *Nova Guinea*, vols. 1 and 2.

Wilson, E.O. 1959. Adaptive shift and dispersal in a tropical ant fauna. *Evolution* 13: 122–144.

Wilson, E.O. 1961. The nature of the taxon cycle in the Melanesian ant fauna. *Am. Nat.* 95: 169–193.

Wiltgen, R.M. 1979. *The Founding of the Roman Catholic Church in Oceania 1825 to 1850*. Australian National University Press, Canberra.

Wolff, T. 1966. The Noona Dan Expedition 1961–1962. General report and lists of stations. *Vidensk. Medd. Dansk naturh. Foren.* 129: 287–336.

Wolff, T. 1969. The fauna of Rennell and Bellona, Solomon Islands. *Phil. Trans. Roy. Soc.* B 255: 321–343.

Wolff, T. 1970. Lake Tegano on Rennell Island, the former lagoon of a raised atoll. *Nat. Hist. Rennell Isl., Br. Solomon Isl.* 6: 7–29.

Wolff, T. 1973. Notes on birds from Rennell and Bellona Islands. *Nat. Hist. Rennell Isl., Br. Solomon Isl.* 7: 7–28.

Woodford, C. M. 1890. *A Naturalist Among the Head-hunters*. Petherick, Melbourne.

Woodford, C. M. 1916. On some little-known Polynesian settlements in the neighborhood of the Solomons. *Geogr. J.* 48: 45–49.

Wurm, S. A. 1982. *Papuan Languages of Oceania*. Günther Narr, Tübingen.

Wurm, S. A. 1983. Linguistic prehistory in the New Guinea area. *J. Human Evolution* 12: 25–35.

Yamashina, Y. & T. Mano. 1981. A new species of rail from Okinawa Island. *J. Yamashina Inst. Orn.* 13: 1–6.

Ziegler, A.C. 1982. An ecological check-list of New Guinea recent mammals. In: J. L. Gressitt (ed.), *Biogeography and Ecology of New Guinea*, pp. 863–894. Junk, The Hague.

Zielinski, G.A., P.A. Mayewski, L.D. Meeker, S. Whitlow, M.S. Twickler, M. Morrison, D.A. Meese, A.J. Gow & R.B. Alley. 1994. Record of volcanism since 7,000 B.C. from the GISP2 Greenland ice core and implications for the volcano-climate system. *Science* 264: 948–952.

Ziswiler, V., H.R. Güttinger & H. Bregulla. 1972. Monographie der Gattung *Erythrura* Swainson, 1837 (Aves, Passeres, Estrildidae). *Bonn. Zool. Monogr.* no. 2: 1–158.

Zohary, D. 1999. Speciation under self-pollination. In: S. P. Wasser (ed.), *Evolutionary Theory and Processes: Modern Perspectives*, pp. 301–307. Kluwer, Dordrecht, the Netherlands.

Zweifel, R.G. 1956. Results of the Archbold Expeditions. No. 72. Microhylid frogs from New Guinea, with descriptions of new species. *Amer. Mus. Novitates* no. 1766.

Zweifel, R.G. 1958. Results of the Archbold Expeditions. No. 78. Frogs of the Papuan hylid genus *Nyctimystes*. *Amer. Mus. Novitates* no. 1896.

Zweifel, R.G. 1960. Results of the 1958–1959 Gilliard New Britain Expedition. 3. Notes on the frogs of New Britain. *Amer. Mus. Novitates* no. 2023.

Zweifel, R.G. 1962. Results of the Archbold Expeditions. No. 83. Frogs of microhylid genus *Cophixalus* from the mountains of New Guinea. *Amer. Mus. Novitates* no. 2087.

Zweifel, R.G. 1966. A new lizard of the genus *Tribolonotus* (Scincidae) from New Britain. *Amer. Mus. Novitates* no. 2264.

Zweifel, R.G. 1969. Frogs of the genus *Platymantis* (Ranidae) in New Guinea, with the description of a new species. *Amer. Mus. Novitates* no. 2374.

Zweifel, R.G. 1972. Results of the Archbold Expeditions. No. 97. A revision of the frogs of the subfamilies Asterophryinae family Microhylidae. *Bull. Amer. Mus. Nat. Hist.* 148: 411–546.

Zweifel, R.G. 1980. Results of the Archbold Expeditions. No. 103. Frogs and lizards from the Huon Peninsula, Papua New Guinea. *Bull Amer. Mus. Nat. Hist.* 165: 387–434.

Zweifel, R.G. & M.J. Tyler. 1982. Amphibia of New Guinea. In: J.L. Gressitt (ed.), *Biogeography and Ecology of New Guinea*, pp. 759–801. Junk, The Hague.

Subject Index

Species Index